MW01643862

Practical Physics

Jerry D. Wilson
Chairman, Department of Science and Mathematics
Lander College, Greenwood, South Carolina

SAUNDERS GOLDEN SUNBURST SERIES
SAUNDERS COLLEGE PUBLISHING
Philadelphia New York Chicago
San Francisco Montreal Toronto
London Sydney Tokyo Mexico City
Rio de Janeiro Madrid

Address orders to:
383 Madison Avenue
New York, NY 10017

Address editorial correspondence to:
West Washington Square
Philadelphia, PA 19105

Text Typeface: 10/12 Times Roman
Compositor: Progressive Typographers
Acquisitions Editor: John Vondeling
Developmental Editor: Lloyd Black
Project Editors: Maureen R. Iannuzzi, Ellen Newman
Copyeditor: Janis Moore
Art Director: Carol Bleistine
Art/Design Assistant: Virginia A. Bollard
Test Design: Emily Harste
Cover Design: Lawrence R. Didona
Text Artwork: J&R Technical Services, Inc.
Production Manager: Tim Frelick
Assistant Production Manager: JoAnn Melody

Cover credit: © Gary Gladstone/THE IMAGE BANK

Library of Congress Cataloging in Publication Data

Wilson, Jerry D.
Practical physics

Includes index.
1. Physics 2. Technology I. Title
QC23.W579 1985 530 85-2510
ISBN 0-03-063512-8

PRACTICAL PHYSICS ISBN 0-03-063512-8

Library of Congress catalog card number 85-2510

3456 032 987654321

CBS COLLEGE PUBLISHING
Saunders College Publishing
Holt, Rinehart and Winston
The Dryden Press

To Joyce and my students

Preface

Practical Physics is intended for a one-semester course in introductory physics for students majoring in the liberal arts or social sciences. Since nonscience majors are often wary of science, it is my purpose in writing this text to make these students aware of and appreciative of their physical environment. The physical universe is a dynamic, exciting place that becomes more meaningful to the person willing to learn the "why" as well as the "what." To help students achieve this goal, I have included the following pedagogical features.

Style. I have endeavored to write in a style that is clear and logical as well as somewhat informal and relaxed to assist students in their learning. New terms are carefully defined, and I have attempted to avoid jargon.

Organization. The book is divided into seven parts: I. Mechanics; II. Properties of Matter; III. Heat; IV. Sound; V. Electricity and Magnetism; VI. Light; and VII. Modern Physics. This is a well-known standard format and was chosen because one topic in physics builds on the next. For example, the concepts of motion, work, and energy (I. Mechanics) are needed to understand the concepts of temperature and heat (III. Heat), which in turn are used in explaining energy losses (joule heating) in electrical circuits (V. Electricity and Magnetism). Similarly, the basic concepts of electrical charge and magnetism are important in the study of light (VI. Light), as well as in identifying particles coming from radioactive isotopes (VII. Modern Physics).

Part Openers. Each opener presents a brief overview of the section that follows. Some historical matter is introduced to give students a perspective on how physics concepts developed through time.

Coverage. The text fully covers the realm of the physical universe. For example, in the mechanics section students will learn of the fundamentals of measurement, microgravity, and moment of inertia. In the section on modern physics the topics include lasers, masers, quasars, quarks, and black holes. Things are explained as simply as possible with little or no mathematics.

Illustrations. The text is highly illustrated with many photographs and skillfully drawn pieces of artwork. Much of the artwork has been rendered in a "realistic" style that places emphasis on accuracy, relative scale, and actual appearance. Students will make the connections between concept and application more easily if objects with which they are familiar really do *look* like those objects. Likewise, new information—whether visual or verbal in nature—should be presented as accurately as possible.

Special Features. These 67 Special Features, of which there are several in every chapter, should be of great interest to most students. They emphasize the practical aspects of the book and help to link the student's everyday experiences with physical principles. Among those included are:

The Automobile Air Bag
Microgravity
A Pioneer in Outer Space
The Greenhouse Effect
Acid Rain
Photochromic Glasses
Blood Pressure
The Microwave Oven
Galloping Gertie: The Tacoma Narrows Bridge Collapse
Noise-Exposure Limits
Ultrasonics
Superconductivity
Personal Safety and Electrical Effects
Xerography and Electrostatic Copiers
Woofers and Tweeters
LCDs—Liquid Crystal Displays
Space-Age General Relativity
Lasers in the Supermarket
The Smoke Alarm
Case History: Three Mile Island
Black Holes

Question/Answer Examples. In these examples a question is posed and an answer is given immediately. These examples provide insight into interesting physical aspects while training students to think about physics in terms of questions as well as answers. These examples also function as models for the student's work with the Exercises.

Summary of Key Terms. Key terms presented within a chapter are defined at the end of each chapter so that students can easily review them.

Exercises. The end-of-chapter Exercises are designed to stimulate student thinking by requiring that students apply what they have learned in the chapter. Most are of the thought question variety, but some do require elementary calculations.

Users of *Practical Physics* will receive an extensive set of ancillary items that I feel can substantially assist the instructor in the presentation of the course as well as motivate the student in his or her learning. These supplements include:

Instructor's Resource Manual. This printed manual contains the following items:

- answers to all Exercises
- a 1200-question Test Bank composed of 40 questions per chapter (10 each of multiple choice, completion, modified true/false, and matching)
- a set of Prepared Tests, consisting of 4 test pages with 10 of the above-type questions per page that may be removed and duplicated directly; this also includes an answer key (these 1200 questions differ from those of the Test Bank, giving a total of 2400 possible test questions)

Computerized Test Bank. The CTB contains the multiple choice, completion, and modified true/false questions found in the Test Bank and is available for both the Apple II and IBM PC microcomputers.

Classroom Demonstration Disk. This program disk contains a series of interesting computer programs that demonstrate or simulate physics phenomena. The disk is available for the Apple II microcomputer.

Home-Study Experiments in Practical Physics. This printed supplement contains simple home experiments that may be assigned by the instructor or done by interested students. The experiments provide a hands-on feeling for some of the physical principles studied in the text, and they are fun to do. This separate softcover Guide is offered free to every student and is packaged with the textbook.

Overhead Transparencies. A selection of 100 figures from the textbook are reproduced on acetate sheets suitable for projection.

All in all, a great deal of time and effort has been spent in writing this book to present the fascinating subjects of physics in an interesting and understandable manner.

The mathematical background of the student taking this course need only include a good knowledge of high school algebra. Although mathematics is an important tool in investigating the physical universe, students should concentrate on using their ability to read, think, and formulate questions as they progress through the course. Once they comprehend the concepts, the use of some basic mathematics will flow naturally from the established principles. Although there is mathematics in the text, its use is kept to a minimum.

Acknowledgments

I would like to acknowledge all the assistance provided by various persons in the preparation of this book. My grateful thanks is given to each of the following for their reviews and many constructive suggestions:

Carl G. Adler, East Carolina University
Angelo Armenti, Jr., Villanova University
Donna A. Berry, Shaker Heights High School
John S. Eck, Kansas State University
Simon George, California State University, Long Beach
Stanley S. Hertzbach, University of Massachusetts
William F. Junkin III, Erskine College
Laurence R. McAneny, Southern Illinois University at Edwardsville
Hans S. Plendl, Florida State University
Lawrence C. Shepley, University of Texas at Austin
Robert F. Simpson, University of New Hampshire
Jon S. Staib, James Madison University

For their help in manuscript preparation, art work, and photography, I thank Ruth Hodges, Jocelyn Sanders, David Williams, Lee Kelly, and Sue Brannon at Lander College. Last but not least, there was the able assistance of the staff at Saunders College Publishing, in particular, John Vondeling, Associate Publisher, and Lloyd Black, Developmental Editor. I salute and thank these people.

Jerry D. Wilson

Lander College

Contents Overview

Contents

Special Features

PART ONE
Mechanics

Mechanics is the study of the motions of material bodies. Historically, it was one of the earliest exact sciences to be developed. Some mechanical principles were known to Greek scientists in the third century B.C. The tremendous growth of physics since the 1600's began with the discovery of the laws of mechanics by Galileo and Newton. Early successes were in predicting the motions of the moon, the Earth, and planets and their satellites (celestial mechanics). Now we apply the same principles to the motions of artificial satellites such as the orbiting Space Shuttle.

In general, the principles of mechanics can be applied to (a) the motions of celestial objects so as to accurately predict events, in some cases many years before they happen, for example, the return of Halley's comet; (b) the motions of ordinary objects on Earth, for example, an automobile or a thrown baseball; and (c) with some success, the behavior of atoms, atomic particles, and subatomic particles. The term ''classical mechanics'' is generally used to differentiate these principles from those of newer physical theories, such as relativistic mechanics and quantum mechanics. (See Part Seven, Modern Physics.)

Mechanics greatly influenced the growth of later sciences such as sound and electricity. It may be said that mechanics furnishes the basic concepts of the whole of physics, so quite naturally, the study of physics begins here.

1 Measurement and the Metric System

What We Measure

How tall are you? How much do you weigh? What time is it? The answers to these and many other such questions require that measurements be made. Rarely does a day go by that we do not make or use a measurement. Many times we are not aware that we are doing this. For example, when you tell someone what time it is, you are stating a measurement. Your clock is the measurement instrument.

Let's think of some of the things we commonly measure. With a little thought, you might say length. (I am 5 feet 8 inches tall, or I live three blocks from school.) Some of us frequently weigh ourselves. (I weigh 160 pounds.) And then there is time. (Class periods are 50 minutes long.)

When buying gas for your car, you may buy 10 gallons of gasoline. So we should add volume or capacity to our list of commonly measured items. Summarizing, we have

Commonly measured items

Length
Weight (Mass)*
Time
Volume or capacity

There are other things we measure. But let's keep the list simple for our discussion. It should be noted that volume really involves length measurements. Recall that the volume (V) of a box is its length (l) times its width (w) times its height (h), or in an equation, $V = l \times w \times h$. However, width and height are lengths too. So volume is a combination of lengths.

* For now, you may consider mass to be the quantity of matter something contains. Weight is the gravitational attraction (force) that the Earth has for an object. Mass and weight are related, and both are used in measurements, as will be learned in the following sections.

In science, things are described as simply as possible. The basic or fundamental properties used to describe things include length, mass, and time. These properties describe the concepts of space, matter, and time, respectively.

Fundamental properties	*Concepts described*
Length	Space
Mass	Matter
Time	Time

There is one other fundamental property associated with electricity, the electric charge. More will be said about this property in a later chapter. A vast majority of what we observe in nature can be measured or described in terms of these four fundamental properties and their various combinations.

Figure 1.1 A measurement is a quantitative description of a fundamental property compared to a standard. The standard units of measurement may be different, but the length of the table is the same in any case.

Measurement Units

Now that we know some of what we measure, how do we do it? It's really a matter of choice. For example, a table has a certain length no matter how we describe it. One person might measure the table in feet, another in yards, and still another in meters. Certainly, the length of the table doesn't change, only the choice of units used to describe it.

The measurement *unit* is the key word. Different people, societies, and civilizations have chosen different units. If a particular unit becomes popular and accepted, it is called a standard unit. That is, *a standard unit has a fixed and reproducible value for the purpose of taking accurate measurements.* For example, the foot and the meter are standard units. To measure something in feet, we compare it to a standard foot ruler or yardstick. To measure something in meters, we compare it to a meterstick standard (Fig. 1.1). Hence,

> A measurement is a quantitative description of one or more fundamental properties compared to a standard.

But who establishes or chooses a standard unit? Traditionally, it has been a head of state or a government. Early standards were referenced to parts of the human body. What could be more convenient? Some units of the British or English system, which is the customary system in the United States, originated from anatomical references. For example, the inch was referenced to the thumb (Fig. 1.2), which, of course, varied from person to person. In the 1300's King Edward II of England decreed the official inch to be equal to three barleycorns taken from the middle of the ear and laid end to end. (Not a great improvement!) He also decreed the foot to be equal to 12 three-barleycorn inches. Perhaps this was the length of his royal foot. Another English monarch, King Henry I, established the yard as the distance from the tip of his royal nose to the end of his fingers with his arm outstretched.

Later, standards became a little more accurate. King Henry VII had an iron bar made that was to be used as the standard yard. Today, most governments have agencies that maintain and establish material standards for common measurements. In the United States, this is the responsibility of the National Bureau of Standards of the Department of Commerce. However, because of the historical development and duplication of many units, they are often quite confusing.

The Metric System

The need for a more uniform and convenient system of units led to the development of the metric system, which is now used in most countries around the world.

(a)

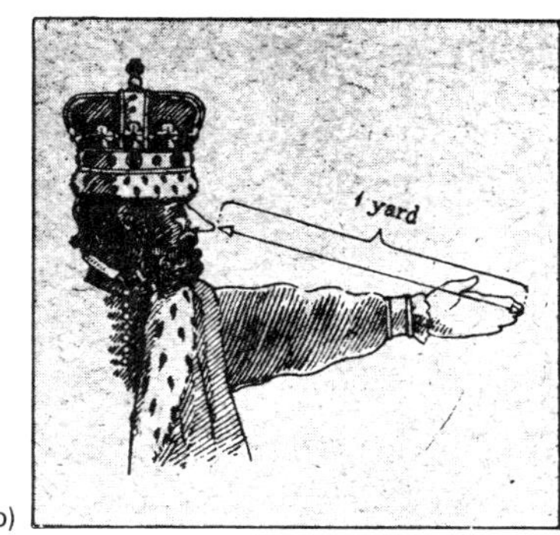

(b)

Figure 1.2 Anatomical units. (a) The inch was once compared to the thumb. (b) The yard was decreed by King Henry I to be the distance from the tip of his royal nose to the end of the fingers of his outstretched arm.

LENGTH

The metric standard length is the meter (see Special Feature 1.1). The meter (abbreviated m) is slightly longer than a yard, in fact, 3.37 inches longer (Fig. 1.4).

With a length standard selected, the next job is to define submultiple and multiple units. (For example, in our British system, 1 foot = 12 inches and 3 feet = 1 yard.) The metric system is a decimal or "base-10" system. That is, larger and smaller units are obtained by

SPECIAL FEATURE 1.1

The Meter

In 1790, in the midst of the French Revolution, the National Assembly of France requested the French Academy of Sciences to "deduce an invariable standard for all the measures and all the weights." The commission appointed by the Academy created a system that was simple and scientific. The name *metre,* which we spell meter, was assigned to the unit of length. This name was derived from the Greek word *metron,* meaning "to measure." The length of the meter was defined as one ten-millionth of the distance along a meridian from the North Pole to the Equator (Fig. 1.3). A portion of a meridian running near Dunkirk in France and Barcelona in Spain was surveyed and the length of a meter determined. Based on these results, a 1-meter bar of platinum was constructed. This bar became the "Meter of the Archives," from which copies were made.

The use of metric weights and measures was legalized in the United States in 1866, and since 1893 the yard has been defined in terms of the meter. Metal bar meter lengths are used for common measurement reference standards, but these lengths are affected by temperature variations. In 1960, the meter was defined in terms of the wavelength of light. In 1983, a new definition was adopted that references the meter to a distance light travels in a vacuum (see Table 1.2).

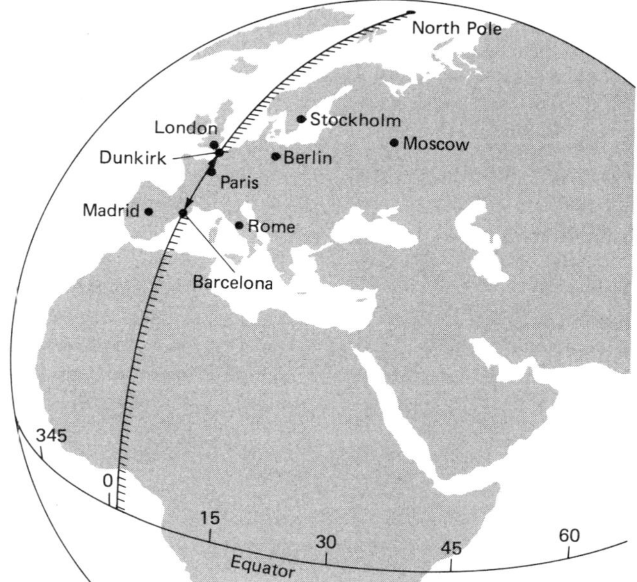

Figure 1.3 Definition of the meter. The meter was originally defined as one ten-millionth of the distance from the North Pole to the Equator along a meridian that ran through France.

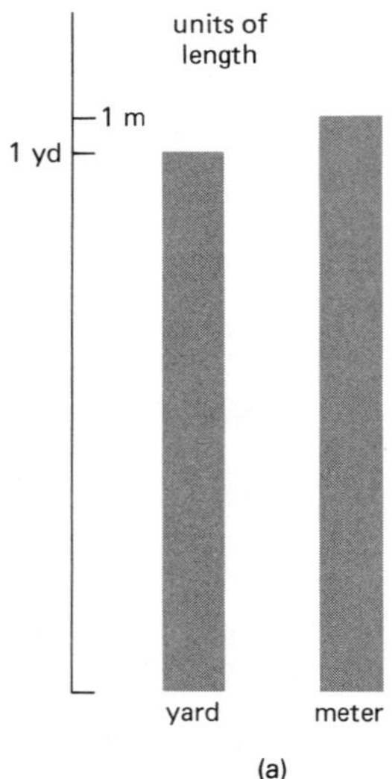

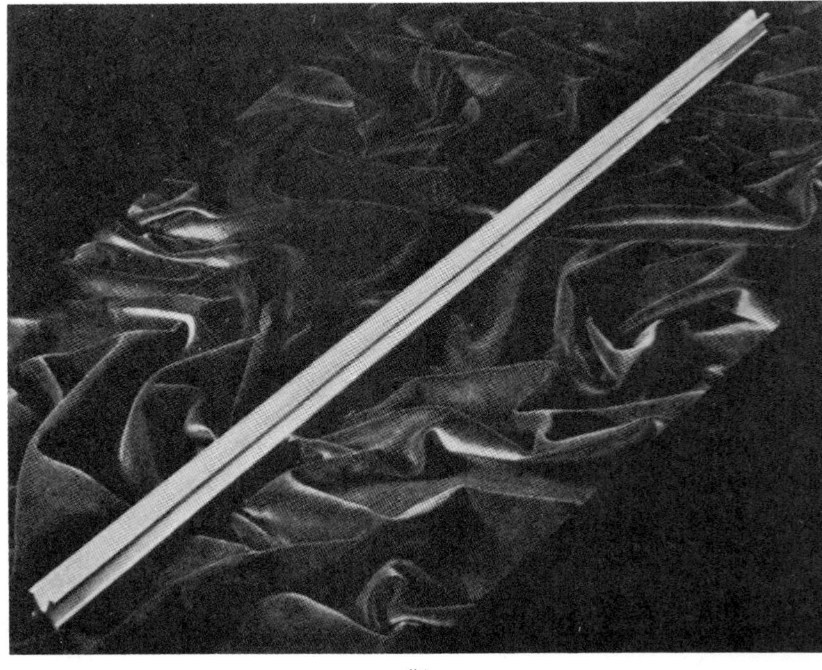

Figure 1.4 The meter and the yard. (a) The meter is slightly longer than a yard (3.37 inches longer). (b) The prototype meter bar that is the United States copy of the Standard Meter (Prototype Meter No. 27). The bar was sent from France in 1890.

multiplying or dividing standard units by factors of ten.

A list of the metric prefixes used to indicate these factors is given in Table 1.1. However, only three prefixes are usually needed to describe everyday measurements of length (see Fig. 1.5):

Milli– 0.001 $\left(\frac{1}{1000}\right)$
1 millimeter (mm) = 0.001 meter
or 1 m = 1000 mm

Centi– 0.01 $\left(\frac{1}{100}\right)$
1 centimeter (cm) = 0.01 meter
or 1 m = 100 cm

Kilo– 1000
1 kilometer (km) = 1000 meters
(pronounced kil-oh-meter)

Let's take a closer look at the decimal base of the metric system, which is one of its great advantages. You are already familiar with a similar decimal system—our money. The dollar is divided into cents, with 100 cents (pennies) making one dollar. If a dollar is comparable to a meter, then a cent or penny is comparable to a centimeter. In fact, we could easily call a penny a "centidollar." For example,

150 cm = 1.50 m and 150 cents = \$1.50 (dollars)

We can carry this comparison one step further. You may have heard how property taxes are assessed in mils. A mil is $\frac{1}{10}$ of a cent, and there are 1000 mils in a dollar. Hence, a millimeter is analogous to a mil or "millidollar."

Notice how much easier it is to convert from one unit to another in the metric decimal system than in the British system. For example, 118 cm can be directly determined to be 1.18 m. However, in our "base-12" (duodecimal) system with 12 inches = 1 ft, finding out the number of feet in 118 inches requires a more involved calculation: 118 in./12 = 9.83 ft. (We divide by 12 since 12 in. = 1 ft.)

Some length conversion factors to go from one system to the other are as follows:

1 inch = 2.540 cm

1 cm = 0.3937 in. (or approximately 0.4 cm)

1 mile = 1.609 km (or approximately 1.6 km)

1 km = 0.6214 mi (or approximately 0.62 mi)

Table 1.1 Metric Prefixes

Prefix (abbreviation)	*Pronounciation**	*Value*	*Meaning*
exa (E)	ex'a (*a* as in *a*bout)	10^{18}	One quintrillion times
peta (P)	as in *peta*l	10^{15}	One quadrillion times
tera (T)	as in *terra*ce	10^{12}	One trillion times
giga (G)	jig' a (*a* as in *a*bout)	10^{9}	One billion times
mega (M)	as in *mega*phone	10^{6}	One million times
kilo (k)	as in *kilo*watt	10^{3}	One thousand times
hecto (h)	heck toe	10^{2}	One hundred times
deka (da)	deck'a (*a* as in *a*bout)	10	Ten times
deci (d)	as in *deci*mal	10^{-1}	One tenth of
centi (c)	as in *senti*ment	10^{-2}	One hundredth of
milli (m)	as in *mili*tary	10^{-3}	One thousandth of
micro (μ)	as in *micro*phone	10^{-6}	One millionth of
nano (n)	nan'oh (*an* as in *an*t)	10^{-9}	One billionth of
pico (p)	peek'oh	10^{-12}	One trillionth of
femto (f)	fem'toe (*fem* as in *fem*inine)	10^{-15}	One quadrillionth of
atto (a)	as in an*ato*my	10^{-18}	One quintrillionth of

* Source: Metric Guide for Educational Materials, American National Metric Council.

The approximate values are convenient for rough calculations.

MASS (WEIGHT) AND VOLUME (CAPACITY)

In the British system, weight units, for example, the pound and the ounce, are commonly used to measure quantities of matter. In the metric system, mass units, such as the kilogram and gram, are used. Mass is the fundamental property. Weight is the gravitational attraction on a quantity of matter but is not a fundamental property.

Near the surface of the Earth, the gravitational attraction on a particular quantity of matter or on "a mass" gives it a certain weight. On the moon, however, where gravity is $\frac{1}{6}$ that on Earth, the *same* mass would have $\frac{1}{6}$ the weight as on Earth. This example serves to point out the difference between weight and mass, and to show why weight is not a fundamental property. The weight of "a mass" is different wherever gravity is different (see Chapter 5), but the mass or quantity of matter is the same.

To avoid confusion between these terms in commercial and everyday use, the American National Metric Council suggests

> . . . the term "weight" nearly always means mass; the use of the word "weight" to mean "mass" and the word "weigh" to

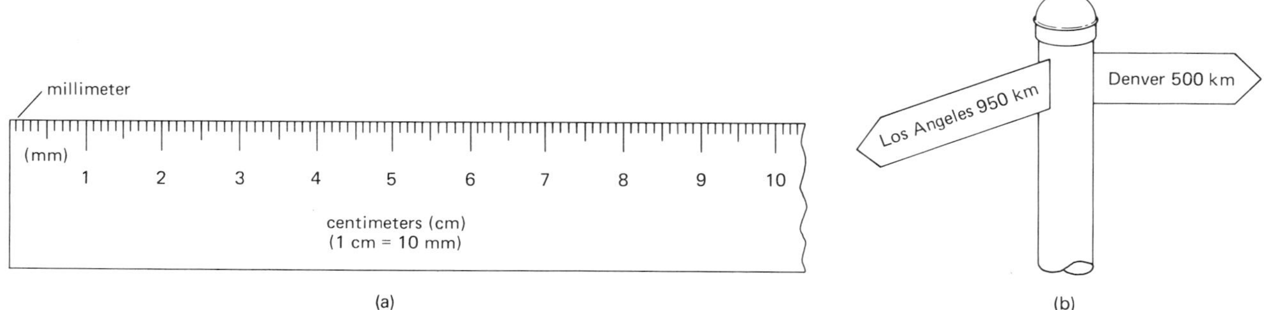

Figure 1.5 Metric prefixes. Only three metric prefixes are needed to describe most everyday measurements: (a) millimeters (mm) and centimeters (cm) and (b) kilometers (km).

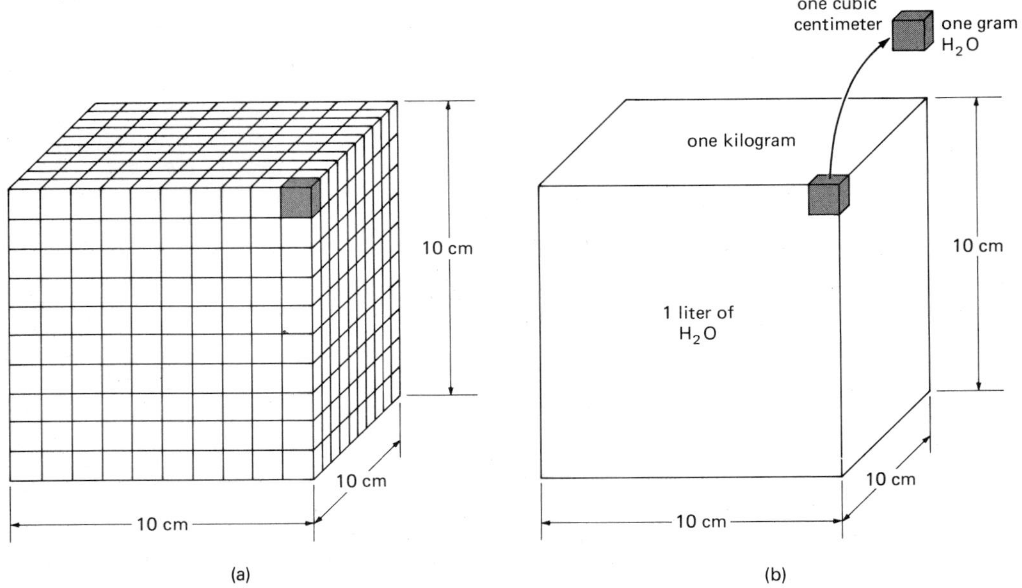

Figure 1.6 Mass units are related to length in the metric system. (a) A cube 10 cm on a side has a volume of 1000 cm^3. (b) The amount of water that fills this volume has a mass of 1 kilogram, and 1 cm^3 of water has a mass of 1 gram. (The volume 1000 cm^3 is defined to be a liter.)

mean "determine the mass of" or "have a mass of" is acceptable. Examples: My weight is 60 kilograms. Weigh the envelope carefully.†

Weight and mass units can be related to volume or capacity. For example, a British Imperial (or Canadian) gallon of water weighs 10 pounds. With 8 pints to the gallon, it follows that one British pint of water weighs 1.25 pounds. However, a smaller U.S. gallon of water weighs 8.327 pounds, and one U.S. pint of water weighs 1.041 pounds.

The quantity of water in a particular metric volume was originally used to define the standard metric mass unit. A container 10 cm on a side has a volume of 10 cm × 10 cm × 10 cm = 1000 cm^3 (cubic centimeter, sometimes abbreviated cc). See Figure 1.6. Filling the container with water, the mass of this quantity of water (1000 cm^3) was defined to be 1 kilogram (kg).‡

Since the metric prefix "kilo" means 1000, it follows that 1 kg = 1000 grams (g), and one cubic centimeter of water has a mass of 1 gram. The gram unit is often divided into milligrams (1 mg = 0.001 g or 1 g = 1000 mg), which is a convenient unit for small quantities. It is interesting to note that the carat unit used for precious stones is really a metric unit, with 1 carat = 200 milligrams. Hence, a 2-carat diamond weighs 400 mg.§ For larger mass quantities, a metric ton is defined to be 1000 kg.

The relationship between the kilogram and pound units is that

> 1 kg mass has a weight of 2.2 pounds (lb) on the surface of the earth.

This relationship is illustrated in Figure 1.7, along with the metal prototype kilogram standard.

The metric unit of volume or capacity is the volume used in defining the kilogram. A volume of 1000 cm^3 is defined to be 1 liter. The abbreviation for liter is either a small "ell" (l) or a capital L. The latter is generally preferred in the United States so as to avoid confusion with the numeral one. A liter is slightly larger than a U.S. quart (Fig. 1.8):

† *Metric Editorial Guide*, 3rd edition, section 7.1. National Metric Council, 1625 Massachusetts Avenue, NW, Washington, D.C., 20036.

‡ At standard atmospheric pressure and at the temperature of the maximum density of water (4°C). See Chapter 10.

§ The carat unit is also "decimalized" into points, and 1 carat = 100 points. A 50-point diamond is then ½ carat, or 100 mg. Another type of karat (with a k) is used to indicate the purity of gold. Pure gold is 24-karat. A piece of 12-karat gold jewelry has 50 percent gold and 50 percent of some other metal. There is also another carrot, but you are familiar with this one.

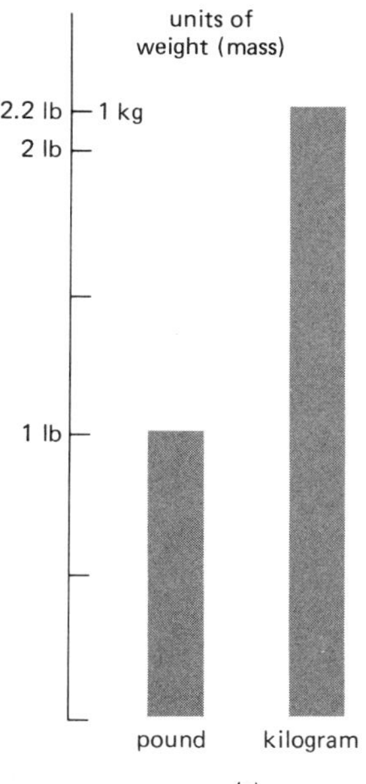

Figure 1.7 The kilogram and the pound. (a) The metric kilogram unit of mass has an equivalent weight of 2.2 lb. (b) The prototype kilogram standard cylinder (Prototype Kilogram No. 20) is the national standard mass for the United States.

$$1 \text{ liter} = 1.056 \text{ quart}$$

A common smaller unit of volume is the milliliter (mL). From the previous discussion, we know that

$$1 \text{ liter} = 1000 \text{ cc} = 1000 \text{ mL}$$

and

$$1 \text{ cc} = 1 \text{ mL}$$

TIME

The standard unit of time may give you some relief. This is the second (s) in both the British and the metric systems. Two events define an interval or unit of time, similar to two marks or positions defining a length interval or unit. (Notice how we often might say, for example, that class periods are 50 minutes *long* or 50 minutes *in length.*)

Prior to 1956, the apparent daily motion of the Sun was used to define the second. A solar day is the time that elapses between two successive crossings of the same meridian by the Sun. A second was defined as $\frac{1}{86,400}$ of this *apparent* solar day. (One day = 24 hours = 1440 minutes = 86,400 seconds.)

However, the Earth's motions vary slightly, and all solar days are not the same. To remedy this, an average or mean solar day was computed from the apparent solar days during a one-year period, and this average was used as the length of a solar day.

Even so, the mean solar day is not exactly the same in each yearly period, owing to minor variations in the Earth's motions plus a steady slowing of the Earth's rate of rotation due to tidal friction. An atomic standard is now used for a more precise reference (see Table 1.2). But at any rate, the second is the standard unit of time for all common measurement systems.

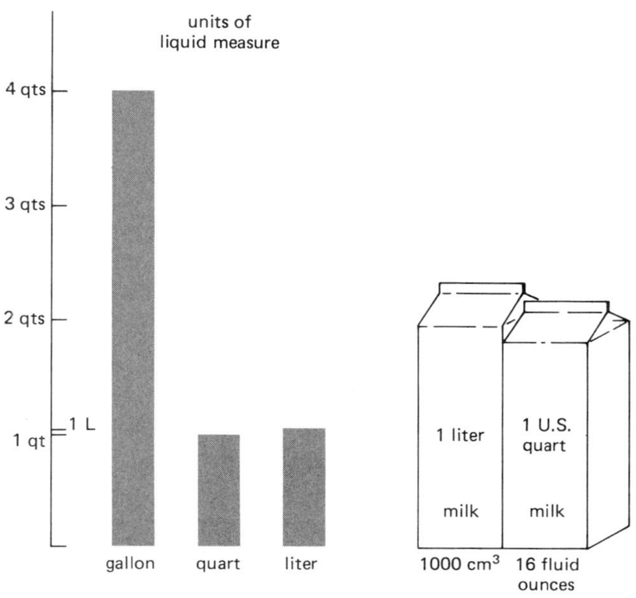

Figure 1.8 The liter and the quart. The liter is slightly larger than the quart (1 liter = 1.056 quart).

Table 1.2 The Modernized Metric System

Name of unit (abbreviation)	*Property measured*	*Definition*	*Further information*
		SEVEN BASE UNITS	
meter (m) 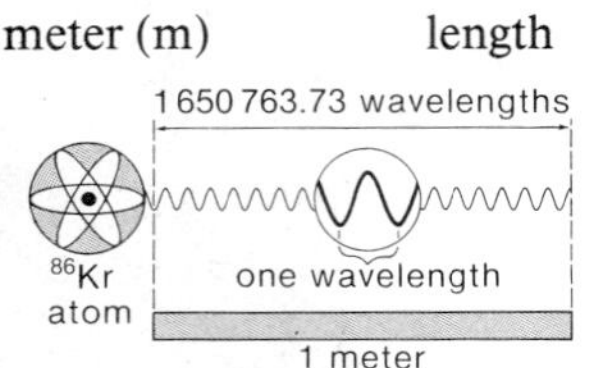 An interferometer is used to measure length by means of light waves	length	The meter is defined as 1 650 763.73 wavelengths in vacuum of the orange-red line of the spectrum of krypton-86. Flash! In October 1983 the General Conference on Weights and Measures meeting in Paris adopted a new definition for the meter. The new method defines the meter as the distance traveled by light in a vacuum during $\frac{1}{299,792,458}$ of a second. Looking at it another way, the new definition establishes the speed of light as 299,792,458 meters per second. Using time, our most accurate measurement, to define length is considered to be a real breakthrough.	The SI unit of area is the **square meter** (m^2). The SI unit of volume is the **cubic meter** (m^3). The liter (0.001 cubic meter), although not an SI unit, is commonly used to measure fluid volume.
kilogram (kg) 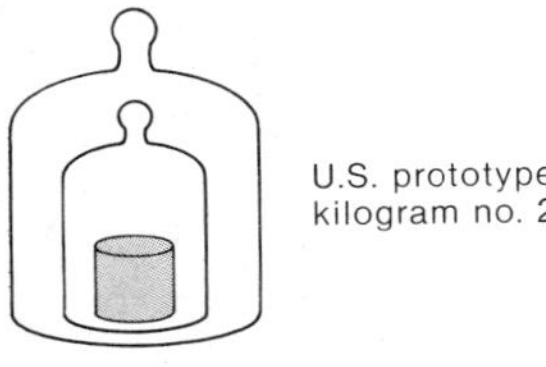 U.S. prototype kilogram no. 20	mass	The standard for the unit of mass,* the kilogram, is a cylinder of platinum-iridium alloy kept by the international Bureau of Weights and Measures at Paris. A duplicate in the custody of the National Bureau of Standards serves as the mass standard for the United States. This is the only base unit still defined by an artifact.	The SI unit of force is the **newton** (N). One newton is the force which, when applied to a 1-kg mass, will give the mass an acceleration of 1 (meter per second) per second. $1\ N = 1\ kg \cdot m/s^2$ 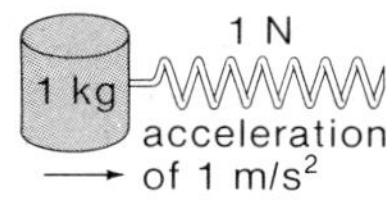 The SI unit for pressure is the **pascal** (Pa). $1\ Pa = 1\ N/m^2$ The SI unit for work and energy of any kind is the **joule** (J). $1\ J = 1\ N \cdot m$ The SI unit for power of any kind is the **watt** (W). $1\ W = 1\ J/s$

* "Weight" is a commonly used term for "mass."

Table 1.2 The Modernized Metric System (*Continued*)

Name of unit (abbreviation)	*Property measured*	*Definition*	*Further information*
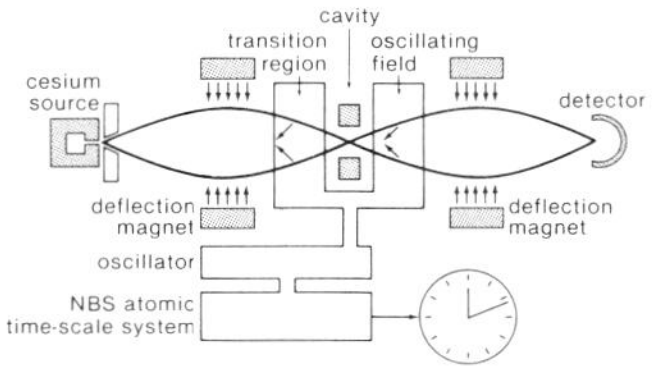 second (s)	time	The second is defined as the duration of 9 192 631 770 cycles of the radiation associated with a specified transition of the cesium-133 atom. It is realized by tuning an oscillator to the resonance frequency of cesium-133 atoms as they pass through a system of magnets and a resonant cavity into a detector.	The number of periods or cycles per second is called *frequency*. The SI unit for frequency is the **hertz** (Hz). One hertz equals one cycle per second. The SI unit speed is the **meter per second** (m/s). The SI unit for acceleration is the **(meter per second) per second** (m/s^2).
Schematic diagram of an atomic beam spectrometer or "clock." Only those atoms whose magnetic moments are "flipped" in the transition region reach the detector. When 9 192 631 770 oscillations have occurred, the clock indicates one second has passed.			
ampere (A) 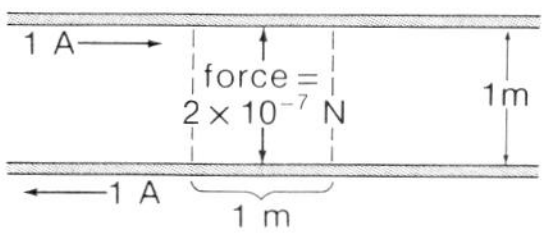	electric current	The ampere is defined as that current which, if maintained in each of two long parallel wires separated by one meter in free space, would produce a force between the two wires (due to their magnetic fields) of 2×10^{-7} newton for each meter of length.	The SI unit of voltage is the **volt** (V). $1\ V = 1\ W/A$ The SI unit of electric resistance is the **ohm** (Ω). $1\Omega = 1\ V/A$
kelvin (K) 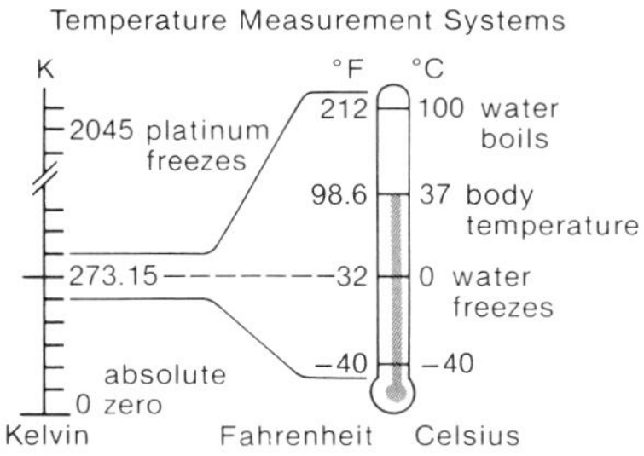	temperature	The kelvin is defined as the fraction 1/273.15 of the thermodynamic temperature of the triple point of water. The temperature 0 K is called "absolute zero."	On the commonly used Celsius temperature scale, water frezes at about 0°C and boils at about 100°C. The °C is defined as an interval of 1 K, and the Celsius temperature 0°C is defined as 273.15 K. 1.8 Fahrenheit degrees are equal to 1.0°C or 1.0 K; the Fahrenheit scale uses 32°F as a temperature corresponding to 0°C.
Temperature Measurement Systems			
mole (mol) 	amount of substance	The mole is the amount of substance of a system that contains as many elementary entities as there are atoms in 0.012 kg of carbon-12.	When the mole is used, the elementary entities must be specified and may be atoms, molecules, ions, electrons, other particles, or specified groups of such particles. The SI unit of concentration (of amount of substance) is the **mole per cubic meter** (mol/m^3).
candela (cd) 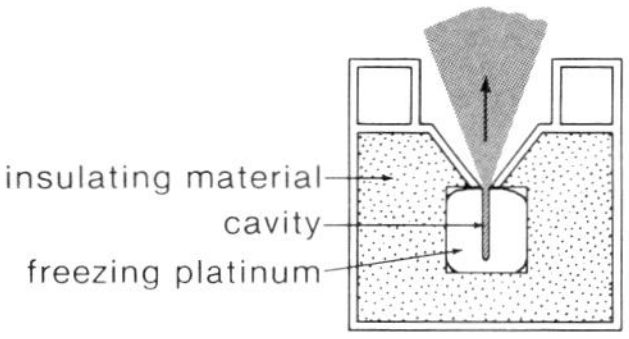	luminous intensity	The candela is defined as the luminous intensity of 1/600 000 of a square meter of a blackbody at the temperature of freezing platinium (2045 K).	The SI unit of light flux is the **lumen** (lm). A source having an intensity of one candela in all directions radiates a light flux of 4π lumens.  A 100-watt light bulb emits about 1700 lumens

(Continued on p. 12)

Table 1.2 The Modernized Metric System (*Continued*)

Name of unit (abbreviation)	*Property measured*	*Definition*	*Further information*
		TWO SUPPLEMENTARY UNITS	
radian (rad)	plane angle	The radian is the plane angle with its vertex at the center of a circle that is subtended by an arc equal in length to the radius.	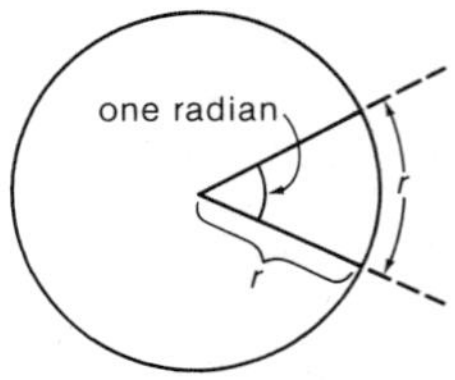 one radian; r; r
steradian (sr)	solid angle	The steradian is the solid angle with its vertex at the center of a sphere that is subtended by an area of the spherical surface equal to that of a square with sides equal in length to the radius.	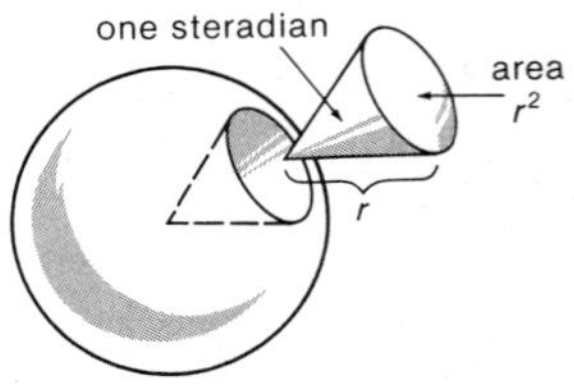 one steradian; area r^2; r

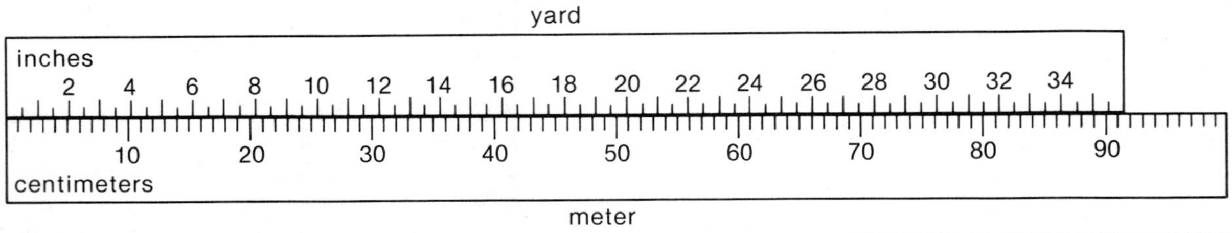

The SI System

You may have heard people refer to the SI system and wondered how this differs from the metric system. The SI system is the *modernized* version of the metric system established by international agreement. (See Special Feature 1.2.)

As may be seen from Table 1.2, the SI system consists of seven base units and two supplementary units. The most common measurements involve length, mass, time, electric current, and temperature. Let's take a closer look at temperature, since this is used every day. The standard SI unit of temperature is the **kelvin** (K, *not* degrees kelvin).‖ The Kelvin temperature scale is a scientific one. It has its zero temperature at *absolute* zero. There can be no lower temperatures, and so there are no negative temperature readings on the Kelvin scale. (See Chapter 13.)

Although the kelvin is the standard SI unit of temperature, ordinary temperature measurements are made on the Celsius scale in **degrees Celsius** (°C). This scale was formerly referred to as the centigrade scale (Latin *centi* meaning, "one hundred," and German *grade* meaning, "degrees"), because there are 100 degrees on this scale between the freezing point (0°C) and the boiling point (100°C) of water.

Figure 1.9 shows a comparison of the common temperature scales, including the Fahrenheit scale in **degrees Fahrenheit** (°F). Notice that a *change* of 1 degree Celsius is equal to a change of 1 kelvin. A degree interval on these scales is almost twice as large as a degree interval on the Fahrenheit scale: 1 K = 1°C = 1.8°F, and 1°F = 0.56°C. Hence, the Fahrenheit scale is a bit finer.

‖ After Lord Kelvin, a British scientist who originated the temperature scale. When a unit is named after a person, it is customary to capitalize the abbrevation, e.g., kelvin (K), but meter (m).

SPECIAL FEATURE 1.2

The SI System

The SI system is the *modernized* version of the metric system established by international agreement. SI stands for the International System of Units (from the French, *Le Système International d'Unités*). It was initiated by a request from the International Union of Pure and Applied Physics and various other scientific organizations for the adoption of a practical international system of units. After a great deal of work, the international General Conference on Weights and Measures (which resulted from the 1875 Treaty of the Meter) adopted in 1960 six base units—meter, kilogram, second, ampere, kelvin, and candela. In 1971, a seventh base unit for the amount of substance, the mole, was added (see Table 1.2). According to our National Bureau of Standards, this system "provides a logical and interconnected framework for all measurements in science, industry, and commerce."

We often hear the temperature given in both degrees Celsius and degrees Fahrenheit on the radio and TV. If we adopt the Celsius scale exclusively, there will be a few important temperatures you should know (see also Fig. 1.10):

Important common temperatures

100°C	(212°F)	Boiling point of water
37°C	(98.6°F)	Normal human body temperature
20°C	(68°F)	Comfortable room temperature
0°C	(32°F)	Freezing point of water

What the Metric System Will Mean to You

As the United States generally goes on the metric system, we will have to become familiar with some new units in our everyday activities. The common temperatures discussed previously are good examples. To help people "think metric," the National Bureau of Standards has published a flyer, *All You Will Need To Know About Metric (For Your Everyday Life),* as shown in Figure 1.11.

To help relate the common metric units to our customary units, one might think:

> A *meter* is a yard plus a little extra (3.37 in.).
> A *kilogram* weighs 2 pounds plus a little extra (0.2 lb).
> A *liter* is a quart plus a little extra (0.056 qt).

Of course, once the official conversion is made, we will only have to think in SI units. In the meantime, some basic conversion factors are listed in Table 1.3.

Let's take a look at what you might encounter in everyday measurements. Length measurements will be made in meters. Carpet will be purchased in square meters rather than in square yards. The length of a

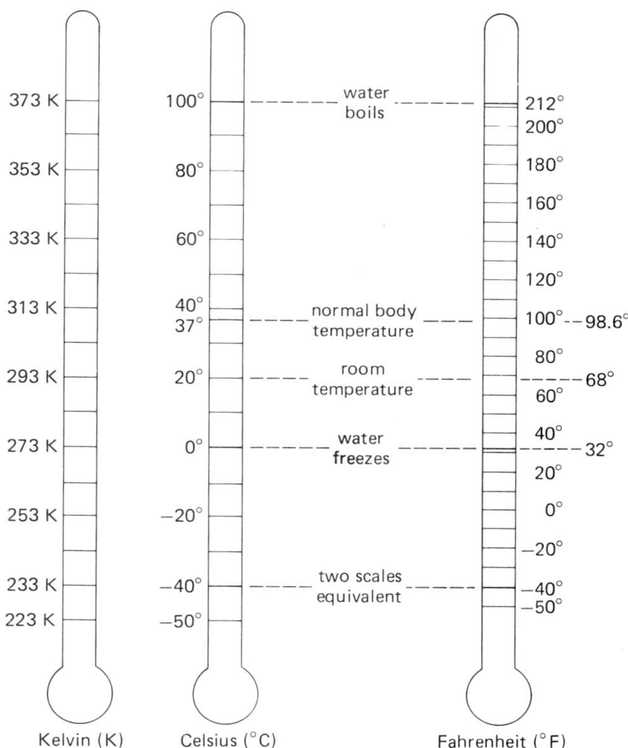

Figure 1.9 A comparison of temperature scales. A change of one degree on the Kelvin or Celsius scales is equivalent to a change of 1.8 degrees on the Fahrenheit scale.

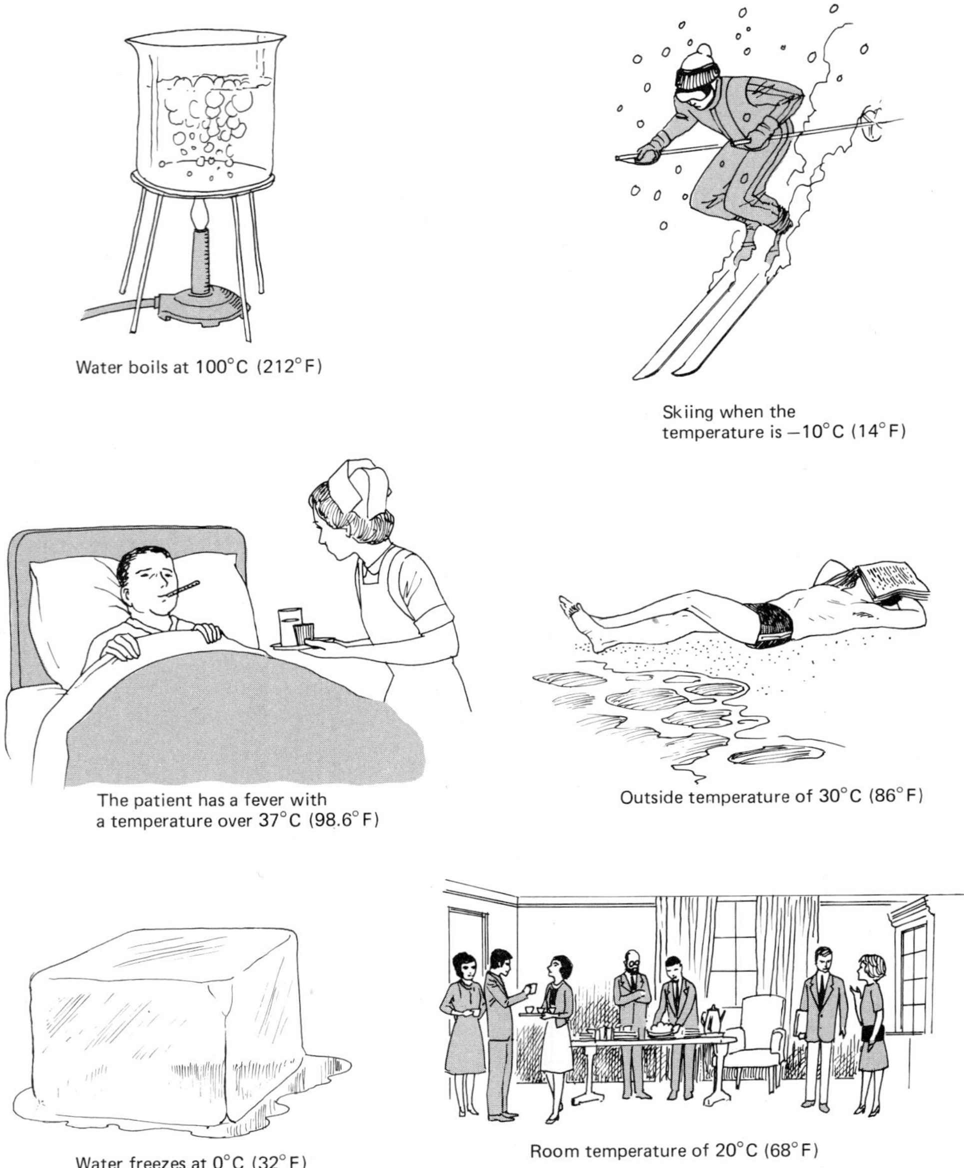

Water boils at 100°C (212°F)

Skiing when the temperature is −10°C (14°F)

The patient has a fever with a temperature over 37°C (98.6°F)

Outside temperature of 30°C (86°F)

Water freezes at 0°C (32°F)

Room temperature of 20°C (68°F)

Figure 1.10 Some important Celsius temperatures.

football field (100 yards) is 91.44 meters. Perhaps this will be changed to 100 meters (109.3 yd). You might then hear, "It's second down on the 40-meter line with 8 meters to go." In some instances, sports lengths are already metric. For example, Olympic measurements in sports are all metric. The Olympic-size pool (50 m long) is now common for swimming competition, and the 100-meter dash has replaced the 100-yard dash.

One of the most noticed conversions would be highway distances and speeds. A road sign would show a town to be 16 km away rather than 10 miles. (See chapter introduction photo.) The speed limit signs and our auto speedometers would be in km/h (kilometer per hour). As illustrated in Figure 1.12, the 55 mi/h (mph) speed limit would probably be posted as 90 kph (55.9 mph). Almost all newer cars in the United States have both mi/h and km/h scales on the speedometers. Check your speed in km/h the next time you drive one.

All You Will Need to Know About Metric
(For Your Everyday Life)

Metric is based on Decimal system

The metric system is simple to learn. For use in your everyday life you will need to know only ten units. You will also need to get used to a few new temperatures. Of course, there are other units which most persons will not need to learn. There are even some metric units with which you are already familiar: those for time and electricity are the same as you use now.

BASIC UNITS

METER: a little longer than a yard (about 1.1 yards)
LITER: a little larger than a quart (about 1.06 quarts)
GRAM: a little more than the weight of a paper clip

(comparative sizes are shown)

1 METER

1 YARD

25 DEGREES FAHRENHEIT

COMMON PREFIXES
(to be used with basic units)

milli: one-thousandth (0.001)
centi: one-hundredth (0.01)
kilo: one-thousand times (1000)

For example:
1000 millimeters = 1 meter
100 centimeters = 1 meter
1000 meters = 1 kilometer

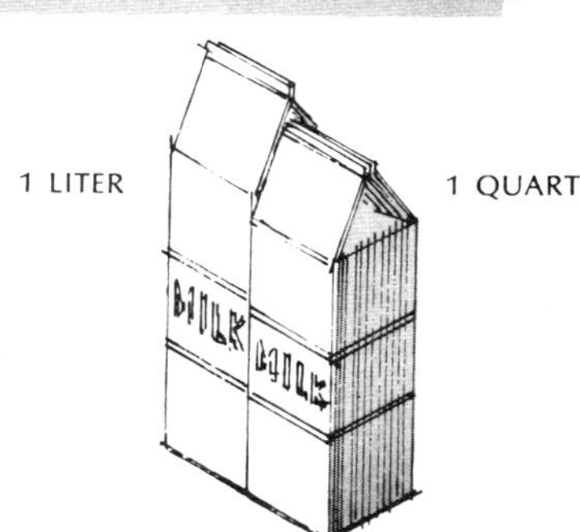

OTHER COMMONLY USED UNITS

millimeter:	0.001 meter	diameter of paper clip wire
centimeter:	0.01 meter	a little more than the width of a paper clip (about 0.4 inch)
kilometer:	1000 meters	somewhat further than ½ mile (about 0.6 mile)
kilogram:	1000 grams	a little more than 2 pounds (about 2.2 pounds)
milliliter:	0.001 liter	five of them make a teaspoon

OTHER USEFUL UNITS

hectare: about 2½ acres
metric ton: about one ton

25 DEGREES CELSIUS

1 POUND

WEATHER UNITS:

FOR TEMPERATURE
degrees Celsius

FOR PRESSURE
kilopascals are used
100 kilopascals = 29.5 inches of Hg (14.5 psi)

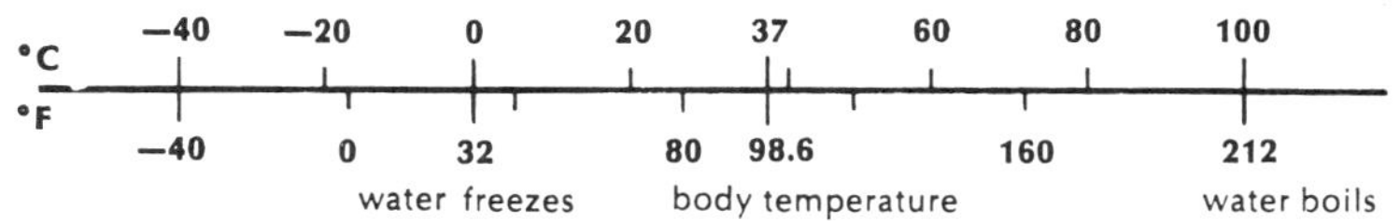

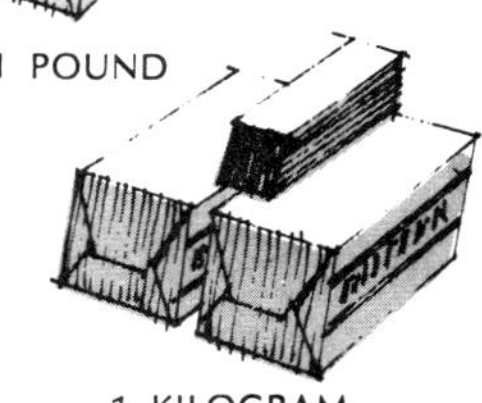

1 KILOGRAM

Figure 1.11 A flyer to help people "think metric."

Table 1.3 Basic Conversion Factors

Metric	*British*
1 millimeter	0.0394 inch
1 centimeter	0.394 inch
1 meter	1.094 yards
1 meter	39.37 inches
1 kilometer	0.62 mile
1 gram (weighs)	0.035 ounce
1 kilogram (weighs)	2.2 pounds
1 liter	1.06 quarts
British	***Metric***
1 inch	25.4 millimeters
1 inch	2.54 centimeters
1 foot	0.3048 meter
1 yard	0.914 meter
1 mile	1.61 kilometers
1 ounce (has a mass of)	28.35 grams
1 pound (has a mass of)	0.45 kilogram
1 quart	0.94 liter

How about weight (mass) and volume? These too would change; many manufacturers are already putting the metric equivalents on their products (Fig. 1.13). We have metric mass advertisements already. You hear commercials for tablets that have "1000 mg of pain reliever." (One thousand milligrams sounds a lot bigger than one gram, doesn't it?)

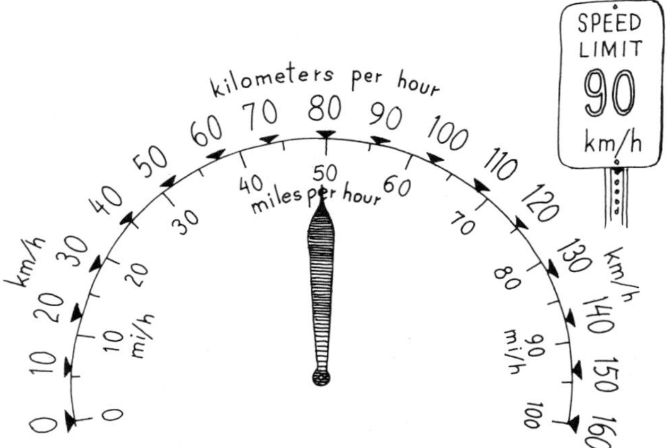

Figure 1.12 Getting used to metric. After conversion to the metric system, our highway signs would be in km/h. The 55 mi/h speed limit would probably be replaced by a 90 km/h (55.9 mi/h) speed limit.

(a)

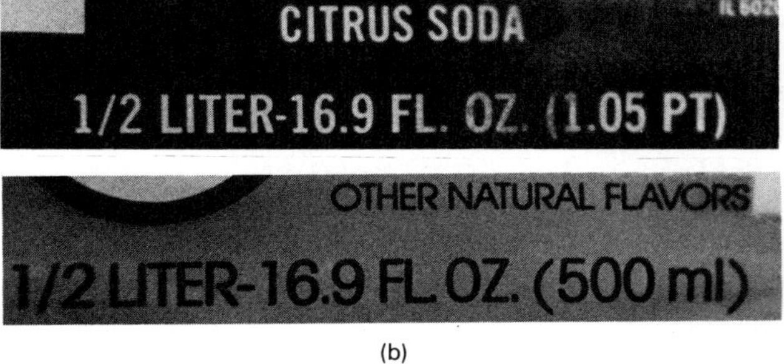

(b)

Figure 1.13 Evidence of a metric change. (a) Weight-mass comparisons on commercial items. (b) Volume comparisons on soft-drink labels. Notice in (b) that the metric unit is listed first.

Figure 1.14 Gasoline would be sold in liters; it already is in some places.

Unit prices would be quoted in price "per 100 grams" or price "per kg or 'kilo'." Something selling at \$1.00/lb would be \$2.20/kg. (Recall that 1 kg of mass weighs 2.2 lb.)

Even our own body weights would be expressed in kilograms. A 160-pound person would weigh (have a mass of) 73 "kilos" in the metric system.

Volume or capacity measurement would be expressed in liters. Some conversion to this unit has already taken place. It is common to buy $\frac{1}{2}$-liter (500 ml) and 2-liter bottles of soda in the store (Fig. 1.13). Notice that the 500-ml bottle is replacing the common 16-oz bottle. We're on our way.

Another use of the liter occurred when the price of gasoline was soaring. Some stations sold gasoline in liters (and some still do). If you are used to buying 10 gallons (37.85 liters) of gasoline, you would probably ask for 40 liters instead (Fig. 1.14).

So, with all of this information, are you ready to go metric? We may be forced to. The United States is becoming an island in a metric world (Fig. 1.15). Even

CANADA
UNITED STATES
Atlantic Ocean
Pacific Ocean
SOUTH AMERICA
EUROPE
ASIA
AFRICA
Indian Ocean
AUSTRALIA
Antarctic Ocean

Metric, Pre World War II

Metric or Committed to Metric, Post World War II

Uncommitted

Barbados	Jamaica	Nauru	Trinidad
Burma	Liberia	Sierra Leone	United States
Gambia	Muscat and Oman	Southern Yemen	
Ghana		Tonga	

Figure 1.15 Islands in a metric world. The map shows the few countries that are uncommitted to, or have not officially adopted, the metric system.

so, there are pros and cons in the metric conversion controversy. These are based on familiarity, general opposition to change, economics and world trade, and practicality. Think about all these attributes carefully. You may someday be asked to decide whether or not we should "go metric" by voting on the issue.

SUMMARY OF KEY TERMS

Customary system the predominant measurement system used in a country. In the United States, the customary system is the British or English system.

Decimal system a "base-10" system in which converting from one unit to the next you either multiply or divide by 10.

Fundamental properties the simplest quantities used to describe nature (length, mass, time, and electric charge).

Measurement a quantitative description of one or more fundamental properties compared to a standard.

SI system the International System of Units, which is the modernized version of the metric system and consists of seven base units and two supplementary units.

Standard unit a popularly or legally accepted unit that has a fixed and reproducible value for the purpose of taking accurate measurements.

EXERCISES

1. The hand is a unit to measure the height of horses. (1 hand = 4 inches). Speculate on the origin of this unit.
2. Some anatomical units of measurement are shown in Figure 1.16. Estimate the equivalent lengths in metric units.
3. A dime would be given what name in the "centidollar" system? (Use a metric prefix.)
4. Which is longer, the 50-yard dash or the 50-meter dash?
5. If a runner ran the 100-yard dash and the 100-meter dash both in the same time, in which event would the runner be running faster?
6. What common object has a mass of about one gram?
7. We say that a class period is 50 minutes *long.* Does this imply a length? Explain.
8. Explain why it is said that originally metric mass units were related to length.
9. The volume of automobile cylinders is sometimes given in cubic inches. What would replace this unit in the metric system?
10. Explain why (a) one cubic centimeter of water has a mass of one gram and (b) why 1 cc = 1 ml.
11. Which is greater, one pound or 0.50 kilogram of steak? Explain.
12. A homemaker ordinarily plans a meal with $\frac{1}{4}$ lb of meat per serving. Twelve people will be at a meal on Sunday. However, the butcher shop just went metric. About how many kilos of meat should be purchased (Fig. 1.17)?
13. Would it be possible to walk 2400 meters in 30 minutes? Explain.
14. (a) Explain why measuring recipe ingredients in the metric system would be similar to the customary system.
    (b) How would baking instructions differ in the metric system? Would there be any difficulty with this?
15. Which is greater and by how much: a metric ton (sometimes written tonne) or a customary ton (2000 lb)?
16. At a metric weight-watchers club, how would each of the following weight losses be reported?
    (a) 22 lb, (b) 11 lb, (c) 2.2 lb, (d) 1.1 lb, (e) 0.55 lb
17. At what temperatures does water freeze and boil in the SI system? If the room temperature were 15°C, how would you feel?
18. On the morning weather report it is announced that the high temperature that day is expected to be 28°C. How would you dress?
19. Give your approximate height and mass in metric units.

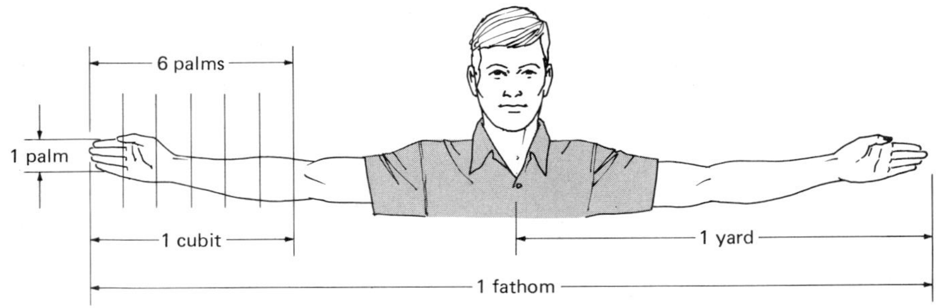

Figure 1.16 See Exercise 2.

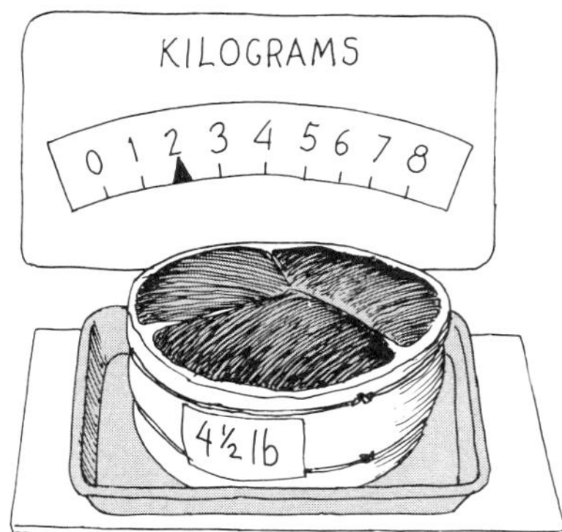

Figure 1.17 See Exercise 12.

Figure 1.18 See Exercise 22.

20. Explain whether each of the following is likely or unlikely:
   (a) One student is 20 cm taller than another student.
   (b) A dog's tail is 300 mm long.
   (c) The height of a room ceiling is 2 meters.
   (d) The normal speed on an interstate highway is 60 kph.
   (e) The floor area of a room in a house is 15 m^2.

21. How would the following popular expressions change when converted to the metric system?
   (a) I wouldn't touch that with a 10-foot pole.
   (b) The cowboy wore a 10-gallon hat.
   (c) Give him an inch and he'll take a mile.
   (d) This thing must weigh a ton.

22. Noah, of Biblical fame, was instructed to build an Ark 300 cubits long, 50 cubits wide, and 30 cubits high (Fig. 1.18). What were the approximate dimensions and volume of the Ark in the metric system? (See Exercise 2.)

2

A Restless World: Motion, Force, and Newton's Laws

Look around you. It is difficult to observe a scene in which something is not moving. Thinking on a bigger scale, recall that you are moving through space as you read this book — the Earth is rotating on its axis and revolving about the Sun.

The study of motion has intrigued scientists since early times. The Greeks, in particular Aristotle, put forth theories on motion that were not always correct. Other great scientists, such as Galileo and Newton, studied motion and helped formulate our present concepts. Isaac Newton set forth three laws of motion, as well as several other principles that will be studied later in the text. (See Special Feature 2.1.)

Motion

You may think that the description of motion would be quite simple. But suppose you were asked, "What is motion?" Could you give an answer? A typical reply is that something is in motion when it is moving. This is true, but does it really define motion?

You are on the right track, however. Moving involves a change of position, and it quickly becomes clear that:

> Motion is the process of a change in position.

SPECIAL FEATURE 2.1

Isaac Newton

On Christmas Day, 1642 Isaac Newton was born in Woolsthrope in Lincolnshire, England. His father had died three months earlier. When he was 14 years old, his mother became widowed for a second time and brought Isaac home from school to help run the family farm. He proved to be a lackadaisical farmer, being more occupied with mathematics than farm chores.

At the age of 18, he entered Trinity College at Cambridge and received his degree four years later in 1665. Later that year, the spread of the Great Plague caused the university to close. Newton returned home, and during the next 18 months he conceived most of the ideas for his famous discoveries in science and mathematics. Chief among these were the development of calculus mathematics and studies on light and color, motion, and gravitation. As Newton later described this period,

> I was in the prime of my age for invention, and minded Mathematics and Philosophy (science) more than at any time since.

After the plague had passed, he returned to Cambridge and at the age of 26 was appointed as professor of mathematics. Newton's early published works were in the field of optics. He developed a new type of telescope, a reflecting telescope, that used a mirror rather than a lens to collect light (see Chapter 25). His most notable book, *Principia Mathematica Philosophiae Naturalis* (Mathematical Principles of Natural Philosophy*), or "Principia" for short, was published in 1687 (Fig. 2.1). The cost of the book was borne by a contemporary, Edmund Halley, who predicted the return of a famous comet that bears his name. In the "Principia" Newton set forth his theories on gravity, tides, and motion.

A bachelor, Newton lived very austerely and is reported to have been the classic absent-minded professor, being so absorbed in his work that he forgot meals and other day-to-day activities. In later life Newton was appointed master of the mint and moved to London in 1701. Queen Anne knighted him in 1705 in recognition of his numerous accomplishments. Early in 1727 he was taken seriously ill and died on March 20 of that year.

An insight into the character of this great scientist is given in one of his statements:

> If I have been able to see farther than some, it is because I have stood on the shoulders of giants.

One of these giants was Galileo. (See Special Feature 2.2.)

* Physics was once called natural philosophy.

(a)

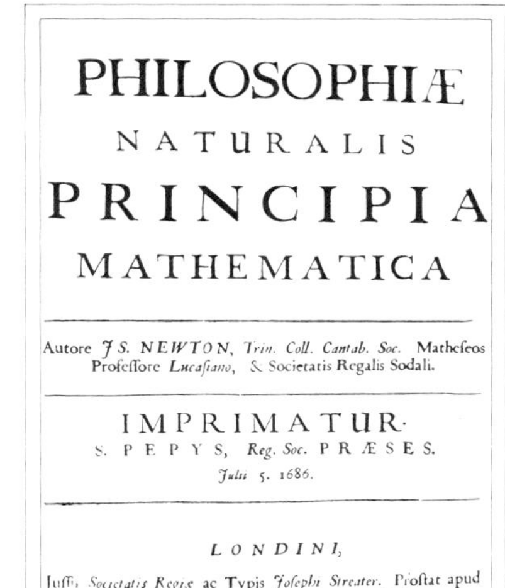
PHILOSOPHIÆ

NATURALIS

PRINCIPIA

MATHEMATICA

Autore *J S.* NEWTON, *Trin. Coll. Cantab. Soc.* Matheſeos Profeſſore *Lucaſiano*, & Societatis Regalis Sodali.

IMPRIMATUR.

S. PEPYS, *Reg. Soc.* PRÆSES.

Julii 5. 1686.

LONDINI,

Juſſu *Societatis Regiæ* ac Typis *Joſephi Streater*. Proſtat apud plures Bibliopolas. *Anno* MDCLXXXVII.

(b)

Figure 2.1 (a) Sir Isaac Newton (1642–1727), one of the greatest physicists of all time. (b) The title page of Newton's *Principia.* Notice that the name of the famous diarist Samuel Pepys appears near the middle of the page. Pepys was president of the Royal Society that sponsored the book. Can you read the date in Roman numerals at the bottom?

Figure 2.2 Motion is the process of a change in position. The book on the table has changed positions, so it was moved or was in motion. The description of motion involves length (distance) and time.

If you put your book on the desk and come back later to find it in another place, you know that motion occurred (Fig. 2.2). To describe motion, we use combinations of the fundamental properties of length and time. There are three quantities: speed, velocity, and acceleration.

Speed

Suppose you are in a car and it is in motion. If someone asks how fast you are moving, you might look at your *speed*ometer and say 55 miles per hour (mi/h), or about 90 kilometers per hour (km/h). This is how fast you are traveling almost instantaneously, so for practical purposes the speedometer expresses what we call instantaneous speed. **Instantaneous speed** is the speed of an object right then or at a particular instant of time.

More generally, speed is expressed by the change in position or distance traveled divided by the time required to make the change or to travel that distance:

$$\text{Speed} = \frac{\text{change in position (distance traveled)}}{\text{time to make change (elapsed time)}}$$

For example, if you drove 100 km in 2 hours, your speed would be 100 km/2 h = 50 km/h (50 kilometers *per* hour).* Speed then is a *rate,* in particular, the time rate of change of position, or how rapidly distance is covered (Fig. 2.3(a)).

Note that this is an **average speed** taken over an appreciable time interval. That is, on the *average* you traveled 50 km each hour. However, on a trip in your car there is usually a variation in speed (stopping, starting, and so on). If, on the other hand, there were no variation, then you would be traveling at a constant or uniform speed. For example, if you drove at a constant speed of 90 km/h on an interstate highway, you would travel 90 km each hour, or cover equal distances in equal intervals of time (Fig. 2.3(b)). Your speedometer would *constantly* read 90 km/h. This is also your *instantaneous* speed—the speed at any particular instant.

* The term *per* essentially means "divided by."

QUESTION: When you are traveling at a constant speed, what is your average speed?

ANSWER: When traveling at a constant speed, the average speed is the same as the constant speed. Let's illustrate this by an analogy. Suppose your class took a test and everyone got a score of 90 points. Then, the class average is 90, the same as the constant or uniform score.

Velocity

We often use the terms *speed* and *velocity* interchangeably. However, there is a difference in physics. Speed tells you how fast you are moving, but not the *direction.* Velocity specifies *both* speed and the direction of the motion.

For example, if you say you are traveling 75 km/h, you are specifying speed. However, if you say you are traveling 75 km/h to the east, then you are specifying velocity. In science, such distinctions are made by referring to *scalar* and *vector* quantities. A **scalar** quantity, like speed, has magnitude or "size" only. A **vector** quantity, like velocity, has both magnitude and direction. Graphically, we often represent a vector quantity by an arrow, where the length of the arrow is proportional to the magnitude and the arrowhead indicates the direction (Fig. 2.4).

QUESTION: If an object travels at a constant speed, does it also have a constant velocity?

ANSWER: Not necessarily. If an object travels with a constant speed in a straight-line path, then

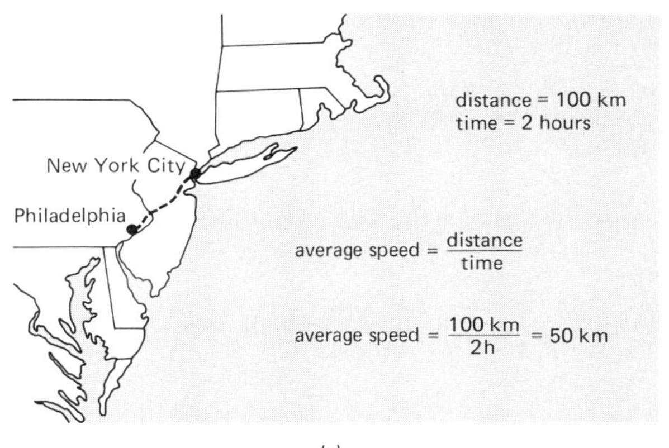

(a)

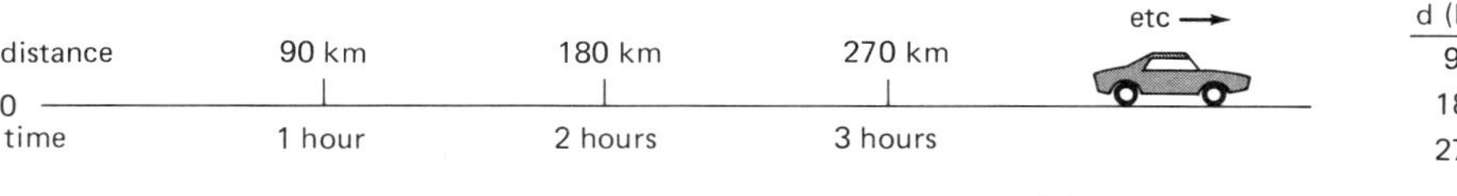

d (km)	t (h)	d/t = speed
90	1	90 km/1 h = 90 km/h
180	2	180 km/2 h = 90 km/h
270	3	270 km/3 h = 90 km/h

(b)

Figure 2.3 Speed. (a) Average speed. The average speed of a trip is the distance traveled divided by the time taken to travel the distance. (b) Constant or uniform speed. When traveling at a constant speed, equal distances are traveled in equal times. A car's speedometer would have a constant reading.

it also has a constant velocity—constant speed and *constant direction.* However, an object may move with a constant speed in a curved path, for example, a car going around a circular track at a constant speed. In this case the velocity isn't constant because the direction of motion is continually changing. As a matter of fact, if the car circles the track once at a constant speed, its average velocity is zero because it has had equal and opposite velocities at each set of points on opposite sides of the track.

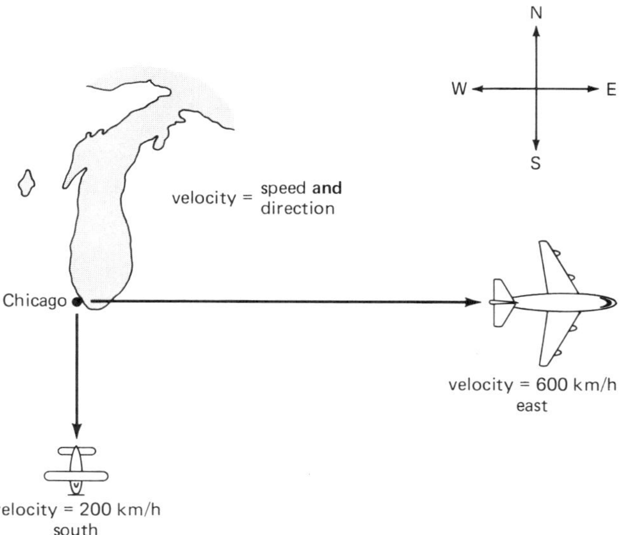

Figure 2.4 A vector quantity has magnitude and direction. It can be represented by an arrow, with the arrowhead indicating the direction and the length of the arrow being proportional to the magnitude, for example, 600 km/h east or 200 km/h south, as illustrated here.

Acceleration

To express a change in velocity, we talk about acceleration. **Acceleration** is defined as the time rate of change of velocity, or

$$\text{Acceleration} = \frac{\text{change in velocity}}{\text{time to make change}} = \frac{\Delta v}{\Delta t}$$

where the Δ symbol means change in or difference, e.g., $\Delta t = t_2 - t_1$. Notice that this is an *average* acceleration taken over an appreciable time interval Δt. As a common example, we refer to the gas pedal on a car as an accelerator. By pushing down on or letting up on the accelerator, the car speeds up or slows down, which is a change in speed and therefore a change in velocity. As a change in velocity, acceleration is also a vector quantity with magnitude and direction.

Suppose you are traveling in your car in a straight-line path at a constant speed of 20 km per hour, and you push on the accelerator so the velocity increases 10 km/h each second—30 km/h, 40 km/h, 50 km/h, and so on. The acceleration, or the rate of change of

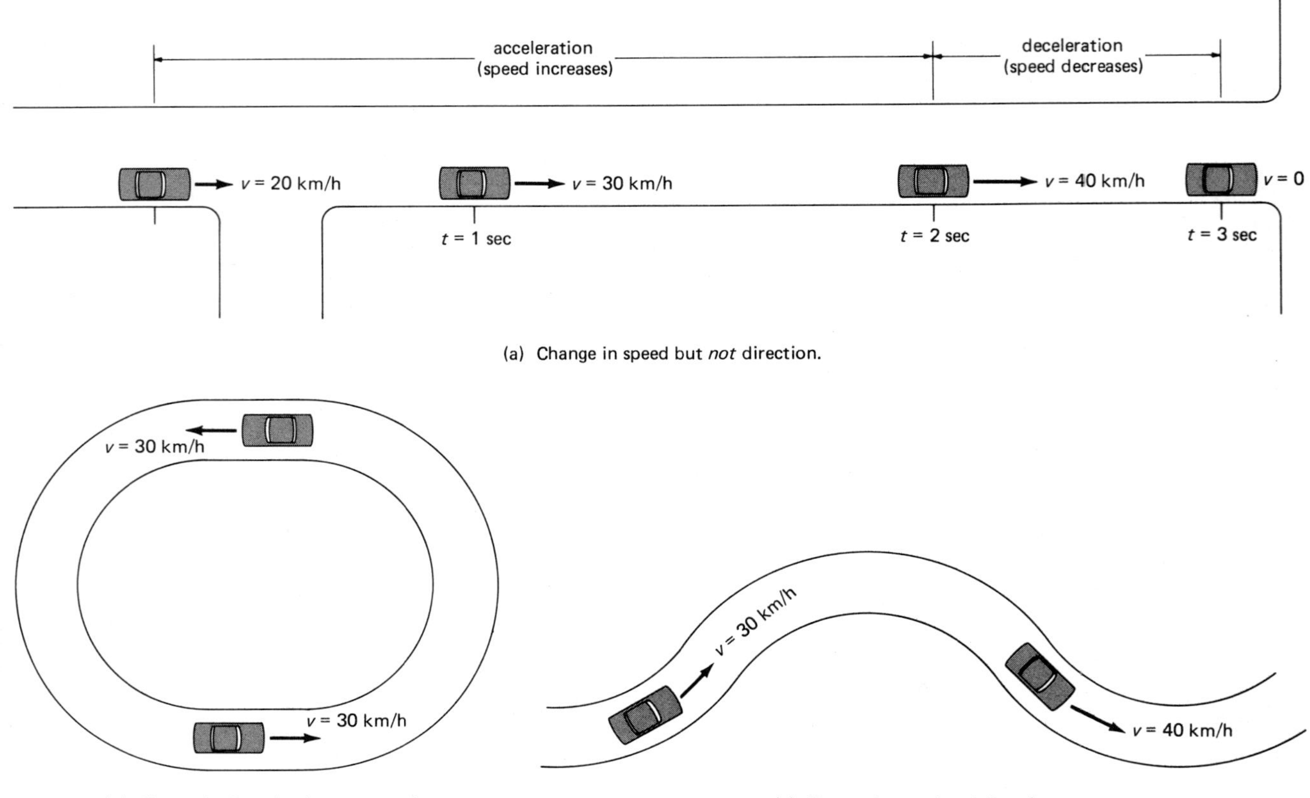

Figure 2.5 Acceleration = time rate of change of velocity (change in velocity/time) due to a change in speed and/or direction.

velocity, is then 10 km/h per second.† This is an example of a constant or uniform acceleration, since the change in velocity is uniform — 10 km/h each second (Fig. 2.5(a)). Here, the speed changes but not the direction of motion. Notice that for a constant acceleration,

$$\text{Change in velocity} = \text{acceleration} \times \text{time}$$

The term acceleration applies to both increases *and* decreases in velocity. When a car slows down, there is a change (decrease) in its velocity. We commonly call this a *deceleration* or a *negative acceleration.* Similar to the gas pedal of a car being called an accelerator, the brake pedal might be called a decelerator.

However, this is not the total story. As was learned from the previous Question and Answer, a change in velocity can also result from a change in the direction of motion, since velocity is a vector quantity. Hence, a change in velocity or an acceleration can result from (a) a change in speed, (b) a change in the direction of motion, or (c) changes in both speed and direction (Fig. 2.5). Perhaps we should also call the steering wheel of a car an accelerator because it can change *direction*!

Force

Closely associated with motion is force. In describing what a force *does,* we know intuitively that a force can produce a change in motion. For example, when an object starts to move (a change in motion) we know a force is acting. Similarly, when a moving object stops (change in motion), we know a force is acting. A change in motion then is *evidence* of a force.

We are commonly familiar with contact forces such as a push or a pull. This is the case of forces resulting from one object being in contact with another. For example, you push or exert a (contact) force on a door

† Acceleration is the change in velocity divided by time, $a = \Delta v/\Delta t$, or $a = v/t$ with the deltas omitted. In standard units, the unit of velocity is meters/second (m/s), so $a = v/t$ is (m/s)/s, and (m/s)/s = m/s-s = m/s^2, which is read as "meters per second squared."

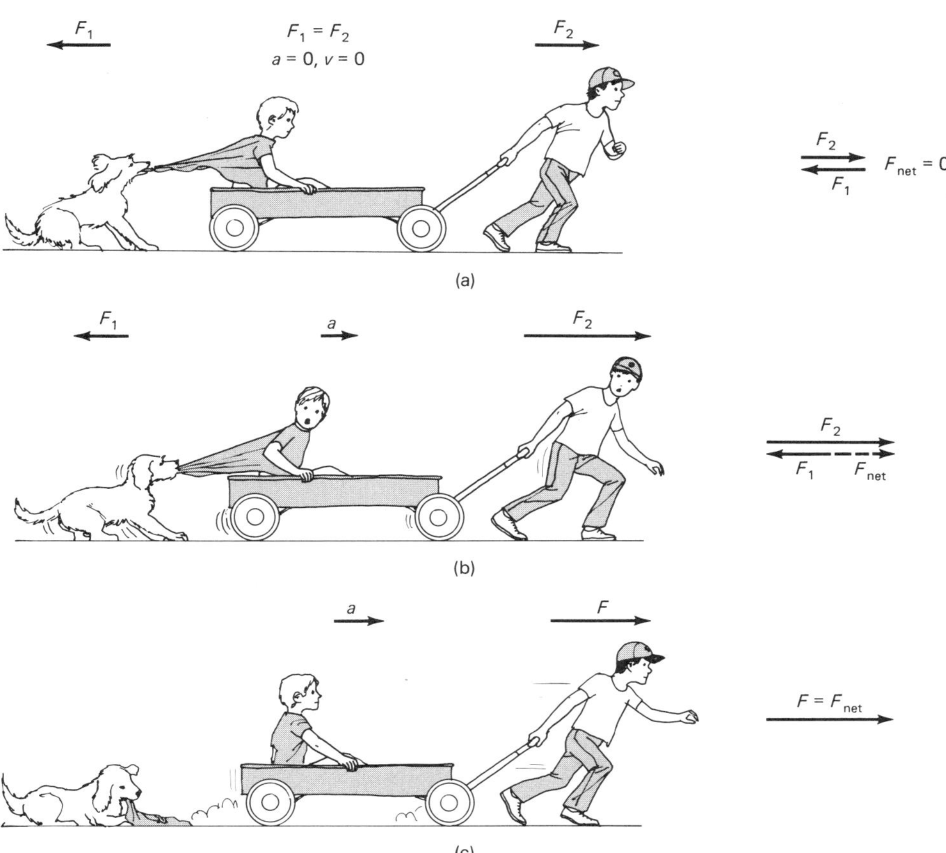

Figure 2.6 Unbalanced or net force. (a) The forces are equal and opposite, so they "cancel," and the net force is zero. (b) The net or unbalanced force is in the direction of F_2 and has a magnitude equal to the difference in the magnitudes of F_2 and F_1. (c) The applied force F is the net force. Note that the acceleration a is proportional to the net force. (Friction is neglected in all cases.)

to open it. There are also action-at-a-distance forces, which will be considered in more detail in later chapters. For example, you know that the gravitational pull or attraction (force) on an object causes it to fall (move) toward the Earth. Also, the gravitational forces of the moon and Sun on the Earth's oceans give rise to tides.

It is common in physics to describe an object's "state of motion," for example, 20 m/s *up* or 20 m/s *west,* and being at rest is also a state of motion (zero speed or velocity). Then,

> A force is something capable of changing an object's state of motion (including starting it from rest).

A key word here is *capable.* A force may not produce motion. You have applied many contact forces to objects and not produced motion. (Try pushing against the side of a building.) As illustrated in Figure 2.6(a), two forces may act against each other (in opposite directions) and effectively cancel each other with regard to producing motion. (Force is a vector quantity, so we may represent it with vector arrows.) In this case, we say that the forces are balanced or that the net force is zero. No change in motion would occur. If an object were initially at rest (no motion), it would stay at rest.

However, if one of the applied forces is greater than the other (case b), then there is a net or unbalanced force. The effect of the two forces would be the same as that of an equivalent net force F, which is in the direc-

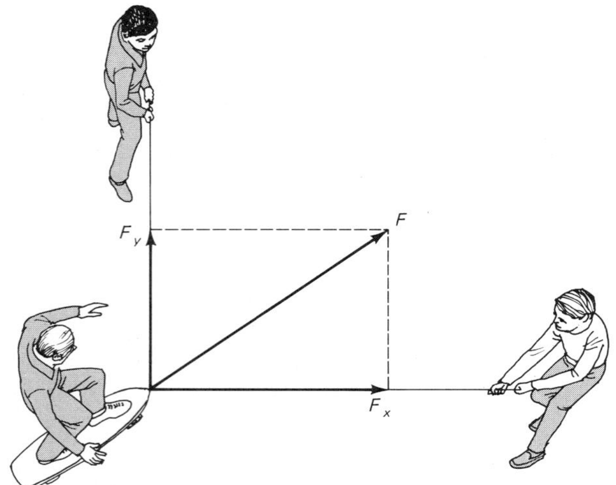

Figure 2.7 Vector addition. The effect of two (or more) forces could be accomplished by a single resultant force equal to the vector sum of the forces.

tion of the larger force and has a magnitude equal to the difference of the magnitudes of the forces.‡

When we speak of a force changing an object's state of motion, it is understood that this may be the resultant net or unbalanced force of two or more individual forces. Usually, we are interested in the *net* motional effect produced by the net force rather than the individual forces.

Two or more forces may act in different directions at various angles to each other. In this case, there is also a net or "resultant" force. For example, in the simple case of two equal forces acting at 90° in the x- and y-directions as in Figure 2.7, the resultant force is in the direction midway between the two forces. The same effect of the two forces could be accomplished with a single force equal to the resultant force. Looking at it another way, a single force F can be thought of as being made up of component forces F_x and F_y in those directions. A vector can be "broken up" or resolved into x and y components to analyze its effects in these directions.

So, knowing that forces can produce changes in motion, the next question is, How are force and motion related? Sir Isaac Newton described that relationship in three laws. Let's look at Newton's laws of motion.

‡ This is a case of vector addition, which takes into account both the magnitude *and* the direction. Opposite directions may be represented as plus (+) and minus (−) so that in vector addition the "sum" of two or more vectors algebraically takes into account this quality to give the net or resultant vector sum.

Newton's First Law of Motion: The Law of Inertia

According to Aristotle's theory of motion, which had prevailed some 1500 years after his death, a body required a force to keep it in motion. That is, the normal state of a body was one of rest, with the exception of celestial bodies, which were naturally in motion. Aristotle observed that moving objects tended to slow down and come to rest (due to friction, we now know), so this conclusion seemed logical to him.

The groundwork for the first law was laid by Galileo (Special Feature 2.2). Rather than rely solely on intuition and general observations, Galileo tested his ideas with experiments. He observed the motions of a ball on inclined planes (Fig. 2.8). When a ball was released and allowed to roll down an incline, it would roll up an adjoining incline to about the same height (a little less, because of friction), being stopped by the retarding forces of gravity and friction. When the angle of incline of the adjoining plane was made less steep, the ball would roll to approximately the same height in each case, *but it rolled farther in the horizontal direction.* Then, when the ball was allowed to roll onto a horizontal plane, it rolled a considerable distance before coming to rest.

Galileo took great pains to make the surfaces of the plane and the rolling ball as smooth as possible to reduce friction, and he found that the smoother the surface, the greater the horizontal distance the ball would roll. So the question arises, How far would the ball travel if friction could be removed completely and the plane made infinitely long? Galileo reasoned that in this ideal case the ball would continue to travel in straight-line, uniform motion forever, since there would be nothing (no force) to alter or change its motion.

Contrary to Aristotle's ideas, Galileo concluded that material bodies exhibited the behavior or property of maintaining a state of motion. Similarly, if a ball is at

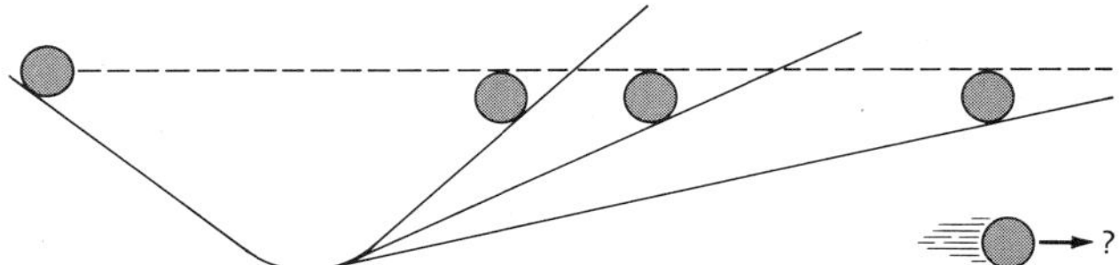

Figure 2.8 An illustration of Galileo's experiment. A ball would roll farther the smoother the surface. In an ideal case of no friction, the ball would continue (slide) indefinitely, since there would be no force to alter its motion.

SPECIAL FEATURE 2.2
Galileo Galilei

Known universally now by his first name, Galileo (Fig. 2.9) was born in Pisa on February 15, 1564. His father was a musician, and Galileo was the eldest of a family of seven children. In 1574 the family moved to Florence, and Galileo was sent to school where he studied Latin, Greek, and mathematics, as well as topics in physics, astronomy, and what is now collectively called the humanities. At the age of 17 he returned to Pisa to study medicine. But this was quickly supplanted by mathematics, a subject in which he excelled. At the age of 25 he received an appointment as professor of mathematics at the University of Pisa, and other prestigious appointments followed.

Throughout his lifetime, Galileo had a variety of scientific pursuits. These included studies of time, motion, floating bodies, and the nature of heat, as well as the construction of telescopes and microscopes (see Chapter 25). Although perhaps best known through the popular stories of his alleged experiments of dropping objects from the Leaning Tower of Pisa, which he probably did not do (see Special Feature 4.1), it was the telescope that played a very critical part in his life. With his telescopes he observed the features of the moon, sunspots, and various planets, including the phases of Venus.

Figure 2.9 Galileo Galilei (1564–1642), the mathematician, astronomer, and physicist who made many contributions to science, including the description of motion. This portrait of Galileo is by Ottavio Leoni, a contemporary.

Galileo lived in an era in which the Aristotelian view of an Earth-centered universe was being challenged. His observations supported the Copernican Sun-centered theory of the solar system, in which the Earth moved. However, Catholic Church dogma at the time held that the Earth was the stationary center of things, as supported by various Biblical references (Joshua 10:12–13, Psalms 19:4–6 and 104:5). To believe and publicly state differently was considered heresy. A former monk, Giordano Bruno, was burned at the stake in 1600 for holding such views.

But Galileo's thought is expressed in one of his favorite quotes: "The Bible shows the way to go to heaven, not the way the heavens go." As early as 1615, he was admonished by the Church about his questionable views. The publication of his book, *The Dialogue of the Two Great World Systems,* in 1632 left little doubt. He was summoned to Rome by the Inquisition, put on trial, and forced to recant his "heretical" ideas.

> I, Galileo, son of the late Vincenzo Galilei, Florentine, aged seventy years . . . have been pronounced by the Holy Office to be vehemently suspected of heresy, that is to say, of having held and believed that the Sun is the center of the world and immovable and that the Earth is not the center and moves. . . . This vehement suspicion justly conceived against me, with sincere heart and unfeigned faith I abjure, curse, and detest the aforesaid errors and heresies . . . and I swear that in future I will never again say or assert, verbally or in writing, anything that might furnish occasion for a similar suspicion. . . .*

Contrary to popular belief, he did not end with the statement, *"Eppur si muove"* ("but it still moves"). At least, there is no record of this, and it is doubtful that Galileo would have been so foolhardy.

Afterward, Galileo returned to Florence to work on more secular things, such as projectile motion. Totally blind the last five years of his life, he died on January 8, 1642, the same year another great scientist, Isaac Newton, was born.

* de Santillana, G., *The Crime of Galileo,* The University of Chicago Press, Chicago, 1955.

Figure 2.10 Inertia "in action" is used to tighten a loose hammer head.

rest, it would remain so, unless something caused it to move. Galileo called this property inertia, and we say,

> Inertia is the property of matter that describes its resistance to changes in motion.

That is, if an object is at rest, it seems to want to remain at rest. If an object is in motion, it seems to want to remain in motion. Here, as in the case of forces and indeed all physical quantities, the property of inertia is defined in terms of behavior or observations. The term simply describes observed effects.

QUESTION: When a hammer head becomes loose on its handle, a person sometimes brings the hammer downward and strikes the butt of the handle sharply on a hard surface (Fig. 2.10). What does this accomplish?

ANSWER: In bringing the hammer downward, the handle and head are in motion. When the handle butt strikes the surface, it stops suddenly because of the large contact force of the surface on the handle. But the massive head continues in motion until it is stopped by the tapered handle. This tightens the hammer head on the handle.

Newton eventually made the idea of inertia quantitative by relating inertia to *mass*. Originally, he thought of mass as a "quantity of matter" but effectively redefined it as the measure of inertia. Today, we know that the mass of a given quantity of matter is not totally independent of conditions. (The mass of an object depends on its speed and even on its temperature, but these changes are commonly too small to measure.) Quantity of matter is still commercially acceptable as a definition of mass, but physicists use this definition:

> Mass is a measure of inertia.

The greater the mass of an object, the greater inertia it exhibits. For example, it is easier to push and get a sports car moving than a more massive van.

Newton summarized these results in his first law of motion, which is also called the law of inertia:

> *Every body preserves its state of rest, or of uniform motion in a right (straight) line unless it is compelled to change that state by forces impressed thereon.*§

Or, in more modern language:

> An object remains at rest or in motion with a constant velocity unless acted upon by an unbalanced force.

Newton's Second Law of Motion: Cause and Effect

Newton's first law says that a body remains at rest or in motion with a constant velocity until it is acted upon by a force. In the absence of an unbalanced force then, the acceleration of a body is zero, since there is no change in velocity. This should lead you, like Newton, to the conclusion that a force acting on a body produces an acceleration. However, Newton recognized that inertia or mass also plays a part. For a given force F, the greater the mass of a body, the less its acceleration a, or change

§ *Principia Mathematica Philosophiae Naturalis,* from Magie, W. F., *A Source Book in Physics,* Harvard University Press, Cambridge, Mass., 1963.

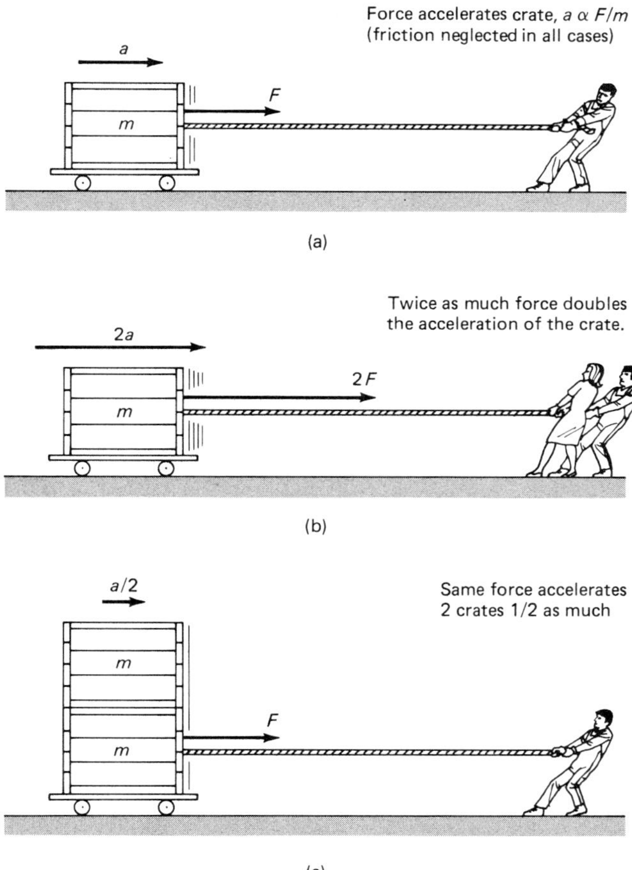

Figure 2.11 Relationships between acceleration, force, and mass expressed in Newton's second law of motion.

in motion. Expressing this as a proportion, ‖

$$a \propto \frac{F}{m}$$

Thus, the acceleration of a body depends *both* on the net force and on the mass of the body. These relationships are illustrated in Figure 2.11.

Newton's second law of motion is commonly expressed in equation form as

‖ The symbol $\propto$ means "proportional to." The expression $a \propto F$ means that a is *directly* proportional to F and they change in the same proportion. For example, if F is doubled, a is doubled. The expression $a \propto 1/m$ means that m is *indirectly* or *inversely* proportional to a and they change in the same inverse proportion. For example, if m is doubled, a is halved. Written together, we have $a \propto F/m$. A proportion gives only relative changes. An equation can be used to calculate the exact values.

$$F = ma$$

$$\text{Force} = \text{mass} \times \text{acceleration}$$

Thus, if the unbalanced force on an object (mass) is zero, its acceleration is zero, and it remains at rest or is in motion with a constant velocity, which is what Newton's first law also tells us.

The metric unit of force using SI quantities is quite appropriately called the newton (N). The second law states that a force of 1 N is the amount of force that gives a mass of 1 kg an acceleration of 1 m/s^2. (One newton is 0.225 or about $\frac{1}{4}$ lb. Recall that the pound is a unit of force.)

Newton's second law relates a force directly to the acceleration it produces—a sort of "cause-and-effect" relationship. Also, the acceleration of a body is always in the direction of the applied net force. If the force is applied in the direction of a body's motion, it will increase the body's velocity. If applied in the direction opposite to the motion, a decrease in velocity will result (negative acceleration). When applied at an angle to the direction of a body's motion, a force will deflect or change the direction of the body's motion. This is also an acceleration or change in velocity (a change in direction certainly *and* possibly a change in magnitude as well).

Another way of looking at this is that an acceleration of a body is evidence of an applied force. Thus, we may say that a force is anything that can accelerate a body.

EXAMPLES OF NEWTON'S SECOND LAW

Friction

The force of friction, which in general resists motion, occurs between contacting media. We usually think of friction in terms of contacting solid surfaces, but it also occurs for liquids and gases.

For solid surfaces, friction arises from the irregularities in the surfaces of objects. All solid surfaces are microscopically rough, no matter how smooth they may appear or feel. Early investigators thought friction to be primarily due to the interlocking or fitting together of surface irregularities. However, modern research shows that the majority of friction between contacting surfaces of ordinary solids (metals in particular) is due to local adhesion or "sticking" between the surface irregularities rather than to their fitting together.

Generally, we represent the overall frictional effect by a friction force f. The force of friction always opposes relative motion of the surfaces in contact. We

SPECIAL FEATURE 2.3

An Accelerometer

A carpenter's air-bubble level can be used as an accelerometer or force meter in detecting an acceleration or a force by observing the motion of the bubble (Fig. 2.12). If a force is applied toward the left to a level at rest, as shown in the figure, which way will the bubble move?

Many people say that the bubble would move to the right, but actually it would move to the left, or in the direction of the acceleration and the force. The incorrect answer arises from the fact that we are used to observing the bubble rather than the liquid. The correct answer is explained by Newton's first law. Because of inertia, the liquid resists the motion and "piles up" toward the rear of the level. This forces the bubble in the other direction, or in the direction of the acceleration. Think of pushing a stationary pan of water. What would happen to the liquid?

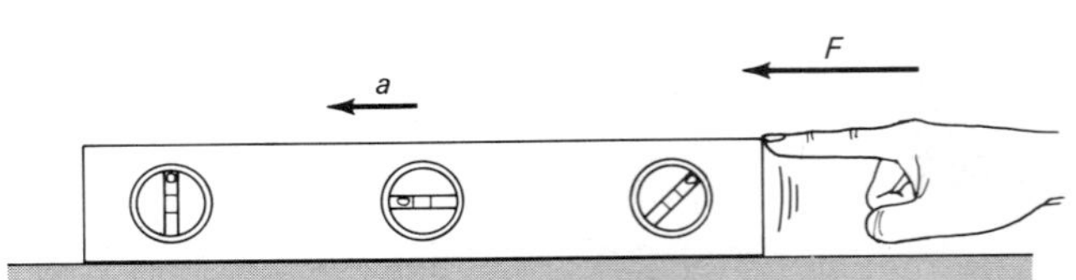

Figure 2.12 Accelerometer. The motion of the bubble indicates the direction of the acceleration of the level.

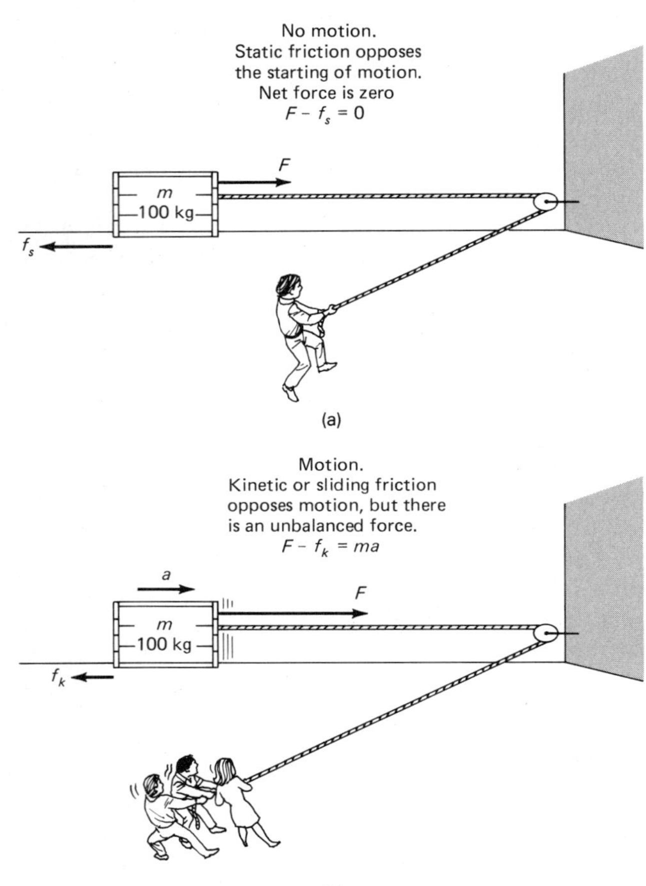

Figure 2.13 (a) Static friction. (b) Kinetic or sliding friction.

have static friction that opposes the starting of motion, and kinetic or sliding friction that acts when the surfaces are in relative motion (Fig. 2.13).

Applying Newton's second law to stationary objects, we have, for static friction,

$$F - f_s = 0 \; (= ma, \text{ but } a = 0)$$

or

$$F = f_s$$

The applied force F and the force of static friction f_s are equal and opposite, so there is no acceleration or motion. This is a case of a net force $(F - f_s)$ being zero.

However, if the applied force is increased so that the object slides, then the frictional force opposing the motion is kinetic, and

$$F - f_k = ma$$

or

$$a = \frac{F - f_k}{m} = \frac{F_{\text{net}}}{m}$$

The acceleration a may or may not be zero, depending on the magnitudes of the applied and frictional forces. It generally takes a smaller applied force to keep an object moving than to get it moving because kinetic friction is less than static friction.

QUESTION: What would happen if the applied force on a sliding object had the same magnitude as the sliding frictional force?

ANSWER: In this case, $F - f_k = F_{\text{net}} = 0 = ma$, so the net force and the acceleration are zero. This means that the object slides with a constant velocity (à la Newton's first law).

Mass and Weight

The weight of a body is the gravitational force acting on it. Usually, this is the gravitational attraction of the Earth, and the force is easily demonstrated. When we drop an object, it falls faster and faster (accelerates) toward the Earth.

We commonly write the formula for weight as $w = mg$, where g is the acceleration due to gravity and has a relatively constant value near the surface of the Earth of 9.8 m/s² (32 ft/s²).

Notice that this is a special form of Newton's second law,

$$w = mg$$

or

$$F = ma$$

where the acceleration due to gravity is given the special symbol g because it is so common.¶ In the metric system, your weight force would be expressed in newtons. However, in the SI system a body's "weight" is expressed in kilograms ("kilos") or in units of mass (Fig. 2.14). Recall that 1 kg of mass has an *equivalent* weight of 2.2 lb or, by the above formula, 9.8 N.

We will consider the motion of falling objects in more detail in the next chapter.

Newton's Third Law of Motion: Action and Reaction

Although we commonly talk of single forces, Newton recognized that it is impossible to have just an individual force. Rather, there is a mutual interaction, and forces always occur in pairs. An example given by Newton was that if you press on a stone with your finger, the finger is also pressed upon by the stone. That is, if one object exerts a force on a second object, then the second object exerts a force on the first. This is like saying that you can't touch something without being touched.

Newton termed these forces *action* and *reaction*, and **Newton's third law** is commonly expressed as follows:

> For every action, there is an equal and opposite reaction.

¶ This should not be confused with the accepted abbreviation for the gram (g).

Figure 2.14 Weight and mass. Weight is force, expressed in units of newtons or pounds. In the SI system, a body's "weight" is expressed in kilograms ("kilos") or in units of mass.

Or, alternatively,

> For every force, there is an equal and opposite force.

In symbol form,

$$F_{\text{action}} = -F_{\text{reaction}}$$

where the negative sign indicates the opposite direction. Which force is the action or reaction is arbitrary and depends on how you look at the situation — it's a relative interaction.

The third law may seem contradictory to the second law. (If you have equal and opposite forces, how can there be an acceleration?) However, the second law is concerned with force(s) acting *on a particular body* and its resulting acceleration. In applying the second law,

Figure 2.15 An example of Newton's third law. For every force there is an equal and opposite force. The forces act on *different bodies.*

we look at only the forces acting on a given body. The force pair of the third law acts on *different bodies* (Fig. 2.15).

Let's take a look at some examples of the third law action-reaction force pair. When you are holding something quite heavy, you supply an upward force (action) *on the object.* After a short time, you may become painfully aware of the reaction force the heavy object exerts *on you* (Fig. 2.16(a), particularly if the cooler is full).

In Figure 2.16(a), there are two sets of force pairs. There is the upward action force *on the cooler handle* by the person, and the downward reaction force *on the person's hand* by the cooler. Also, there is a downward action force *on the cooler* due to gravity (its weight). Although not obvious, there is an upward reaction force *on the Earth.* Notice that there are equal and opposite forces acting *on* the held cooler, so there is no net force acting on it and it is stationary (Newton's first law).

But, suppose the cooler is dropped (Fig. 2.16(b)). Now there is a net force on it, and it falls or accelerates downward (Newton's second law). The third law action-reaction force pair between the cooler and the Earth is still there (as it always is since we can't turn off gravity). The Earth is so massive (6×10^{24} kg), however, that its reaction motion (acceleration) is negligible.

Let's look at a less massive example that involves acceleration by applying the third law to the firing of a rifle (Fig. 2.17). When the charge explodes, the bullet is accelerated down the barrel. It is acted upon by a force (an action), as evidenced by its acceleration. The reaction force acts on the rifle. This force accelerates it in the opposite direction, and gives rise to the backward recoil or "kick" that is commonly experienced with a large-caliber rifle or shotgun.

By Newton's third law, the magnitudes of the action-reaction forces are equal, or $F_b = F_r$ (b for bullet and r for rifle). Then,

$$A_b = \frac{F_b}{m_b}$$

and

$$a_r = \frac{F_r}{M_r}$$

where the symbol size is meant to emphasize relative magnitudes. Hence, the acceleration of the rifle a_r is much smaller than the acceleration of the bullet A_b (as one would hope), owing to the difference in masses.

A similar situation occurs in jet propulsion. Exhaust gases from burned fuel are accelerated out the back of a rocket or jet engine, and the rocket or aircraft is accelerated forward by the reactive force. Similarly, astronauts use hand rockets to maneuver on "space walks" (Fig. 2.18).

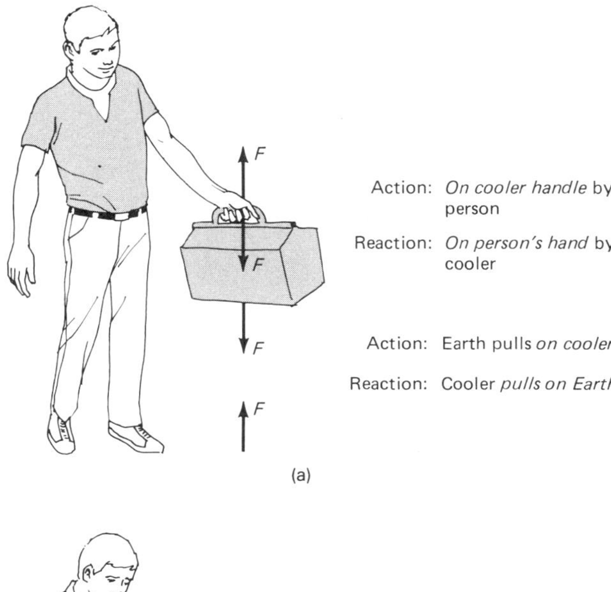

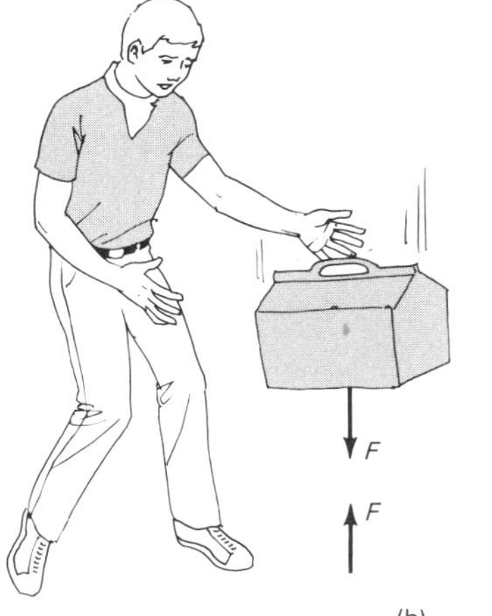

Figure 2.16 Newton's third law force pairs. (a) There are two force pairs associated with the cooler when it is held stationary. Notice that the net force on the cooler is zero. (b) When the cooler is dropped and falling, there is still a force pair, but the net force acting on the cooler is not zero, since it accelerates toward the Earth.

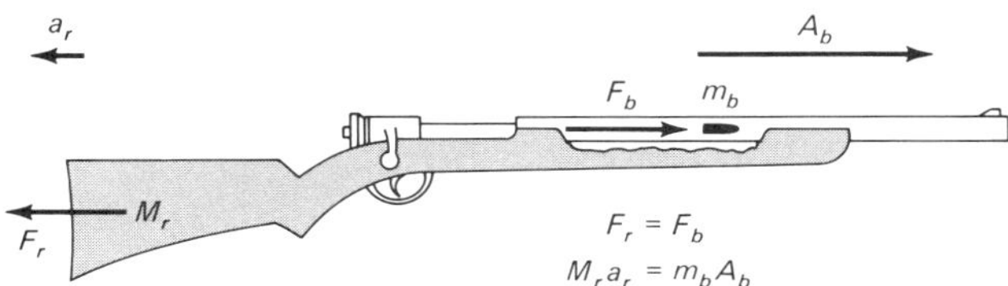

Figure 2.17 Newton's third law force pair for an isolated rifle and bullet. There are equal and opposite forces on the rifle and bullet. Because the bullet has less mass, its acceleration is much greater than that of the rifle.

(a)

(b)

Figure 2.18 Examples of Newton's third law. Rockets operate on the action-reaction forces of the rocket and exhaust gases. (a) An Apollo/Saturn V rocket blasting off. (b) An astronaut on a space walk using a hand-held, self-maneuvering unit, or hand rocket.

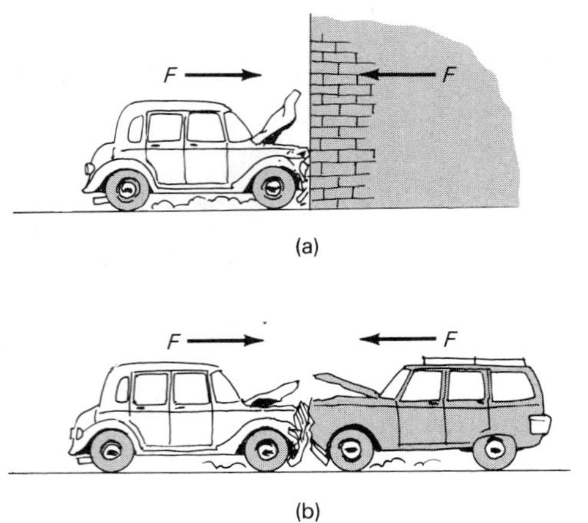

Figure 2.19 Undesirable action-reaction pairs. The reaction force of the wall on the car (a) is recognized when a substitution is made (b).

QUESTION: When a rocket blasts off, what causes it to lift off the launching pad? Is it the exhaust gases "pushing against" the pad?

ANSWER: The launching pad is just there as a launching place. It is actually the reactive force of the gases pushing on the rocket that causes the rocket to accelerate upward. If this were not the case, there would be no space travel, since there is nothing to "push against" in space.

Sometimes a not-so-obvious reaction force can be better understood by substituting another force in its place. For example, the reaction force of a wall on a car produces the same effect as the force applied by another car (Fig. 2.19). As another example, consider a 10-kg mass suspended by a rope fastened to a wall, as shown in Figure 2.20. A spring scale is used to measure the force. This would be the weight of the mass, $w = mg = (10 \text{ kg})(9.8 \text{ m/s}^2) = 98 \text{ N}$. The force is transmitted through the rope and the scale to the wall.

To illustrate that the wall "pulls" on the rope with a reaction force of 98 N, suppose you unfastened the rope and held the mass stationary (like the wall). You

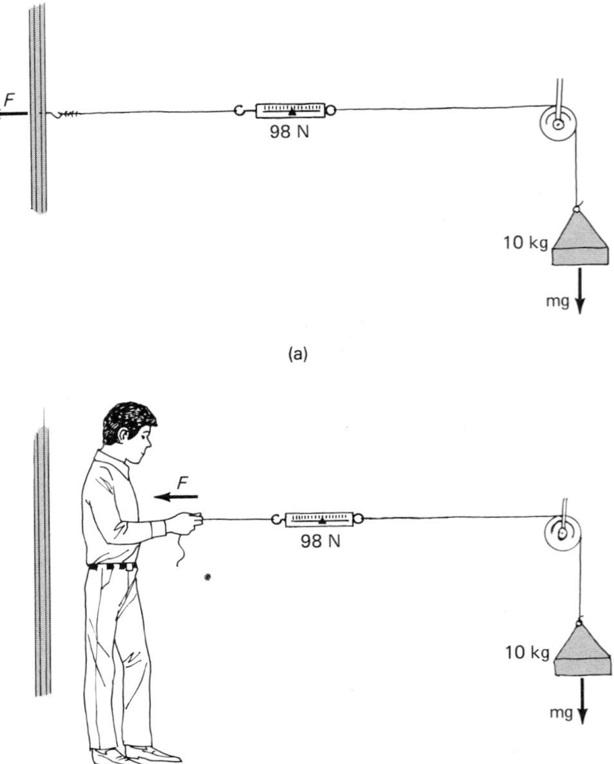

Figure 2.20 Newton's third law one more time. (a) The weight of the mass exerts a force of 98 N on the wall, and the wall "pulls" back with an equal force. (b) This can easily be seen by replacing the wall with yourself or another person.

would have to pull with a force of 98 N, so the wall must be doing the same.

Hence, we see by Newton's third law that forces occur in pairs acting on different things. A clue to this is that an object, or a person, cannot exert a force on nothing. There has to be a reaction force, and there is a simple way to describe it. If you describe a given force as exerted *on* object X *by* object Y, then if you switch the two prepositions "on" and "by" and reverse the direction of the force, you have now described the reaction to that force.

SUMMARY OF KEY TERMS

Motion the process of a change in position.

Force a quantity capable of producing motion or a change in motion.

Newton's first law of motion (law of inertia) an object remains at rest or in uniform motion with a constant velocity unless acted upon by an unbalanced force.

Inertia the property of matter that resists changes in motion. Mass is a measure of inertia.
Speed time rate of change of position, or how rapidly the distance is covered: Speed = change in distance/change in time.
Velocity the rate of change of position *and* the direction of the motion. Velocity is a vector quantity.
Acceleration time rate of change of velocity: Acceleration = change in velocity/time it takes for change.
Unbalanced or net force the equivalent or resultant force F of two or more forces.
Newton's second law of motion $F = ma$. Relates force to acceleration.
Friction the force that opposes the relative motion of contacting media.
Weight the gravitational force acting on a body: $w = mg$.
Newton's third law of motion for every force (action) there is an equal and opposite force (reaction).

EXERCISES

1. We often refer to a moving van or a moving company. Is the adjective "moving" descriptive of what they do? Why?
2. When playing checkers or chess, a player's turn is called his or her move. Why is this?
3. Is it possible to be in the same place and to have moved? Explain.
4. When observing an object, how can you tell if there is a force acting on it? Can you always "see" what is applying the force? Explain.
5. Is it possible to have motion without a force? Explain.
6. How did Galileo's and Aristotle's ideas of motion differ?
7. Can you actually isolate or find the inertia of a body? Explain.
8. An old parlor trick involves suddenly pulling a tablecloth from underneath a setting of plates and glasses. Rather than falling to the floor and breaking, the plates and glasses remain on the table (Fig. 2.21, which also shows a modern variation that is a bit hard on the tableware). Explain the "magic" of this trick.
9. Suppose you are standing on an icy surface (very little friction) and a heavy object and a light object of the same

(a)

(b)

Figure 2.21 See Exercise 8.

size and shape are also on the ice. How could you tell which object is heavier without picking them up?

10. An astronaut in a spaceship and in a "weightless" condition wants to know which of two identical containers used to store equipment is empty. How can the astronaut tell which container is empty without opening them?
11. When a paper towel is torn from a roll on a rack, a jerking motion tears the towel better than a slow pull. Why is this? Does this method work better when the roll is large or when it is small and near the end of the roll? Explain.
12. A passenger in a car struck from behind sometimes experiences "whiplash," which results from the upper vertebrae of the spine being bent backward. What causes this?
13. Common safety devices in automobiles are seat belts and shoulder straps. Explain why these devices are needed in terms of Newton's first law.
14. When a police officer using radar gives someone a speeding ticket, is the officer concerned with the driver's long-term average speed or the approximate instantaneous speed? Explain.
15. Two cars, A and B, travel the same 100-km distance when driven by students going home on a spring break. Car A makes the trip in 2 hours and car B in $2\frac{1}{2}$ hours. (a) Which car has the greater average speed? (b) Calculate the average speed of each car.
16. An automobile travels 50 km in 30 minutes, and a train travels 90 km in 1 hour. (a) Which has the greater average speed? (b) What is the average speed of each?
17. A person walks with a constant speed of 0.5 m/s. How far does he or she walk in (a) 1 second, (b) 20 seconds, (c) 1 minute?
18. An airplane flies eastward with a constant velocity of 300 km/h. Given the plane's position at a particular time, where is the plane 2 hours later?
19. A car travels along a straight road with a constant speed of 75 km/h. What is its acceleration?
20. A motorist in a moving car applies the brakes. Does the car accelerate? Explain. Discuss what could be called "accelerators" and "decelerators" on a car. Could a steering wheel be a "decelerator"? How about "gearing down"?
21. A race car travels around a racetrack with a constant speed. Is the car accelerating? Explain.
22. An air-bubble level is pushed along a table surface. Where is the bubble or in which direction does it move (a) when the level is moved with a constant velocity? (b) When the applied force is removed and the level comes to a stop?
23. A child sitting in a stationary car holds a helium balloon by a string. Explain what happens to the balloon when the car starts to move forward.
24. A person riding a bus cannot find a seat, so he stands in the aisle. Explain what happens to the person when (a) the bus starts up from a stop, (b) the moving bus accelerates, (c) the bus travels at a constant velocity, (d) the bus slows down, and (e) the bus turns a corner.
25. If an object at rest is acted upon by a force and the object doesn't move, what can be concluded?
26. On packaged items, such as breakfast cereals, the net weight is listed. What does this mean?
27. The amount of money ($) one receives when returning returnable bottles to a store is proportional to the number (n) of bottles returned. (a) Write the symbol relationship for this proportionality. (b) Does the relationship tell you how much money you would actually get for a certain number of bottles? (c) If you received 5¢ for each bottle returned, write the equation relationship. (d) Does the equation allow you to calculate the amount of money you would receive for a certain number of bottles, say $n = 10$?
28. Student study time is inversely proportional to extracurricular activities. Write and explain the symbol relationship for this proportionality.
29. Newton's first law of motion may be derived from his second law of motion. Explain this statement.
30. By Newton's second law of motion, a net force of 1 newton acting on a mass of 1 kilogram produces an acceleration of 1 m/s^2. What would be the acceleration if (a) the force is doubled (same mass)? (b) if the mass is halved (same force, 1 N)? (c) if both the force and mass are doubled (2 N and 2 kg)?
31. What is the weight in newtons of a 1-kilogram mass? Can

TIGER

Figure 2.22
See Exercise 33.

Figure 2.23 See Exercise 34.

you write a conversion factor for newtons and pounds?

32. In a tug-of-war, two teams pull on the rope with equal and opposite forces. Analyze the situation (forces) in terms of (a) Newton's second law and (b) Newton's third law.

33. What suggestion would you give Suzy in Figure 2.22? Give a detailed suggestion and explanation as Isaac Newton might have given her.

34. Identify the action and reaction forces of Newton's third law for each of the following cases:
(a) A person pushing on a compressed spring.
(b) A swimmer changing directions in starting another lap at the end of the pool.
(c) A person jumping onto a bank from an untied canoe (Fig. 2.23). Why does the canoe move backwards or opposite the direction the person jumps? What might happen to the person and why?

35. When a person pushes on a wall, the wall pushes on the person à la Newton's third law. Suppose the person put a block of wood between his or her hand and the wall. Analyze the forces acting on the block of wood. Why doesn't the block move?

36. (a) A car sits stationary on a level road. Identify the action-reaction force pair(s). (b) Suppose the car starts to move down the road. What force pairs are acting now? What force causes it to accelerate?

37. What causes a rotary lawn sprinkler to rotate?

38. What force causes a helicopter to begin to rise?

39. Two people pull with equal forces of 100 N on the ends of ropes which have the other ends tied to each end of a spring scale. (a) What does the scale read? (b) If one person tied his end around a post, what would the scale read?

40. Two masses are attached to a spring scale, as shown in Figure 2.24. (a) If both masses were 1 kg, what force in newtons would the scale read? (*Hint:* See Figure 2.20.)

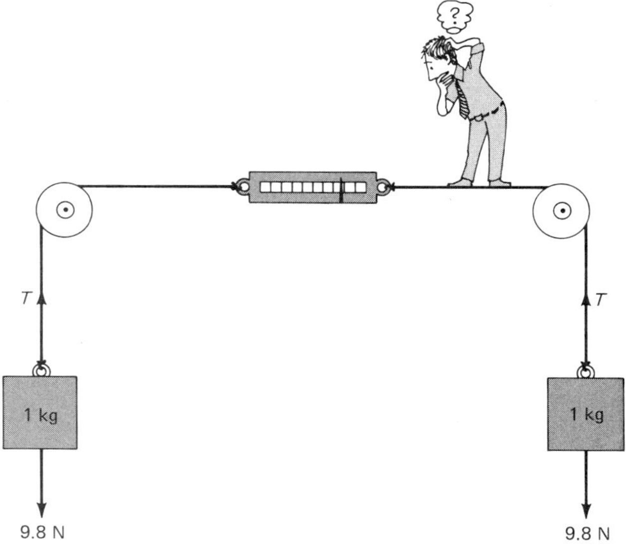

Figure 2.24 See Exercise 40.

(b) We say that the force is transmitted undiminished by the rope. What would happen if this were not the case, that is, if the tension were different in different parts of the rope?

41. In approaching (falling toward) the moon, astronauts in a spaceship slow their spacecraft down by firing retro-rockets in order to go into orbit and not crash into the moon. How does firing a rocket slow down the spacecraft?

42. (a) A person places a bathroom scale (not the digital type) in the center of the floor and stands on the scale with his arms at his sides (Fig. 2.25). Keeping his arms *rigid* and quickly raising his arms over his head, he notices the scale reading increases as he brings his arms upward. Why? (b) Then, with his arms over his head, he brings them quickly to his side. How does the scale reading change and why? (Try this experiment yourself.)

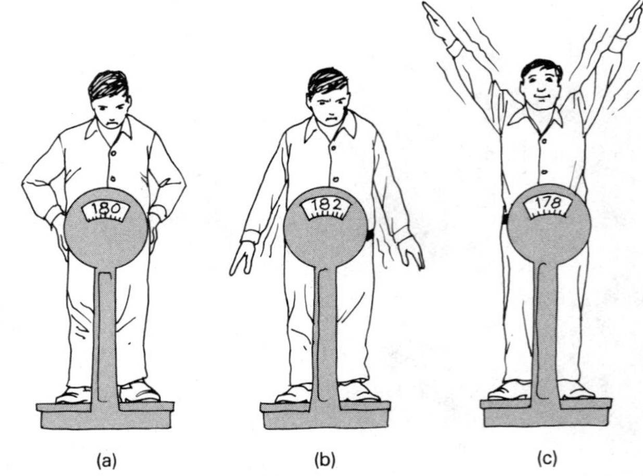

Figure 2.25 See Exercise 42.

3 Momentum

Impulse and Momentum

Let's take a look at force and motion from another point of view. According to Newton's second law, $a = F/m$, an object or mass accelerates when a net force acts on it. This force is often applied through a contact collision of objects, such as hitting a ball with a bat or "shooting" a billiard ball with a cue stick. The resulting motion or change in motion of an object, such as a billiard ball, depends not only on its mass and the applied force but also on the *time* of the contact or application of the force. We are all aware of this from pushing a stalled car. You have to push for some time to get it rolling well.

This fact is expressed in terms of a quantity called **impulse.**

$$\text{Impulse} = \text{force} \times \text{time}$$

or, in symbols,

$$\text{Impulse} = F\Delta t$$

where Δt is the contact or application time of the force (Fig. 3.1).

The force involved in an impulse is not usually a steady force, but varies with time. For example, when a bat hits a ball, the force on the ball increases rapidly from an initial zero value as the ball is deformed (Fig. 3.2). The force decreases as the ball recoils, and it returns to its original shape on leaving the bat. (This is an example of elasticity. All solid materials, even steel, are elastic to some degree.)

The force changes with time in such cases, and the impulse is difficult to determine. However, impulse is related to the resulting change of motion or velocity of an object that receives the impulse. This can be seen

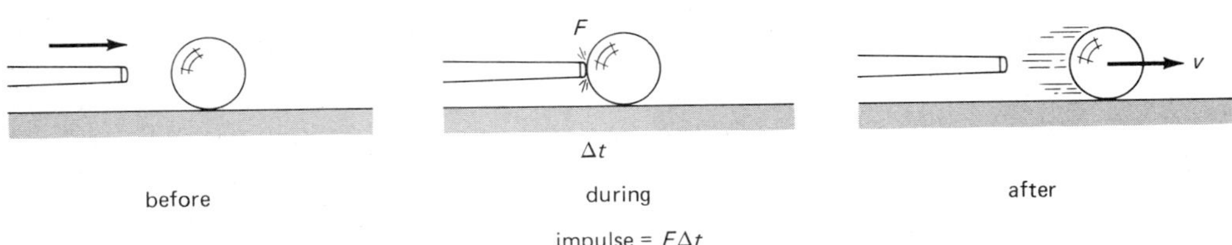

Figure 3.1 An example of impulse. The change in motion of an object depends not only on the applied force, but also on the contact time of the applied force.

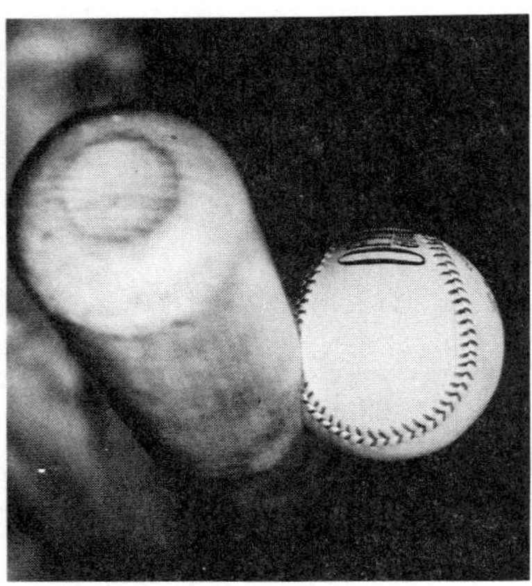

Figure 3.2 Objects usually deform in contact collisions.

directly from Newton's second law, $F = ma$, where a = change in velocity/time or $a = \Delta v/\Delta t$, and

$$F = ma = \frac{m\Delta v}{\Delta t}$$

Rearranging the equation (multiplying both sides by Δt and cancelling $\Delta t/\Delta t$ on the right side) gives

$$F\Delta t = m\Delta v = \Delta(mv), \text{ or change in } mv$$

since m is usually constant and doesn't change. Then,

$$\text{Impulse} = F\Delta t = \Delta(mv)$$

The quantity mv, or the product of the mass of an object and its velocity, is called its **momentum.** That is,

$$\text{Momentum} = \text{mass} \times \text{velocity}$$

or

$$\text{Momentum} = mv$$

Momentum is a vector pointing in the same direction as the velocity. An impulse produces a change in momentum, since impulse = $\Delta(mv)$.

Newton called momentum a "quantity of motion." It takes into account not only the inertia of an object but also its motion. We sometimes say that a football player running down the field has a lot of momentum because he is difficult to stop. This is usually attributed to the football player's large mass, although his velocity also contributes to the momentum. However, a bullet (small mass) with a large velocity can have an appreciable momentum (Fig. 3.3) and is also difficult to stop. A stationary object has zero momentum since it has zero velocity.

In most instances, a change in momentum involves a change in velocity, since the mass is constant. That is, an object either slows down, speeds up, and/or its direction of motion changes. This is easily observed or measured and gives us a "handle" on the impulse, or the product of force and time. Let's see how this works in some common applications and manipulations of the impulse quantities—force and time.

Suppose someone throws you a hard ball with a velocity v and you catch it with your arms rigidly extended. As we all know, the ball "stings" your hands.

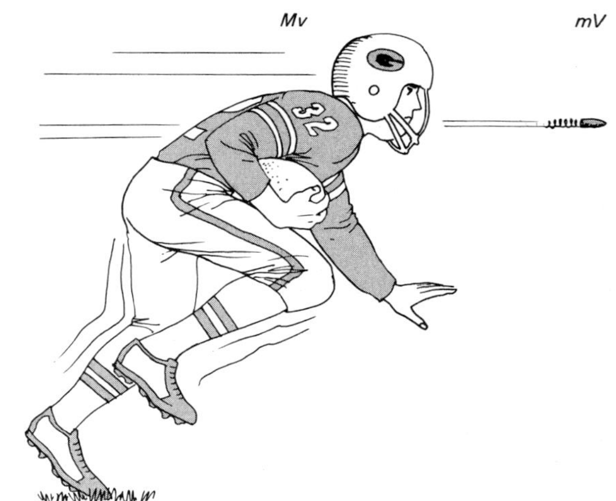

Figure 3.3 Momentum is mass times velocity, mv. It is common to think of moving, massive objects as having a lot of momentum, but a small object moving at a high velocity also has appreciable momentum.

The change in momentum is just mv, since when the ball stops $v = 0$ and its momentum is zero. (Change in momentum $= mv - 0 = mv$.) Then,

$$F\Delta t = mv$$

where the small Δt indicates that you stopped the ball in a short time. This makes the impulse force large, which stings your hands (Fig. 3.4).

(a)

(b)

Figure 3.4 Impulse in action. In stopping (catching) a ball, the change in momentum is mv, which is equal to the impulse. (a) If the contact time is small, F is large and the ball "stings" the hands. (b) If the contact time is increased by moving the hands along with the ball, the force is smaller, and there is little or no sting.

After a few stinging catches, you get smart and learn to move your hands backward while catching the ball; that is, you learn to manipulate the impulse (Fig. 3.4). Assuming the ball is thrown with the same velocity, the change in momentum is the same for each catch. However, when you move your hands in the direction of the motion of the ball, the contact time is increased, which lessens the impulse force and the sting. That is, in symbol form,

$$F\Delta t = mv$$

Another example of manipulating the contact time to control the impulse force is in jumping from a high place on to a solid surface or floor. If you were to land stiff-legged, you would stop suddenly (small Δt), and the large impulse force might hurt your knees, legs, or spine. You quickly learn to bend your knees when landing so as to increase the contact time and reduce the force.

Impulse plays an important role in automobile safety. For example, cars were once built without padded dashboards, but now padded "dashes" are the rule. If you were not wearing your seat belt or shoulder strap and the car came to a sudden stop, you would continue in motion according to Newton's first law. (The external force on the seat of your pants would not be enough to stop you.) In this case, you would probably hit and be stopped by the dashboard. The padding of the dash would increase your contact time and reduce the impact force, possibly preventing injury. See Special Feature 3.1 for another application of the same principle.

Impulse and momentum also are involved in sports. For example, in baseball or golf when it is desired to hit the ball a long distance, the player "follows through" with the swing so as to increase the contact time. With a fixed force or maximum swing, the ball receives a larger impulse and a larger change in momentum or a greater velocity.

Another example of a contact force occurs when karate experts break bricks or boards with their bare hands. There is an impulse here, but another factor comes into play. This is pressure, which is force/area (F/A). When the karate expert strikes the boards with the edge of the hand (small area), the downward momentum is suddenly reduced and the impulse force is great. With a small contact area, the pressure is great, and the boards break.

By the same token, the karate expert could hit the boards with the palm of the hand (large area) and probably not break the boards. In this case, the change in

SPECIAL FEATURE 3.1

The Automobile Air Bag

Air bags are scheduled to be installed in new automobiles as a safety feature; however, this has been delayed for some time for economic and manufacturing reasons. When they *are* installed, air bags in cars will prevent many injuries in accidents, particularly for people in the front seat who now do not "buckle up" for safety with seat belts and shoulder straps. The principle involves the automatic inflation of an air bag on the hard impact of an automobile so as to prevent the driver (and passenger) from hitting the steering wheel, the dashboard, and/or the windshield in accordance with Newton's first law (Fig. 3.5).

In terms of impulse, the air bag increases the stopping contact time, thereby reducing the impact force and preventing injury.

Figure 3.5 Automobile air bag. The air bag increases the collision contact time in stopping a person, thereby reducing the impulse force and possible injury.

momentum may be greater (final $v = 0$) and the impulse greater. But the impulse force is spread over a larger area, the contact area of the palm, so the pressure is less. Also, the "expert's" hand would probably sting a bit in this case.

Conservation of Momentum

Recall that Newton's first law tells us that a body remains at rest or in motion with a constant velocity unless it is acted upon by an unbalanced force. Then, if the velocity is constant (even zero), the momentum of a body is also constant (if its mass is also constant). Hence we have the condition necessary for the **conservation of momentum** of a body:

> In the absence of an unbalanced force, the momentum of a body is conserved.

By conserved, we mean unchanged or constant with time. That is, each time we measured the momentum we would obtain the same value.* Because nature is so difficult to analyze, scientists look for conserved quantities and treasure them. When something is conserved or constant, it makes the job and life a bit easier for scientists.

The beauty of the conservation of momentum is that it also applies to more than one body, in fact, to any number of bodies. We call a collection of objects a system. In the case of more than one body, we speak of the total momentum of the system. Since velocity is a vector, so is momentum. As such, the total momentum is the vector sum (the net or unbalanced momentum) of the momentum vectors of the individual bodies. Like the individual forces that produce net or unbalanced forces (which we saw in Chapter 2), the individual momentum vectors may add and/or cancel, depending on their directions.

* Newton expressed his second law in terms of the time rate of change of momentum, $F = ma = m\Delta v/\Delta t = (mv_2 - m_1v_1)/\Delta t = p_2 - p_1/\Delta t = \Delta p/\Delta t$, or $F = \Delta p/\Delta t$. (The symbol p is commonly used for momentum; $p = mv$.) So, if $F = 0$, $\Delta p/\Delta t = 0$, and $\Delta p = 0 = p_2 - p_1$, or $p_2 = p_1$, and momentum is constant or conserved.

So we may extend the conservation of momentum to a system of bodies.

> In the absence of an unbalanced, *external* force, the total momentum of a system is conserved.

Notice that we distinguish between *external* or outside forces that act on the system and *internal* forces that act within a system. An external force or impulse is required to change the momentum of a body or a system. Internal forces, for example, the internal molecular forces of a billiard ball, have no effect on the ball's momentum. Or if you are a passenger in a moving car and push on the floorboard with your foot (as some passengers do to apply the "brakes"), there is no change in the car's velocity or momentum. This is because these forces are internal ones. By Newton's third law, they are equal and opposite, and the net force or impulse within the system is zero, so there is no change in momentum.

Hence, the bodies of the system may move around and bump into each other, but nothing that happens within a system can change the total momentum of the system. The total momentum of the system is conserved—in the absence of an unbalanced, *external* force.

Let's take a look at some examples of this "before and after" conservation effect. Consider two skaters standing on a frictionless, icy surface (Fig. 3.6). Initially, the total momentum of the system (the two skaters) is zero, since they are not moving. Now, suppose they "push off" each other (internal forces). They move apart, but the total momentum is still zero—the momentum vectors cancel. Because the skaters have different masses, their speeds are different. However, the product of mv is the same for each—but in opposite directions, which makes the net momentum equal to zero. (How would their speeds compare if they were both the same mass?)

Figure 3.6 Conservation of total momentum. (a) The total momentum P of the skaters' system is zero. (b) With no external forces ("pushing off" involves internal forces), the total momentum is conserved and remains zero.

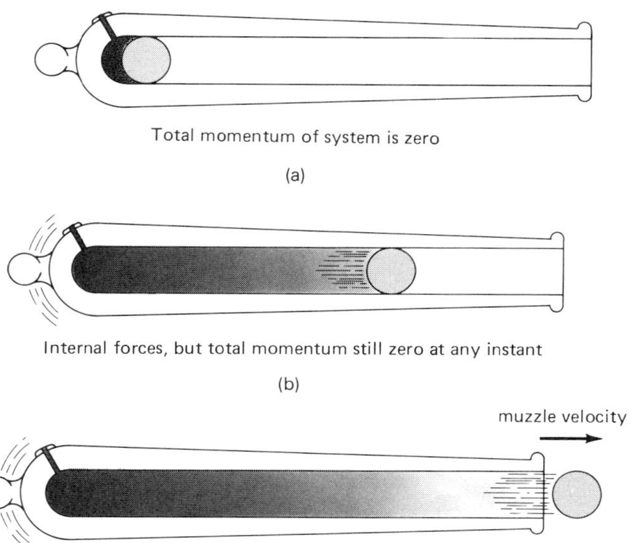

Figure 3.7 Conservation of momentum. The total momentum of the system is zero in each idealized case.

QUESTION: In reality, there was some friction between the skaters' skates and the ice in the previous example. Is momentum conserved?

ANSWER: No. Friction would be an external force, so the momentum would not be conserved. Each skater would gradually slow down and lose momentum.

In the last chapter we used Newton's third law to explain the recoil or "kick" of a rifle (see Fig. 2.17). We can also look at this in terms of momentum. Let's think big this time and consider a cannon (Fig. 3.7). When the powder charge explodes (an internal force), the cannon ball accelerates down the barrel. The total instantaneous momentum of the system is still zero. This is easier to see after the cannon ball leaves the barrel. We call the velocity with which a projectile leaves the barrel of a rifle or cannon its muzzle velocity. The speed of the cannon ball is much greater than that of the cannon, but the total momentum is still zero. Why?

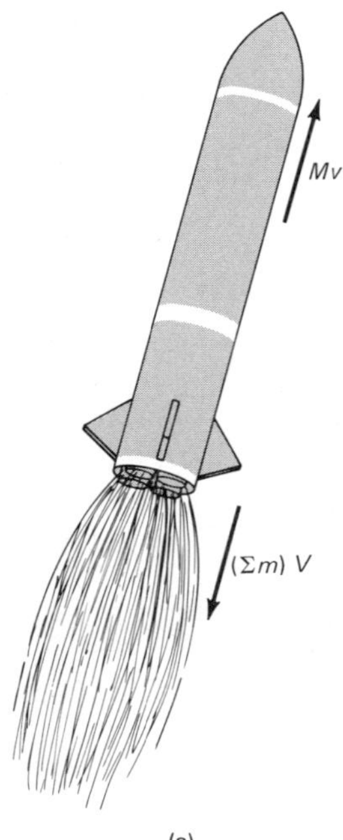

Figure 3.8 Jet propulsion and conservation of momentum. If originally zero, the total momentum is zero at any instant, and the magnitude of the momentum of the exhaust gas molecules $(\Sigma m)V$ is equal to the magnitude of the momentum of the rocket, Mv. (The symbol Σ means "sum of" and Σm is the sum of the masses of the molecules, or the total mass of the exhausted gas.)

In a similar manner, the conservation of momentum can be applied to a rocket or jet propulsion (Fig. 3.8). Assuming the rocket blasted off from rest, the total momentum of the molecules of the exhaust gases (assuming they don't strike anything) *at any instant* is equal and opposite to the momentum of the rocket. This is a more difficult case to analyze completely because one body, the rocket, is continually losing mass, and the other body, the expelled gases, is continually gaining mass. Your instructor may wish to explain this case more fully. A simple example is given in Special Feature 3.3 at the end of the chapter.

Another application of momentum is given in Special Feature 3.2.

Collisions

Things are always bumping into each other or colliding—sometimes purposefully, like billiard balls, and sometimes not so purposefully, like automobiles. We ordinarily think of objects coming into contact during a collision with contact forces involved. However, this is not always the case. Action-at-a-distance forces can be involved. For example, a meteor may "collide" with the Earth (gravity acting) and miss it completely.

Because there is always an interaction between colliding objects, momentum is involved. Therefore, in a broad sense, a **collision** is any interaction in which momentum is exchanged or transferred. Since collision forces are internal forces of the system, we can analyze collisions in terms of the conservation of momentum, and if there are no external forces,

Total momentum before collision
= total momentum after collision

Let's take a look at some simple, head-on, contact collisions to see how momentum helps us understand what happens. Usually, we know or can measure the masses and velocities of the colliding objects before collision, so the conservation of momentum allows us to predict the motions of the objects after collision. We distinguish between elastic and inelastic collisions.

ELASTIC COLLISIONS

In elastic collisions the colliding objects rebound without lasting deformation. In addition, no heat (a form of energy) is generated and lost during collision.† This is an ideal case, but many hard objects, such as billiard

† In an elastic collision, the kinetic energy, or energy of motion, is conserved. Only a descriptive treatment is presented here to avoid mathematical details. Kinetic energy will be discussed in Chapter 6.

SPECIAL FEATURE 3.2

Reverse Thrust

After touching down when landing a jet plane, the pilot "revs" up the engines in order to apply braking action. This is called applying reverse thrust, and one method is to use clamshell doors that deflect the exhaust gases from the jet engines in the forward direction (Fig. 3.9). With the doors in the in-flight open position (a), the exhaust gases go out the back of the engine, and the plane receives momentum in the forward direction, or a forward thrust. With the clamshell doors closed (b), the exhaust gases are deflected toward the forward direction, and there is a (vector) component of momentum in that direction. By the conservation of momentum, the engine and the airplane receive a momentum in the reverse direction, which gives an impulse or reverse thrust in that direction, thus helping to slow and stop the plane. (The airplane brakes would never do it alone.)

The deflector doors may also be exterior to the engine and form part of the engine casing. Smaller commercial jet aircraft use reverse thrust to "back away" from loading docks. Larger planes are pushed away with two motors. (Why?) The next time you are at a large airport, watch to see if you can see the deflector doors when a plane backs out.

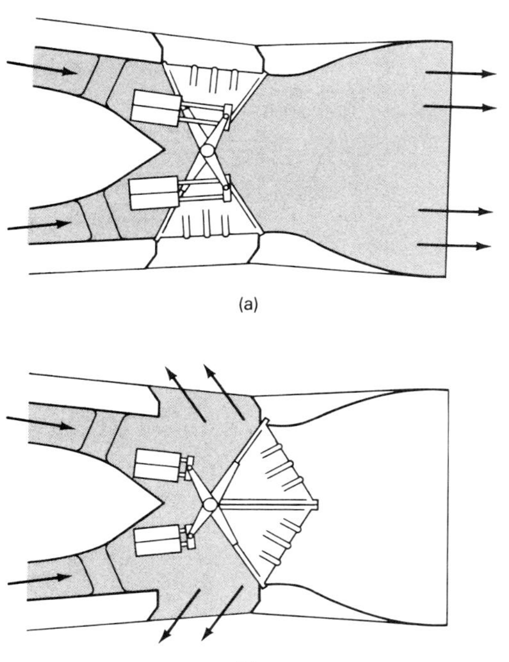

(c)

Figure 3.9 Thrust. (a) With the exhaust gases going out the back of the engine, there is a forward thrust (momentum) on the engine. (b) When the clamshell doors are closed and the gases are deflected forward, there is a reverse thrust that slows a jet plane. (c) Actual jet engine clamshell doors.

balls, steel balls, bowling balls, and marbles, have nearly elastic collisions. We find in actuality that only atoms and subatomic particles can have truly elastic collisions.

Consider the elastic collisions of balls of equal mass, as shown in Figure 3.10. During a collision only internal forces are involved, so the momentum is conserved. There is an exchange or transfer of momentum during collision, but the total momentum of the system before collision equals the total momentum after collision.

This is the principle of the novelty item shown in Figure 3.11. When one ball swings in, one swings out; when two balls swing in, two swing out; and so on. You can now understand why this happens.

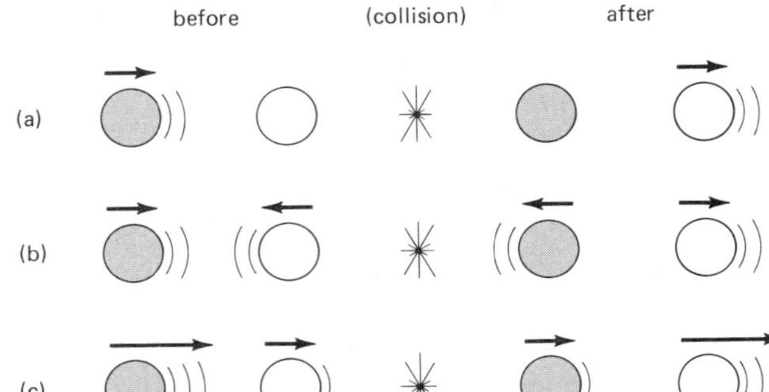

Figure 3.10 Collisions and conservation of momentum. For collisions of balls of equal mass, there is simply an exchange or complete transfer of momentum—each ball after collision has the same momentum as the other ball had before collision. The arrows represent the momenta of the balls.

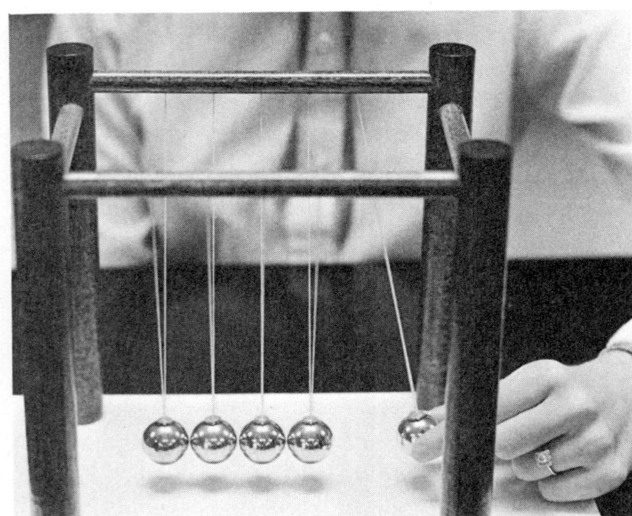

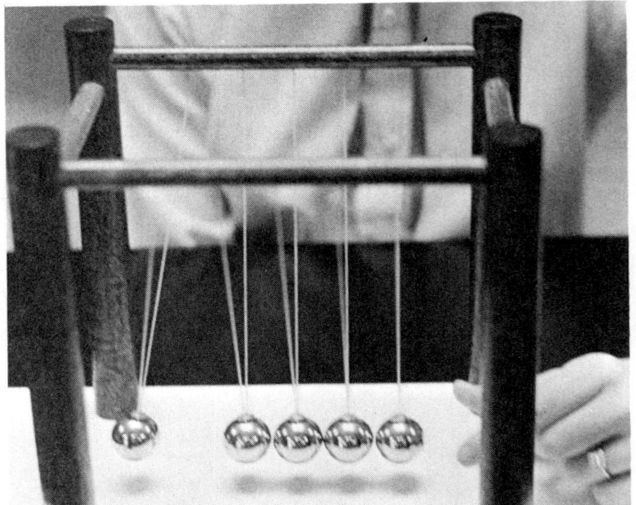

(a)

(b)

Figure 3.11 A case of momentum transfer in elastic collisions. Momentum is transferred down the line of balls, and when one ball swings in, one swings out. When two balls swing in, two swing out; and so on. The line drawing (b) illustrates the collision transfer process when a single ball swings in.

Figure 3.12 A simple apparatus for studying collisions. All you need is a grooved ruler and some marbles.

QUESTION: Why does one ball swing out when one ball swings in?

ANSWER: During the collisions, momentum is conserved in the horizontal direction. Since the balls are of equal mass, there is, ideally, a complete transfer of momentum. When one ball swings in, it stops as it transfers all of its momentum to the first ball in the row. This ball collides with the next ball in the row, and so on down the line. Because the balls in the row are in contact or almost in contact, we do not "see" the collisions, but they occur just as though the balls were separated and the motions before and after could be observed (see Fig. 3.10(a)). In the absence of a commercial apparatus, you can study collisions with the simple arrangement shown in Figure 3.12.

So, momentum is passed down the row by the collisions, and the final ball flies out with the same momentum of the ball that swung in (assuming perfectly elastic collisions). How about two balls in? Here, there is a double shot of momentum passed down the row.

When the two balls swing in, the first incoming ball collides with the first stationary ball in the row. Transferring all of its momentum, the first incoming ball stops, and its momentum is transferred down the row. Instantaneously, the second incoming ball strikes the first, now-stationary incoming ball, which now acts as part of the stationary row. The momentum of the second ball is transferred down the row after that of the first. The first momentum transfer causes the far-end ball to swing out, and the second momentum transfer causes the next-to-the-end ball to swing out. Because the collisions of the second momentum transfer instantaneously follow the first, the balls swing out together.

Returning now to the elastic collision of only two balls, suppose the masses of the balls aren't equal. The momentum is still conserved. Two extreme cases are illustrated in Figure 3.13. In the first case (a), the incoming ball is much less massive than the stationary second ball. After collision, the incoming ball rebounds in the opposite direction, while the big (massive) ball moves slowly to the right. It has a large momentum, chiefly because of its mass, which conserves the total momentum, but it has a small velocity.

In the second case (b), the incoming ball is much more massive than the stationary second ball. After collision in this case, the incoming ball continues in the same direction after having transferred a small amount of momentum to the little ball. However, because of its small mass, the little ball has a relatively larger velocity and so speeds ahead of the larger, more massive ball.

Not all collisions are head-on collisions. There are glancing collisions where the objects go off at angles to the direction of the initial motion (Fig. 3.14). Here, we

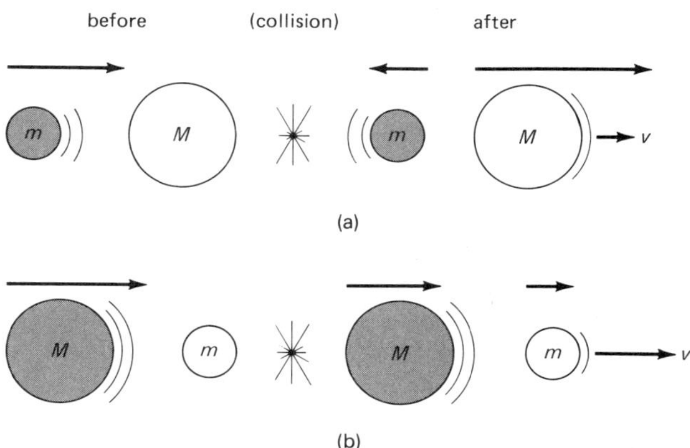

Figure 3.13 Collision of balls of unequal masses. (a) When a small mass collides with a much more massive one, the small ball rebounds and the massive ball moves slowly to the right with a large momentum because of its mass. The arrows over the balls represent momenta. (b) A massive ball colliding with a ball of small mass transfers a small amount of momentum, but the small ball has a relatively large velocity and so speeds ahead of the larger, more massive ball.

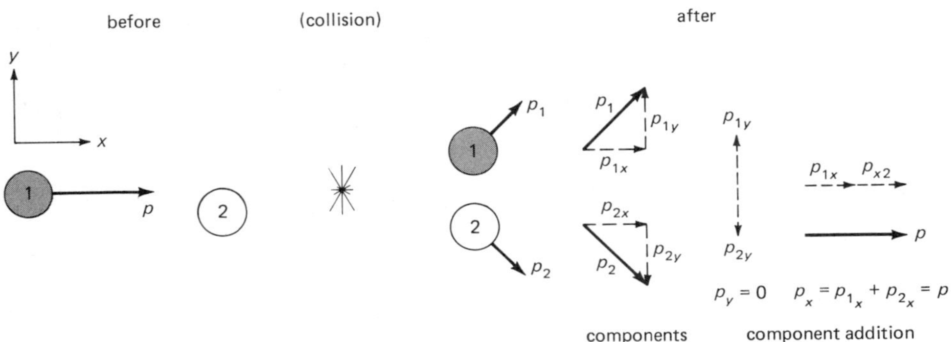

Figure 3.14 A glancing collision. The total momentum is conserved before and after collision in the component (x and y) directions.

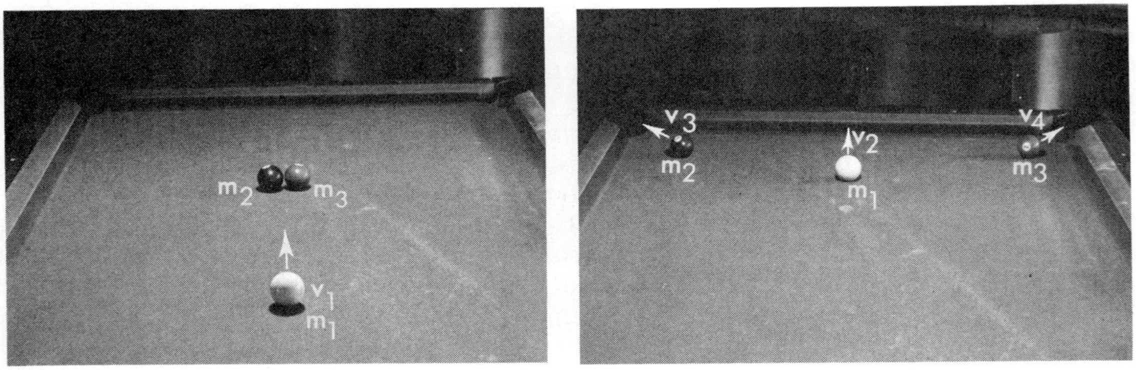

(a)

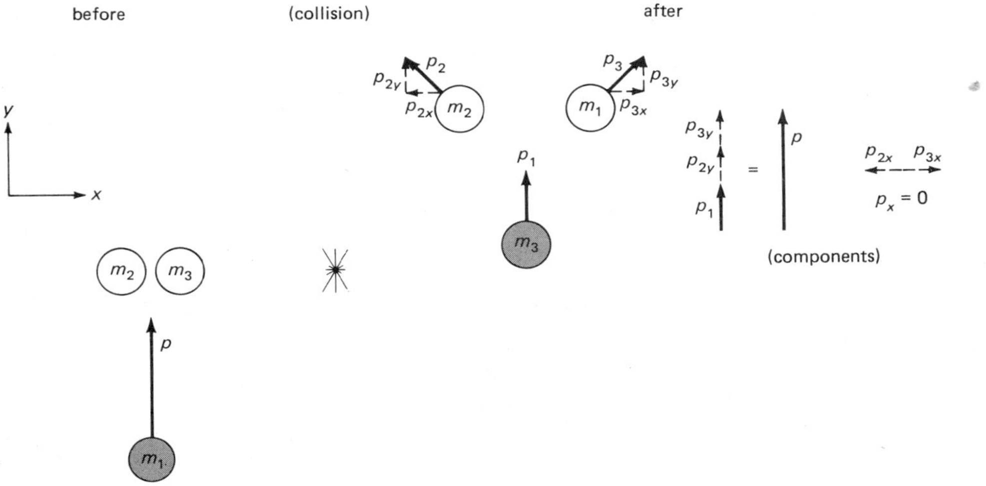

(b)

Figure 3.15 Three-body collision. The total momentum is conserved before and after collision. This requires that the momentum in the component (x and y) directions also be conserved.

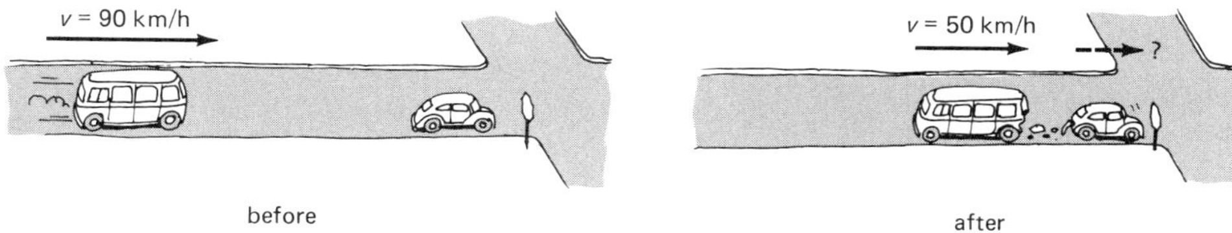

Figure 3.16 In an inelastic collision, energy is lost, but even so, the total momentum is conserved if there is no net external force. See text for description.

look at the conservation of momentum in different directions through the vector components. Recall from Chapter 2 how a vector can be resolved or "broken up" into x- and y-components. The symbol p is used for momentum, and at any time $p = mv$. Notice that before the collision all of the momentum is in the x-direction. We say that the momentum in the y-direction is zero, since there is no motion in that direction.

After collision, the momentum must be the same as before in both the x- and y-directions. Looking at the components of the momentum vectors after collision, we see that the y-components are equal and opposite. These component vectors cancel each other, and the momentum in the y-direction is still zero. The two components of momentum in the x-direction add, and their sum is equal to the momentum before collision.

Collisions can involve more than two objects. A collision of three pool balls is shown in Figure 3.15. As the component vector drawings show, the total momentum is conserved. Also, it was a pretty good shot.

INELASTIC COLLISIONS

The collisions we observe in everyday life are generally inelastic collisions. In an inelastic collision, there is some deformation of the object(s), or heat is generated (energy is lost). Even so, momentum is conserved, provided there is no net external force. The internal forces of the collision process may be quite complicated, but the total momentum before and after collision remains unchanged.

Suppose a van of mass $2m$ going 90 km/h hits a stationary auto of mass m, as illustrated in Figure 3.16. After the collision, the velocity of the van is 50 km/h. What would be the velocity of the car? Ignoring friction, we have by the conservation of momentum

$$\text{Total momentum before} = \text{total momentum after}$$

$$\text{or} \quad 2m \times 90 \text{ km/h} = (2m \times 50 \text{ km/h}) + (m \times v_{car})$$

Cancelling the m's on each side of the equation (or dividing both sides of the equation by m), we have

$$2 \times 90 \text{ km/h} = (2 \times 50 \text{ km/h}) + v_{car}$$

$$180 \text{ km/h} = 100 \text{ km/h} + v_{car}$$

and it can be seen that the velocity of the car after collision must be 80 km/h. (Maybe there will be a case of "whiplash." Do you remember why from Chapter 2?)

When colliding objects stick together after collision, it is called a "completely inelastic" collision. The conservation of momentum comes to our aid in analyzing these collisions, too. Suppose two railroad cars roll toward each other and become coupled during collision (Fig. 3.17). What happens after the collision?

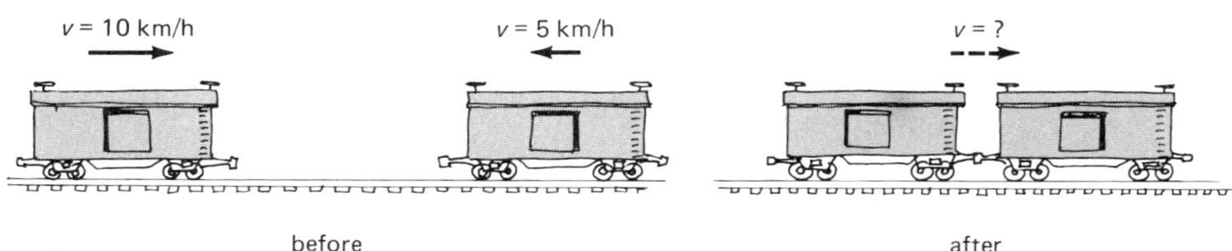

Figure 3.17 Completely inelastic collision. In a completely inelastic collision, the objects stick together. Here, two railroad cars are coupled during collision. See text for description.

Again applying the conservation of momentum,

Total momentum before = total momentum after

$$(m \times 20 \text{ km/h}) - (m \times 10 \text{ km/h}) = (m + m) \times v = 2m \times v$$

Here, the negative sign indicates that the second car is moving in the opposite direction to the first (positive direction to the right). Then, cancelling the m's,

$$20 \text{ km/h} - 10 \text{ km/h} = 2 \times v$$

or

$$10 \text{ km/h} = 2 \times v$$

and v can be seen to be 5 km/h. Since this velocity is positive, the coupled cars move to the right with a speed of 5 km/h. (If the velocity had come out to be negative, the motion would be in the designated negative direction or to the left.)

Can you tell what would happen if the railroad cars approached each other with equal speeds?

SUMMARY OF KEY TERMS

Impulse the product of the force and contact time. Impulse = $F\Delta t$.

Momentum the product of a body's mass and velocity. Momentum $p = mv$.

Conservation of momentum the momentum of a body or system remains constant or unchanged. Condition: there is no unbalanced external force acting on the body or system.

System a collection of masses or objects.

Total momentum the vector sum of the individual momenta of all of the objects in a system.

External force an outside force that acts on a body or system.

Internal force a force that acts within a body or a system. By Newton's third law, internal forces occur in equal and opposite pairs.

Collision any interaction in which momentum is exchanged or transferred.

Elastic collision a collision in which the colliding objects are not deformed and/or heat is not generated (no energy lost).

Inelastic collision a collision in which there is a deformation of one or more of the colliding objects and/or heat is generated (energy lost).

EXERCISES

1. Pole vaulters and gymnasts use padded mats for landings. What is the purpose of these mats?
2. Prize fighting was once done with bare knuckles. Why are padded boxing gloves now used? (Some people believe there should be more padding in the gloves or that boxing should be banned completely.)
3. A boxer quickly learns to move his head backward when he sees he is going to receive a jab to the head. What does the head motion accomplish?
4. Also in boxing, sharp jabs are used as "punishing" blows, whereas a boxer usually follows through with a "KO" punch. Explain the difference.
5. A golfer using a 9-iron to chip onto the green or a wedge to get out of a sand trap often uses a short "chopping" swing. Why is this?
6. Most new automobiles are equipped with bumpers that collapse under a large impact. What is the purpose of such bumpers?
7. Guard rails along roadsides collapse (bend and crumple) when a car runs into them. Wouldn't it be better to install stronger guard rails so they wouldn't have to be replaced so often?
8. In the science fiction series Star Trek, when the warp drive goes out on the Starship *Enterprise* the slower impulse engines are used for propulsion. Describe how an impulse engine might work.
9. While you are driving along, another car pulls out in front of you so you can't avoid hitting it. Would it be better to hit the car head-on or try to sideswipe it? Explain.
10. On September 8, 1974, the daredevil motorcycle rider Evel Kneivel tried to jump across the Snake River Canyon on a jet-powered motorcycle (Fig. 3.18). The nose cone of the motorcycle rocket was purposely constructed so that it would crumple on impact. What was the purpose of this? The rocket malfunctioned, and Evel crashed to the bottom of the canyon—with only minor injuries.
11. When a balloon is blown up and released, it flies around in a zig-zag fashion. What causes this?
12. When a rifle or gun is fired, there is generally more recoil or "kick" the greater the caliber (the more massive the bullet) of the gun. Why is this? Would you want to fire a rifle that was only a few times heavier than the bullet?
13. A bazooka or rocket launcher used as a weapon against

Figure 3.18
See Exercise 10.

tanks is essentially a tube open at both ends that a soldier holds on the shoulder when firing a heavy shell. There is very little recoil on the person firing the rocket launcher. Why is this? Also, it is very important that another person loading the launcher not stand behind it when it is fired. Why is this?

14. The Army uses large-caliber "recoilless" rifles. Explain how a firearm might be made recoilless. (Actually, the rifles are not completely recoilless, but the recoil is greatly reduced.)

15. Suppose a person in a sailboat is becalmed on a lake and has a large battery-operated fan on board. Getting a bright idea, the person aims the fan at the sail in order to blow the boat to the shore (Fig. 3.19). However, he finds

Figure 3.19 See Exercise 15.

that the boat goes nowhere. Getting an even brighter idea (and applying physics principles he learned in college), he takes the sail down and turns the fan around. The boat then moves forward. Explain the motion (or lack of it) in each case.

16. Explain why it is difficult for a fire fighter to hold a high-pressure hose when trying to put out a fire. What must the fire fighter do in holding the hose stationary?

17. An astronaut approaching the moon must fire retro-rockets to slow the craft down so it will not crash (fall) into the moon. The exhaust gases of the retrorockets are expelled toward the moon. Explain how firing these rockets slows down the spacecraft.

18. A rubber ball is thrown horizontally against a wall, and the ball rebounds in the opposite direction with essentially the same speed it had just before hitting the wall. Is momentum conserved? Explain.

19. Analyze the maneuverability of an astronaut using a hand rocket in space in terms of momentum (see Fig. 2.18).

20. In applying a reverse thrust in a jet engine, suppose a gas molecule of mass m and velocity v is deflected directly opposite to its original direction. The change in momentum (impulse) of the molecule is then $2mv$. Show that this is correct. (*Hint:* the change in momentum is a vector addition or subtraction, and direction is a consideration.)

21. A person standing on the Earth jumps vertically upward. According to the conservation of momentum, what happens to the Earth? (You can move the Earth!) Can we observe this effect on the Earth? Explain.

22. Two masses on a horizontal frictionless surface with a spring between them are pushed together, and a string is attached to hold them together (Fig. 3.20). If the string is then burned and the $3M$ mass moves to the right with a velocity of 2 m/s, what is the velocity of the other mass? What kind of force does the spring apply to the system?

23. Suppose you are standing in the middle of a frozen lake and that the ice is perfectly frictionless. How could you get to shore?

24. In a collision of balls of equal masses, the momentum is conserved. We might also say that the velocity is conserved. Why is this? If the balls were not of equal mass, would the velocity be conserved? Would the momentum be conserved?

25. A ball of mass m and velocity v collides with a stationary ball of unknown mass. What is the mass of the stationary ball if after collision the incoming ball is stationary and the other ball has a velocity?

26. In a suspended ball set with five balls (cf. Fig. 3.11), if three balls are allowed to swing in toward two stationary balls, explain why three balls swing out.

27. A person who weighs 80 kilos skates with a constant velocity of 6 m/s and catches a child who has fallen from a window. If the child weighs 20 kilos, what is their velocity after the catch?

28. Two Volkswagen "bugs" collide in a completely inelastic collision, as illustrated in Figure 3.21. What is the velocity of the cars immediately after collision?

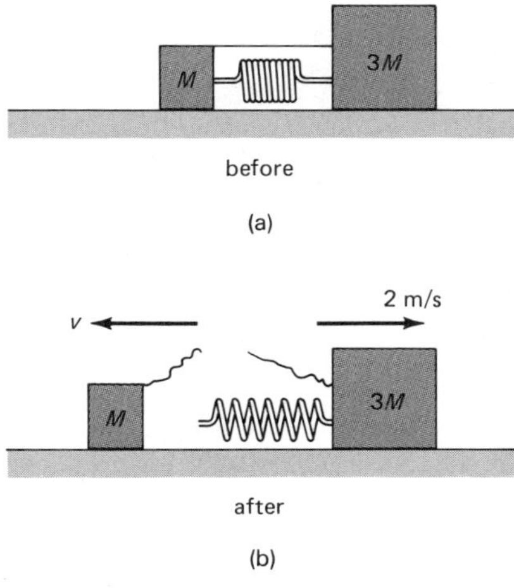

Figure 3.20 See Exercise 22.

Figure 3.21 See Exercise 28.

29. A cannon ball is fired horizontally from a cannon with a constant muzzle velocity. Is the horizontal momentum conserved thereafter? Is the vertical momentum conserved? (*Hint:* Think in terms of components of momentum.)

30. In testing explosive charges in space, an astronaut leaves a bomb there that is later detonated by remote control. The bomb explodes into many fragments that fly off in various directions. What is the total momentum of the bomb fragments?

31. Two cars of the same model and year (same mass) approach an intersection traveling at the same speed. One is traveling north and the other east. At the intersection, one runs a stop sign and they collide in a completely inelastic collision. In what direction do the cars move after collision? (*Hint:* use a vector diagram.)

32. Two railroad cars of equal mass roll toward each other on a level track at the same speed. What happens after collision?

33. A single railroad car with a velocity v rolls toward two coupled railroad cars moving with a velocity of $v/2$ in the opposite direction. If the track is level and all the cars have equal masses, what happens after collision?

34. A railroad car with a speed of 20 km/h rolls on a level track in the same direction as another car of the same mass that has a speed of 10 km/h. The faster car overtakes the slower one, and they couple together on collision. What is the velocity of the cars after collision?

SPECIAL FEATURE 3.3

The Rocket*

A rocket or spaceship consists of two primary parts: (1) the payload—the rocket hull, astronauts, instruments, and so on that we wish to propel somewhere—and (2) the fuel and anything else that gets ejected in propelling the rocket.

To simplify the situation, let's consider a rather primitive type of spacecraft of mass 1001 kg far out in space, where no forces are acting on the craft (Fig. 3.22). Let m_2 be the mass of the rocketship *at any time* and m_1 be the mass of a 1-kg "fuel pellet."

Suppose an astronaut throws one of the fuel pellets out the back of the rocket with a velocity of 20 m/s. We take the position where the pellet was thrown out as our zero position (see figure). Then, by the conservation of momentum,

$$m_1 v_1 = -m_2 v_2$$

or $$(1 \text{ kg})(20 \text{ m/s}) = -(1000 \text{ kg})v_2$$

and $$v_2 = -0.02 \text{ m/s}$$

where the minus sign indicates that the change in the velocity of the rocket is in the opposite direction to that of the fuel pellet, or to the right in the figure.

Hence, the rocket has changed its velocity by a small amount relative to the position in space where the fuel pellet was thrown out, and the rocket has 1 kg less mass.

* Adapted from Highsmith, P. E. and A. S. Howard, *Adventures in Physics*, W. B. Saunders Co., Philadelphia, 1972.

Figure 3.22 An "impulse" rocket propelled by the ejection of "fuel pellets." See text for description.

The astronaut then throws out another 1-kg chunk of fuel with the same velocity, and

$$m_1 v_1 = -m_2 v_2$$

or $$(1 \text{ kg})(20 \text{ m/s}) = -(999 \text{ kg})v_2$$

and $$v_2 = -0.02002 \text{ m/s}$$

The change in the velocity of the rocket this time is slightly greater because the rocket lost mass in the first throw. Relative to the original zero position, the change in the rocket's velocity is now $-0.02 + (-0.02002) = -0.04002$ m/s.

If the astronaut continued to throw out fuel pellets, the rocket's velocity would change slightly each time and it would continue to lose mass, which would give the rocket a slightly greater velocity change each time. For example, by the time the rocket had a mass of 401 kg, the next-thrown fuel pellet would give the rocket a change in velocity of

$$m_1 v_1 = -m_2 v_2$$

$$(1 \text{ kg})(20 \text{ m/s}) = -(400 \text{ kg})v_2$$

and $$v_2 = -0.05 \text{ m/s}$$

The change in velocity for this throw is $2\frac{1}{2}$ times greater than for the first fuel pellet.

If the astronaut continued to throw out fuel pellets, the rocket would continue to lose mass, and the change in its velocity would increase each time. If the fuel pellets were thrown out often enough, the change in velocity would appear smooth, which the astronauts would interpret as an increasing acceleration.

For a real rocket during a "burn," the exhaust gas molecules are ejected out the rear of the rocket at speeds of several thousand meters per second. The average velocity of the exhaust gases and the amount of material ejected is relatively constant. Therefore, the rocket thrust and the force on the rocket are constant. The increasing acceleration is due to the rocket's decreasing mass.

Of course, the masses of the exhaust gas molecules are much less than that of the 1-kg fuel pellets of our simplified rocketship, but there are a lot of them. To show the effect of high exhaust velocities, suppose the astronaut had thrown out the first fuel pellet with a speed of 20,000 m/s. Then,

Figure 3.23 In a multistage rocket, the burned-out stages are jettisoned for an in-flight reduction of mass. The photo shows the interstage section that connects the first and second stages of the Saturn V rocket falling away after the separation of the two stages.

$$m_1 v_1 = -m_2 v_2$$

$$(1 \text{ kg})(20{,}000 \text{ m/s}) = -(1000 \text{ kg})v_2$$

and $$v_2 = -200 \text{ m/s}$$

A large velocity change indeed!

To further reduce the in-flight mass of a rocketship, multistage rockets are used. The burnt-out stages are jettisoned for an in-flight reduction in mass (Fig. 3.23).

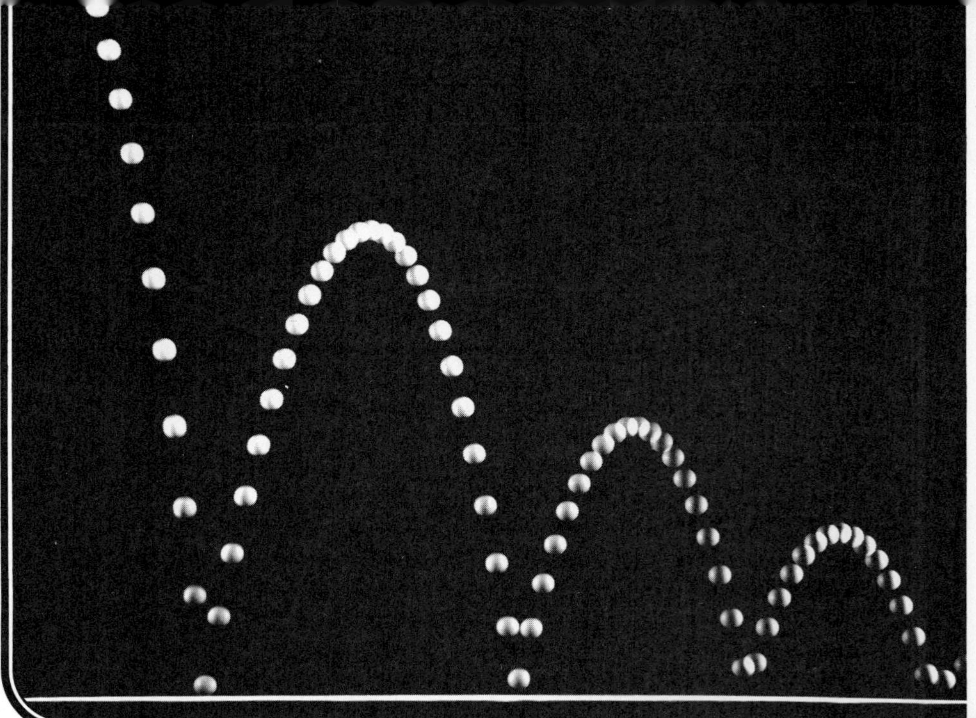

4

Projectile and Circular Motions

Projectiles

Now that you have the basics of force and motion under your belt, we can consider some special, but common, cases of objects in motion. One such case is projectile motion.

When you throw a ball or shoot a rifle, you put the ball or bullet in motion — projectile motion. A projectile is just an object that has been thrown or projected by some means. We do it all the time.

In describing projectile motion, we usually consider only the motion of the projectile after it has been thrown or launched. That is, a projected object has an initial velocity in some direction. For example, the initial velocity of a fired bullet would be its muzzle velocity on leaving the barrel. Often we are not concerned with the force that caused the initial motion, only what happens afterward.

Other forces may act on a projectile after it is launched. In particular, here on Earth there is gravity. Normally, projected objects fall to the ground. (Space launches and Earth satellites are topics in Chapter 5.) Let's see how we describe this projectile motion. First, we will consider objects that are moving only directly up and down (in one dimension) and then expand our analysis to two dimensions.

Falling Objects

In a sense, a falling object is a projectile. When you drop something, *you* do not apply a force to get it moving. It is projected with an initial velocity of zero. Gravity takes over, and the object accelerates toward the Earth. Of course, you could always throw or project an object downward so that its initial velocity would not be zero.

FREE FALL

When an object falls toward the Earth with only gravity acting on it, we say it is in **free fall.** Air resistance and anything else are neglected. This makes life easier and is a good approximation in some cases, for example, for a heavy object falling a short distance.

All objects in free fall near the Earth have the same acceleration—the acceleration due to gravity. Recall that the gravitational force acting on an object near the Earth is its weight force, $w = mg$ ($F = ma$), where the acceleration due to gravity g, is equal to 9.8 m/s^2 (or 32 ft/s^2).

It might be thought that heavy objects would fall faster than lighter objects. After all, if one object weighs twice as much as another, or the gravitational force is twice as great, why shouldn't it fall twice as fast? Aristotle thought that this was the case in his theory of motion, but Galileo showed, or at least believed, otherwise. See Special Feature 4.1.

SPECIAL FEATURE 4.1

Galileo and the Leaning Tower of Pisa

There is a popular and well-known story that Galileo performed experiments with falling bodies by dropping objects from the Leaning Tower of Pisa (Fig. 4.1). However, the authenticity of this story is seriously questionable. The original question as addressed by Aristotle was, Why do bodies fall to the ground? According to the

(a)

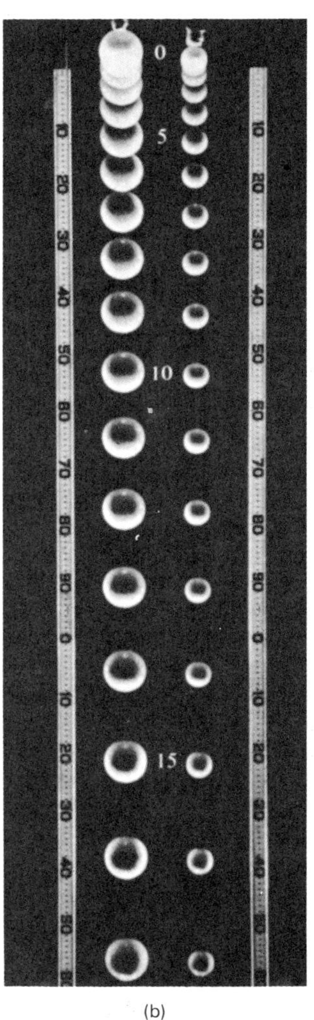

(b)

Figure 4.1 Free fall. Objects in free fall have the same acceleration. (a) Legend has it that Galileo demonstrated this by dropping objects of different mass or weight from the top of the Leaning Tower of Pisa. (b) A photograph of a baseball and a golf ball released simultaneously showing that they fall together, or have the same acceleration.

(Continued)

Aristotelian view, they were seeking their natural place at the center of the Earth, which itself was the center of the universe. How this occurred supposedly depended on the "earthiness" of a body; that is, the heavier a body, the faster it would fall in seeking its natural place. There is little doubt that Galileo questioned this view. In an early writing from about 1590 when Galileo was living in Pisa (see Special Feature 2.2) he stated:

> How ridiculous is this opinion of Aristotle is clearer than light. Who ever would believe, for example, that . . . if two stones were flung at the same moment from a *high tower,* one stone twice the size of the other, . . . that when the smaller was half-way down the larger had already reached the ground?*

And from a later passage concerning the falling of wood and lead objects,

> . . . but a little later the motion of the lead is so accelerated that it leaves the wood behind, and if they are let go from a *high tower,* precedes it by a long space; and I have often made this test.

Here there were probably some air resistance considerations.

But in a 1638 work we find,

> Aristotle says that 'an iron ball of one hundred pounds falling from a height of one hundred cubits reaches the ground before a one-pound ball has fallen a single cubit.' I say that they arrive at the same time.

The first account of a Tower of Pisa experiment came a dozen years after Galileo's death and was written by Vincenzo Viviani, his last pupil and first biographer. Viviani relates that the falling bodies

> . . . all moved at the same speed; demonstrating this with repeated experiments from the height of the Campanile (Tower) of Pisa in the presence of the other teachers and philosophers, and the whole assembly of students. . .

Yet there is no independent record of this from the time, and it is not mentioned in Galileo's writings. Did Galileo relate this to Viviani in his declining years, or did Viviani use his imagination in describing his teacher's experiments from a "high tower"? In debating this, many lose sight of the point. Whether or not Galileo in myth or in fact dropped objects from the Tower of Pisa, the acceleration due to gravity is independent of an object's mass.

* This and following quotations from Cooper, L., *Aristotle, Galileo, and The Tower of Pisa,* Cornell University Press, Ithaca, N.Y., 1935.

Galileo gave no reason why objects in free fall have equal accelerations, but Newton's second law does. Recall that the acceleration of an object depends not only on the applied force, but *also* on its mass ($a = F/m$). If one object has twice the mass of another object, then it also has twice as much inertia. Hence, it needs twice as much force to accelerate it at an equal rate. This result of Newton's law is illustrated in Figure 4.2.

Thus, we say that the *acceleration* of an object in free fall is independent of its mass or weight. This means that all objects in free fall accelerate at the same rate—the acceleration due to gravity—regardless of their masses.

When an object is accelerated, its velocity changes with time. For an object in free fall, its velocity changes (increases) 9.8 m/s each second. ($g = 9.8$ m/s per second, and acceleration = change in velocity per second.) This means that the distance fallen by the object increases during each time interval or second because its velocity increases each second (Fig. 4.3).

Figure 4.2 A heavier object (mass $2m$) in free fall has twice the (gravitational) force acting on it as a lighter object (mass m); but having twice as much mass and inertia, it falls with the same acceleration.

QUESTION: A body is dropped from rest. How fast is it falling at the end of 3 seconds?

ANSWER: The body's instantaneous velocity at the end of 3 seconds can be found from the definition of acceleration (Chapter 2): acceleration = change in velocity/elapsed time, or $a = \Delta v/\Delta t$. The change in velocity, Δv, can be written $\Delta v = v - v_o$, where v is the instantaneous velocity at the end of a certain time interval Δt and v_o is its initial velocity (at $t_o = 0$ and $\Delta t = t - t_o = t$). In this case, $v_o = 0$, since the dropped body is initially at rest. Then we may write,

$$a = \frac{v}{t}$$

or

$$v = at = gt$$

where, as we have learned for a free-falling body, $a = g$.

Let's use customary units in this case for familiarity and to make a point ($g = 32$ ft/s²). Then, with $t = 3$ s,

$$v = gt = (32 \text{ ft/s}^2)(3 \text{ s}) = 96 \text{ ft/s}$$

So the body has a speed of 96 ft/s in the downward direction. We don't commonly express speeds in the standard units of ft/s. To better appreciate how fast a freely falling body is traveling at the end of 3 seconds, there's a convenient conversion factor, 60 mi/h = 88 ft/s. Hence, at the end of 3 seconds, the body is falling with a speed in excess of 60 mi/h! (About 65.5 mi/h or 105 km/h.)

Think about this when a less serious student than yourself drops something like a water balloon out of a dormitory window to hit a fellow student or, heaven forbid, a professor. (Of course, when a water balloon hits someone, there are impulse considerations, and you already know about these. The impulse force on the balloon and its contents would probably cause the balloon to burst as you would want, particularly for a [physics] professor.*)

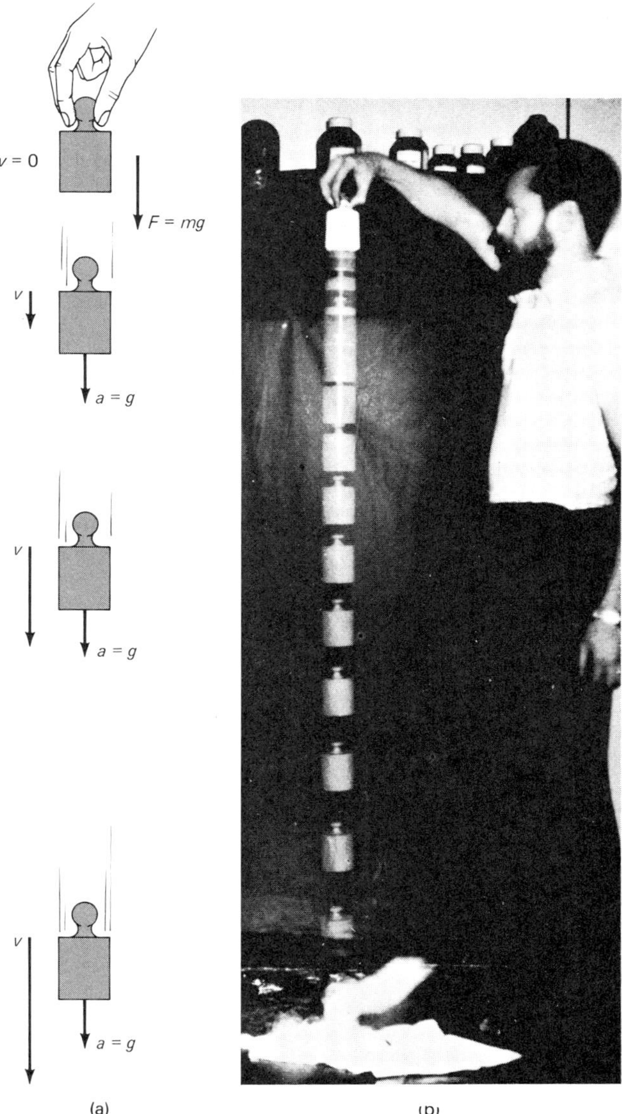

Figure 4.3 Constant acceleration, changing velocity. The distance traveled by an object increases during each time interval or second, since its velocity increases each second. With a constant acceleration, the velocity increases uniformly with time.

NONFREE FALL WITH AIR RESISTANCE

Technically, free fall would occur only in a vacuum because air resistance normally acts on a falling body. The resistance opposes the motion, so an object falls at a slower rate than in free fall. In some cases, the air resistance is appreciable and drastically affects the mo-

* If you are interested in how far a free-falling object dropped from rest falls with time, this distance is given by $d = \frac{1}{2}gt^2$. In the case of $t = 3$ seconds, $d = \frac{1}{2}gt^2 = \frac{1}{2}(32 \text{ ft/s}^2)(3 \text{ s})^2 = \frac{1}{2}(32 \text{ ft/s}^2)(9 \text{ s}^2) = 144$ ft. So for a water balloon to fall for 3 seconds, you'd have to be on the 15th floor of a high-rise dormitory (assuming 10 ft per story or floor).

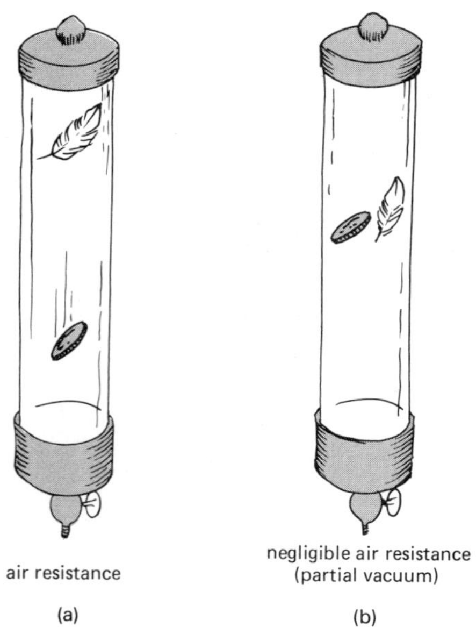

Figure 4.4 Air resistance. (a) When falling in air, the motion of a feather is retarded more than that of a coin. (b) If most of the air is removed from the tube, the coin and feather fall together.

tion. For example, a feather "floats" or falls slowly because of the upward air resistance. A common lecture demonstration is illustrated in Figure 4.4. With air in the tube, the heavier coin always beats the feather to the bottom. However, if enough air is evacuated from the tube, giving a partial vacuum, the air resistance is negligible and the feather and coin fall together at the same rate.

A similar experiment with a hammer and feather was performed on the moon in 1971 by astronaut David Scott. The moon has no atmosphere, and hence there is no air resistance. When released simultaneously, the hammer and feather fell together at the same rate and hit the lunar surface at the same time. Of course, they fell quite a bit slower than they would in a vacuum chamber on Earth, since the acceleration due to gravity on the moon is one-sixth that on Earth.

The air resistance on a falling body depends on its (1) shape, (2) size (exposed area), and (3) speed.† The amount of air a body "catches" depends on its shape. For example, a skydiver with an unopened parachute falls quite rapidly. But when the chute opens, the shape and size of the falling "body" change, so the air resistance is increased and the descent is slowed, which makes skydiving a pleasure, or at least repeatable.

† The air density is also a factor, but we will assume this to be relatively constant near the Earth's surface.

This effect can be observed by moving your hand through water. With the palm of the hand in the direction of motion, there is more resistance than if you have the tips of the fingers in the direction of motion. Automobiles are now "streamlined" in shape to reduce air resistance and improve fuel consumption.

Air resistance is also velocity- or speed-dependent. That is, the greater the speed of a falling object, the greater the air resistance. This is because the faster an object falls, the greater the number of air molecules it hits or collides with per second. This gives a greater total resistive impulse force, or air resistance. In many cases, the air resistance is proportional to the square of the speed.

So as an object falls, it gains speed or accelerates due to gravity, and the retarding force of air resistance increases with speed (Fig. 4.5). This continues until the force of air resistance equals the weight force of the falling object. The net force is then zero (downward weight force balanced by the upward force of air resistance), and the object no longer accelerates but falls with a constant speed. We then say that the object has reached *terminal velocity*. For a sky diver with an unopened parachute, the terminal velocity is about 200

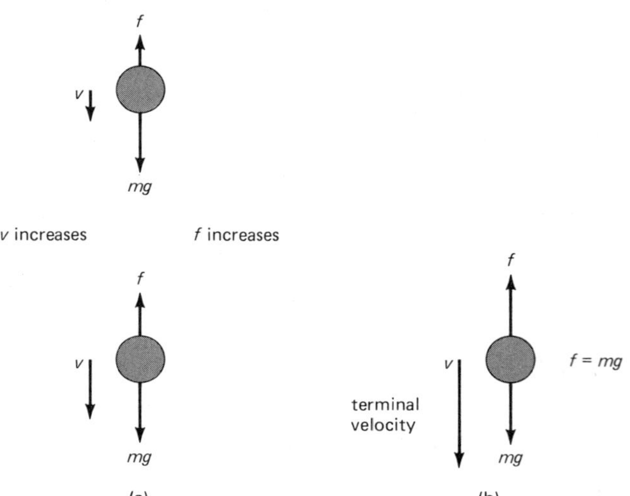

Figure 4.5 Air resistance is velocity-dependent. (a) As the velocity of a falling object increases, the force of air resistance f increases. (b) At a certain velocity, called the terminal velocity, the force of air resistance is equal to the weight of the object. Then there is no longer a net acceleration, and the object falls with a constant velocity.

Figure 4.6 Air resistance in action. Skydivers assume a "spread-eagle" position to increase the air resistance and prolong the time of fall.

km/h (125 mi/h). Sky divers are aware of the shape-dependence of air resistance. In falling they use a "spread-eagle" position to increase the air resistance and prolong the time of fall (Fig. 4.6). When the parachute is opened, the fall is slowed by the additional resistive force to a terminal velocity of about 40 km/h (25 mi/h). See Special Feature 4.2 for the effect of air resistance on falling objects of different weights.

Vertical Projections

We often throw or toss things directly upward. This is a vertical projection. There is an initial velocity v_o upward, but the acceleration (due to gravity) is downward, as always. This means that the acceleration due to gravity for the upward motion is a deceleration, or tends to slow down a vertically projected object. Vectorially, the velocity and acceleration vectors are in opposite directions on the way up (Fig. 4.8).

As we all know, the deceleration slows the object to an (instantaneous) stop at its maximum height. Hence, at its maximum height a vertically projected object stops instantaneously ($v = 0$), as it must to change directions. However, the acceleration is not zero here. Gravity is still acting on the object as always, and $a = g$.

SPECIAL FEATURE 4.2

Terminal Velocity and Falling Objects

The effect of air resistance on falling objects is to retard their motion. The magnitude of the air resistance on an object depends on the object's shape, size, and speed. Ordinarily, heavy objects of similar shape and size dropped short distances appear to accelerate at the same rate and strike the ground at the same time. This is the essence of Galileo's alleged Leaning Tower of Pisa experiment (see Special Feature 4.1). However, suppose the objects were dropped from a very high altitude so that speed and air resistance were appreciable factors. For example, if two metal balls made of different materials, one much more massive (heavier) than the other, of the same size and shape were dropped from a high-flying aircraft (Fig. 4.7). Which would strike the ground first?

Here, air resistance is a factor, and the heavier ball would hit the ground first. As the falling balls gain speed and the air resistance increases, the weight of the lighter ball would be the first to be balanced by the force of air resistance. The heavier ball would continue to accelerate downward until it reached its faster terminal velocity. So the heavier ball would be ahead of the lighter ball and would be falling faster, thereby reaching the ground first.

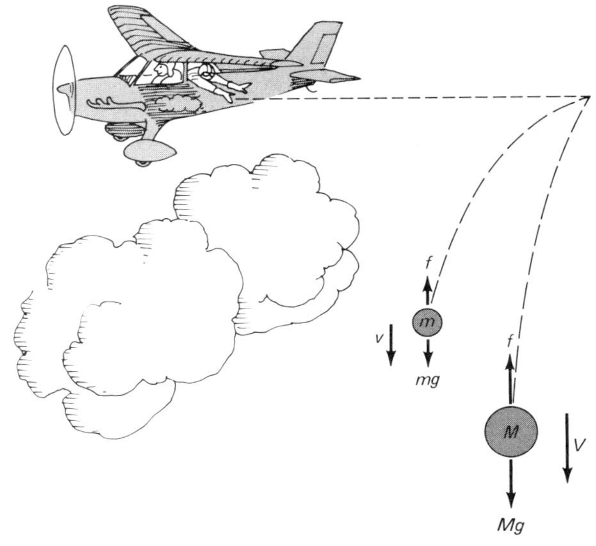

Figure 4.7 Air resistance and terminal velocity. Because of air resistance a lighter (less massive) ball will reach its slower terminal velocity before a heavier ball of similar shape and size. As a result, the heavier ball will strike the ground first.

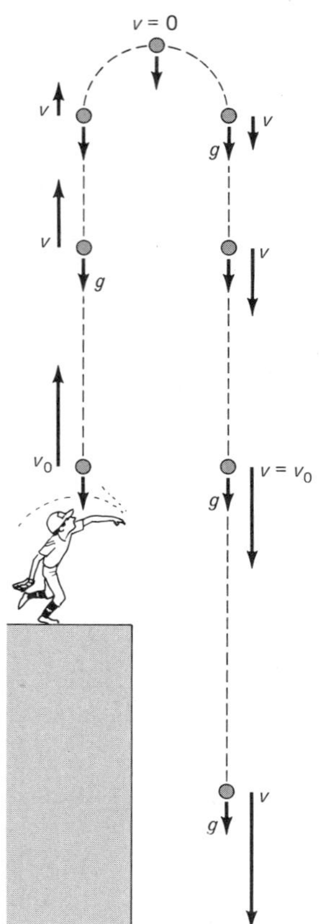

Figure 4.8 Vertical projection. The vector arrows illustrate how the acceleration remains constant and the velocity varies.

Thereafter, it is as though someone dropped the object from its maximum height (initial velocity downward is zero). At the instant a vertically projected object reaches its maximum height, it essentially becomes a dropped object in free fall.

An interesting fact for a vertical projection without air resistance is that the object returns to its starting point with the same speed as it had initially (directions and hence velocities are different—equal in magnitude but opposite in direction). Hence, in the absence of air resistance, if you shot a rifle bullet or something else vertically upward, it would return with the same speed, and you might shoot yourself.

Horizontal Projections

Another common type of projection is one in a horizontal direction. Such projections include a ball rolling off a level table, a horizontally thrown ball, or a bullet fired from a rifle held horizontally. For this type of projection, there is an initial velocity in the horizontal, or x-direction, but none in the vertical, or y-direction, $v_{y_o} = 0$ (Fig. 4.9). However, there is an acceleration in the downward direction due to our old friend gravity.

A horizontal projectile follows an arc-like path until it finally hits the ground. To analyze the motion, it is convenient to consider the individual components of motion in the two directions—much the same as we did with components of momentum in the last chapter. The actual motion is a combination or the resultant of the individual component motions.

Since there is no acceleration or force in the x-direction after it is projected, the projectile moves in this direction with a constant speed v_{x_o} (Fig. 4.9). As the object moves horizontally, it also falls (accelerates) downward owing to the force of gravity at a rate of 9.8 m/s^2 (acceleration due to gravity). In the downward direction, the motion is the same as that of a dropped object.

So the combination or resultant of these component motions is the actual arc motion of the projectile. To project an object a greater distance, we only need to give it a greater horizontal velocity and/or start at a greater distance above the ground. The latter gives the object a longer time to fall, and it travels farther horizontally during this longer time. The horizontal distance a projectile travels is called its *range.*

Projections at an Angle

The most general type of projection is at some angle to the horizontal (Fig. 4.10). Here the initial velocity v_o has components in both the horizontal directions (v_{x_o}) and the upward vertical direction (v_{y_o}). You can probably guess what happens in the component directions.

In the x-direction with no acceleration, the object moves along with a constant speed. Viewed from above, it would appear that the projectile travels in a straight line.

In the y-direction, the motion is the same as that of a vertically projected object. If a person moved along under the projectile at the same horizontal speed, that's what the person would see. Notice that at its maximum height the vertical velocity of the projectile is zero.

The combination of the component motions causes the projectile to describe a curved path called a parabola. We say that the path of the projectile is a parabolic arc. Notice that the second half of the motion is the same as that of a horizontal projection.

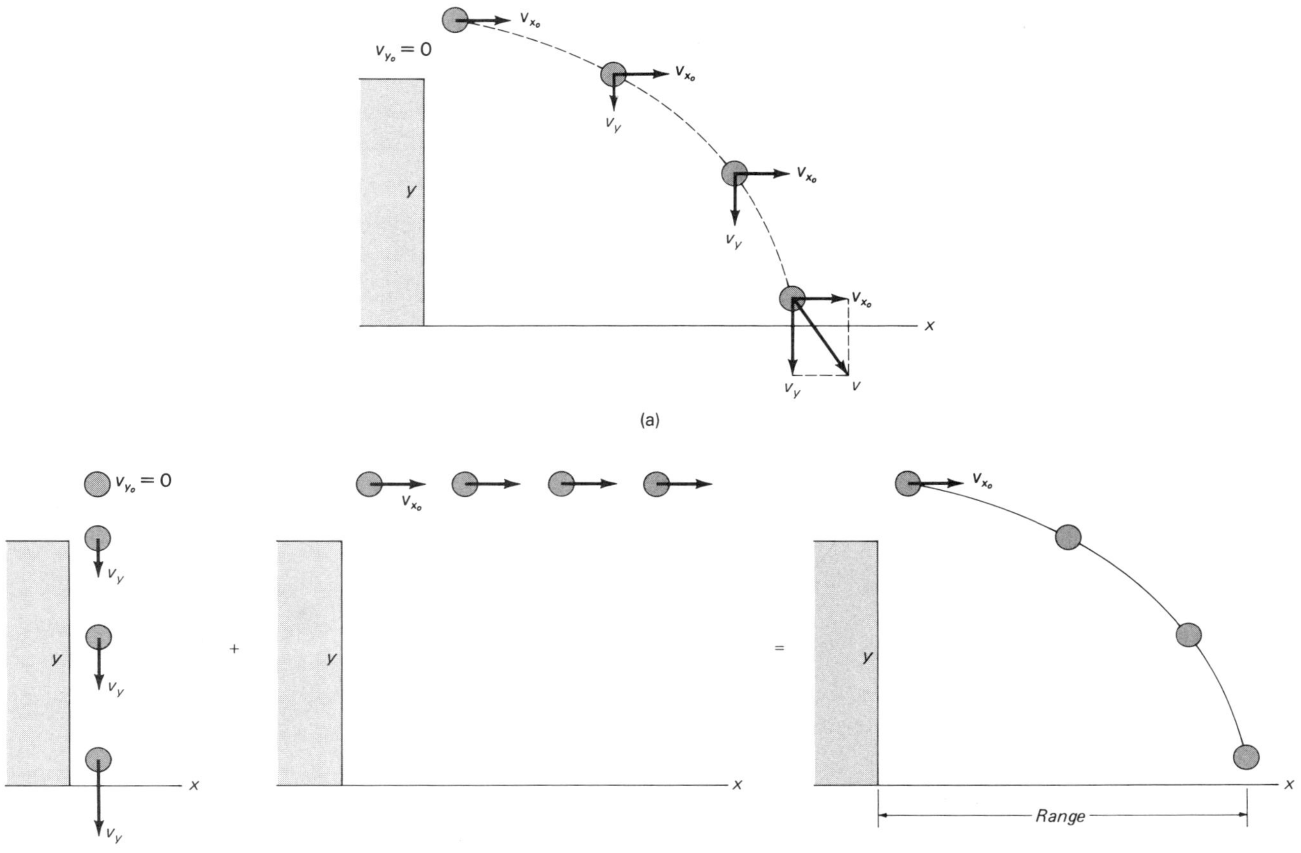

(a)

(b)

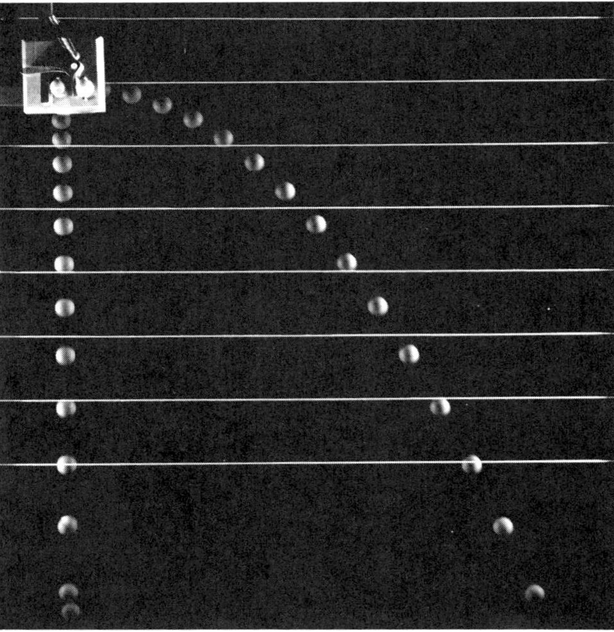

(c)

Figure 4.9 Horizontal projection. (a) The velocity in the horizontal direction is constant, since there is no acceleration in that direction. (b) The projectile motion is a combination of an object falling from rest and moving horizontally with a constant velocity at the same time. (c) A photograph of dropped and projected balls. Notice that the downward motion of the balls is the same in each case.

We are often interested in getting the maximum range (horizontal distance) for a projectile. A moment's thought should tell you that this depends on the initial launch speed and the angle of projection. Suppose the launch speed were fixed. Which angle would give you the maximum range? As you might guess, it is 45° (Fig. 4.11). This is a consideration in several sports—for example, shotput, baseball, javelin throwing, and golf—when maximum ranges are desired. In actual practice, other considerations such as air resistance and spin come into play. The projection angle for maximum range is less than 45° with air resistance. A reverse spin on a driven golf ball provides lift, and a

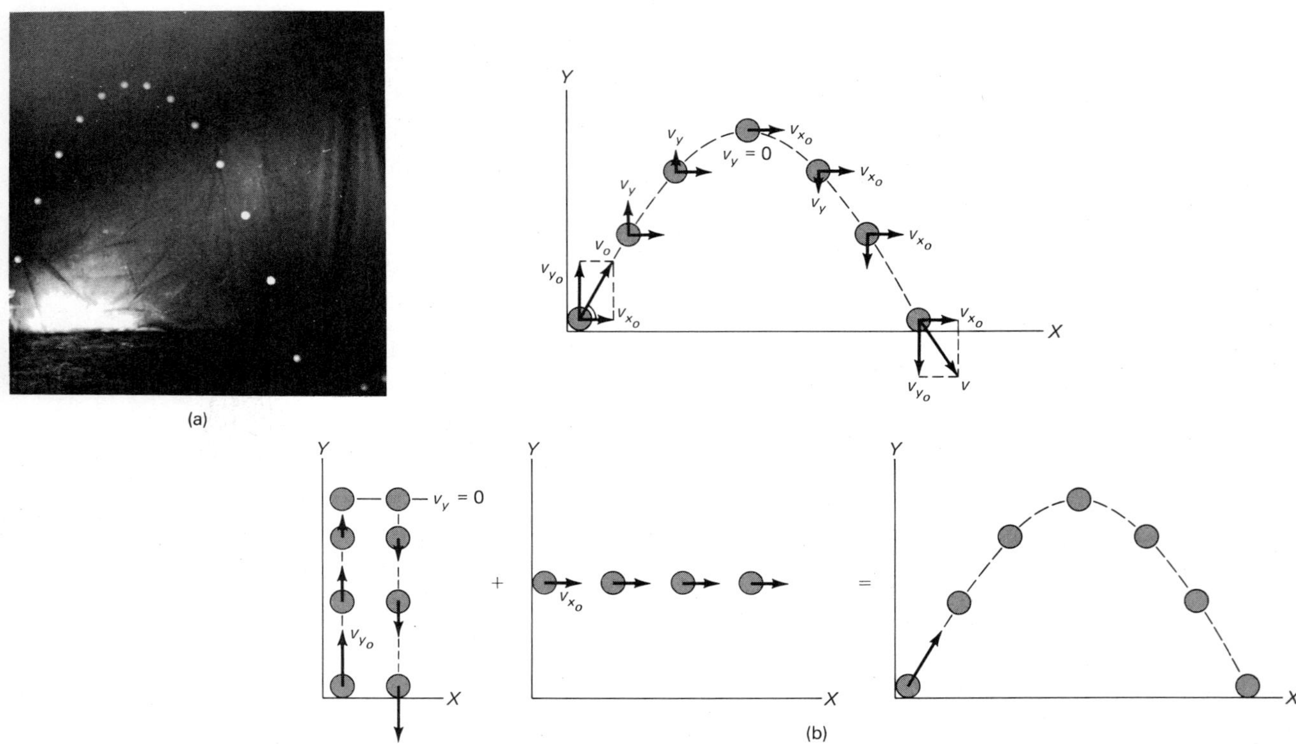

Figure 4.10 Projection at an angle (a). The motion is a combination of vertical and horizontal motions (b). The object describes a parabolic path.

typical launch angle is considerably less than 45°. Also, when the launch height is above the ground or the landing height, the angle of projection for the maximum range is less than 45°.

Circular Motion and Centripetal Force

Another common type of motion is circular motion, for example, a car traveling around a circular track or yourself on the rotating Earth—you have a circular path in space. In analyzing circular motion, we will generally consider uniform circular motion. An object traveling in uniform circular motion travels in a circular path with a constant or uniform speed.

As was learned, an acceleration produces a change in velocity due to a change in speed and/*or* a change in direction. For an object in uniform circular motion, its speed is constant, but its velocity is *not,* because the direction of motion is continually changing (Fig. 4.12). Where is the acceleration and thus the force that produces this change in velocity?

If we add the velocity vectors to find the direction of the change in velocity Δv (Fig. 4.12), it is found to be instantaneously directed *toward* the center of the circle. This "center-seeking" acceleration is called **centripetal acceleration** (a_c). Hence,

> A centripetal acceleration, and so a centripetal force, is required for circular motion.

The centripetal force that supplies the centripetal acceleration for circular motion comes from various sources, depending on the situation. For an automobile on a circular track or rounding a corner, the centripetal force is provided by the friction force between the tires and the road. For you on the rotating Earth, the force is gravity. For a ball on a string being swung in a circle (Fig. 4.13(a)), the force is the string tension supplied by the person. A centripetal force then is just any force that produces circular motion. The force is directed toward the center of the circular path or at right angles (90°) to the instantaneous velocity. Notice that the object is accelerated toward the center of the circle, but never gets any closer to the center.

It is easy to show that a center-seeking or centripetal force is required for circular motion. Suppose you were swinging a ball on a string as in Figure 4.13 and the

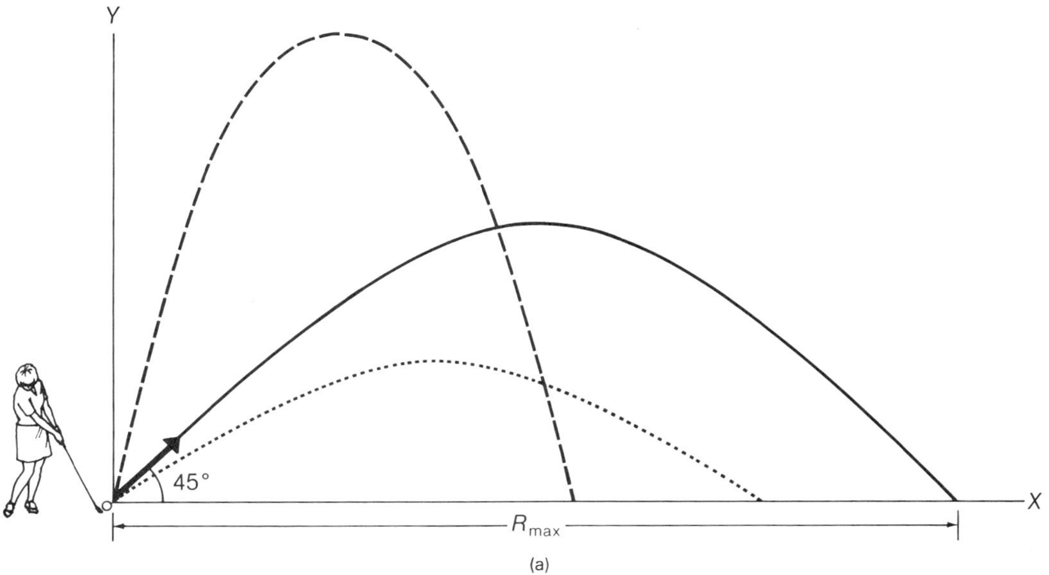

Figure 4.11 The projection angle for the maximum range is 45°. This is important in several sports.

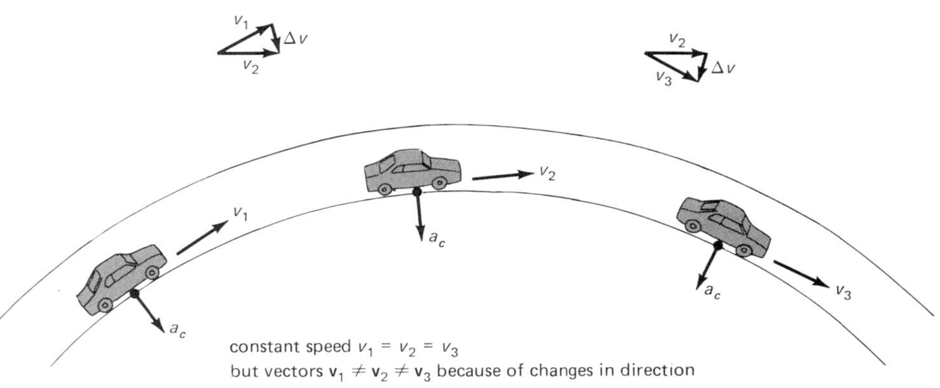

Figure 4.12 Centripetal "center-seeking" acceleration. The speed of an object in circular motion may be constant, but because of changes in direction, the velocity is not. Circular motion requires a centripetal force or acceleration (a_c).

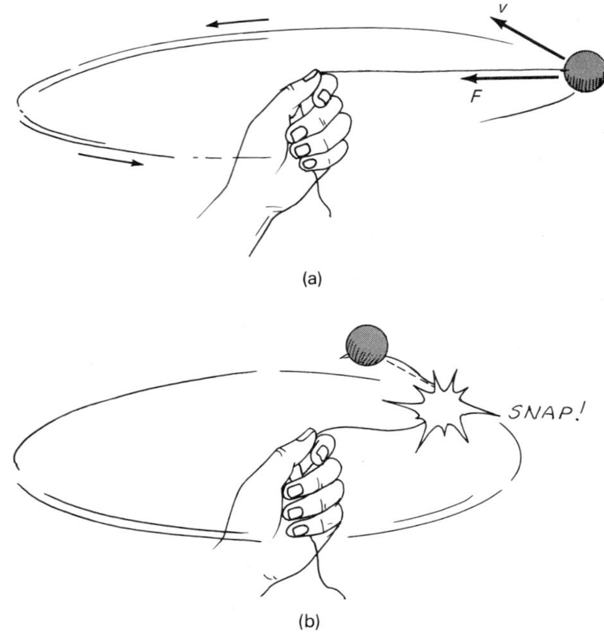

Figure 4.13 (a) The centripetal force for the circular motion of a ball on a string is supplied by the person "pulling inward." The velocity of the ball at any point is called the tangential velocity. (b) If the string breaks (centripetal force removed or goes to zero), the ball flies off tangentially (at a right angle to the radius of the circle).

string broke or you let go of it (centripetal force and acceleration on the ball become zero). What happens? The ball would no longer follow a circular path but would fly off in the direction of the instantaneous velocity. This direction is tangential to the circle (at a right angle to the radius of the circle at that point), and we refer to this velocity as the **tangential velocity.**

QUESTION: A car going around a circular curve at high speed hits a patch of ice on the road. What happens?

ANSWER: When the car hits the icy surface, the friction supplying the centripetal force is reduced. It can be shown that the magnitude of the centripetal acceleration of an object in uniform circular motion is given by the simple formula $a_c = v^2/r$, where v is the circular speed and r is the radius of the circular path. The centripetal force is then $F = ma_c = mv^2/r$.

So the centripetal force required for the car to negotiate the curve depends on its speed and the "radius of curvature" of the circular curve. At a high speed, the required centripetal force is large. If the frictional force supplying this is reduced below the required value on the icy surface, then the car "slides outward," because the reduced frictional force is not enough to change its direction so as to go around the curve. This is like swinging a ball horizontally on a string and letting the string slip through your fingers, rather than releasing it completely.

Also, notice that the centripetal force depends on the square of the speed ($F = mv^2/r$). Hence, if the speed of an object in uniform circular motion is doubled (from v to $2v$), then the centripetal force required to keep the object traveling in the same circular path increases by a factor of four.

Centripetal force, or the lack of it, is used in various practical applications. For example, in a washing machine's spin cycle, water is separated from the clothes (Fig. 4.14). The tub of the washer rotates rapidly, but the force exerted on the water by the clothes is not great enough to make the water travel in a circle with the clothes. The water flies off, leaving the clothes less wet. This is similar to the lack of centripetal force in the case of a car traveling too fast on an icy curved roadway, as discussed in the preceding Question and Answer.

A less desirable case of lack of centripetal force is when the rear wheel of an automobile spins in mud.

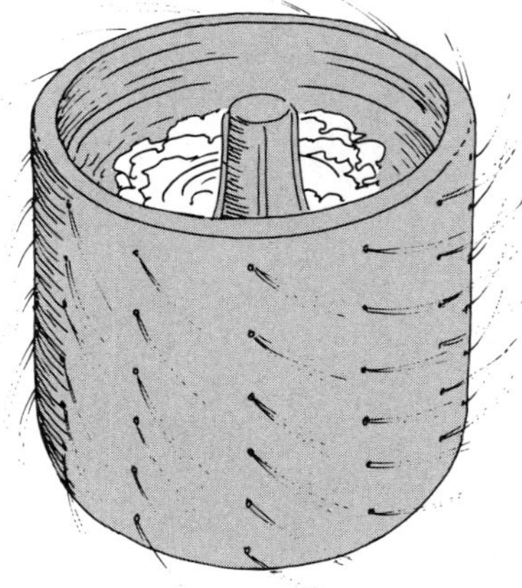

Figure 4.14 Application of centripetal force. In a washing machine, water is separated from the clothes by a spinning action. The (centripetal) force exerted on the water by the clothes is not enough to make the water travel in a circle with the clothes, so the water flies off.

Have you ever been pushing a stuck car and gotten a bit muddy? Stay away from behind the spinning wheel(s). The adhesion of the mud to the wheel (centripetal force) is not great enough to hold the mud on the tire, so off it comes tangentially to the tire's circular motion.

Centrifugal Force

You have probably heard the term "centrifugal force" or "centrifugal acceleration." When in a fast-moving car rounding a sharp curve or on a rotating ride in an amusement park, we "feel" a force "pushing" us outwardly or away from the center of curvature. This is the centrifugal or "center-fleeing" force.

Some people avoid using the term *centrifugal force* because they say it is a false or pseudo-force that doesn't exist according to Newton's laws. You may argue that it is "real" enough for you. The problem is a matter of definition and distinction between frames of reference, or how one looks at the situation.

To illustrate this, consider a car rounding a corner after having traveled along a straight roadway with a constant velocity (Fig. 4.15). An outside, stationary observer (such as yourself) would describe the situation as follows: The car and its occupants were moving along with a constant velocity. The driver turned the wheels and the frictional force between the tires and the road supplies the necessary centripetal force for the car to round the corner.

Figure 4.15 Centrifugal force is a pseudo-force. A person in a car rounding a curve experiences what is believed to be an outward force, but to a stationary observer, the passenger in the car is moving in accordance with Newton's first law of motion, and there is no "center-fleeing" force.

The passenger, who is not wearing his seat belt, continues to move in a relatively straight line in accordance with Newton's first law. (The friction on the seat of his pants is not great enough to supply the necessary centripetal force.) However, the car turns "in front" of the passenger, and the door pushes against him. This supplies the centripetal force needed for him to accelerate inwardly and around the corner with the car. Notice that in this description the passenger has no outward force acting on him.

Now, imagine yourself to be the passenger in the car. Your immediate frame of reference is the car. Then the situation may be described like this: When moving along the straightaway with a constant velocity, you are at rest with respect to your reference frame (the car). As the car rounds the corner, you notice yourself moving with respect to the car—outwardly toward the door. From your knowledge of Newton's law, you think that this requires a force to be acting on you, which could conveniently be called a centrifugal or center-fleeing force. Finally, as a result of this "force" you push up against the door and it pushes back, stopping your outward motion. You are now in equilibrium, with your outward centrifugal force being balanced by the inward door force acting on you.

Centrifugal force then is something that is "invented" to make $F = ma$ work in an accelerating reference frame (the car changing direction or velocity). An inertial or nonaccelerating reference frame is one for which Newton's first law holds. In such a reference frame, forces described by Newton's second law are real forces, that is, the interactions of a body with its environment.

If a reference frame is accelerating, it is said to be noninertial. Here, the observer is moving along with the accelerating body, for example, in circular or rotational motion. A noninertial observer must introduce fictitious or pseudo-forces to make Newton's second law work in his or her accelerating reference frame. These forces, such as the centrifugal force, *appear* to be real in the noninertial frame. However, the pseudo-forces *do not* exist when the motion is observed from an inertial frame. Hence, only to an observer in an accelerating or noninertial reference frame does the term "centrifugal force" have any significance. Also, notice that such fictitious forces are not exerted by another object, so there is no third-law reaction force.

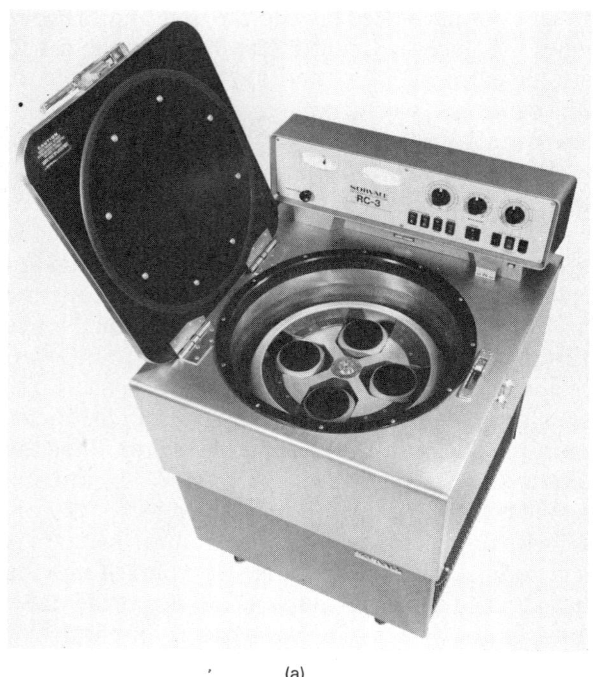

(a)

(b)

Figure 4.16 Centrifuge applications. (a) A general-purpose refrigerated centrifuge capable of separating blood plasma and blood cells. (b) A dairy centrifuge used to separate cream and milk.

It is preferable to describe situations from the point of view of an outside, inertial observer, for whom "centrifugal force" is a meaningless term. However, it is sometimes helpful to use a noninertial reference frame. We build machines called *centrifuges* to separate materials of different weights or densities by spinning action. The spinning drum in a washing machine used to separate water from clothes is a centrifuge. From a noninertial reference frame point of view, it would be said that "the centrifugal force throws the water outward" as the drum rotates with wet clothes. (The inertial frame description with centripetal force has already been given. See Fig. 4.14.)

Other applications of centrifuges are in separating blood cells from plasma and cream from milk in dairy separators (Fig. 4.16). Ultracentrifuges with speeds of the order of 500,000 rpm (revolutions per minute) are capable of concentrating viruses in solutions.

In a centrifuge, the denser or heavier materials migrate toward the outer end of a spinning horizontal container such as a test tube. (The centrifuge container holder is pivoted so that the container is horizontal when the centrifuge spins rapidly.) For example, when blood samples are sufficiently centrifuged, the red cells are bottom-most in the tube, with the lighter white cells on top. To an noninertial observer, this is because the heavier cells are thrown outward farther by the centrifugal force (much like heavier and lighter objects in nonfree fall). To an inertial observer, this occurs because of the greater inertia of the heavier particles, which respond less to the real center-seeking centripetal force. To the inertial observer, a centrifuge should really be called a "centripuge."

Let's look at another example of circular motion you may have experienced. There is an amusement park ride called the "Rotor" or some other name, where riders are spun in a large drum-like room (Fig. 4.17). When the Rotor reaches a certain speed, the floor drops out. But the riders do not fall! A centrifugal force pushes a Rotor rider against the wall, and the reaction force of the wall balances this force so he is at rest with respect to the rotating reference frame. To an outside inertial observer, a rider standing away from the wall cannot maintain the circular motion (lack of centripetal force) and flies off tangentially up against the wall, where the reaction force supplies the centripetal force needed for circular motion.

But what keeps the people from falling? The "centrifugal force" wouldn't do this because it is directed outward, and the centripetal force is directed inward.

(a)

(b)

Figure 4.17 An amusement park Roto ride. The reaction force of the wall on an object supplies the centripetal force needed for circular motion. At a certain critical speed, the upward friction force equals the downward weight force, and the floor is dropped out. Riders feel as though they are "stuck" against the wall.

The only force to balance the downward gravitational weight force is the upward frictional force between a rider and the wall. This frictional force is proportional to the reaction force the wall exerts on a person. The reaction force in both reference frames increases with speed (recall that $F = ma_c = mv^2/r$), so the reaction force and the friction force build up as the Rotor speed increases. At a certain critical speed, the friction force equals the gravitational force in magnitude, and the floor may be dropped without causing the person to fall. If the wall were smooth, a person would fall even in a fast-spinning Rotor.

QUESTION: Since the frictional force is proportional to the reaction force of the Rotor wall on a person, what would happen if the Rotor speed was greater than the critical speed? Would the friction force become greater and the person move up the wall?

ANSWER: No. At a speed above the critical speed, the centrifugal force in the rotating system would be greater and would "throw" the person harder against the wall. The centripetal reaction force of the wall on the person would be greater too, as is necessary to keep the person in circular motion at a greater speed.

However, remember that a frictional force always *opposes* motion and does not produce motion. The force of static friction increases as the Rotor gains speed, until it balances or fully opposes the gravitation weight force of *any* particular rider at the critical rotational speed, but no greater.

It is interesting to note that analysis shows the critical speed is independent of the masses of the riders. (Your instructor may wish to show you this. It is similar to the acceleration due to gravity g being independent of the mass of a falling object.) So, at this minimum critical speed, the weights of all of the riders are balanced by equal and opposite frictional forces, regardless of their masses. A small child would have a small balancing frictional force and a large adult would have a large balancing frictional force. This is because the frictional force *is* dependent on mass, as well as the rotational speed (similar to the weight force being mass dependent, $w = mg$).

The frictional force also depends on surface properties. This would be an important consideration in selecting a material for the Rotor wall—you'd want a rough wall so the critical speed wouldn't have to be excessively large.

SIMULATED GRAVITY

A potential use for centripetal force or centrifugal force, depending on how you look at it, is to provide simulated gravity in space. Have you ever wondered how the crews of spaceships in science-fiction movies are able to walk about normally when on missions in free space?

Figure 4.18 Rotating space colony. The reaction force on the people's feet would provide simulated gravity. Rotating the structure at the right speed would give a gravity environment similar to that on Earth.

(So has the author.) We Earthlings live in a gravitational environment, and many of our normal activities and functions depend on gravity. Think of how things would be without it.

Our awareness of gravity comes primarily from falling objects and weight. The reaction force to your weight force acts on the Earth when you are falling or not falling (see Fig. 2.16). To avoid a "weightless" feeling, you must *experience* the reaction to the force of your feet (or bottom) pushing against the floor (or chair). In an ideal gravity-free environment, your feet would not push against the floor and there would be no reaction force. Of course, nowhere is gravity-free (see Chapter 5). In free fall, however, one seems weightless.

To stimulate gravity, we might have a rotating spacecraft (Fig. 4.18). When rotated at the proper speed so that the centripetal reaction force of the spacecraft on a person was equal to that experienced under normal conditions of gravity, the effect would be the same as standing on Earth. Objects would also "fall" toward the person's feet (due to a centrifugal force?). Don't be concerned about standing on the "wall." This would be the floor, since downward is defined by the direction of gravity or the direction in which things fall. (Think about two people standing on Earth, one at the North Pole and one at the South Pole. "Downward" to them would be in opposite directions.)

Space colonies of the future may have a circular or wheel design so as to provide simulated gravity by this means (Fig. 4.18). Of course, such space stations would have to be quite large. Otherwise, they would have to spin so rapidly to simulate normal gravity conditions that the inhabitants would feel dizzy because of the effects on the inner ear.

SUMMARY OF KEY TERMS

Projectile a body that has been thrown or projected by some means.

Free fall an object falling with only gravity acting on it.

Nonfree fall an object falling under the influence of gravity with another force or forces (e.g., air resistance) acting on it.

Terminal velocity the maximum constant velocity a non-freely falling object reaches when a retarding force balances the gravitational force and the object's acceleration is zero.

Vertical projection an object thrown or projected vertically upward.

Horizontal projection an object thrown or projected horizontally to the Earth's surface.

Range the horizontal distance traveled by a projectile.

Projection at an angle an object projected at an angle to the horizontal.

Uniform circular motion the motion of an object traveling in a circle with a constant speed.

Centripetal force (acceleration) a center-seeking force (acceleration) necessary for an object to travel in circular motion.

Tangential velocity the instantaneous velocity of an object in circular motion, which is at right angles to the radius of the circular path.

Centrifugal force the fictitious center-fleeing force in a rotating or accelerating system that is "invented" to describe the motion in this system in terms of Newton's laws.

Inertial (nonaccelerating) reference frame one for which Newton's first law holds.

Noninertial reference frame an accelerating reference frame, for example, one moving with an object in circular or rotational motion.

EXERCISES

1. Near the surface of the Earth, the acceleration due to gravity is measured to have a relatively constant value of 9.8 m/s^2. Does this mean that a freely falling object travels 9.8 m each second of fall? Explain.
2. Two rocks, one with a mass m and the other with mass $4m$, are in free fall. Show that both rocks have the same acceleration.
3. In the legendary Leaning Tower of Pisa experiment, suppose that Galileo had used a cannon ball and a wooden ball of equal size. What would have been the result?
4. What is the speed in m/s of a freely falling object dropped from rest at the end of (a) 2 seconds and (b) 5 seconds? (Recall $v = gt$.)
5. A sky diver steps out of a hovering helicopter. Assuming the diver to be in free fall, between what seconds of fall would a speed of 120 mi/h be reached? ($v = gt$ or $t = v/g$, and $g = 32$ ft/s^2. Also, 88 ft/s = 60 mi/h.)
6. A freely falling body falls with a constant acceleration. Is the distance traveled by the body constant for each time interval? Explain.
7. The distance traveled by an object in free fall dropped from rest is given by $d = \frac{1}{2}gt^2$, where t is the time of fall. How far would an object fall in (a) 1 second, (b) 2 seconds, (c) 3 seconds?
8. Considering the equation given in Exercise 7, could you measure the height of a tall building using a stop watch? Explain. What would be the sources of error?
9. Two sky divers jump at the same time from an airplane, after having agreed to open their parachutes at a particular altitude. One diver falls with his arms and legs pulled in (fetal position) and the other in a spread-eagle position. Which diver would open his parachute first?
10. Explain what effect an updraft of air would have on the terminal velocity of a falling object.
11. Snowflakes fall more slowly than sleet. Why?
12. Is it possible for a heavy object and a lighter object to have the same terminal velocity? Explain.
13. A sky diver in a spread-eagle position reaches her terminal velocity. She then decides to fall feet first. What happens?
14. Suppose you live in the future, when moon vacations become common. Someone asks you to join a parachute club they are going to form for jumping on the moon. Would you sign up?

Figure 4.19 See Exercise 15.

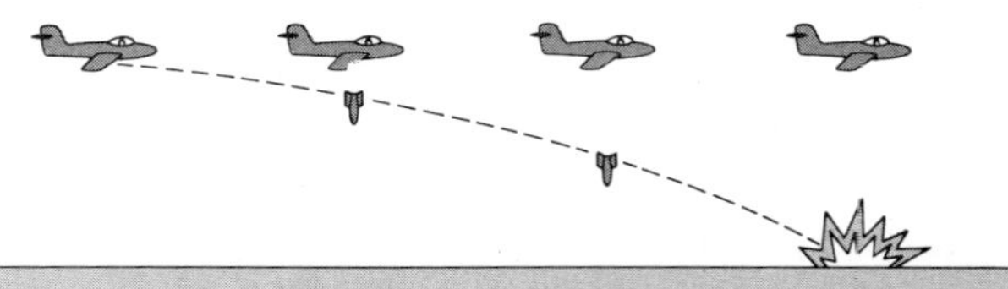

Figure 4.20 See Exercise 20.

15. A group of sky divers form a chain, as shown in Figure 4.19. Is the group's terminal velocity less than that of a single diver in a spread-eagle position? Explain.
16. A sky diver falling at terminal velocity opens his parachute. Describe what happens. Is the force due to air resistance constant? Is terminal velocity reached again?
17. If a marble and a golf ball were projected vertically with the same initial velocity, (a) which would have the greater maximum height? (b) Which would return to the starting point first? (Neglect air resistance.)
18. While sitting in a stationary automobile, you flip a coin upward and it falls back into your hand. Suppose the car was traveling on a level road with a speed of 90 km/h and you flipped the coin in the same manner. What would happen? Explain.
19. What is the angle of projection for a horizontal projection?
20. An object is dropped from an airplane, as shown in Figure 4.20. Explain why the object hits the ground directly below the plane. What must be assumed about air resistance for this to happen?
21. The coyote is trying to catch the road runner again, this time wearing a pair of Acme jet-powered skates (Fig. 4.21). If the road runner makes a sudden turn at the cliff, but the coyote doesn't, sketch the path of the coyote's descent to the floor of the canyon below when (a) his skates fall off at the edge of the cliff and (b) his skates are still on and in operation when he is in "flight."
22. Discuss the effect (make a sketch) of air resistance on (a) a vertical projection, (b) a horizontal projection, (c) a projection at an angle.
23. Several years ago, a pilot of a jet plane fired the plane's cannon at an imaginary target, then put the plane into an evasive dive. Rather embarrassingly, the pilot shot himself down. How did this happen?
24. Is it possible for a vertical projection and a projection at an angle to have the same maximum height? Explain.
25. In throwing a baseball to home plate from the outfield, a fielder never throws the ball horizontally. Why?
26. Suppose you and another student are playing catch with a ball. He throws the ball at an angle of 35° and you catch it while standing still at the same height as it was thrown. Could you throw the ball back with the same initial speed but at another angle so he could catch it as you did? If so, at what angle, and which throw would have the greater maximum height?

Figure 4.21 See Exercise 21.

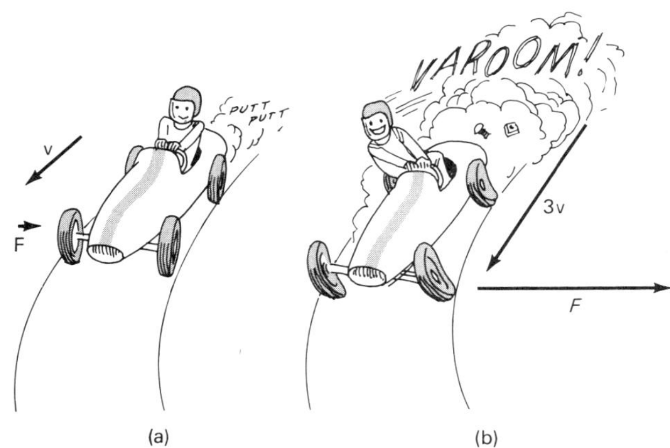

Figure 4.22 See Exercise 30.

27. A projection angle of 45° is important to a shot-putt thrower. Give some examples of other situations in which this is a critical angle.
28. Show that $a_c = v^2/r$ has the units of acceleration.
29. In a washing machine during a spin cycle, the clothes stay against the drum in circular motion. However, in a dryer the clothes are tumbled and do not go in complete circular motion. Why is this? Isn't there enough centripetal force?
30. A race car travels around a circular race track with a constant speed v (Fig. 4.22). If the driver increases the speed to $3v$, how many times greater is the centripetal force on the car? What supplies the centripetal force?
31. Curved roadways and race-track curves are often "banked" so that cars can travel around the curves at higher speeds than on level curves without sliding outward. (The bank slope is inward toward the center of curvature.) How does banking increase the centripetal force? (*Hint:* Consider the components of the force of gravity acting on the car.)
32. Do Newton's laws hold on Earth? Explain.
33. On a Rotor ride, suppose the Rotor slowed down below its critical speed before the floor came back up. What would happen to the people on the ride?
34. On the rotating Earth a person at the equator weighs slightly less than a person of equal mass at the North Pole. Explain why.
35. In a dairy centrifuge used to separate the cream from the milk, which is on the bottom?
36. The *Coriolis force* is a fictitious or pseudo-force that we use here on Earth. It "deflects" moving objects to the right (viewed along the direction of motion) in the Northern Hemisphere and to the left in the Southern Hemisphere. As a result, air circulation around low-pressure areas ("lows") is counterclockwise and around high-pressure areas ("highs") is clockwise in the Northern Hemisphere (opposite circulations in the Southern Hemisphere). The next time you see a satellite weather picture on TV, notice the rotation around a low, which is outlined by clouds. (A hurricane is a low.) Suppose you are at the North Pole and fire a high-speed, long-range projectile toward the equator along a meridian. Prove that the projectile would hit the Earth to the west or to the right of the meridian, and explain why. What is required to cause this "deflection" to the right? Which way would the projectile be deflected if you were at the South Pole and shot it toward the equator?
37. The inhabitants of a rotating space colony have become overweight, and the person in charge wants to put everyone on a diet. As an alternative, one clever fellow suggests a sure-fire way that everyone can lose "weight" without going on a diet. How might this be done?

5

Gravitation, Planetary Motion, and Earth Satellites

Gravitation

Gravity, or gravitation, is one of the fundamental forces of nature. This force acting between all masses holds the universe together, so to speak, on a large scale. There are three other fundamental forces — the *electromagnetic* force and the *strong* and *weak* nuclear forces. The electromagnetic force is dominant on the atomic level and is responsible for holding the atoms together that make up molecules. It also binds the electrons of an atom to the nucleus. Strong and weak nuclear forces act only *within* the nucleus.* More will be said about these forces in later chapters. Here we will be concerned with gravity, intrinsically the weakest of the fundamental forces. But even though it is the weakest force, it causes objects to fall, determines the motions of celestial bodies, and gives rise to the ocean tides.

Isaac Newton, who gave us the laws of motion, also formulated an expression for the fundamental interaction that takes place between all objects or masses. This law of gravitation was the result of Newton's study of the motions of celestial objects.

In Aristotle's theory of motion, which was the dominant theory for 1800 years afterward, the Earth was considered to be the center of the universe (geocentric theory). The apparent motions about the Earth of celestial objects such as the Sun, other stars, planets, and the moon were considered to be *natural* motions. They just "did their thing" without the benefit or need of forces. However, Newton and others recognized that this could not be the case if the laws of motion were valid.

* Recent research suggests that the electromagnetic force, the weak nuclear force, and possibly the strong nuclear force may all be the same type of interaction or a unified (fundamental) force. Scientists continually look for ways to unify their theories, thereby simplifying the description of nature.

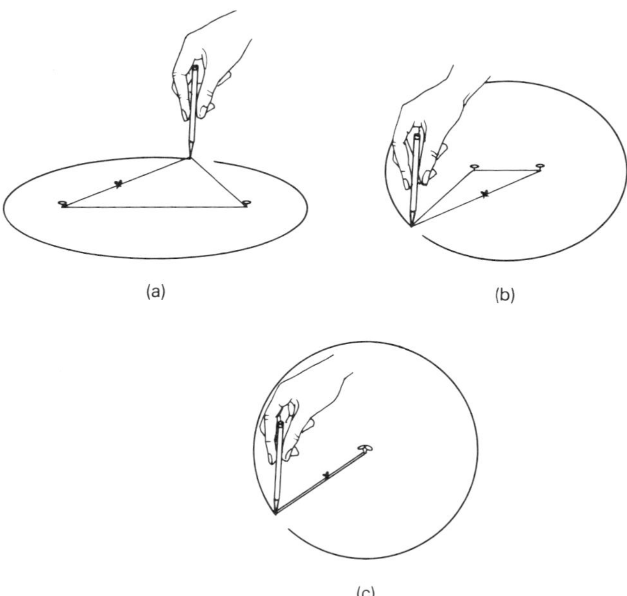

Figure 5.1 How to draw an ellipse. (a) A loop of string over two thumb tacks allows an ellipse to be easily drawn. The locations of the thumb tacks are the foci of the ellipse. (b) As the foci are brought closer together, the ellipse approaches a circle, and when they are at the same point (c), they are at the center of a circle.

Kepler's Laws

During the 16th and 17th centuries much attention was focused on the motions of celestial objects, in particular the planets. Tycho Brahe (1546–1601), a Danish astronomer, spent most of his life studying the planets and stars and is considered to be one of the greatest practical astronomers since the Greeks. His measurements of the planets and stars, all made with the unaided eye (the telescope had not yet been invented), proved to be more accurate than any previously made.

Brahe's data were edited by his colleague Johannes Kepler (1571–1630), a German mathematician and astronomer, who had joined Brahe during the last years of Brahe's life. After Brahe's death, his lifetime of observations were at Kepler's disposal, and they proved invaluable in the formulation of the empirical "laws" we know today as **Kepler's laws of planetary motion.**†

† "Empirical" means derived from experiment or experience rather than from theory. An empirical law gives no explanation of why things occur in a particular manner, only that they do. It is interesting to note that the three laws of planetary motion, which Kepler spent a great deal of his life formulating by trial and error, can be derived on a couple of sheets of paper using advanced mathematics.

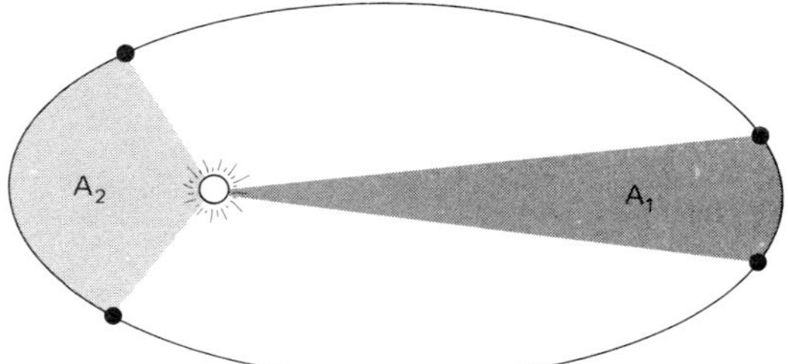

Figure 5.2 Equal areas. A planet sweeps out equal areas A_1 and A_2 in equal times; hence, it must move faster or have a greater orbital speed when it is near the Sun.

Kepler found that Brahe's data fit three simple rules or laws that describe planetary motion in a heliocentric or "sun-centered" theory.

1. First law, or **the law of elliptical orbits.** All planets move in elliptical orbits about the Sun, with the Sun at one focus of the ellipse.

   An ellipse has the appearance of an oval or a flattened circle. One can be drawn using string and thumb tacks, as shown in Figure 5.1. The tacks are at the positions of the foci (singular: focus) of the ellipse. As the foci are moved closer together, the ellipse approaches a circle, which is a special case of an ellipse.

   Although the orbits of the planets are ellipses, they are almost circular, with the exception of those of Mercury and Pluto.
2. Second law, or **the law of equal areas.** Kepler also found that the planets do not have uniform orbital speeds. They move faster as they orbit closer to the Sun than when they are farther away from the Sun. The Earth has its greatest orbital speed during the winter, when it is closest to the Sun (about January 4).‡ This was stated in terms of equal orbital areas: An imaginary line from a planet to the Sun sweeps out equal areas in equal periods of time (Fig. 5.2).
3. Third law, or **the law of periods.** It was also found that there is a common relationship between the average radii of the orbits of the planets and their orbital periods. (A period is the time it takes for a planet to make one revolution about the Sun.) This relationship is that the ratio of the square of the

‡ We are closer to the Sun during our winter, but because of the inclination of the Earth's axis, the Northern Hemisphere is tilted away from the Sun at this time. The Southern Hemisphere, tilted toward the Sun, receives the more direct rays, so it is summer there.

period of a planet (T^2) and the cube of the radius (R^3) is a constant and is the same for all planets ($T^2/R^3 =$ a constant).

Newton's Law of Gravitation

In his study of planetary motions, Newton started a bit closer to home and focused his attention on the motion of the moon in its orbit about the Earth. Obviously, the motions of the moon and the known planets could not be "natural" or without a force of some kind. If there were no force exerted on them, then their motions would be straight lines by Newton's own first law.

As you know, and Newton knew, circular motion requires a centripetal force. But what supplies this centripetal force? Allegedly, Newton's insight was sparked by observing an apple falling to the ground. If gravity attracts an apple toward the Earth, perhaps it also attracts the moon, and the moon is "falling" or centripetally accelerating toward the Earth under the influence of gravity (Fig. 5.3).

Actually, in its nearly circular orbit the moon is continually "falling" toward the Earth, since it is accelerated toward the center of its orbit. But it never gets any closer because it is also moving tangentially, which by itself would cause the moon to move farther from the Earth (Fig. 5.4). The combination of these component motions is a circular orbit. The planets are also "falling" toward the Sun in their orbits.

Using his knowledge of centripetal force and Kepler's third law, along with amazing insight, Newton formulated an expression for the attractive gravitational force that acts between the Earth and the moon, the Sun and the planets, and indeed, between any two spherical or point masses m_1 and m_2. This is known as **Newton's law of gravitation**

$$F = \frac{Gm_1 m_2}{r^2}$$

where G is a constant (the universal gravitational constant, called "big" G to distinguish it from "little" g, the acceleration due to gravity) and r is the distance between the masses.

Gravity is always an attractive force. Thus, Newton's law of gravitation states that every particle in the universe is attracted toward every other particle. The gravitational interactions between two particles are equal and opposite attractive forces. Notice that the law of gravitation is an "inverse-square law." That is, the force falls off (decreases) as the reciprocal of the square of the distance between the masses. For spherical objects like the Earth, we measure the distance from the center of the Earth just as if all of the mass inside this radical distance were concentrated there as a particle (Fig. 5.5).

At the surface of the Earth, the force of gravity on an object is equal to its weight, and if dropped, the object

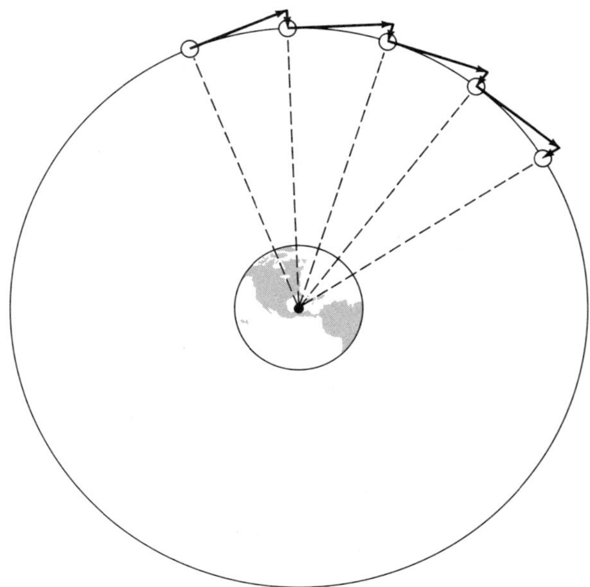

Figure 5.4 Circular motion is a combination of a tangential component and one toward the center of the circle. Because of the latter, the moon is continually "falling" toward Earth in its nearly circular orbit.

Figure 5.3 Newton and the apple. Legend has it that Newton's insight into gravity came from observing a falling apple. Did this force extend to the moon and provide the centripetal force to keep it in orbit?

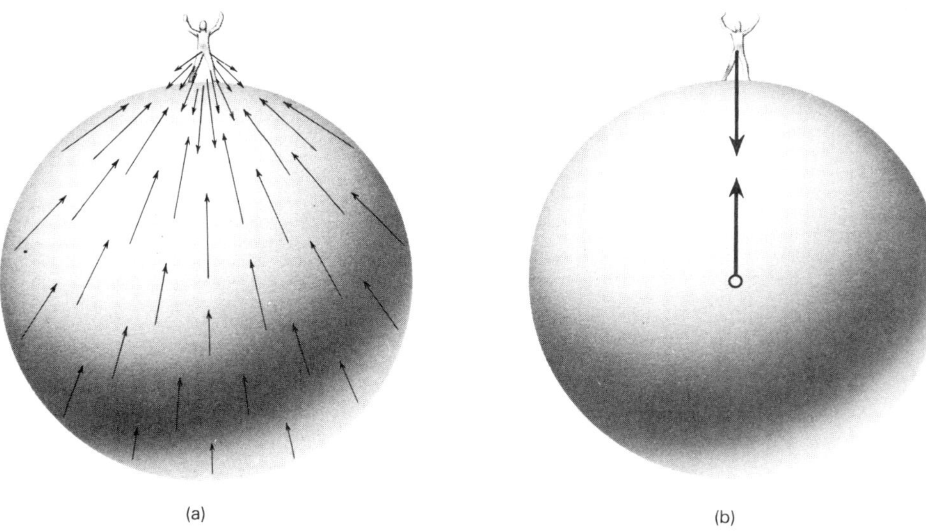

(a) (b)

Figure 5.5 Gravity. (a) Mutual gravitational attraction exists between any two particles. (b) For a uniform spherical object, the effect is as though all the mass of the sphere were concentrated at its center.

would fall to the Earth with an acceleration due to gravity g (9.8 m/s^2). The gravitational force of the Earth also acts on the moon, which is at a distance from the Earth equal to about 60 Earth radii. The acceleration due to gravity falls off with an inverse-square relationship. Hence, the acceleration due to the Earth's gravity on the moon is $1/r^2 = 1/(60)^2 = 1/3600$ less than on the surface of the Earth, or 0.0027 m/s^2 instead of 9.8 m/s^2. This is the centripetal acceleration needed for the moon's nearly circular motion about the Earth.

UNIVERSAL GRAVITATIONAL CONSTANT

Newton could not use his expression to calculate the gravitational force between two masses because he didn't know the value of G, the universal gravitational constant. Nor was he able to determine its value experimentally with the equipment of his day.

It was not until 1798 (71 years after Newton's death) that the gravitational constant was experimentally determined by Henry Cavendish, an English physicist. He used a sensitive balance that could measure the force between two masses. The value of G is 6.67×10^{-11} N-m^2/kg^2.

Since it is believed that Newton's law of gravitation applies to all masses everywhere, G is called the ***universal*** **gravitational constant.** Newton's law of gravitation is sometimes referred to as the universal law of gravitation.

QUESTION: Why do you not experience an attraction (gravitational) toward a textbook or another person?

ANSWER: The gravitational force is too small for you to detect. Consider a couple of 100-kg (220-lb) football players standing 1 meter apart. Then, by Newton's law with the appropriate numbers,

$$F = \frac{Gm_1m_2}{r^2} \approx 10^{-7}\text{ N (or 0.0000001 N)}$$

A force on the order of 10^{-7} Newtons is very small (much less than the weight of a flea). It would be even smaller between you and a textbook ($m \approx 1$ kg). Because G is so small, you must interact with a very large mass, for example the Earth, at a relatively close distance to detect the gravitational force.

Ocean Tides

If you have visited an ocean beach, you probably noticed the periodic rise and fall of the water, or tides—two high tides and two low tides per day. Tides have long been associated with the moon, and it is easy to see how the moon would gravitationally attract the ocean water toward it to give a tidal bulge on the side of the

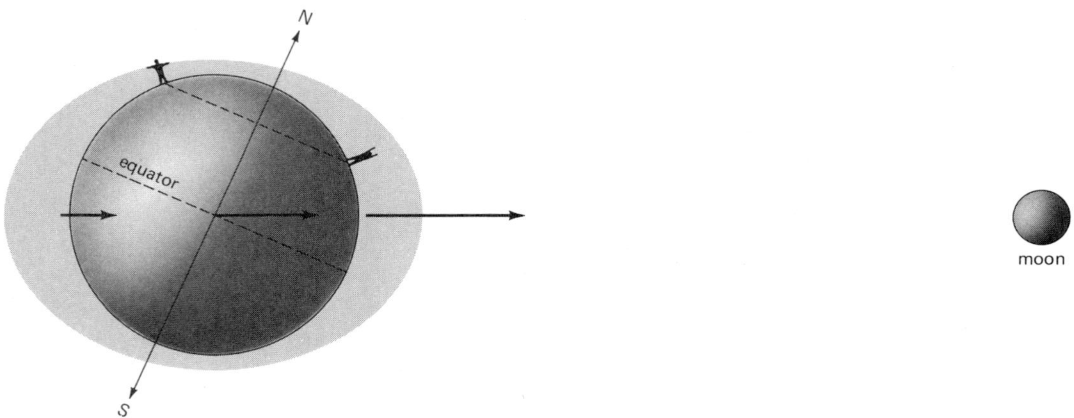

Figure 5.6 Tides. There are two tidal bulges, or two high tides, per day because of the difference in the gravitational attraction of the moon at different distances.

Earth nearest the moon. However, as the Earth rotates beneath it, this would seem to give only one high tide and one low tide (the water depression on the opposite side of the Earth) per day. So what causes the other bulge for the second daily high tide?

Newton showed that the two high tides per day are caused by the differences in the gravitational pulls on the opposite sides of the Earth (Fig. 5.6). Keep in mind that the gravitational pull of the moon acts on the Earth itself, as well as on the water on both sides of the Earth, with the attraction getting weaker the greater the distance from the moon (inverse-square law).

The water nearest the moon has the greatest attraction, and this pull forms one tidal bulge. The Earth itself is attracted toward the moon with more force than the water on the side opposite the moon. As a result, the Earth is effectively pulled away from the water on the distant side, which gives a second and opposite tidal bulge. So, as the Earth rotates, two high tides "travel" around the Earth daily. The intervening water depressions are the two low tides.

Notice in Figure 5.6 that the two daily high tides at a particular location need not be equally "high." The tidal bulges can be above and below the equator because of the inclination of the moon's orbit.

QUESTION: The gravitational pull on the Earth by the massive Sun is about 200 times greater than that of the moon. Why doesn't this produce mammoth tides?

ANSWER: Although the Sun's gravitational attraction is greater than the moon's, the *difference* between the Sun's pull on opposite sides of the Earth is appreciably less. This is because the Earth's diameter is a relatively small fraction of the distance from the Sun as compared to that of the moon. The Sun's gravitational pull on each side of the Earth is relatively large, but the difference is relatively small. To illustrate this idea, the difference between two large numbers like 20,000 and 20,002 is relatively small compared with the difference between two small numbers like 100 and 150.

The gravitational pull of the Sun does have a "helping" effect at certain times to produce higher-than-usual and lower-than-usual tides. When the Sun and moon are on the same or opposite sides of the Earth, the added pull of the Sun on the bulges makes them higher than usual. This gives rise to higher high tides and lower low tides, which we call spring tides. These occur at the times of the full and new moons and have nothing to do with the season of spring.

Similarly, when the Sun and moon are at right angles (90°) relative to the Earth, or at the times of the first-quarter and third-quarter moons, the Sun's pull tends to attract water to the tidal depressions and decreases the bulges. These lower high tides and higher low tides are called neap tides.

Actual tides involve other complicating factors than those presented in the preceding simple description which assumed the Earth to be completely surrounded by water. The presence of land masses that stop the flow of water, "tidal" friction between the oceans and the ocean floors (which slows down the Earth's rotation about 1/1000 of a second per century), and the variable depths of the oceans are some of these.

Generally, the tidal range (the difference in height between high and low tides) in an ideal open ocean would be a meter or two. However, as this tidal bulge comes "ashore" at certain places, the tidal range can be quite different. A classic example occurs in the funnel-like Bay of Fundy between New Brunswick and Nova Scotia, Canada. The highest tides on Earth occur at the head of the Bay of Fundy, where under favorable circumstances the tidal range can be as much as 16 meters (Fig. 5.7).

It is interesting to note that the tidal forces of the moon and the Sun also produce *Earth tides* as well as water tides. The Earth is not totally rigid; its surface rises and falls in the same manner as water tides. However, these tides generally go unnoticed. It is believed that Earth tides may be a factor in initiating the relief of stress along cracks or faults in the Earth, which results in earthquakes.

Discovery of Planets — Neptune, Pluto, and Planet X(?)

Newton's law of gravitation also assisted in the discovery of the planets Neptune and Pluto. Not only does the Sun's gravity act on the planets, but there are also mutual interactions among the planets themselves. These lead to slight departures or perturbations from smooth elliptical orbits.

Uranus, the seventh planet from the Sun, was discovered in 1781 by telescopic observation. Its motion was studied, and after allowing for perturbation effects of the known planets, scientists determined that something was affecting Uranus' orbit. In the 1840's Urbain Leverrier in France and John C. Adams in England, using Newton's law, independently calculated where an unknown eighth planet should be located to account for the perturbation.

Both Adams and Leverrier had little success in convincing the astronomers in their countries to look for the new planet. However, Leverrier wrote to a colleague, Johann Galle, at the Berlin Observatory. Galle trained a telescope on the position predicted by Lever-

Figure 5.7 A 16-meter (52-foot) tidal range at Hantsport, Nova Scotia, at the head of the Bay of Fundy. The objects floating between the posts in the high-tide photo (right) are pieces of ice — not the top of the car. (The car was removed.)

rier on the very night of the day he received the letter and discovered Neptune!

Even after the discovery of Neptune, the perturbation of Uranus was not fully explained. Perhaps there was another planet. Percival Lowell, an American astronomer, calculated the position of such a planet. He searched for it until his death in 1916. With the development of a more advanced telescope, Pluto was discovered by Clyde Tombaugh at the Lowell Observatory in Arizona in 1930.

But Pluto's mass is not sufficient to account completely for Uranus's perturbation. Is there a "Planet X" beyond Pluto? This was a subject of discussion at a recent American Astronomical Society meeting. With calculations using Newton's law of gravitation, the search goes on.

As long as we are looking at history, we might mention Edmund Halley, a contemporary of Newton's. Halley began a friendship with Newton that resulted in the publication, at Halley's expense, of Newton's *Principia* in 1686. The *Principia* contained Newton's laws of motion and the law of gravitation.

Halley is perhaps most famous for the calculation of the orbit of a comet he observed in 1682, which has been named in his honor—Halley's comet. (Kepler wrote about the comet in 1607.) Halley's correct prediction of its return in 1758 was one of the first application of Newton's laws.

Halley's comet is in a large elliptical orbit about the Sun, with a period of about 76 years (Fig. 5.8). In this century it was visibly seen (near the Sun) in 1910 and, as you may be aware, near the beginning of 1986. You may have been fortunate enough to see this perhaps once-in-a-lifetime occurrence. A comet (from the Latin, *cometes,* meaning "longhaired") is generally characterized by a tail that extends away from the Sun. As a comet passes near the Sun, evaporated particles driven away from the Sun reflect light, giving rise to a long "tail." Notice that a comet's tail is not behind it or in a direction opposite to its motion.

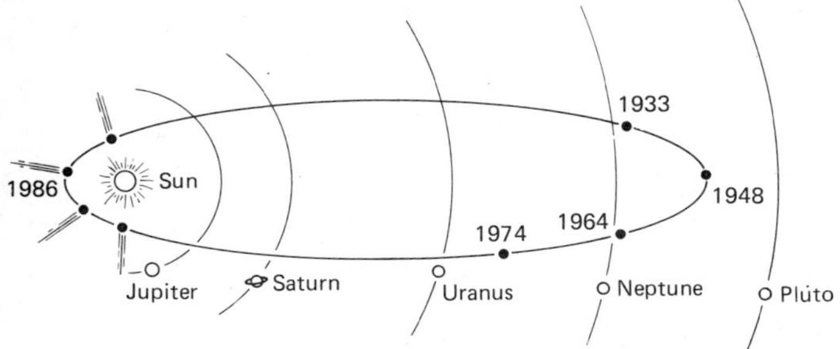

Figure 5.8 Halley's comet. With an orbital period of about 76 years, in this century Halley's comet passed near the Sun in 1910 and 1986. An observed comet's tail is away from the Sun, not behind it or in the direction opposite its motion.

A Closer Look at *g*

We have talked about the acceleration due to gravity quite a bit. Let's take a closer look at g with the help of Newton's law of gravitation.

It has been stated several times that the acceleration due to gravity is constant near the Earth's surface. Why is this, when we now know that the force due to gravity, and hence the acceleration due to gravity, varies with distance? Also, it was learned that the acceleration of a freely falling body is independent of its mass. Newton's law explains.

The gravitational force on an object on Earth is its weight force, $F = mg$. But this must also be the same as Newton's law of gravitation for the special case of two interacting masses, m the mass of an object and M_e the mass of the Earth. Hence, we may equate the two equations

$$F = mg = \frac{GmM_e}{R_e^2}$$

where the separation distance is the radius of the Earth, R_e. Then, cancelling the m's,

$$g = \frac{GM_e}{R_e^2}$$

So, g is independent of the mass m of the object (m does not appear in the equation), and it is also constant since G, M_e, and R_e are constants.

QUESTION: To determine your mass, you weigh yourself and divide your weight by the acceleration due to gravity, $w = mg$ and $m = w/g$. The mass of the Earth is listed in many references. How is the mass of the Earth determined?

ANSWER: Determining the mass of the Earth might appear to be a sizable task. One thing for sure, we do not weigh it. Instead, the preceding expression for g is used. Rearranging the equation, we have

$$M_e = \frac{gR_e^2}{G}$$

The values of all of the quantities on the right side of the equation are known from experimental measurements. Plugging in these values, the mass of the Earth can be calculated to be about 6.0×10^{24} kg.

We consider g to be constant near the Earth's surface. However, it does vary with altitude, $g = GM_e/r^2$, where r is the distance from the center of the Earth. But this variation is not great percentagewise over the heights of our normal activities.

At an altitude of 1.6 km (1 mile), the acceleration due to gravity and hence the gravitational force or weight (mg) is only 0.05 percent less than g at the Earth's surface. At an altitude of 160 km (100 miles) g, and therefore weight, is still 95 percent of its value on Earth. No matter how high we go, gravity still acts, and we still have weight (Fig. 5.9).

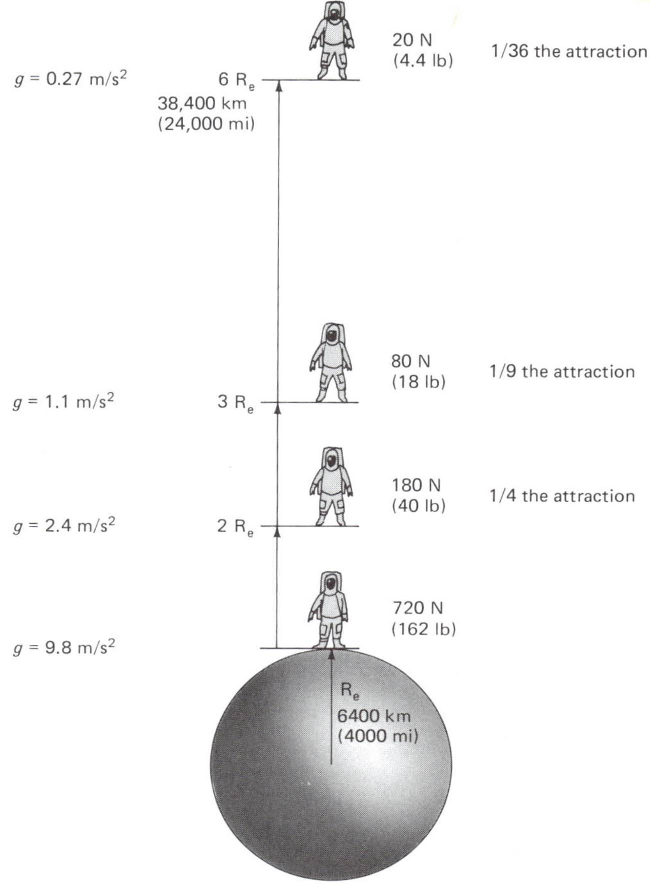

Figure 5.9 Weight and gravity. The force of gravity and g decrease as $1/r^2$, which results in small values, but an object has weight at any height. (Surface weight equivalent values are approximate for illustration.)

Gravitational Field

Another way of looking at the gravitational interaction is to consider only the effect rather than the cause. This is done in terms of a gravitational field. Essentially, we map the gravitational force in space due to some body like the Earth, then forget the body itself and consider only the field. That is, we think of an object interacting with the gravitational field rather than with the Earth or some other body responsible for it.

A gravitational field can be represented pictorially by vector arrows (Fig. 5.10). We compute the **gravitational force per unit mass** at points in space. Note that this is really g, since $F = mg$ or $g = F/m$ (force/mass).

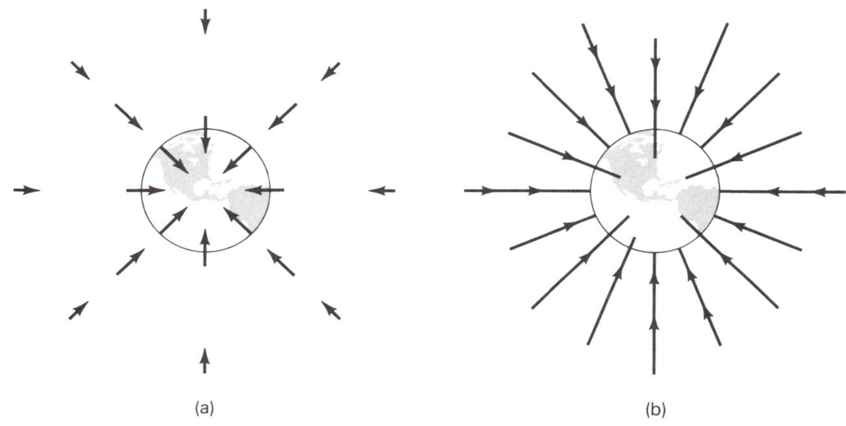

Figure 5.10 Gravitational field. (a) Around a mass, the gravitational force per mass ($g = F/m$) can be represented by vector arrows. (b) Joining the vector arrows forms lines of force.

As shown by the lengths of the vector arrows in the figure, the magnitude of g decreases with increasing distances from the Earth.

We commonly join the arrows together to form "lines of force" [Fig. 5.10(b)]. The closer together the lines of force are, the stronger the gravitational field or force in that region. Thus, a body in a field would experience an acceleration in the direction indicated by the arrowheads on the lines of force.

Einstein extended this idea in his model of gravity in his general theory of relativity, which was quite different from Newton's description of gravity. He perceived a gravitational field as a warping of four-dimensional space and time. Large bodies like the Earth put dents in this space.

To help visualize this effect, consider a heavy object on a loosely stretched rubber sheet (Fig. 5.11). The object puts a dent or depression in the sheet. If a marble were rolled on the sheet, several things could happen. It would roll in a straight line until it came sufficiently close to the "space warp" in the sheet. Depending on the direction of the marble's motion and speed, it may curve around the outer side of the dent and continue on, or it may follow a closed path and orbit the dent in an elliptical or circular path. In a similar manner, the curvature of space determines the gravitational effects.

You may have heard about the incredibly strong gravitational field in the vicinity of a black hole in space—so strong that even light that comes sufficiently close cannot escape its gravitational attraction. Black holes are believed to be burned-out stars that have gravitationally collapsed until they have incredibly high densities. The gravitational attraction near the black hole is so enormous that nothing can escape.

In terms of our space-time warp analogy of the rubber sheet, it would be as though the heavy object sank an enormous distance into the sheet (Fig. 5.12). If a marble rolled into the dent, it would be gone. Similarly,

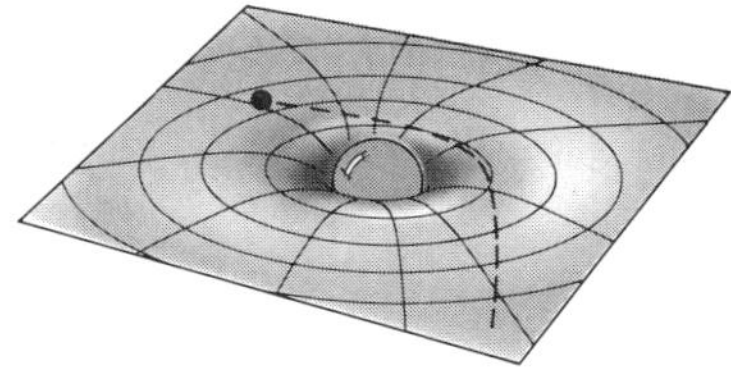

Figure 5.11 An analogy of gravity as a warping of space and time. A large mass warps space-time in a manner analogous to a heavy object denting a stretched rubber sheet. An object rolling on the sheet is deflected as it would be in a "space warp" in a gravitational field.

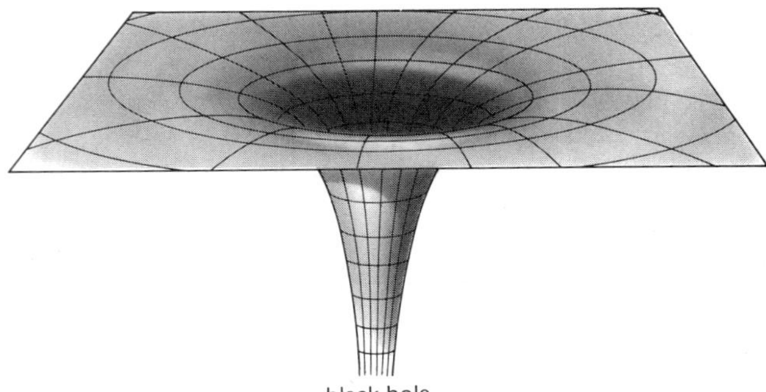

Figure 5.12 A black hole would be represented in the rubber-sheet analogy of a space-time warp as an object that sinks an enormous distance into the sheet.

anything that comes too close to a black hole disappers from the observable universe—even light. It has been suggested, particularly in popular literature, that matter going into a black hole may come out in a parallel universe through a "white hole." (More on black holes in Chapter 30 on astrophysics.)

QUESTION: If black holes are completely invisible or can't be seen, how do we know they exist?

ANSWER: If we were lucky, we might be able to observe the bending of light that passed near a black hole, which would indicate its presence. But we have never been so lucky. However, their presence can be detected by their gravitational influence on other stars. A majority of the stars are binary stars—two stars orbiting about each other. One star may have gravitationally shrunk into a black hole, while the other is still luminous. The luminous star would then appear to be orbiting about an empty point in space.

But this is not enough. The black hole candidate may simply be a faint star that cannot be seen next to a bright companion. Near an object of high density (like a black hole), matter falling toward it or into it is accelerated to a high speed. Charged particles accelerated toward a black hole can emit X-rays, which are energetic enough to escape.

Thus, we must look for X-ray sources associated with binary stars with invisible companions. Some candidates have been discovered. The first was in the constellation Cygnus, called Cygnus X-1. Of course, we cannot be certain that Cygnus X-1 is a black hole, but many astronomers think it is.

Figure 5.13 In orbiting Skylab, astronauts "float" in space, apparently weightless. However, gravity supplies the necessary centripetal force to keep them in orbit along with Skylab, so they do have weight.

Apparent Weightlessness

The terms *weightlessness,* or *zero gravity,* and *g's of force* have become common as a result of the space program. Let's investigate the meaning of these terms.

Weightlessness, or the condition of "zero gravity," generally refers to the situation when astronauts "float" in space, apparently because they have no weight (Fig. 5.13). This would be the case if $g = 0$. But a more correct term for this condition is *apparent weightlessness,* because gravity does act on astronauts in space, and therefore they do have weight. (Recall the definition of weight.)

In the case of an astronaut in a space craft in circular orbit about the Earth, we know that gravity provides the necessary centripetal force for the circular motion. The astronaut therefore has weight. The "weightless" feeling arises because the upward reaction force normally provided by the floor or a chair is missing. This is because the floor or the space craft is "falling" toward the Earth just as fast as the astronaut. In the astronaut's frame of reference, he "floats" relative to the space craft.

To help understand this, consider a person standing on a scale in an elevator (Fig. 5.14). When stationary, the scale reads the person's weight as indicated by the reaction force R. When the elevator accelerates downward, the reaction force is less, and the scale reads less weight. Finally, if the elevator cable broke and the elevator (and the person) were in free fall, the reaction force or the scale reading would be zero. The scale is falling as fast as the person. One might call this a "weightless" condition because the scale reads zero—but is it really?

A relatively new term used to describe the apparent weightless conditions experienced by astronauts orbiting the Earth is "microgravity." See Special Feature 5.1.

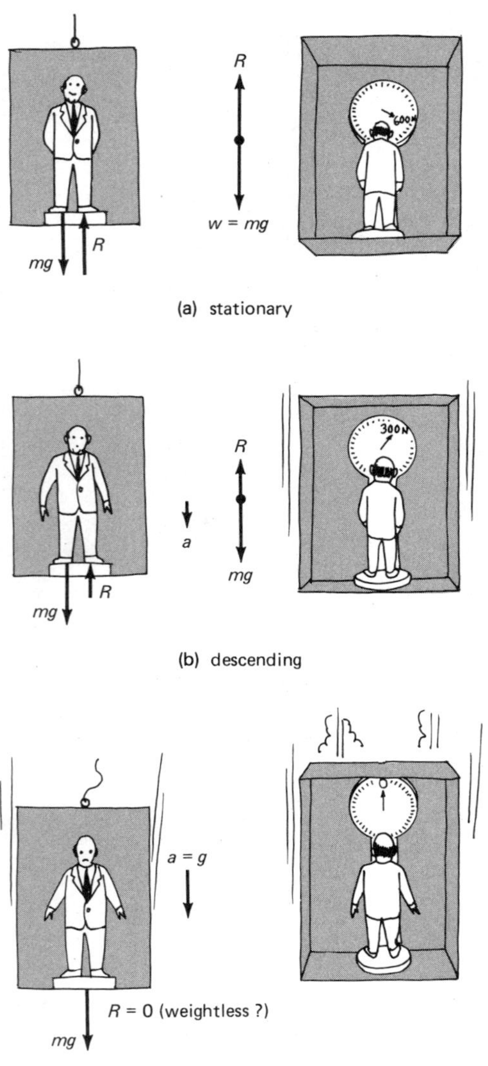

Figure 5.14 Weight in an elevator. (a) In a stationary elevator, the reaction force of the scales indicates the person's weight. (b) In a descending (accelerating) elevator, the reaction force is less, and a person appears to weigh less. (c) In an elevator in free fall, the scales would indicate zero. Is the person weightless?

g's OF FORCE

The constant value of g near the Earth's surface ($9.8\ m/s^2$) is sometimes used as a standard unit of acceleration. For example, an acceleration of 3 g's would be an acceleration three times that of g. We also speak of "g's of force." On the surface of the Earth, you experience a gravitational force of 1 g, which is your weight. A force of 2 g's would simply be some force equal to twice your weight. During the take-off of a commercial jet aircraft, you experience an average horizontal force of about $\frac{1}{5}$ g. This means that as the plane accelerates down the runway, the seat back exerts a force of about $\frac{1}{5}$ your weight against you to overcome your inertia. The reaction force (Newton's third law) is the force you exert against the seat, which gives you the feeling that you are being pushed back into the seat. On take-off at an angle of 30°, this increases to about $\frac{7}{10}$ g with a component of normal gravity helping to push you into the seat.

g's of force are also well known to pilots flying in an arc or loop. Suppose a plane is put into a circular vertical loop with just enough speed to go around the loop (Fig. 5.15). At the top of the loop, the pilot has a downward force of 1 g or his weight, which supplies the needed centripetal force. However, at that instant, the pilot would feel "weightless" because there is no reaction force as ordinarily exerted by the seat. Both he and the aircraft are "falling" toward the center of the loop (as an astronaut in orbit "falls" toward the Earth).

This condition is (incorrectly) called a state of "zero gravity," as discussed previously. When astronauts are trained, the apparent weightlessness condition in space is simulated by flying a plane in an arc, which gives a greater time duration of the "weightless" condition.

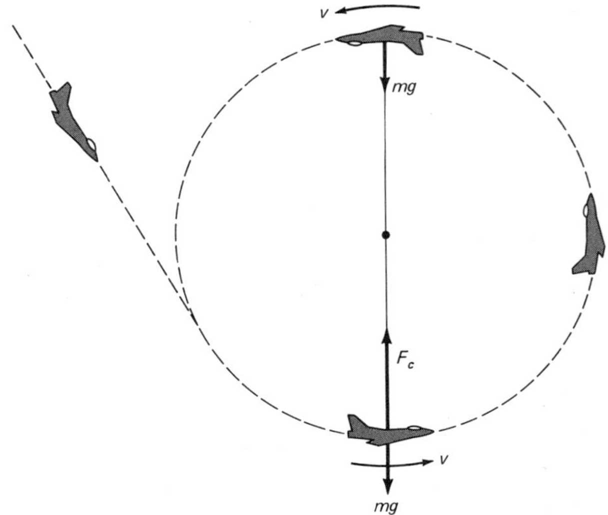

Figure 5.15 Loop-the-loop. In an airplane flying in a vertical circular loop at minimum speed, at the top of the loop there would be a force of 1 g on the pilot and at the bottom of the loop, a force of 2 g's.

SPECIAL FEATURE 5.1

Microgravity*

Because the trajectory of a spacecraft, such as the Space Shuttle, compensates for the force of Earth's gravity, the craft approaches a state of free fall. In a state of free fall, all objects in a craft are "weightless." However, gases venting from the Shuttle and the minute drag exerted by a tenuous atmosphere at obital altitude create nearly imperceptible forces, which are collectively called microgravity.

Once you adjust to it, microgravity can be intriguing. You can perform slow-motion somersaults and handspins (Fig. 5.16). You can float with the greatest of ease. You can push off from one side of your craft's interior and drift to the other side. You can lift or move normally heavy objects, which are nearly "weightless" in microgravity. You need never worry about dropping anything. Whatever slips out of your hand will float.

Living in space calls for special design technology. Here are some problems and solutions:

A beverage or water in a normally open container would leave the container, if shaken, as free-floating droplets. The droplets are not merely annoying; they could inconvenience people and pose a hazard to equipment. A beverage is served in a flexible dispensing unit with a plastic straw, which is clamped when the drink is not being sipped. Sponge baths rather than showers or regular baths are available. Because water adheres to the skin in microgravity, a little of it can go a long way. Water and other wastes are directed by air flow to a drain or are flushed into a sealed container.

Perspiration can be annoying in space. In the absence of proper air circulation, perspiration can accumulate layer by layer on the skin. It doesn't "roll" off.

Shaving could cause problems if whiskers ended up floating around. The whiskers could damage delicate equipment or irritate eyes or lungs. The solution: shaving cream and a safety razor. The whiskers adhere to the cream until wiped off with a disposable towel. Also available is an electric razor with a built-in vacuum pump.

* Adapted from NASA Educational Briefs for the Classroom, an Educational Publication of the National Aeronautics and Space Administration, (U.S. Government Printing Office, publ. no. 1982-361-570/3255) which deals with apparent weightlessness conditions in space, which NASA calls microgravity.

Figure 5.16

Microgravity calls for special considerations in the handling of solid foods. Crumbly foods are provided only in bite sizes to prevent the crumbs from floating around the cabin. The food trays are equipped with magnets, clamps, and double-adhesive tape to hold metal, plastic, and other utensils.

Furniture is bolted in place. Tether lines, belts, and handholds enable people to move around and keep themselves and other objects where they are supposed to be. If you try tightening a screw or turning a nut without a restraint, you rotate in the opposite direction.

A microgravity environment causes a variety of bodily changes. For example, people in space seem to have smaller eyes because their faces become puffy. Many of the effects of microgravity are apparently due to body fluid shifts from the lower to upper parts of the body. Some people in space suffer temporarily from a condition resembling motion sickness on Earth. Prolonged periods in space may also result in changes in the rates of blood cell and bone tissue formation. The good point about most of these effects of microgravity is that they seem to level off eventually in space and to reverse themselves after return to Earth.

At the bottom of the loop the situation is different. Here the pilot's weight is opposite to the direction of the centripetal force (supplied by the reaction of air against the air foil surfaces of the plane). The upward force exerted on the pilot by the seat must support the pilot's weight *and* supply the necessary centripetal force for circular motion (which is still 1 g), so a force of 2 g's is experienced.

You can get an idea of the g's of force in a similar situation by swinging a pail of water in a vertical circle. Here, you are supplying the centripetal force. When you swing the pail at the minimum speed needed for the water to go in a circle, the pail would seem "weightless" to you at the top of the swing, but at the bottom it would be a bit different. Why doesn't the water fall out of the pail at the top of the loop? What would happen if you were to swing the pail more slowly than the minimum speed?

In the preceding example, the centripetal force acting on the pilot and the plane is not excessive because of the minimum speed. However, this may be critical in the case in which an airplane diving at a high speed makes a "tight" circular turn to pull out of the dive (large v, small r, and $a = v^2/r$). The excessive g's of force may cause structural damage to the plane, to the point that the wings may actually shear off. In addition, the pilot may "black out." During the dive, the pilot's blood is accelerated downward, and in the turn it tends to continue in that direction (Newton's first law), draining from the pilot's head. Unconsciousness and, in extreme cases, death can result.

Earth Satellites

In our space age, many artificial satellites orbit the Earth—the moon is the only natural satellite. Most of us have watched (usually on TV) rocket "blast-offs" that send satellite payloads toward their orbital positions. Manned satellites are now common. There was Skylab and currently there is the Space Shuttle—a recyclable satellite.

Putting a satellite into orbit about the Earth involves gravitational interaction. Gravity must be overcome to get the satellite to the planned orbital altitude, and the satellite must be given sufficient tangential velocity so it will not fall back to Earth. Once this is done, the gravitational force provides the needed centripetal force to maintain the satellite in orbit.

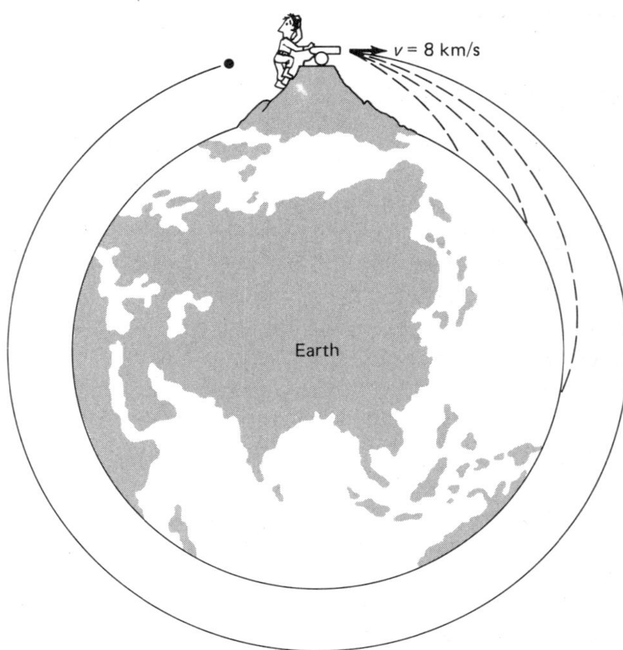

Figure 5.17 ". . . the greater the velocity . . . with which (an object) is projected, the farther it goes before it falls to the Earth. We may therefore suppose the velocity to be so increased, that it would describe an arc of 1, 2, 5, 10, 100, 1000 miles before it arrived at the Earth, till at last, exceeding the limits of the Earth it should pass into space without touching it." (From Newton's *System of the World.*)

CIRCULAR ORBITS

It is interesting to note that more than 300 years ago Sir Isaac Newton considered the possibility of putting an object into orbit about the Earth. He theorized about firing a cannon from the top of a mountain. If the cannon ball had sufficient tangential velocity, it would not fall back to Earth, only toward it (Fig. 5.17).

Newton had the right idea, but he also realized that a cannon ball could never be put into orbit like this. A cannon could never do the job of supplying enough tangential velocity. How fast do you think an object would have to be projected horizontally for it to go into a circular orbit just above the Earth's surface?

Near the Earth's surface, the object's weight force (mg) would supply the centripetal force ($F = mv^2/r$), and

$$mg = \frac{mv^2}{R_e}$$

where the radius of the orbit is the radius of the Earth, R_e, and air resistance is neglected. Using known values of g and R_e and solving for v, we find (as Newton did) that the required tangential speed is about 8 km/s (5 mi/s or 18,000 mi/h).

If the cannon ball were fired with a smaller velocity, it would fall back to Earth in a parabolic path (see Chapter 4). This is really part of a Kepler ellipse with one of the foci at the Earth's center. Over the relatively short flight path of the cannon ball, a parabolic path and an elliptical path are indistinguishable.

Similarly, if a satellite is given a tangential velocity of about 8 km/s, it will go into circular orbit about the Earth. A rocket with a satellite payload might be sent upward, then turned horizontally so there is a speed of 8 km/s at burn-out or when the rocket engines are shut off.

ELLIPTICAL ORBITS

What would happen if a satellite projectile were given a horizontal speed greater than 8 km/s? In this case, the projectile would curve around the Earth and start to move away from it. But gravity slows it down and attracts it back toward Earth. The satellite then would gain speed and come back to its starting point in an elliptical orbit (Fig. 5.18).

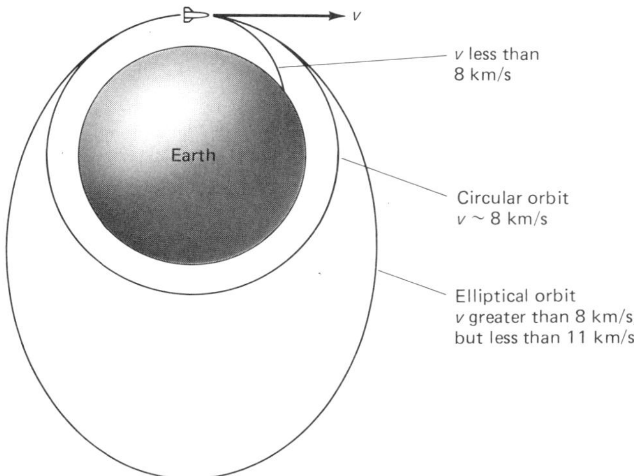

Figure 5.18 Earth orbits. The critical tangential speed for a circular orbit is about 8 km/s. At greater speeds, the orbit is elliptical.

ORBIT DECAY

Although satellites orbit for some time, they eventually return toward the Earth. Even at altitudes of several hundred miles, there is *some* atmosphere that supplies frictional "drag." This causes the satellite orbit to "decay," and it eventually spirals toward the Earth if the perturbation of its orbit is not corrected (by on-board rockets). A descending satellite may either burn up in the atmosphere (like a meteor) or fall to Earth (like a meteorite), depending on its size.

We have had several notable cases of the latter. In 1978 a Soviet *Cosmos* satellite with a nuclear reactor on board fell to Earth, spreading radioactive material over parts of Canada. The U.S. satellite *Skylab* came down in 1979. When astronauts returned from it in 1974, they left it in an orbit that was supposed to be stable until the 1980's. However, unexpected solar activity caused the Earth's atmosphere to heat up slightly and expand. This put more atmosphere into Skylab's orbital altitude and increased the frictional drag. Orbital maneuvering was attempted so as to reduce the drag, but to no avail. *Skylab* came tumbling down. Then, early in 1983, another *Cosmos* satellite came down. This time the nuclear reactor portion fell into the Atlantic Ocean.

ESCAPE VELOCITY

The next logical question is, How much velocity must a projectile be given so it does not go into orbit but escapes the Earth completely? We know that ordinary vertically upward projections are decelerated by gravity to a stop, then are accelerated downward so they return to Earth.

But as the distance from the Earth increases, the force of gravity becomes less. Although a projected object would never escape the Earth's gravitational field, there is a critical launching speed whereby the object would never quite be slowed to a stop. This turns out to be about 11 km/s and is called the **escape velocity.**

Hence, if a projectile had an initial speed of 11 km/s or greater, it would escape from the Earth. Of course, in practical applications a spacecraft or space probe eventually comes under the gravitational influence of another body, such as the moon or another planet, and would not return to Earth. (The process is reversed in getting astronauts back from the moon.)

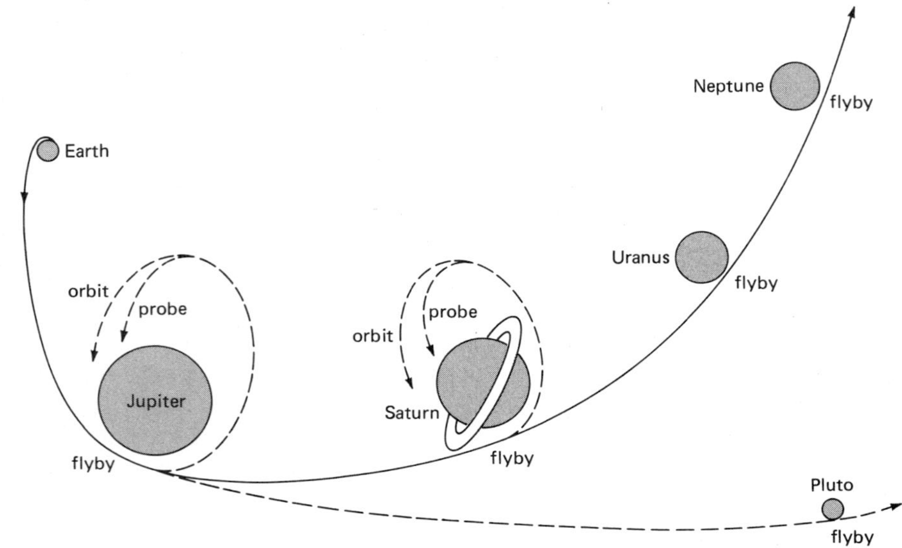

Figure 5.19 A gravitational "slingshot" for interplanetary flybys. The large gravitational field of Jupiter can be used to deflect probes and flybys of the outer planets when they are properly aligned.

Using the gravitational forces of other planets is the principle of interplanetary flybys (Fig. 5.19). These are launched at a time when the outer planets are lined up in such a way that the large gravitational field of Jupiter can be used to deflect the probe in a flyby path toward other planets. The probe may also be slowed and deflected so that it goes into orbit about a large planet.

A probe may eventually go into orbit about the Sun or leave the solar system and travel into interstellar space. For example, as a probe approaches and whips around Jupiter on its way to a Pluto flyby, it gains speed. If its speed is then equal to or greater than the escape velocity from the Sun and from Jupiter at this distance, the probe will trip out into outer space—and this has happened. See Special Feature 5.2.

MULTISTAGE ROCKETS

The escape velocity of an object from the Earth is the *initial* velocity it must be given, for example, the muzzle velocity of a cannon. However, satellites are not launched by cannons. Rockets are used. The burning rocket engines continually accelerate the rocket itself and the payload. Nor does the rocket take the satellite

SPECIAL FEATURE 5.2

A Pioneer in Outer Space

On June, 13, 1983, Pioneer 10 crossed the orbit of Neptune—over 4.5 billion kilometers from the Sun—into interstellar space, becoming the first spacecraft to leave the solar system.* This unique event in human history climaxed an 11-year series of impressive achievements.

Launched on March 3, 1972, and designed for a 21-month lifetime, the flyby of Jupiter in December 1973 was Pioneer 10's primary mission. It was the first spacecraft to fly beyond Mars and to cross the asteroid belt, to fly by Jupiter, to chart that planet's intense radiation belt, to measure the mass and density of its four planet-sized moons, to find that it is a liquid planet, to take closeup pictures of its atmosphere and Great Red Spot, and the first to cross the orbits of Uranus and Pluto before reaching Neptune's.

Scientists predict that Pioneer will travel among the stars almost indefinitely because interstellar space is so empty. Some 10,500 years from now it will pass a star, and in 32,000 years it will have its nearest encounter with one—3.2 light-years distant. Pioneer may outlast the solar system itself.

* Although Pluto is usually the outermost known planet, its highly elliptical orbit actually goes inside Neptune's orbit. Currently, Pluto is inside Neptune's orbit, which now defines the boundary of the known solar system or "outer" space.

Figure 5.20 Multistage rockets are used to put satellites into orbit.

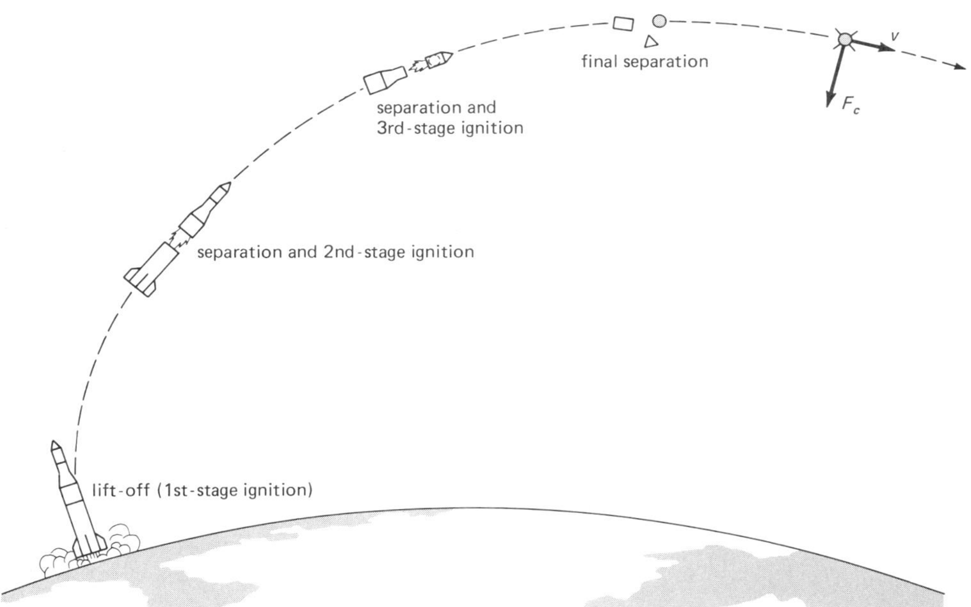

vertically upward to its orbital altitude and then give it the required tangential velocity. After blast-off, the rocket veers away from the vertical (Fig. 5.20). This gives a component of the rocket force in the horizontal direction and an increasing velocity.

So that the necessary speeds can be achieved efficiently, multistage rockets are used. This permits an in-flight reduction in the mass of the rocketship by jettisoning the burnt-out stages. The next stage then has less mass to accelerate. (See Special Feature 3.3.)

SUMMARY OF KEY TERMS

Gravitation (gravity) the attractive interaction or force between two masses.

Kepler's laws empirically derived laws that describe planetary motion.

1. Law of elliptical orbits: All planets move in elliptical orbits about the Sun, with the Sun at one foci of the ellipse.
2. Law of equal areas: An imaginary line from a planet to the Sun sweeps out equal areas in equal times; i.e., planets do not have uniform orbital speeds and have greatest speed when closest to the Sun.
3. Law of periods: The period T and the radius R of a planet have the relationship $T^2/R^3 = \text{a constant}$, where the constant is the same for all planets.

Newton's law of gravitation an expression for the magnitude of the force of gravity between two masses separated by a distance r: $F = Gm_1m_2/r^2$.

Universal gravitational constant, G the constant in Newton's law of gravitation, which is believed to be constant throughout the universe.

Ocean tides the periodic rising and lowering of the ocean's level due to the gravitational attraction of the moon (and Sun) and the Earth's rotation.

Spring tides higher high tides and lower low tides caused by the added pull of the Sun at the times of new and full moons.

Neap tides lower high tides and higher low tides caused by the cancelling effect of the Sun's pull at the times of first-quarter and third-quarter moons.

Gravitational field a mapping of space that represents the gravitational force per mass due to some body or bodies.

Weightlessness (zero gravity) the *apparent* weightlessness experience in an accelerated or "falling" system in the absence of a reaction force.

g's of force a unit of force using a body's weight on Earth as a standard; for example, 2 g's of force acting on a body is a force with a magnitude equal to twice the weight of the body.

Escape velocity the initial or launching velocity an object would have to be given to escape from a large body like the Earth.

EXERCISES

1. There are four fundamental forces. Why are they "fundamental"? (*Hint:* Compare with fundamental properties, Chapter 1.)
2. Evaluate the constant of Kepler's third law using Earth data. (*Hint:* A convenient unit for R is the astronomical unit (A.U.), which is defined as the distance from the Earth to the Sun, i.e., 1 A.U. What would be a convenient unit for T, days or years?)
3. If the moon is "falling" toward the Earth, why doesn't it reach it?
4. If the force due to gravity on you due to the Earth is mg, what is the force due to gravity on the Earth due to you? Explain.
5. According to Newton's law of gravitation, how is the force of gravity affected when the distance between two masses is doubled?
6. G is a *universal* constant. Cite another universal constant. (*Hint:* Consider the ratio of the circumference and the diameter of a circle.)
7. Being an ultra-conscientious consumer, would you rather buy items by mass or by weight at the Dead Sea Supermarket? Explain why. How about at the Top–of–Mt. Everest Supermarket?
8. Suppose a hole is drilled through the center of the Earth to the other side. If you dropped a stone down the hole, what would happen? Remember, an object is attracted (gravitational force) toward the center of the Earth.
9. If you drive down a street past a large skyscraper, does the force of gravity due to the skyscraper accelerate your car on approach and decelerate it after you go past?
10. The Sun is much more massive than the Earth (about a million times more). Why doesn't the gravitational attraction of the Sun pull us off the Earth?
11. If the law of gravitation were somehow repealed, what would happen to the solar system?
12. The law of gravitation is an inverse-square law. Suppose it varied as (a) $1/r$ or (b) $1/r^3$. What effect would this have on the ocean tides in each case? (*Hint:* The tides are caused by the *difference* in the gravitational pulls on opposite sides of the Earth. How would differences vary in these cases, say between r and $2r$?)
13. The high tide on the side of the Earth nearest the moon occurs sometime later than when the moon is highest in the sky. What causes this time lag?
14. Would you expect to weigh more at full moon or at new moon? Why is this effect not detected?
15. Why are the twice-daily Earth tides unnoticed?
16. It was once speculated that the Earth had a sister planet at the same distance from but on the opposite side of the Sun, so it could never be seen. How could it be proved that there is no such planet, both directly and indirectly?
17. If perturbation calculations indicate the possibility of a Planet X beyond Pluto, why hasn't it been discovered?
18. If the Earth shrank to one-half its present size but kept its

Figure 5.21 See Exercise 21.

mass constant, what effect would this have on the acceleration due to gravity on the Earth's surface?

19. What is the acceleration due to gravity at an altitude of 3 R_e (three Earth radii) above the Earth's surface?
20. Why is the acceleration due to gravity on the moon one-sixth of that on Earth?
21. An astronaut on a "moon walk" easily picks up a lot of equipment that would be too heavy for him on Earth (backpacks of 300 lb; Fig. 5.21). Explain how this is possible.
22. Is the acceleration due to gravity different on the surfaces of different planets? Explain.
23. The planet Saturn is 94 times more massive than Earth, but the acceleration due to gravity on the surface of Saturn is about the same as g on Earth. What does that tell you in general about the size of Saturn? Can you give an estimate of the radius of Saturn compared to Earth's?
24. Is it correct to speak about the weight of a planet?
25. Sketch the gravitational field around two equal masses separated by a short distance.
26. Sketch the gravitational field for the Earth-moon system in terms of (a) a Newtonian field and (b) an Einsteinian field.
27. The gravitational field of a black hole and the original star would be identical in a certain region. Identify this region.
28. A black hole is no more massive than the original star. Why is the gravity of a black hole so great near the "hole" (within original star radius)?
29. How far from the Earth would you have to go to be truly weightless?
30. Suppose you are standing on a chair and jump off. Are you weightless until you hit the ground? Explain.
31. If you were in an elevator standing on a scale and the elevator accelerated upward, what would the scale read, and why?
32. Describe the sensations you feel in an elevator when it suddenly accelerates (a) upward and (b) downward. Compare these to how an astronaut feels (a) on blast-off and (b) in a spacecraft in orbit about the Earth.
33. Considering only the Earth and the moon, is it possible for an astronaut on a moon trip to have zero gravity? Explain.
34. The plans for a new super Skylab suggest that a handball or racquetball court be included so the astronauts can get exercise. Describe such a game with the Skylab in orbit about the Earth.
35. When a pail of water is swung in a vertical circle, why doesn't the water fall out of the pail at the top of the loop above a certain minimum speed? What would happen below this speed?
36. In putting a satellite into circular orbit about the Earth after it reached an altitude of several hundred kilometers, would a minimum tangential velocity of 8 km/s be required? Explain.
37. Would a tangential velocity of 8 km/s be required to put a spacecraft in circular orbit near the surface of the moon? Would air resistance be a consideration?
38. Do all planets have the same escape velocity? Explain.
39. Why does the moon have no atmosphere, while the Earth does?

6 Work and Energy

What Is Energy?

Energy. How many times have you read this word recently in a newspaper or magazine, heard it on radio or TV, or even discussed it yourself? We now speak of an "energy crisis" and of ways to conserve energy. But what *is* energy?

If someone asked you to define energy, what would you say? If your instructor asked you to go out and bring back a bucket of energy, with what would you return?

Although energy may be difficult to explain, it is familiar to all of us. The energy needed for life processes is supplied by food. We say we are "full of energy" when ready (and willing) to do something. Fuels supply the energy needed to run factories, to bake pizzas, to provide transportation, and for many other purposes.

From experience we might expect energy to be in some way related to force—something that does things or gets things done or moving. Like force, energy is a *concept.* It is defined in terms of what it does or is capable of doing. To explain this critical concept, we begin with a related concept—work. Are you ready for a little brain "work"?

Work

Work is done when you pick up your physics book. The heavier the object you pick up, the more work you do. Holding a heavy object like a cement block is commonly considered work because it is physically tiring. However, there is no work done on the block in this case, in the scientific sense. What, then, is this seemingly unfair scientific definition of work?

Obviously there must be a force involved, but there is another ingredient—motion. For the simplest case of a constant force and motion in a straight line parallel to the direction of the force, we define work as follows:

> The work done by a force acting on an object is defined as the product of the force and the distance through which the object moves.

That is,

Work = force × distance moved in the direction parallel to the force

or

$$W = Fd$$

So we see why no work is done when holding a heavy object or pushing against a wall. There is no motion ($d = 0$). Also, when a constant force is not parallel to the direction of motion, only the part or *component* of force in the direction of the motion does work. This is illustrated in Figure 6.1. The component of force parallel to the direction of motion does work. The vertical force component does no work because the lawn mower does not move in that direction.

The units of work are force times distance (Fd) or in the SI system newton-meter (N-m) and, in the British system, pound-foot, customarily reversed to foot-pound (ft-lb). The N-m is given the special name of *joule* (J) (pronounced "jewel") in honor of the English scientist James Prescott Joule, and 1 J = 1 N-m.

(a)

(b)

Figure 6.1 Work is force times distance. (a) Work requires motion. When there is no motion (or no distance travelled), no work is done. The vertical component of the applied force ($F_{\perp}$), or the component perpendicular to the direction of motion, does no work since the lawnmower does not move in this direction. (b) Only the parallel component of force ($F_{\parallel}$) in the direction of motion does work.

$$(Fd = W)$$

(SI) newton-meter = joule (J)

(British) pound-foot = ft-lb

Work is a *scalar* quantity and has no direction.

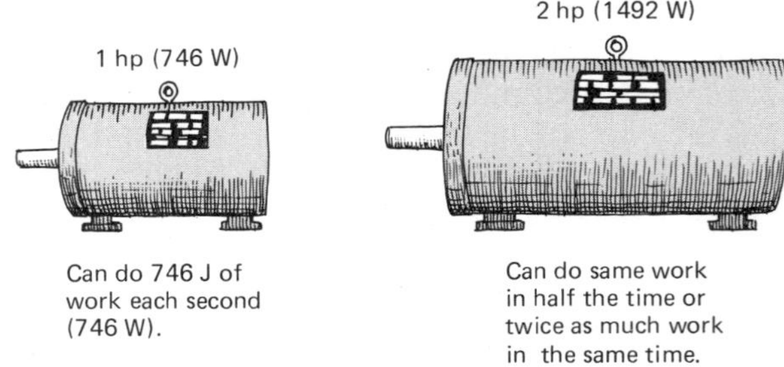

Figure 6.2 Power is work per time. The power ratings of motors tell how much work they can do per time.

Power

Suppose your instructor asks you to move a stack of boxes up to the second floor. You might do this *work* in 10 minutes. Then the instructor asks another student to move the boxes up to the third floor. Not being too enthusiastic, your classmate takes 20 minutes to move the boxes.

The same amount of work is done in each instance, but there is a difference—time, or the time rate of doing the work. To distinguish between such cases, we talk about power. **Power is the time rate of doing work,** or simply the work divided by the time it takes to do the work.

$$\text{Power} = \frac{\text{work}}{\text{time}}$$

or

$$P = \frac{W}{t}$$

In the example of transporting the boxes, since you did the same work in half the time, you expended twice as much power as your classmate.

The units of power can be seen from the defining equation. In the SI system the units are joule/second (J/s). The J/s is given a special name, the watt (W), in honor of James Watt, the Scottish engineer who invented the modern steam engine. In the British system, the units of power are foot-pound/second (ft-lb/s).

$$(W/t = P)$$

(SI) joule/second = watt (W)

(British) foot-pound/second = ft-lb/s

You may have never heard of a ft-lb/s. A more common unit used to express power in the British system is the horsepower (hp), and

$$1 \text{ hp} = 550 \text{ ft-lb/s} = 746 \text{ W}$$

Oddly enough, the horsepower unit was originated by James Watt, after whom the SI unit is named. The value of the horsepower was based on his experiments on the work a horse could do in a working day. Perhaps he was trying to sell his steam engines to replace horses, which were commonly used to bring coal out of mines. The horsepower unit would be a good sales comparison of what the engine could do.

The horsepower unit is still the common customary unit used to rate motors and engines. A 2-hp motor can do twice as much work in the same time as a 1-hp motor, or the same amount of work in half the time (Fig. 6.2).

Energy

So what is the difference between work and energy? Work is not energy itself, but a transfer of energy. Thus, if a body has energy it can do work through a transfer of energy. Energy appears in many forms. We will focus on two basic forms of **mechanical energy**—potential energy and kinetic energy.

POTENTIAL ENERGY

One might ask, when work is done on an object, where does the work "go"? For example, you do work in compressing a spring or lifting an object (Fig. 6.3). It might be thought that the work goes into the spring or the object. However, work is not something a body *has,* but is something *done on* (or by) a body. After being "worked on," the spring and the object do not have work, but do possess the *ability* to do work, or energy. Hence, **a body has energy when it is capable of doing work.** Energy was transferred to the spring and the object in the process of doing work.

In the case of a compressed spring or a raised object, this energy is called potential energy. It was brought about by a change in position.

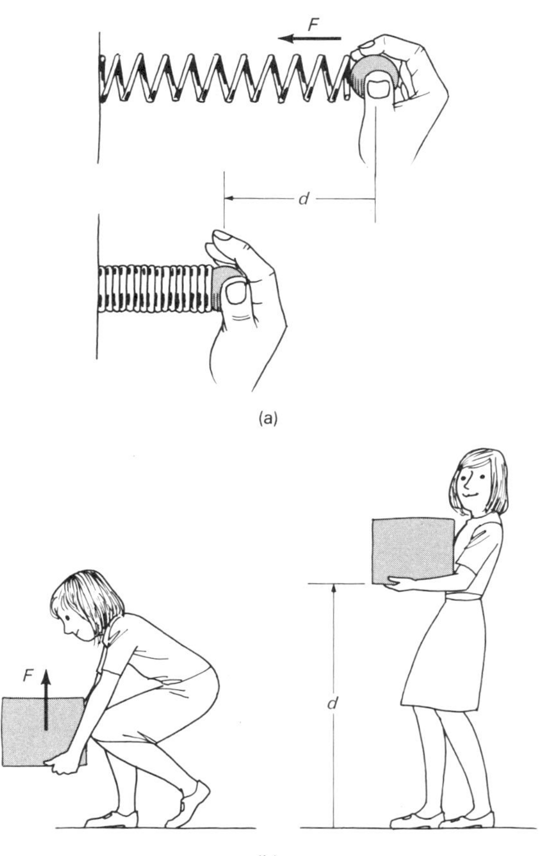

Figure 6.3 Potential energy. Work is done in compressing a spring (a) or in lifting an object (b), giving rise to a change in potential energy, or energy of position.

Potential energy is the energy of position.

That is, an object has more energy at one position relative to another position. Practical examples of spring potential energy and gravitational potential energy are shown in Figure 6.4. The potential "position" energy of a spring is due to a change in its length or a change from its equilibrium position, (where it is neither compressed nor stretched).

Let's take a closer look at gravitational potential energy, since it is so common. The potential energy of a raised object is equal to the work done against gravity in lifting it. The upward force required is equal to (initially, slightly greater than) the weight of the object (mg). The distance through which the object is moved or raised is the height h (Fig. 6.3). The work (Fd) is then equal to mgh, and

(a)

(b)

Figure 6.4 Examples of applications of (a) spring potential energy and (b) gravitational potential energy.

$$\text{Gravitational potential energy} = \text{weight} \times \text{height}$$

$$\text{P.E.} = mgh$$

The units of energy are the same as those of work, for

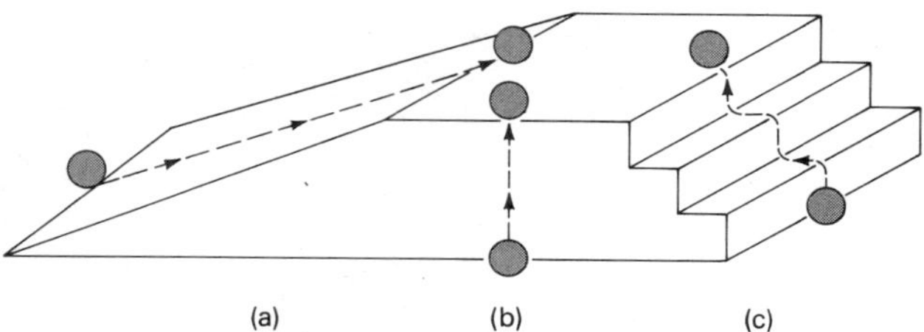

Figure 6.5 Potential energy is independent of path. All the balls on the platform have the same potential energy relative to the floor, regardless of the path used to get there.

example, joules of energy, and energy is also a scalar quantity.

The height h is measured relative to its original position. The potential energy is "independent of path" or how the object gets to a height (Fig. 6.5). It depends only on the vertical height between the initial and final positions.

The initial position, usually taken as $h = 0$, is arbitrary. That is, you can put the zero reference position wherever you like (Fig. 6.6). This may give rise to negative potential energy and the concept of a potential energy "well." Like a real well, energy or work is required to get higher in the energy well. On Earth, we are near the bottom of a potential energy well. Work or energy is required to put satellites into orbit at a higher energy level in the well.

KINETIC ENERGY

How about when a force is applied to an object and it is moved without friction along a horizontal surface? Work is done, but there is no change in potential energy because the height doesn't change. Where does the work "go" in this case? It goes into moving or setting the object in motion, and an object in motion has kinetic energy.

Kinetic energy is the energy of motion.

Hence, when an object is in motion (in any direction), it has the ability to do work. For example, a moving car may run into another one and expend its kinetic energy in doing the work of a "fender bender."

The kinetic energy of a moving object is equal to one half the product of its mass and the square of its velocity:

$$\text{Kinetic energy (K.E.)} = \tfrac{1}{2}mv^2$$

Figure 6.6 Negative potential energy. Relative to the zero reference point, the potential energy of a body may be negative. This gives rise to the concept of a potential energy "well."

Notice that unlike momentum (mv), another "quantity" of motion, kinetic energy is proportional to the square of the velocity (v^2). If the velocity of an object is doubled, its momentum is doubled, but its kinetic energy is quadrupled ($2^2 = 4$). This means that the object could provide twice the impulse but do four times as much work.*

If the work done on a stationary or moving object

* This leads to an important distinction between elastic and inelastic collisions (Chapter 3). Momentum is conserved in both, but kinetic energy is conserved only in elastic collisions. In an inelastic collision, some of the kintic energy goes into the work of deforming the object(s) or is lost as heat.

Figure 6.7 The change in kinetic energy depends on the square of the velocity, K.E. $= \frac{1}{2}mv^2$. If the speed is doubled, the kinetic energy increases by a factor of four. This is an important factor in the stopping or braking distance of a car.

goes only into changing its motion, then the *change* in kinetic energy is equal to the work done on the object. Similarly, because of its kinetic energy, a moving object can do work (Fd). If the object is brought to rest in doing the work, then we may write

$$\text{Work} = \text{kinetic energy}$$

$$Fd = \tfrac{1}{2}mv^2$$

where F is the "stopping" force and d is the stopping distance.

This equation shows that speed or velocity is an important factor in stopping a car. For example, a car traveling 90 km/h has four times as much kinetic energy as when it is traveling 45 km/h (v^2 dependence). Then, assuming the stopping (braking) force F to be constant, it would take *four* times the distance to stop when traveling at 90 km/h than when traveling at 45 km/h (Fig. 6.7). Say it took you 50 m to stop when going 45 km/h. It would take 200 m to stop when going 90 km/h. Add your reaction time in applying the brakes, which would increase the stopping distance, and you can see why we are required to drive 35 km/h (20 mi/h) in school zones.

QUESTION: The potential energy of an object can be negative, owing to the selection of the zero reference position. Can there be negative kinetic energy?

ANSWER: No. Since K.E. $= \frac{1}{2}mv^2$, even if the velocity of an object is negative (by direction), $-v$, the squaring always makes the kinetic energy positive $(-v)(-v) = +v^2$. The "squaring" of the velocity vectors gives a scalar quantity, which kinetic energy or any form of energy is.

FORMS OF ENERGY

We say that the potential energy and kinetic energy discussed thus far are forms of "mechanical" energy. This is because their uses involve the mechanical activities of work with forces and motions. In fact, if you think about it, you might come to the conclusion that *energy is the result of work being done.*

In general, all forms of energy are the result of work being done by forces or against forces. This work may be taking place right now with the "release" of energy, or may have taken place sometime in the past with the "storing" of energy. On the bottom line, the forces involved are the fundamental forces of nature.

We have already seen an example of this with the gravitational force. Work is done against gravity in lifting an object, and the result is potential energy. When an object falls, the gravitational force does work, and kinetic energy results. Let's take a quick look at some other forms of energy and the forces involved. Specific forms and forces will be discussed in greater detail in later chapters.

Electrical energy is a big one. The basis of electrical energy lies in the fundamental force interactions between electrically charged atomic and subatomic particles. Electrical energy is associated with the motion of electric charges, or electric currents. It is this form of electrical energy that operates common electrical appliances (which do work for us).

QUESTION: What do we "buy" from an electric power company?

ANSWER: If you have ever paid an electric bill, you know we pay for electricity in units of kilowatt-hours (kW-h). These are multiple units of power (watts) times time (seconds). Since power = work/time ($P = W/t$), we have work = power × time ($W = Pt$). Hence, the kilowatt-hour is a unit of work or energy. So we pay for or buy electrical energy (to do work) from an electric "power" company. Perhaps we should really call it an energy company.

Electrical forces also hold together the atoms and molecules that make up matter. We might think of atoms as being held together by electrical "springs," and hence there is potential energy. The elastic energy of an actual spring can be traced to the work done against the intermolecular electrical forces.

During a chemical reaction, such as the burning of fuel, electrons and atoms are rearranged (work is done), and energy is released. The result of the combination is another chemical compound in which the particles have less energy. We refer to this form of energy, the energy released, as chemical energy.

Heat is also commonly considered a form of energy.† Heat energy is associated with the random motional (kinetic) energy of the atoms and molecules in matter. When heat (energy) makes a body hotter, its molecules move faster, on the average. In a solid, the molecules vibrate faster, whereas in a gas the molecules are free to move and collide with other molecules. (Collisions and momentum transfer again.)

Light is a form of electromagnetic energy. We sometimes call this radiant energy. Something "radiates" as the result of the acceleration of electrically charged particles. Other forms of electromagnetic energy include radio and television waves. Electromagnetic energy can travel through vacuum. This is how energy gets to us from the Sun, our major source of energy. The Sun's energy enters into the life cycles of plants and animals and is the basis of our chemical energy in fossil fuels (coal, oil, and gas). It also evaporates water from the oceans, which through winds, clouds, and rain ends up inland behind dams with gravitational potential energy for hydroelectric energy.

Nuclear energy is the source of the Sun's and other stars' energy. This involves the fundamental nuclear forces (strong and weak). These forces do work on nuclear particles. The result is a rearrangement of nuclear particles to form different nuclei with the release of energy.

In energy-releasing nuclear processes, part of the masses of the nuclei is transformed into energy. Hence, we now consider mass to be a form of energy. Einstein showed that the amount of energy released in a mass transformation is given by his famous equation $E = mc^2$, where E is the amount of energy released, m is the part of the nuclear mass that is transformed, and c is the speed of light (a constant). This transformation can work the other way, too—energy can be transformed into mass. More about this later in Chapter 29.

Mass-energy transformation is the principle of nuclear weapons and also of nuclear reactors, which supply an appreciable part of the energy used to generate electricity. Nuclear mass energy (or fossil fuel chemical energy) is transformed into heat, which is used to produce steam that turns a turbine. The mechanical energy of the turbine is transformed by a generator into electrical energy, which can again be transformed into heat, light, or the mechanical energy (by motors) that we use to do work. Indeed, energy transformations are all around (and inside) us.

CONSERVATION OF ENERGY

Whenever energy is transformed from one form to another, or transferred from one object to another, we find that no energy is lost in the process. This is expressed in one of the most important conservation laws—the law of **conservation of energy.**

> Energy cannot be created or destroyed. It may be transformed from one form to another, or transferred from one object to another, but the total amount of energy is constant or conserved.

Another way of saying this is that the total energy of the universe is conserved. It is always there somewhere in some form. Of course, the universe is the largest system we can think of, and its energy content might be expected to be constant. Let's apply the conservation of energy to some smaller, more practical systems.

To simplify the understanding of the conservation of energy, we often use *ideal* systems in which the energy is only in two forms—mechanical potential and kinetic energies. Then we talk about the conservation of mechanical energy. Consider the diver in Figure 6.8. On the platform, the diver's total mechanical energy is all potential energy. During the dive, potential energy is converted to kinetic energy, but at any point the total mechanical energy (K.E. + P.E.) is constant. Just before hitting the bucket, the diver's total mechanical energy is all kinetic energy. Of course, when he hits the bucket, it's another story. The total mechanical energy is no longer conserved, since the bucket is not part of our *ideal* system.

Another common example used to illustrate the conservation of mechanical energy is an ideal pendulum (no air friction or friction in the support). As illustrated in Figure 6.9, there is a continual conversion

† More correctly, heat is "energy in motion," or the energy transferred from one body to another because of a temperature difference. See Chapter 13.

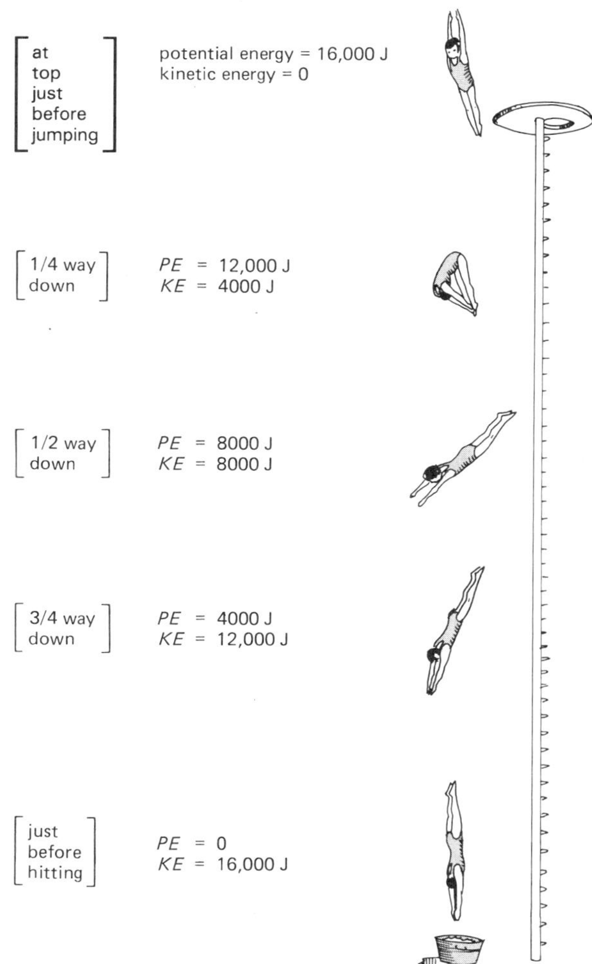

Figure 6.8 Conservation of energy. The total mechanical energy, or the sum of the kinetic and potential energies, is constant at any point during the dive.

between kinetic and potential energies. At the maximum heights of the swing (on each side), the pendulum bob stops instantaneously ($v = 0$), and the total energy is all potential. At the bottom of the swing ($h = 0$), the total energy is all kinetic. In between, the kinetic energy and potential energy always add up to equal the constant total energy, so the total mechanical energy is conserved.

We call such an ideal system in which the total mechanical energy is conserved a conservative system. But what about in the real world? On Earth there is always some energy lost as a result of friction or some other cause. In this case, we have nonconservative systems.

A water analogy of a conservative system and a nonconservative system is shown in Figure 6.10. In the

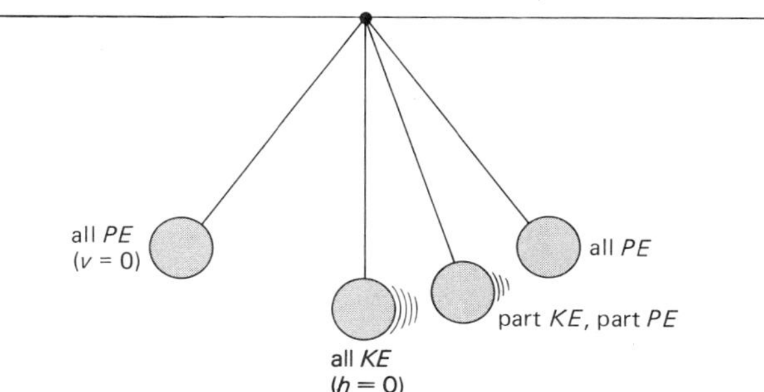

Figure 6.9 Another example of the conservation of energy. The energy of a swinging pendulum exchanges back and forth between kinetic and potential energies.

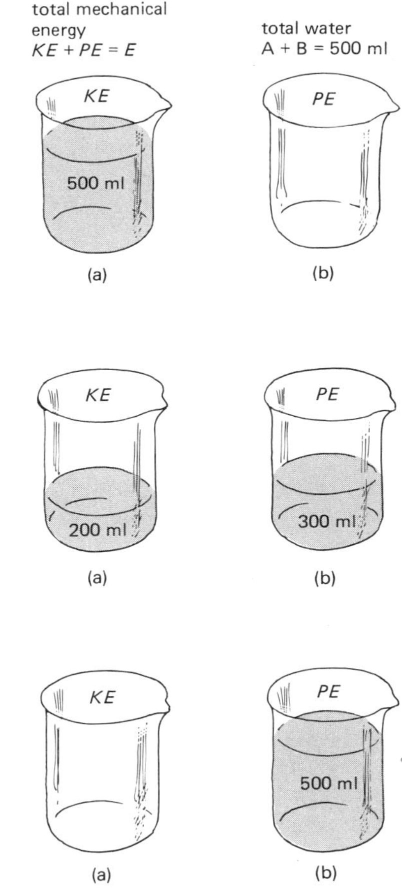

Figure 6.10 A water analogy of energy conservation. The total amount of water remains constant in a "conservative" beaker system, just as the total mechanical energy of a conservative system remains constant. However, if some water is spilled, the system is "nonconservative," just as energy is lost in a nonconservative system.

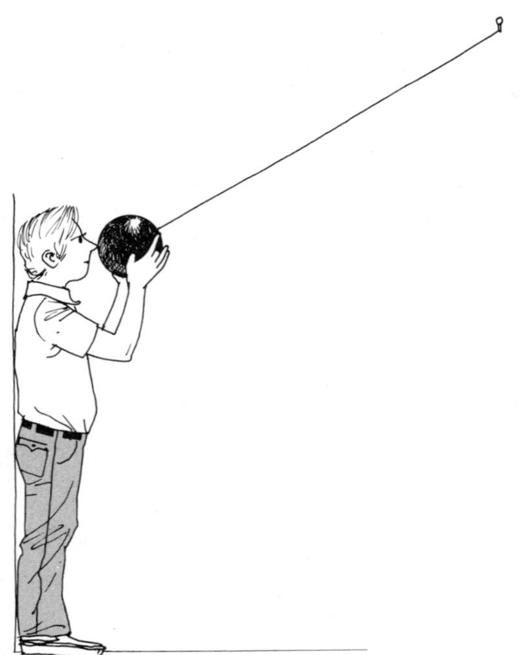

Figure 6.11 Will the ball hit him in the nose on the backswing?

conservative system, the total amount of "mechanical" water in the kinetic-energy and potential-energy beakers is conserved. However, if some of the water is spilled in pouring it back and forth, the beaker system is then nonconservative. The spilled water is analogous to the pendulum energy lost as a result of friction. But even for a nonconservative system, the *total energy* is conserved. All the energy (like the water) is around somewhere in some form.

We say that the lost or wasted energy is no longer available to do *useful work.* In a conservative system work can be done and stored as energy, and the same amount of work can be gotten back. In a real, nonconservative system some energy is always lost. A demonstration of the conservation of energy (and one's belief in it) is to have a person hold a suspended bowling ball to the tip of his nose and then release it (Fig. 6.11). The person stands perfectly still and lets the ball swing back toward him. Many are reluctant to try this. However, by the conservation of energy, the bowling ball will swing back no farther (or to no greater height) than its point of release. Since the system is slightly nonconservative, the ball will not quite reach his nose. Of course, he should be advised simply to release the ball and be careful not to give it a push. Why?

Machines and the Conservation of Energy

How much energy is lost is an important consideration in machine "systems" that are used to do useful work. If you ask someone what a machine is, a typical reply might be that a machine is something that does work. This answer is prompted by the experience of using machines to perform tasks, some of which require more strength than our muscles alone can supply. A person cannot lift an automobile, but with a jack (a machine) this is easily accomplished.

However, machines *do not* actually save us work. That is, we do not get something for nothing. Nature doesn't operate that way—no free lunches. Indeed, in practical applications we always get less work out of a machine that we put in.

In the mechanical sense, **a machine is any device used to change the magnitude or direction of a force.** Let's see how this makes doing work "easier." Consider a simple machine called a lever (Fig. 6.12). Neglecting any frictional losses, we have, by the conservation of energy in terms of work, that the work output will equal the work input, or

$$\text{Work in} = \text{work out}$$

or, since $W = Fd$,

$$(Fd)_{\text{in}} = (Fd)_{\text{out}}$$

But notice in the figure that by applying a small force through a large distance, a large force is exerted through a small distance. Thus, the lever increases or multiplies the force, but at the expense of distance. The force multiplication factor is called the **mechanical advantage** of a machine.

You use the lever principle when you pull a nail with a claw hammer, use a bottle opener, or jack up a car (Fig. 6.13). Suppose you exerted a 200 N force (about

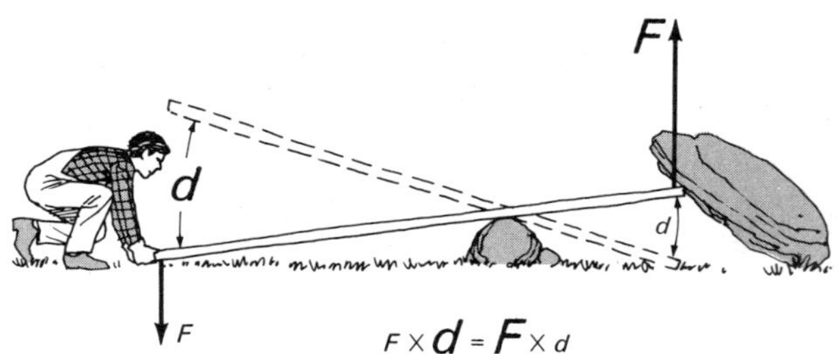

Figure 6.12 The lever. A simple machine that multiplies force (greater force output than input), but at the expense of distance.

Figure 6.13 Mechanical advantage. With the mechanical advantage provided by a lever, a person can lift a heavy car.

Figure 6.14 The pulley. (a) A single fixed pulley is a "direction changer" and not a force multiplier. (b) In a system with a movable pulley, or block and tackle, the mechanical advantage is equal to the number of strands or ropes supporting the movable pulley.

45 lb) to jack up a car weighing 10,000 N (more than a ton). Then your force multiplication factor or mechanical advantage would be 50. However, the output distance (the distance the car moves) compared with the input distance (the distance you move the jack handle) would be reduced by a factor of $\frac{1}{50}$. If you moved the jack handle through 50 cm, the car would be raised only 1 cm.

Another type of simple machine is the pulley (Fig. 6.14). A single fixed pulley doesn't give you any force multiplication (mechanical advantage of one). It simply gives a change in the direction of the force. However, if you use a system of pulleys, called a block and tackle, there is a mechanical advantage. For example, if the suspended movable pulley has three supporting ropes, to make the load rise 50 cm you would have to pull the rope down 150 cm (to shorten each support strand by 50 cm). A moment's thought should tell you that the mechanical advantage of the system is three; that is, ideally you could lift 600 N by pulling down on the rope with a force of slightly more than 200 N.

Efficiency

In the previous ideal examples, friction was neglected, but machines do have friction. By the conservation of total energy we should really write:

$$\text{Work in} = \text{work out} + \text{work "lost"}$$

where the work "lost" is due to friction. The work output is the "useful" work done by the machine.

To express how much useful work we get out for the work we put in, we talk about the **efficiency** of a machine:

$$\text{Efficiency} = \text{work out/work (or energy) in}$$

The efficiency ratio is a decimal fraction, but it is usually expressed as a percent. For example, if 300 J is the useful work output of a machine for a 500-J input, the efficiency is $\frac{300}{500} = 0.60$ or 60 percent. Sixty percent of the work input goes into useful work, which means that 40 percent, or 200 J, of work input is lost.

Some typical efficiencies of complex energy-work conversion machines are given in Table 6.1. Obviously, it pays to have highly efficient machines so as to save work and energy. We'll look at efficiency again in the next chapter, where energy conservation is a major consideration in terms of energy resources and dollars and cents. For now, take a look at Special Feature 6.1 to see where the energy losses are in the not-so-efficient automobile.

Table 6.1 Typical Percent Efficiencies of Some Complex Machines

Automobile	12–15
Electric motor	Up to 95
Steam engine	50–75
Steam locomotive	5–10
Turbine	Up to 40

As a final thought for this chapter, what would be the efficiency of an *ideal* machine, such as a motor or engine, with the work output equal to the work (or energy) input? The answer is 1.0 or 100 percent. But this is not possible, because in practice there is always some work or energy lost.

If it were possible, we could build a *perpetual motion machine.* By feeding the machine's output back into its input, it would run forever!

Even better would be a machine with greater than 100 percent efficiency. In this case energy would be created—more work out than put in. Not only could it run perpetually, but we could remove the extra energy and never have to worry about an energy shortage. Sound too good to be true? It is.

SPECIAL FEATURE 6.1

Energy and the Automobile§

Automobiles powered by gasoline engines are known to be very inefficient machines. Even under ideal conditions, less than 15 percent of the available energy in the fuel is used to power the vehicle. This situation is much worse in stop-and-go driving in the city.

There are many mechanisms that contribute to the energy losses in a typical automobile. About two thirds of the energy available from the fuel is lost in the engine. Part of this energy ends up in the atmosphere via the exhaust system, and part is used in the engine's cooling system. About 10 percent of the available energy is lost in the automobile's drive-train mechanism. This loss includes friction in the transmission, drive shaft, wheel and axle bearings, and differential. Friction in other moving parts accounts for about 6 percent of the energy loss. Approximately 4 percent of the available energy is used to operate fuel and oil pumps and such accessories as power steering, air conditioning, power brakes, and electrical components. Finally, about 14 percent of the available energy is used to propel the automobile ("useful" work). This energy is used mainly to overcome road friction and air resistance. These energy losses are summarized in Table 6.2.

§ From Serway, R. A. *Physics for Scientists and Engineers,* Saunders College Publishing, Philadelphia, 1983.

Table 6.2 Energy Losses in a Typical Automobile

Mechanism	*Power loss (%)*
Exhaust (heat)	33
Cooling system	33
Drive train	10
Internal friction	6
Accessories	4
Propulsion of vehicle ("useful" work)	14

SUMMARY OF KEY TERMS

Work the product of the force acting on an object and the parallel distance through which the object moves. $W = Fd$

Energy the ability or capability to do work.

Joule (J) the SI unit of work and energy. 1 J = 1 N-m

Power the time rate of doing work. $P = W/t$

Watt (W) the SI unit of power.

Horsepower (hp) the British system unit of power. 1 hp = 550 ft-lb/s = 746 W

Potential energy the energy of position. Gravitational potential energy, P.E. = mgh

Kinetic energy the energy of motion. K.E. = $\frac{1}{2}mv^2$

Conservation of energy energy cannot be created or destroyed. It may be transformed from one form to another, or transferred from one object to another, but the total amount of energy is constant or conserved.

Conservation of mechanical energy the sum of the kinetic energy and potential energy (K.E. + P.E.), the total mechanical energy, is constant or conserved.

Conservative system a system in which the total mechanical energy is conserved.

Machine (mechanical) any device used to change the magnitude or direction of a force.

Mechanical advantage the force multiplication factor of a machine.

Efficiency the ratio of work out/work (or energy) in for a machine that expresses the fraction (or percent) of useful work done by a machine.

Perpetual motion machine an ideal machine with 100 percent efficiency (or greater) that could run forever after an initial work input.

EXERCISES

1. Can we detect energy with our senses? That is, can we feel, smell, or taste energy? Explain.
2. Explain the concept of energy.
3. A weightlifter holds a set of weights over his head. Is he doing work? Has he done work? Explain.
4. (a) A short weightlifter and a tall weightlifter both lift 1000 N (Fig. 6.15). Who does more work? (b) How could they do the same work?
5. Is it possible to have negative work? (*Hint:* Although work is a scalar quantity, force and distance are vectors with directions.) What does this mean in terms of the motion of an object on which the work is done?
6. You and another student are late to class and race to your classroom on the second floor by different routes. (a) Who does more work against gravity? (b) If you beat your classmate there, who had the greater power output?
7. How much work is required in drawing a bucket of water from a well?
8. Why are water towers tall structures and often placed on high elevations?
9. Explain how a sling shot and an archery bow can have potential energy. What happens to the P.E. when a stone or an arrow is shot?
10. Could a $\frac{1}{4}$-horsepower motor and a $\frac{3}{4}$-horsepower motor be used to do the same amount of work? Explain. Does either motor have an advantage over the other?
11. Some factory workers do "piece" work; that is, they are paid according to the number of pieces or items they process or produce. Others are paid by the hour. Is there a power consideration involved in these two methods of remuneration? Explain.
12. An 8-hp riding tractor that is used to mow lawns cannot generally be used to plow a garden. Explain why not.
13. A constant applied force F does work in moving an initially stationary object through a distance. If a constant force of $2F$ moves the object through the same distance,

Figure 6.15 See Exercise 4.

(a) are the power outputs the same? (b) If not, does the $2F$ force have twice the power output? (*Hint:* Consider acceleration and time. You can justify your answer mathematically by recalling that $d = \frac{1}{2}at^2$.)

14. Suppose you are at the bottom of a deep hole in the ground of depth h. How much initial kinetic energy would you have to give a stone you threw for it to reach the top of the hole? (The stone's initial velocity would be its "escape" velocity.)
15. A ball lies on the first floor of a house. A person on the first floor says the ball has zero potential energy. A person on the second floor says the ball has −20 J of potential energy. A person in the basement says it has +20 J of potential energy. Can they all be right? Explain.
16. When a moving car comes to a stop, what happens to its kinetic energy?
17. A car traveling with a constant velocity on a level road coasts up a hill and comes to a stop. What happened to its kinetic energy?
18. How much work is required to bring a moving object to rest?
19. A baseball player slides into home plate. What happens to his kinetic energy?
20. An automobile traveling at a velocity v accelerates until its velocity is tripled. How does this affect (a) its momentum and (b) its kinetic energy?
21. A state patrol officer investigating the head-on collision of two automobiles of the same make and model notes that the skid marks caused by braking before collision of one car are four times the length of the skid marks of the other. A witness testifies that both drivers locked their brakes at the same time before collision. (a) What can be inferred from this? (b) Was the collision elastic or inelastic? (c) Would the automobiles be stationary immediately after collision? Explain.
22. A person sits in a moving car. Does the person have kinetic energy? Explain.
23. A moving object has work done on it by a force parallel to its motion. How does this affect its kinetic energy?
24. Explain the difference between the formulas $Ft = \Delta p$ and $Fd = \Delta\text{K.E.}$
25. Explain the work and energy considerations in lifting an object and letting it fall. What happens to the energy when it hits the ground?
26. Why is it easier to drive a large nail with a carpenter's hammer rather than with a small tack hammer?
27. A person on a trampoline can go higher with each bounce. Explain the energy considerations of a trampoline and how this is possible. Is there a maximum height the person can go? Explain.
28. (a) A pile driver is used to drive a pile into the ground (Fig. 6.16). Explain the principle of the pile driver. (b) If the "stopping" force of the pile was the same with each strike, how would the distance the pile is driven into the ground be affected with each strike? Explain.

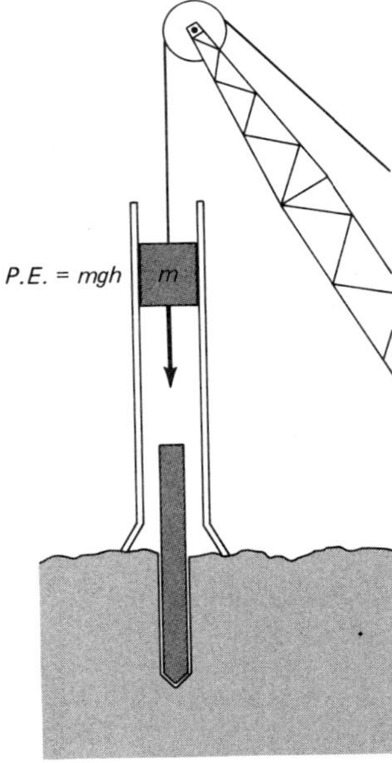

Figure 6.16 See Exercise 28.

29. On a cold day you rub your hands together to warm them. How does this work?
30. A basketball player running down the floor stumbles and gets a "floor burn" on his knee. Explain why. Could a floor burn be worse in one case than another?
31. Explain on the atomic level how friction generates heat.
32. What happens to the energy supplied to electrical devices?
33. Draw an energy flow diagram for the generation and expenditure of electrical energy.
34. Sound is considered to be a form of energy. Explain why.
35. The law of conservation of mass says that matter cannot be created or destroyed. Is this true in light of nuclear reactions?
36. A pendulum released from rest swings inward and its string comes in contact with a rod, as shown in Figure 6.17. Is this a trick photograph, or does the bob really swing to the same height as shown in the photo? Explain.
37. A pendulum is taken to the moon, and its bob is released from the same height above the ground as it was on Earth. Would there be any difference(s)? (Consider work, speed, and maximum heights of swing.)
38. In an acrobatic act, two people jump onto a seesaw to propel another acrobat, as shown in Figure 6.18. (a) Why must two people jump on the seesaw? (b) Are the weights of the acrobats a consideration? Explain. (c) Suppose the board support were moved so that one quarter of the board's length was on the single acrobat's side of the support. What effect would this have?
39. Sketch a block and tackle that has a mechanical advantage of five.
40. An inclined plane, such as a loading ramp, is a simple machine. Explain how an inclined plane has a mechani-

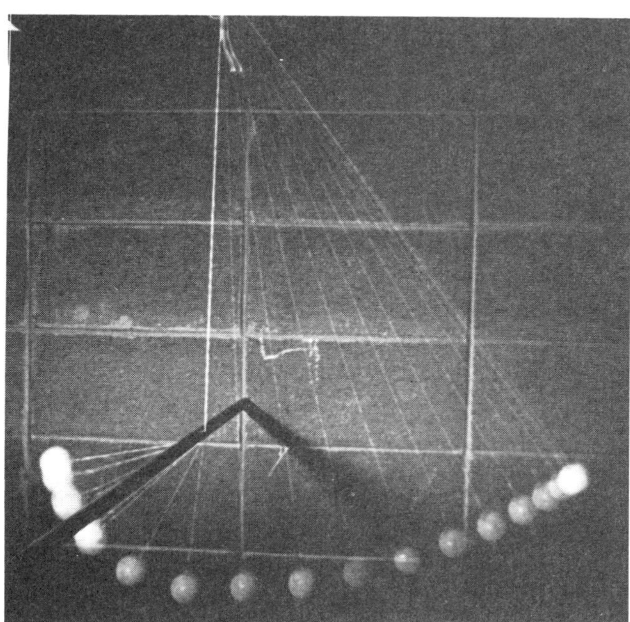

Figure 6.17 See Exercise 36.

Figure 6.18 See Exercise 38.

cal advantage. Does the angle of incline affect the mechanical advantage?

41. Which is more efficient, machine A that has a work output of 500 J for a work input of 700 J, or machine B that does 900 J of useful work with an input 1400 J?

42. Is the horsepower rating of a motor its power input or its power output? [*Hint:* Would the power (work/time) rating tell you about the useful work the motor could do if you did not know its efficiency rating?]

43. Many amusement-park roller coasters now have loops (Fig. 6.19). (a) Does the roller coaster have potential energy at the top of the loop? (b) Is the initial height of the roller coaster critical? Explain.

Figure 6.19 See Exercise 43.

7

Energy and the Environment

An Energy and Environmental Crisis?

Now that the concept and scientific principles of energy have been discussed, we'll take a look at the "practical" aspects of energy — the demand, resources, and related environmental problems. No doubt you have heard, and will continue to hear, a great deal about these topics. Do we have enough energy, and more important, will we have enough energy in the future? What about environmental problems?

Depending on whom you talk to or what you read (or hear or watch, not forgetting radio and TV), you'll get conflicting views. The purpose of this chapter is to present the facts and to get you thinking. Some of the political and social implications will be considered. Keep in mind that when a particular issue gets "hot," the voters are called upon to make decisions. For example, after the nuclear incident at Three Mile Island in 1979, there were several voter decisions on the use of nuclear energy in various parts of the country. Perhaps there will be similar decisions on the use of (imported) oil or coal and on pollution levels. Such decisions should be made by well-informed persons.

Energy Consumption and Demand

Certainly there is a demand for energy, and our present supply seems to be adequate. When we switch on a light, it comes on — but in some cities in the summer during "brownouts," lights come on less brightly. Here, the demand exceeds the supply. We have a large appetite for energy. Statistics show that the United States, with about 5 or 6 percent of the world's population, accounts for about one third of the world's energy consumption. An example of the difference in electrical energy consumption is shown in the introductory chapter photograph. Satellite photos taken at midnight show many areas of the United States all lit up, while most of Europe is dark.

Where did it begin? Well, if you like stories, perhaps it began with the Greek god Prometheus, who according to myth stole fire from the gods and gave it to mankind. Zeus, the father of the gods, punished Prometheus (which means "forethinker") by having him bound to a rock while an eagle continually devoured his immortal liver. In addition, Zeus sent Pandora to Earth, where she removed the lid from a box and released evil, *work*, and disease among mankind "as a price for fire."

However we got fire (probably as a result of lightning), it has been one of the major agents for energy conversion for centuries. Until the latter part of the 1800's, wood was the major fuel. Today, wood seems to be making a minor comeback as a fuel with many people using wood stoves for supplemental home heating. Coal took over and held sway for the first part of the 20th century (1900–1950). Since then, gas and oil have dominated. Figure 7.1 shows recent energy production by type in the United States and the consumption by

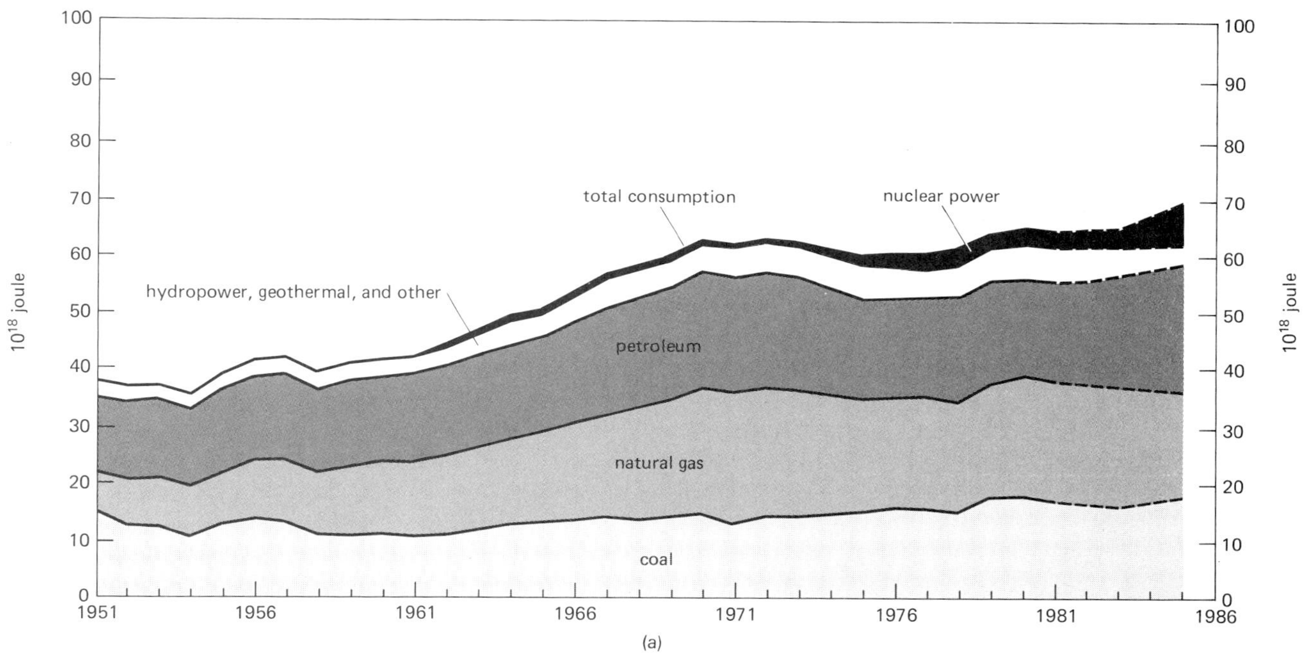

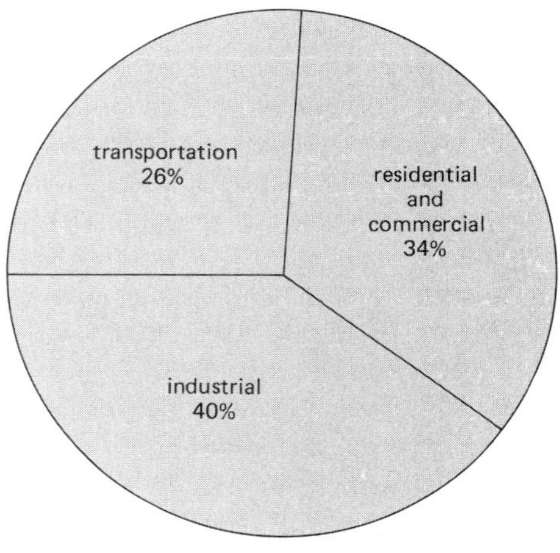

Figure 7.1 (a) Energy consumption in the United States, by type. (b) Energy consumption by sector.

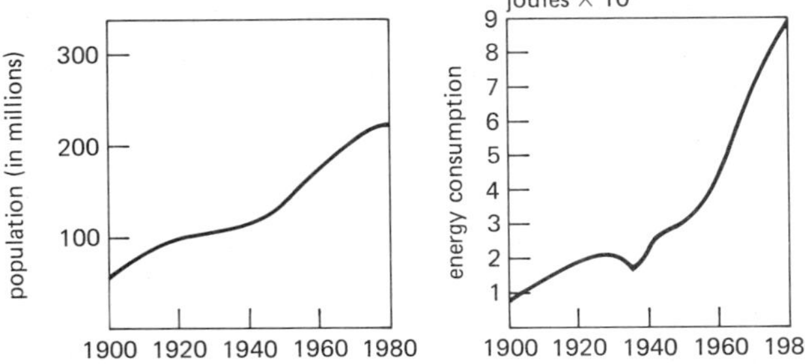

Figure 7.2 A comparison of population growth and energy consumption in the United States for eight decades since 1900.

sector. The energy sources of choice after wood were to some extent a matter of convenience. But supply and demand also played a part. In some countries in rural areas, wood is still the major domestic energy source, and it is in short supply.

Some major factors in energy demand in the United States and the world are population, industrialization, and standard of living. Figure 7.2 compares population growth and energy consumption in the United States for eight decades since 1900. The effect of population growth on energy demand and consumption cannot be overemphasized. Note that per-capita energy consumption has also risen significantly in recent years.

Between 1970 and 1980 the U.S. population increased by about 25 million people. All of these *new* people will need homes and food, and they will want automobiles and other things that are part of our standard of living. All of this will require additional energy. Do we have the energy resources for the future? That's the big question.

Energy Resources

Overall, the Earth has three main energy sources: (1) *solar energy* from the Sun, (2) *tidal energy* from the gravitational pulls of the Earth-moon-Sun system, and (3) *terrestrial or geothermal energy* from inside the Earth (includes nuclear energy). As you might imagine, the Sun is by far the major source of our energy—both past and present.

The Sun is an enormous ball of hot gases some 150 million kilometers (93 million miles) from Earth. The Sun's energy comes from nuclear fusion. In the interior of the Sun, nuclei of hydrogen are fused together to form nuclei of helium. In the process, mass is converted to energy. The energy produced radiates from the Sun, mostly in the form of light. (We shall consider the details of nuclear fusion in Chapter 29.)

The Earth intercepts a very small part of this energy. About one third of the incident energy is reflected back into space, and the other two thirds goes into terrestrial processes, such as maintaining life and warming the atmosphere, the oceans, and the ground. The fraction or percentage of incident sunlight reflected by a celestial body is called its albedo. The Earth then has an albedo of 0.33, or 33 percent. The moon's albedo is 0.07, or 7 percent; that is, it reflects only 7 percent of the incident sunlight. (See Special Feature 7.1.)

The Earth was receiving solar energy long before the dawn of civilization. Fortunately, some of this energy was stored away for use at a later time. Solar energy is captured by green plants and stored chemically in a process called **photosynthesis.** Plants contain carbon dioxide and water, as well as a chlorophyll pigment that gives them a green color (Fig. 7.3).

The chlorophyll acts as an agent which allows sunlight to cause a chemical reaction between the carbon dioxide and water that converts them to glucose—a sugar—and oxygen, which is released. The glucose molecules form cellulose, and the plants grow. Some of these plants are used as food for animals, including humans. (Incidentally, the oxygen product of photosynthesis maintains the oxygen balance of the atmosphere. Recall that animals breathe in oxygen and expel

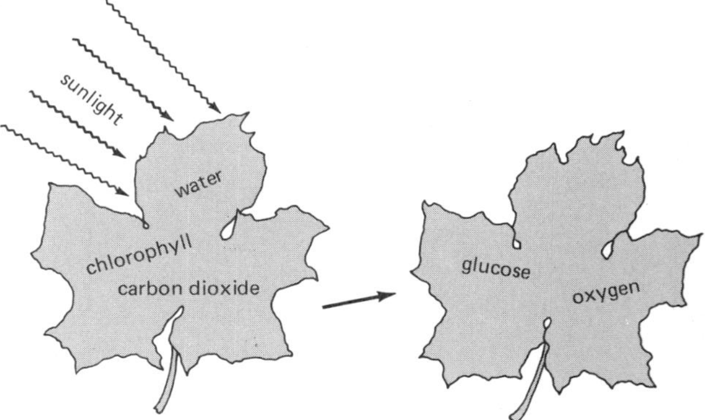

Figure 7.3 Photosynthesis. Carbon dioxide and water in the presence of sunlight and chlorophyll react to produce glucose (a sugar) and oxygen.

SPECIAL FEATURE 7.1

The Greenhouse Effect

If the Earth continually receives energy from the Sun, you may wonder why the Earth doesn't get hotter or why its average temperature does not increase. To maintain a relatively constant long-term average temperature, the Earth must lose energy, which on the average is equal to the amount it receives. This is accomplished through the re-radiation of energy back into space. Atmospheric effects on how this is done are important in preventing large daily temperature variations. On the moon, which has no atmosphere, daily temperature variations range from about 100°C (212°F, the boiling point of water) on the day or Sun side to −173°C (−280°F) on the dark side.

Incoming solar radiation warms the atmosphere and the surface of the Earth, and the warm Earth re-radiates energy in the form of infrared radiation, which is nonvisible (see Chapter 14). The gases of the atmosphere, in particular water vapor and carbon dioxide (CO_2), are "selective absorbers." That is, they allow the visible incoming sunlight to pass through, but they absorb or trap certain infrared radiations. This atmospheric absorption helps to retain the Earth's energy so we don't have the daily temperature fluctuations as on the moon. Clouds (water vapor) also assist in maintaining the Earth's warmth. In the absence of cloud coverage, nights are "*cold* and clear."

Hence, the atmospheric gases have a "thermostatic" effect in maintaining the Earth's daily temperature variations. We call this process the **greenhouse effect.** Glass has absorption properties similar to those of the atmospheric gases. As used in a greenhouse (Fig. 7.4), the glass allows the visible sunlight to pass through, then blocks or absorbs the infrared radiation. Actually, in this case the warmth is primarily due to the prevention of the escape of warm air heated by the ground within the glass enclosure. The temperature of a greenhouse in the summer is controlled by painting the glass panels white, which reflects the sunlight, and opening panels to allow the hot air to escape. The interior of a closed greenhouse is quite warm, even on a cold day. You have probably experienced the "greenhouse" effect in an automobile on a cold, sunny day.

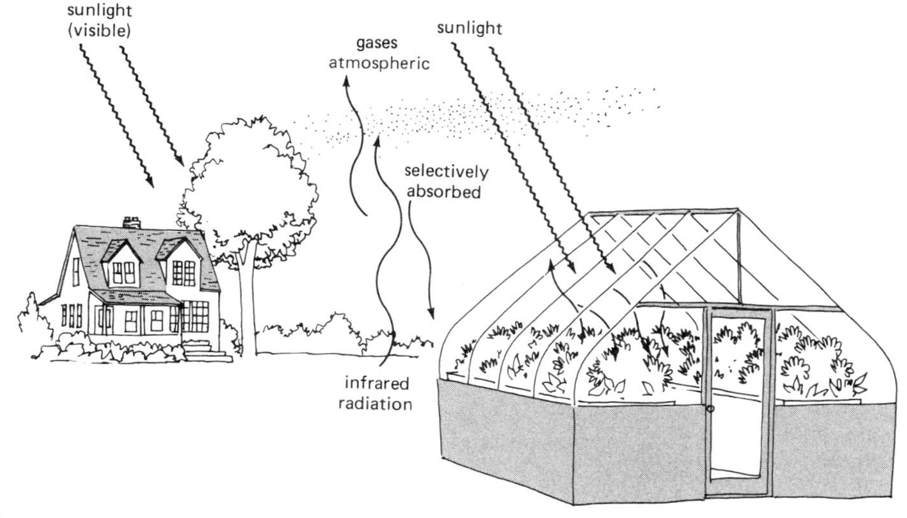

Figure 7.4 The greenhouse effect. The gases of the atmosphere, particularly water vapor and carbon dioxide, are selective absorbers of radiation with absorption properties similar to those of glass as used in a greenhouse. Visible light is transmitted, and infrared radiation is absorbed.

CO_2. Were it not for photosynthesis, animals would deplete the atmosphere's oxygen supply. Most photosynthesis occurs in the oceans.)

The energy stored in the cellulose of plants can be released under the proper conditions. For example, at high enough temperatures and in the presence of oxygen, a piece of wood (cellulose) burns, and energy is released. The products of complete combustion are CO_2 and water, so we are chemically back where we started.

Nearly all plant and animal materials eventually decay, and the stored energy is released. However, if deposits of dead plants are removed from oxygen, the decay process will not occur and the energy is not released. This is what happened millions of years ago when dense vegetation and forests covered large portions of the Earth's surface. Climatic conditions changed and the vegetation in some regions was covered by layers of eroded silt. The pressure of layers of silt caused the vegetation to be pressed into thick layers, or seams, which formed into coal. Similarly, oil (petroleum) and natural gas were formed from layers of small marine plants and animals that accumulated on sea floors. Now you can understand why we call coal, gas, and oil **"fossil" fuels.**

The carbon compounds stored in fossil fuels for millions of years are now being released relatively suddenly through combustion in the form of carbon dioxide. With increased energy consumption and demand, there is concern that the vast amounts of CO_2 being released to the atmosphere may cause an appreciable increase in the normal CO_2 content of the air. This could affect the greenhouse effect such that the Earth would warm up. Part of the polar ice caps would melt, raising the level of the oceans. This would cause severe problems in coastal areas. A recent study predicts that the level of atmospheric carbon dioxide will double by the year 2065, with an increase of 1.5 to 4.5°C in the Earth's surface temperature.

On the other hand, the particulate matter released from fuel combustion might affect the Earth's albedo. When a large, natural volcanic eruption puts volumes of particulate matter (volcanic ash) into the atmosphere, more incoming sunlight is reflected back into space (increased albedo), and climatic changes can occur; for example, winters can become colder than normal. Supersonic transports (SSTs) or aircraft, like the Concorde, emit large amounts of exhaust gases and particulate matter. Many SSTs flying at high altitudes, where there is no precipitation to wash the particles back to Earth, could contribute to an increased albedo. If the Earth cooled enough, perhaps another Ice Age would be triggered.

But, more important, the exhaust emissions of the SSTs could react chemically to deplete the ozone layer.* The ozone layer, at an altitude of about 30–40 km, acts as a shield, absorbing about 99 percent of the deadly ultraviolet radiation from the incoming sunlight. This radiation would otherwise burn our skin and kill plants. (The small amount of ultraviolet radiation that gets through is responsible for sunburns and suntans.) You may remember the concern for the ozone layer in the 1970's associated with the gases used as propellants in aerosol spray cans. Chemical reactions of these gases in the atmosphere could deplete the ozone layer. Different gases are now used in spray cans to avoid this problem.

So some processes warm the Earth and others cool it. It's a dangerous environmental game to play. In any case, let's take a look at how our reserves of fossil fuels stack up for the future and at some of the associated environmental problems.

COAL

The United States has an abundance of coal—an estimated several hundred years' supply at current energy consumption rates. As pointed out earlier, fuel use has shifted from coal to oil and natural gas. Anyone who had or has a coal stove or furnance knows why. Oil and gas are more convenient and cleaner to use—they are easier to handle and leave no ashes. More recent considerations are the environmental pollution problems associated with coal impurities and coal mining. Some of the major problems are sulfur impurities and acid rain. (See Special Feature 7.2.)

Some coal deposits contain more sulfur than others. Much of the coal in the eastern United States is high-sulfur coal (>2 percent).† Western states, such as Wyoming and Montana, have low-sulfur (<1 percent), but generally lower-quality, coal deposits. These deposits are surface mined, which has associated land and water pollution problems.

There are various types of coal with various subclassifications. These are generally based on carbon content, or heat value, and moisture content—the higher the carbon content and the lower the moisture content, the better the coal. In general, there are three major types of coal.

ANTHRACITE. Anthracite coal is sometimes called "hard" coal. It has a brilliant luster and can be polished and made into costume jewelry. With a carbon content of 90 percent or greater, it burns with a pale blue flame and is a high-quality coal.

BITUMINOUS. Bituminous coal is the most abundant variety and the most widely used. It is sometimes called

* Ozone (O_3) is triatomic oxygen.

† The symbol > means "greater than," and < means "less than."

SPECIAL FEATURE 7.2

Acid Rain

Rain is normally slightly acidic as a result of carbon dioxide (CO_2) in the air. Water vapor and carbon dioxide combine to form carbonic acid (H_2CO_3), a mild acid we all drink in the form of carbonated beverages (carbonated water). However, the term *acid rain* refers to a more serious and abnormal source.

Coal generally contains sulfur compounds. When coal is burned, the sulfur combines with oxygen, forming sulfur oxides (SO_x). Vented to the atmosphere, the sulfur oxides SO_2 (sulfur dioxide) and SO_3 (sulfur trioxide) combine with water vapor in the air to form H_2SO_3 and H_2SO_4, sulfurous acid and sulfuric acid, respectively (Fig. 7.5). Also, nitrogen oxides (NO_x) formed from the normal nitrogen and oxygen in the air in the presence of high-temperature combustion can form nitric acid. The exhaust gases from automobiles and trucks are a major source of nitrogen oxides.

Precipitation from contaminated clouds is then acidic—"acid rain" (and also acid snow, sleet, fog, and hail). Sulfur and nitrogen emissions from the industrialized midwest and northeast United States are believed to be responsible for acid rain in New England and eastern Canada. The acid levels in lakes and ponds in these regions are increasing. Acid levels are measured on a pH scale, on which pure water has a pH of 7.0. The lower the pH, the more acidic a solution. Rainfall with a pH of 1.4 has been recorded in the United States. This is more acidic than lemon juice (pH 2.2). Sufficient concentrations of acids kill fish and aquatic plants, and lakes and ponds "die."

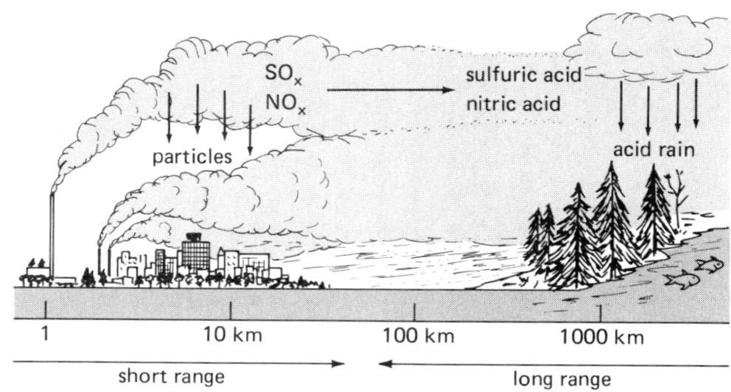

Figure 7.5 Acid rain. Acid rain arises from sulfur oxides (SO_x) and nitrogen oxides (NO_x) emissions combining with water vapor in the air to form acids.

In addition to acid rain, there are acid snows. Over the course of a winter, acid precipitations build up in snowpacks. During the spring thaw and runoff, the sudden release of these acids gives an "acid shock" to streams and lakes.

Nor is acid rain confined to the eastern United States. Acid fogs with pHs as low as 1.7 have been measured along the west coast near Los Angeles. The culprit here is believed to be nitrogen oxides coming primarily from auto emissions.

"soft" coal, being quite brittle and easily crushed. The carbon content is about 70 percent.

LIGNITE. Lignite or "brown" coal is a low-ranking coal with a relatively low heat value and high moisture content. It disintegrates rapidly in air and is subject to spontaneous combustion. Although there are thousands of square miles of lignite deposits in the United States, little is mined because of its poor qualities.

Because of coal's inconvenience in handling and pollution problems, processes of coal gasification and coal liquefication are being considered. These processes convert coal to gaseous and liquid products. The idea is to produce synthetic fuels or "synfuels" from coal. The processes are mostly experimental in this country owing to economic considerations. Underground coal gasification has been practiced on a commercial scale in the Soviet Union. The gasification is done by means of underground explosions. During World War II, Germany produced significant amounts of liquid fuels from coal. Some countries now have commercial facilities in operation for coal liquefication.

OIL

We have come a long way since oil seeping from the ground near Titusville, Pennsylvania aroused interest, and the first oil well in history was drilled in 1859. This 70-ft (21-m) well changed the course of many things.

Figure 7.6 An off-shore oil rig.

Crude oil, as it is pumped out of the ground, is a thick, gooey, dark liquid. It is refined to produce many different products, including gasoline, heating oil, motor oil, and tar. Some oil products are used as medicines, and our now ever-present plastics come indirectly from oil. If all the oil-based products suddenly disappeared, things would look quite different.

How much oil do we have? Domestic proven and potential reserves may last about 40 years or so, depending on the rate of consumption. A major oil find in Alaska in 1968, with subsequent construction of the Alaskan pipeline, helped increase our reserves. But even this major find is only a five-year supply at our current total consumption rate.

Because oil was formed in sea beds, we now look for it off-shore (Fig. 7.6). To supplement our reserves, about 30 percent of the oil we use is imported.‡ This has a drastic effect on our balance of payments.

Other sources of oil in the forms of oil shale and tar or oil sands have received consideration. Oil shale is a fine-grain sedimentary rock containing significant amounts of organic material (not oil). When heated to about 480°C (900°F), the shale yields a viscous liquid that can be transformed into synthetic crude oil. But one ton of high-grade shale will yield only about half a barrel of oil. There are large deposits of oil shale in the western United States, particularly in Colorado. However, the low yield gives you some idea of the problems involved both technically and environmentally.

Tar or oil sands contain a heavy crude oil that is too thick to flow into wells. The largest known deposits are in Alberta, Canada. There are also substantial deposits in the United States, in Utah. Here too there are problems with extraction, economics, and the environment.

‡ Department of Energy 1981 Annual Report to Congress. This 1985 figure is a large percentage, but it is better than 1977, when almost 55 percent of the oil consumed was imported.

NATURAL GAS

Natural gas, which is mostly methane gas, is the cleanest-burning of the fossil fuels and is in the shortest supply. Natural gas is found with or near deposits of crude oil. It was once "flared" or burned off as an unwanted by-product of oil production. We have possibly 30 to 50 years worth of domestic reserves. It is hoped that new deep-drilling techniques will enable the tapping of new deposits.

There are large deposits of natural gas in the world, with more than 65 percent of the reserves in the Soviet Union and Iran. A large gas pipeline is now being built to bring natural gas from Siberia to Europe. We import natural gas in liquid form. The gas that is produced from fields relatively near deep-water ports is refrigerated and liquefied. The liquefied natural gas (LNG) is then shipped in refrigerated tankers. Liquefication allows 600 volume units of gas to be reduced to 1 volume unit of liquid.

Liquefication also eliminates the need for pressurized storage containment of gas near the place of use. However, it was recently pointed out that tankers loaded with LNG docked in coastal ports, such as New York City, are potential bombs. Should one explode, heaven help New York City and its fire department.

There you have it: A synopsis of our principal energy resources—fossil fuels. The point to keep in mind is that fossil fuels are nonrenewable, and once gone, they're gone forever. It took millions of years for fossil fuels to form. We have the potential and ability to use them up in a century or two.

NUCLEAR ENERGY

Without a doubt, the most controversial of our "conventional" energy sources is nuclear energy. Some might even argue whether it is conventional or not. But nuclear energy accounts for about 12 percent of the electrical energy production in the United States, so nuclear energy as an energy source has arrived.

The way it arrived is a large part of the problem. The uncontrolled release of nuclear energy over Hiroshima, Japan, in 1945 during World War II announced to the world the awesome power of the nucleus. In 1959 the controlled release of nuclear energy in a reactor in Shippingsport, Pennsylvania began the commercial production of electrical energy from the nucleus in the United States.

Fission

The nuclear energy we use to generate electricity and to propel ships comes from a fission process. In this process, a nucleus, usually uranium, is "split" into two lighter nuclei with the emission of two or more neutrons (hence the origin of the phrase "splitting the atom"). The lighter nuclei are about half the size of the original nucleus, and a neutron is an electrically neutral particle.

After the fission process, the fission fragments have less mass than the original nucleus. So, in accordance with Einstein's $E = mc^2$ relationship, energy is released and goes into the kinetic energy of the fission fragments.

The fission process is started by a neutron striking a large fissionable nucleus. The emitted neutrons then strike other fissionable nuclei, and a cascading or chain reaction takes place. This reaction can be self-sustaining. If it is controlled, we have a continuous source of nuclear energy. (See Chapter 29 for fission reaction details.)

Controlled nuclear fission takes place in a nuclear reactor. The nuclear fuel most commonly used is uranium. Natural uranium is predominantly uranium-238 (U-238, uranium nuclei with 92 protons and 146 neutrons). However, U-238 does not undergo fission easily. The fissionable uranium nucleus is U-235 (92 protons and 143 neutrons).§ Only about 0.7 percent of natural uranium is U-235. The U-235 is concentrated to about 3 percent so there will be enough U-235 nuclei for the chain reaction to proceed.

This *enriched* uranium is put into the fuel rods of a reactor, where the chain reaction takes place (Fig. 7.7). To control the chain reaction, neutron-absorbing control rods (usually cadmium) are inserted between the fuel rods. By removal of neutrons, the growth of the chain reaction can be controlled. If more energy is needed, the control rods are partially pulled out of the reactor core so that more neutrons are available for the fission process. The reactor can be shut down by full insertion of the control rods.

A major problem associated with fission nuclear energy production is nuclear wastes. The resulting nuclear fragments from the fission reaction are radioactive and dangerous. Some remain so for thousands of years. Currently, they are stored in underground tanks. (More details on radioactivity and the operation of a nuclear reactor will be given in Chapter 29.)

QUESTION: If the control rod mechanism in a nuclear reactor malfunctioned and the chain reaction proceeded uncontrolled, wouldn't the reactor explode like a bomb?

ANSWER: No. A nuclear reactor cannot explode like a bomb. Reactor-grade uranium contains only about 3 percent U-235, whereas weapons-grade material is more than 90 percent. The high-grade material must be held together briefly to achieve an explosive release of energy.

If control is lost in a reactor and energy is not removed, the fuel rods would fuse together, and a core "meltdown" could occur. However, the energy release is not concentrated enough to create an explosion. The fissioning mass might melt through the floor of the reactor containment building, and the environment would be contaminated with dangerous radioactive material.

The energy release from fissionable material is enormous. One pound (less than 0.5 kg) of fissionable uranium, the amount in a cube 1.1 inches (2.8 cm) on a side, has the energy equivalent of 1000 tons of coal!

We have about a 50-year domestic reserve of uranium. Like oil and gas, our domestic supply will run out in the not-too-distant future. However, it is possible to extend the supply of fissionable material through breeder reactors (see Chapter 29).

Fusion

There is another source of nuclear energy through a fusion reaction, which is the fusion of two light nuclei, such as hydrogen, into a heavier nucleus (helium) with the release of energy. As discussed previously, this is the source of the Sun's energy, and we would like to imitate

§ U-238 and U-235 are called isotopes of uranium. Both are uranium nuclei (each has 92 protons), but isotopes have different numbers of neutrons (Chapter 28).

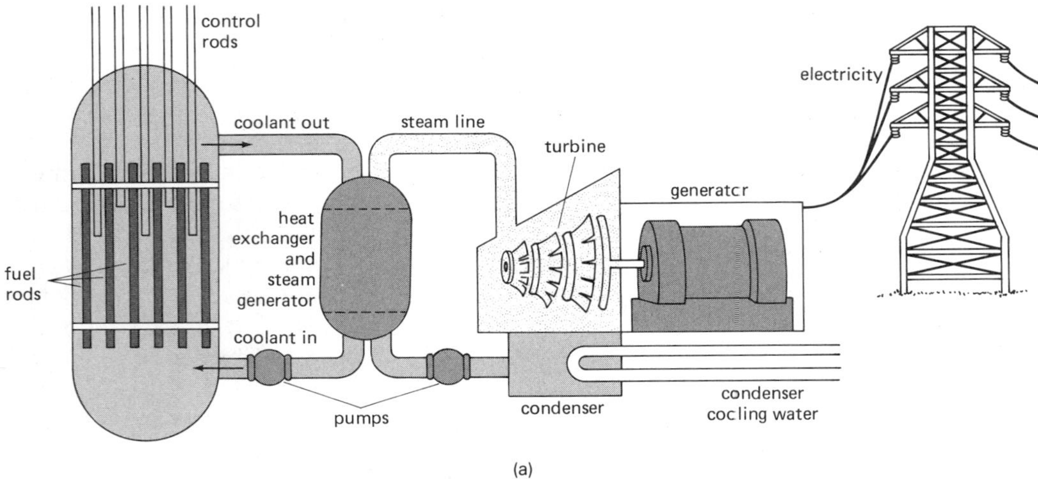

(a)

(b)

Figure 7.7 (a) A schematic diagram of a nuclear reactor and electrical generating plant. (b) An actual nuclear power plant with three nuclear reactors (in cylindrical containment buildings).

it on a smaller scale on Earth. We have achieved the uncontrolled release of energy from nuclear fusion in the form of the "hydrogen" bomb. However, the controlled release of fusion energy in a sustained manner has not been accomplished.

Fusion is called a *thermonuclear* reaction because it takes enormous temperatures to initiate the reaction —on the order of 100 million degrees. The hydrogen bomb is set off by a small atomic or fission bomb. As can be imagined, there are many technological problems in making a practical fusion reactor.‖ Research is going on, but it will probably be well into the 21st century before fusion becomes a practical energy source. When this does occur, the energy available to us will be enormous.

‖ The development of the fusion reactor will be discussed further in Chapter 29.

HYDROGEN ENERGY

Hydrogen is a chemical fuel as well as a nuclear fuel. It could be used in place of natural gas, and like natural gas it can be conveniently liquefied. The nice thing about hydrogen is that when it is burned or oxidized, the product is plain old water (H_2O), so there would be no pollution problem.

Hydrogen can be obtained from natural gas or from a really abundant source—water. When an electric current is passed through water in a process called electrolysis, the hydrogen and oxygen of the water molecule can be disassociated, and hydrogen (and oxygen) can be generated. Of course, this takes energy, and using hydrogen as a fuel is not efficient. It is basically a secondary energy source.

As a practical convenience, hydrogen energy is currently being used in special applications. Fuel cells use reverse electrolysis and combine hydrogen and oxygen to generate electricity. New York City and Tokyo have pilot fuel-cell projects designed to supply electricity in quick response to fluctuating power demands.

Renewable Energy Sources

Unlike fossil fuels, which will be gone when they are used up, there are some energy sources that are renewable. For example, wood is a renewable fuel, since we can grow more trees.

Some renewable energy sources were once primary energy sources before the major use of fossil fuels. But today these sources are, by and large, a "drop in the (energy) bucket." Almost all these sources have limitations. Even so, they should be kept in mind because they may become important in the future, when our conventional fuel supplies begin to run low.

HYDRO ENERGY

Hydro energy or water power is an old familiar standby. It has been used for centuries and still is, to some extent. Atmospheric processes evaporate water from the oceans and carry it inland, where it falls as precipitation. With gravitational potential energy, the water flows back to the sea. On its way, its energy was once commonly used to turn water wheels. Now it is stored by means of dams and used to generate electricity (Fig. 7.8). Hydroelectricity accounts for about 10 to 11 percent of the total production in the United States.

Unfortunately, not all locations are suitable for hydro energy, and most have already been developed in this country. Then, too, droughts can occur, which reduces the reliability of hydro energy.

WIND ENERGY

Wind energy has been used since ancient times for propelling sailing ships and through wind mills. Some modern ships have again been equipped with sails to harness the wind and reduce oil consumption (Fig. 7.9). Wind generators for the production of electricity have been built and installed on mountain tops. However, electricity can be produced more cheaply by conventional means, and what happens if the wind doesn't blow?

Figure 7.8 Hydro energy. Water flowing from a higher level to a lower level is used to generate electricity at dam sites such as the Hoover Dam, shown here.

TIDAL ENERGY

The energy sources discussed thus far (with the exception of nuclear energy) can be traced back to the Sun. But the tidal rise and fall of the oceans is primarily due to the gravitational attraction of the moon (with a little help from old Sol, see Chapter 5).

The movements of the tides are a source of energy in some coastal areas, particularly where the tides are especially high. Tidal energy from water stored in tidal

(a)

(b)

Figure 7.9 Wind energy. (a) An oil tanker built by Nippon Ko Kan Co. of Japan uses sails to assist the conventional diesel engines to reduce fuel consumption. The sails are set and furled by remote control from the bridge. (b) Wind turbine generator near Boone, NC. The blades are as long as the wing span of a Boeing 707. At wind speeds of 25 mph and above, the generator can produce 2 megawatts of electrical power, enough to supply the needs of about 500 homes.

dams was once used for sawmills along the coast of Maine. There is a commercial tidal electrical generating facility operating in Brittany, France, in the tidal basin of the Rance River off the Gulf of St. Malo (Fig. 7.10). Its power output is equivalent to that of a small fossil-fuel plant and is about one fourth of the hydroelectric output of Hoover Dam.

GEOTHERMAL ENERGY

The Earth itself is a storehouse of energy. The outer core of the Earth is hot and molten. This is thought to be due to the energy release of radioactive nuclear processes. The temperature at the center of the Earth is estimated to be about 3000°C (5400°F).

(a)

(b)

Figure 7.10 Rance tidal power station. The tide on the Rance estuary is one of the highest in the world, reaching 13.5 meters (44 ft). When the tide rises, water passes through the dam gates; the dam is then closed, and the trapped water is used to turn 24 turbines when the tide goes out.

Toward the surface of the Earth, the temperature decreases, but it is still quite high. Imperfections in the Earth's crust allow some of this energy to reach the surface in certain places, particularly in volcanic regions. This is observed in the form of hot springs, geysers such as Old Faithful in Yellowstone Park, and steam vents. Hot springs have been used as baths or spas down through history. In Iceland, hot water and steam from geyser fields are used for domestic heating.

A significant commercial application of geothermal energy occurred in 1905 with the building of the first geothermal electrical generating plant in Larderello, Italy. The first geothermal generation facility in the United States was built by the Pacific Gas and Electric Company in 1960 at The Geysers, California, 90 miles north of San Francisco (Fig. 7.11). The Earth supplies the steam, and the electrical output is about equivalent to that of Hoover Dam.

Further tapping of this energy source will involve new drilling techniques to reach the hot rocks and steam beneath the Earth's surface. However, there is a pollution problem. Underground steam and hot water are often contaminated with sulfur compounds, such as hydrogen sulfide ("rotten egg" gas, so called because of its smell).

SOLAR ENERGY

A natural energy source without pollution is the Sun. If we could use an appreciable part of the sunlight that falls on the Earth, our energy problems would be solved. Here the major problem is concentrating or collecting ample amounts.

Solar energy systems have been used to heat homes and buildings in various parts of the country for some time. The systems are divided into two categories: pas-

(a)

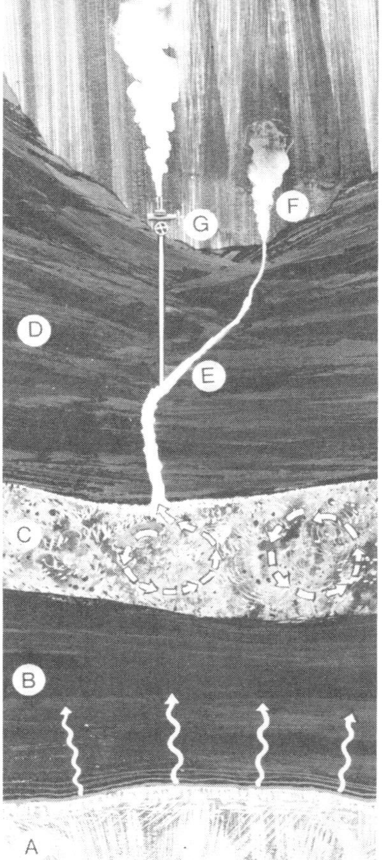

(b)

Figure 7.11 (a) A geothermal power plant in Sonoma County, California. (b) Diagram of a geothermal field.

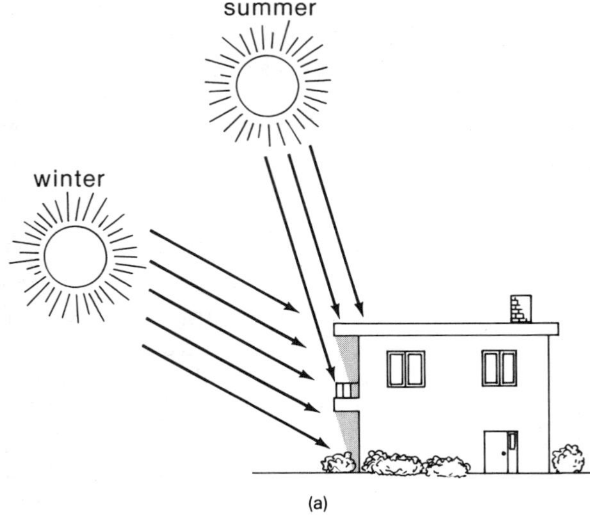

(a)

(b)

Figure 7.12 Passive solar heating. (a) Typical designs have large windows facing south, with overhangs to protect them from summer radiation. (b) A solar-heated home with a water-filled storage wall. The array at the right is a solar water heater. The windmill in the background pumps the water.

sive systems and active systems. In **passive solar systems,** the building itself is designed as the solar collector. The heat energy is then distributed by natural means, without the use of any mechanical or electrical equipment.

One of the simplest passive designs is to have large windows facing south (Fig. 7.12). In some instances, thermal storage walls are built inside the south-facing windows. The wall blocks and stores the solar energy transmitted through the windows. Some designs use containers of water as a wall to improve the energy storage capacity. Still others use a rooftop pond system in which a shallow pond of water on the roof is the collector and provides thermal storage. In any case, the stored thermal energy is used to heat the house at night when the Sun goes down.

Active solar systems make use of special collectors, storage tanks, and pumps and fans to collect, store, and distribute the solar energy. As in the case of passive designs, there are many active system designs. A general schematic diagram is shown in Figure 7.13. The collector may use water or air ducts to collect and transfer the energy to a storage tank. Storage tanks may use water, rocks (in the case of air heat transfer), or some chemical to store the energy. The type of storage material depends on the particular application and economy.

Both passive and active systems generally have an auxiliary heat supply, usually electric. Keep in mind that the reception of sunlight is variable and intermittent. It varies seasonally, and a series of cloudy or rainy days can interrupt the solar heat supply.

Incidentally, solar energy can also be used for cooling or air conditioning. How do you get cold from heat? Well, one such process, once commonly used in (natural) gas refrigerators, will be discussed in Chapter 15 on thermodynamics.

Thermal-Electric Conversion

Solar thermal-electric conversion systems collect solar radiation, converting it first to thermal energy and then to electrical energy. The latter part of the operation is carried out by conventional steam-turbine generators. Hence, such systems must collect enough solar energy to heat a transfer fluid to a temperature of several hundred degrees ($\sim 500\,^{\circ}\mathrm{C}$).

There are two types of thermal-electric conversion systems under consideration: distributive collector systems and central receiver systems. In the distributive collector system, a large number of solar collectors collect and concentrate the solar energy (Fig. 7.14). The energy is then transported by a fluid, such as molten sodium or potassium salts, to an electrical generating facility. It has been suggested that such systems could be used in desert areas, such as in the southwestern United States. Because of the large land areas that are required to collect sufficient solar energy for practical electricity production, such facilities are called solar farms.

In a central receiver system, a large number of sun-tracking mirrors, called heliostats, concentrate the solar energy on a single receiver atop a solar energy or

Figure 7.13 Active solar heating. (a) A schematic diagram of an active home-heating system and an active solar home. (b) Solar collectors fit right into the front, south-facing roof of a McDonalds restaurant in Ontario, Canada. The solar system heats water for the kitchen and rest rooms.

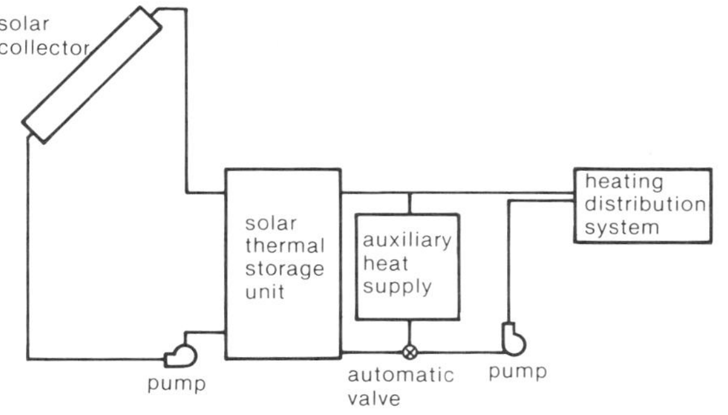

(a)

(b)

Figure 7.14 Distributive collector system. In this test facility, a parabolic trough collector collects and concentrates the solar radiation on a fluid-transfer system.

power tower (Fig. 7.15). Here again, a large mirror array or many small installations would be needed to produce commercial amounts of electric power. Then, too, the electric power would have to be transmitted over long distances to populated consumer areas.

Solar Cells

You might be wondering if solar energy can be converted directly into electrical energy. The answer is yes. Solar cells allow this direct conversion. These cells are solid-state devices; silicon solar cells are the most common.

(a)

(b)

Figure 7.15 Central receiver system. (a) A solar tower at Sandia Laboratories, New Mexico, receives sunlight from 1775 mirrors. (b) When the sunlight is concentrated on a steel target on the "power tower," there is enough energy to melt the steel, which can be seen dropping down from the tower.

Solar cells, sometimes called photocells, are used in light meters in photography. Some wrist watches and hand calculators are powered by solar cells (Fig. 7.16). Spacecrafts use arrays of solar cells or solar panels to provide electrical energy.

There is a limiting factor, however, namely efficiency. Conventional solar cells have a maximum efficiency of about 20 percent. It would take a very large array at a large capital cost to produce commercial amounts of electricity. Applications are limited to small scales, such as electricity for remote, sparcely populated areas. It has been suggested that large arrays of solar panels could be put into orbit about the Earth and the energy beamed back to a ground station by microwaves, a form of electromagnetic energy (Fig. 7.17). In this case, there would be no interruption of sunlight by cloudy or rainy days, but only when the orbiting array passed through the Earth's shadow.

BIOMASS—PLANTS, GARBAGE, AND MANURE

Another renewable energy source is plants and other organic materials—generally referred to as biomass. Wood is a common example. Other high-yield plants,

(a)

(b)

Figure 7.16 Solar cells convert solar energy to electrical energy. (a) A solar-cell calculator that also works on indoor "nonsolar" light, which allows its use in the classroom. (b) Satellite solar panels. Solar panels on the Skylab space station photographed against a black-sky background from a command module on a "fly-around" inspection prior to docking.

such as swamp grasses, have been suggested as an energy source.

Agricultural products and by-products are also a possible and real energy source. If corn is fermented, alcohol can be extracted, which is a good fuel. Mix it with gasoline and we have the automotive fuel called gasohol (Fig. 7.18).

Trash and garbage are a problem in some areas. However, they are also an energy source. In the United States, a few facilities burn trash as a fuel, but usually it

Figure 7.17 Solar panels in space. Artist's conception of an orbiting array of solar panels. Energy would be transmitted to Earth in the form of microwaves for conversion to electricity.

Figure 7.18 A gasohol pump that dispenses a mixture of gasoline and alcohol.

Table 7.1 Comparison of Various Energy Sources

Energy source	*Current use*	*Future availability of energy supply*	*Relative cost*	*Environmental problems*
Oil	42%	Shortages by year 2000	Cheap	Considerable
Gas	27%	Shortages by year 1985	Cheap	Cleanest fossil fuel
Coal	21%	Two- to three-hundred-year supply	Cheap	Serious problems
Nuclear	4%	Unknown	More expensive than fossil fuel	Serious problems
Solar	Less than 1%	Excellent	Slightly more expensive than oil	Negligible
Hydro power	4%	Small expansion	Cheap	Some problems
Geothermal	Less than 1%	Small expansion	Cheap	Some problems
Waves and tides	Negligible	Excellent	Slightly more expensive than oil	Some problems
Wind	Negligible	Excellent in some regions	Slightly more expensive than fossil fuels	Negligible
Wood, plant matter, and garbage	1%	Small but continuous	Cheap	Some problems
Hydrogen fusion	0	Excellent (if reactors can be built)	Unknown	Serious thermal pollution problems

is buried in a landfill. In France, the burning of such wastes is used as a domestic energy source. A large facility in Paris produces electrical energy and 20 percent of the total steam used for heating the city.

Other, more exotic wastes can be used to produce fuel. Methane gas is released from the decomposition of cow manure, which is in ample supply around cattle feed lots. You may have read about a methane-powered automobile that derives its fuel from chicken manure.

All of these are potential sources of fuel, but they do not compete with conventional fossil fuels at present. Maybe someday they will. A summary and comparison of the various energy sources is given in Table 7.1.

Energy Conservation

Energy conservation can have a couple of different meanings. In the last chapter, the conservation of energy was concerned with the total accounting of the energy in any form. In this chapter, energy conservation is concerned with reducing energy consumption and conserving energy resources.

A great deal is heard about energy conservation, and we are encouraged to practice it. Basically, there are two methods of conserving energy: (1) a reduction in use and (2) increased efficiency. The first involves making do with less, and the second involves doing more with the same amount or getting more from the energy we use. The elements of both methods may not be desirable and could involve sacrifices. But either we conserve energy now voluntarily or we may be forced to do so in the future.

Let's take a look at how energy is and might be conserved. A reduction in energy use is the easiest to think of. For example, lights can be turned off when not needed (Fig. 7.19). The temperatures of our homes can be adjusted. Until recently, many homes in the United States were heated to 72°F (22°C) in the winter and cooled to 68°F (20°C) in the summer. Now, federal guidelines call for 65°F (18.3°C) in winter and 78°F (25.6°C) in summer. Current energy prices encourage this.

A simple way to conserve gasoline and energy is not to drive our cars so much. Sometimes the price of gas helps in this respect; for example, people take vacations closer to home. During oil shortages, filling stations

(a)

(b)

Figure 7.19 Conservation of energy. (a) A reminder to turn off the lights when not needed. (b) Federal guidelines and laws in federal and some state buildings call for 65°F thermostat settings in winter. (No cheating here. Photo was taken on a warm day.)

closed on Sundays to conserve our reserves. The author once suggested to a high school class that energy could be conserved if all students walked or rode the bus to school, rather than driving personal cars. The idea didn't go over too well.

Methods of energy conservation through increased efficiency usually involve some expense. For example, we are told to keep our cars "tuned up" for better mileage or gasoline efficiency. Also, existing energy consumption systems can be modified or replaced. We are encouraged to put additional insulation in the walls and attics of our homes to save heat and energy. Conventional home heating and cooling systems are now being replaced by heat pumps, which are more efficient in the heating mode. (See Chapter 15 for heat pump operation.) When buying home appliances, you should consider energy conservation and efficiency (Fig. 7.20). This includes lights or lamps. Fluorescent lamps are

(a)

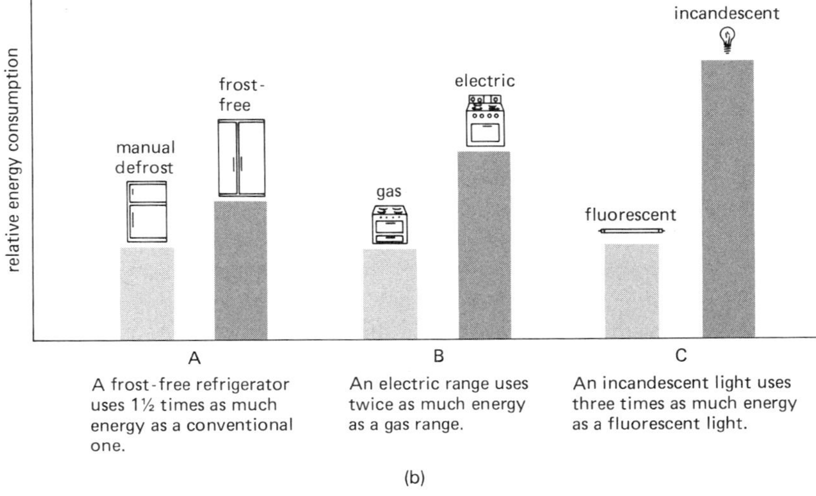

(b)

Figure 7.20 (a) Federal law now requires appliances to have an Energy Guide label that compares energy costs. (b) A comparison of the energy consumption of some types of appliances.

more efficient than incandescent lamps, and new sodium-vapor lamps are even more efficient.

Transportation can be made more efficient by using plastics and lighter materials. (See Special Feature 10.3.) This has been done in automobiles and airplanes to improve gas mileage. Many more smaller cars have replaced the energy-consuming "gas hogs."

Industry has installed waste-heat recovery systems and now uses previously discarded energy for the space heating of factories and for other purposes. Heat recovery is taken advantage of in some office buildings where the heat from lights is used to heat the building. Many buildings now have computerized heating and cooling systems and lighting controls so as to conserve energy when the building is closed or not in use.

These are just a few of the ways to conserve energy. You can probably think of others, and you may have to someday.

SUMMARY OF KEY TERMS

Main energy sources of the Earth solar energy, tidal energy, and terrestrial or geothermal energy.

Photosynthesis the process by which plants use sunlight and chlorophyll to convert water and carbon dioxide to sugar and oxygen.

Greenhouse effect the absorption or trapping of infrared radiation by atmospheric gases, such as CO_2 and water vapor, which prevents large daily temperature variations.

Fossil fuels fuels, specifically coal, gas, and oil, that formed from the remains of plant and animal matter.

Acid rain rain with acid concentrations resulting, in particular, from atmospheric sulfur pollution.

Anthracite ("hard") coal a high-quality coal with a carbon content of 90 percent or greater.

Bituminous ("soft") coal the most abundant coal with a carbon content on the order of 70 percent.

Lignite ("brown") coal a low-ranking coal with a relatively low heat value and high moisture content.

Synfuels synthetic fuels, such as gaseous and liquid products produced from coal gasification and liquefication.

Fission a nuclear process in which a large nucleus "splits" into two lighter nuclei, with the emission of two or more neutrons and the release of energy.

Chain reaction a cascading fission reaction used to produce nuclear energy.

Fusion a nuclear process in which two light nuclei, such as hydrogen, are fused together into a heavier nucleus, with the release of energy.

Renewable energy resources energy sources that are continually available from renewable processes. These include hydro, wind, tidal, geothermal, solar, and biomass energy sources.

Passive solar systems systems in which the energy is distributed by natural means, without the use of any mechanical or electrical equipment.

Active solar systems systems that use special collectors, storage tanks, pumps, and fans to collect, store, and distribute solar energy.

Thermal-electric conversion systems systems that collect solar radiation, converting it first to thermal energy and then to electrical energy.

Solar cells solid-state devices that convert sunlight directly into electrical energy.

Biomass a renewable energy source consisting of plants and other organic materials.

Energy conservation the reduction of energy consumption and the conservation of energy resources through less use and increased efficiency.

EXERCISES

1. Using the graph of energy fuel consumption by type in Figure 7.1, make a circular percentage representation for fuels for the current year (similar to the other circular display in the same figure).
2. (a) Using Figure 7.2, compute the average energy consumption per capita in the United States for the major years listed on the horizontal scale.
   (b) Using your calculations, plot a graph of energy consumption versus time (years).
3. In Figure 7.2, explain (a) the "dip" in the energy consumption curve in the 1930's and (b) the rapid increase in energy consumption between 1960 and 1980.
4. List five ways a serious energy shortage might affect you and your family.
5. Do you believe there is an energy crisis? Defend your position.
6. Estimate the approximate "albedo" of a mirror and of a blackboard.
7. After a sunny day, if there is a cloud cover at night the temperature does not drop a great deal. But on a cloudless night, it may be "cold and clear." Explain the difference.
8. In 1815 the volcano Tamboro, located in the East Indies, had the most violent volcanic explosion in recorded his-

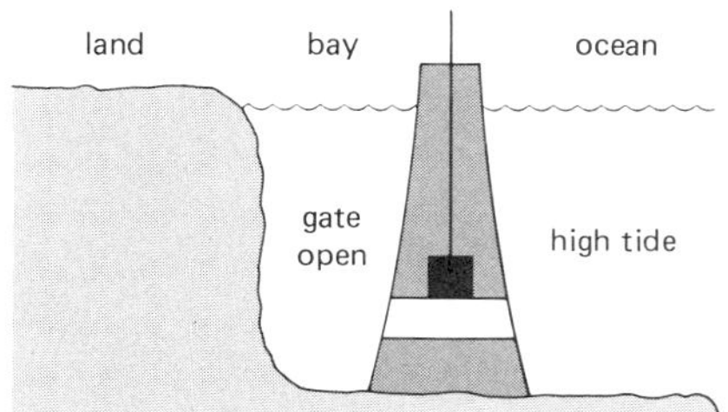

Figure 7.21 See Exercise 14.

Figure 7.22
See Exercise 17.

tory. It blew as much as 35 cubic miles of particulate debris into the air. The next year was uncommonly cold. Explain why.

9. Is it correct to say that energy from fossil fuels is really solar energy? Explain. How about hydro energy and wind?
10. The industrialized northeastern and midwestern United States have large deposits of high-sulfur coal, while sparcely populated western states have low-sulfur, but generally lower-quality (sub-bituminous and lignite), coal deposits. Discuss some of the environmental and economic implications of this situation.
11. Discuss some of the environmental and economic problems associated with the production of fuels from oil shale and oil sands.
12. Explain how a nuclear reactor produces energy and how the energy production is controlled. What are the environmental problems?
13. Choose one of the renewable energy resources discussed in the text that you think holds the most hope for the future, and explain the reasons for your choice.
14. Tidal energy can be extracted by constructing a large dam across a channel between a bay and the ocean (Fig. 7.21). Gates located at the lower portion of the dam allow the water to flow through or be blocked. The flowing water is used to drive turbines, which generate electricity. The greatest potential for electrical generation is at low tide and at high tide. Show this with a series of drawings, starting with the situation shown in Figure 7.21.
15. Design a solar-heated classroom building for your campus. Consider both passive and active systems.
16. Solar collectors using fluid ducts are covered with glass or plastic to make use of the greenhouse effect. Explain how this improves the efficiency of a solar collector.
17. Most "18-wheelers" are now equipped with wind deflectors on top of the cabs (Fig. 7.22). What is their purpose?
18. Many street lights are now turned on and off by using solar or photocells. How does this save energy?
19. In most if not all states, right turns are permitted at red lights after stopping (unless otherwise indicated). Was this instituted to speed up traffic or to conserve energy? Explain.
20. Discuss energy conservation in terms of use and efficiency for the transportation modes shown in Figure 7.23. (Mopeds can get more than 100 miles per gallon, and "bike-pooling" is pretty good too.)

Figure 7.23
See Exercise 20.

21. The greatest electrical energy demands occur in "peak periods" during the day. In hydroelectric generation, some facilities use electricity during non-peak periods to pump water back behind the dam. The process is called pumped storage. When do peak periods occur, and is pumped storage an efficient procedure?
22. Discuss the energy conservation involved in (a) car pooling and (b) recycling.
23. Give some possible examples of waste-heat recovery that would conserve energy in your community.
24. Would eating less or eating a selective diet be a method of conserving energy? Explain.
25. For a quick cup of instant coffee, some people fill a kettle and then use only enough hot water for one or two cups. Discuss the efficiency of this practice.
26. Discuss some examples of energy conservation and the related economic impact. For example, if cars were built out of aluminum and plastics, how would this affect the steel industry? Here's another: Some states have legislated deposits on all beverage containers, including "throw-aways." Glass companies generally oppose this. Why the opposition, and why the deposit?
27. Give a couple of examples of what might be thought of as "far-out" methods of energy conservation.

8

Rotational Motion

Before leaving Part One on Mechanics, there is another important topic we should discuss — rotational motion. Special cases of rotational motion — circular motion and orbiting satellites — were considered in Chapters 4 and 5. But there are a lot more, for example, the spinning (rotating) Earth and many other common things that rotate: wheels, helicopter rotor blades, and door knobs. This chapter will require a little rotational thinking, so get your head and mind spinning.

Particles, Center of Mass, and Center of Gravity

In previous chapters, physical "things" were commonly referred to as objects, bodies, or masses. In general, when we studied the motion of objects no mention was made of possible rotational motion. However, it is difficult to throw a baseball or football without rotation or spin.

Sometimes rotational considerations are ignored by talking about the motion of "particles." A particle is the physical counterpart of a "point" in mathematics. It has no physical dimensions and can be accurately localized in space. We can also say a particle has mass. But since a particle has no physical dimensions, we don't have to worry about it spinning. There is no such thing as a particle (as thus defined) in nature.

However, the imaginary particle is a very useful concept. We say that a solid object or a **rigid body** is a system of particles in which the particles are fixed, constant distances apart.* A quantity of water would ob-

* This definition of a "rigid body" is an idealized one because the atoms and molecules of a solid undergo thermal vibrations and because of elasticity. Even so, most solids macroscopically approximate the idealized case.

viously not satisfy this criterion. But if the water were frozen, the ice would be a rigid body.

Another important particle concept of a rigid body is the **center of mass.** This can be thought of as the average location of all the mass particles that make up the body. That is, it is the point at which all the mass of a body can be considered to be concentrated. For example, a uniform, symmetric object such as a sphere can be thought of as having all of its mass concentrated at its center. However, an unsymmetric object such as a hammer has more mass toward one end. The center of mass of a hammer is therefore toward the head of the hammer.

Since weight and mass are related by a constant for objects near the surface of the Earth ($w = mg$), the center of mass is also the center of gravity for such objects. The **center of gravity** is simply the average location of the weight distribution of a body. That is, it is the point where all the weight of a body can be considered to be concentrated.

You have located the center of gravity (and center of mass) of an object when you have balanced it on your finger. For example, when you balance a stick (Fig. 8.1). If the mass of the stick is uniformly distributed, the center of gravity is at the center of the stick. You balance or support the entire weight of the stick by pushing upward through the center of gravity. It's as though all the weight of the stick were concentrated at its center of gravity. Suppose you were blindfolded and the stick was replaced by a small object of equal mass or weight, say a cube. As far as weight goes, you would not be able to tell the difference. For a nonuniform body such as a hammer, the center of gravity is nearer the heavier end.

The center of gravity of a freely suspended body lies directly below the point of suspension (when it is not

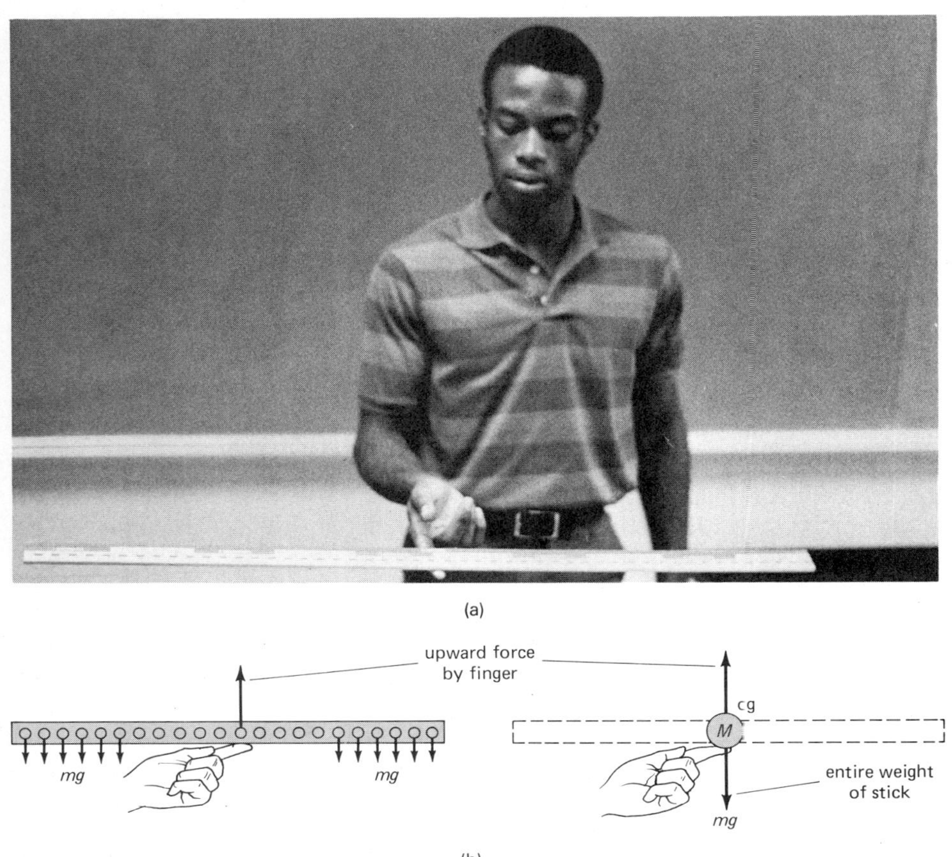

Figure 8.1 Center of gravity. A uniform stick's center of gravity is at its center where it can be balanced. It is as though the entire weight (mass) of the stick is at the center point.

(a)

(b)

Figure 8.2 Location of the center of gravity by suspension. (a) The intersection of two vertical lines of suspension locates the position of the center of gravity of a flat, irregularly shaped object. The object could be balanced on the student's finger at the point of intersection. (b) Location of the "center of state" (South Carolina). Of course, states aren't flat objects, but it makes a nice demonstration.

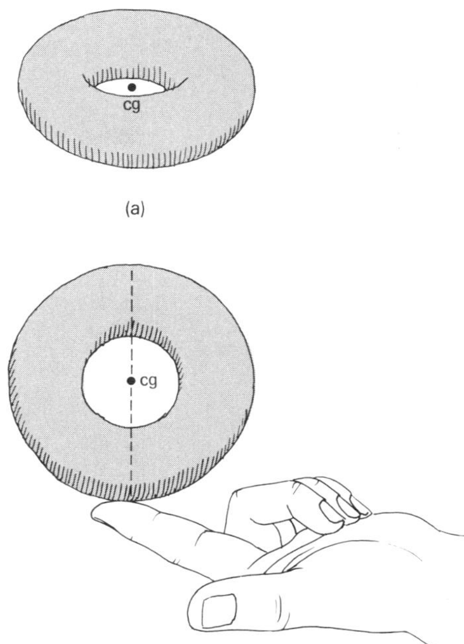

Figure 8.3 Example of center of gravity not located within the body of an object. (a) The center of gravity of a uniform donut is at a point in space at the center of the donut (hole). (b) This can be shown by suspension or support methods.

suspended at its center of gravity) when two points of suspension are used (Fig. 8.2), the intersection of the two vertical lines of suspension locates the position of the center of gravity. This method is useful in finding the center of gravity of a flat, irregularly shaped object.

QUESTION: Does the center of gravity always lie within a body?

ANSWER: No. In some cases the center of gravity is outside the body itself. For example, consider a uniform donut (Fig. 8.3). Here, the center of gravity is at the center of the donut hole. You could prove this by using two points of suspension or by balancing the donut at a couple of places.

Can you think of other bodies in which the center of gravity lies outside the body? How about a horseshoe?

Rotational and Translational Motions

In general, we have previously considered objects in translational motion and ignored spinning rotations. In **pure translational motion,** every particle of a body has the same instantaneous velocity. As illustrated in Figure 8.4(a) for a thrown football that is not spinning or turning end over end, all the particles, including the center of gravity, have the same instantaneous velocity.

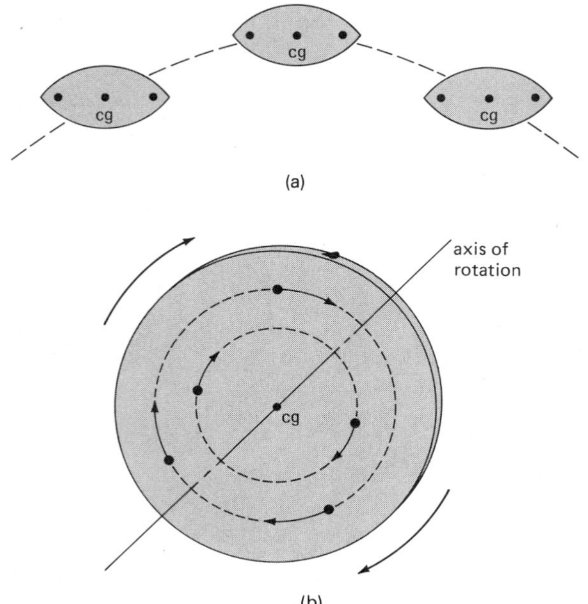

Figure 8.4 Translational and rotational motions. (a) In pure translational motion, all the particles have the same instantaneous velocity, so there is no rotation. (b) In pure rotational motion, the particles of a body move in circles about a line called the axis of rotation. The rotational speed of the particles on each circular path is different.

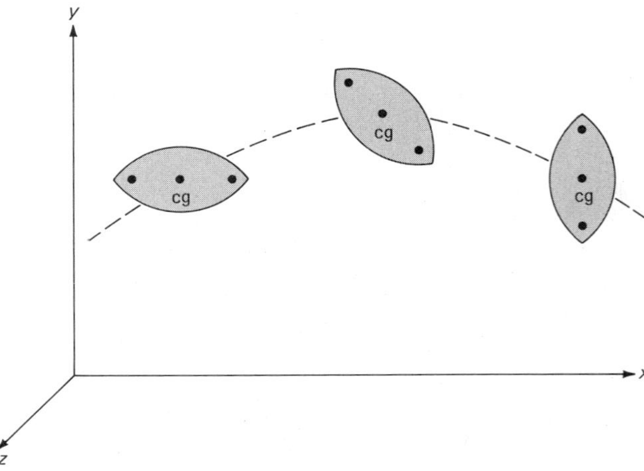

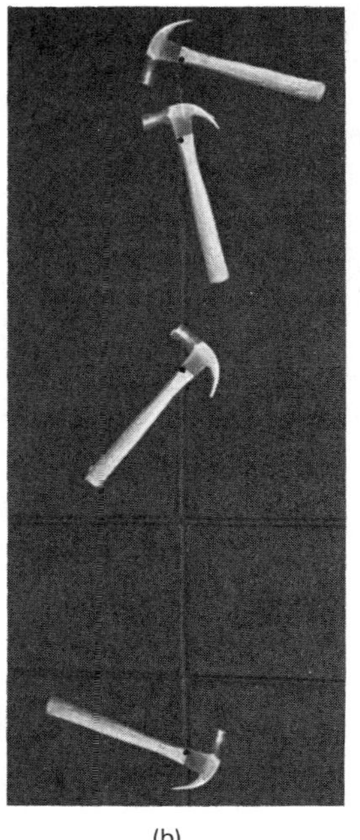

Figure 8.5 The general motion of a rigid body is a combination of translational and rotational motions. Notice that the centers of gravity of the football and the hammer follow paths as though they were "particles."

But there is also the possibility of rotational motion. In **pure rotational motion,** the particles of a body move in circles about a line called the axis of rotation. The instantaneous velocity (and speed) of the particles on each circular path is different. This is illustrated in Figure 8.4(b) for a wheel or disk. The axis of rotation is through the center of mass, but it need not be.

The axis of rotation may be either inside or outside of the body. Also, a body may rotate about more than one axis at a time. For example, the Earth rotates daily about an axis through its center. It also rotates yearly about an axis through the Sun. We commonly say that a body "rotates" when the axis of rotation is inside or passes through the body. When the axis of rotation is outside the body, we commonly say the body "revolves," as the Earth revolves about (an axis through) the Sun.

Hence, the general motion of a rigid body is a combination of translational and rotational motions. Examples of this general motion are shown in Figure 8.5 for a tumbling football and a dropped hammer. Notice that the center of gravity of the football follows a parabolic projectile path and that the center of gravity of the hammer falls in a straight line as though they were "particles" representing the total weight of the objects.

Another example of this combination of motions is a rolling object (Fig. 8.6). If you put a pure translational motion and a pure rotational motion together without slipping or spinning, you have rolling motion. Look at the velocity vector addition in the figure. A rolling object has an instantaneous axis of rotation at its point or line of contact (for example, a sphere or cylinder, respectively). The particles at this point or along a line are

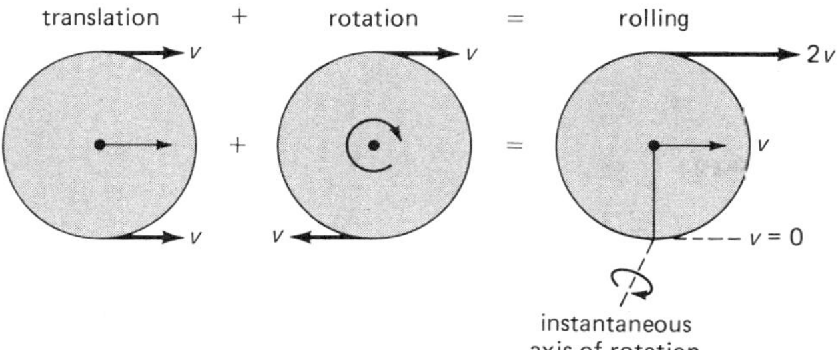

Figure 8.6 Rolling is a combination of translational and rotational motion.

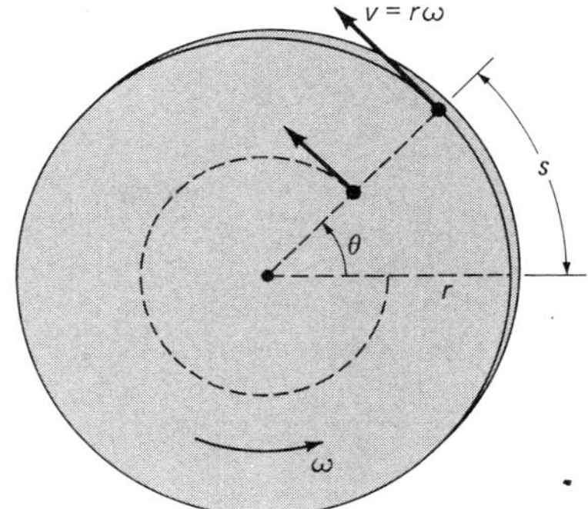

Figure 8.7 A particle in circular motion traveling an angular distance θ in a time t has an angular speed, $\omega = \theta/t$. The tangential speed of a particular particle is $v = r\omega$.

instantaneously at rest. Above this, the particles have increasingly greater speeds, such that the object "falls over itself" and rolls. Instantaneously, there is no translational motion. Notice also that the center of gravity of the rolling object (assumed to be uniform) moves with a velocity v, as if it were a particle in uniform motion according to Newton's first law (Chapter 2).

DESCRIPTION OF ROTATIONAL MOTION

The description of rotational motion is analogous to that of translational motion. We have angular distance, angular speed and velocity, and angular acceleration. There are also rotational analogs of force, energy, and momentum, as we will see.

The particle nature of rigid bodies is helpful in analyzing rotational motion. For example, you might think of yourself as a rotating "particle" because you are on a rotating Earth. As part of a rotating body, you and other "particles" on and in the Earth move in circles about the axis of rotation.

Consider the particles of a rotating disk, as shown in Figure 8.7. All of the particles move in circles around the axis of rotation and move an **angular distance** θ (Greek theta) in a certain time. The angular distance may be measured in degrees (360° in one rotation or a complete circular path). Scientists often use another unit for angular distance called the radian (rad)†. There are 2π radians in one rotation or circle, and 2π rad = 360°.

Since the particles move through an angular distance θ in a time t, they have an **angular speed** ω, and $\omega = \theta/t$ (analogous to regular speed, $v = d/t$). The units of angular speed are radians per second (rad/s). Other common units are revolutions per second (rev/s or rps) and revolutions per minute (rev/min or rpm). For example, there are several types of phonograph records, one of which is called $33\frac{1}{3}$. This is the record's angular speed on the turntable—$33\frac{1}{3}$ rpm.

Notice that all of the particles in a rotating rigid body have the same *angular* speed. They all travel through one circle or 2π radians in the same time. (Think about the objects on a rotating merry-go-round.) However, the particles farther from the axis of rotation travel in greater circular paths in the same time. So the farther the particles are from the axis of rotation, the greater their tangential speeds. The tangential speed of a particle is related to its angular speed by the relationship $v = r\omega$, where r is the radial distance from the axis of rotation.

There is also angular velocity, which is a vector, so a direction has to be specified. We might say that an object rotates either clockwise or counterclockwise. But this is a sense and not really a direction. If you looked at one side of a rotating disk it might be rotating clockwise, but if you looked at the other side, it would be rotating counterclockwise. (Try rotating a sheet of paper in your hands and prove this to yourself.)

To specify the direction of the angular velocity, a right-hand rule is used (Fig. 8.8). If the fingers of the right hand are curled in the circular direction of the angular motion, the thumb points in the direction of the angular velocity. This eliminates any confusion.

Since the angular velocity can change, we can have an **angular acceleration** α, and $\alpha = \Delta\omega/\Delta t$ (analogous to regular acceleration $a = \Delta v/\Delta t$). The direction of the angular acceleration vector is in the same direction as

† When a particle moves through an arc length s that is equal to the radius r of its circular path, it moves an angular distance θ of one radian (57.3°).

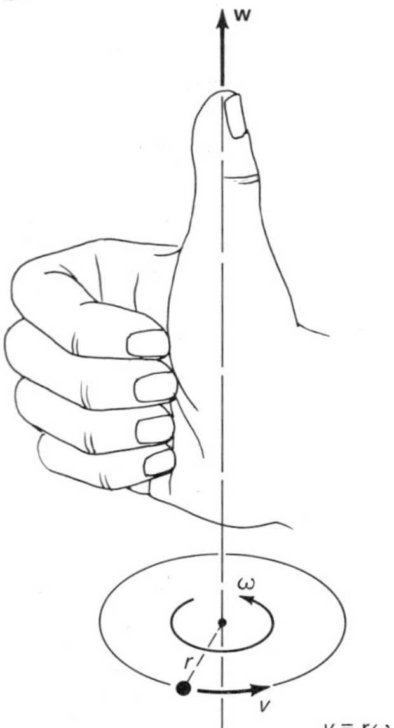

Figure 8.8 The direction of the angular velocity vector is given by a right-hand rule. Curl the fingers of your right hand in the circular direction of a particle's motion and your thumb will point in the direction of the angular velocity vector.

(a) (b)

Figure 8.9 Rotational inertia depends on mass distribution. (a) When the weights are close together or near the axis of rotation, the bar is easy to twist or rotate. (b) When the weights are farther apart, getting the bar to rotate is more difficult because there is greater rotational inertia.

that of the angular velocity vector if the acceleration speeds up the rotation. If the angular acceleration vector is in the opposite direction to that of the angular velocity, then this is a "negative" acceleration or deceleration, and the rotating object would be slowed down. Friction commonly gives rise to negative angular accelerations that cause rotating objects to slow down and stop. Tidal friction causes the Earth's rotational speed to decrease somewhat. This lengthens the rotational period or days by about 1/1000 second per century.

ROTATIONAL INERTIA

Just as the property of inertia resists changes in translational motion, there is also rotational inertia that resists changes in rotational motion. Rotational inertia, like inertia in the translational case, depends on the mass of the body. But rotational inertia also depends on the mass distribution about the axis of rotation.

Consider a barbell on which the weights can be moved to different positions (Fig. 8.9). If the weights are close to the axis of rotation, the bar is relatively easy to rotate. However, if the weights are moved farther apart, the bar is more difficult to rotate. The barbell has the same mass in both cases, but the rotational inertia is different. When the mass is distributed farther from the axis of rotation, there is greater rotational inertia. We say that there is a greater moment of inertia, I. Moment of inertia is the rotational analog of mass and a measure of rotational inertia.

The mass distribution dependence of rotational inertia or moment of inertia is the principle of the flywheel. Flywheels generally have most of their mass concentrated near the rim to give a greater moment of inertia (Fig. 8.10). Once a flywheel is rotating, there is a greater tendency for it to continue rotating because there is a greater rotational inertia. Automobiles have flywheels to promote an even power output between the firings of the cylinders. The rotational inertia of flywheels will also figure in the proposed electric cars of the future (Fig. 8.10).‡

QUESTION: Why is it easier to balance yourself with your arms outstretched when walking on a narrow wall, as in Figure 8.11, or on the rail of a railroad track?

ANSWER: In extending your arms, you increase your rotational inertia by distributing more mass

‡ In some cases of ultra-fast spinning flywheels, the mass is concentrated near the axis because of the strength of materials. Recall that in a rotating system, centripetal force acts on the particles, and the farther from the axis, the greater the force. ($F = v^2/r = r\omega^2$, since $v = r\omega$.)

(a)

(b)

Figure 8.10 Flywheels have most of their mass concentrated near the rim to give a greater moment of inertia.

farther from the axis of rotation (along the rail). When you start to "rotate," the increased inertia resists the change and gives you more time to regain your balance.

It would be wiser to try this on a railroad rail. If you rotated too far around the top of a narrow wall or fence, gravity would take over and you'd become a falling object. As we know, the vertical motion (acceleration) of your center of gravity is independent of mass or inertia in this case — until you hit the ground.

Torque

Now that we know how to describe rotational motion, the next obvious question is, What produces it? As we know, a (net) force produces translational motion. The rotational counterpart of a force is a torque (pronounced tork).

As you might guess, a torque involves a force, but there is something else. The distance from the axis of rotation is always an important factor in rotational motion. In the case of a torque, this distance from the

Figure 8.11 Increased rotational inertia helps one's balancing act. See Question and Answer.

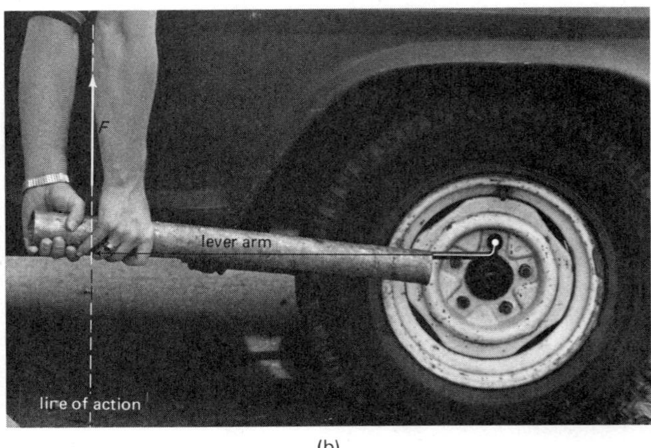

(a) (b)

Figure 8.12 Torque = lever arm × force. The lever arm is the perpendicular distance from the axis of rotation to the line of action of the force. With the force the same, the torque in (b) is greater than in (a) because the lever arm is greater. To loosen a really stubborn lug, the lever arm is extended with a section of pipe.

axis of rotation to the "line of action" of an applied force is called the lever arm (Fig. 8.12). A **torque** is defined as the product of the lever arm and the applied force that tends to produce a rotation about an axis.

$$\text{Torque} = \text{lever arm} \times \text{force}$$

Notice that the lever arm is the perpendicular distance from the axis of rotation to a line along which the force acts. The forces in the figures may be the same in each case, but the torques would be different because the lever arms are different. Every mechanic knows that it is easier to get a tight bolt or nut to rotate by applying more torque. This can be done by extending the lever arm of the wrench by putting a section of pipe on the wrench handle. Also, notice that if a force acts through the axis of rotation, the torque is zero. Why?

Just as the production of or a change in translational motion requires an *unbalanced* force, the production of or a change in rotational motion requires an *unbalanced* or net torque. In the situation shown in Figure 8.13, the net torque is zero, because the torques are equal and opposite. That is, one torque tends to produce a clockwise rotation, and the other a counterclockwise rotation. Since the torques have equal magnitudes, the net effect is zero, or no net torque.§

§ The vector direction of a torque is found by a right-hand rule similar to that for angular velocity. Curl the fingers in the direction the torque force would produce rotational motion and the thumb will be in the direction of the torque.

So, we have a version of Newton's first law for rotational motion:

> A rigid body remains at rest or in motion with a constant angular velocity unless acted upon by an unbalanced (net) torque.

There is also a rotational form of Newton's second law ($F = ma$). Using the symbol notation for rotational quantities, we have

$$\tau = I\alpha$$

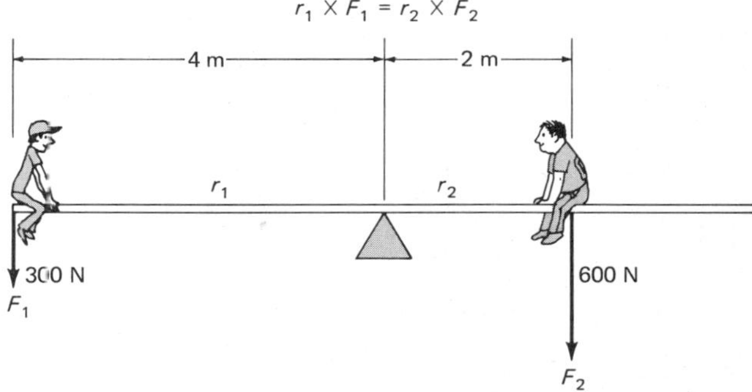

Figure 8.13 Zero net torque. The net torque is zero because the torque tending to produce a clockwise rotation is equal to or balanced by the torque tending to produce a counterclockwise rotation.

or

Torque = moment of inertia × angular acceleration

Notice the almost one-to-one correspondence between translational and rotational quantities. Of course, they are different, but the analogies make rotational effects easier to understand. We will consider energy and momentum shortly. Here too there are similarities and many interesting rotational effects.

QUESTION: Why is it easier to open a door when pushing on it farther away from the hinges?

ANSWER: We have all experienced this, particularly when mistakenly pushing on the hinge side of a glass door. The axis of rotation of the door is along the hinges, and the closer to the hinges, the shorter the lever arm.

Suppose the same pushing force was used in both cases (Fig. 8.14). Near the hinges, the torque would be $\tau = r \times F$, whereas the torque with a greater lever arm (farther from the hinges) would be greater, $T = R \times F$. With the moments of inertia of the doors the same, the door with the larger torque would swing open faster (greater angular acceleration), as can be seen from the equation $\tau = I\alpha$.

If the doors swing open at the same rate in each case, then the torques would be the same, but a smaller force would be required with a greater lever arm. Hence, the larger the lever arm, the "easier" it is to open the door. Check this out on a door for yourself and become familiar with torques and lever arms.

Stability and Equilibrium

Closely associated with torque and center of gravity is the stability of a rigid body. By **stability,** we mean whether an object will fall over or topple when it is slightly moved or tipped about an axis of rotation.

Stability depends on the location of the center of gravity. Consider the L-shaped object in Figure 8.15(a). The center of gravity lies outside of the body in this case. When sitting on one of its long sides, the body is stable. But when sitting on an end side, it will topple over (rotate). Hence, we see that **a body is stable when its center of gravity lies vertically above and inside its**

Figure 8.14 The lever arm makes a difference.

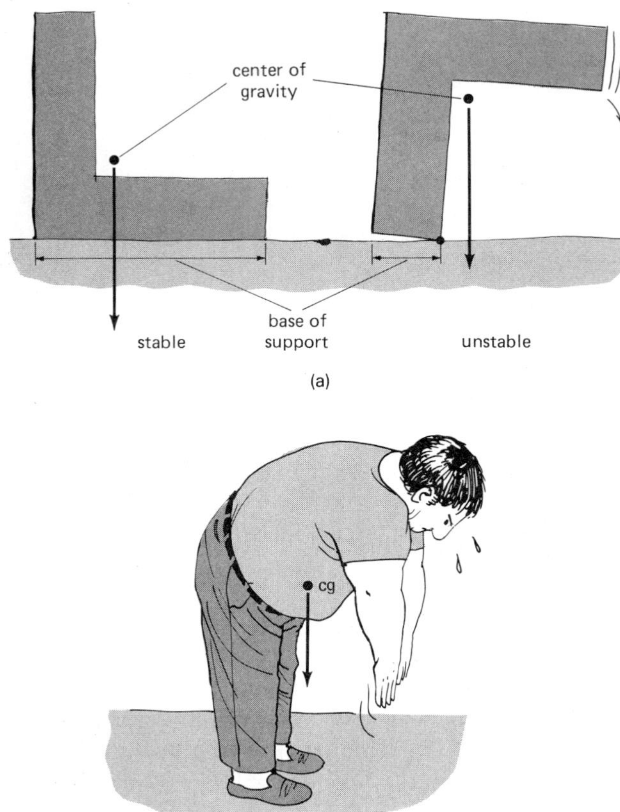

Figure 8.15 Stability. (a) A body is stable when its center of gravity lies vertically above and inside its base of support. (b) Can he touch his toes?

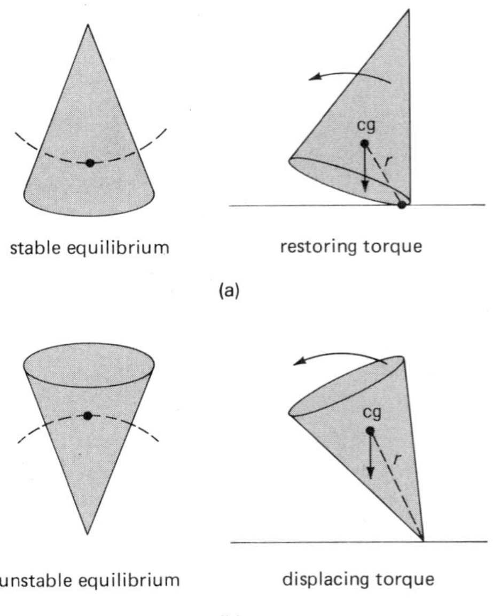

Figure 8.16 Equilibrium. (a) If a body in stable equilibrium is slightly displaced, a restoring torque tends to bring it back to its equilibrium position. Notice that the center of gravity acts like a particle in a bowl. (b) If a body is in unstable equilibrium, a slight displacement will cause it to topple, owing to a displacing torque.

base of support but is unstable when the center of gravity falls outside its base of support. When you are standing, your base of support is the area under you bounded by the edges of your feet. An example of human stability, and perhaps instability, is shown in Figure 8.15(b). Do you think that the fellow could touch his toes? Probably not. His center of gravity is a bit forward, owing to an ample diet. If he bends over and his center of gravity falls outside his base of support, over he'll go. Maybe he would then touch his toes, but not in the desired manner.

When watching someone touch his toes, notice that his hind parts are extended backward somewhat so that the center of gravity is adjusted to be over the base of support. Ask him to try touching his toes while standing with his heels against a wall and see what happens.

We describe an object as being in stable equilibrium or unstable equilibrium (Fig. 8.16). If an object in **stable equilibrium** is slightly displaced, its center of gravity is still vertically above and inside its base of support. Because of the weight acting through the center of gravity, a restoring torque tends to bring the body back to its equilibrium position. Notice that the center of gravity behaves like a particle in a bowl.

As can be seen in the figure, if an object is in **unstable equilibrium** its center of gravity is still vertically above its base of support, but this is a point or an edge. A slight displacement will cause it to topple, owing to a displacing torque that rotates the object about an axis through its base. In this case, the center of gravity is like a particle sitting on top of a bowl or inverted dome.

Hence, we see that objects with wide bases and lower centers of gravity are more stable and less likely to be tipped over. These conditions are evident in the design of high-speed racing cars, which have wide wheel bases and are very close to the ground.

QUESTION: Suppose you and your classmates get together and stack your textbooks as shown in Figure 8.17, with each book displaced 4 cm from the one beneath it. How many books could be stacked before they topple over?

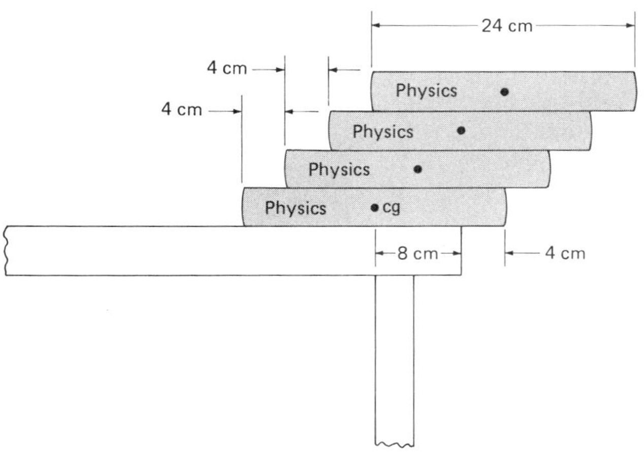

Figure 8.17 How many books can be stacked before they topple? See Question and Answer on p. 136.

ANSWER: The stack will fall over when the center of gravity of the stack is not above the base of support (the table). The textboks are fairly uniform, so their centers of gravity are at or near their centers. (How could you determine this for sure?) Hence, when the second text is placed on the first one, their centers of gravity are 4 cm apart. The center of gravity of the combination of both books is midway between, or displaced 2 cm to the right.

The third book will displace the combined center of gravity another 2 cm to the right, and so on. The edge of the table is 8 cm from the center of the bottom book, so when four books are stacked on the table, the center of gravity is still above the edge of the base of support (the table), having been displaced 6 cm to the right. But the fifth book will put the stack in unstable equilibrium. The stack may not fall if the fifth book is positioned *very* carefully, but it is doubtful if this could be done in practice. Just placing the fifth book on the stack would probably cause it to fall. Try it and see.

When you stand erect, your center of gravity lies within your body. The center of gravity of an average woman is lower than that of an average man of the same height. This is because women generally have larger pelvises and men have broader shoulders (different mass distributions). As a result, women have greater stability. For example, women can often touch their toes more easily than men. This center of gravity property is used in a common classroom demonstration and party game. First you stand with your feet together two foot lengths (length of one's feet) from a wall with your head against the wall, and then you pick up a chair (Fig. 8.18). The challenge is to exert yourself and stand up straight.

(a)

(b)

Figure 8.18 A test of stability. (a) In general, men cannot pick up a chair as shown and stand up, but women can. For a man, the center of gravity falls outside his base of support (feet). (b) However, by swinging the chair behind his back and shifting the center of gravity backward, he too can stand up.

In general, women can do this and men cannot. When bent over with the head against the wall, a man's center of gravity is more forward than a woman's. When the chair is picked up, it is enough to shift the combined center of gravity outside the feet base of support for most men, but not quite or only slightly for women. Hence, by exerting backward (shifting their muscles), women can stand up and men cannot. Try this with someone of the opposite sex and prove it to yourself.

As shown in Figure 8.18(b), a male student not wishing to seem less agile than his female friend can use scientific principles to show that he can "stand up" to the challenge. The instructions are usually to "put your head against the wall, pick up the chair, and stand up." So, this student picked up the chair and swung it behind his back. This shifts the center of gravity of the combination backward, and he too can stand up. Next time the rules would probably be more specific.

Rotational Energy

When we talked about kinetic energy in Chapter 6, we ignored rotational energy. We were generally focusing on the translational motion of the center of mass "particle." But a rotating object has rotational kinetic energy. Consider, for example, the pure rotational motion of a disk rotating about a fixed axis through its center (of mass). There is no translational motion, but there is motion, and hence kinetic energy—the energy of motion. The **rotational kinetic energy** is given by $\frac{1}{2}I\omega^2$, where I is the moment of inertia of the body and ω its angular velocity. (Notice the similarity to the expression for translational kinetic energy, $\frac{1}{2}mv^2$.)

A body may have both translational and rotational kinetic energies. For example, when a ball or cylinder rolls down a hill or incline, the center of mass "particle" moves translationally and the other particles of the rigid body are in rotation.

The conservation of energy also applies. The potential energy of the body at the top of the incline (mgh) goes into both translational and rotational kinetic energies (Fig. 8.19). If a cylinder slides down an incline (negligible friction), it moves down the incline faster than a rolling cylinder because the potential energy goes only into translational kinetic energy. For a rolling cylinder, part of the potential energy goes into the rotation of the body. A solid cylinder rolls faster than a hollow one. See Special Feature 8.1.

Angular Momentum

Translational or linear momentum (mv) involves objects moving in straight lines. In the case of circular motion, there is rotational or angular momentum. For a particle of mass m moving in a circle, the **angular momentum** is defined as follows:

$$\text{Angular momentum} = mvr$$

where r is the radius of the circular path and v is the instantaneous tangential velocity. For a rigid body, you could probably guess by analogy that the angular momentum is $I\omega$ (compare with mv).

Like the conservation of linear momentum (in the absence of an unbalanced force), the **conservation of angular momentum** is helpful in explaining many things. The conservation of rotational or angular momentum is stated as follows:

> In the absence of an unbalanced torque, the total angular momentum of a system is conserved.[‖]

There are many interesting and important examples of the use of conservation of angular momentum. You have no doubt used it yourself. Angular momentum, like its linear counterpart, is a vector. Hence, if angular momentum is conserved, the magnitude *and* direction do not change. That is, the direction of the angular momentum vector remains fixed in space. This is the principle used in spirally passing a football and in giving a rifle bullet a spiral rotation (by the rifling in the rifle barrel) to prevent them from tumbling in flight. In the absence of a torque, the angular momentum vector is constant in direction, which gives greater accuracy in throwing a football or shooting a rifle bullet. This is also the principle of the gyrocompass used to navigate ships and airplanes. If you've got a known, constant direction, you've got it made.

Notice that if the $I\omega$ expression for angular momentum is constant, then I and ω can change and the product will still be constant. For example, the products $3 \times 4 = 2 \times 6 = 1 \times 12 = 12$ are all constant (equal to 12). Varying I and ω is the basis of several applications of the conservation of angular momentum, as discussed in Special Feature 8.2.

[‖] Since $\tau = I\alpha$ and $\alpha = \Delta\omega/\Delta t$, $\tau = I\Delta\omega/\Delta t$ or $\tau\Delta t = I\Delta\omega$ = change in angular momentum, where $\tau\Delta t$ is the rotational impulse. Hence, if the unbalanced torque τ is zero, there is no change in the angular momentum, and it is constant or conserved.

SPECIAL FEATURE 8.1

The Great Cylinder Race

The race is between two cylinders, one solid and one hollow (Fig. 8.19). They are released at the same time from rest at the top of the incline, and the race is on. Which is your favorite? Which rolling cylinder will reach the bottom of the incline first?

Actually, the race is "fixed" (in the scientific sense). The solid cylinder will roll faster. In fact, any solid cylinder *always* beats any hollow one to the bottom. It makes no difference whether they have the same mass or outer diameter.

You may be quick to point out that a hollow cylinder has its mass concentrated farther from its axis (of symmetry) and therefore has greater rotational inertia. This is a good point, but it is possible for a solid cylinder to have a greater moment of inertia than a hollow cylinder by selecting appropriate masses and radii. It would seem that such a solid cylinder would have more rotational inertia and lose the race, but it doesn't. A solid cylinder may have a greater moment of inertia than a hollow one; however, the moment of inertia *per unit mass* of a solid cylinder is smaller.

There is no quandary if the conservation of energy is analyzed. It turns out that for a solid cylinder, $\frac{1}{3}$ of the potential energy always goes into rotational motion and $\frac{2}{3}$ into translational motion. More than $\frac{1}{3}$ of the potential energy goes into the rotational motion of the hollow cylinder. (Your instructor may wish to show you this with some simple mathematics.) You can experiment with some solid and hollow cylinders yourself and prove that solid cylinders will always beat the hollow ones to the bottom of an incline. Food cans make good solid cylinders, and the cans with the food removed and the ends cut out are hollow cylinders, as are napkin rings, and so on. Keep in mind that this is a rigid-body experiment. A can of juice or soda isn't a rigid body. Why?

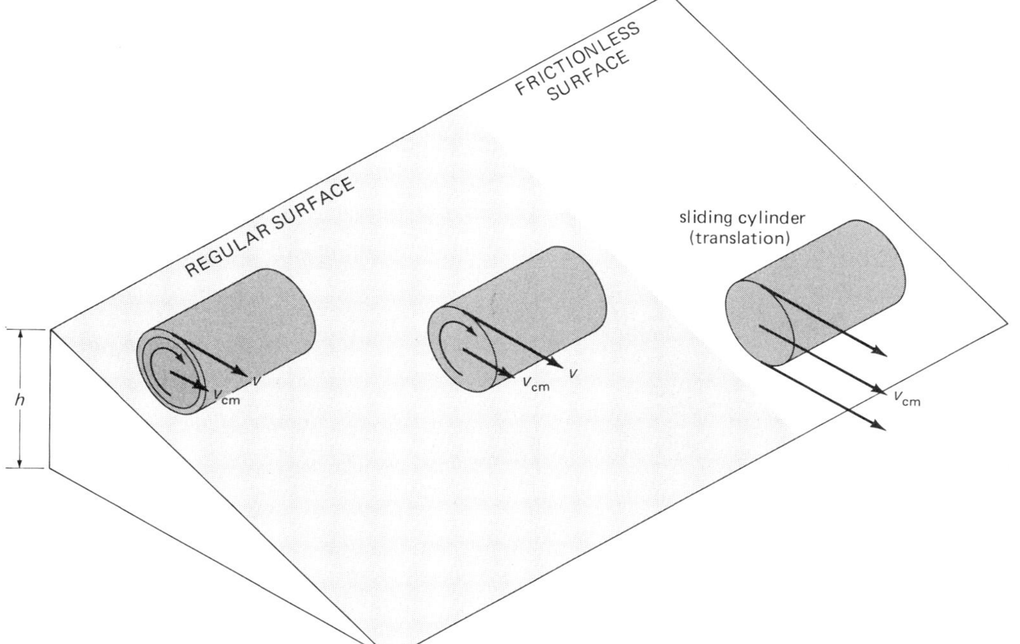

Figure 8.19 Rotational energy. An object at rest at the top of the incline has potential energy *(mgh)*. In rolling down the incline, the potential energy goes into both translational and rotational kinetic energies, K.E. $= \frac{1}{2}mv^2_{cm} + \frac{1}{2}I\omega^2$. If an object slides down a frictionless surface, the potential energy goes only into translational kinetic energy.

(a) (b)

Figure 8.20 Conservation of angular momentum. (a) A diver spins by "tucking" and decreasing the moment of inertia. By the conservation of angular momentum, this causes the angular speed to increase. (b) An ice skater uses the same principle in toe spins.

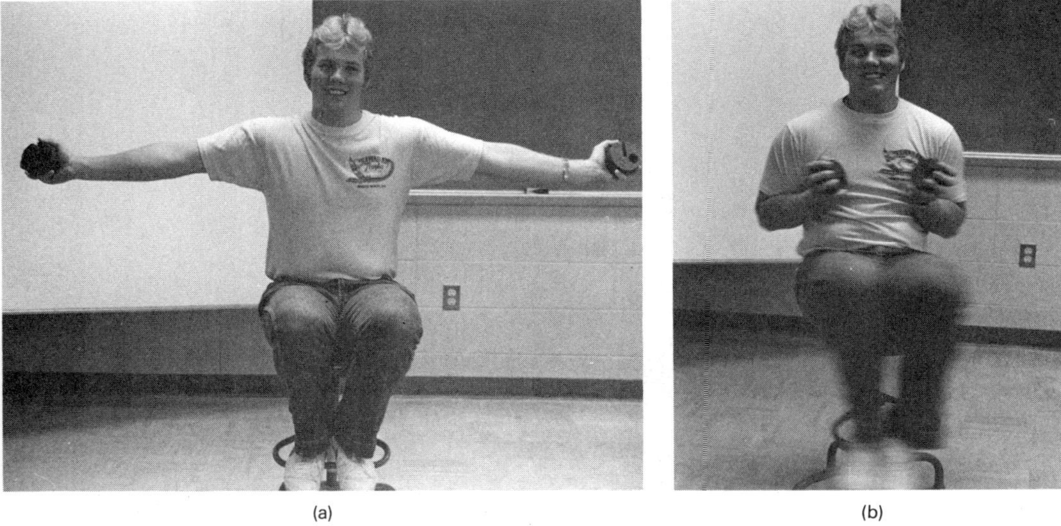

(a) (b)

Figure 8.21 Conservation of angular momentum. (a) When the person's arms are outstretched, there is a large moment of inertia and the rotation (angular speed) of the stool is slow. (b) When the arms and weights are brought inward, the moment of inertia decreases and the angular speed increases.

SPECIAL FEATURE 8.2

Conservation of Angular Momentum in Action

When angular momentum is conserved, $I\omega$ is constant. I and ω can change, but the product is still constant. In particular, the angular speed of an object can be changed by changing I by rearranging the mass distribution of the object. If this is done within or *internal* to the system, the angular momentum is still conserved (no *external* torques). This is what a diver does in a "tuck" (smaller I) to rotate faster during a dive (Fig. 8.20a). To reduce the rotation, the diver may "pike" (greater I), then "layout" (even greater I) to break the water cleanly. Gymnasts use similar techniques for twisting somersaults, for example, in dismounts from the horizontal bar and parallel rings where landing on the feet is essential.

Cats make use of a similar technique in always landing on their feet when dropped from a sufficient height. While falling, a cat changes its moment of inertia by instinctively reorienting its legs and tail. By the proper twisting reorientation it is possible for the head to rotate one way and the feet the other, so that the feet are downward or underneath when the cat lands.

The conservation of angular momentum is also used by ice skaters in doing spins (Fig. 8.20b) and by ballet dancers in doing pirouettes. Sweeping motions of the arms and legs start the body in rotational motion. When pivoting on the tip of the skate blade or on the tip of the toe, the skater or dancer is virtually free of external torques. The arms are then "tucked in" or raised vertically overhead to reduce the moments of inertia of their bodies, and the skater and dancer spin rapidly to conserve angular momentum.

Should you lack these talents, you can still get in the act (of conservation of angular momentum). All you need is a rotating stool or turntable and some weights (Fig. 8.21). This is a popular lecture demonstration. With the feet off the floor, you'll need someone (an external torque) to get you started. Try to get yourself rotating when isolated on the stool, as shown in the figure (or by standing on a turntable). You'll learn to appreciate Newton's rotational first law. Don't start rotating too fast at first. When you bring in the weights toward your body (reducing I), you'll spin quite fast. You don't want to get dizzy, go into unstable equilibrium, and fall off the stool.

Another very practical application of conservation of angular momentum is in helicopters (Fig. 8.22). Suppose a helicopter had only one set of rotor blades. When the helicopter is sitting on the ground with the engine off, its angular momentum is zero. The helicopter engine supplies the (internal) torque to start and maintain the rotor blades in motion. If the helicopter lifts off the ground, the helicopter body would rotate in the opposite direction to that of the rotor blades to conserve angular momentum (equal angular momen-

(a)

(b)

Figure 8.22 Conservation of angular momentum. (a) Large helicopters have two rotors that rotate in opposite directions to balance the angular momentum. With a single rotor, the helicopter body would rotate to conserve angular momentum. (b) On small helicopters, an "anti-torque" tail rotor provides a torque to prevent the rotation of the helicopter body.

tum vectors in opposite directions), which is a bit undesirable.

As a result, large helicopters are equipped with two sets of rotor blades that rotate in opposite directions. On smaller helicopters with a single set of overhead rotor blades, an "anti-torque" tail rotor provides a torque to prevent the rotation of the helicopter body.

SUMMARY OF KEY TERMS

Particle a physical point concept that has no physical dimensions but may have mass and can be accurately localized in space.

Rigid body a system of particles in which the particles are fixed, constant distances apart.

Center of mass the average location of all the mass particles of a body, or the point at which all the mass of a body can be considered to be concentrated.

Center of gravity the average location of the weight distribution of a body, or the point at which all the weight of a body can be considered to be concentrated. In a uniform gravitational field the location of the centers of gravity and mass is the same.

Pure translational motion motion in which every particle of a body has the same instantaneous velocity, and the path of each particle is the same as that of the body as a whole.

Pure rotational motion motion in which every particle of a body moves in a circle about an axis of rotation.

Angular speed the time rate of change of angular position, $\omega = \theta/t$.

Angular acceleration the time rate of change of angular velocity, $\alpha = \Delta\omega/\Delta t$.

Rotational inertia the property of matter that resists changes in rotational motion.

Moment of inertia a measure of the rotational inertia of a body.

Lever arm the perpendicular distance from the axis of rotation to a line along which a force acts.

Torque the product of the lever arm and force. Torques are capable of producing rotational motion and changes therein.

Newton's first law for rotational motion a rigid body remains at rest or in motion with a constant angular velocity unless acted upon by an unbalanced torque.

Newton's second law for rotational motion torque = moment of inertia × angular acceleration, $\tau = I\alpha$.

Stable equilibrium the condition of a body with its center of gravity vertically above and inside its base of support.

Unstable equilibrium the condition of a body when its center of gravity is vertically above an edge or point such that a slight displacement will cause the body to topple.

Rotational kinetic energy energy of rotational motion, $\text{K.E.} = \frac{1}{2}I\omega^2$.

Angular momentum (particle in circular motion) mvr, (rigid body) $I\omega$.

Conservation of angular momentum in the absence of an unbalanced torque, the total angular momentum of a system is conserved.

EXERCISES

1. Is it possible for a rigid body to have a center of mass and not a center of gravity? Explain.
2. Where is the center of gravity of a boomerang located? How could you determine the location of a boomerang's center of gravity?
3. Is it possible for a person's center of gravity to lie outside of his body? Explain.
4. When a pole vaulter goes over the bar (Fig. 8.23), his or her center of gravity may actually pass beneath the bar. (a) How is this possible? (b) Why is this an advantage compared with having the center of gravity go over the bar? (*Hint:* Explain in terms of energy.)
5. In finding the center of gravity of an object by suspending it from different points, is there any advantage in suspending it from three or four points? Explain.
6. Where is the center of gravity of the mobile shown in Figure 8.24?

Figure 8.23 See Exercise 4.

Figure 8.24 See Exercise 6.

7. Suppose someone in your class said that it was possible for a body to have pure translational motion and pure rotational motion at the same time. Would you agree with them? Why?
8. Describe the motion(s) of a frisbee in flight (Fig. 8.25). Also, discuss its energy and angular momentum.
9. How many radians are there in (a) 360°, (b) 180°, (c) 90°?
10. What are the angular speeds of (a) the second hand, (b) the minute hand, and (c) the hour hand of a clock?
11. What is the direction of the angular velocity vector of a phonograph turntable? (In case you don't remember, turntables rotate clockwise, looking down from above.)
12. The Earth rotates from west to east. What is the direction of the angular velocity vector in (a) the Northern Hemisphere, and (b) the Southern Hemisphere?
13. Considering the relative motion of a phonograph needle in the grooves of a record, does the needle have a greater angular speed near the outer portion of the record or near the center of the record? How about tangential speed?
14. Two songs of equal length are put on a phonograph album, one at the beginning or outside of the record and the other near the center of the record. Will the band widths of the selections be the same? Explain.
15. A person was given a speeding ticket because a patrolman's radar showed her car to be exceeding the 55 mi/h speed limit. At the hearing, three witnesses in the car testified that the speedometer read exactly 55 mi/h. Yet the judge imposed a fine when he learned that new, oversized tires had recently been installed on the car. Was this a fair ruling? Explain. Would the new tires have any effect on the odometer (mileage meter) reading?
16. A linear acceleration can change the magnitude and/or the direction of the linear velocity. Can an angular acceleration have the same effects on the rotational velocity of a body? Describe what the effects would be.
17. A type of potter's wheel consists of a heavy "kickwheel" connected by a shaft to a smaller and lighter working wheel on which the clay is spun (Fig. 8.26). The kickwheel is rotated by foot, leaving both of the potter's hands free to work the clay. The working wheel rotates evenly, even though the energy from kicking is delivered periodically. Explain why.
18. Compare the rotational inertia of a baseball bat about axes through (a) the length of the bat, (b) the handle end, and (c) the opposite (larger) end.

Figure 8.25 See Exercise 8.

Figure 8.26 See Exercise 17.

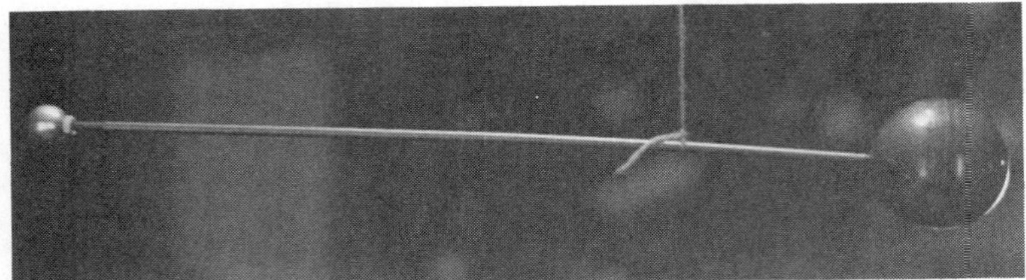

Figure 8.27 See Exercise 26.

19. When a front-engine automobile is stopped suddenly, why does the front end go down and the back end come up? How would a rear-engine car behave?
20. Is it possible to have forces applied to a body such that the net force is zero, but the net torque is not? (*Hint:* Consider a rod that can be rotated about a fixed axis at its center and forces applied perpendicularly to the ends of the rod.)
21. When pedaling a bicycle, when do you apply the maximum and minimum torques? (Assume the downward force of your foot is constant.) What is the value of the minimum torque?
22. A stubborn screw won't come loose when you try to remove it with a screw driver. If you had two other screw drivers available, one with the same diameter metal shaft and a bigger handle and one with same diameter but longer metal shaft and the same size handle, which one would you use to give the screw another try and why?
23. Why is the steering wheel of a tractor trailer rig bigger than that of an automobile? Wouldn't a smaller steering wheel give the driver more room?
24. The center of gravity of a rod or stick can be located by balancing as in Fig. 8.1. Explain this in terms of torque. (*Hint:* Note that a balanced stick does not rotate about your finger axis.)
25. Explain the principles of a double-pan balance in terms of torque. Such balances are used to weigh gold and precious stones.
26. A rod with end weights is balanced as shown in Figure 8.27. If the big ball weighs 4 N and the little ball weighs 1 N, how far is the center of gravity of the system from the little ball? (Assume the rod is 1 meter long, and neglect its mass.)
27. Where would you say the centers of mass of the following systems would be? (a) the Earth and moon, (b) the Sun and the Earth, (c) the solar system. (Check with your instructor to see how close you are.)
28. When an object, such as a ball or cylinder, rolls down an incline from rest, what produces the rotational motion? (*Hint:* Discuss in terms of center of gravity.)
29. Would a rotational form of Newton's third law be valid? (For every torque there is an equal and opposite torque.)
30. Why does a lower center of gravity give an object greater stable equilibrium?
31. We have all balanced a vertical pole or stick on our fingers, and quickly learned that a longer pole is easier to balance than a shorter one (for example, a meter stick and a pencil). Why is this so when a shorter pole has a lower center of gravity? Also, explain how you keep the pole balanced.
32. A tightrope walker often carries a long, droopy pole. As was learned in the chapter, the pole increases the rotational inertia. Does the droopiness of the pole make a difference?
33. Circus performers ride bicycles across tightwires. Some use long poles and others do not. How do the latter maintain their (unstable) equilibrium?
34. A common child's toy with a rounded, weighted bottom always rights itself when knocked over. Explain why this toy can't be knocked down to stay.
35. Given a bunch of bricks 21 cm long, how many bricks could be stacked on the bottom brick before the stack falls if each stacked brick is displaced 2 cm from the edge of the one beneath it?
36. (a) What is the most stable position you can assume? (b) Why is it easier to do sit-ups when your arms are extended in front of you than when your hands are behind your head?
37. How does standing with your legs apart, for example, on a moving bus, increase your stability? Is the stability increased in all directions? (Better hang on.)
38. Is a tall person with big feet more stable than a shorter person with smaller feet? (Don't forget height and center of gravity.)
39. Does it make a difference how the trailer of a tractor-trailer rig is loaded? (Consider both sideways and vertical load distributions for a truck that will go around some banked curves on a trip.)
40. The Leaning Tower of Pisa leans because of settling under one side of its base (see Fig. 4.1). (a) If this continues, when will the tower fall? (b) Corrective actions have been taken.¶ What else might be done to keep the tower from toppling?
41. How many different positions are there of stable and

¶ The Leaning Tower (56.2 m high) was begun in 1174 and completed in the 14th century. The settling began in the early stages of construction, and the third, fifth, and top (eighth) stories were straightened somewhat to compensate for the lean. Its foundation is only as wide as the circumference of the tower. The foundation has been injected with cement, but the tower is still threatened with collapse. The top of the tower leans about 5 m from the vertical.

Figure 8.28 See Exercise 42.

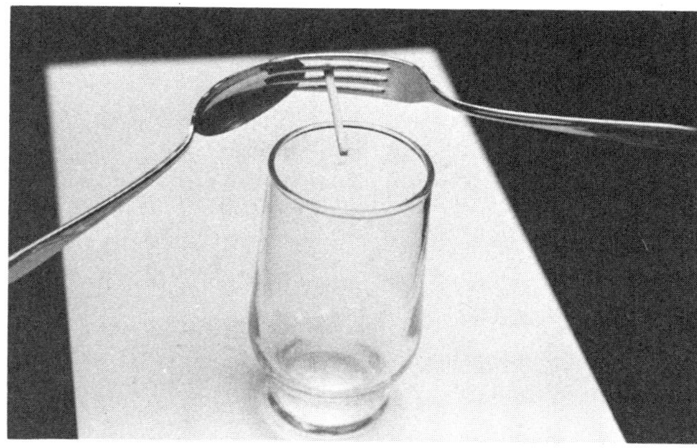

Figure 8.29 See Exercise 43.

unstable equilibrium for a cube? (Consider that resting on each different side is a different position.)

42. Does putting "big wheels" on a pick-up truck make it more stable (Fig. 8.28)? Explain. (*Hint:* What is the purpose of the bar behind the cab in the photo?)

43. Explain the balancing act shown in Figure 8.29.

44. Suppose you are standing on a large turntable near the outside edge and you start to walk around the outer circumference of the table. How would you move relative to someone watching you who is not on the turntable? (Consider the turntable bearings to be frictionless.)

45. Unlike helicopters, single-engine aircraft have only one propeller (or set of rotor blades). What keeps the body of the airplane from rotating? (*Hint:* Think in terms of torques from air foils.)

PART TWO
The Properties of Matter

The structure and properties of matter have been major topics of science from the earliest times. We may simplistically define matter as a fundamental property or anything that occupies space. However, the questions of what is the basic constituent of matter and why different substances behave differently still remain.

The early Greeks addressed the problem of the makeup of matter in terms of atoms. According to Democritus, who lived about 450 B.C., "The only existing things are the atoms and empty space; all else is mere opinion." As scientific focus expanded to include the atomic microcosm as well as the celestial macrocosm, theories and new discoveries began to fall into place. Today, the general properties of matter are described and understood in terms of atomic theory. This involves not the indivisible atom of the Greeks, but an atom with structure and particle makeup.

The ultimate constituent of matter is still a question of modern physics. However, atomic theory provides us with an insight of the general makeup of substances. That is, how the various atoms of matter are combined together generally determines the properties of the solids, liquids, and gases we experience in everyday life.

9

Atoms, Molecules, and Matter

Atomic Theory

We have discussed the concept of matter, which is generally defined as anything that occupies space. You, the kitchen sink, your physics book, the stars—everything we see is matter, and so are some things we don't see, like air. But what is matter made of?

Suppose you cut a piece of matter in half, then cut one of the halves in half, and so on into smaller and smaller pieces. Would you ever reach a "tiniest" piece of matter, or is matter a continuous thing? This was a major question of science down through the centuries.

The notion that matter consists of discrete particles is an old one. The idea was expressed by Democritus, a Greek philosopher, around 400 B.C. According to Democritus, there was a limit to the subdivision of matter. He called the ultimate particles atoms, after the Greek word *atomos,* meaning indivisible.

The atoms of Democritus were in perpetual motion and could not be created or destroyed. He implied that all of our experience is the result of the rearrangements of atoms. All these ideas have a ring of truth according to our modern concepts, as we shall see. (Democritus also thought that atoms could not fuse or combine together, which wasn't such a good idea.)

However, Democritus's ideas were rejected by Plato and Aristotle, who believed matter to be continuous. Because of their influence, the atomic theory wasn't accepted at the time and lay dormant for about 20 centuries. In the 1600's the concept of atoms was suggested again by the Italian physicist Gassendi (Fig. 9.1). This time there was some influential support for the idea. Sir Isaac Newton wrote:

> It seems probable to me that God, in the Beginning, formed Matter in solid, massy, hard, impenetrable, movable Particles, of such Sizes and Figures, and with such other Proper-

Figure 9.1 Big names in atomic theory.

ties, and in such Proportions to Space, as most conduced to the End for which He formed them. . . .

Prior to the 1800's the concept of the atomic nature of matter was basically speculative. Then in 1808 John Dalton, an English chemist and school teacher, developed explanations of several of the laws of chemistry using atomic theory. Some of his ideas had to be discarded, but the essentials of atoms have withstood the test of time and are with us today. In his atomic theory, Dalton referred to the chemical combination of two or more atoms as a molecule.

More direct evidence for the atomic theory was accidentally discovered in 1827 by the Scottish botanist Robert Brown. While observing pollen grains suspended in water under his microscope, he noticed that the tiny grains zig-zagged about in an agitated manner (Fig. 9.2). Dust particles in water were also observed to do a similar dance.

Figure 9.2 Brownian motion. Pollen grains suspended in water being knocked about by a barrage of randomly moving water molecules give evidence of the atomic "particle" theory.

This phenomenon is called **Brownian motion,** in honor of the observant Scot, and is readily explained by atomic "particle" theory. The pollen grains are knocked about by a barrage of randomly moving water molecules (collisions and momentum transfer as in Chapter 3). Oddly enough, it was Albert Einstein who first explained Brownian motion in 1905, the same year he published another paper on his theory of special relativity (Chapter 26). Perhaps the great man relaxed by thinking of slower-moving objects such as pollen grains.

Brownian motion and other evidence gathered in the last century have established the atomic theory of matter. Today, we even have pictures of atoms (Fig. 9.3). These are not ordinary pictures taken using visible light. The dimensions of the light waves are too big to be affected by individual atoms. To be "visible" or detected, a particle must be large enough to influence the

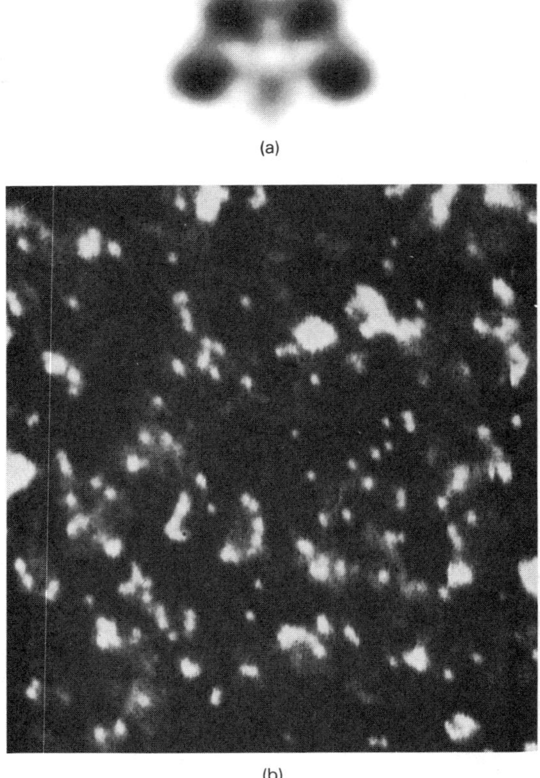

Figure 9.3 Atoms. (a) An electron microscopy image of a complex organic molecule. Four mercury atoms (dark spots) are clearly visible. (b) Single uranium atoms shown as small bright spots magnified more than 5 million times by an electron microscope. The large white spots are clusters of atoms.

wave or disturbance. To "see" atoms, the high-energy electron beam in an electron microscope is used (Chapter 27). Here, the electron beam disturbances or waves are over a thousand times smaller than those of visible light, and atoms can be detected.

So, considering the atomic theory, typical questions might be: How big are atoms and molecules? How many are there in a quantity of matter? Do atoms have structure or are they "billiard ball" particles? Let's take a look at the answers to these questions.

SIZES AND NUMBERS OF ATOMS AND MOLECULES

Atoms are extremely small. From quantitative studies of Brownian motion, Einstein estimated atoms to be on the order of 0.00000001 cm (1×10^{-8} cm) in diameter. This was a good "ball-park" figure. More modern measurement methods have shown nitrogen to be the smallest atom, with a diameter of about 0.0000000106 cm (1.06×10^{-8} cm). The largest atom is cesium, a real monster in size with a diameter of about 0.000000054 cm (5.4×10^{-8} cm).*

The size of a molecule of course depends on the sizes of its constituent atoms and how they are put together. For example, a water molecule (H_2O) has a length of about 3×10^{-8} cm (3 Å). This is a curved length. The hydrogen atoms hook on the oxygen atom at an angle, so the molecule has a bent crescent-moon shape (see Fig. 9.7). There are also huge macromolecules found in plastics and living organisms. Here, the molecules are long chains of atoms (Fig. 9.4). For example, if one of the DNA molecules in the human cell were stretched out, it would be several centimeters long.

How many atoms are there in the world? So many that the number would be difficult to comprehend. Let's think on a smaller scale. Every time you breathe, you take in about a half liter of air. In this quantity of air there are about 10,000 billion billion gas molecules — 10^{22} molecules and about twice that many atoms, since the gas molecules of the air are mostly diatomic (two-atom) molecules. On a hot day, you may easily drink a liter of water. (Recall that this is a little more than a quart.) How many molecules are you consuming? About 3.3×10^{25}.† Being a sharp student, you might be quick to ask, How do you know? Who counted the molecules? No one, of course. The numbers are computed from mass considerations and chemistry principles. Another aspect of atoms is considered in Special Feature 9.1.

* To avoid these small decimals, scientists often use a smaller unit of length, the angstrom (Å). 1 Å = 10^{-10} m = 10^{-8} cm. Hence, the nitrogen atom has a diameter of 1.06 Å and the cesium atom a diameter of 5.4 Å.

† Written out so as to look impressive, this is 33,000,000,000,000,000,000,000,000 molecules.

Figure 9.4 DNA model showing a two-turn helix. The DNA molecule is a macromolecule with a spiraling helix form. The DNA molecules in the human cells would be several centimeters long if stretched out.

ATOMIC STRUCTURE

Atoms are not the "indivisible" particles of Democritus and Dalton. Atoms have structure and are made up of subatomic particles—electrons, protons, and neutrons. A common, over-simplified model of the atom resulting from an early modern theory (see Chapter 27) is often referred to as the solar-system model (Fig. 9.5). Similar to the planets orbiting the Sun, the electrically negative (−) electrons orbit about the much more massive, electrically positive (+) charged protons and neutral neutrons, which are grouped very closely together in the center nucleus of the atom. The electrons orbit the nucleus in three dimensions, with the electrical force between an electron and the nuclear protons supplying the necessary centripetal force to keep each electron in orbit.

SPECIAL FEATURE 9.1

Recycled Atoms

Atoms are ageless. They just get recycled around, usually in molecules, but they never change (except in relatively infrequent nuclear processes). Suppose you were a carbon atom in a plant that lived millions of years ago. Then you got buried and layers of dirt piled up on top of you over the years in geologic processes. Its weight drove away some of your water molecule friends and neighbors, and you found yourself in a bed of coal. In the 20th century, someone dug you up, threw your lump into a fire, and the next thing you know you find yourself rubbing shoulders and combining with a couple of oxygen atoms in the atmosphere (becoming CO_2).

Then, some plant doing its photosynthesis thing takes you in, and you wind up in a sugar molecule that the plant uses for growth (see Chapter 7). But a cow comes along and eats the plant. After trips through two stomachs, you eventually end up as a bovine carbon atom, but not for long. The next thing you know, you're sitting on a plate as a part of a USDA Choice cut of meat and are about to be consumed by a strange-looking creature with a knife and fork in hand. After a hard day or two of chemical processes, you proudly become an integral part of the highest order of animal life on Earth. Who knows, if you're lucky you may make it to the top and become part of a brain cell.

Where does our carbon atom character go from there? You can write the next scenario. There are many possibilities.

Unlike the Sun and the Earth satellites, the electrons can only be in certain orbits at particular distances from the nucleus and cannot orbit at intermediate distances. These specific orbits, sometimes called shells, can contain only a certain number of electrons—the innermost shell, two electrons; the next shell, eight electrons; the next, 18 electrons, and so on. Most of the volume of the atom is empty space—again, much like the solar system. The electrons contribute very little to the mass of the atom. A proton or neutron is almost 2000 times more massive than an electron, so most of the atomic mass is in the nucleus (over 99.9 percent).

But how is one kind of atom distinguished from another kind, for example, a carbon atom from a nitrogen atom? Or, put another way, what makes an atom that of a particular species? For instance, what makes an oxygen atom an oxygen atom? The species or kind of atom is determined by the atomic number or proton number, which is simply the number of protons in the nucleus of the atom. For example, an oxygen atom is

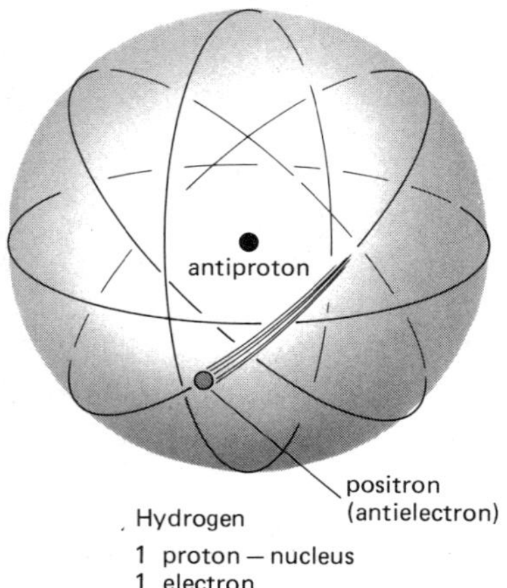

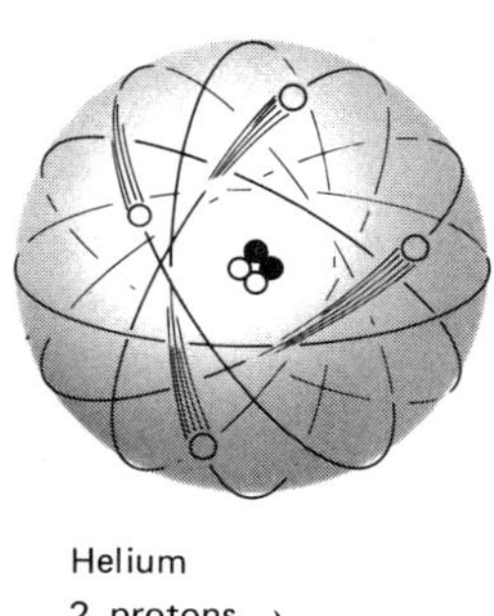

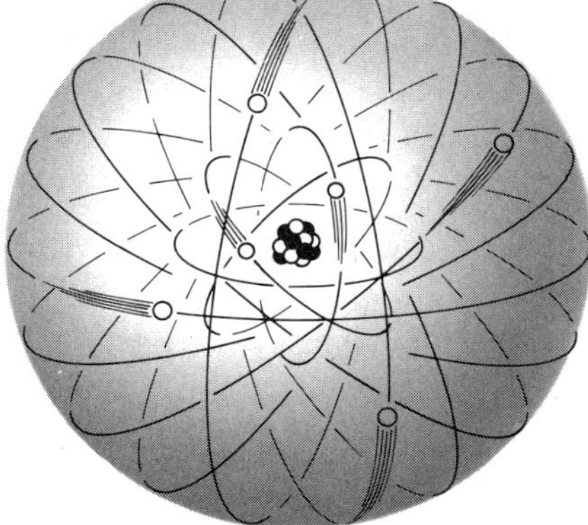

Figure 9.5 Atomic structure. The "solar-system" model of the atom views the electrons to be in orbits about a central nucleus.

oxygen because it has eight nuclear protons. All atoms with eight protons are oxygen atoms. Similarly, a carbon atom can be distinguished from a nitrogen atom because the carbon atom has six protons and the nitrogen atom has seven protons, by definition.

You might think that you could use the number of electrons to determine the atomic species, since there are the same number of electrons as protons in a neutral atom. This is true, but an atom may be ionized by removing an electron. Even so, it is still the same atomic species, for example, a nitrogen ion (N^+) with seven protons and six electrons. However, if the proton (atomic) number of an atom somehow changed by a nuclear process, the atomic species would change.‡

Then there are the neutrons. One species of atom may have different numbers of neutrons. These are called isotopes; they have the same number of protons, but different neutron numbers. Recall from Chapter 7 that natural uranium is 99.3 percent U-238 (92 protons and 146 neutrons) and 0.07 percent U-235 (92 protons and 143 neutrons). These atoms are (nuclear) isotopes of uranium.

Elements and the Periodic Table

Elements are a special class of substances. Chemically, an **element** cannot be decomposed into a simpler substance. From the previous discussion, the definition of an element is easy. An element is a substance in which all of the atoms have the same number of protons.

The ancient Greeks thought that all matter was composed of four "elements"—earth, air, fire, and water. Just four elements would have made science a lot easier but a lot less exciting. Of course, this notion was inadequate and incorrect. Even in the very early civilizations, nine of what we now call elements were isolated: gold, silver, copper, iron, lead, tin, mercury, sulfur, and carbon.

Until about the 17th century, chemistry developed rather slowly. Prior to that period, it was chiefly a by-product of alchemy. The main goal of the alchemists was to change common metals into gold. There also was a sideline of trying to develop a substance that would restore youth. They didn't have much luck with either goal, but their chemical experiments added five new elements to the known list: arsenic, antimony, bismuth, phosphorus, and zinc.

‡ Such nuclear transformations are possible and will be discussed in Chapter 28 on nuclear physics.

Today, the list of elements numbers around 106. Table 9.1 gives an alphabetical listing of the elements along with their chemical symbols, atomic (proton) numbers, and atomic masses. The atomic mass is a relative measure based on the common carbon atom (six protons and six neutrons), which is given a value of 12.0000.§ You will notice that carbon (and other elements) do not have whole-number atomic masses. This is because of isotopes. For example, natural carbon has three isotopes, but it is mainly composed of the common isotope with six neutrons. The other two isotopes, with seven and eight neutrons, contribute to the atomic mass so that the average is not a whole number.

Notice the convenient symbol notation for the elements. In general, the first letter of an element's name is used, along with the next or another letter when there are more than one element beginning with the same letter, for example, C, Ce, and Cs for carbon, cerium, and cesium, respectively. But we have Cl and Cr for chlorine and chromium, and there is no element with the symbol Ch. There is really no fixed method for assigning symbols. The official symbols are adopted by an international chemistry organization.

QUESTION: If the first letter of an element's name generally appears in its symbol, why are K, Na, and Pb used for potassium, sodium, and lead, respectively?

ANSWER: Because the symbols were taken from their original Latin names, *kalium, natrium,* and *plumbum,* respectively.

By 1869 some 63 elements had been discovered. Chemists tried to categorize or classify the growing list of elements, for example, as metals and nonmetals. However, this classification was much too broad, and a better method was desired. It had been noticed in the early 1800's that the elements could be listed in such a way that similar chemical properties reoccurred periodically throughout the list. In 1869 the Russian chemist Mendeleev (pronounced "Men-duh-lay-eff") formulated a table of elements based on this periodic property (Fig. 9.6). His periodic table of the elements in its modern version is still used today and can be seen on the walls of every chemistry or science building, as well as in Table 9.2.

§ A small unit for mass is thus defined: $\frac{1}{12}$ of a common carbon atom is 1 atomic mass unit (amu). A common carbon atom has a mass of 12 amu, and 1 amu = 1.66×10^{-27} kg.

Table 9.1 The Chemical Elements

	Symbol	Atomic no.	Atomic mass*		Symbol	Atomic no.	Atomic mass*
Actinium	Ac	89	227.0278	Mercury	Hg	80	200.59
Aluminum	Al	13	26.98154	Molybdenum	Mo	42	95.94
Americium	Am	95	[243]†	Neodymium	Nd	60	144.24
Antimony	Sb	51	121.75	Neon	Ne	10	20.179
Argon	Ar	18	39.948	Neptunium	Np	93	237.0482
Arsenic	As	33	74.9216	Nickel	Ni	28	58.70
Astatine	At	85	[210]	Niobium	Nb	41	92.9064
Barium	Ba	56	137.33	Nitrogen	N	7	14.0067
Berkelium	Bk	97	[247]	Nobelium	No	102	[259]
Beryllium	Be	4	9.01218	Osmium	Os	76	190.2
Bismuth	Bi	83	208.9804	Oxygen	O	8	15.9994
Boron	B	5	10.81	Palladium	Pd	46	106.4
Bromine	Br	35	79.904	Phosphorus	P	15	30.97376
Cadmium	Cd	48	112.41	Platinum	Pt	78	195.09
Calcium	Ca	20	40.08	Plutonium	Pu	94	[244]
Californium	Cf	98	[251]	Polonium	Po	84	[209]
Carbon	C	6	12.011	Potassium	K	19	39.0983
Cerium	Ce	58	140.12	Praseodymium	Pr	59	140.9077
Cesium	Cs	55	132.9054	Promethium	Pm	61	[145]
Chlorine	Cl	17	35.453	Protactinium	Pa	91	231.0359
Chromium	Cr	24	51.996	Radium	Ra	88	226.0254
Cobalt	Co	27	58.9332	Radon	Rn	86	[222]
Copper	Cu	29	63.546	Rhenium	Re	75	186.207
Curium	Cm	96	[247]	Rhodium	Rh	45	102.9055
Dysprosium	Dy	66	162.50	Rubidium	Rb	37	85.4678
Einsteinium	Es	99	[252]	Ruthenium	Ru	44	101.07
Erbium	Er	68	167.26	Samarium	Sm	62	150.4
Europium	Eu	63	151.96	Scandium	Sc	21	44.9559
Fermium	Fm	100	[257]	Selenium	Se	34	78.96
Fluorine	F	9	18.998403	Silicon	Si	14	28.0855
Francium	Fr	87	[223]	Silver	Ag	47	107.868
Gadolinium	Gd	64	157.25	Sodium	Na	11	22.98977
Gallium	Ga	31	69.72	Strontium	Sr	38	87.62
Germanium	Ge	32	72.59	Sulfur	S	16	32.06
Gold	Au	79	196.9665	Tantalum	Ta	73	180.9479
Hafnium	Hf	72	178.49	Technetium	Tc	43	[98]
Helium	He	2	4.00260	Tellurium	Te	52	127.60
Holmium	Ho	67	164.9304	Terbium	Tb	65	158.9254
Hydrogen	H	1	1.0079	Thallium	Tl	81	204.37
Indium	In	49	114.82	Thorium	Th	90	232.0381
Iodine	I	53	126.9045	Thulium	Tm	69	168.9342
Iridium	Ir	77	192.22	Tin	Sn	50	118.69
Iron	Fe	26	55.847	Titanium	Ti	22	47.90
Krypton	Kr	36	83.80	Tungsten	W	74	183.85
Lanthanum	La	57	138.9055	Uranium	U	92	238.029
Lawrencium	Lr	103	[260]	Vanadium	V	23	50.9415
Lead	Pb	82	207.2	Xenon	Xe	54	131.30
Lithium	Li	3	6.941	Ytterbium	Yb	70	173.04
Lutetium	Lu	71	174.967	Yttrium	Y	39	88.9059
Magnesium	Mg	12	24.305	Zinc	Zn	30	65.38
Manganese	Mn	25	54.9380	Zirconium	Zr	40	91.22
Mendelevium	Md	101	[258]				

* Atomic masses given here are 1977 IUPAC values based on Carbon-12.
† A value given in brackets denotes the mass number of the longest-lived or best-known isotope.

Table 9.2 Periodic Table of the Elements

IA	IIA	IIIB	IVB	VB	VIB	VIIB	VIII	VIII	VIII	IB	IIB	IIIA	IVA	VA	VIA	VIIA	0
1 H 1.0079																1 H 1.0079	2 He 4.00260
3 Li 6.941	4 Be 9.01218											5 B 10.81	6 C 12.011	7 N 14.0067	8 O 15.9994	9 F 18.998403	10 Ne 20.179
11 Na 22.98977	12 Mg 24.305											13 Al 26.98154	14 Si 28.0855	15 P 30.97376	16 S 32.06	17 Cl 35.453	18 Ar 39.948
19 K 39.0983	20 Ca 40.08	21 Sc 44.9559	22 Ti 47.90	23 V 50.9415	24 Cr 51.996	25 Mn 54.9380	26 Fe 55.847	27 Co 58.9332	28 Ni 58.70	29 Cu 63.546	30 Zn 65.38	31 Ga 69.72	32 Ge 72.59	33 As 74.9216	34 Se 78.96	35 Br 79.904	36 Kr 83.80
37 Rb 85.4678	38 Sr 87.62	39 Y 88.9059	40 Zr 91.22	41 Nb 92.9064	42 Mo 95.94	43 Tc (98)	44 Ru 101.07	45 Rh 102.9055	46 Pd 106.4	47 Ag 107.868	48 Cd 112.41	49 In 114.82	50 Sn 118.69	51 Sb 121.75	52 Te 127.60	53 I 126.9045	54 Xe 131.30
55 Cs 132.9054	56 Ba 137.33	57 ★La 138.9055	72 Hf 178.49	73 Ta 180.9479	74 W 183.85	75 Re 186.207	76 Os 190.2	77 Ir 192.22	78 Pt 195.09	79 Au 196.9665	80 Hg 200.59	81 Tl 204.37	82 Pb 207.2	83 Bi 208.9804	84 Po (209)	85 At (210)	86 Rn (222)
87 Fr (223)	88 Ra 226.0254	89 ᵞAc 227.0278	104 Unq (261)	105 Unp (262)	106 Unh (263)	107 Uns		109									

★ Lathanide Series

58 Ce 140.12	59 Pr 140.9077	60 Nd 144.24	61 Pm (145)	62 Sm 150.4	63 Eu 151.96	64 Gd 157.25	65 Tb 158.9254	66 Dy 162.50	67 Ho 164.9304	68 Er 167.26	69 Tm 168.9342	70 Yb 173.04	71 Lu 174.967

ᵞ Actinide Series

90 Th 232.0381	91 Pa 231.0359	92 U 238.029	93 Np 237.0482	94 Pu (244)	95 Am (243)	96 Cm (247)	97 Bk (247)	98 Cf (251)	99 Es (252)	100 Fm (257)	101 Md (258)	102 No (259)	103 Lr (260)

Note: Atomic masses shown here are 1977 IUPAC values.

Figure 9.6 Mendeleev formed the periodic table by arranging the elements based on the periodic reoccurrence of chemical properties. The elements in a vertical column have similar properties.

Mendeleev arranged the known elements in rows, which are called periods, in order of increasing atomic masses. When he came to an element with chemical properties similar to one of the previous ones, he went back and put this element below the similar, lighter one. As he did this, the columns formed groups or families of elements with similar properties. The table was later rearranged in order of increasing atomic or proton number (the numbers at the top of the boxes in Table 9.2) because of some inconsistencies. Notice the atomic masses of cobalt and nickel (atomic numbers 27 and 28).

When Mendeleev made his table with 63 elements, there were vacant spaces in it, which he reasoned to be due to missing or undiscovered elements. Since the missing elements had properties similar to those of the other elements in a group, Mendeleev was able to predict their masses and properties. Three of the missing elements were discovered within 20 years after Mendeleev formulated his table.

There is a great deal of interesting and useful information in the periodic table. (Your instructor may want to go into it a bit deeper.) For example, all the elements above 92 are artificial; they do not occur naturally. These elements are made by nuclear processes in particle accelerators and nuclear reactors. They are radioactive, and most have very short "lives" before they disintegrate or "decay" into something else (see Chapter 28). The elements with atomic numbers greater than 92 are called "transuranic" elements. Notice the next two elements above uranium—remind you of the planets of the solar system? You'll also find that the names of the rest of these elements have a very modern ring. Check Table 9.1 and see if you can associate a person or place with their names.

So, is it safe to say that we have 92 naturally occurring elements? Well, many textbooks say that there are only 90 such elements. This is because technetium (43) and promethium (61) are not currently found in nature. They can be prepared artificially but are radioactive and decay. Some scientists believe that these elements did occur naturally along with the other 90 when the Earth was formed, but are long since gone. Incidentally, the name technetium comes from the Greek word meaning artificial, and it was the first unknown element to be created by artificial means.

Notice in Table 9.3 that two elements, oxygen and silicon, make up 75 percent of the Earth's crust by weight. You pick up these elements in compound form when you scoop up a handful of sand (SiO_2). Many of the common elements, such as mercury and sulfur, are relatively scarce.

Molecules and Compounds

Atoms combine to form molecules, or on a larger scale, elements combine to form compounds. A **molecule** is a group of two or more atoms held together by forces called chemical bonds (Fig. 9.7). A pure substance made up of more than one element is called a **compound.** Some elements occur naturally as molecules, for example, H_2, O_2, and N_2 (diatomic molecules of hydrogen, oxygen, and nitrogen).

Table 9.3 Percentages of Elements in the Earth's Crust

Element	*Percentage (by weight)*
Oxygen	47
Silicon	28
Aluminum	8
Iron	5
Calcium	3.6
Sodium	2.8
Potassium	2.6
Magnesium	2
All others	1

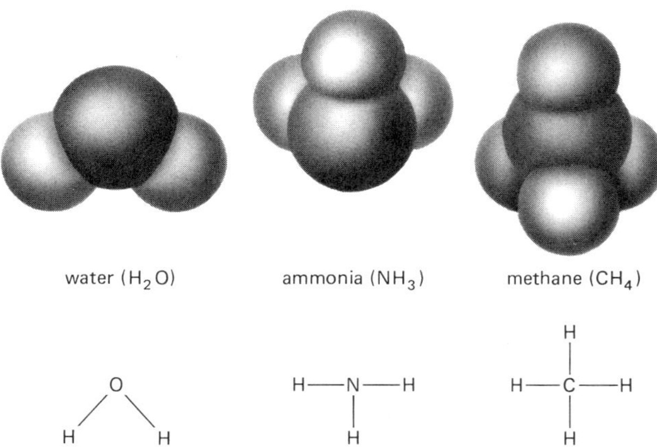

Figure 9.7 Molecules are groups of two or more atoms held together by forces called chemical bonds.

The chemical bonding is associated with the electrons of the atoms of a molecule. In fact, all ordinary chemical reactions involve only the outermost electrons in atoms. The main reason that the elements in a vertical group in the periodic table have similar chemical properties is that they have the same number of electrons in their outermost shells. Molecules are formed by atoms that combine so as to have filled shells (or subshells, not discussed here).

There are two general types of chemical bonds: ionic bonds and covalent bonds.

In **ionic bonds,** electrons are transferred between atoms (Fig. 9.8). The resulting ions are then held together by attractive electrical forces. In general, we cannot isolate an ionic "atom" such as NaCl because the ions are in an array and each ion is associated with several others (see Chapter 10). The ionic chemical bond is quite strong. This is evidenced by the relatively high melting-point temperatures of ionic compounds.

It is interesting to note how a simple electron transfer makes such a big difference in properties. For example, sodium (Na) is a soft, gray metal that is so reactive that it is kept stored under oil. Otherwise, it would react with the moisture in the air. A piece of sodium placed on your hand would burn you quite

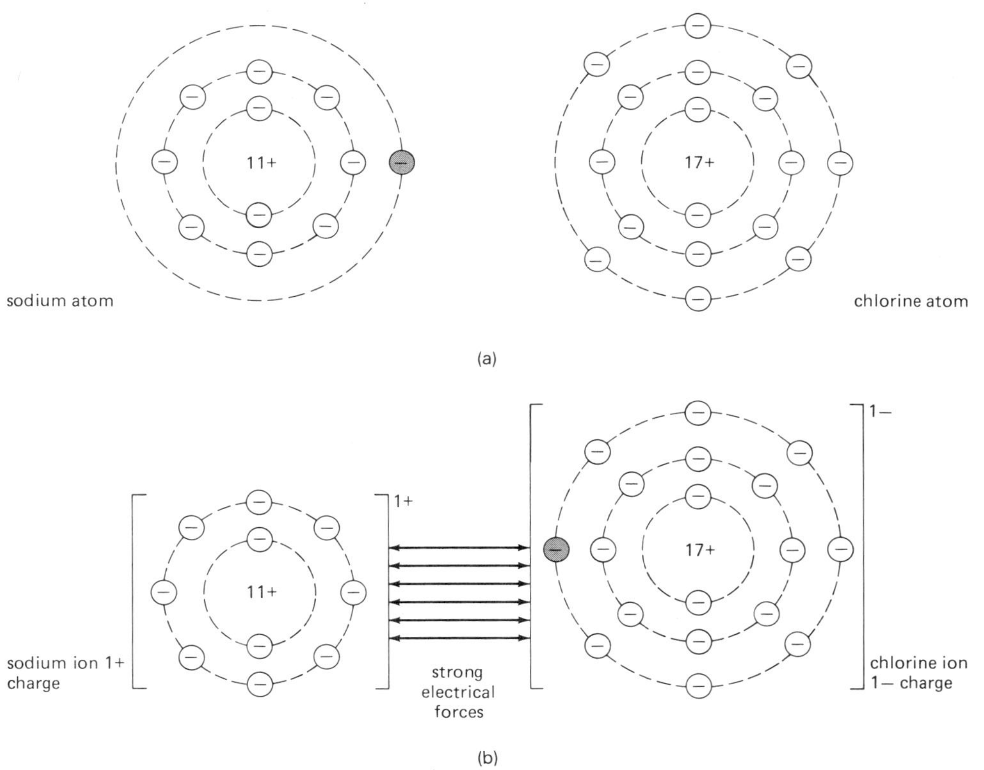

Figure 9.8 Ionic bonding. An ionic bond is formed when an electron is transferred between atoms, forming oppositely charged ions.

badly. Chlorine (Cl), on the other hand, is a greenish-yellow gas and is equally injurious. It was used as a poison gas in World War I. We have all smelled a hint of chlorine from bleaching solutions used in the washing of clothes or from city drinking water in which it is used to kill bacteria. Yet we can cause sodium and chlorine to react together chemically, and with tiny electron transfers between the atoms we get NaCl or common table salt, which most of us eat every day.

In **covalent bonds,** the electrons of atoms are mutually shared rather than permanently transferred (Fig. 9.9). Covalent bonds are not as strong as ionic bonds. The sharing of the electrons between atoms may or may not be equal. That is, the electrons may spend more time on the average with one atom than the other. As a result, such molecules have unsymmetric charge distributions, and we refer to them as having (covalent) **polar bonds** or being polar molecules (Fig. 9.10). The water molecule is polar, with regions of electric charge. The electrons spend more time with the oxygen atom, making this region of the molecule negatively charged and the hydrogen atom regions positively charged. This can be demonstrated as shown in Figure 9.10.

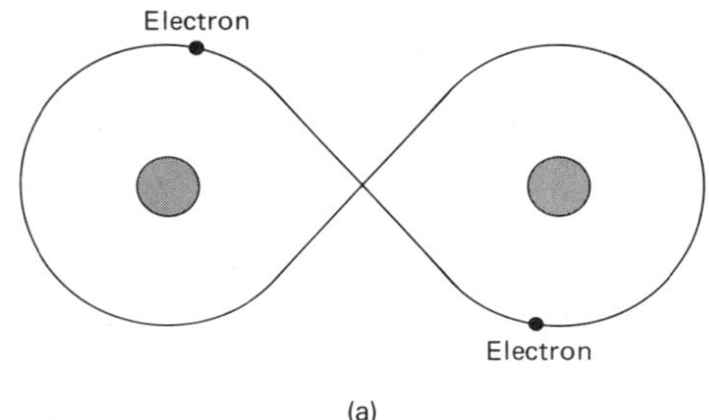

(a)

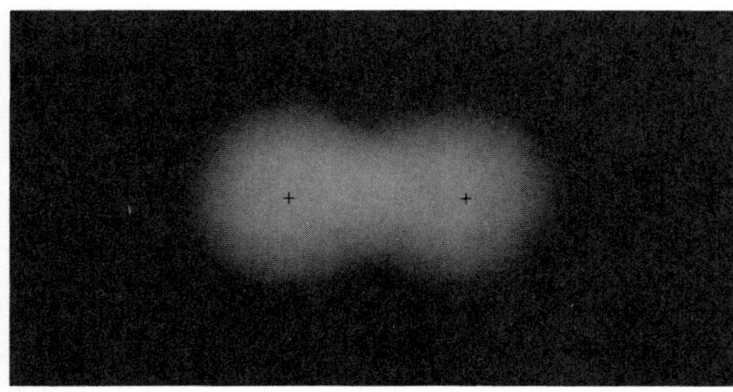

(b)

Figure 9.9 Covalent bonding. (a) In covalent bonds, the electrons of atoms are mutually shared rather than permanently transferred. (b) Equal sharing gives rise to a symmetric charge distribution.

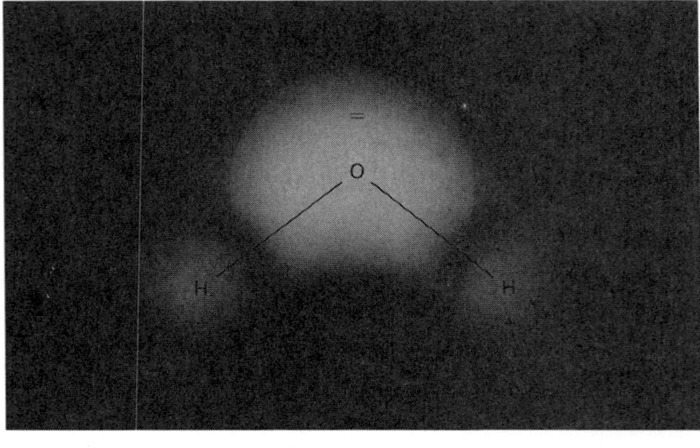

(a)

(b)

Figure 9.10 Polar molecules. (a) The water molecule is a polar molecule. Because of unequal sharing of electrons between atoms, there are regions of charge. (b) This can be demonstrated by using electrical forces. The rod, having been rubbed on fur, is negatively charged and repels the water molecules. (See Chapter 18.)

Since compounds are composed of elements, they can be decomposed chemically into their component elements. For example, ordinary sugar (sucrose) has the chemical formula $C_{12}H_{22}O_{11}$. By decomposition, the hydrogen and oxygen can be removed, and carbon is left (Fig. 9.11). Many elements are obtained from naturally occurring compounds, for example, metal ores.

Figure 9.11 Chemical decomposition. Sugar is composed of carbon, hydrogen, and oxygen. When the hydrogen and oxygen are chemically removed, carbon is left.

States of Matter

Everyone learned in grade school that at ordinary temperatures matter exists in three (physical) states or phases: solid, liquid, and gas. We may not be familiar with the states of some substances as they exist at various temperatures and pressures, for example, liquid nitrogen, which boils at a temperature of $-196°C$. However, we are all familiar with the three states of water (Fig. 9.12). Note that the "steam" seen coming from the teakettle is condensed water droplets (liquid state) and not steam (invisible gas near the spout).

Figure 9.12 Examples of the three states of matter: solid, liquid, and gas.

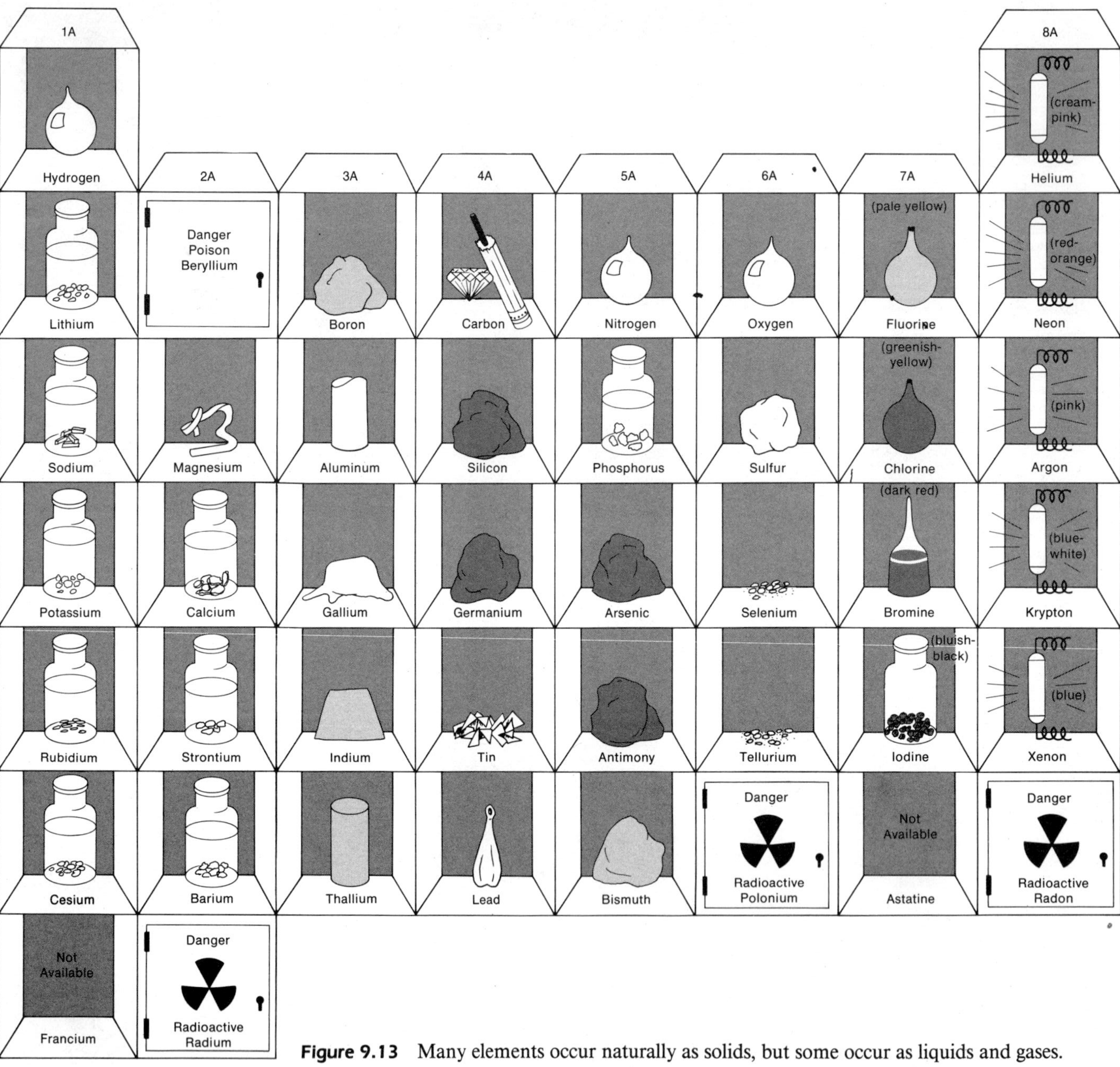

Figure 9.13 Many elements occur naturally as solids, but some occur as liquids and gases.

In general, we describe the states of matter as follows. Solids have definite shapes and definite volumes. Their molecules are held together by comparatively strong forces. Many elements occur naturally as solids (Fig. 9.13).

Liquids have definite volumes and their molecules have significant attractive forces between each other. However, a liquid has no definite shape. It assumes the shape of its container.

Gases spread out and fill the entire volumes of their containers. Therefore, a gas has no definite shape or volume. The gas molecules are relatively far apart, so the force between them is very small (except during molecular collisions).

If enough energy is added to a molecular substance, the molecules will separate into atoms, and the atoms themselves will come apart, giving a "gas" of free electrons and ions or nuclei. We call this a plasma, and plasma is considered to be a fourth state of matter.‖ The plasma state is not as common as the other three states of matter on Earth, but it exists in various applications and natural phenomena. For example, the ionized gas in fluorescent lamps and the matter of the Sun is in a plasma state.

More will be said about the various states of matter in the following chapters.

Antiparticles and Antimatter

Before leaving this chapter on matter, let's mention briefly something you may have heard of—antimatter. In 1932 a particle was discovered that had the same mass as an electron and behaved like an electron, except that it had a *positive* electric charge. This "positive" electron is called a positron and is said to be the antiparticle of an electron.

All subatomic particles have been found to have antiparticles, which are observed in cosmic rays from outer space and/or are produced in nuclear processes. There is the antiproton, with the same mass as a proton but a negative charge. There are also antineutrons.

However, the antiparticles do not get along well with their particle counterparts. When a particle and its antiparticle collide, they vanish or "annihilate" each other, and their mass is converted *completely* into energy or into other particles (Fig. 9.14). This is another example of mass-energy conversion and conservation according to Einstein's $E = mc^2$ relationship, with a possible 100 percent conversion. In the uranium fission reaction used in nuclear reactors for energy production for electrical generation, there is only about a 0.1 percent mass-to-energy conversion.

‖ This plasma is pronounced and spelled the same as the liquid part of blood (blood plasma), but it is entirely different.

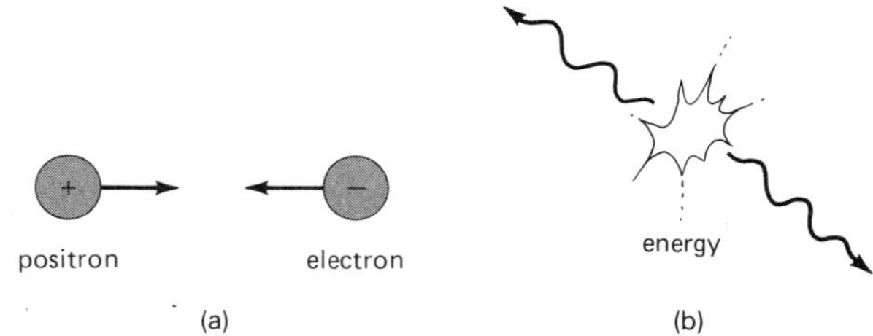

Figure 9.14 Annihilation. A particle and its antiparticle coming together are "annihilated," and their mass is converted to energy.

Since antiparticles exist, it could be possible that somewhere in the universe there are quantities of antimatter, whose antiatoms are made up of antiparticles (Fig. 9.15). This might be the substance of stars or even entire galaxies (antistars and antigalaxies, to be more exact). Science-fiction writers have written about antiworlds composed of antimatter, but no one really knows. One thing for sure is that we couldn't visit our antineighbors, nor they us. There would be annihilation when the matter and antimatter came into contact.

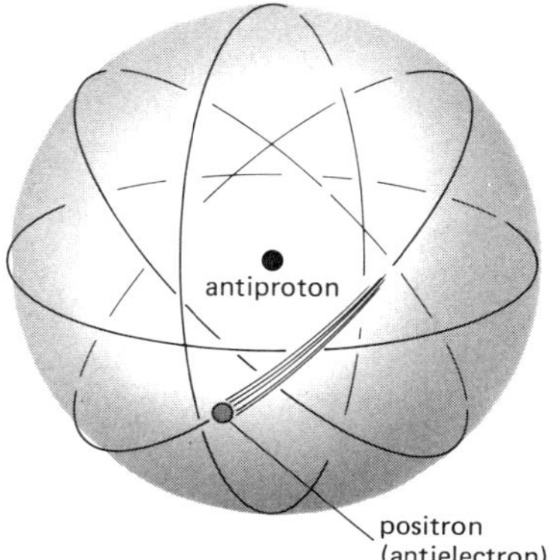

Figure 9.15 Antiatoms would make up antimatter. A hydrogen antiatom would have a nuclear antiproton and an orbiting positron.

SUMMARY OF KEY TERMS

Atom the smallest unit of an element that can exist alone or in combination with other atoms.

Element a substance in which all the atoms have the same number of protons.

Brownian motion The erratic motion of small particles (pollen grains, dust particles, etc.) in suspension due to collisions with the molecules of the suspension medium.

Solar-system model the atomic model that pictures the protons and neutrons to be in a central core or nucleus with the electrons in orbits (or shells) about the nucleus.

Atomic (proton) number the number of protons in a nucleus or atom. The atomic or proton number determines the species of an atom.

Ion an atom with a net electrical charge due to the transfer (loss or gain) of one or more electrons.

Isotopes atoms or nuclei of the same species (same number of nuclear protons) having different numbers of neutrons.

Periodic table of the elements an arrangement of elements based on atomic number and recurring chemical properties that are periodically repeated.

Period a horizontal row in the periodic table.

Group a vertical column in the periodic table. The elements of a group or "family" have similar chemical properties.

Molecule a group of two or more atoms held together by forces called chemical bonds.

Compound a pure substance made up of one or more elements.

Ionic bond a bond formed by a transfer of electrons between atoms.

Covalent bond a bond formed by a sharing of electrons between atoms.

Polar bond a covalent bond in which there is an unequal sharing of electrons between atoms such that there is an unsymmetric charge distribution or molecular regions of net charge.

States of matter the forms of matter in terms of being a solid, liquid, gas, or plasma.

Plasma an ionized gas of free electrons and ions, which is considered to be a fourth state of matter.

Antiparticle a particle in which the electrical (and magnetic) properties are reversed from those of ordinary particles.

Annihilation the complete conversion of mass into energy that occurs in the collision of a particle and antiparticle.

Antimatter matter made up of antiparticles.

EXERCISES

1. Why doesn't a small piece of wood floating in water exhibit Brownian motion?
2. The Austrian physicist Ernst Mach (1838–1916) was one of the last holdouts against the atomic theory. He thought that it was meaningless to believe in atoms since they could not be sensed directly. If Mach were alive today, how would you convince him of the existence of atoms?
3. Compare the populations of (a) the United States and (b) the world to the number of molecules in a half liter of air. (*Hint:* See Chapter 7 for populations.)
4. On the average, a person breathes in about 6 liters of air per minute. (a) What is the molecular intake per minute? (b) Are the same molecules expelled? Explain.
5. Compare the magnitude of the national debt (about 1.5 trillion dollars) with the number of molecules in a milliliter of water.
6. Are the common atomic model and the solar system similar or different with respect to (a) mass distribution, (b) empty space, (c) the nature of centripetal force, (d) three-dimensional orbital shells, (e) the distance of orbiting "particles" from the center of the system?
7. Does an atom always have as many electrons as protons? Explain.
8. Sketch the electron shell arrangements for the atoms of Be and Mg, and explain why these elements would have similar chemical properties.
9. The elements sodium (Na) and fluorine (F) form the ionic compound NaF (sodium fluoride). Sketch the electron shell arrangements for the atoms of these elements and explain why an ionic bond is readily formed.
10. What elements do the following chemical symbols represent? (a) Ag, (b) Au, (c) Hg, (d) Fe.
11. Which one of the elements in each of the following sets has the greatest number of protons in its atomic nucleus? (a) sulfur, copper, and calcium, (b) thorium, krypton, and strontium, (c) einsteinium, mendelevium, and nobelium.
12. Nickel and platinum are metals. What would you predict the lesser-known element palladium to be, and why?
13. Suppose a U-238 atom lost two protons, two neutrons, and two electrons. What would it be then?
14. Webster's Dictionary defines a molecule as "the smallest quantity of an element or compound which can exist separately and still retain the properties and character of the element or compound." One of the characteristics of sugar is its sweetness, but one molecule or even several molecules of sugar cannot be detected by taste. Is there something wrong with Webster's definition? Explain.
15. Suppose common sugar or sucrose were decomposed

into carbon and water. How many carbon atoms and water molecules would be obtained from a dozen sugar molecules?

16. A quantity of table salt (NaCl) is dissolved in a volume of water. Is this a chemical change or is the solution a chemical compound? Explain. (*Hint:* A mixture is a combination of substances that can be separated by physical means rather than chemical means, as is required for the separation [decomposition] of a compound.)
17. If positrons are created by cosmic rays and nuclear processes, why are they not commonly found in nature?
18. Sketch the solar-system model for hydrogen and carbon antiatoms.
19. Could the problem of handling the antimatter for an antimatter engine be solved by having the antimatter created directly by the engine itself? (*Hint:* Consider the efficiency of such an engine.)
20. If a spaceship came to Earth from an antimatter universe, what would happen if it landed? Would it even be able to land? Explain.

10

Solids

We say that solids have definite shapes and volumes. It is tempting to think of a solid as an ideal rigid body in which all the particles are fixed distances apart, but the particles in a solid actually move or vibrate around their equilibrium positions.

From atomic theory we know that the particles (atoms, ions, or molecules) are held together by attractive electrical forces. It is something like the particles being held together by springs (Fig. 10.1). The degree of vibration depends on the internal energy of the material. For example, when heat is added, the particles vibrate more. If enough energy is added, the particles may no longer be held in place by the interparticle forces. Macroscopically, the solid would no longer retain its definite shape and would be observed to melt. Hence, the melting-point temperature of a solid depends on the strength of the interparticle forces.

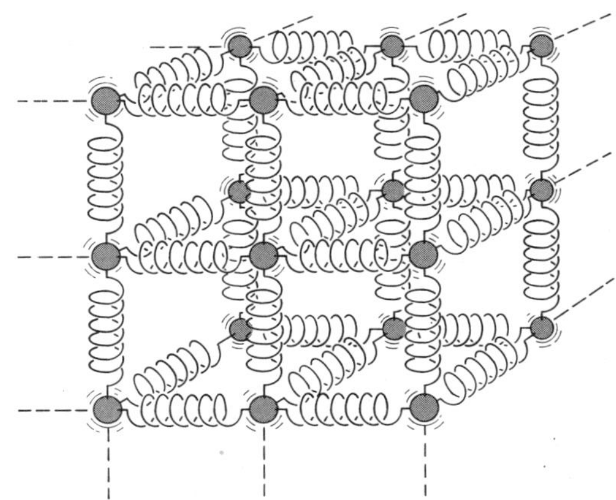

Figure 10.1 Spring model of a solid. The particles of a solid are held together by attractive electrical forces and vibrate as though there were interconnecting springs between the particles.

Solids can be described as either crystalline or amorphous. Crystalline solids have a regular arrangement of particles; amorphous solids have a completely random particle arrangement.

Crystalline Solids

Most substances exist as solids in some characteristic crystalline form. For example, the "spring" solid in Figure 10.1 has a simple cubic crystalline lattice. Each particle is at the corner of a cube. The orderly arrangement of the particles in a solid defines a pattern or lattice structure. You might think of the fixed-chair seating arrangement in a full classroom as being "crystalline," since there is an "orderly" array of student "particles."

The crystalline nature of some solids is evident from their external appearance. The crystalline structure of other solids may not be so evident. Scientists use X-rays as a tool to study the crystalline structure of solids. When a crystalline solid is illuminated with X-rays, distinct patterns appear on photographs (Fig. 10.2). These are called X-ray diffraction patterns, and each pattern is characteristic of a particular crystalline lattice. (Diffraction is discussed in more detail in Chapter 23.)

Figure 10.2 An X-ray diffraction pattern of ice. The pattern is characteristic of the crystalline structure of ice.

Figure 10.3 Cleavage. Some crystalline solids can be split or cleaved along certain planes in the crystalline lattice, as shown here for a cubic lattice of NaCl.

Because of the orderly particle arrangement, some crystalline solids can be split or cleaved along certain planes in the crystal lattice. For example, a solid with a simple cubic lattice can be cleaved along the molecular planes of the crystal (Fig. 10.3). Sodium chloride (table salt) has a simple cubic lattice. If you tried to cleave the crystal along a plane diagonal to the cubic structure, things wouldn't work so well. The crystal would shatter or fracture, and you'd end up with a lot of little broken pieces.

In various crystalline substances, the particles have different lattice structures, which give rise to different properties. For example, diamond is one form of carbon. The other common form of carbon is graphite. Both diamond and graphite consist only of carbon atoms, but in different crystalline structures (Fig. 10.4).

Diamond is very hard and has complicated cleavage planes. Diamond cutters must study the structure before cutting or cleaving a diamond. Not cutting along a cleavage plane results in fracture, which isn't too desirable. The fractured pieces are no longer good for gemstones, but could be used for industrial grinding or etching. (Diamond is one of the hardest substances known.)

In graphite, the carbon atoms are arranged in lattice layers. The atoms within any one layer are strongly bonded to each other. However, the bonding *between* layers is much weaker, and the layers can readily slide past each other. As a result, graphite is soft and slippery

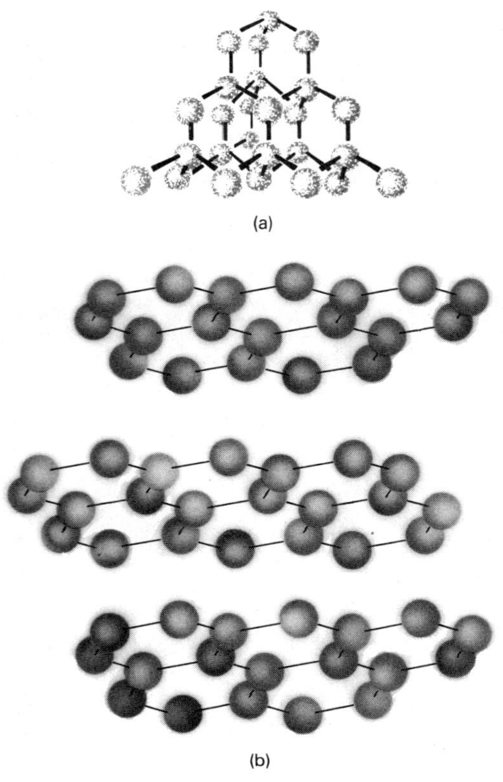

(a)

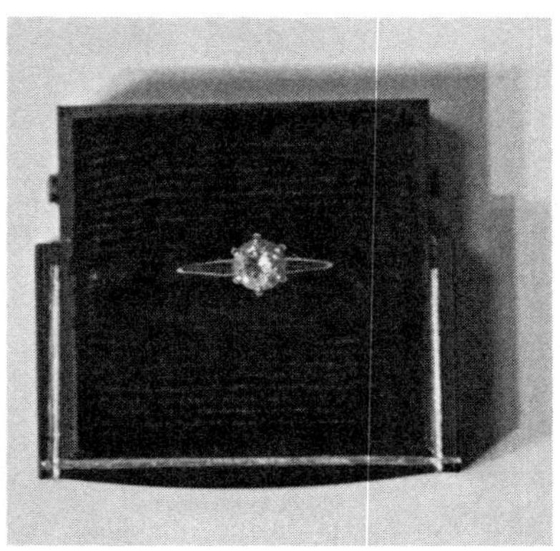

(c)

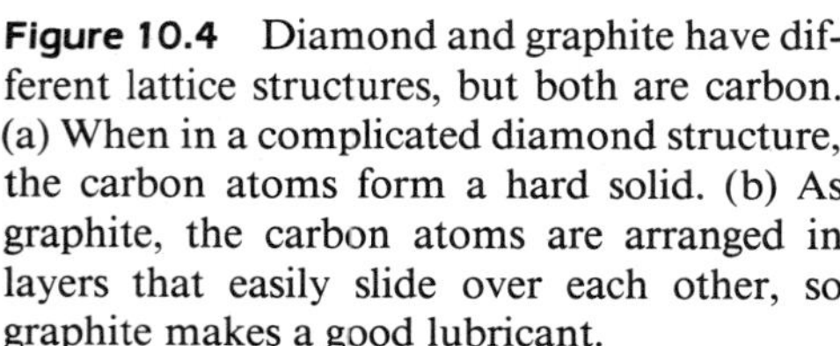

(b)

(d)

Figure 10.4 Diamond and graphite have different lattice structures, but both are carbon. (a) When in a complicated diamond structure, the carbon atoms form a hard solid. (b) As graphite, the carbon atoms are arranged in layers that easily slide over each other, so graphite makes a good lubricant.

and can be used as a lubricant. A common application is in the lubrication of latches on car doors. The "lead" in a lead pencil is actually graphite that has been mixed with other materials to make it more firm. When you write with a pencil, you are spreading graphite out on the paper.

Pure, finely divided graphite powder is called carbon black. It is used as the pigment in India ink and makes up about 30 percent of the weight of a typical auto tire.

Amorphous Solids

Amorphous solids have a random particle arrangement. Unlike the orderly "crystalline" seating arrangement in a classroom, you might think of a crowd sitting on a grassy slope at an outdoor rock concert as an "amorphous" array.

Truly amorphous solids are rare. Many solids that were once thought to be amorphous have been found to have a partially crystalline structure. However, materials like glass and paraffin may be considered amorphous (Fig. 10.5). These materials have the properties of solids but lack the sharply defined melting points of crystalline solids. In many respects, they resemble liquids and flow very, very slowly at normal temperatures. As a result, the bottoms of old window panes (as in cathedrals) are somewhat thicker than at the tops.

Common applications of glass include mirrors and lenses in eyeglasses that many of us wear (Chapter 24). A special eyeglass application is discussed in Special Feature 10.1.

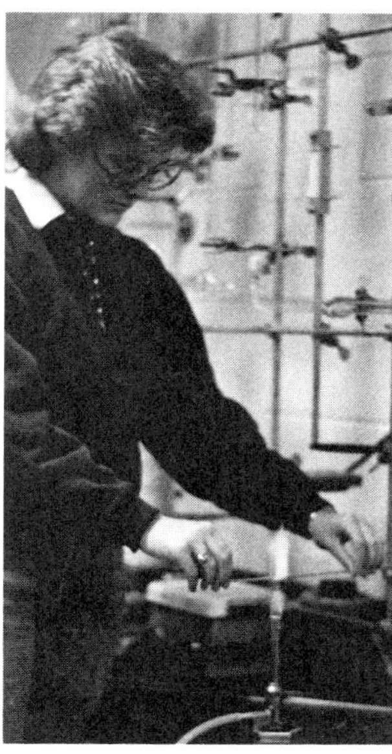

Figure 10.5 Glass is an amorphous substance with a random particle makeup.

Liquid Crystals

In general, when a crystalline substance melts, the resulting liquid no longer has an orderly particle arrangement. However, in certain cases, there is a carry-over of the crystalline properties of the solid state to the liquid state, where the liquid shows some degree of molecular order. This is an in-between state known as the liquid crystal state. A liquid crystal behaves like a liquid, inasmuch as it can flow and assume the shape of its container. However, the optical properties of the liquid crystal depend on the order of its molecules. For example, certain liquid crystals with orderly molecular arrangements are transparent. But this crystalline structure is very weak and can be scrambled or disordered by applied forces (electrical, magnetic, and so on). The disordered liquid crystal then scatters incident light.

These properties are used in LCDs (*L*iquid *C*rystal *D*isplays), which are now quite commonly used in cal-

SPECIAL FEATURE 10.1

Photochromic Glasses

Small amounts of various substances are often added to glass to produce different-colored glasses such as those used in sunglasses. For example, Cr_2O_3 (chromium oxide) gives green glass, and MnO_2 (manganese dioxide) gives a violet glass. This shading or coloring is permanent. However, photochromic glasses that darken and lighten automatically are now commonly available (Fig. 10.6). Photochromic glass darkens on exposure to ultraviolet light (Chapter 21) and becomes lighter again when removed from exposure. Eyeglasses made of photochromic glass darken in sunlight, which contains ultraviolet light, and so act as sunglasses. They become lighter again when the wearer goes indoors.

This special glass contains finely dispersed crystals of silver chloride (AgCl), which is an ionic compound. When exposed to ultraviolet light, some of the silver ions (Ag^+) are converted to atoms of metallic silver (Ag), which is dark and almost black. This process is similar to that used in the exposure of a photographic film that contains silver chloride in the photographic emulsion. However, the silver crystals formed in photochromic glass are many times smaller than those in a photographic emulsion, and in glass the chlorine atoms are not removed by side reactions and do not migrate away from the reaction zone, so the reaction is reversible. When removed from ultraviolet light, the silver atoms and chlorine atoms in photochromic glass recombine to form silver chloride again, and the glass becomes lighter in color.

Figure 10.6 Photochromic eyeglasses darken automatically in bright sunlight.

(a)

(b)

Figure 10.7 Common LCD (liquid crystal display) applications. (See also Figure 7.16.)

culators and digital watch readouts (Fig. 10.7). You can also see LCDs on gas pumps. More will be said about how these displays work in Chapter 23.

Density

The masses of particles and the spacing between them determine the density of a material. Density is a measure of the compactness of the particles in a material (either solid, liquid, or gas). That is, density expresses the amount of mass per unit volume:

$$\text{Density} = \text{mass/volume}$$

Density may be expressed in gram/cubic centimeter (g/cm^3) or kilogram/cubic meter (kg/m^3). Recall from Chapter 1 that the mass of water in a cube 10 cm on a side (volume of 1000 cm^3) is 1 kilogram, or 1000 grams. Hence, water has a density (mass/volume) of 1000 g/1000 cm^3 = 1 g/cm^3. The densities of some substances are listed in Table 10.1.*

The densest solid on Earth is the metal osmium. Another metal, iridium, runs a close second. This is because the atoms in these solids are closely packed. Atoms of gold, lead, and uranium are all more massive than those of osmium and iridium (see periodic table, Chapter 9). However, in the crystalline structures of these substances, the atoms are not as close together as in osmium and iridium, and their densities are smaller —less mass in a given volume.

* Since mass and weight are related ($w = mg$), density can also be expressed in terms of weight, or weight density = weight/volume. Weight densities are not uncommon in the customary British system. For example, the weight density of fresh water is 62.4 lb/ft^3 (or 9812 N/m^3). In the SI system, mass density is commonly used. The SI density units are kg/m^3, but the smaller units of g/cm^3 generally have more practical applications.

Types of Solids

Diamond is much less dense than osmium because its atoms are lighter and are packed less compactly in its lattice. Yet, diamond is harder and melts at a higher

Table 10.1 Densities of Some Substances

	Density		
Material	*g/cm³*	*kg/m³*	
Air	0.00129	1.29	
Aluminum	2.7	2,700	
Brass (70% Cu)	8.5	8,500	
Copper	8.9	8,900	
Glass (general)	2.6	2,600	
Gold	19.3	19,300	
Ice	0.92	920	
Iridium	22.42	22,420	
Iron	7.9	7,900	
Lead	11.5	11,500	
Mercury	13.6	13,600	
Osmium	22.5	22,500	
Water	1.00	1,000	(62.4 lb/ft³)
sea water	1.03	1,030	(64.3 lb/ft³)
Wood (oak, general)	0.70	700	
Zinc	4.3	4,300	

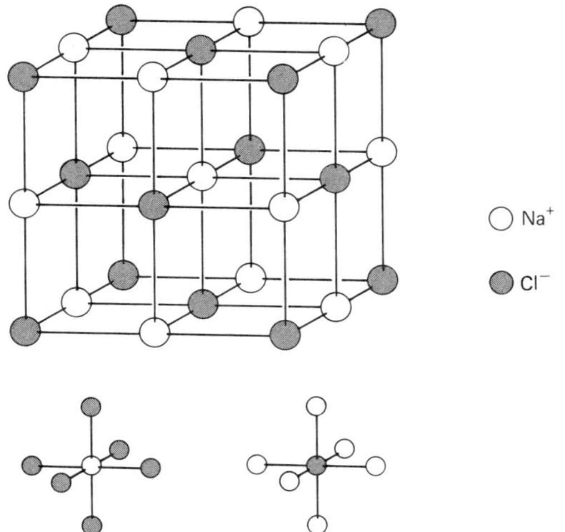

Figure 10.8 NaCl lattice. Each ion (Na^+ or Cl^-) is bonded to six other ions.

temperature. Solid materials have a variety of properties. Steel is strong, lead bends easily, quartz is brittle, and talc is soft.

All of these properties are in some way related to the nature of the particles (atoms, ions, or molecules) of a solid and the forces between the particles.

In ionic solids, sodium chloride for example (NaCl, table salt), oppositely charged ions are held together by the strong electrical forces of ionic bonds (electron transfer). In an ionic solid with a simple cubic lattice, such as NaCl, each ion (Na^+ and Cl^-) is bonded to six other ions (Fig. 10.8). There are ionic lattice structures other than simple cubic, for example, face-centered cubic and body-centered cubic. (Can you imagine how the various ions would be arranged in these structures?)

Because of the strong ionic bonding, ionic solids have relatively high melting points (typically 600–2000°C). The ionic bond must be broken to melt the solid by separating oppositely charged ions from each other. Only at high temperatures do the ions acquire enough kinetic energy for this to happen.

Molecular solids are made up of molecules rather than ions. Intermolecular forces between molecules bond them together as a solid. Generally there is an unequal sharing of electrons within the molecules. This gives rise to regions of electrical charge (polar molecules) and attractive forces (Fig. 10.9). The strength of the intermolecular forces depends on several factors which will not be discussed here. (We'll leave this for an introductory chemistry course.)

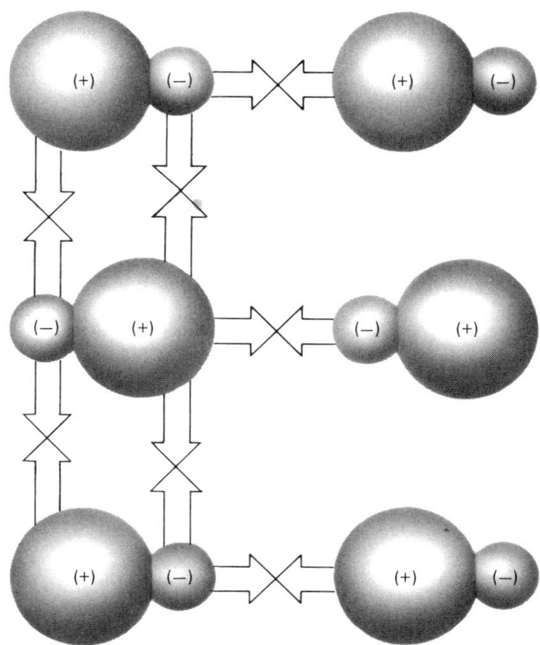

Figure 10.9 Molecular solids are generally held together by intermolecular forces between polar molecules.

The intermolecular forces of molecular solids are not as strong as ionic bonds. As a result, molecular solids in general have low melting points, usually below 300°C. Also, they tend to be volatile and lose molecules as a vapor. Many molecular solids can sublime, or pass directly from a solid to a vapor upon heating. Some sublime at room temperature, for example, solid air fresheners and mothballs. Solid carbon dioxide (dry ice) is another molecular solid that sublimes readily. At room temperature, dry ice "disappears" by vaporizing.

Probably the most common molecular solid is ice. In one sense, ice is very unusual. It is almost unique in having a density less than that of the liquid from which it solidifies (density of ice = 0.918 g/cm^3, density of water = 1.000 g/cm^3). This is why ice floats in water (see Chapter 11). When water freezes to ice, an open hexagonal (six-sided) lattice pattern results (Fig. 10.10). This structure is externally evident from snowflakes. The large empty spaces in the ice structure explain why ice is less dense than water. Also, the variation of the density of water with temperature explains why water freezes from the top down rather than from the bottom up. See Special Feature 10.2.

In some solids, each atom in a structure is covalently bonded to its neighbors. The resulting crystals are com-

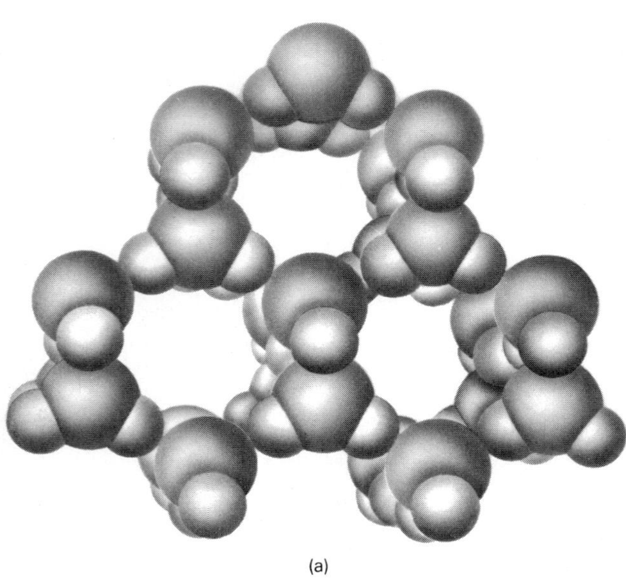

(a)

(b)

Figure 10.10 Ice lattice. (a) When water freezes, an open hexagonal (six-sided) structure results. (b) This six-sided structure is evident in snowflakes.

SPECIAL FEATURE 10.2

Water Density and Freezing at the Top

The densities of most liquids increase as the liquids are cooled and their temperatures reduced. This is the case for water as it is cooled to 4 °C (Fig. 10.11). However, when water is cooled below 4 °C to its freezing point, its density decreases. This implies that the formation of an open lattice structure occurs over the temperature range of 4 °C to 0 °C, rather than taking place solely at the freezing point.

This unique property accounts for the fact that open containers of water freeze at the top first. Most of the cooling takes place at the open surface. As the temperature of the top layer of water is lowered toward 4 °C, this cooler, denser water sinks to the bottom. However, below 4 °C, the water at the top is less dense than the water below and remains at the top, where it freezes when the freezing point is reached.

Think of the environmental effects if this were not the case. Otherwise, lakes, ponds, and rivers would freeze from the bottom up, and much of the aquatic animal and plant life would be destroyed—not to mention what it would do for ice skating.

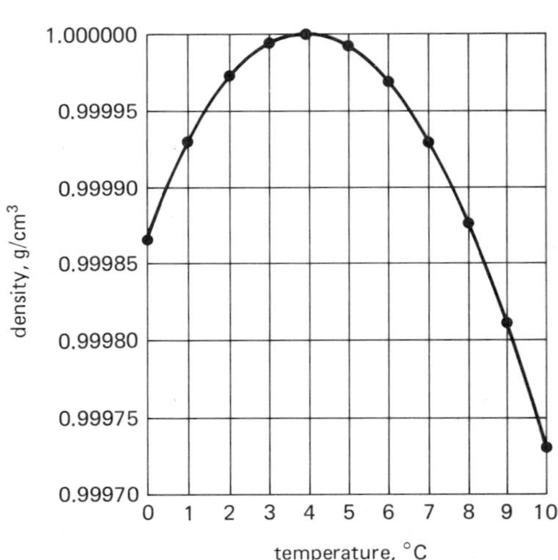

Figure 10.11 Density of water versus temperature. The maximum density of water occurs near 4 °C. Below 4 °C, water is less dense due to the formation of an open hexagonal structure of the molecular units.

pact, interlocking, covalent network structures. Substances of this type are called macromolecular solid. In effect, the entire solid consists of one huge molecule. Common examples of macromolecular solids are diamond and silicon dioxide (SiO_2, the mineral quartz and the main component of sand). Macromolecular solids have relatively high melting points, often above 1000°C.

POLYMERS

Hardly a day goes by when you don't use a dozen or more materials made of synthetic **polymers,** which are macromolecular substances. Many of these materials are plastics of one sort or another. Examples of these include dishes and cups, combs, telephones, pens, false eyelashes, and fibers in clothing. A **plastic** is a polymer substance that will flow under heat and pressure and hence can be molded into various shapes. All plastics are polymers, but not all polymers are plastics.

Unlike macromolecular solids, which are in effect one large molecule, polymer substances are made up of giant or macromolecules. These are formed by the combination of small molecular units called **monomers** (Greek *mono,* meaning "one"). The monomer units form giant molecules, which make up a polymer (Greek *poly,* "many" and *meros,* "parts"). An illustration of a linear polyethylene molecule is shown in Figure 10.12. A typical polymer molecule may contain a chain of monomers several thousand units long.

There are many natural polymers produced by plants and animals that are essential to life. Many of these fall into the class of organic polymers known as proteins. Natural polymers are common in daily life, for example, wood, paper, and clothes made of cotton, wool, silk, and leather.

Some of our most useful synthetic polymers have resulted from copying giant molecules found in nature. For example, synthetic rubber is copied from natural latex rubber. Some synthetic polymers, however, do not have duplicates in nature, such as nylon, Dacron (polyester), and Teflon.

It is interesting to note that a typical radial auto tire is only 39 percent rubber, of which about 70 percent is synthetic rubber. Only in special-use tires that get very hot does the percentage of natural rubber increase because of its superior heat resistance. The rubber in a high-performance aircraft tire, for example, is almost all natural for this reason. However, the tread of a regular aircraft tire is about 50 percent synthetic rubber because of its better resistance to wear and abrasion. This characteristic is mainly responsible for synthetic rubber's wide use in auto tires.

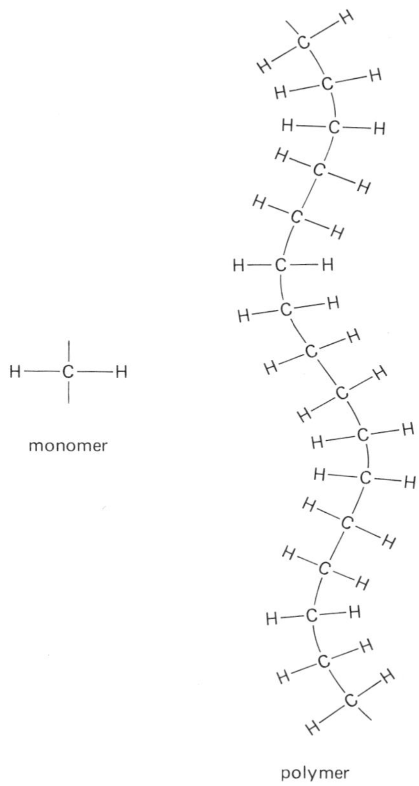

Figure 10.12 A polymer molecule is made up of a series of monomer units.

The properties of a polymer depend on its molecular structures. Polymers in which the monomer units are arranged in nearly parallel chains form strong and flexible fibers. If the chains are tangled or linked together, the polymer may form a strong film or a rigid solid. The melting points of polymers vary over a range of several hundred degrees.

Additives and chemical processes are used to give polymers desired properties. For example, many plastics are used where they are exposed to sunlight. The ultraviolet (uv) portion of the sunlight has sufficient energy to break the chemical bonds found in polymers. If enough bonds are broken, the polymer becomes brittle and will break under a force (Fig. 10.13). Compounds called uv stabilizers, which absorb uv light, can be added to reduce the degradation.

Most of the raw materials for synthetic polymers come from petroleum and some from coal. Only a small amount of our petroleum consumption goes into making polymer products. See Special Feature 10.3.

Figure 10.13 The ultraviolet portion of sunlight has sufficient energy to break the chemical bonds in polymers. If enough bonds are broken, the polymer becomes brittle and will break under force.

SPECIAL FEATURE 10.3

Petrochemicals and Plastics*

Many people are unaware of how many of the products we use every day come from petroleum and natural gas. In the United States each family of four uses more than *two tons* of petroleum products annually. That's almost 1200 pounds of chemicals each year for every man, woman, and child in the United States—a staggering total of 225 billion pounds of chemicals from petroleum, and to a lesser extent, natural gas.

Of the vast amount of petroleum and natural gas we consume, over 90 percent is burned as fuels. Only about 5.5 percent is used for the manufacture of petrochemicals by the chemical industry. These petrochemicals vary widely in their functions and include such products as drugs, detergents, rubber, paints, fertilizers, dyes, perfumes, explosives, food preservatives, artificial sweeteners, and agricultural chemicals. Finally, about 1.5 percent of the oil and natural gas is used as raw material for plastics. This small percentage translates into the production of billions of pounds of polymers that yield many different and useful products.

In the post-World War II years, the United States was flooded with domestic and imported items of extremely low cost, low quality, and limited lifetime. This led to the image of "cheap plastics" with low durability. Today, however, the image of plastics has changed. Plastics perform an extremely broad range of functions, from heart valves and artificial kidneys to ski boots, nonstick surfaces, super glues, and spacecraft parts, and they compete with natural products in durability. No other materials except plastics could perform all these different functions.

Plastics are replacing more and more parts of your car. The use of 1 pound of plastic can replace an average of 3.5 pounds of metal in an automobile. An automobile with 400 pounds of plastic substituted for metal will weigh about 1000 pounds less, which increases its gas mileage by about 3 miles per gallon. The fuel savings are estimated to be about 160 million barrels of oil annually. That's more than the total amount used by the chemical industry as raw materials to make the polymers. As another example, synthetic polymer fibers are commonly used in fabrics, for both economical and practical reasons. If the world's synthetic fibers were replaced by cotton, this would require an additional 40 million acres of farm land.

Certainly the use of polymer plastics will increase. One can expect to find more applications in home construction and furniture because of the unlimited design freedom of plastics. Plastics will be used more in drink containers and food packaging. The 700 billion gallons of liquids consumed each year in the United States will find their way to the consumer more and more in plastic bottles. (Beer and champagne in plastic bottles?) Diseased or malfunctioning parts of the body will be replaced by specialized plastic components to a greater degree. We are indeed becoming a plastic society.

* Courtesy of Dr. Peter A. Vahjen, Professor of Chemistry, Lander College.

METALLIC SOLIDS

Of the known elements, about 80 can be classified as metals. Of the four major types of solids, only metals are good electrical conductors in the solid state. All metals are solids at 25 °C (77 °F), with the exception of mercury.

The metallic lattice consists of positive ions surrounded by an electron "gas" or "sea." The electrons are donated by the atoms of the metal and belong to the crystal as a whole. The positive ions are anchored in position like bell buoys in a mobile "sea" of electrons. The electrons are free to migrate or wander throughout the lattice, somewhat like gas molecules in a closed container.

This simple model of metallic bonding explains why metals are good electrical conductors. However, metallic solids vary widely in hardness and melting-point temperature.

Alloys

While on the topic of metals, we should mention alloys. **Alloys** are blends of two or more metallic elements or metallic and nonmetallic elements that give materials with properties different from those of the individual elements. For example, bronze is stronger and harder than either of its constituents, copper and tin. Some common alloys are listed in Table 10.2.

Alloys have been known for a long time. The rise of some civilizations is credited to their knowledge of and work with metal alloys (for example, those of the Bronze Age). We are familiar with alloys in the form of coins. Light aluminum and magnesium alloys are used in aircraft. However, the strengths of these alloys decrease rapidly between 300 and 400 °C. Air friction on very-high-speed jet aircraft necessitates the use of titanium alloys. (The melting point of titanium is 1660 °C, as compared with 660 °C and 650 °C for aluminum and magnesium, respectively.)

Table 10.2 Some Common Alloys

Alloy	*Constituents (Percentages by Weight)*
Brass	Copper (60–85%) and zinc
Bronze	Copper and tin (5–10%)
Cast iron	Iron and carbon (2–5%)
Steel	Iron and carbon (<1.3%)
Stainless steel	Steel, chromium (18%), and nickel (8%)

Alloys may be thought of as solid solutions. They are generally prepared by melting and mixing together the constituent metals. Not all metals will form alloys, however. The fitting together of different metallic atoms (ions) in a common lattice is possible only if their sizes do not differ greatly. Some alloys are difficult to make because the heavier element tends to settle out in the melt before solidification. An example is aluminum antimonide, which is an alloy of aluminum and antimony. Antimony is more than three times denser and tends to settle faster than aluminum when mixed in the molten state.

One attribute of aluminum antimonide is that it may be 30 to 50 percent more efficient than silicon in solar cells and computer-circuit chips. The settling effect causes a lack of uniformity in the alloy, which severely limits its use in these applications. However, this effect does not take place in the "microgravity" of space (see Chapter 5). Experiments in Skylab, Apollo-Soyuz, and Space Shuttle orbital missions have achieved largely uniform mixtures in alloys that are used extensively in computers and other electronic products.

Mechanical Properties of Solids

A wide variety of materials is used in various applications. In many cases, the use of a particular material depends on its mechanical properties. In some instances it is desirable for a material to withstand an applied force or torque, for example, as on the axle of an automobile. In other instances, a material that can be stretched or easily deformed is required, for example, rubber in a rubber band or automobile tire.

The mechanical properties of solids depend on the composition and internal forces of the materials, as you might expect. The internal processes that occur in materials when they are subjected to forces are complex. However, these processes are manifested as changes in a material's external properties, which are described by common terms.

ELASTICITY

Solids have definite shapes because the forces between the particles resist changes. However, some solids, such as rubber, are easily deformed but return fairly well to their original shapes when external forces are removed. In such cases, the interparticle forces allow for stretching and twisting. When the distorting force is removed,

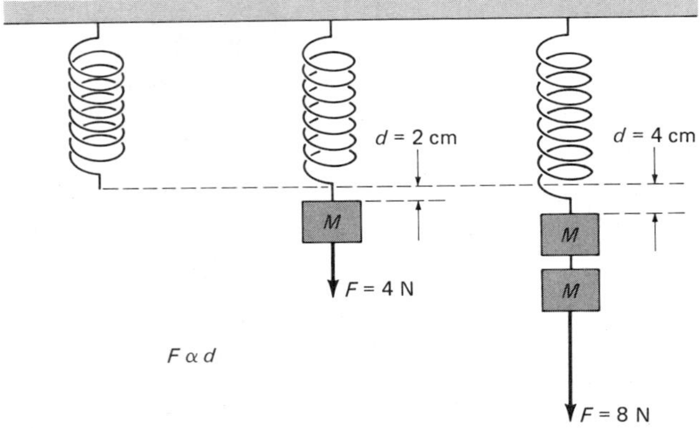

Figure 10.14 Hooke's law, $F \propto d$. If a force of 4 N stretches a spring 2 cm, an 8 N force will stretch it 4 cm.

Figure 10.15 Not a stress test. Winds caused this bridge to oscillate and be stressed until it collapsed. (See Chapter 16, Figure 16.19.)

the attractive forces between the particles bring them back to their original positions and the solid as a whole back to its original shape. This property is called **elasticity.**

All materials are elastic to some extent, even steel. However, some materials are easily deformed and have little tendency to return to their original shapes. Such materials as putty and clay are generally said to be inelastic.

To better understand the terms *elastic* and *inelastic,* consider an "elastic" spring, as illustrated in Figure 10.14. When force (weight) is applied to a spring, it is stretched or deformed. In general, it is found that the stretch distance d is proportional to the applied force F, or $F \propto d$. For example, if a force of 4 N stretches a spring 2 cm, an additional force of 4 N will stretch it another 2 cm (total force 8 N and total stretch distance 4 cm). This relationship is known as Hooke's law.† Hooke's law also applies to a compressive force and compressed spring distance.

If the force is not too great, the spring will return to its original shape and length when the weights are removed. However, if too much force is applied, the elastic limit of the spring will be exceeded. When this happens, the spring will be stretched or bent out of shape. The deformation is then permanent, and the spring will not return to its original shape.

A straight steel wire also has elasticity, but this is not generally observed visually. Large forces and sensitive detectors are usually required to observe this behavior. Occasionally this elastic behavior can be observed for what are usually considered inelastic materials, as shown in Figure 10.15. This is the famous Tacoma Narrows bridge, which collapsed in 1940. More will be said about what caused the vibrations in the bridge in Chapter 16.

If enough force is applied to a spring or a steel wire (or any material), it will break or fracture. This is a case of the external applied force exceeding the internal particle forces, causing the bonds to break. In applying a force to a solid, it is common to speak of **stress** and the resulting **strain** or deformational change in an object's dimensions.

After the elastic limit is reached, metals and most polymers exhibit what is known as **plasticity** or **plastic deformation.** Unlike elastic deformation, plastic deformation is permanent; the material does not recover when the force or stress is removed. For example, a dent in an automobile fender is the result of a force that produces a plastic deformation. Plastic deformation was also required in forming the fender from sheet metal, but elastic deformations are usually desirable thereafter.

If the force is continually increased, the material finally breaks or fractures. Some materials, such as ceramics, fracture without noticeable plastic deformation.

QUESTION: When a piece of wire, such as a straightened paper clip, is bent, it shows permanent plastic deformation. Yet when a bending force of the same magnitude is applied back and forth, the wire eventually breaks. Why is this?

† Not all elastic materials follow Hooke's law (named in honor of Robert Hooke, an English physicist who first described the relationship). A rubber band is a good example.

ANSWER: Some metals and polymers, especially crystalline polymers, strengthen as a result of plastic deformation. This is called work-hardening or stress-hardening. When a wire is bent or flexed (stressed) repeatedly, it becomes harder and more brittle as a result of work-hardening in the bending area. The more brittle a material, the less stress it can stand before fracturing.

When metals are worked, a metal may have to be heated and cooled slowly (annealed) to remove the excessive hardening caused by work-hardening.

SUMMARY OF KEY TERMS

Crystalline solid a solid with a regular or orderly particle arrangement in a lattice structure.

Lattice the geometric pattern or arrangement of particles in a crystalline solid.

Amorphous solid a solid with random particle arrangement.

Liquid crystal an in-between state in which a substance with liquid properties shows some degree of molecular order as in a crystalline solid.

Density a measure of the compactness of the particles in a material, which is expressed as mass per unit volume: density = mass/volume.

Ionic solid a solid consisting of oppositely charged ions, e.g., NaCl.

Molecular solid a solid consisting of molecules with intermolecular bonding, e.g., ice.

Macromolecular solid a solid consisting of covalently bonded atoms, such that in effect the solid consists of one huge molecule.

Polymer a substance consisting of giant or macromolecules made up of repeating monomer units.

Plastic a polymer substance that will flow under heat and pressure and hence can be molded into various shapes. All plastics are polymers, but not all polymers are plastics.

Metallic solid a solid consisting of a lattice of positive ions surrounded by a "sea" of electrons, which gives rise to good electrical conduction.

Alloy a blend of two or more metallic elements or metallic and nonmetallic elements.

Elasticity the property of a solid whereby it returns to its original shape after a distorting force or stress is removed.

Plasticity the property of a solid characterized by permanent deformation when a force or stress is removed.

EXERCISES

1. Describe how an amorphous "spring" solid might look.
2. The simple cubic lattice of two different particles (ions) is illustrated in Figure 10.8. Two other types of cubic lattices are (a) face-centered cubic and (b) body-centered cubic. Using light and darkened circles to indicate different particles, sketch how these lattices would look. (*Hint:* The names are descriptive of the lattices.)
3. Wood is a cellulose polymer. Explain why wood can be split "with the grain" but has to be sawed into pieces going "across the grain."
4. Diamond and graphite are both carbon. Why are the properties of two carbon substances so different?
5. The density of uranium is 18.7 g/cm^3, and the density of gold is 19.3 g/cm^3. Yet an atom of uranium is more massive (heavier) than an atom of gold (238 vs. 197; see periodic table, Chapter 9). Why does uranium have a smaller density?
6. When you squeeze and compress something, such as a sponge or a balloon, do you change its density? Explain.
7. The density of liquid mercury is 13.6 g/cm^3 and the density of solid mercury is 14.2 g/cm^3 (freezing point of $-39\,^\circ$C). What does this tell you?
8. Explain why sea water has a greater density (1.03 g/cm^3 or 64.3 lb/ft^3) than fresh water (1.00 g/cm^3 or 62.4 lb/ft^3).
9. Density = mass/volume, so mass = density $\times$ volume. (a) What are the masses of gold and lead bricks with dimensions of 5 cm $\times$ 10 cm $\times$ 20 cm? (b) How much would these bricks weigh in pounds?
10. Why isn't the density of water in SI units 1.000 kg/m^3?
11. Other than hardness, what is a basic difference between "hard" wood and "soft" wood?

‡12. To help understand the empty space in a lattice structure, consider the two-dimensional close-packed circles in Figure 10.16 and compute the percentage of empty space in the square not occupied by the circles. *Hint:* (a) calculate the area of the square, (b) calculate the total area of the circles (area of a circle = πr^2), (c) calculate

‡ Exercises 12 and 29 from Turk, J., and A. Turk, *Physical Science*, Second Edition, Saunders College Publishing, Philadelphia, 1981.

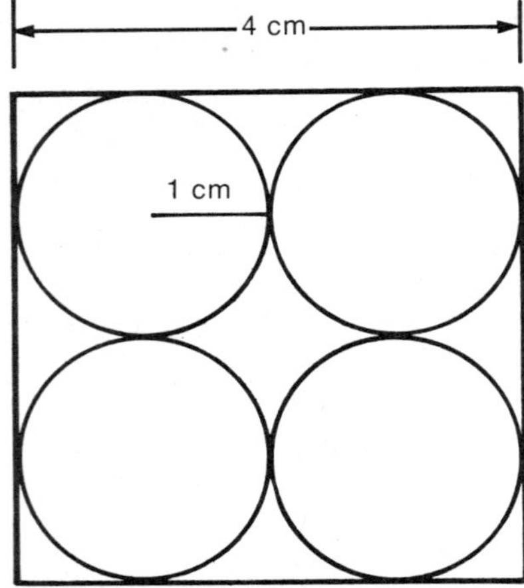

Figure 10.16 See Exercise 12.

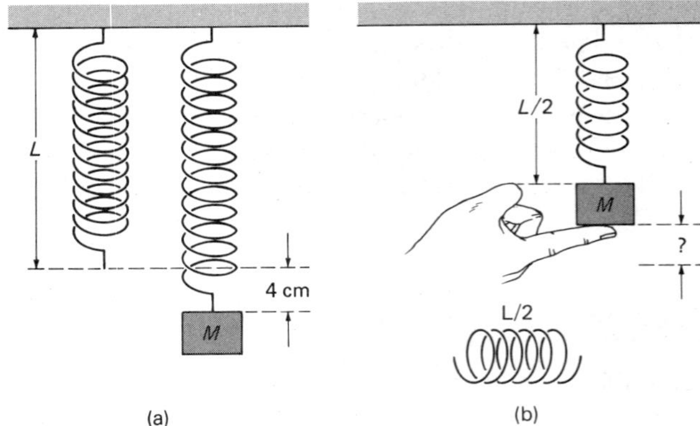

Figure 10.17 See Exercise 24.

the area of empty space = (a) − (b), and find percentage.

13. Compare the similarities and differences between ionic and macromolecular solids.
14. Compare the general melting-point temperatures of various solids, and explain the differences.
15. Why are moth balls and air fresheners volatile solids? Is there a practical advantage to this? Explain.
16. Ice cubes in a freezer or freezer tray are observed to be smaller after having been in the freezer for a long time. Explain why.
17. Why is ice less dense than water? Water below 4°C is less dense than water at 4°C. What does this imply?
18. Many synthetic polymers are ultraviolet degradable but are not biodegradable (not broken down by bacterial action, as are paper and wood). What does this mean in terms of environmental litter, for example, plastic soda bottles and carry-out soft drink lids? Discuss how this might be helped by the addition of uv *un*stabilizers to these products.
19. Suppose all the synthetic polymers suddenly disappeared from your classroom. How would things (and you) look?
20. Why are some metals better electrical conductors than others?
21. Magnesium and sodium are both metals, but magnesium has a much higher melting point than sodium (650°C vs. 98°C). Explain why.
22. Is the quarter (25¢ piece) made with "sandwiched" copper an alloy? Explain.
23. Explain why the density of brass is less than the density of copper.
24. A long spring is stretched 4 cm when a mass is suspended from it (Fig. 10.17). If the spring is cut in half and the same mass is suspended from one of the halves, how far would the mass descend?
25. Two identical springs stretch the same length (2 cm) when a 0.5-kg mass is suspended from either of them. If the springs are hooked together and the mass is suspended from the lower spring, what will be the stretch of each spring?
26. A wire is not very elastic, but if the wire is formed into a coil spring, there is a great deal of elasticity. What does this tell you about the interparticle forces of the wire?
27. Why do hammers generally not have metal handles, in particular, sledge hammers? (Some carpenter's hammers do have metal handles, but with rubber grips.) Why do we use rubber mallets in some instances of hammering?
28. Automobile tires were once made of solid rubber. Why did we switch to inflated tires?
‡29. Enter each of the following four solids into the appropriate boxes shown in Figure 10.18: lead, aluminum, diamond, and iridium (a metal used for penpoints). If you are not sure of the properties of these substances, make your best guess, or ask someone who works in a machine shop. What factors account for the hardness and densities of solids?
30. Compare the advantages and disadvantages of ceramic and plastic (dinner) plates in terms of mechanical properties.

	harder than iron	softer than iron
denser than iron		
less dense than iron		

Figure 10.18 See Exercise 29.

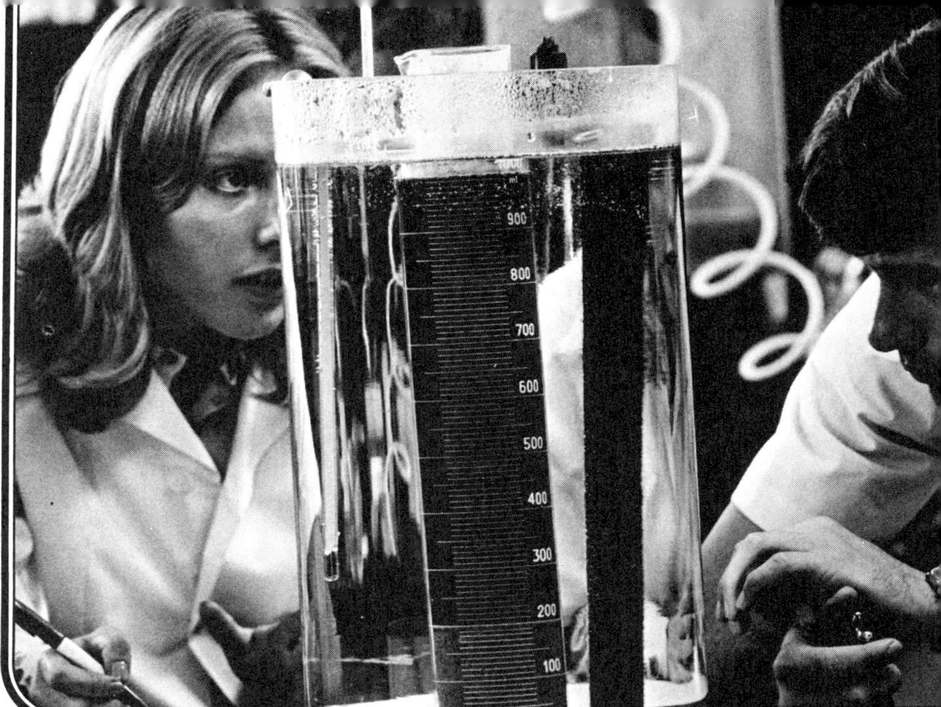

11

Liquids

Fluids

Liquids have definite volumes, but no definite shapes. The molecules of a liquid can move around, allowing the liquid to take on the shape of its container. We often use the term "fluid" to mean liquid. A **fluid** is a substance that can flow. Hence, gases as well as liquids are fluids (Fig. 11.1). Because of this common property of being able to flow, gases and liquids have many common properties, and they are often treated together in a chapter on fluids. However, there are many important differences, so we will treat them separately in different chapters (see Chapter 12, Gases).

A liquid can flow, but its molecules stay about the same distance apart. This means that liquids are essentially (but not perfectly) *incompressible.* As a result, a liquid keeps the same volume except for minute changes. The molecules of a gas, on the other hand, do not stay the same distance apart. This allows a gas to fill up the complete volume of any container, so a gas has no definite shape or volume. A gas is also *compressible,* with its volume being affected by pressures upon it. For example, air (a mixture of gases) is compressible, but water is relatively incompressible.

A basic mechanical difference between solids and fluids is that a solid can support a shear force or stress. For example, a solid can support the flat side or even the cutting edge of a knife blade without being cut or sheared if the force on the knife is not too great. But a liquid such as water (or a gas) could not support the shearing stress of the knife and would flow. A fluid flows under the application of a shear stress instead of being elastically deformed.

Keep in mind that although gases and liquids are different, many of the (fluid) principles discussed in this chapter apply to both.

Figure 11.1 A fluid is a substance that can flow. Liquids and gases are fluids, but solids are not.

Pressure

It is easy to apply a force to a solid. In fact, we didn't think much about it in previous chapters. A vector force arrow could easily represent a force applied with the flat of the hand, a finger, or even a pencil point. But since a liquid (or a gas) cannot support a shear stress, an object supplying a force might cut or plow right through it.

Hence, we have to think of applying a force to a liquid in a different way—in terms of pressure. Pressure takes into account not only the force but also the *area* of application. Pressure is defined as the force divided by area, or force per unit area:

$$\text{Pressure} = \text{force/area}$$

$$p = F/A$$

The SI unit of pressure is newton/per square meter (N/m^2), which is called a pascal (Pa) in honor of the French scientist Blaise Pascal (1623–1662). In the British system, the common unit is pound per square inch (lb/in^2). For example, we might inflate automobile tires to a pressure of 30 lb/in^2 (30 "pounds" pressure), or about 200 kPa (kilopascals).

To get a feel for the difference between force and pressure, try holding a pin as shown in Figure 11.2. If you push on the pin, the thumb and finger experience the same force (Newton's laws), but the finger holding the pin point hurts! This is because there is a smaller area of contact, or greater pressure. The force is the same, but the point concentrates it over a much smaller area:

$$\underset{\text{Head end}}{\mathrm{p} = \mathrm{F}/\mathbf{A}} \qquad \underset{\text{Point end}}{\mathbf{p} = \mathrm{F}/A}$$

In order to apply pressure on a liquid, we commonly use a surface area of its container. For example, when you squeeze a liquid-filled plastic bottle, you apply pressure. Another common method is to put the liquid in a cylinder and apply a force to a piston or plunger, which puts pressure on the liquid. A medical syringe used to give shots is a good example.

QUESTION: Why do spike heels on women's shoes dent tile floors?

ANSWER: This was a problem in the 1960's when spike heels were popular. A woman weighing 500 N (112 lb) applies the same weight force to the floor when wearing regular shoes or even when barefoot. The difference is pressure.

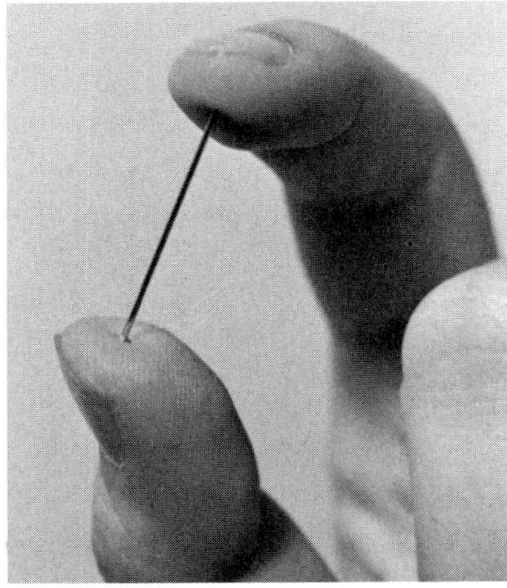

Figure 11.2 Pressure versus force. When holding a pin as shown, the thumb and finger experience equal forces (Newton's third law), but the pressure on the finger with the smaller area of the pin point is greater, and it can hurt. (Ouch!)

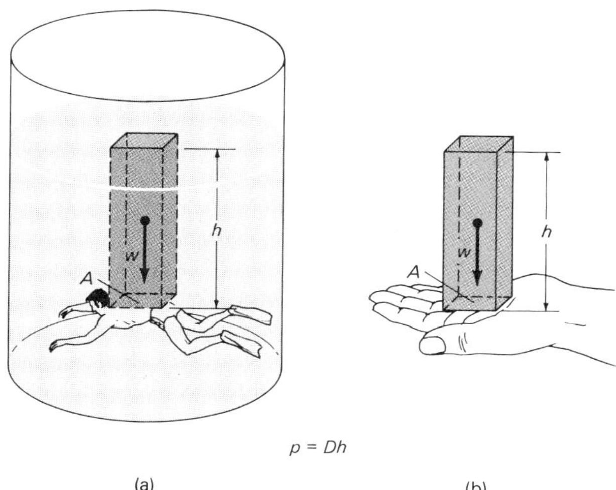

Figure 11.3 Pressure in a liquid. The pressure at any depth h in a liquid due to the weight of the column of liquid above is $p = Dh$, where D is the weight density of the liquid ($D = w/V$).

When spike heels are worn, a large part of the person's weight is distributed over the small heel area, and there is a greater pressure. This can cause the floor to be dented (plastic deformation).

PRESSURE IN A LIQUID

Inside a liquid without any externally applied pressure there is pressure due to the weight of the liquid itself. Most of us have experienced this when swimming underwater. The deeper one goes, the more pressure there is from the weight of the water above. For an open container of liquid, there is an additional pressure due to the weight of the atmospheric gases above the container. We'll ignore this for now and consider the effects of atmospheric pressure in the next chapter.

Consider a column of liquid, as shown in Figure 11.3. The pressure in the liquid at the bottom of the column is the weight force of the liquid column divided by its area. This can conveniently be expressed in terms of the weight density D of the liquid and the height or depth h of the column (sometimes called the **pressure-depth relationship**):*

$$p = Dh$$

Pressure = weight density × depth

The pressure on the bottom of the container due to the liquid may be found by using the same expression, taking h to be the total depth of the liquid. This is like considering the whole liquid as a column. The liquid pressure on the bottom of a container does not depend on the shape of the container, only on the depth of the liquid (and its weight density).

Pascal's Principle

When you apply a force to one end of a rigid rod with the other end against something, the same pressure (force/area) is transmitted by the rod to the other end. This is because the rod is incompressible. The same is true for a liquid, since liquids are practically incompressible. However, there is an important difference. In a confined liquid, the pressure is transmitted undiminished to all the walls of its containers.

This fact was first stated by Blaise Pascal, after whom the SI unit of pressure is named, and is known as **Pascal's principle:**

> Pressure applied to an enclosed liquid (fluid†) is transmitted undiminished to every other part of the liquid and to the walls of its container.

As a demonstration of Pascal's principle, suppose some small holes are poked in a water-filled balloon—small enough so the water would not run out. Then, if pressure were applied to the balloon, the water would be forced out of all the holes. This demonstrates that the pressure is transmitted (equally) throughout the liquid.

A common application of Pascal's principle is shown in Figure 11.4. When you push on the brake pedal of your car, the brake fluid transmits the pressure to the brake cylinders that apply the brake shoes to the drums—then friction slows you down.

Pascal's principle is used in the hydraulic press, which is a simple machine. Recall from Chapter 6 that a machine is basically a force multiplier that gives a mechanical advantage. Consider the systems shown in

* Weight density = weight/volume, $D = w/V$. The weight is given by $w = DV$. The volume of the column is $V = Ah$ (area times height). Hence, $p = F/A = w/A = DV/A = DAh/A = Dh$, or $p = Dh$.

† Pascal's principle holds for a gas, but the pressure in a gas is not transmitted as quickly as in a liquid.

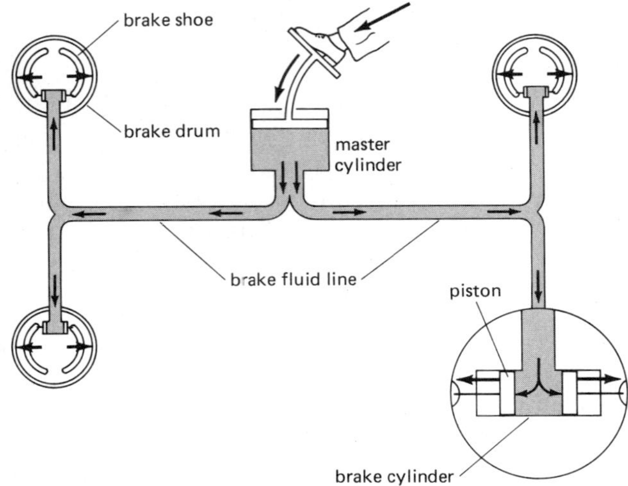

Figure 11.4 Pascal's principle. A practical application in the hydraulic (fluid) brake system of an automobile.

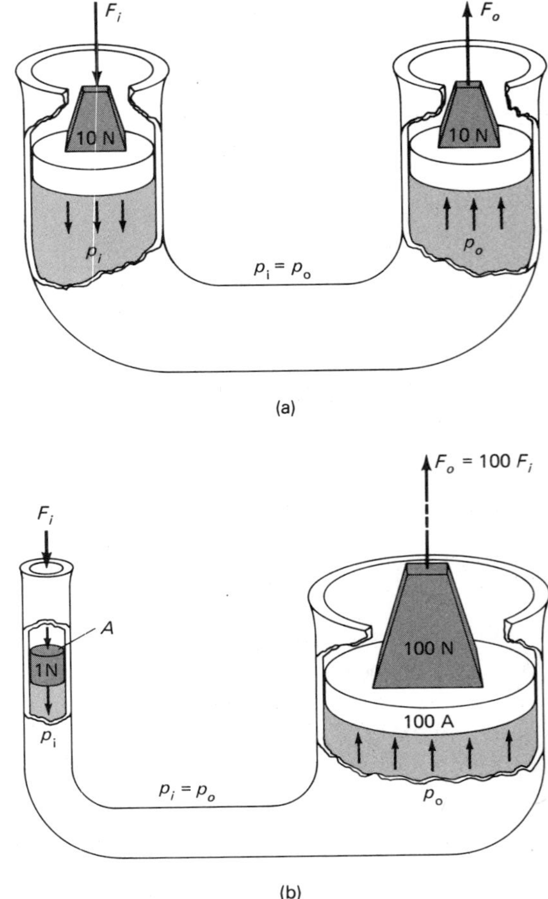

Figure 11.5 Mechanical advantage and Pascal's principle. (a) If the areas of the pistons are equal, there is no force multiplication, or a mechanical advantage of one. (b) The pressure of the small piston is transmitted to the larger one (Pascal's principle), and with the large piston's area 100 times greater, the output force is 100 times greater than the input force.

Figure 11.5. With equal forces (weights) on pistons of equal area, the pressure of one piston is transmitted undiminished to the other by the liquid (Pascal's principle). Since the areas are equal, the input force F_i is equal to the output force F_o, and there is no force multiplication.

Suppose the areas of the pistons aren't equal. Even so, the pressures on them are still equal. That is, the pressure on the input piston is the same as the pressure on the output piston, $p_i = F/A$ and $p_o = 100\ F/100\ A = F/A$. Hence, there is a force multiplication of 100, or the output force is 100 times the input force, $F_o = 100\ F_i$. This means that an input force of 9.8 N can produce an output force of 980 N. See an application of hydraulic force multiplication in Special Feature 11.1.

SPECIAL FEATURE 11.1

The Hydraulic Garage Lift

A common application of the hydraulic press is in the garage lift used to raise heavy automobiles and trucks (Fig. 11.6). Compressed air on the surface of an oil reservoir transmits pressure to the lift piston. The compressed air pressure doesn't have to be too great. Suppose it were 200 kPa, which is about the pressure in your automobile tires (30 lb/in^2). The pressure on the liquid in the reservoir is transmitted to the lift piston. If the diameter of the piston is 0.30 m, then its area is 0.07 m^2, and the lift force would be 14,000 N ($F = pA$), or about 3150 lb. So the next time you see a car on a lift, you'll know that Pascal's principle is in operation.

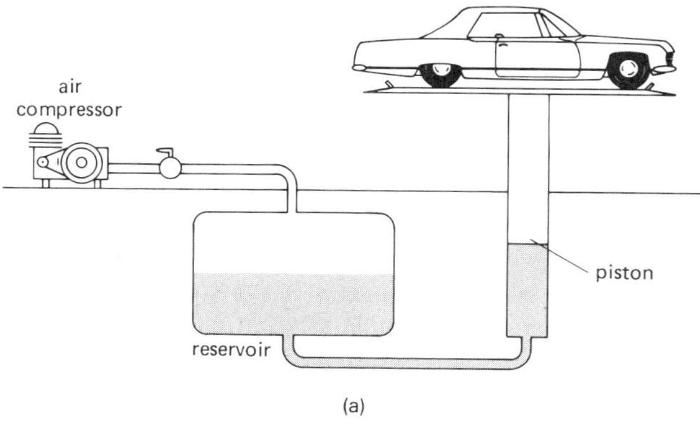

Figure 11.6 An application of Pascal's principle in the form of a hydraulic lift. The air pressure on the surface of the oil in the reservoir is transmitted undiminished to the lift piston, giving a large output force.

Buoyancy and Archimedes' Principle

When lying in a tub of water or standing neck-deep in a swimming pool, you have no doubt noticed that your submerged arms (or legs) feel lighter. What causes this? Certainly your arms have their regular mass and weight. When you pick up something, you have to exert an upward force slightly greater than the object's weight. If someone helped you and supplied part of the force without your knowing it, you might think the object to be "lighter." So, there must be an upward "helping" force with you in the bathtub.

This phenomenon is called **buoyancy,** and the upward-acting force is called the **buoyant force.** To help you understand this buoyant force, consider an imaginary "block" of liquid, as shown in Figure 11.7. The water block exerts a downward force equal to its weight on the liquid surface at the bottom of the block, and the liquid exerts an equal upward force to keep it there. If the water block is removed and replaced with an object of the same weight and rectangular base, as shown in the figure, it would float in place.

Hence, the upward "buoyant" force must be equal to the weight of the block of water removed. This is the same as the volume of water the object would displace when normally placed in the water. The upward buoyant force then depends only on the weight of the volume of liquid an object displaces. This idea was expressed by the Greek scientist Archimedes (287–212 B.C.) and is known as **Archimedes' principle:**

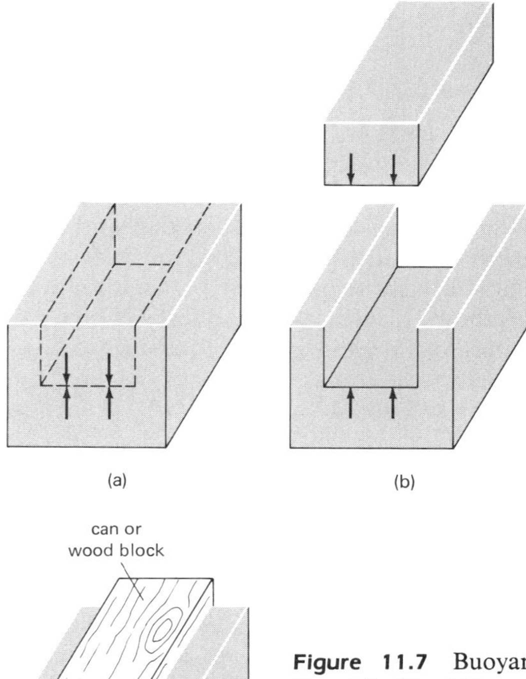

Figure 11.7 Buoyancy. If a "block" of liquid is removed and replaced by a material block of the same weight, it would float because of the upward buoyant force, which must be equal to the weight of the liquid displaced (the removed imaginary water "block" in this case).

> An object in a liquid (fluid‡) is buoyed up by a force equal to the weight of the volume of liquid it displaces.

Archimedes' principle can be easily demonstrated, as shown in Figure 11.8(a). When a heavy object suspended on a scale is submerged in a liquid, it weighs less and displaces a volume of liquid equal to its own volume. The weight of the displaced liquid is equal to the decrease in the measured weight of the submerged object, or the buoyant force.

You may be wondering what causes the buoyant force. Think in terms of the pressure in a liquid or fluid. As illustrated in Figure 11.8(b), the liquid pressure on the bottom of the block (via Pascal's principle) is greater than that on the top of the block, since the bottom portion is at a greater depth. The pressures on the sides of the block are equal and opposite, so they cancel each other. Hence, there is a net upward pressure or net upward (buoyant) force on the object.

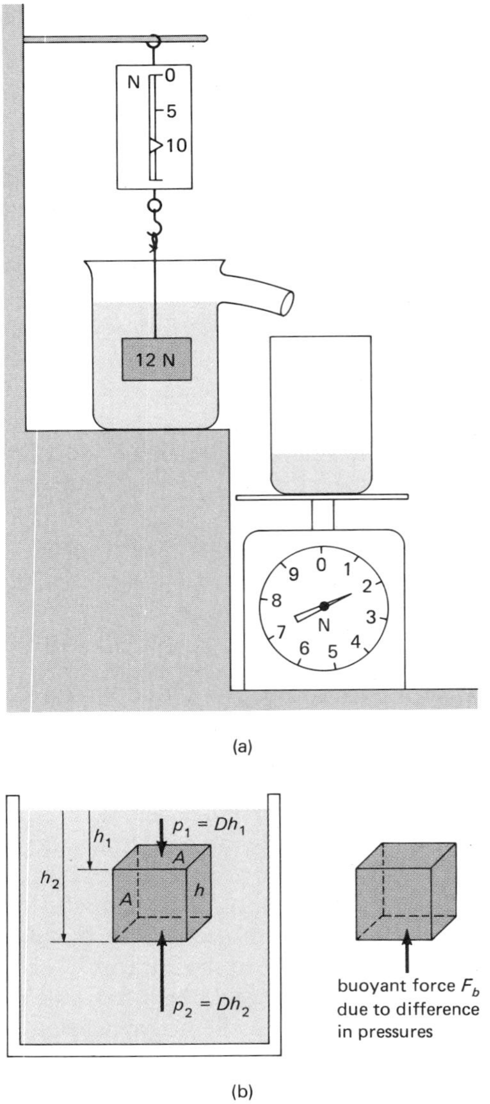

(a)

$p_1 = Dh_1$ h_1 h_2 A h A $p_2 = Dh_2$

buoyant force F_b due to difference in pressures

(b)

Figure 11.8 Archimedes' principle. (a) A demonstration of the principle. The weight of the volume of liquid displaced is equal to the buoyant force, which causes the submerged object to weigh less. (b) The source of the buoyant force is the difference in pressures—more pressure on the bottom of the block than on the top.

FLOATING AND SINKING

In terms of buoyancy we see that a submerged object will sink if the buoyant force is less than the object's weight. If the buoyant force is greater than a submerged object's weight, the object will be buoyed up and rise to the surface. When the object reaches the surface and some of its volume is exposed, the buoyant force decreases (less liquid displaced). The object will float partially submerged when the weight of the displaced liquid or buoyant force equals the object's weight.

Rather than talk about "heavy" and "light" objects and liquids, it is convenient to consider densities. Recall that density is the amount of mass or weight in a given (unit) volume. It easily follows that

> 1. An object will float in a fluid if the density of the object is less than the density of the fluid.
> 2. An object will sink in a fluid if the density of the object is greater than the density of the fluid.
> 3. An object will be in equilibrium at any submerged depth if the density of the object and the density of the fluid are equal.

For example, the density of water is 1.00 g/cm³, and the density of wood is 0.70 g/cm³ (Table 10.1).§ Thus, a

‡ This works for gases, too.

§ The density units g/cm³ will be used instead of the SI units (kg/m³) because of the convenient smaller values. This is one of the advantages of the metric system and its submultiple units. (g/cm³ × 1000 = kg/m³)

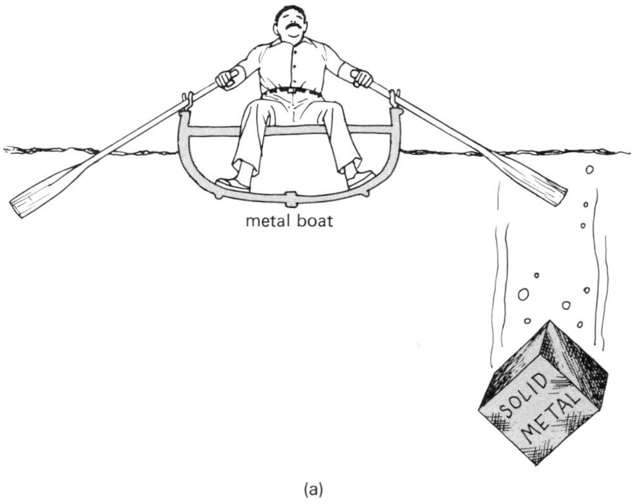

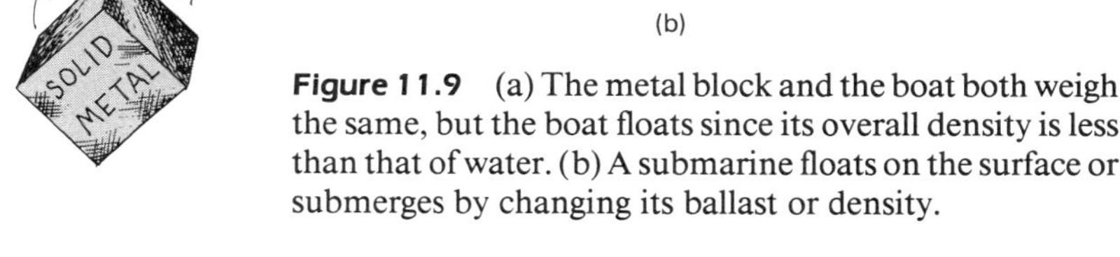

Figure 11.9 (a) The metal block and the boat both weigh the same, but the boat floats since its overall density is less than that of water. (b) A submarine floats on the surface or submerges by changing its ballast or density.

block of wood will float in water. However, the density of iron is 7.9 g/cm^3, so a block of iron would sink in water. If a block of some material had a density of 1.00 g/cm^3 (the same as water), it would stay at any submerged depth at which it was placed.

QUESTION: A block of iron will sink in water. But ships are made of iron. Why don't they sink?

ANSWER: In applying the above rules, you have to consider the density of an object as determined by its total volume. A block of iron will sink because in this form the density of the block is greater than that of water (Fig. 11.9). But if the iron is fashioned into sheets and constructed so that the density of the structure as a whole is less than that of water, it will float. (The volume includes the hollow internal space.) In this case, water is displaced by the submerged hollow portion of the structure, and the buoyant force equals the weight force of an iron ship.

A submarine can float or run submerged. To submerge, the ballast tanks of the sub are flooded with sea water until its total density is greater than that of sea water. The submerged depth can also be adjusted by means of engine power and fins. To surface, water is pumped out of the ballast tanks, thereby decreasing the sub's density.

It can be shown that the ratio of the density of a floating object and the density of the liquid (density of object/density of liquid) gives the fraction of the volume of the object that will be submerged. For example, if the density of a wooden block is 0.70 g/cm^3, then in water the density ratio is 0.70 g/cm^3/1.00 g/cm^3 = 0.70, or 70 percent of the block's volume will be below the surface when floating in water.‖ See Special Feature 11.2 for another example.

Surface Tension

If you drop a razor blade or a needle into a glass of water, it will sink. But if you lay a razor blade very gently on the water surface with its flat side (long dimension of a needle) parallel to the surface, it will float (Fig. 11.11). Is there something wrong with Archimedes' principle? No, there is another effect here called **surface tension.**

Consider a molecule within a liquid (Fig. 11.12). It is attracted in every direction by the neighboring molecules surrounding it. However, a molecule on the surface of a liquid is attracted to the sides and downward, but not upward. These molecular attractions on a surface molecule tend to pull the molecule into the liquid. The net effect is that the surface becomes as small as

‖ The relative density of a liquid is sometimes expressed in terms of specific gravity, which is the ratio of the weight of a volume of liquid and the weight of an equal volume of water. Since the volumes are equal, this is also the ratio of their densities: sp.gr. = ρ_ℓ/ρ_{H_2O} (where ρ is the mass density). The density of water is ρ_{H_2O} = 1.0 g/cm^3, so the specific gravity is just equal to the magnitude of a liquid's mass density in these units (sp.gr. = ρ_ℓ/1.0). Being a ratio, specific gravity has no units.

SPECIAL FEATURE 11.2
The Tip of the Iceberg

Icebergs are huge floating masses of ice that have broken off the ends of glaciers or polar ice sheets. These floating "islands" of ice range in size up to 30 km (19 miles) long and 60 m (200 ft) thick. See Figure 11.10.

The density of ice is 0.92 g/cm³, and the density of sea water is 1.03 g/cm³ (Table 10.1). Then, forming the density ratio, an iceberg floating in the ocean has $0.92/1.03 = 0.89$, or 89 percent of its volume below the surface. We see only the "tip of the iceberg," or 11 percent (1/9) of its volume.

Icebergs pose a serious threat to shipping. After the *Titanic* struck an iceberg and sank in 1912, an international ice patrol was established by joint action of all nations with shipping interests in the North Atlantic. The ice patrol is the responsibility of the U.S. Coast Guard, which locates icebergs, predicts their drift, and issues warnings.

Figure 11.10 The tip of an iceberg. Almost 90 percent of an iceberg lies below the surface.

Figure 11.11 Surface tension allows a razor blade to float on water.

possible. The surface is "stretched," or in a state of tension, hence the name surface tension. The surface behaves much like a thin elastic film. A razor blade or needle will sink if an edge penetrates this surface-tension "film."

The contracting surface force of a liquid accounts for our being able to carefully add more water to a glass of water filled to the rim. The surface of the water can be raised above the rim of the glass because the tendency for the surface to contract keeps the water from spilling.

Surface tension is the reason why drops of liquids are almost spherical, for example, rain drops and falling

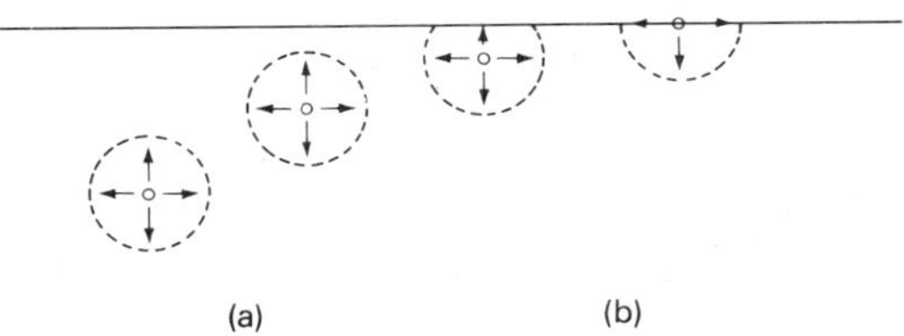

Figure 11.12 Intermolecular forces. (a) Within a liquid, a molecule has intermolecular forces in all directions due to surrounding molecules, so there is no net force. (b) At or near the surface, there is a net inward force that gives rise to surface tension.

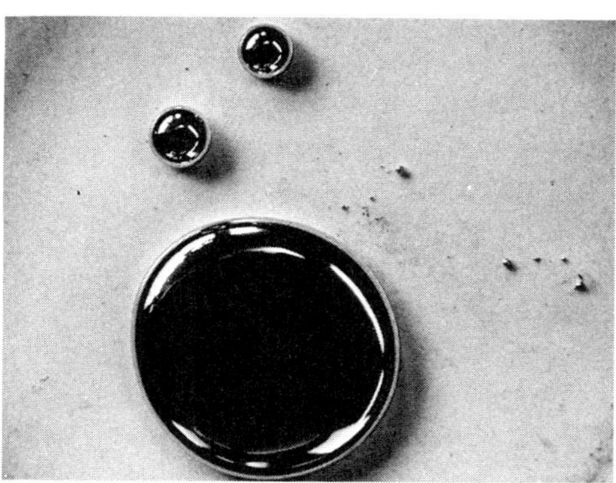

Figure 11.13 Surface tension causes a free liquid to form spherical drops.

drops of molten metal. (Lead shot is made by dropping molten metal from a tower, and the drops solidify into spherical shapes on the way down.) The contracting surface tension causes the free liquid to form into the shape with the smallest surface area. A sphere is the geometrical shape that has the smallest surface area for a given volume (Fig. 11.13).

DETERGENTS

Back to the floating razor blade for a moment. Suppose you put a few drops of liquid detergent in the water. You'd find that the razor blade would sink, because the detergent reduces the surface tension. This is also a detergent's job in cleaning. Surface tension keeps the water's surface smooth, and it does not "wet" particles of dirt or get under a particle on the fabric of cloth or on a dish. Detergents reduce the surface tension and make the water "wetter" so cleaning is more efficient. Ordinary soap is a detergent.

QUESTION: Why does hot water clean better than cold water?

ANSWER: Hot water has less surface tension than cold water and so washes or "wets" better. This is because the molecules in hot water have more energy and move around more.

Modern detergents are more effective in reducing the surface tension in cold water than are ordinary soaps. So now you can do your washing with cold "power" and save energy.

Capillary Action

Some liquids wet surfaces and others do not. For example, in Figure 11.13 the mercury drops do not wet the surface. This is because the contractive *cohesive* force (surface tension) of the mercury is greater than the *adhesive* force between the mercury and the surface.¶ A liquid that wets a solid has greater adhesion than cohesion.

If you examine the surface of water in a glass container, you will find it is not exactly level. It is slightly bowed downward at the center or has a concave shape (curved inward like a "cave"). The water in contact with the glass is lifted above the normal level because adhesion between the water and the glass is greater than the cohesion between the water molecules (Fig. 11.14). If the container is filled with mercury instead of water, the surface of the mercury will be bowed upward at the center or will have a convex shape. In this case, the cohesion between the mercury molecules is greater than the adhesion between the mercury and glass.

The crescent shape (concave or convex) of the surface of a liquid is called the meniscus (pronounced meh-NISS-kus). This comes from the Greek word meaning "crescent moon."

When tubes of very small diameter, called capillary tubes, are inserted in water and in mercury, we find that the water rises in the tubes while the mercury level is depressed (Fig. 11.15). This is called **capillary action** or capillarity. The rising of a liquid in a capillary tube depends on both adhesion and surface tension. Adhesion between the liquid and the tube causes the liquid to creep up the tube walls and produce a concave meniscus surface. The surface tension tends to flatten the surface by contraction. The combined action of these two forces raises the liquid column until the upward force is balanced by the weight of the liquid column. Notice that the liquid column is taller in a smaller capillary tube.

For liquids that do not wet the tube (cohesion greater than adhesion), the surface tension causes the liquid in the tube to be depressed. The meniscus is convex in this case.

It is capillary action that causes liquids to be absorbed by paper towels and oil to rise in lampwicks. An

¶ Molecular attraction between *like* molecules is called a cohesive force. The molecular force of attraction between *unlike* molecules, for example, between a liquid and its container, is called an adhesive force.

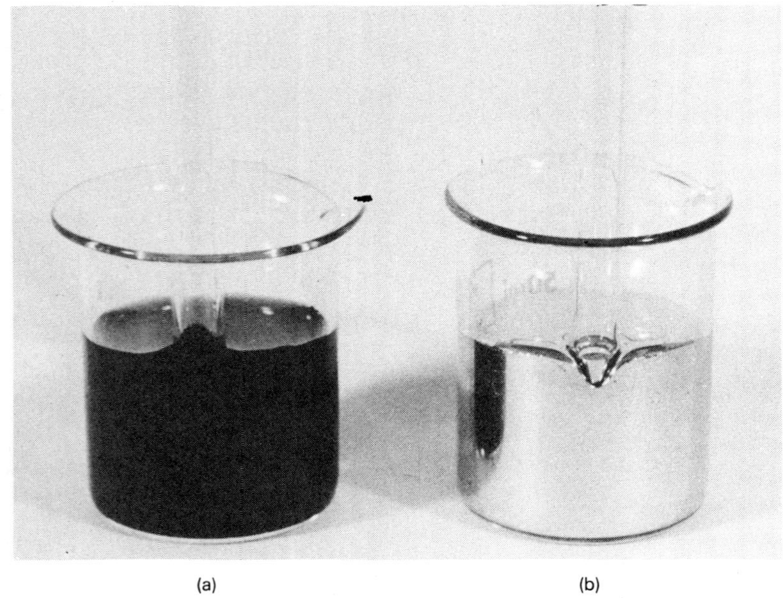

Figure 11.14 Adhesion and cohesion. (a) Because of adhesive forces between the water and glass, the water wets the glass. (Water darkened with ink so effect can be seen.) Notice that the overall water surface is bowed in or concave in shape. (b) The internal cohesive force of the mercury is greater than the adhesive force with the glass, so it does not wet the glass. The overall surface of the mercury is bowed upward or convex in shape.

important environmental capillary action is the rise of water in the soil.

Bernoulli's Principle

In the previous discussions, the liquids were generally at rest or in equilibrium. The study of liquid (fluid) flow is also an important topic, but it is a difficult one. Unlike in rigid-body solids, the molecules of a fluid can move around, and their motions within a fluid can get pretty mixed up.

To simplify matters for an introductory study of fluid flow, it is convenient to consider **streamline flow.** In streamline flow, the fluid particles follow parallel paths and don't get mixed up, so to speak. This idea is illustrated in Figure 11.16. The initial rising smoke approaches streamline flow (which is really an ideal

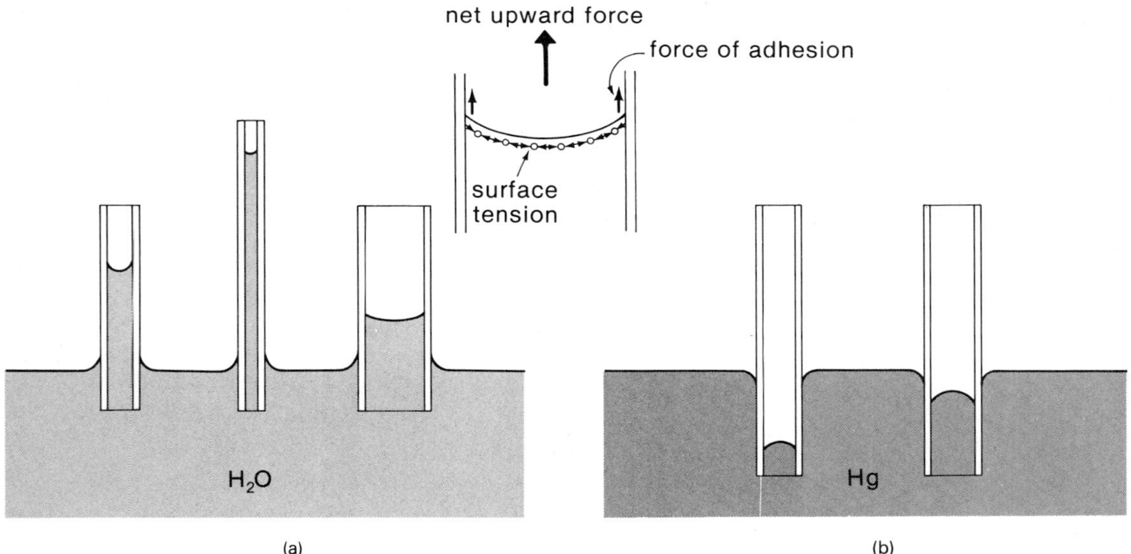

Figure 11.15 Capillary action depends on adhesion and surface tension. (a) Water wets the glass and rises in a capillary tube. (b) Mercury does not wet the glass and is depressed in a capillary tube.

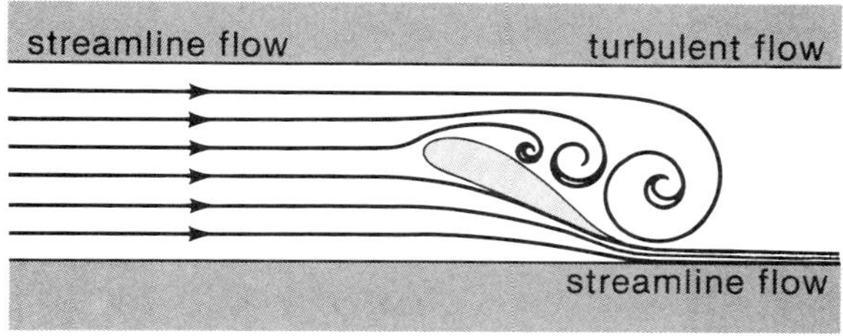

(a)

(b)

Figure 11.16 Streamline and turbulent flows. (a) When fluid particles flow in parallel paths, the flow is said to be streamlined. Mixing causes turbulent flow. (b) Smoke particle "flow," illustrating these conditions.

case). Then mixing causes the flow of the smoke particles to become *turbulent*. You can see how complicated it would be to describe this motion.

Figure 11.17 Water comes out of a hose with a nozzle at a greater speed because the nozzle constricts or decreases the cross-sectional area. (The "loop of flow" doesn't really exist. Why?)

Streamline flow is represented graphically by parallel streamlines. Notice that in streamline flow the streamlines never cross. If they do, this gives the particles the option of two or more paths, and the flow becomes mixed or turbulent. Turbulent flow is characterized by little whirlpools or eddies.

A group or bundle of parallel streamlines is often referred to as a tube of flow. In a uniform "tube" all the fluid particles have the same speed. But what would happen if the tube or pipe became constricted or decreased in diameter or cross-sectional area? You have no doubt experienced this effect when a nozzle is put on a garden hose (Fig. 11.17). Without the nozzle, the water "particles" coming out of the hose have a low speed and are not "shot" very far. However, with a nozzle the water speed is increased, as evidenced by how far it is projected. (Recall from Chapter 4 that the range of a projectile depends on its initial speed.)

Streamline flow, along with some other physical principles, allows us to explain this and other general cases of fluid flow. In terms of streamlines (Fig. 11.18), a constriction in a pipe forces the streamlines closer

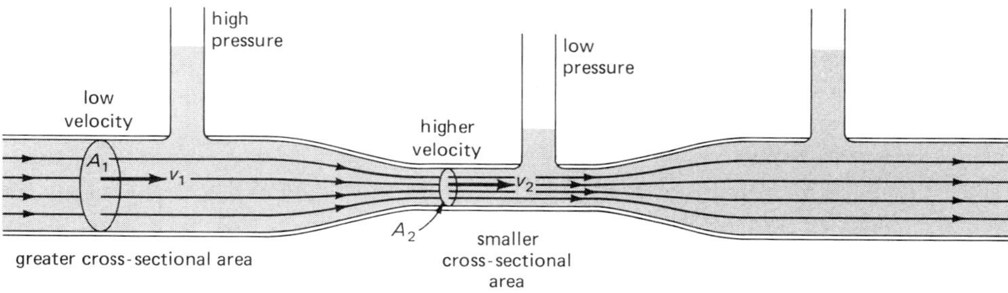

Figure 11.18 Velocity and pressure in streamline flow. The "bunching" together of steamlines increases the flow velocity and decreases the pressure as work is done in increasing the velocity.

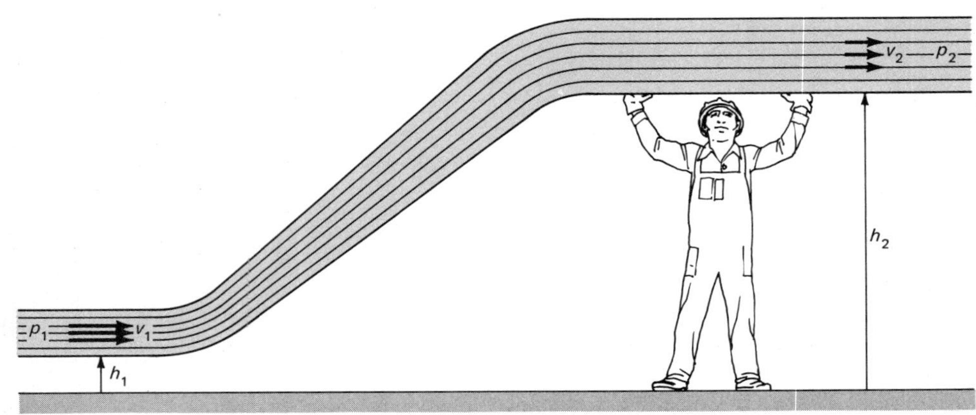

Figure 11.19 Bernoulli's principle takes into account changes in height or potential energy in fluid flow.

together, which indicates that the speed of the liquid is greater. This can be understood in a couple of ways. Since liquids are practically incompressible, if a certain mass of liquid goes into one end of the tube in a given time, an equal mass must come out of the other end (conservation of mass). The only way for this to occur is to have the liquid speed up in the smaller part of the tube.

Notice that the pressure in the liquid also changes, as indicated by the standpipes, which act as pressure gauges. The greater the pressure, the greater the height of the liquid in the standpipe. Why? A change in pressure is a measure of the work done. As can be seen from Figure 11.18, the pressure decreases in the constricted portion of the tube, which means work was done (in forcing the liquid into the smaller tube). By the work-energy theorem ($W = \Delta$K.E.), the kinetic energy or speed of the liquid increases.

Daniel Bernoulli (1700–1782), a Swiss mathematician, expressed these work-energy considerations for ideal fluid flow in what we call **Bernoulli's principle.** He also took into account potential energy, which involves a change in height along the length of fluid flow (Fig. 11.19). In this case we have $W = \Delta$K.E. + ΔP.E., and work can go into gravitational potential energy of the fluid. In terms of work-energy conservation, this can be written

$$p + \tfrac{1}{2}\rho v^2 + \rho gh = \text{a constant}$$

where the work is proportional to a change in p, and the mass density ρ takes the place of mass in the kinetic and potential energy expressions. We commonly use pressure to pump water into water towers, where it has potential energy. The fluid flow can be analyzed in terms of the conservation of energy, which is what Bernoulli's principle represents. Some common examples and applications of Bernoulli's principle will be given in the next chapter on gases (fluids).

Viscosity

Streamline flow is an ideal case but quite helpful. There are many other factors in real fluid flow. One of these is viscosity, which is the internal resistance of a fluid to flow—internal "friction," so to speak. The greater the viscosity of a fluid, the greater its resistance or the slower its fluid flow (less fluidity).

Viscosity is temperature-dependent. The viscosity generally decreases with increasing temperature. For example, a liquid flows more readily when it is heated (decreased viscosity). Viscosity is the physical principle behind the old saying "as slow as molasses in January." It is also an important feature in proper lubrication. See Special Feature 11.3.

SUMMARY OF KEY TERMS

Fluid a substance that can flow. Both liquids and gases are fluids.

Pressure force divided by area: $p = F/A$.

Pressure in a liquid $p = Dh$, where D = weight density and h = depth in liquid.

Pascal's principle pressure applied to an enclosed fluid is

SPECIAL FEATURE 11.3

Viscosity and Motor Oils

Viscosity is a prime consideration in the type of oil you use in your car. In the winter you want a low-viscosity or "thin" oil so it will flow more easily and lubricate the car engine properly, particularly when the engine is cold at startup. In the summer a higher-viscosity or "thick" oil is used because the temperature is generally higher and a "cold" motor is warmer.

Figure 11.20 Multigrade or multi-viscosity motor oil. Polymer additives or "viscosity improvers" facilitate oil flow at low temperatures and provide greater viscosity at high temperatures so the oil can be used year-round.

Motor oils are rated in SAE (Society of Automotive Engineers) viscosity numbers. Common SAE numbers are 10, 20, 30, and 40. The greater the number, the greater the viscosity of the oil for a given temperature. Oils of SAE 10W and 20W are used in automobiles in the winter. The W denotes oils primarily for winter service—below 0°C (32°F). In summer, SAE 30 and 40 oils are used. SAE numbers are commonly referred to as "weights," for example, "30-weight" oil.

You may have given all this up and now use a "multigrade" or multi-viscosity oil year-round (Fig. 11.20). These oils are a development made possible by a new group of polymer additives called viscosity improvers. Viscosity improvers facilitate oil flow at low temperatures and provide greater viscosity at high temperatures. As shown in the figure, such oils are multi-rated, for example, SAE 10W-20W-40 (or 10W-40, for short). Multigrade oils span the range of given SAE numbers, so you don't have to change oils seasonally.

transmitted undiminished to every other part of the fluid and to the walls of its container.

Buoyant force the upward force on an object when submerged in a fluid.

Archimedes' principle an object in a fluid is buoyed up by a force equal to the weight of the fluid it displaces.

Surface tension the contracting force on the surface of a liquid due to unbalanced molecular attraction on the surface molecules.

Capillary action the rise of a liquid in a small tube due to adhesion and surface tension.

Meniscus the crescent-shaped surface of a liquid.

Streamline flow flow in which fluid particles follow parallel paths.

Bernoulli's principle the application of the conservation of work-energy to fluid flow.

Viscosity the internal resistance of a fluid to flow.

EXERCISES

1. Given an ordinary brick, how many ways could you place the brick on a table so that the pressure on the table would be different? Which position would give the greatest pressure? Is the force on the table different in each case?
2. In case you didn't think of it, suppose you could place the brick in Exercise 1 on an edge or corner (using a slight *horizontal* force to prop it up so that the vertical weight force of the brick is unaffected). How many ways could you position the brick in this manner to get different pressures? Which position would give the greatest pressure?

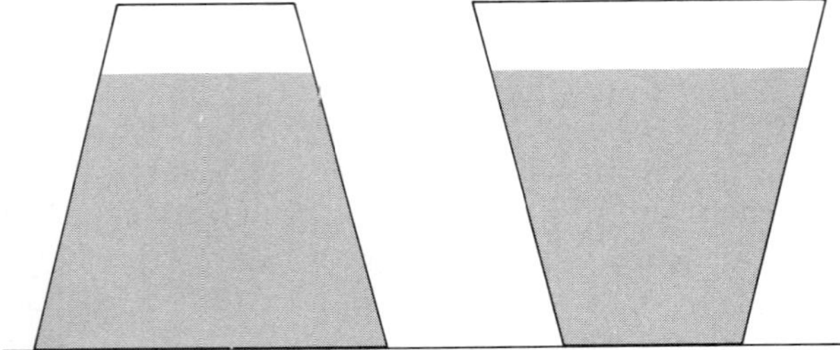

Figure 11.21 See Exercise 6.

3. The formula for pressure in a liquid is sometimes written $p = \rho gh$, where ρ is the mass density and g the acceleration due to gravity. Is this the same as $p = Dh$? Explain.
4. A cylindrical container of liquid has two holes in its side plugged by corks. One hole is near the base of the container and the other is midway down the side. If the corks are removed from the holes (same size), from which hole would the liquid flow faster, and why? Could you have the same flow rate from each hole by varying the sizes of the holes? Explain.
5. Suppose your waterbed had a leak from a small hole. Would the water run out faster when you were "in" bed? Explain. (Water beds are used in some hospitals to help prevent bedsores, which bedridden patients can get from regular mattresses. How do waterbeds help?)
6. Two containers, as shown in Figure 11.21, are filled to equal levels with a liquid. Which container has the greater pressure on its bottom surface?
7. Dams are built with wide bases. Why is this? Might it be necessary for a dam on a river to have a wider base than an ocean dam or dike? (The little Dutch boy is said to have saved the day by sticking his finger in a leaking dike. How could he have held back the whole ocean with his finger?)
8. A vertical cylinder 1.0 meter in height is filled with water. If a piston applies a pressure of 2000 kPa at the top of the cylinder, what is the total pressure on the bottom of the cylinder?
9. In a hydraulic system such as in Figure 11.5(b), suppose that no energy is lost when the input piston is moved downward a distance of 50 cm. How far would the output piston be raised? (See Chapter 6 if necessary.) Explain this in terms of liquid displacement.
10. A hydraulic hand jack can be used to raise a car, a mobile home, or even a house. Explain how this can be done by hand. Why does the jack handle have to be moved many times to raise such a heavy object by only a short distance?
11. Suppose in the hydraulic system shown in Figure 11.5(b) that the input force is the force applied to the larger piston. What would be the mechanical advantage in this case? How far would the large piston have to be pushed down in order to raise the output piston 10 cm?
12. Some hospital patients receive fluids intravenously from bottles hung on tall stands. Why are the bottles elevated? What might happen if the bottles were hung too low?
13. Cruise ships weigh on the order of 70,000 tons (600 million newtons). Explain how such heavy boats can float. How much water does a 70,000-ton ship displace? (Take the density of sea water to be 10,000 N/m^3.) Does the ship float higher or lower in the water when it is loaded? Explain.
14. An ocean-going freighter loads up at a Great Lakes port and then sails out into the Atlantic. Will it float higher or lower in the ocean?
15. If a Ping-Pong ball is held well below the surface of a body of water and then released, it rises and "pops" out of the water. What causes the ball to rise above the surface?
16. Why can humans float in water? Can you affect your floating depth in any way? Explain.
17. Why can people float easily when swimming in the Great Salt Lake in Utah?
18. A steel (iron) ball floats in a liquid (Fig. 11.22). What can you say about the density of the liquid? (See Table 10.1.) What do you think the liquid might be? (It's a metal, too.)
19. Do ice cubes float higher in water or in mixed drinks? (Explain your answer. The density of alcohol is 0.79 g/cm^3.) How about an ice float in a party punch that has had fruit frozen into the ice? (*Hint:* Does fruit float in water?)
20. How could you use a plastic milk jug in a swimming pool to illustrate the density floating and sinking rules given in the chapter?
21. The setup shown in Figure 11.8 is used to illustrate Ar-

Figure 11.22 See Exercise 18.

chimedes' principle. Could you also use it to determine the density of an irregularly shaped heavy (sinking) object like a rock? How about finding the density of an irregularly shaped floating object? (*Hint:* Use a sinker.)

22. A hydrometer is a weighted glass instrument used to measure specific gravity (Fig. 11.23). The stem is calibrated to read 1.0 when the hydrometer floats in water. (a) How do the markings run on the stem? That is, are the readings greater or smaller on the upper part of the stem (above the 1.0 mark)? (b) Hydrometers are used to measure the "strengths" of battery acid and antifreeze in your car. How does the specific gravity give an indication of these "strengths"? (*Hint:* Battery acid is denser than water, and antifreeze is less dense.)
23. Would the buoyant force on an object floating in a container of water be increased if some of the water were displaced by scooping it out of the container?
24. A container of water sitting on a scale has a certain weight. If a block of wood is placed in the water and it floats due to the upward buoyant force, would the scale reading change? Explain.
25. What is the principle of a life preserver?
26. Suppose there is a hole in the bottom of a boat (Fig. 11.24). Explain why the boat sinks in terms of water displacement. How does bailing help?
27. Why can water bugs "walk on water"?
28. Why are drops of dew on blades of grass or hanging from twigs semi-spherical? Why aren't they completely spherical?
29. Why are soap bubbles easier to blow and longer-lasting than "water bubbles"?
30. Discuss the drops shown in Figure 11.13 and drops of dew hanging from a twig in terms of adhesive and cohesive forces.
31. Why is glue called an adhesive? Is its adhesive or cohesive

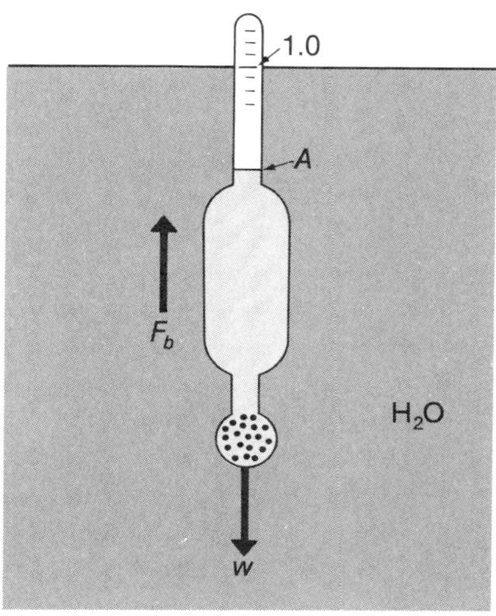

(a)

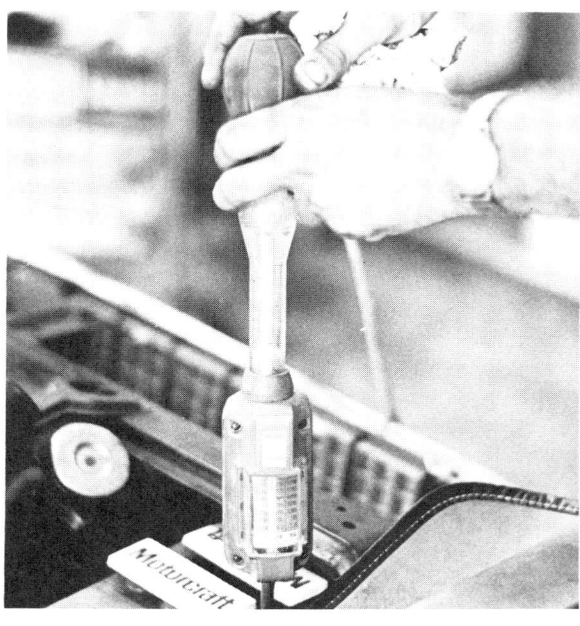

(b)

Figure 11.23 See Exercise 22.

Figure 11.24 See Exercise 26.

force stronger? Why won't glue stick to some surfaces, for example, a greasy surface?

32. What gives rise to a "belly smacker" when you dive into a swimming pool improperly?
33. To demonstrate the wetting action of modern detergents, an experimenter smeared the bottom side of a duck with detergent and then placed the duck in water; the duck sank. (It was saved, of course.) Why did the duck sink?
34. At the beach, sand can be observed to be wet above the water line (even at high tide). Why is this? Why is it easier to walk on wet sand than on dry sand?
35. A light liquid and a heavier (more dense) liquid both have the same surface tension and adhesion to the walls of similar capillary tubes. Would the liquids both rise to the same height in the tubes?
36. Would a detergent added to a liquid increase its capillary action?
37. Why are you able to shoot water farther from a hose with a nozzle than from one without? How can you do this with a hose without a nozzle by using your finger?
38. When you are taking a shower and someone flushes a toilet (on the same floor), you may notice a decrease in the flow of the shower water, particularly the cold water. Why is this, when the water pressure coming into the house is still the same?
39. (a) A hypodermic needle and syringe are used to give "shots" into muscles and tissue. The relatively small force applied to the plunger of the syringe alone would not be able to do this. How does the needle help accomplish this? (b) What is the principle of the new "air-injected" shots that do not use needles? (What does hypodermic mean?)
40. Explain what happens in terms of pressure and flow speed if the liquid flow in Figure 11.19 is reversed.
41. In Figure 11.19, the flow speed is decreased at the top of the elevation. Devise a way you could keep the flow speed constant.
42. What causes viscosity in fluids (liquids and gases)?
43. Would the capillary action of a liquid be increased if you heated it so as to reduce its viscosity? Don't forget surface tension.
44. Does milk or cream have a greater viscosity? Which has the greater density?
45. What would happen if you used SAE 40 oil in your automobile in the winter?

12

Gases

Now for a look at the other side of the fluid "coin"—gases. Gases are fluids since they can flow, and as such they can be described by many of the same principles that apply to liquids (Chapter 11). But an important difference between liquids and gases is that gases are easily compressible. The distances between gas molecules are relatively large, so molecules can be pushed or forced together.

Since little or no cohesive force exists between gas molecules, a gas has no definite shape. A gas can expand to fill the volume of any container and takes on the container's shape and size. On a big scale, this might be a "gravitational container." For example, the expansion of the Earth's atmosphere or that of a star is limited by gravitational forces, which determine the size and shape of a very large mass of gas.

The Perfect Gas Law

As with a liquid, pressure is used to express the force (per area) exerted on or by a gas. However, it may be difficult to see how pressure comes about in a confined gas. In order to get a picture of this, we use molecular theory.

Molecular theory views the molecules of a gas to be moving about at high speeds. They not only collide with each other, but also with the walls of the container (Fig. 12.1). Each collision with a wall exerts a tiny force on it. Taken alone, this would be insignificant. But the frequent collisions of billions of gas molecules exert a fairly steady force (per area) or pressure on the walls.

The pressure of a gas depends on several factors. One is the volume of the gas or the size of its container.

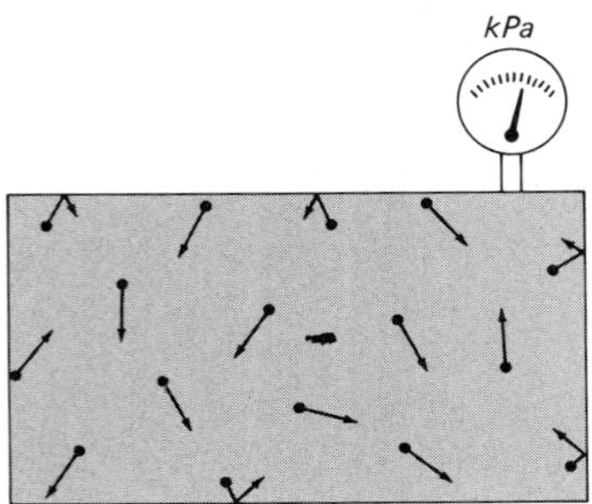

Figure 12.1 Gas molecules colliding with the walls of a container exert, on the average, a steady force (per area) or pressure on the walls or on a pressure gauge.

Consider a mass of gas in a cylinder as shown in Figure 12.2. If the volume is decreased by one half, the pressure doubles. This is because the gas molecules occupy half the space and so make twice as many collisions with the walls. Similarly, if the volume is doubled, the pressure decreases (by one half, why?).

The relationship between the pressure and volume of a given quantity of gas (at constant mass and temperature) is given by

$$p_1V_1 = p_2V_2$$

This is known as Boyle's law, after Robert Boyle (1627–1691), an English chemist who is credited with its discovery. The relationship holds for all gases (except when the pressures are very large).

Using modern kinetic molecular theory, another relationship can be derived for the product pV. It is found that pV is proportional to the average kinetic energy of the gas molecules,

$$pV \propto \text{average kinetic energy of gas molecules}$$

This makes sense because the more kinetic energy the gas molecules in a container have, the faster they move, giving rise to more collisions and a greater pressure.

The kinetic energy of a perfect gas is its internal energy. By adding or removing heat energy, we can therefore affect the pressure and/or volume (Fig. 12.3). When heat is added and the gas is kept at a constant volume, the pressure increases. However, there are cases in which both the pressure and the volume can change.

Temperature (T) is a relative indication of heat energy, and in terms of temperature we may write

$$pV \propto T$$

For example, on a warm day an automobile tire may be properly inflated. But if it turns cold overnight, you may notice that the tire looks a little flat in the morning. The temperature decrease (reduction in internal energy) gives rise to the volume decrease.

Another factor that affects the pressure of a gas is its mass, or how many molecules there are. The more molecules, the more collisions and the greater the pressure. For example, suppose an automobile tire has a gauge pressure of 165 kPa (25 lb/in^2) and you *add* more

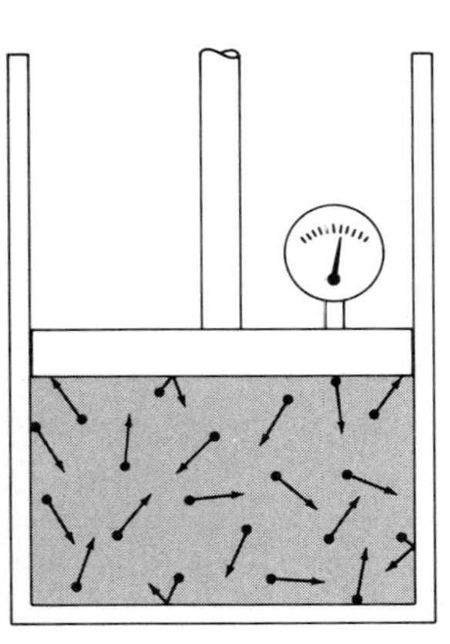

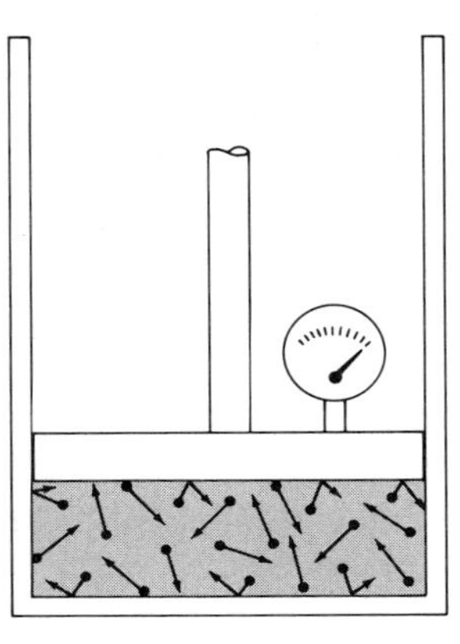

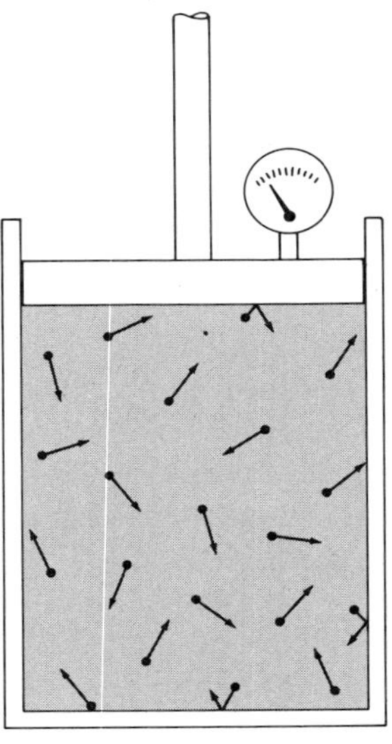

Figure 12.2 Pressure and volume. If the volume of a gas (a) is decreased by one half, the pressure is doubled (b). If the volume is doubled (c), the pressure is decreased by one half.

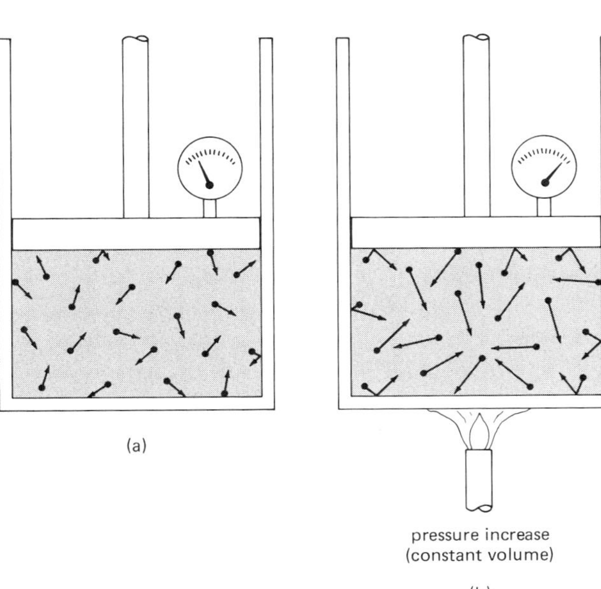

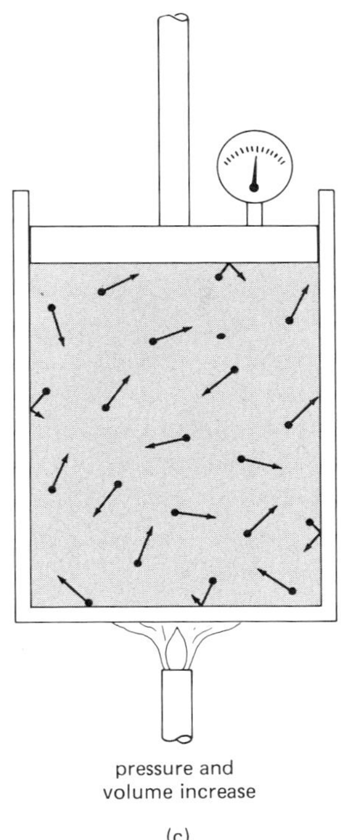

Figure 12.3 Pressure, volume, and temperature. When heat is added to a gas kept at a constant volume, the pressure increases (b). However, both the pressure and the volume can change (c).

air so that the pressure becomes 200 kPa (30 lb/in^2). The pressure increases because there are more molecules in the tire and more molecular collisions with the walls. Another way of looking at this is as a density increase. Notice in Figure 12.2(b) that the density and pressure were increased by making the volume smaller.

All of these factors are expressed in a relationship called the perfect gas law (an older name is the ideal gas law):

$$pV = NkT$$

where N is the number of molecules in the quantity of gas and k is a constant (called Boltzmann's constant).*

The Atmosphere

Now that we know some of the factors that affect gases, let's take a look at some of the properties and phenomena of the mixture of gases in which we live—the atmosphere. Our atmosphere is the gaseous shell or envelope of air that surrounds the Earth. It is composed chiefly of two gases, nitrogen (78%) and oxygen (21%).

Just as certain sea creatures live at the bottom of the ocean, humans live at the bottom of a vast atmospheric sea of gases, which is on the order of 600 km (400 miles) deep. The Earth's gravitational attraction retains our atmosphere. The moon and the planet Mercury do not have atmospheres. Having relatively small masses and weak gravitational attractions, they lost any original atmosphere they might have had.

The downward attraction of the Earth's gravity causes our atmosphere to be concentrated or more dense near the Earth. One half of the entire atmosphere lies below 11 km (7 miles), and 99 percent lies below an altitude of 30 km (19 miles). So the air is quite "thin" in the upper part of the atmosphere.

Living at the bottom of the atmosphere, we support the weight of the air above us (Fig. 12.4). This gives rise to a pressure at sea level of about 100 kPa or 10^5 N/m^2, which in customary-system units is about 15 lb/in^2 (actually 14.7 lb/in^2). This means that the roof of a house with an area of 10 m $\times$ 25 m = 250 m^2 has a downward force of 25 million newtons (or about 2800 tons!) on it due to the weight of the atmospheric column of

* The temperature in the perfect gas law is *absolute* temperature, which is measured in kelvins (SI unit). More will be said about this temperature scale in Chapter 13, where the perfect gas law is used to define *absolute* zero. The perfect gas law is sometimes written in another form. Since $pV/T = Nk$ = a constant,

$$p_1V_1/T_1 = p_2V_2/T_2$$

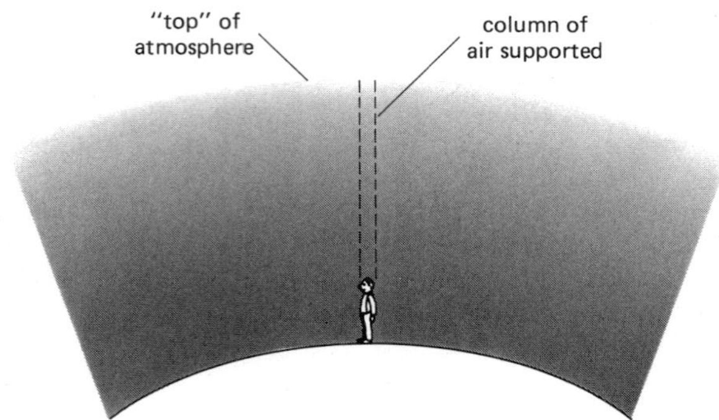

Figure 12.4 At the bottom of the atmosphere, we support the weight of a column of air which gives a pressure of about 100 kPa (15 lb/in^2) at the Earth's surface.

Figure 12.5 Atmospheric pressure demonstration. A small amount of water in a metal can is boiled so that the steam drives most of the air from the can. When the heat is removed and the can capped, the steam condenses, leaving a partial vacuum in the can, and atmospheric pressure crushes it.

air. ($p = F/A$ so $F = pA$.) Of course, this is balanced by the air pressure inside the house (Pascal's principle, Chapter 11).

An effective demonstration of atmospheric pressure can be made with a large metal can and a rubber stopper (Fig. 12.5). A small amount of water is placed in the open can, which is then heated over a bunsen burner. When the water boils, as evidenced by water vapor seen coming out of the can, the fire is extinguished, and the stopper is pushed securely into the can opening.† After a short time, the can begins to make stressing noises and is slowly crushed as if by an invisible force. Can you explain why?

The steam from the boiling water drives the air from the can. After the heat is removed and the stoppered can cools, the steam condenses, leaving a partial vacuum in the can. The greater external atmospheric pressure on the can does the rest.

Of course, we don't notice the weight of the atmospheric gases because it is balanced by internal pressures. However, we sometimes notice variations or changes in pressure. A relatively small, quick change in altitude may cause one's ears to "pop." This is because the pressure in the inner ear does not equalize as quickly, which puts a force on the eardrum. When the pressure equalizes (swallowing helps), the ears pop. Airplanes are equipped with pressurized cabins that maintain normal atmospheric pressure at high altitudes. Our bodies are accustomed to an external pressure of 100 kPa. Should this be reduced, the excess internal pressure may be evidenced in the form of a nose bleed and other discomforts.

† A rubber stopper is preferred to the usual gasketed screw cap that comes with the can, for safety reasons. Should one forget to turn off the burner, the pressure build-up will simply "pop" the stopper. With a screw cap in this case, the can might explode.

THE BAROMETER

On TV weather reports the announcer gives you the temperature and so on, along with the barometric pressure (see Fig. 14.10). This is the atmospheric air pressure that is measured with a barometer.

The barometer was invented by Evangelista Torricelli (1608–1647), a contemporary of Galileo. The principle of the mercury barometer is shown in Figure 12.6, along with a modern instrument. If a long tube full of mercury is inverted in a container of mercury, some of it runs out, but a column of mercury remains in the tube. The atmospheric pressure on the mercury surface supports the column (via Pascal's principle). The pressure due to the weight of the mercury column is equal to the atmospheric pressure.

The pressure of a liquid column is $p = Dh$ (Chapter 11). At sea level, the atmosphere supports a column of mercury 760 mm tall (30 inches). Instead of reporting the pressure in pascals or N/m^2, it is common practice to give the pressure in terms of the height of the mercury column. One millimeter of mercury (Hg) is called

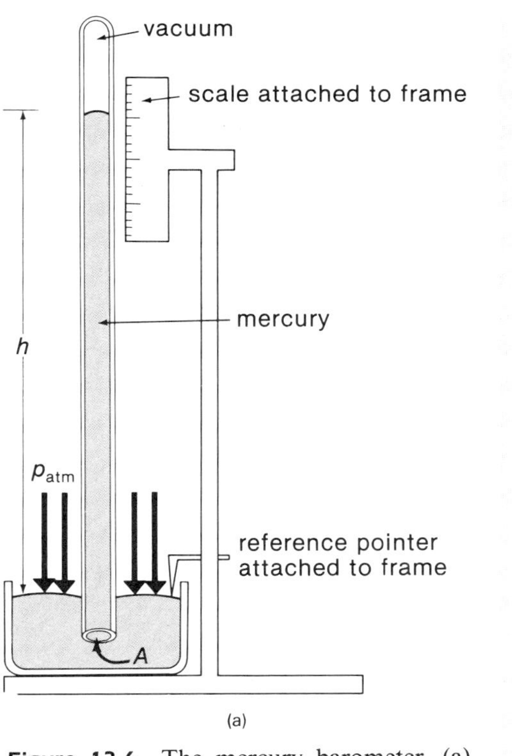

Figure 12.6 The mercury barometer. (a) The height of a mercury column depends on the atmospheric pressure. (b) An actual mercury barometer.

a torr, in honor of Torricelli. Hence,

$$1 \text{ atmosphere}\ddagger = 760 \text{ mm Hg} = 760 \text{ torr} = 30 \text{ inches Hg}$$

The barometric pressure on weather reports is usually given in inches of mercury. The variations in pressure as high- and low-pressure air masses move across the country are important in weather forecasting.

Any liquid could be used in a barometer. Other liquids would be less expensive. Why not use water? That's a good question.

QUESTION: Why isn't water used in barometers?

ANSWER: The density of mercury (13.6 g/cm^3) is 13.6 times greater than that of water (1.00 g/cm^3). Therefore, a column of water with the same weight as a 760-mm column of mercury would be 13.6 times taller, or 10.3 *meters* (34 ft).

This would be a water column about as tall as a three-story buiiding. Water barometers are a bit impractical.

Another type of barometer in common use is the aneroid barometer (aneroid means "without fluid"). This is a mechanical device having a sealed metal container with a diaphragm "lid" or cover that is sensitive to small pressure changes (Fig. 12.7). The diaphragm is much like a drumhead responding to pressure. A pointer is used to indicate changes in pressure.

Figure 12.7 An anaeroid barometer. The pressure is measured by an evacuated metal container sensitive to small pressure changes (seen in the background of the central opening). The lid or diaphragm of the container acts like a drumhead responding to pressure.

‡ Atmospheres (atm) are used as a unit for very large pressures. Another pressure unit used by meteorologists is the millibar. 1 millibar (mb) = 100 N/m^2 (pascals), so 1 atm $\approx$ 1000 mb.

(a)

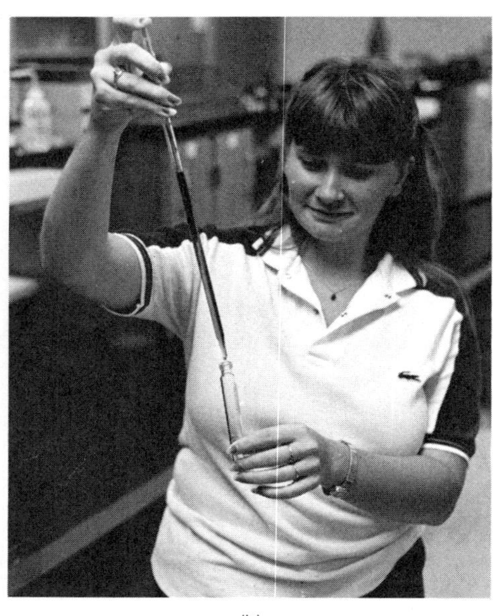

(b)

Figure 12.8 Atmospheric pressure in use. (a) When drinking with a straw, the person drawing on it reduces the pressure in the straw, and atmospheric pressure on the surface of the liquid pushes it up the straw. (b) When a liquid is transferred with a pipette, the net upward pressure supports the liquid column. When the finger is removed, the pressures are equalized, and the liquid drains from the pipette.

Aneroid barometers with dial faces are common in homes. Notice in Figure 12.7 that fair weather is associated with high barometric pressures and rain with low barometric pressures. Do you know why?§

The altimeters used in airplanes and by skydivers to indicate altitude are really aneroid barometers. The pressure decreases rather uniformly with height in the lower atmosphere. If the barometer dial face is replaced by one calibrated inversely in height, the barometer becomes an altimeter. As you go up in altitude, the pressure becomes less, and this is read as increasing height on the barometer altimeter.

ATMOSPHERIC EFFECTS

There are a lot of common effects associated with the atmosphere and its pressure. You see or use some of them quite often. For example, a frequent use of atmospheric pressure is in drinking through a straw. It is generally believed that sucking on the straw draws the liquid up the straw. Actually, the sucking action reduces the pressure in the straw. The atmospheric pressure on the liquid's surface then pushes the liquid up the straw (Fig. 12.8).

Liquids are often transferred and dispensed by medicine droppers and pipettes. A liquid goes up such glass or plastic tubes by the same principle as for a straw. The liquid remains in a pipette stoppered with a finger over one end because the external air pressure on the open end is greater than the pressure in the upper part of the tube. This pressure difference supports the weight of the column. When the finger is removed, the pressures are equalized, and the liquid flows from the pipette under the influence of gravity.

Another method of transferring a liquid is by siphoning action. A **siphon** is a tube filled with liquid, which is siphoned from one level over a small elevation to a lower level (Fig. 12.9). Atmospheric pressure is involved, but it really doesn't do the job. Notice in the figure that the upward pressure at B, which is atmospheric pressure, p_a, and the upward pressure at A are essentially the same. Thus, atmospheric pressure tends

§ When the atmospheric pressure is high, air heated near the Earth's surface does not expand greatly. Being only slightly buoyant (Archimedes' principle), it rises a relatively short distance before it cools off. Clouds are formed when air rises high enough for water vapor to condense into droplets (Chapter 14). This does not generally occur with a high atmospheric pressure, and without clouds, there is no precipitation but rather fair weather.

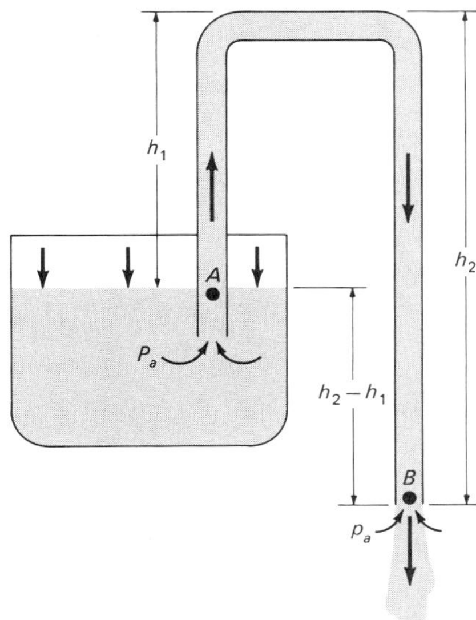

Figure 12.9 A siphon is another method of transferring liquids. See text for description.

to push the liquid up each arm of the tube with equal force (uniform tube diameter).

However, the downward pressure at A due to the weight of the liquid in the shorter arm is less than the downward pressure at B, which is due to the weight of the liquid in the longer arm. As a result, there is a pressure difference between A and B, and the liquid flows through the tube. The pressure difference is given by $p = D(h_2 - h_1)$. The flow will stop if the liquid levels are equal, $h_2 = h_1$, or if the liquid level at A falls below the tube opening, allowing air to break the liquid column in the tube.

Buoyancy

Being a fluid, the atmosphere also has buoyancy, with the buoyant force given by Archimedes' principle (Chapter 11). We don't feel this buoyant force because our bodies don't displace much air, so the force is quite small. Having a greater density than air, we "sink" in the atmosphere. However, if enough air is displaced, objects will float. This is observed for helium and hot-air balloons (Fig. 12.10).

We commonly say that helium is "lighter" (less dense) than air and that "hot air rises." The perfect gas law shows why hot air is less dense than cooler air. Hot air has a greater temperature, and by $pV \propto T$, its volume increases. When heated air expands and increases in volume, its density is lowered ($\rho = m/V$). Hence, it is less dense than the surrounding cooler air, and it is buoyed up and rises.

(a)

(b)

Figure 12.10 Buoyancy. When filled with enough hot air from a propane burner (a), balloons become buoyant and rise (b).

Pumps

Fluids (liquids and gases) can flow, but they must be given the energy to do so. This is the job of pumps. Basically, a pump is a machine in which mechanical energy is transferred to a fluid. For example, a fan may be thought of as a pump. Pumps are used to deliver water to homes, to pump oil and natural gas to furnaces and hot and cold air throughout our homes, to create partial vacuums, and to circulate blood in our bodies. There are many types of pumps. Let's consider here one common basic type—the force pump.

THE FORCE PUMP

A force pump mechanically displaces a volume of fluid by some means. For example, a cylinder-piston arrangement might be used, as shown in Figure 12.11. When the piston is brought forward, the liquid is forced from the cylinder. The air chamber on the outlet pipe is filled on the forward stroke, which compresses the air above and increases the pressure ($p \propto 1/V$, perfect gas law). The compressed air forces the water from the chamber during the backstroke of the piston and maintains a steadier flow. During the piston backstroke, a partial vacuum is created in the cylinder, and the atmospheric pressure on the liquid in the reservoir forces more liquid into the cylinder. The cycle is then repeated.

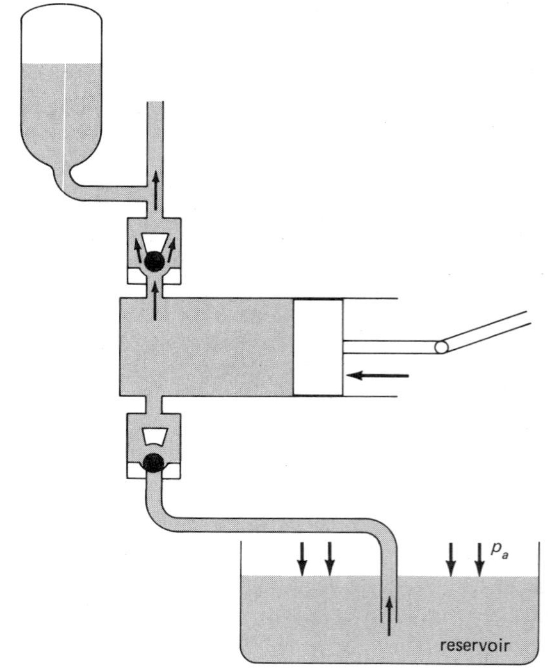

Figure 12.11 A cylinder-piston mechanical force pump. Variations in the air pressure in the chamber on the outlet pipe allow for a steadier flow between pumping strokes.

Suppose the fluid in the force pump is air and it is pumped into a tank. The pressure in the tank is gradu-

SPECIAL FEATURE 12.1

The Vacuum Cleaner

With more and more carpets in homes (wall-to-wall), the vacuum cleaner is an indispensable cleaning tool. Its operation involves a difference in pressure. Within a vacuum cleaner, the motor and blower create a partial vacuum (reduced pressure) in the cleaner chamber and hose. A vacuum cleaner "picks up" dirt because atmospheric pressure forces it into the partial vacuum (Fig. 12.12). Remember, it is pressure or a pressure difference that "pushes" a fluid. A vacuum does not "pull." Perhaps we should call a vacuum cleaner an "air-pressure" cleaner. (In some cases, we use the exhaust of a vacuum cleaner as an air blower.)

Figure 12.12 Pressure at work.

ally built up, and the pump acts as a compressor. The compressed air can then be used in a garage lift, a spray painter, or some other application.

In some cases, gases are pumped to create partial vacuums—vacuum pumps. Here, the air or gas is pumped from a volume. Gases can be pumped mechanically and by moving vapors that carry the gas molecules along with them (diffusion). Mechanical methods can produce partial vacuums of the order of 10^{-3} torr (mm Hg). Diffusion pumps can produce partial vacuums on the order of 10^{-6} torr, which is about the atmospheric pressure at an altitude of 270 km (170 mi).

Space is a good partial vacuum. Outside a satellite in orbit at 500 km (300 mi) the pressure is about 10^{-10} torr. In the vacuum of "deep" space, there are only a few molecules per cubic meter and virtually no pressure. A more down-to-earth partial vacuum and its practical application is discussed in Special Feature 12.1

BREATHING

In a manner of speaking, air is pumped in and out of our lungs—at a rate of about 5 liters per minute. The intake and output of our respiratory pump are controlled by the diaphragm, the muscular partition between the chest and abdominal cavities. An apparatus used to demonstrate the breathing process is shown in Figure 12.13.

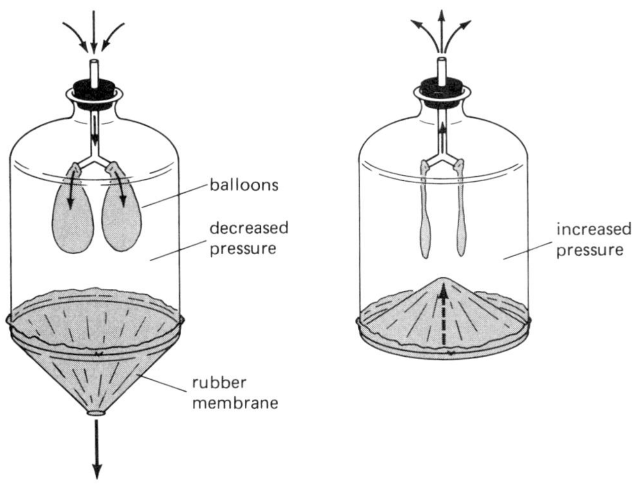

Figure 12.13 A demonstration to show the mechanics of breathing. A rubber membrane acts as the diaphragm and balloons as lungs.

When we inhale, the diaphragm moves downward, thereby increasing the volume of the chest cavity. This decreases the pressure inside the chest ($p \propto 1/V$) to below the external atmospheric pressure. The atmosphere then forces air into the lungs. In exhaling, the diaphragm relaxes and moves upward. The internal pressure increases, and air is expelled from the lungs.

THE HEART

You may have been thinking about pumps and have realized that there is an important pump in our bodies—the heart. The heart is basically a muscular force pump that pumps blood through the body's network of arteries, capillaries, and veins in order to bring nourishment and oxygen to the cells and to carry away wastes.

Rather than having a piston, the heart has muscular walls that contract and relax, thereby changing the volumes of the heart's chambers as it "beats," that is, as it forces blood through the circulatory network (Fig. 12.14). The heart pumps or beats 60 to 90 times a minute in the average adult. With each pumping cycle, the heart distends and takes in freshly oxygenated blood arriving from the lungs. On contraction, the blood is forced through the aorta into the arterial network.

The working of the circulatory system is externally monitored by taking one's pulse and blood pressure. The pulse is commonly taken with one's fingers on a person's wrist. The pulsating beat is monitored at about 60 pulses per minute, on the average, for a calm, resting person. Blood "pressure" is measured with a special instrument. See Special Feature 12.2.

Bernoulli Effects

Bernoulli's principle, as discussed in Chapter 11, applies to fluids and so also to gases. There are many common effects of gas flow that can be explained by Bernoulli's principle, for example, why airplanes can fly and why baseballs can be made to curve.

In gaseous flow, the potential energy term in Bernoulli's equation is relatively unimportant. The elevation of the gas flow may change, but because of the relatively small mass or density of a gas (as compared to a liquid) the change in potential energy (ρgh) is small. See Chapter 11. Hence we need only to consider the work and kinetic energy terms in Bernoulli's principle ($p + \frac{1}{2}\rho v^2 =$ a constant) for a good approximation.

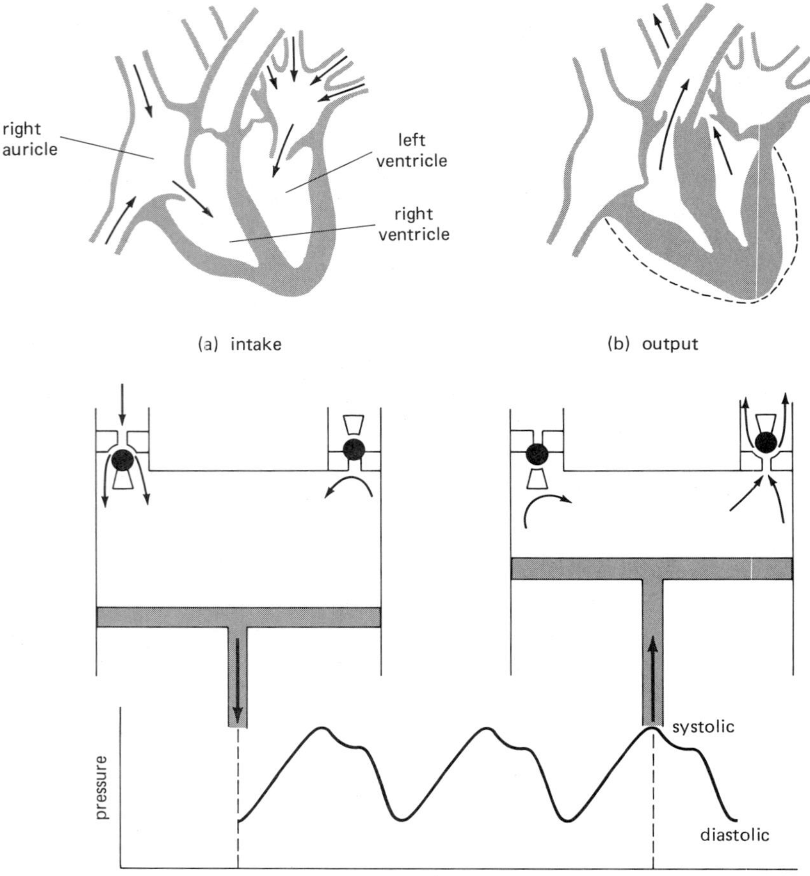

Figure 12.14 The heart as a force pump. The pumping action or "beating" causes the blood pressure to vary between a high (systolic) pressure and a low (diastolic) pressure.

This means that in regions where the gas speed is greater (streamlines closer together), the pressure is less, and vice versa. This is like the liquid flow in Figure 11.19. Keeping this in mind, let's look at airplanes and curve balls.

Airplanes are heavier (denser) than air, so by Archimedes' principle they would sink (or crash) or not even get off the ground by a buoyant force. So what upward force allows airplanes to take off and remain in flight? This "lift" force in terms of Bernoulli's principle is due to the air flow over the wings (Fig. 12.15). Because the upper wing surface is curved, the speed of the air flowing over this surface is greater than the flow speed past the bottom surface. As a result, there is a greater pressure on the bottom of the wing than on the top. This pressure difference provides the upward lift force. A lift force due to air passing over a surface can be demonstrated, as shown in Figure 12.17.

Some people prefer to analyze airplane wing lift in terms of momentum and Newton's second law (see Chapter 3). Notice in Figure 12.15 that the streamlines have a downward component or a downward component of momentum in that direction behind the airfoil. The rate of change of the vertical component of momentum (or upward force) on the airfoil passing

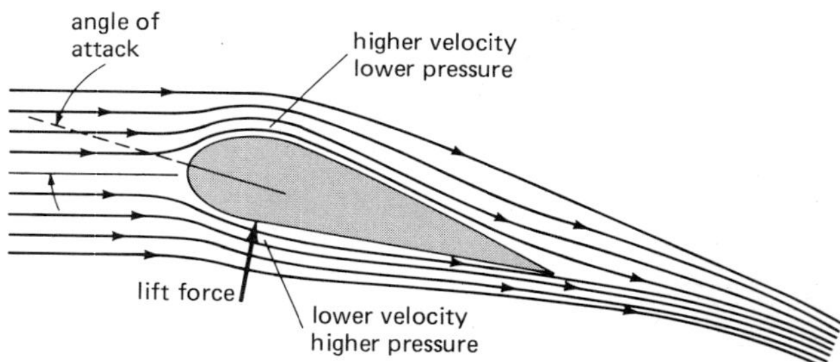

Figure 12.15 Airplane lift. Because of the curved surface of an airfoil or wing, there is an air speed and pressure difference between the upper and lower surfaces that gives a lift force. The lift can also be analyzed in terms of momentum and Newton's second law.

SPECIAL FEATURE 12.2

Blood Pressure

Taking one's blood pressure monitors the pumping action of the heart. When the heart contracts, the blood pressure in the arteries increases. When the heart relaxes, the pressure goes down. The "upper" or maximum pressure is called the systolic pressure, and the "lower" or minimum pressure is called the diastolic pressure (Fig. 12.14).

When taking your blood pressure, a doctor or nurse (or even a coin-operated machine) is measuring the pressure of the blood on the arterial walls. This is done by an instrument called a sphygmomanometer (Greek *sphygmo,* pulse). See Figure 12.16. This tongue twister is pronounced *ss-fig-mom-an-om-it-er.* The dial pressure gauge is calibrated in mm Hg or torr. (Older types of meters use a mercury column to measure the pressure.)

An inflatable cuff is placed snugly around the upper arm. When inflated to a sufficient pressure, the cuff shuts off the blood flow in the large arteries of the arm. Air is then slowly released from the cuff, and the doctor or nurse listens with a stethoscope over the artery just below the cuff in the bend of the arm. When the cuff pressure is slightly lower than the blood pressure, blood begins to flow through the artery with each beat of the heart. The rhythmic escape of blood beneath the cuff produces a distinct sound that can be heard through the stethoscope. As soon as the sound is heard, the pressure is noted, for example, 120 mm Hg (systolic or "upper" pressure).

As more air escapes from the cuff, blood flows more steadily through the artery. At a specific lower pressure the distinct beating sound is no longer heard, for example, at 80 mm Hg (diastolic pressure). Blood pressure readings are commonly reported in a ratio form, such as 120/80 (a reading of "120 over 80"). Normally, the pressure in the arteries ranges between 100 and 140 systolic (contraction pressure) and between 70 and 90 diastolic (relaxation pressure).

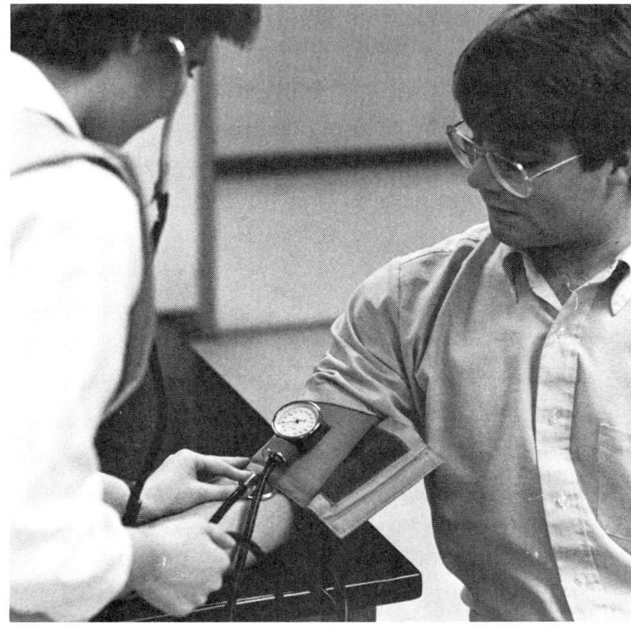

Figure 12.16 Blood pressure is measured with a sphygmomanometer, as shown on the "patient's" arm. A student nurse listens with a stethoscope placed over an artery.

through the air must equal the lifting force on the wing.

The airplane engines supply the forward motion that gets the air flowing over the wings. An airplane cannot "lift off" until it has gained sufficient air speed. Hang gliders must "take off" from heights and use gravitational potential energy to gain speed and lift. Updrafts may also provide some lifting force.

Another upward component of force also results from the air pressure on the leading edge of the wing when it is tilted at an angle to the horizontal—the so-called angle of attack. If the angle of attack is too great, turbulence develops over the upper wing surface, and the lift is lost. The airplane then "stalls" (loss of lift, not an engine failure) and falls.

A similar Bernoulli effect causes a baseball to curve. A pitcher causes the ball to curve by giving it an appropriate spin when thrown. If the air were an ideal nonviscous fluid, the spin would have no effect in changing the direction of the ball. However, since the air is viscous, friction between the ball and the surrounding air causes a thin layer of air to be dragged around the spinning ball (Fig. 12.18).

Figure 12.17 Bernoulli's effect. Because of the increased air flow over the top of the strip of paper, there is a pressure difference that causes the strip to rise.

The actual velocity of the air at any point is then the resultant or vector sum of the velocities. Notice on one side of the ball that the spin direction and velocity of the air layer is opposite to that of the relative air velocity, and on the other side they are in the same direction. So the air speed is increased on one side of the ball and retarded on the other side (note the streamlines). By Bernoulli's principle, the low-speed side has a greater pressure than the high-speed side. The ball then experiences a net force toward the low-pressure (high-velocity) side and is deflected, or curves.

The effect of moving air and pressure is important in home fireplaces. Winds passing over a chimney reduce the pressure and cause the chimney to "draw" or to have a draft up the chimney. Tall chimneys get more air flow over the top and draw better than shorter chimneys. Without good draft, fireplaces often "smoke."

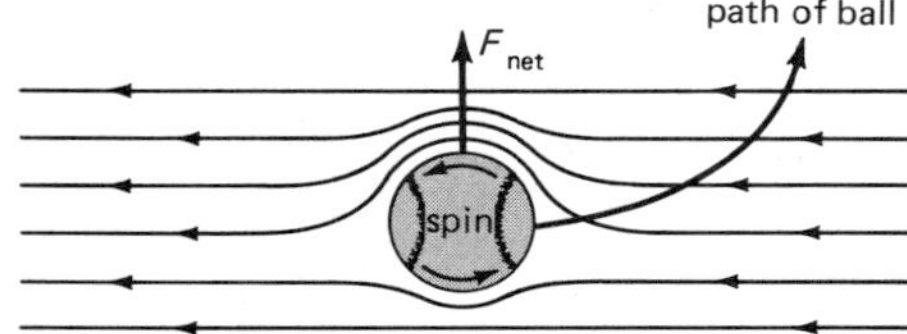

Figure 12.18 Bernoulli curve ball. The difference in the air velocities gives rise to a pressure difference that causes the ball to curve.

Another novel application of Bernoulli's principle is suspending a light ball, such as a Ping-Pong ball, in a stream of air (Fig. 12.19). You may have seen a similar demonstration in a store with a beach ball suspended above a vertical fan. The stream of air pushes the ball upward and it moves about. But what causes the ball to stay in the air stream?

Bernoulli's principle explains. As the ball moves to one side of the center of the diverging air stream, the air speed on the outer side of the ball is less than the air speed on the inner side nearer the center of the air stream. Hence, the pressure is higher on the outer side of the ball, and the pressure difference forces the ball back toward the center of the stream—a "Bernoulli-trapped" ball, so to speak.

A Bernoulli effect used in an application to harness the wind's energy is discussed in Special Feature 12.3.

(a)

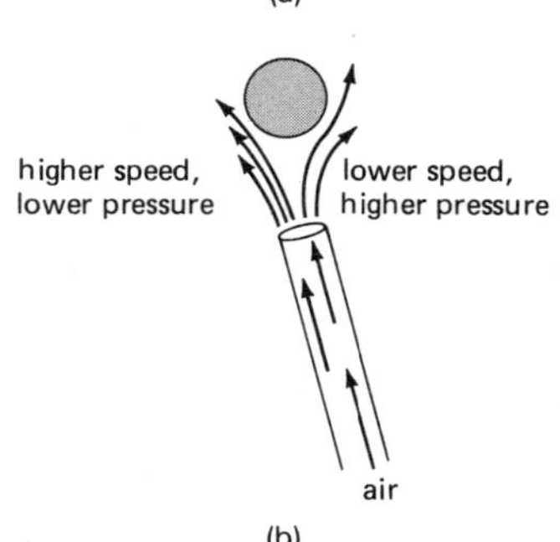

(b)

Figure 12.19 Bernoulli-trapped ball in an air stream.

SPECIAL FEATURE 12.3

Harnessing the Wind's Energy with the Bernoulli Effect

Jacques Cousteau, the ecologist and environmentalist, is reportedly looking for a replacement vessel for his famed exploration ship *Calypso* (built in 1942 as a mine sweeper). Under serious consideration is a prototype candidate called the *Moulin à Vent* ("Windmill"), shown in Figure 12.20. It is a wind-powered boat with no sails. Instead, this innovative craft—a 20-meter (65-ft) catamaran—has a 13.5-meter- (44-ft)-high cylinder that looks like a smokestack with a flap down the side. The fuel in this case is nonpolluting, abundant, renewable, and inexpensive—the wind.

The concept is a new application of an old idea. A stationary cylinder "sail" or column in the wind would produce no "lift" or lateral (sideways) force. This would be analogous to a cylindrical air foil (see Fig. 12.15). However, if the cylinder were rotated, there would be a lateral propelling force, similar to that experienced by a rotating baseball (see Fig. 12.18). Although this is a Bernoulli effect, it is sometimes called the Magnus effect after the 19th-century German scientist Heinrich Magnus, who explained the effect using a rotating cylinder.

Aton Flettner, a German engineer, first applied the Magnus effect to ship propulsion in 1824. He replaced the masts of an old sailing ship with two vertical cylinders that rotated at 100–150 rpm. It worked as a complementary energy source, but there were disadvantages. The propelling force was lateral to the wind direction, with the lateral direction depending on which way the cylinder was rotated. But maybe you don't want to go in either of these directions, so you "tack" into the wind, which is slow. Flettner's large cylinders also offered a great deal of air resistance.

Figure 12.20 The *Moulin à Vent* ("Windmill"). A prototype vessel that uses a smokestack-type "sail" to harness the wind.

The *Moulin à Vent* has a nonrotating, orientable cylinder with a "shutter-flap" that is automatically positioned by an on-board computer to cover one of two vent columns that run the length of the cylinder. A fan draws wind through the uncovered cylinder top. The orientation of the flap deflects the air such that the air flow behind the cylinder gives an increased "lift" force and propels the boat in the desired direction. If all goes well with the prototype, an 80-meter (260-ft) single-hull *Calypso II* with two cylinders is planned. It is estimated that on the average there would be a 30 to 40 percent savings on fuel costs by harnessing the wind in this manner.

SUMMARY OF KEY TERMS

Boyle's law $p_1V_1 = p_2V_2$ or $pV =$ a constant (at a given temperature).

Perfect gas law $p_1V_1/T_1 = p_2V_2/T_2$ or $pV = NkT$.

Barometer an instrument used to measure atmospheric pressure.

$$1 \text{ atmosphere} = 760 \text{ mm Hg} = 760 \text{ torr} = 30 \text{ in Hg} = 100 \text{ kPa} = 14.7 \text{ lb/in}^2$$

Pump a machine that transfers mechanical energy to a fluid, causing it to flow.

EXERCISES

1. (a) What happens to the volume of a quantity of gas if its temperature is decreased without changing the pressure? Why? (b) What happens to the temperature of the gas if its volume is decreased without changing the pressure?
2. If a quantity of gas is transferred from one container to another container half the size, how would this affect (a) the density and (b) the pressure of the gas? Would work be required to make the transfer? Explain.
3. When you pump up a tire with a hand pump, the lower part of the body of the pump becomes warm. Neglecting the friction of the plunger, why does this happen? Why does it get harder to pump as the tire becomes inflated?
4. A cup of water leaks through a hole, as shown in Figure 12.21. What would happen if you placed the palm of your hand tightly over the mouth of the cup, and why? (Explain in terms of the air trapped in the cup between the water surface and palm.)
5. Bubbles rising in a liquid, for example in boiling water, get larger as they approach the surface. Why is this? What must the pressure in the bubbles be for them to break the surface?
6. If you inflated an automobile tire to proper pressure on a very cold day, what would happen if the weather suddenly turned warm?
7. Would an astronaut be able to drink with a straw on the moon? Would you take a job selling vacuum cleaners to "moon people"?
8. Would it be easier to drink with a straw on top of Mt. Everest or in Death Valley?
9. Is there a limit to how long (tall) a drinking straw could be? Explain. (*Big hint:* The answer is 10.3 meters.)
10. How do suction cups work? Would they work on the outside of a space ship on the moon?
11. A "plumber's helper" or plunger is used to unplug drains. When a plunger is used on a sink, is the plug pushed or pulled in the drain?
12. A glass is filled with water by submerging, then lifted up, as shown in Figure 12.22. How far could the glass be raised before the water runs out? The glass is in a salt-water aquarium. A curious sea urchin looks on in the background.
13. What is the total force due to the atmosphere on the body of a person with a total body area of 0.20 m^2?
14. A table top measures 1 m × 1 m. What is the force on the table top due to the atmosphere in (a) newtons and (b) pounds? Why can such a table be easily picked up with such a force on its top?
15. When a soda bottle full of water (or soda) is turned over, the water does not run out steadily, but comes out in spurts with a gurgling sound. Why is this? Why does the water flow out steadily when the bottle is slowly tilted?
16. Gasoline cans with spouts commonly have a plastic-capped vent hole on the top. What is the purpose of the vent and what happens if you forget to remove the cap when pouring? (Some cans have a screw-top cap.)
17. Cakes and bread "rise" when baked because of the gas formed by the chemical reaction of baking soda in the batter. Cake mixes prepared for high-altitude locations have less baking soda. Why?
18. Automobile tires are inflated to pressures on the order of 200 kPa (30 lb/in^2). Yet bicycle tires are inflated to pressures more than twice this. Why such a greater pressure for a lighter vehicle? (How many more times greater is the typical automobile pressure than atmospheric pressure?)
19. Would a siphon work on the moon? Explain.
20. What would be the height of a column of a barometer that contained a liquid (a) one half as dense as mercury and (b) twice as dense as mercury?
21. In a mercury or liquid barometer, does the height of the column depend on the diameter or cross-sectional area of the tube? Explain.
22. Explain in terms of fluid principles (a) how a medicine dropper works, both in filling and dispensing, and (b) how a hypodermic syringe works. When a syringe is filled

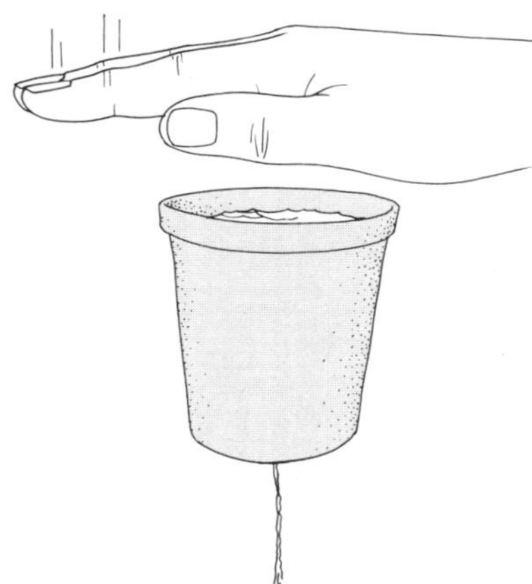

Figure 12.21 See Exercise 4.

Figure 12.22 See Exercise 12.

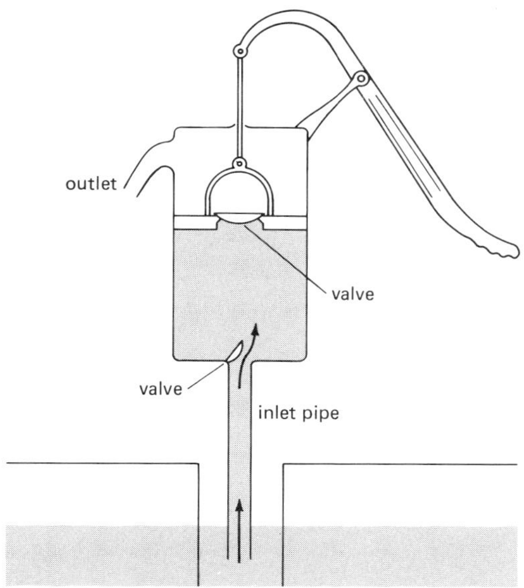

Figure 12.23 See Exercise 25.

from a sealed, rubber-topped medicine vial, why is air first injected into the vial?

23. When a released helium-filled balloon rises in the atmosphere, it gets larger. Why? With increasing volume, does the buoyant force increase so the balloon will rise indefinitely?
24. Hot-air balloonists continually give the balloon a shot of hot air (from a propane burner) while in flight. Why is this necessary?
25. Explain the principles involved in a "pitcher" or "lift" pump, as shown in Figure 12.23. It is common to have to "prime" such pumps by pouring water into the top if the pump has not been used for some time. Why is this necessary?
26. When you open your mouth and breathe in quickly, are you sucking in air? Explain.
27. How many times a minute do you breathe? If you take in 5 liters of air per minute on the average, how much air do you take in with each breath?
28. Explain the difference between a pulse rate of 60 and a blood pressure of 120/80.
29. In blowing up an inflatable mattress by mouth, what would determine the maximum pressure to which the mattress could be inflated?
30. High blood pressure is associated with arteriosclerosis or "hardening of the arteries." The arterial walls are normally elastic in young people, but the elasticity diminishes with age. Fatty deposits can narrow the arterial passageways (which would decrease the pressure), but they also roughen artery surfaces, which slows down the blood flow. Explain how the latter condition gives rise to high blood pressure. Would the heart have to work harder to maintain a normal blood flow?
31. Why do big jet aircraft require longer runways for takeoffs and landings?
32. If a high-flying jet airliner lost a window, would things in the aircraft be blown out the window or away from the window?
33. (a) Birds soar or glide using Bernoulli's principle, but how do they gain altitude? What principle is involved here? (b) How do rockets fly without wings?
34. How do paper airplanes work? Is there any lift involved?
35. How do helicopters lift off vertically and fly without wings?
36. How does a Frisbee "fly"? Why does it fall?
37. Why does the top on a convertible (automobile) "pop up" or bulge when the car goes at a fast speed?
38. When you drive your car on an interstate and a large truck with "the hammer down" goes by you in the passing lane, the car sometimes seems to sway or veer toward the truck. What causes this?
39. What is the principle of an atomizer that is commonly used to spray perfume?
40. What produces an updraft in a fireplace chimney, in addition to Bernoulli's principle, when there is a fire in the fireplace? (Suppose there were no wind passing over the chimney.)
41. Is the draft of a fireplace chimney good in terms of energy conservation? Explain.
42. A student in a class states that a nonspinning baseball will curve. Is this correct?
43. A glass is filled with water and an index card is placed on the top. If you hold the card and turn the glass over, you can remove your hand without causing the card to fall off and the water to drain out (Fig. 12.24). Explain why. Could you turn the glass sideways or horizontally with same result?

Figure 12.24 See Exercise 43.

PART THREE
Heat

This portion of physics having to do with heat may in a broad sense be called thermal physics. The effects of heat on the properties of matter are commonplace and have many practical applications in everyday life. With our present knowledge, it is quite natural to look toward the atomic structure of matter in describing these effects, and this provides great insight. However, it was not always so.

The major study of heat began with the start of the industrial revolution when it became important to understand the effects of heat and its conversion to mechanical energy. This study, sometimes called classical thermodynamics, concerned itself solely with the macroscopic properties of matter, such as temperature and pressure, without reference to the underlying atomic structure. Indeed, the atomic nature of matter was not fully understood at the time.

In the latter half of the nineteenth century, when the atomic nature of matter began to be understood, efforts were made to learn how the macroscopic or bulk properties of matter depended on the assumed behavior of atoms. One of the first successes of this study was with gases in terms of the average values of properties, such as the kinetic energy, of atoms and molecules. This part of the study of heat became known as kinetic theory.

In this section you will be treated to an overview of how macroscopic experimental observations are complemented by theoretical atomic theory in the understanding of the nature of heat and its effects.

13

Temperature and Heat

The Difference Between Temperature and Heat

The terms *temperature* and *heat* are used frequently, and we all understand their general meanings. We turn up the "heat" in the house to make it warmer or raise the temperature. We put things in the refrigerator to cool them off or lower the temperature. However, most people find it difficult to give precise definitions of temperature and heat. (Can you define them?)

It is evident from experience that temperature is associated with how hot or cold something is. For example, if you had two bowls of water at sufficiently different temperatures, you could tell which was hotter, or had a higher temperature, by comparing the water in the bowls with your hands. Notice that this is a comparison, or *relative* measure. You are comparing the hotness and coldness of the water in the bowls relative to the (body) temperature of your hands. Hence, we can say

> Temperature is a relative measure or indication of hotness or coldness.

With regard to heat, we know from previous chapters that heat is associated with energy transfer. When you put your hand in a bowl of water, your hand feels warmer or cooler because heat is transferred to or from it. Heat energy will always "flow" from a substance with a higher temperature to one with a lower temperature.* Thus, we can say

* An early theory of heat considered it to be a fluid-like substance called caloric (Latin *calor,* meaning "heat") that could be made to flow in and out of a body. Even though this theory has long since been abandoned, we still commonly say that heat "flows" from one object to another.

> Heat is energy transferred from one body to another because of a temperature difference.

It is a common observation that when a hot object (or fluid) is brought into contact with a colder object, the objects eventually come to the same temperature, or to thermal equilibrium in scientific jargon. For example, thinking of the two bowls of water again, if you dumped the water from one bowl into the other, all the water would eventually come to the same temperature. This would be somewhere between the temperatures of the individual bowls of water.

The preceding statements are really operational definitions. To obtain a better insight into temperature and heat, let's look at molecular theory. Recall that all matter is made up of molecules—either single atoms or combinations of atoms. The molecules are continuously jiggling around in motion. In a gas at ordinary pressures, the molecules move around freely, interacting through collisions. In liquids, the molecules also move around but are bound together by relatively weak forces. In solids, the molecules move back and forth relative to each other, as if held together by tiny "springs" (see Chapter 10). The energy associated with these random translational motions is often called thermal energy.†

However, in diatomic and other complicated gases, liquids, and solids there may also be rotational and/or vibrational motions of the atoms within the molecules (Fig. 13.1). Then too, there is potential energy associated with the molecular "springs." The *total* energy (kinetic plus potential) contained within a body is called its internal energy.

In terms of these considerations, temperature is associated with the thermal energy or random motions of the molecules of a substance. Thermal energy is the "temperature" energy or the energy that gives rise to temperature. When heat is added to a substance, it can go into the thermal or random translational kinetic energy of the molecules, which increases temperature. It can also go into the internal vibrational and rotational energies, which do not raise the temperature. Generally, there is a combination of both. Because the molecules have different translational speeds or kinetic energies, we think in terms of the *average kinetic energy* of the molecules, and

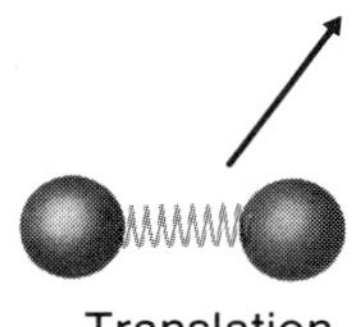

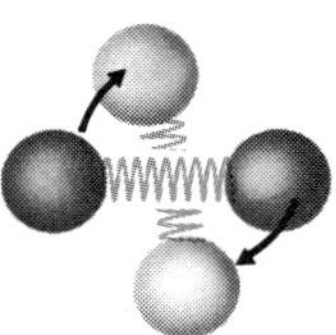

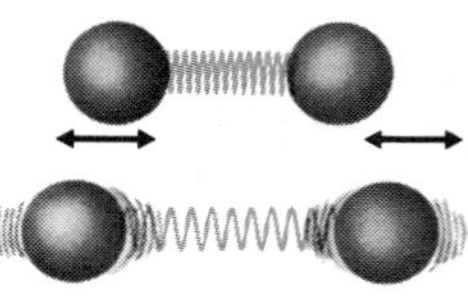

Figure 13.1 Molecular motions. In translational motion the molecule moves as a whole. There may also be intramolecular rotational and vibrational motions of the atoms.

> Temperature is a measure of the average random translational kinetic energy per molecule of a substance.

If a body has a high temperature, the average translational kinetic energy of each of its molecules is relatively high. If a body has a low temperature, the average translational kinetic energy of each of its molecules is relatively low.

You may have already guessed the molecular definition of heat.

> Heat is internal energy that is added to or removed from a substance.

For example, when heat is added to a body and its internal energy increases, some of the transferred energy generally goes into thermal energy or the transla-

† *Thermal energy* is a general term rather than a scientific one. It is used in different contexts by different people. We will use it here in association with the *random* translational motion of molecules, which, as will be seen, has to do with temperature. An object moving as a whole, much as a thrown baseball, has translational motion, but this is ordered motion, *not* random motion.

tional kinetic energy of its molecules.‡ This increases the average kinetic energy of the molecules of the body, raising its temperature and making it hotter.

Measuring Temperature

Now that we know what temperature is, how do we measure or express it? Humans have a temperature sense, but it is somewhat unreliable, varying from person to person, and its range is too limited for scientific and even practical purposes. (Who wants to check the temperature of almost-boiling water?) Also, our temperature sense of touch does not provide a means for measuring temperature quantitatively.

THERMAL EXPANSION

To measure temperature, we use some physical property of matter that depends on temperature. We construct a thermometer, a device that measures relative hotness or coldness by a change in some physical property. Fortunately, there are many physical properties that can be used (see Special Feature 13.3 at the end of this chapter). Probaby the most obvious and by far the most commonly used property is thermal expansion (and contraction, i.e., a negative expansion). Thermal expansion refers to the changes in the dimensions of substances that occur with changes in temperature.

Almost all substances expand with increasing temperature, but by different amounts. Conversely, most substances contract with decreasing temperature. Metals are a good example. The thermal expansion is small, but it can be made evident by using two metals in the form of a bimetallic strip (Fig. 13.2). Two metal strips, such as brass and iron, are bonded together. Because the metals expand differently, the bimetallic strip bends toward the strip with the smaller linear expansion.

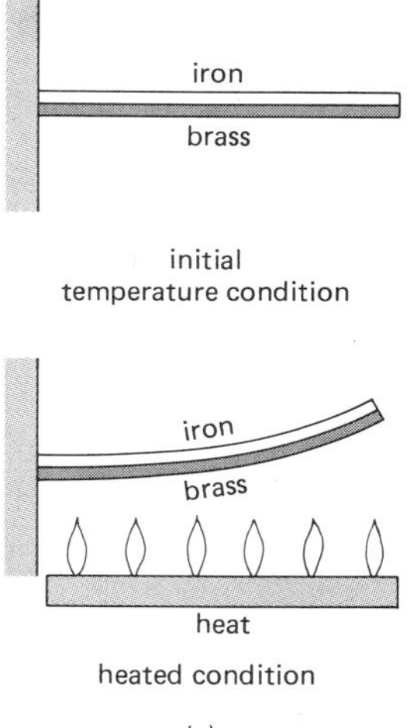

(a)

(b)

Figure 13.2 Bimetallic strip. Because the metals expand differently, the bimetallic strip bends toward the metal with the smaller linear expansion.

A bimetallic strip could be used as a thermometer by calibrating its deflection, since each position of the deflecting end corresponds to a different temperature. A more convenient form of a bimetallic thermometer is a strip wound in a coil or helix. The deflection is indicated by a dial on a calibrated scale. Such dial thermometers are commonly used as cooking thermometers in ovens (Fig. 13.3). Some have long probes that can be stuck into meat roasts. A common application of a bimetallic coil for temperature control is given in Special Feature 13.1.

‡ An exception occurs during a change of state, for example, going from a solid to a liquid or from a liquid to a gas. This will be discussed in the next chapter.

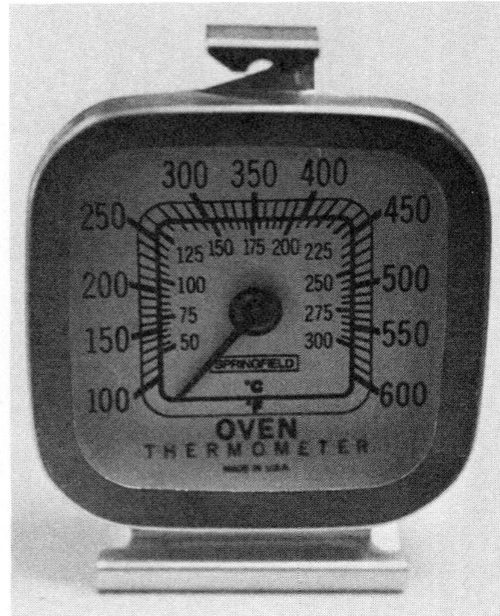

Figure 13.3 Bimetallic thermometer. A bimetallic strip wound in a coil or helix is used in a dial thermometer.

SPECIAL FEATURE 13.1

The Thermostat

When we want to adjust the temperature in our homes, we usually turn the thermostat up or down. This common wall device controls the temperature by turning heating and cooling systems on and off.

If you pull the cover off a thermostat, you will ordinarily see a mechanism such as that shown in Figure 13.4. The temperature sensor is a bimetallic coil to which a glass vial containing mercury is attached. As the coil expands or contracts with temperature changes, the vial is tilted, and the mercury moves from one end to the other. Within the vial are electrical contacts. In moving to different ends of the vial, the mercury makes or breaks electrical contact to turn heating and cooling systems on or off, depending on the temperature for which the thermostat is set.

Setting a thermostat for a desired temperature tilts the vial to the appropriate position. Remove the cover of a thermostat set in the "off" position and observe the tilting of the vial and movement of the mercury when you adjust the temperature control.

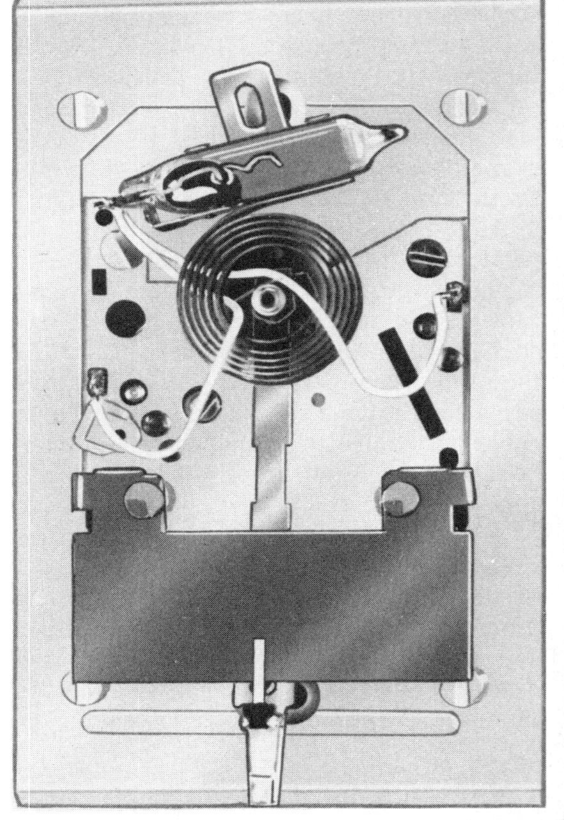

Figure 13.4 Thermostat. The expansion of a bimetallic coil causes a vial of mercury to tip and make or break electrical contact, which turns heating and cooling systems on and off.

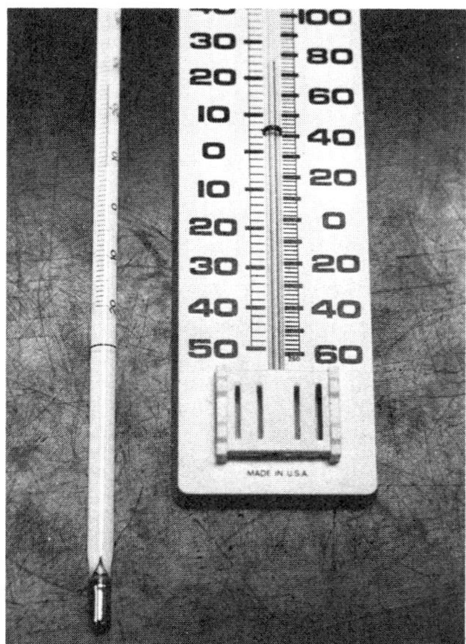

Figure 13.5 Liquid-in-glass thermometers. The liquid in a thermometer expanding and contracting up and down in a capillary bore in a glass tube indicates temperature changes. Commonly used liquids are mercury (left) and colored alcohol (right).

By far the most common temperature-measuring device is the liquid-in-glass thermometer (Fig. 13.5). This makes use of the thermal volume expansion of a liquid. It consists of a glass bulb, usually containing mercury or alcohol (colored with a dye to make it more visible), which is connected to a glass tube with a small capillary bore. An increase in the temperature of the liquid in the bulb causes it to expand up the bore, and the change in the height of the liquid in the tube provides a means of measuring a change in temperature.

Thermal expansion is an important consideration in many applications. It can cause bonded or glued joints to break because of "mismatch" expansion between the adhesive and bonded pieces with temperature changes. The different expansions of the materials can produce large stresses that break the bond. Engineers are well aware of the expansion of materials over seasonal temperature ranges. For example, you may have noticed the expansion joints and rocker supports on bridges (Fig. 13.6). Without these to allow for thermal expansion, damage could occur.

QUESTION: If a piece of metal sheeting with a hole in it, as shown in Figure 13.7, were put in an oven and heated, would the hole get larger or smaller?

ANSWER: Most people think that metal would expand so as to make the hole smaller—but not so. The metal expands, but the hole gets larger. One way of seeing this is to think about the piece of metal that was cut out to make the hole. If it was heated, it would certainly expand and get larger. The metal in the sheet with the hole behaves the same as if the cut-out piece were still there, so the hole gets larger. (For a more conclusive proof, see Exercise 13.) As a practical example, when a cannon is repeatedly fired, it gets hot, and the diameter of its bore increases.

(a)

(b)

Figure 13.6 Allowances for thermal expansion. (a) Bridge expansion joint and (b) rocker support. Serious damage could occur without these.

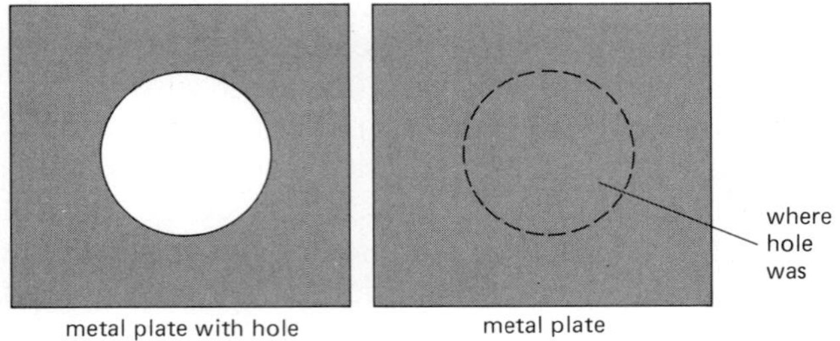

Figure 13.7 When a metal plate with a hole in it is heated, does the hole get larger or smaller? See Question and Answer on previous page.

TEMPERATURE SCALES

Thermometers are calibrated so that a numerical value may be assigned to a particular temperature. This is done by defining a temperature scale and a standard unit of temperature. As with any measurement or scale, we can use two reference points or marks. (Recall that the meter and the yard are practically defined by two reference marks on metal bars.)

To establish a temperature scale, we may take two reference or "fixed" points defined by physical phenomena that always occur at the same temperatures, for example, the ice point and steam point of water. These fixed points are convenient and readily available.§ More commonly called the freezing and boiling points, the ice and steam points are the temperatures at which pure water freezes and boils under a pressure of one atmosphere (standard pressure).

The Kelvin temperature scale and the kelvin unit are the official SI system standards. But before considering these, let's first discuss the more common Celsius and Fahrenheit temperature scales (Fig. 13.8). The **Celsius temperature scale** is named after its originator, A. C. Celsius (1701–1744), a Swedish astronomer. On this scale, the interval between the ice and steam points is divided into 100 equal subintervals or degrees (much like dividing a meter into centimeters). The temperature is read in degrees Celsius (°C), with the ice point having a temperature of 0°C and the steam point a temperature of 100°C.‖ Room temperature is 20°C.

The **Fahrenheit temperature scale** is named after G. D. Fahrenheit (1686–1736), a German physicist

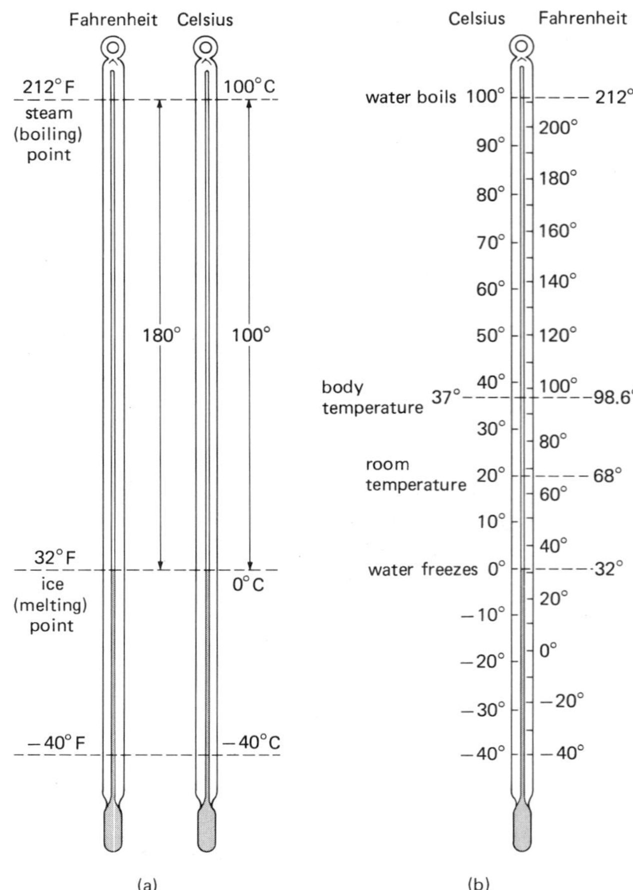

Figure 13.8 The Fahrenheit and Celsius temperature scales. See also Figure 13.5.

who originated it and who invented the liquid-in-glass thermometer. (His liquid was linseed oil.) On this common scale, the interval between the ice and steam points is divided into 180 equal degrees, and the temperature is read in degrees Fahrenheit (°F). As we all know, the ice (freezing) point of water is 32°F and the steam (boiling) point is 212°F. Room temperature is 68°F (20°C).

Notice that the temperature *units* on both scales are degrees. However, the degree unit on the Celsius scale is larger than the degree unit on the Fahrenheit scale—almost twice as large (1 Celsius degree = 1.8 Fahrenheit degrees, since 100 Celsius degrees equals 180 Fahrenheit degrees). A comparison of these temperature scales is shown in Figure 13.8. We now commonly hear or see the temperatures on weather reports given in both degrees Fahrenheit and degrees Celsius so we can become used to Celsius temperatures. The Celsius scale

§ A single fixed point and an interval unit can also be used to define a scale. The technical fixed point and standard SI temperature unit are referenced to a special phenomenon. See Table 1.2 and Chapter 14.

‖ This scale is commonly and incorrectly referred to as the centigrade scale (Latin *centi,* meaning "one hundred" and German *grade,* meaning "degrees").

is used throughout most of the world for general temperature measurements, the most notable exception being the United States.

QUESTION: Could the temperature reading given on the weather report ever be the same on both the Celsius and Fahrenheit scales?

ANSWER: The temperatures on the Celsius and Fahrenheit scales always seem to be numerically different, for example, 20°C and 68°F. However, it is possible to have the same reading on both scales. This occurs at −40° (−40°C = −40°F).

The equation relationship between the Fahrenheit temperature T_F and the Celsius temperature T_C is $T_F = \frac{9}{5}T_C + 32$, or inversely, $T_C = \frac{5}{9}(T_F - 32)$. Let $T_F = -40°$ or $T_C = -40°$, then plug this value in and see what you get. Whether you ever hear this temperature reading on a local weather report depends on where you live.

You may wonder why Fahrenheit chose 32° and 212° for the temperatures of the ice and steam points of water. Actually, he didn't. Other fixed points were used on his original scale. Fahrenheit chose for the zero reading on his scale the lowest temperature then obtainable in the laboratory—that of an ice-salt mixture. The upper fixed point was taken to be normal human body temperature, which was designated as 96°. (More exact measurements showed this to be 98.6°F.) When the ice and steam points of water were adopted as fixed points, these corresponded to 32° and 212° on the Fahrenheit scale.¶

ABSOLUTE ZERO AND THE KELVIN TEMPERATURE SCALE

The Celsius and Fahrenheit scales are used for everyday temperature measurements, and negative temperatures (below zero) are common. Thinking about this, you may wonder if there is a lower limit of temperature or an absolute zero. The answer is yes. It is determined by using the perfect gas law.

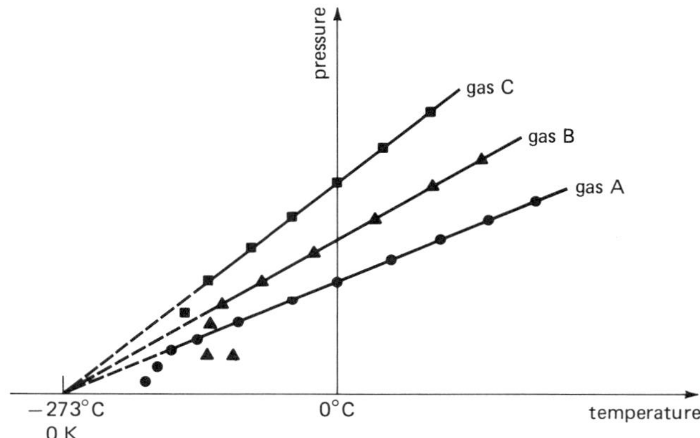

Figure 13.9 Absolute zero. The pressures of gases follow a linear relationship with temperature. Extrapolating to zero pressure gives the value for "absolute" zero temperature, 0 K (or −273°C).

Recall from Chapter 12 that the perfect gas law is $pV = NkT$. Then for a gas in a rigid container (constant volume V), the pressure is directed proportional to the temperature. If the pressures of some gases are measured at various temperatures and the data plotted, the points follow straight lines down to relatively low temperatures (Fig. 13.9).

At lower temperatures there are deviations from the straight-line relationship of the perfect gas law. This is because real gases eventually condense into liquids. Since there are no molecular interactions other than collisions in a "perfect" or "ideal" gas, it would theoretically remain a gas at any temperature. The pressure of a perfect gas would then be zero when the temperature is zero—absolute zero.

This absolute minimum temperature is implied by extrapolation of the straight lines in Figure 13.9 to zero pressure. It is found that they all converge to −273°C (actually −273.16°C or −459.69°F) on the temperature axis. This temperature is taken as absolute zero of the **Kelvin temperature scale.**** The Kelvin or absolute temperature (T_K) is related to the Celsius temperature by

$$T_K = T_C + 273$$

Notice that at $T_C = -273°\text{C}$, $T_K = 0$ K (see Fig. 13.10). The individual intervals on the Kelvin scale are the same size as those on the Celsius scale. This temperature unit is called a kelvin (*not* degree kelvin) and is abbreviated with the symbol K (*not* °K). The kelvin is

¶ ". . . the degree 48, which in my thermometers holds the middle place between the most intense cold artificially in a mixture of water, of ice and sal ammoniac (ammonium chloride, a chemical 'salt') or even sea-salt, and the limit of the heat which is found in the blood of a healthy man." Fahrenheit, D. G., *Philosophical Transactions,* Vol. 33, 1724, from Magie, W., *A Source Book in Physics,* Harvard University Press, Cambridge, Mass., 1963. In this article Fahrenheit reports the temperature of boiling rainwater to be 212° on his temperature scale.

** Named after Lord Kelvin (William Thomson, 1824–1907), a British physicist.

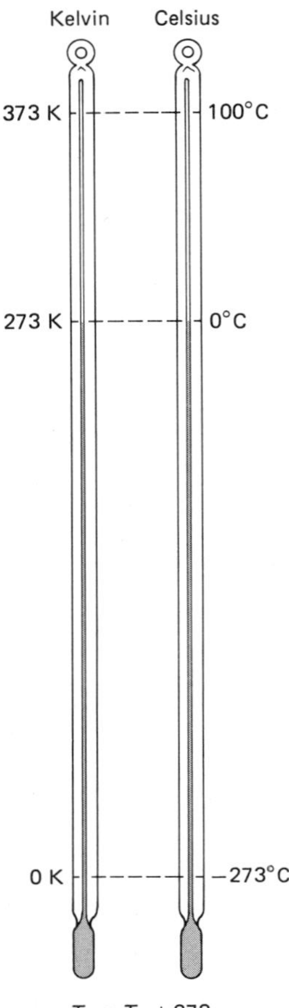

Figure 13.10 A comparison of the Kelvin and Celsius temperature scales. The individual intervals on these scales are the same size; a kelvin is the same size as a Celsius degree.

the SI base unit of temperature and is used primarily in scientific work. When temperature intervals are involved, the kelvin may be used interchangeably with the degree Celsius (°C).

However, in the perfect gas law, $pV = NkT$, the absolute temperature must be used. Since the internal energy of an ideal gas is proportional to the absolute temperature, this means that if the absolute temperature is doubled, the internal energy is also doubled, and vice versa. (The internal and thermal energies are the same for a perfect gas. Why?) Such is not the case for the Celsius (or Fahrenheit) temperature. To illustrate this, suppose that a gas were at a temperature of 0°C (or 0°F) and the internal energy were doubled by adding heat. Certainly the Celsius (or Fahrenheit) temperature is not doubled. Doubling zero (degrees) still gives zero. Also what if the temperature of the gas were −10°C and you doubled the internal energy? These problems do not occur with absolute temperatures.

Absolute zero (0 K) is believed to be the lower limit of temperature, although we have never physically been able to obtain this temperature. (In fact, there is a physical law that states that absolute zero cannot be reached, as will be discussed in Chapter 15.)

Kinetic theory predicts that at absolute zero all molecular motion would cease. This includes translational, vibrational, and rotational motions in a real gas. However, by modern concepts there would still exist some "zero-point motion." The zero-point energy of this motion is associated with the energy of the atoms of the gas molecules or of the atoms of a solid because they are confined.

No upper limit of temperature is known. In terms of a measuring stick analogy, we have a temperature "measuring stick" with a zero end that extends indefinitely, or at least with no upper end (limit) in sight. The interior temperatures of stars are on the order of 100 million kelvins or degrees Celsius. A difference of 273 degrees between the temperature scales doesn't make much difference at such temperatures. Various methods of measuring temperatures here on Earth are discussed in Special Feature 13.3 at the end of the chapter.

Measuring Heat

In the strict sense, a body does not contain heat, but contains internal energy. Heat is internal energy *transferred* from one body to another as a result of a temperature difference. The amount of heat or energy transferred is measured by some change that occurs in the transfer process, usually a change in temperature.

The SI unit of all forms of energy, transferred or otherwise, is the joule. So in this system the unit of heat is the joule. However, until the SI system is universally adopted, you will often hear heat energy expressed in a widely used unit called the **calorie.** It is defined as follows (Fig. 13.11):

> A calorie (cal) is the amount of heat required to change the temperature of 1 gram of water 1 degree Celsius.

This is sometimes called a gram calorie to distinguish it from the larger Calorie associated with the energy value of foods. See Special Feature 13.2.

Another common unit of heat energy is the **British thermal unit** (Btu). A Btu is the amount of heat energy

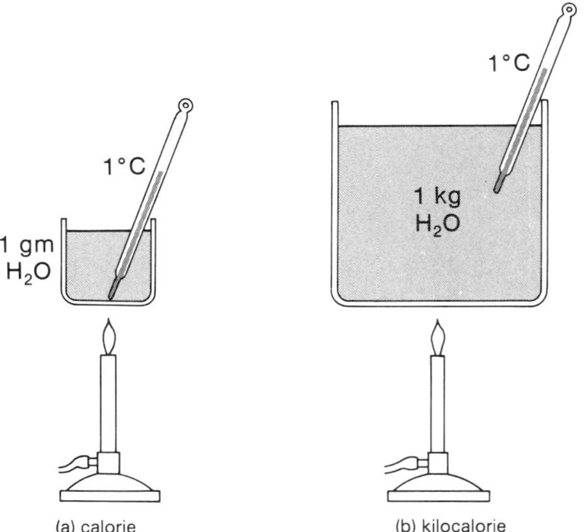

Figure 13.11 Units of heat energy. (a) A calorie raises the temperature of 1 gram of water 1°C. (b) A kilocalorie raises the temperature of 1 kg of water 1°C. (c) A Btu raises the temperature of one pound of water 1°F.

required to change the temperature of 1 pound of water 1 degree Fahrenheit (1 Btu = 252 cal).

HEAT OF COMBUSTION

We burn fossil fuels to obtain energy for heating purposes. In a similar sense, foods are body fuels. These processes involve the conversion of chemical energy to heat.

The intrinsic energy value of a fuel or food is expressed in terms of **heat of combustion.** This is the heat produced per unit mass of a substance when burned in oxygen. The units of heat of combustion are calorie/gram (cal/g) or kilocalorie/kilogram (kcal/kg).

Some heats of combustion for typical substances are given in Table 13.1. Notice that gasoline has a heat of combustion of 11,400 cal/g, so one gram of gasoline releases 11,400 calories of heat when burned. Compared to alcohol, which has a heat of combustion of 6400 cal/g, gasoline has a much greater intrinsic energy value. From the table you can see that you would gain more weight from eating scrambled eggs, 2100 kcal/kg, than from eating boiled eggs, 1600 kcal/kg. Remember, the kilocalorie is the food "Calorie." Of course, you probably wouldn't eat a kilogram of eggs, at least not at one sitting. The Calories listed for foods in diet tables are given for average amounts or portions, for example, an average-sized egg.

SPECIAL FEATURE 13.2

Counting Calories

If you are a weight watcher, you no doubt count Calories which are used to specify the energy values of foods. The Calorie contents of food servings are now commonly listed on labels. However, this is not the same as the calorie (cal) defined as the amount of heat required to change the temperature of 1 gram of water 1 degree Celsius. The food Calorie (Cal) is really a kilocalorie (1 Cal = 1000 cal = 1 kcal). This amount of energy would change the temperature of 1 kilogram of water 1°C (Fig. 13.11).

To distinguish between the kilogram Calorie and the gram calorie, the bigger unit is written with a capital C and is sometimes referred to as a "big calorie." For example, a piece of pie containing 300 Calories or "big calories" has 300 kcal or 300,000 "little calories" (gram calories). Perhaps if food values were listed in gram calories there would be a greater incentive for one to stay on a diet. At least it looks and sounds as though you get more.

Table 13.1 Typical Values of Heats of Combustion

Substance	*cal/g or kcal/kg*
Fuels	
Alcohol	6400
Coal	
Anthracite (hard coal)	8000
Bituminous (soft coal)	7500
Diesel oil	10,500
Fuel oil	10,300
Gasoline	11,400
Natural gas	10,000
Wood (pine)	4500
Foods	
Bread (white)	2000
Butter	8000
Eggs	
Boiled	1600
Scrambled	2100
Ice cream	2100
Meat (lean)	1200
Milk	700
Potatoes (white, boiled)	970
Sugar (white)	4000

Specific Heat

Suppose you had equal masses of copper and aluminum at the same temperature, and you added equal amounts of heat to each. Would the temperature increase be the same for each? You might be surprised to find that if the temperature of the copper increased by 100°C, the corresponding temperature change in the aluminum would be only 44°C. To get the temperature of the aluminum to change 100°C, you'd have to add more than twice the amount of heat than was added to the copper.

If you tried the same experiment with mercury and water, the amount of heat required to give a 100°C temperature change in the mercury would produce only a 3.3°C temperature change in the water. The water would require more than 30 times as much heat for a 100°C temperature change. Why the big difference?

This shows that different substances have different capacities for storing internal energy. When heat is added to a substance, it can go into thermal or random kinetic energy that will increase the temperature. It can also go into the internal vibrational and rotational energies, which do not raise the temperature. Generally, there is a combination of both. For example, it takes more heat to raise the temperature of a mass of hydrogen (a diatomic gas) by 1°C than for a degree change in an equal mass of helium (a monatomic gas). This is because in the hydrogen a portion of the internal energy is stored as atomic rotational or vibrational energy and is not effective in raising the temperature.

To express this difference, we say that hydrogen has a greater specific heat (or specific heat capacity) than helium. Each substance has a **specific heat** value, which is generally defined as follows:

> The specific heat is the amount of heat required to raise the temperature of a unit mass of a substance by one degree.

The specific heat is descriptive of how much heat energy a substance will "hold," or its "capacity," per unit mass for a given temperature change (one degree). For example, by definition of the calorie or kilocalorie, water has a specific heat of 1.0 cal/g/°C (calorie per gram per degree Celsius). By definition, 1 calorie raises the temperature of 1 gram of water by 1°C. The specific heat of water is also 1.0 kcal/kg/K. Why? The kelvin could be replaced with °C, since the temperature interval is the same. A list of specific heats of some substances is given in Table 13.2.

Table 13.2 Specific Heats of Various Substances

Substance	*cal/g-°C or kcal/kg-°C*
Solids	
Aluminum	0.22
Brass	0.094
Copper	0.093
Glass (typical value)	0.16
Ice	0.50
Iron	0.113
Lead	0.031
Soil (typical value)	0.25
Wood (typical value)	0.40
Liquids	
Ethyl alcohol	0.60
Benzene	0.41
Gasoline	0.50
Mercury	0.033
Water	1.00
Gases	
Air	0.17
Steam	0.48

We often say that certain materials "hold more heat." This is because such materials have relatively large specific heats. Since it takes more heat per unit mass to raise their temperatures or "heat them up," they have more stored energy. This is sometimes painfully evident when eating a baked potato or cheese on a pizza. A large portion of potato and cheese is water, which has one of the highest specific heats.

The high specific heat of water arises because much of the added heat goes into the rotational and vibrational energies within the H_2O molecules. Because of its high specific heat, along with its availability and cheapness, water is used to store energy in solar homes (see Chapter 7). Energy stored in large bodies of water such as the oceans or the Great Lakes affects climates. Lakes warm up in the summer, and ocean currents, such as the Gulf Stream, bring warm water to various regions. In the winter the release of the stored energy helps moderate the climate in the surrounding regions.

SPECIAL FEATURE 13.3

Thermometry — Temperature Measurement

Temperature is measured by reproducible changes in the physical properties of materials. As we have seen in this chapter, thermal expansion is a commonly used property in bimetallic and liquid-in-glass thermometers, but it is not the only one. For example, in addition to thermal expansion, electrical and radiation properties are also used.

Various physical properties have different ranges over which they are applicable for temperature measurement or thermometry. Some of the common practical temperature-measuring instruments and their ranges are listed in Table 13.3.

In this feature, some of the not-so-common temperature-measuring devices will be discussed. Although some of the electrical and radiation principles involved will be discussed in later chapters, you should be able to grasp the general ideas from your practical experience.

ELECTRICAL RESISTANCE THERMOMETERS

The electrical resistance of most metals increases with temperature. (Electrical resistance is the opposition of a material to the flow of electric current. See Chapter 19.) These changes are rather large. For example, for platinum a 39 percent change occurs between 0°C and 100°C. Hence, changes in the electrical resistance of a metal can be calibrated to temperature and used as a resistance thermometer. Platinum, which can be used to measure over a temperature range of −260°C to 1100°C, is especially suitable.

If the metal in a circuit is connected to the appropriate meters, the temperature can be read. The metal temperature probe is convenient in many applications (Fig. 13.12).

THERMOCOUPLE THERMOMETERS

The thermocouple thermometer is based on an electrical effect. If two different metal wires, such as copper and iron, are joined together to form a closed loop, an electrical current flows in the loop if one junction is kept at a different temperature than the other (Fig. 13.13).

The electrical current is commonly measured in terms of voltage with a millivoltmeter. The voltage is proportional to the difference in the temperatures of the junctions. If the temperature of one of the junctions (reference junction) is known, the temperature of the other junction can be measured. For example, a cold (reference) junction in ice and water is at 0°C, since ice and water co-exist at this temperature.

Table 13.3 Temperature-Measuring Instruments

Instrument	*Approximate range*
Liquid-in-glass thermometer	
Alcohol	−80°C–100°C
Mercury	−38°C–350°C
Bimetallic thermometer	−40°C–500°C
Electrical resistance thermometer	−272°C–1600°C
Thermocouple	−260°C–1600°C
Optical pyrometer	600°C upward
Infrared pyrometer	−20°C–1700°C

(Continued)

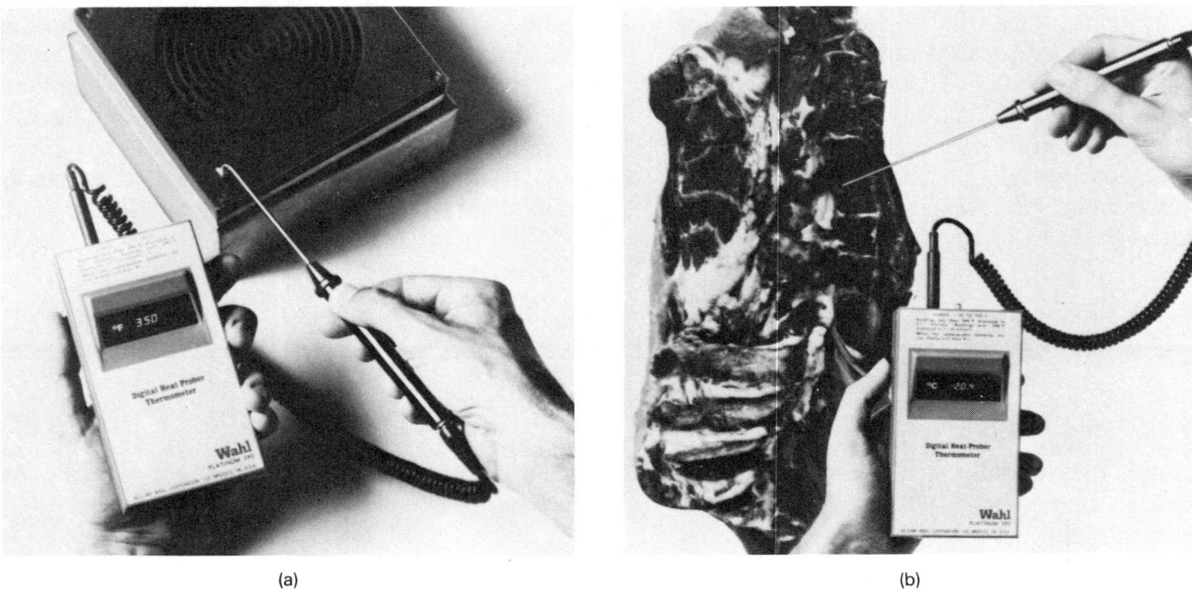

Figure 13.12 Resistance thermometers. Based on the change of electrical resistance of metals with temperature, resistance thermometers (with digital readouts here) are versatile temperature-takers, such as measurement of (a) surface temperatures and (b) interior temperatures of solids and liquids.

The thermocouple is particularly convenient for many applications because of the small size of the metal wire junction. It can be inserted wherever a temperature measurement is desired. The reference junction and electrical meter can be some distance away.

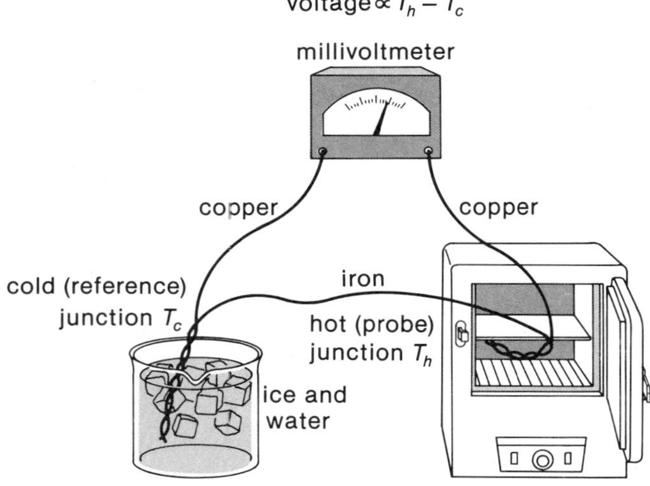

Figure 13.13 The thermocouple thermometer. When the junctions of a thermocouple are at different temperatures, a voltage develops that is proportional to the temperature difference. One of the junctions is used for a reference at a known temperature.

PYROMETERS

Pyrometers make use of the radiation or light emitted by an object. The wavelength or color of the light is proportional to the temperature of the body. For exam-

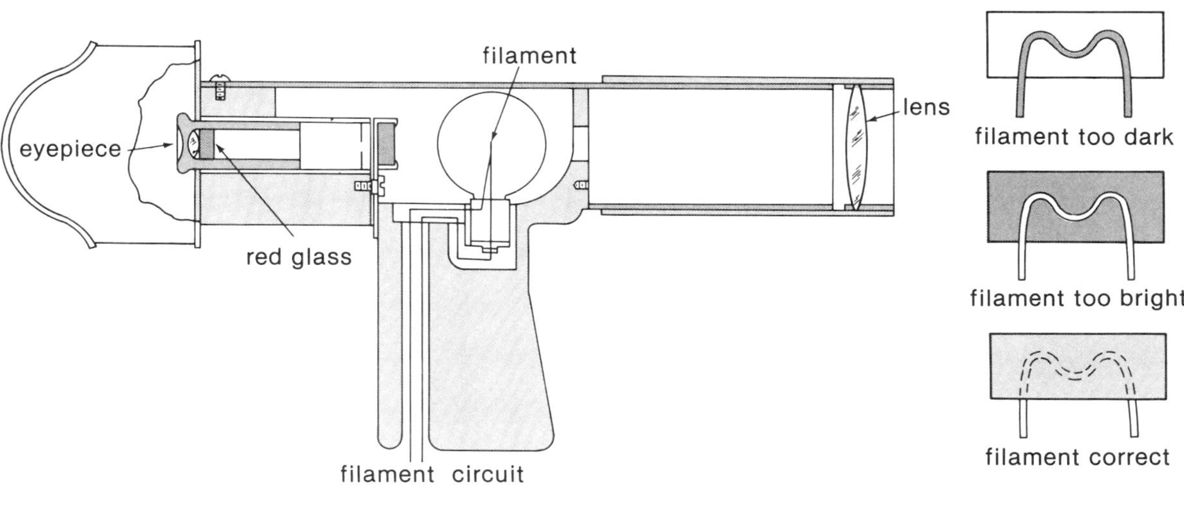

(a)

(b)

Figure 13.14 Optical pyrometer. (a) The filament provides a visual brightness comparison of a temperature "background" that is matched. The filament current is proportional to the temperature. (b) An optical pyrometer being used to measure the temperature of molten steel.

ple, when metals are heated to high temperatures they glow with different colors (emit different radiations) with increasing temperature. The approximate relationships are

Minimum visible red	475°C
Dull red	600°C
Cherry red	700°C
Light red	850°C
Orange	900°C
Yellow	1000°C
Blue-white	1150°C and higher

Thus, the temperatures of very hot bodies can be judged visually by their color and brightness (intensity). Direct visual estimation of temperature is subjective and may be quite inaccurate. An optical pyrometer is an instrument designed to improve visual estimates of temperature by providing the eye with a source of brightness for comparison (Fig. 13.14). The comparison source is a lamp filament. Varying the current in the filament varies its brightness so as to match the brightness of the image of a hot body as seen through the pyrometer. The filament current is calibrated in terms of temperature.

(Continued)

All warm bodies emit infrared radiation. The characteristics of this radiation depend on a body's temperature. Infrared radiation cannot be seen with the human eye, so photocells and other special detectors must be used. An infrared pyrometer is used to detect the infrared radiation from a body, and the radiation is then related to its temperature. These instruments have more precision and accuracy and a greater range than optical pyrometers. Infrared "thermometers" are now used quite extensively (Fig. 13.15). You don't even have to be close to take a body's temperature; you can just point the pyrometer at it.

(b)

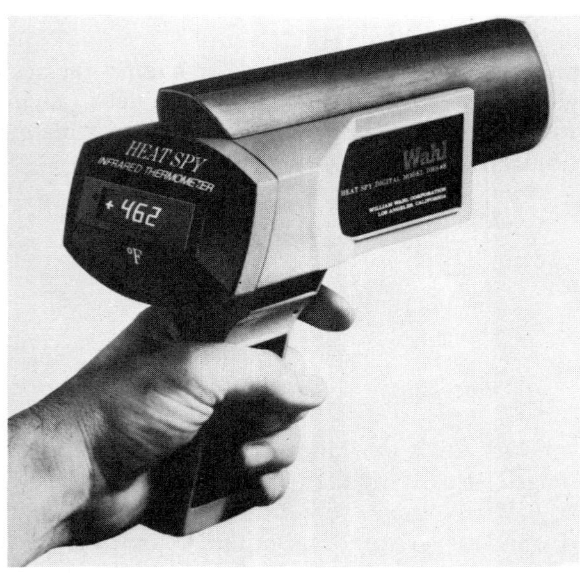

(a)

Figure 13.15 An infrared thermometer (a) and some applications (b) and (c). The temperature of a surface can be determined from the emitted infrared radiation merely by pointing the instrument toward the surface.

(c)

SUMMARY OF KEY TERMS

Temperature a relative measure or indication of hotness or coldness or, more specifically, a measure of the average random translational kinetic energy per molecule of a substance.

Thermal energy the energy associated with the random translational motions of the molecules.

Internal energy the *total* energy (kinetic and potential) contained within a body.

Heat the internal energy that is added to or removed from a substance owing to a difference in temperature.

Thermal expansion the changes in dimensions of substances that occur with changes in temperature.

Thermometer a device that measures changes in temperature through a change in some physical property.

Celsius temperature scale the scale with designations of 0° and 100°, respectively, for the ice and steam points of water. The temperature unit is the degree Celsius.

Fahrenheit temperature scale the scale with designations of 32° and 212°, respectively, for the ice and steam points of water. The temperature unit is the degree Fahrenheit.

Absolute zero the absolute minimum or lower limit of temperature, which is −273°C (−460°F).

Kelvin or absolute temperature scale the scale based on absolute zero; its temperature unit is the kelvin. The ice and steam points of water are 273 K and 373 K, respectively.

calorie the amount of heat required to change the temperature of 1 gram of water 1 degree Celsius.

Calorie a kilocalorie (1000 cal) and the unit associated with the heat energies of foods.

Heat of combustion the heat produced per unit mass of a substance when burned in oxygen.

Specific heat the amount of heat required to raise the temperature of a unit mass of a substance by one degree.

EXERCISES

1. Heat always flows from a substance at a higher temperature into one with a lower temperature. Does it always flow from a substance with more thermal energy into one with less thermal energy? How about internal energy? (*Hint:* Consider dropping a hot BB into a tub of water at room temperature.)
2. Heat might be thought of as one of the "middle men" of energy. Why is this?
3. When you leave an outside door open on a cold day, does the cold come in or the heat go out?
4. What is the difference between the thermal energy and internal energy of a perfect gas?
5. If equal masses of helium and oxygen gases are at the same temperature, do they have equal internal energies? Explain.
6. A piece of ice is placed on a bimetallic strip, as shown in Figure 13.16. What happens?

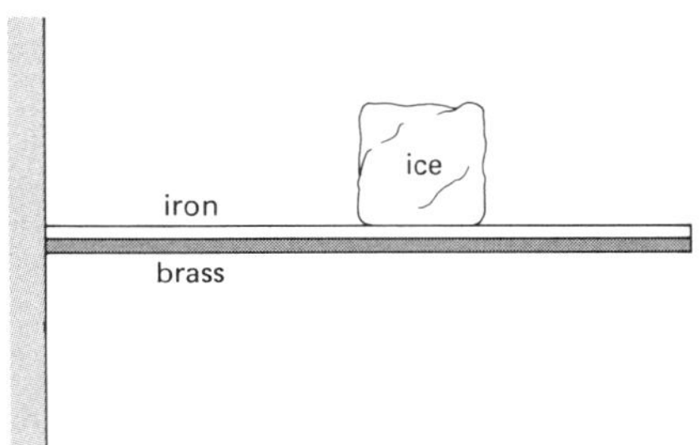

Figure 13.16 See Exercise 6.

7. On a hot afternoon you may hear creaking noises in a house's attic. Creaking may also be heard at night. What causes this?
8. Why are concrete highways and driveways poured in sections rather than in one long strip? Sidewalks are sometimes poured in strips, but groove joints are made. What does this do?
9. (a) Railroad tracks are laid with some space in between. What would happen if the rails were butted up against each other when they were laid on a cold day? (b) On a hot afternoon you may observe gasoline dripping from a recently filled automobile gas tank. Is the owner trying to waste fuel and money? Explain.
10. When cold water is run on a hot plate or a hot liquid is poured in a glass, the plate or glass often cracks. Why?
11. Certain glass and ceramic cooking dishes can be used in ovens for baking, while others cannot. What is the difference?
12. When one drinking glass is stuck inside another an old trick to unstick them is to put water in one of them and run water at a different temperature over the outside of the other. Which water should be hot and which should be cold?
13. In the demonstration shown in Figure 13.17(a), at room temperature the ball fits easily through the ring. However, when the ball is heated, it will not fit through the ring, as shown in Figure 13.17(b). If both the ball and ring are heated, the ball fits through the ring as in the first figure. Explain why.
14. If you turn on a hot-water faucet or the hot water in a shower to a moderate flow, you may observe that the

(a)

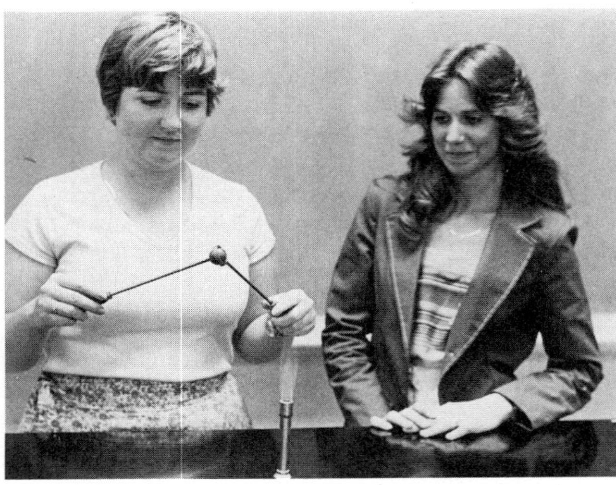

(b)

Figure 13.17 See Exercise 13.

flow decreases after a while, but this is not observed for cold water. What causes the flow of hot water to change?

15. When a thermometer is inserted into hot water, the mercury or alcohol is sometimes observed to fall slightly before it rises. Why?
16. (a) Most substances expand with increasing temperature. Explain this expansion in terms of molecular theory.
(b) A thermometer initially at room temperature is used to measure the temperature of a hot substance. Explain how this is done in terms of molecular theory. Is it possible that the thermometer might introduce error in the temperature measurement?
17. Water has unique thermal expansion properties (Fig. 13.18). As the temperature increases between 0°C and 4°C, the volume *decreases*. Compare this graph with Figure 10.11 and explain the differences. Explain why the volume of water increases in going from 4°C to 0°C. (*Hint:* See Chapter 10.)

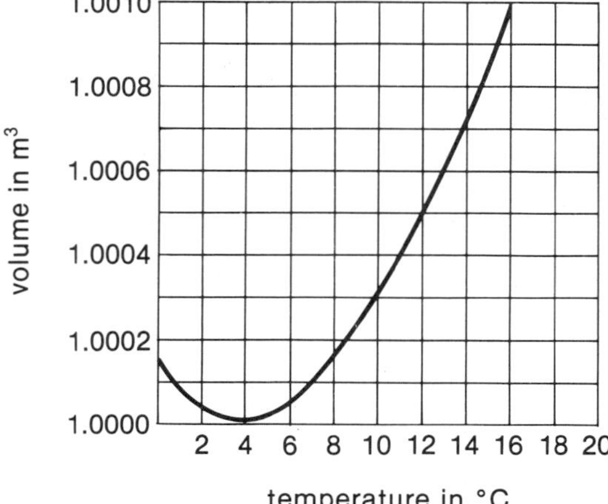

Figure 13.18 See Exercise 17.

18. In light of the preceding exercise, would water be a good liquid for a liquid-in-glass thermometer? How would such a thermometer behave near the ice point?
19. Why are thermal expansion properties an important consideration for materials used in tooth fillings?
20. What is the most commonly used temperature scale in the world?
21. Which unit is larger, a degree Fahrenheit or a kelvin, and by how much?
22. On what temperature scale would a temperature change of 10 units be the largest?
23. What is the interval between the ice and steam points of water on the Kelvin scale? What is room temperature on the Kelvin scale?
24. A dilute (perfect) gas at constant pressure decreases in volume as the temperature is lowered. What does the perfect gas law predict the volume to be at absolute zero? Does this actually happen? Explain.
25. A quantity of ideal gas is at room temperature (20°C). If enough heat is added to double its internal energy, what would be the temperature of the gas? (*Hint:* Change to kelvins and then change back.)
26. The calorie is defined in terms of the Celsius temperature unit. How would the definition differ if it were defined in terms of the kelvin?
27. One calorie is equal to 4.2 J. Give a definition of a joule in terms of heat measurement.
28. In Table 13.1, the heats of combustion are given in cal/g or kcal/kg. Are these ratios the same? Explain.
29. Why would scrambled eggs have a greater heat of combustion than boiled eggs? (See Table 13.1.)

30. Gasohol is a mixture of gasoline and alcohol. Does gasohol have a greater intrinsic energy value than gasoline?
31. When you eat a hot apple pie, you may find that the crust is only warm, but you may burn your mouth on the apple filling. Why is this?
32. If equal amounts of heat were added to two continers of water and the temperature change of the water in one of the containers was twice that of the other water, what could you say about the quantities of water in the containers?
33. Given equal masses of copper and lead at the same temperature, to which heat is added so each undergoes the same temperature change. To which metal was more heat added and how much more? (*Hint:* See Table 13.2.)
34. At a beach, the sand may be very hot to your feet, but the water is relatively cool. What does this tell you about the specific heats of sand and water?
35. Blocks of metal with equal masses are heated to the same temperature and placed on a block of paraffin (Fig. 13.19). Explain why they melt to different depths when they are initially at the same temperature. Give an estimate of the specific heat of zinc.

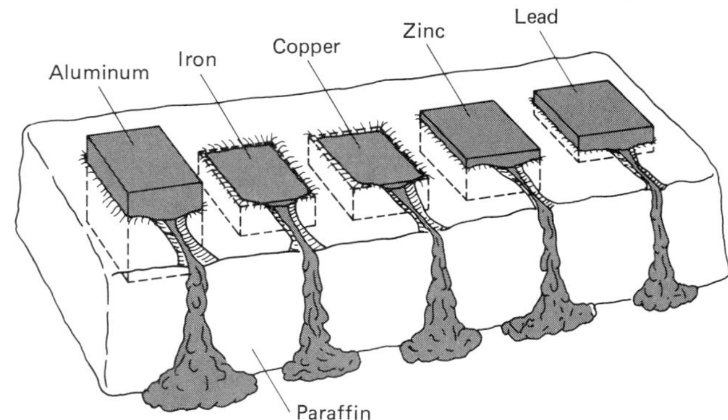

Figure 13.19 See Exercise 35.

14

Heat Transfer and Change of State

Heat Transfer

Heat is the transfer of energy caused by a temperature difference. It goes on continuously, and in many instances we try to control and channel the flow of heat from one place to another for our benefit. For example, the Earth continuously receives heat from the Sun, and we use heat transfer in cooking foods and in heating (and cooling) our homes. Hence, heat transfer is an important practical consideration.

How is heat transferred? We know this requires a temperature difference, but what are the transfer mechanisms? There are three ways to transfer heat from one place to another. These methods are called conduction, convection, and radiation. Let's take a look at them.

CONDUCTION

We can boil a pot of coffee on a stove because heat is *conducted* through the coffee pot from the hot stove burner. Heat transfer by conduction takes place through molecular collisions. To visualize heat conduction over a longer distance than through the bottom of a coffee pot, consider holding one end of a metal rod or poker in a fire. There is a temperature difference between the ends of the rod, so conditions are right for heat transfer.

The molecules and electrons in the heated end of the rod become agitated and move rapidly. These molecules and free electrons collide with their less-energetic neighbors toward the cooler end of the rod and pass some of their extra energy to them. These molecules in turn collide with and pass energy on to their cooler neighbors, and so on. As a result, energy is passed down the rod, in a "fire-bucket brigade" fashion, and heat is *conducted* through the rod. Energy arriving at the cooler end of the rod causes the molecules there to vibrate faster, and the temperature of this end rises. This may be painfully evident. Notice that there is no *net* mass movement in the heat conduction process.

How well a substance conducts heat depends on the electrical bonding of its molecular structure. Solids are generally the best thermal conductors, with metals taking first prize. In addition to collisions between bound molecules, there are also a large number of electrons in a metal that are "free" to move around (not permanently bound to a particular molecule or atom [see Chapter 10]). These free electrons contribute significantly to heat transfer or thermal conductivity. (As will be learned in Chapter 18, these free electrons of a metal are also responsible for electrical conductivity.)

Nonmetal solids have relatively few free electrons and are poorer conductors. Cooking utensils are made of metals because of their good thermal conductivity, but potholders are made of cloth and pot handles are made of wood or plastic because they are good *thermal insulators* (poor conductors).

Liquids and gases are, in general, poor conductors. Liquids are better thermal conductors than gases because their molecules are closer together. Gases are poor conductors since their molecules are relatively far apart and collision interactions are relatively infrequent. (The mobility of the molecules in liquids and gases gives rise to another form of heat transfer called convection, which will be discussed below.)

Many solids, such as cloth, wood, fiber glass, and Styrofoam, are porous and have a large number of minute air pockets or spaces that add to their poor conductivity. We use such materials as thermal insulators. Examples include fiber glass insulation in our homes and Styrofoam "coolers" (Fig. 14.1). Insulating materials with air pockets are used instead of free air spaces because the air could move around in free spaces and promote heat transfer. (See the following discussion on convection.)

QUESTION: People commonly say that an ice chest or cooler "keeps the cold in" or that home insulation "keeps the cold out." Is this correct?

ANSWER: No. Poor conducting materials or insulators are used in these cases to "keep heat out of the cooler" and to "keep heat in" the house. (In the case of air conditioning, home insulation helps keep the heat out.) Remember that heat is what is transferred, *not* cold. Cold is the lack of heat.

The conductive heat-transfer capability of a substance is characterized by a quantity called thermal conductivity. Some typical values of thermal conductivities of materials are given in Table 14.1. Notice from the units of thermal conductivity that it gives the rate of energy flow per temperature difference ΔT. (W (watt) = joule/s, and W/K = J/s/K.) The length unit (m) arises from dimensional considerations of the material (area and thickness). Your instructor may wish to explain this more fully. We will use the values of the thermal conductivities only for comparison purposes.

Figure 14.1 Poor thermal conductivity. Materials such as Styrofoam, which have a lot of air spaces, are poor thermal conductors. This property helps keep objects inside a Styrofoam cooler cold for long periods of time. Notice the 2-liter (metric) bottles.

Silver is the best conductor, with copper a close second (both are metals). Air is way down on the list, as you might expect. You might not expect snow to be a good insulator. In the far north, dogs and other animals often burrow in the snow to sleep. There are a lot of air spaces in the snow, as well as in their thick coats. As a poor conductor, the snow slows the loss of body heat. Also notice from the table that the conductivity of a vacuum is zero. Why?

QUESTION: In a bedroom with a tile floor (common in dorms), when you rise and shine with both feet hitting the floor at the same time, you might remark, "Oh, that floor's cold!" A throw rug is usually obtained to avoid this discomfort (Fig. 14.2). How does this help when the rug and floor are in contact and in thermal equilibrium at the same temperature?

ANSWER: The rug and the floor are at the same temperature, but their thermal conductivities are

Table 14.1 Some Typical Values of Thermal Conductivities

Material	Thermal conductivity (W/m · K)
Good conductors	
Silver	425
Copper	390
Aluminum	235
Iron	80
Average conductors	
Ice	3.5
Brick	1.0
Concrete	0.8
Floor tile	0.7
Water	0.6
Glass	0.4
Poor conductors (good insulators)	
Wood	0.2
Snow	0.16
Fiber board	0.1
Cotton	0.08
Glass wool	0.04
Styrofoam	0.033
Air	0.026
Vacuum	0

different (see Table 14.1 and assume the rug is cotton). The tile floor only *feels* colder because it conducts heat from your feet faster. You should more correctly say, "Oh, that floor has a high thermal conductivity!"

Figure 14.2 Is a tile floor colder than a rug? See Question and Answer on previous page.

CONVECTION

Compared to solids, liquids and gases are not good thermal conductors, but the mobility of the molecules in fluids permits heat transfer by another method—convection. Unlike conduction, convection involves mass transfer. Heat is not transferred by molecular collisions, but rather the heat is carried by the moving fluid en masse. For example, when you turn on the hot-water faucet, heat is transferred from the water heater to the kitchen or bathroom with or by the moving water.

Heat transfer can occur within a liquid or gas by convection (mass movement). When a fluid is heated at a surface or in one region, it expands. Becoming less dense, it rises or is buoyed upward in the surrounding cooler fluid (Archimedes' principle, Chapter 11). You may have noticed these rising "currents" in water being heated in a glass coffee pot or in air near a heating unit (Fig. 14.3).* When a portion of the fluid rises, colder, more dense fluid descends to take its place, and a convection cycle is set up. This circulation mixes the fluid and distributes the heat.

On a larger scale, thermal convection cycles stir the atmosphere and result in surface winds, particularly near large bodies of water. During the day, the land (soil) temperature increases faster than the water temperature. The land has a lower specific heat than water. The water also has mixing currents that serve to distribute the solar energy. The air near the hot land surface is heated by conduction. The hot air rises and a convection cycle is set up. The bottom part of this cycle gives rise to a sea breeze. You may have noticed this continuous "on-shore" sea breeze at the beach on a sunny day, which makes it more pleasant.

At night, the situation is reversed. The water cools off more slowly than the land. As a result, the air over

* These currents in a fluid are seen because of the optical effect of the bending of light caused by temperature and density differences in the fluid. Another example is "seeing" hot air rising from a blacktop road in the summer. See Chapter 24.

Figure 14.3 Convection. Heat transfer by convection involves mass transfer. Rising hot air sets up a convection cycle that distributes heat around the room.

the water is warmer and the convection cycle is reversed, giving rise to a land or "off-shore" breeze.

A larger, but similar, effect occurs for summer and winter (compared to the heating effects of day and night), particularly in southeast Asia. The on-shore summer winds, called summer monsoons, bring moist air to the land, giving rise to heavy rains. In the winter (also warm in this area), the off-shore winds or winter monsoons bring cool, dry air from the mountains.

Not all convection cycles are natural. In some cases the medium of heat transfer is moved by mechanical means. Common examples are the cooling systems of automobile engines and the forced-air heating of homes and buildings. In homes with forced-air heating, a blower fan aids the convectional transfer of heated air (Fig. 14.4). Air ducts allow the return of cooler air. Older homes used natural convection heating.

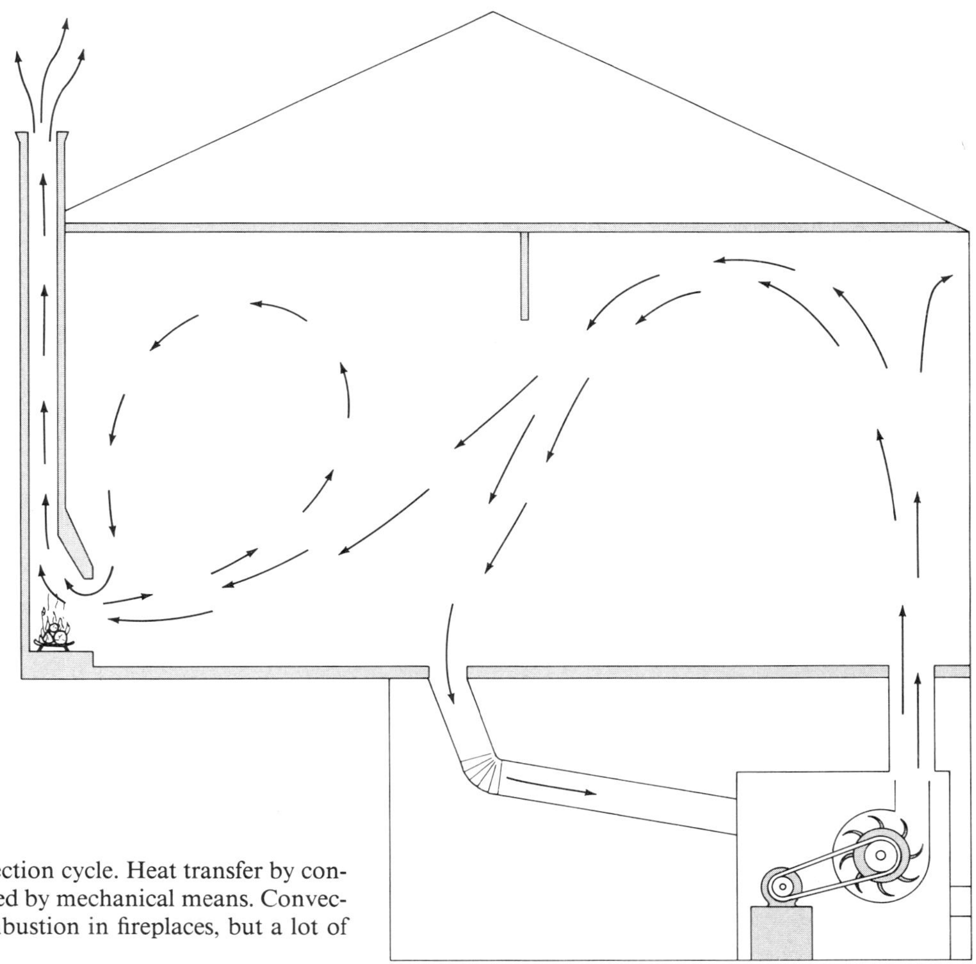

Figure 14.4 Forced-air convection cycle. Heat transfer by convection in most homes is assisted by mechanical means. Convection provides fresh air for combustion in fireplaces, but a lot of energy is lost up the chimney.

Convection currents are also important to the burning of a fire in a fireplace. The fire gets fresh air needed for combustion from convection currents, which are set up by the rising hot air in the chimney and aided by the "draft" of the Bernoulli effect (Chapter 12). The air flow can come from other portions of the house and may take heated air up the chimney, or from the air in a closed room, that is replaced by colder air coming through cracks around windows or under doors. (A fireplace would not operate in a completely leakproof house.) In either case, a fireplace is not very efficient. The efficiency is increased in fireplaces with "heatolators," in which air pipes in the chimney are heated and the warmed air is circulated back into the room.

RADIATION

Both conduction and convection require a medium or matter for energy transport. Yet energy is transmitted to us from the Sun through the void of space. Also, part of this solar energy is transmitted through the atmosphere directly to the Earth's surface. This couldn't be by conduction, since air is a very poor conductor. Convection is out too. Convection cycles begin at the Earth's warm surface and generally remove heat. Another heat transfer process, called radiation, is responsible here.

Radiant energy from the Sun is transmitted through space by means of electromagnetic waves, generally called **radiation,** which require no transport medium. These waves originate from accelerated charged particles (see Chapter 21). Electrons vibrating in matter have acceleration, so all matter, whether hot or cold, radiates energy. The molecular vibration rate depends on the thermal energy and temperature of an object, so objects at different temperatures radiate different types of electromagnetic waves. By analogy, waves with different characteristics are generated in a rope when you shake the end of the rope up and down at different rates (Fig. 14.5).

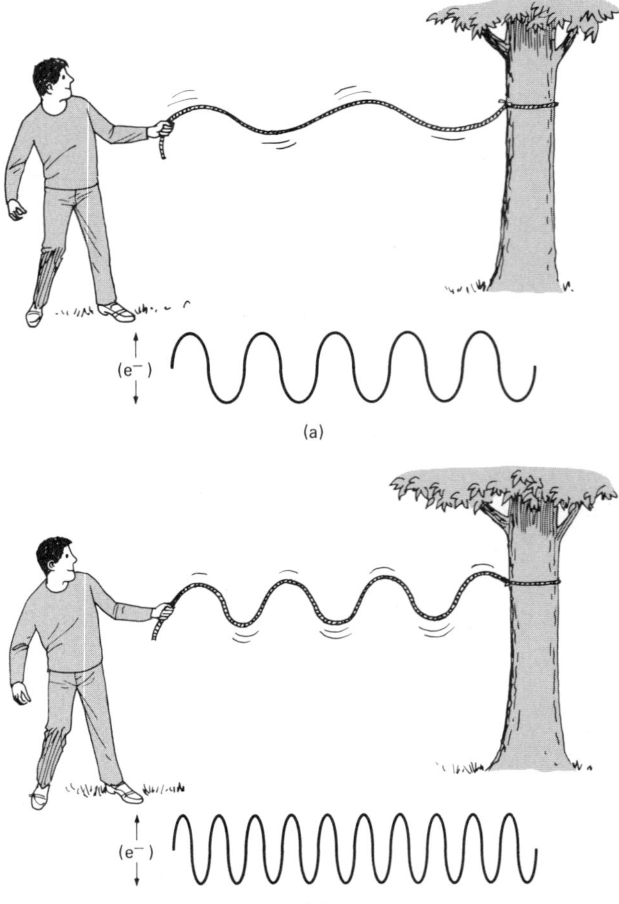

Figure 14.5 Electromagnetic radiation. Electromagnetic waves originate from accelerated charged particles. Electron vibrations in matter depend on the temperature, so objects at different temperatures radiate different types, or electromagnetic waves of different frequencies. By analogy, different waves are generated in a rope when the shaking rate is varied. These are not electromagnetic waves.

Waves and electromagnetic radiation will be discussed more fully in later chapters. For the present discussion, we will think generally of *radiant energy* that travels through space. One body emits radiant energy, and it may be absorbed by another body. Recall the selective absorption by the atmospheric gases of the infrared radiation emitted by the Earth in the greenhouse effect (Chapter 7).

Another example is observed for visible radiation. A dark object is dark because it absorbs most of the visible radiation that hits it, whereas a light-colored object is light because it reflects radiation. You have no doubt noticed that a black asphalt driveway on a sunny summer day is much hotter on your bare feet than a concrete driveway or sidewalk. A good absorber is also a good emitter of radiation of the same type, so a dark object loses absorbed heat faster by radiation than a light-colored object does.

We have all experienced heat transfer by radiation when near an open fire. At an appreciable distance, you feel the heat on your face and hands from the fire. This cannot be by convection, since in general the air movement is toward the fire as part of the convection cycle initiated by it. Also, conduction is out. (Why?) Visible radiation is emitted by the burning material. But most of the heating effect comes from the invisible infrared radiation emitted by glowing embers or coals. We feel this radiation because it is absorbed by the water molecules of our skin. (Tissue is about 85 percent water.)

The water molecule has a natural vibration that is the same as an infrared radiation vibration, so the radiation is absorbed. This energy absorption causes the molecules to vibrate faster, which increases the kinetic energy and the temperature. Because of this, infrared radiation is commonly called "heat rays," and infrared lamps are used to keep foods warm in cafeterias and to warm aching joints in medical therapy. You may have noticed another example of this infrared radiation heating when the radiation was "turned off." Have you ever been "catching a few rays" (sunbathing) on a hot, sunny day when the Sun goes behind a cloud? It suddenly feels cooler. This is primarily because the infrared portion of the sunlight is absorbed by the cloud and not by you. A practical application of heat transfer by radiation is discussed in Special Feature 14.1.

Change of State

In general, when heat is added to or removed from a substance its temperature increases or decreases. However, at certain temperatures a substance will undergo changes of *state* or changes of *phase,* for example, solid to liquid or liquid to gas. While this is going on, say with heat being added, the temperature does not change. So where does the energy go?

Let's look at the common examples of the phase changes of water. Suppose you have a chunk of ice at $-10°C$. If heat is added, the temperature of the ice increases until it reaches $0°C$, the melting point. At this point, additional heat causes the ice to melt, and the ice and water coexist at $0°C$ until all of the ice is melted. Once the ice is melted, adding more heat causes the temperature of the water to increase. A similar situation occurs for the phase change from liquid to gas (water to steam) at the boiling point, $100°C$. Additional heat causes the temperature of the steam to rise. The graph in Figure 14.7 shows how the temperature of water varies with heat.

As you know, if the heat doesn't go into the random thermal energy associated with temperature, it must go

SPECIAL FEATURE 14.1

The Microwave Oven

Heat transfer by radiation is a relatively new boon to cooking in the form of the microwave oven. Microwaves are a type of electromagnetic radiation that are also absorbed by water molecules. In a microwave oven (Fig. 14.6), the microwaves are distributed by reflection from a metal stirrer and the metal walls. Because they reflect the radiation, the walls of the oven do not get hot. Microwaves pass through plastic wrap and glass or ceramic cooking dishes but are absorbed by water molecules in the food, causing it to be heated. The microwaves do not penetrate completely through the food, but are absorbed in the outer layers. Heat is then conducted to the interior. You are advised to let large food items sit for a while after the oven has shut off so they will be heated or cooked throughout.

An important safety feature of a microwave oven is the automatic shut-off when the door is opened. If you were able to reach into the oven while it was running, you could be injured, since the water molecules in your hand would absorb the radiation.

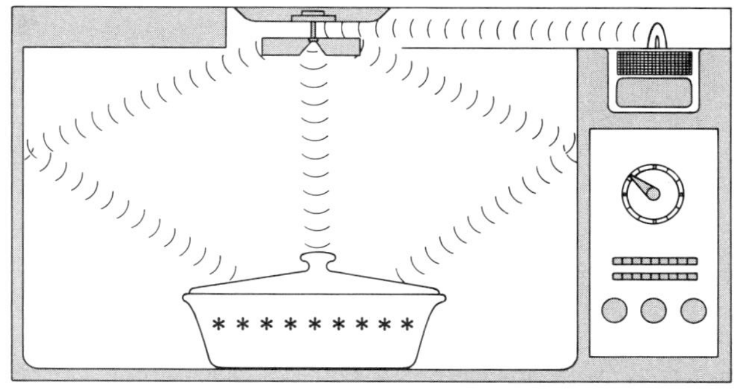

Figure 14.6 Heat transfer by radiation. In a microwave oven, microwaves (a type of electromagnetic radiation) are absorbed by water molecules in the food, causing it to become hot.

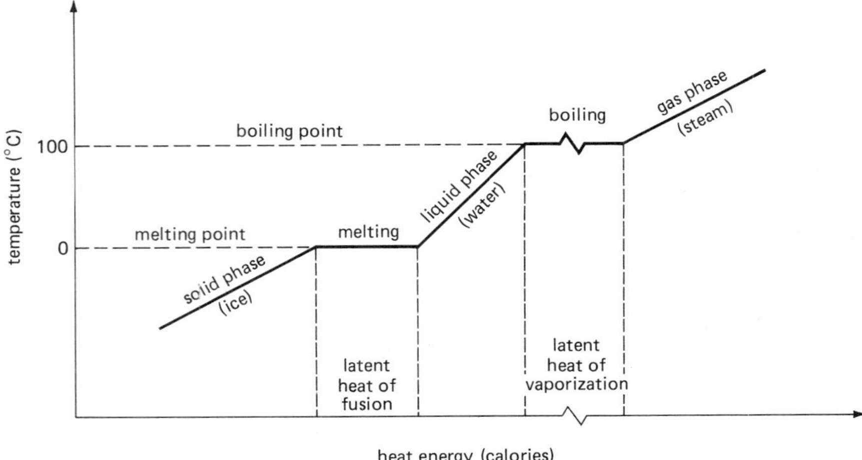

Figure 14.7 Change of state. A graph of how the temperature of water varies with heat. Phase changes or changes of state occur along the horizontal segments of the graph.

into other parts of the internal energy of the substance. This is what happens during a phase change. The energy goes into the work of breaking up the lattice of a solid in becoming a liquid or in separating the liquid molecules in becoming a gas. We call the heat energy involved in a phase change latent heat (latent means "hidden").

The latent heat involved in the solid-liquid phase change is called the latent heat of fusion, and for the liquid-gas phase change it is called the latent heat of vaporization. For water, the latent heat of fusion at 0°C is 80 cal/g (or kcal/kg), and the latent heat of vaporization at 100°C is 540 cal/g. That is, 80 calories are required to melt each gram of ice at 0°C, and 540 calories are needed to convert each gram of water to steam at 100°C (Fig. 14.8).

The term *latent* or *hidden* heat may be more descriptively appreciated for the reverse processes of condensing and freezing. By the conservation of energy, when 1 gram of steam condenses to water, 540 calories of heat are given up. This is why steam burns are more serious than burns from boiling water, even though both are at 100°C. When a gram of steam condenses on the cooler skin, an additional 540 calories is released that was seemingly "hidden."

When water freezes, for example, when rain changes to sleet or water vapor forms ice crystals, latent heat is given up. This is why sleeting and snowing are warming processes in the atmosphere. It was once common practice to keep large containers of water in cellars where canned goods were stored in the winter. If the temperature dropped below freezing (0°C), the water, which had a higher freezing point than the liquids in the canned goods, would freeze first and release latent heat. The released energy would moderate the cellar temperature and help prevent it from dropping to the point at

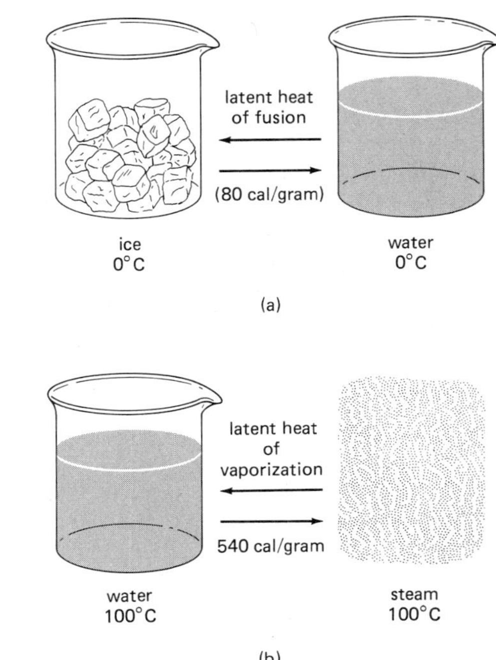

Figure 14.8 Latent heat. (a) The latent heat of fusion for water is 80 cal/gram. (b) The latent heat of vaporization for water is 540 cal/gram. These amounts of heat must be added or removed for each gram of water to change the phase at the respective phase change temperatures.

which the jars of fruit and vegetables would freeze and burst.

There is another common change of state — solid to gas. Some solids, such as dry ice (solid carbon dioxide) and mothballs (*p*-dichlorobenzene) vaporize directly from the solid form. This direct change from the solid to gas phase is called sublimation. It requires latent heat of sublimation.

Figure 14.9 Air from a fan feels cool because it promotes the cooling rate and evaporation.

EVAPORATION

You don't have to be at the boiling point of a liquid to have a change of phase. The slow evaporation of water from wet clothes hanging on a line or from an open container of water at room temperature becomes evident after a time. The molecules in a liquid are moving around at many different speeds. (The temperature of the liquid is associated with an average of the speeds in terms of the average kinetic energy.) Some of the particularly fast molecules at or near the surface of the liquid may have enough energy to overcome the surface tension forces and break free of the liquid, or become "evaporated." Since the evaporated molecules take energy with them, the average kinetic energy of the liquid and its temperature are lowered. Hence, evaporation is a cooling process. See Special Feature 14.2.

Evaporation plays an important role in the cooling of our bodies. When hot, we perspire. The evaporation of perspiration takes heat from our bodies, and we feel cooler. How about rubbing alcohol? You really feel cool when it is rubbed on your skin. This is because alcohol evaporates more readily and cools more quickly. Air circulation also makes us feel cooler. Have you ever stood in front of a fan and thought how cool the circulating air feels? In a closed room, the air temperature is pretty much the same everywhere, and the fan is just blowing hot air from one side of the room to the other (Fig. 14.9). So why do we feel so much cooler?

SPECIAL FEATURE 14.2

Does Hot Water Freeze Before Cold Water?

This is a common question, and the answer depends on the conditions and principles discussed in this chapter. Suppose that two *covered* identical pans with equal amounts of water at temperatures of 20°C and 40°C, respectively, are placed in a freezer at 0°C. All other things being equal (except the cooling rates), the pan with the lower initial temperature (20°C) would freeze first. The pan at the higher temperature would lose heat at a faster rate (greater temperature difference), but it has more heat to lose. It would cool down to 20°C, then cool at the same rate as the other pan did during this time.

Think of it this way: Suppose you were going to run 20 m and another person was going to run 40 m on the same track. The other runner might run at a faster rate to your 20 m mark, but then if he slowed down and ran the last 20 m at the same rate as you, he would never be able to catch up, and you'd get to the finish line first.

It might be possible for the hot water to freeze first under some special conditions. Suppose the pans were *uncovered* and the hot water were very hot. In cooling down, more of the hot water would evaporate, which would leave less water (mass) in the pan to cool to freezing. Also, a hot pan may melt into the frost layer on the freezer shelf and with better contact would have greater conductivity and a faster cooling rate.

There are a couple of reasons. Your body is usually warmer than the surrounding air, so it loses heat to the air by conduction. In still air, the air around your body is warmed, and the temperature difference becomes less. This reduces your cooling rate. Air circulated by a fan carries the warmed air away, and your cooling rate increases. The air circulation also promotes cooling by evaporization. In still air, the evaporated water molecules from perspiration can collide with the molecules of the air. Being knocked backward, they may condense on you, which is a warming process. (You probably have had water condense on you in a moist bathroom after drying off from a shower and felt a bit warm.) Much of the cooling by evaporation can be cancelled out by the warming of condensation. By moving the air with a fan, you remove the evaporated molecules and promote cooling by evaporation.

So, air circulation or wind plays an important part in how cool or cold we feel. At moderately low temperatures in a brisk wind, we may feel extremely cold. The wind promotes the loss of body heat, which adds to the chilling effect. The effect of wind on how cold we feel is expressed in terms of a wind chill index or wind chill factor. This is the temperature in still air that would have the same cooling effect as that of wind on exposed flesh for a given air temperature. For example, if the air temperature is 10°F and the wind speed is 20 mi/h, the chill factor is −24°F; that's how cold it would *feel* on bare skin (see Table 14.2, where customary units are used by the U.S. Weather Service).

HUMIDITY

Another factor that affects our comfort is humidity. Humidity refers to the amount of water vapor in the air. The air always contains water vapor as a result of evaporation from oceans, lakes, the ground, and so on, and the amount varies.

The humidity may be expressed in terms of *absolute* humidity, which is simply the actual amount of water vapor in a given volume of air, say mg/m^3. The atmospheric water vapor can be thought of as being in solution with the air, similar to table salt being dissolved in or in solution with water. The amount of water vapor or table salt in solution depends on temperature. The greater the temperature, the more vapor a given volume of air will hold. At a particular temperature, a volume of air or water can hold only so much water vapor or table salt. When this amount is reached, we say the solution or air is saturated.

It is more common to express humidity in terms of relative humidity. (You hear and see this on the weather report; see Fig. 14.10). **Relative humidity** is the ratio of how much water vapor a volume of air actually holds to what it could hold (saturated) at that temperature. This fraction is usually expressed as a percent. For

Table 14.2 Wind Chill Index (or Factor)*

		Air temperature (°F)					
Calm	*30*	*20*	*10*	*0*	*−10*	*−20*	*−30*
Wind speed (mi/h)							
5	27	16	7	−6	−15	−26	−35
10	16	2	−9	−22	−31	−45	−58
		Very cold					
15	11	−6	−18	−33	−45	−60	−70
20	3	−9	−24	−40	−52	−68	−81
			Bitter cold				
25	0	−15	−29	−45	−58	−75	−89
30	−2	−18	−33	−49	−63	−78	−94
				Extreme cold			
40	−4	−22	−36	−54	−69	−87	−101

* Find the air temperature in the top row and the wind speed in the left column. Read the chill factor where the corresponding row and column intersect. For example, a calm air temperature of 10°F and a wind speed of 20 mi/h together are equivalent in cooling effect to a temperature of −24°F.

Figure 14.10 Relative humidity. The humidity on weather reports is given in terms of relative humidity. At 44%, this means the air holds 44% of the water vapor it could hold at the current temperature. (The barometer reading is in inches of mercury and the winds are from the south at 8 mi/h.)

Figure 14.11 Cumulus cloud. Clouds are formed in rising air that cools below the dew point. Rising air commonly forms big, billowy cumulus clouds in the summer.

example, if the relative humidity is 50 percent, a volume of air is "half full" or contains half the water vapor it could hold at that temperature. When the relative humidity is 100 percent, the volume of air is saturated, or it holds all the water vapor it can.

Since the amount of water vapor the air can hold depends on temperature, as the air temperature decreases the relative humidity increases (the same amount of water vapor in a volume of air that can hold less). The temperature may be decreased until the relative humidity is 100 percent. The temperature at which this occurs is called the dew-point temperature, or just the dew point for short.

At the dew point and below, the water vapor condenses. This is analogous to cooling a table salt–water solution to the temperature at which the salt crystallizes out of solution. The water molecules do not coalesce or stick together in mid-air collisions but condense on blades of grass as dew or on particles in the air to form visible droplets. Masses of these visible droplets form clouds. Clouds are formed in rising air that cools below the dew point. Convection cycles with rising air are a major method of cloud formation. You can watch big, billowy cumulus clouds form on a hot summer day (Fig. 14.11). Under the right conditions, some of the cloud droplets grow by condensing water vapor. These larger droplets then coalesce and form rain drops. Then, we may have a dark cumulonimbus cloud or "thunderhead."

If the relative humidity is high, along with the temperature, then it is hot and "muggy." Since the air has most of the water vapor it can hold, the evaporation from our bodies is small, and we feel hot. We feel most comfortable at room temperature (20 °C) when the relative humidity is about 50 or 60 percent. At very low relative humidities, the rapid evaporation can cause drying of the nose and throat. We may feel cool, but uncomfortable. To prevent these conditions by controlling the humidity, many people use humidifiers in their homes in the winter and dehumidifiers in the summer. (A humidifier may be a pan of water.)

Boiling and Freezing

The boiling and freezing (melting) points of substances vary widely. This is due to differences in the molecular bonding of the substances. For example, the boiling and freezing points of water are 100 °C and 0 °C, respectively; for silver, 2212 °C and 960 °C, respectively; and for nitrogen, −196 °C and −210 °C, respectively. All these phase-change temperatures are at a pressure of one atmosphere (standard pressure).

Pressure makes a difference on the boiling and freezing points. Let's first take a look at boiling. Why does a liquid boil? As heat is added to a liquid and its temperature increases, the average kinetic energy of its molecules increases. With more highly energetic mole-

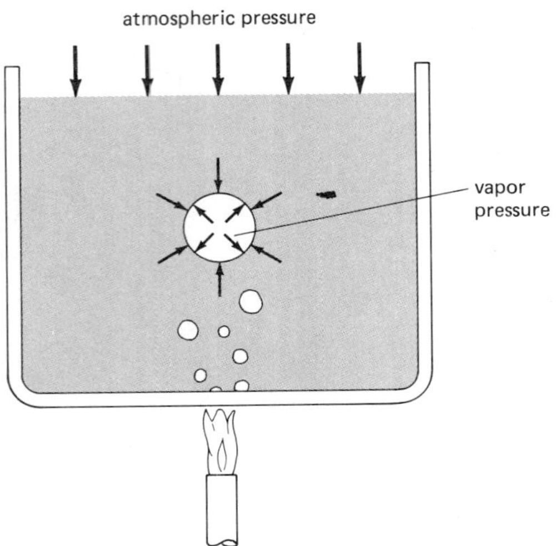

Figure 14.12 Boiling. When the vapor pressure in bubbles is greater than the external pressure, boiling starts. The bubbles rise and break through the surface, and the gas escapes.

cules, the evaporation rate also increases. As the temperature is further increased, a point is reached (the boiling point) at which the average molecular energy is equal to the work necessary to change the liquid into a gas (100 °C at one atmosphere of pressure for water). Unless the molecules in the liquid are moving fast enough, the atmospheric pressure on the surface of the liquid keeps the gas (bubbles) from forming and the phase change from occurring.

More heat produces vapor bubbles in the liquid, usually near the bottom of the container where it is being heated and is hottest (Fig. 14.12). If the vapor (gas) pressure in the bubbles is greater than the pressure above due to the atmosphere (and the weight of the liquid†), boiling starts. The bubbles rise and break through the surface, and the gas escapes.

Thus, we can keep a liquid from boiling, or raise its boiling point, by increasing the pressure on the liquid. A practical application of this is given in Special Feature 14.3.

At reduced pressures, the boiling points of liquids are lower. For example, at high altitudes, where the atmospheric pressure is less (than at sea level), the boiling point of water is less than 100 °C. At Pike's Peak the atmospheric pressure is about 600 torr (mm Hg), and water boils at about 94 °C rather than at 100 °C. This has an effect on cooking times.

† This is negligible compared with atmospheric pressure, in most cases.

By reducing the pressure on a liquid, you can have boiling without external heating. This is a common demonstration using water in a vacuum jar. As air is evacuated from the jar, the pressure and hence the boiling point of the water are lowered. Eventually, the boiling point is lowered to the temperature of the water, and it boils "without heating." If this pressure is maintained, the boiling quickly stops because the latent heat of vaporization is taken from the internal energy of the water. This cools the water and lowers its temperature to below the boiling point.

If the pressure is reduced further, the boiling and cooling continues until the temperature is reduced to the freezing point and ice forms over the surface of the boiling water—boiling and freezing at the same time! Here we have water coexisting in all three phases—solid, liquid, and gas. This is known as the triple point, and it occurs at a pressure of 4.6 torr and a temperature of 0.01 °C. This fixed point is used in the definition of the known temperature unit and scale (see Table 1.2).

The boiling points of all liquids generally increase with increasing pressure. But how about the freezing points? Most liquids contract on freezing or solidifying. Increased pressure helps this process along and thereby raises the freezing point. However, there are some exceptions, most notably water. The open molecular structure of freezing water causes an expansion or increase in volume, so it is less dense as a solid (Chapter 10).‡

Because the volume of a given amount of water is less than that of its ice in solid phase, it would seem reasonable that increased pressure on ice would tend to lower the freezing point temperature and make it melt or have a smaller volume. This is the case, and increased pressure lowers the freezing (melting) point of ice. The melting point isn't lowered much, only about 0.0075 °C per atmosphere of pressure increase.

It is sometimes believed that the melting-point reduction due to pressure is the principle of ice skating. An ice skater is able to skate or glide on the skates because of a lubricating thin film of water between the skate blades and the ice. The pressure ($p = F/A$) due to the weight of a person on the small area of a skate blade is large, perhaps 10 atmospheres. However, it would take a pressure increase of 100 atmospheres to reduce

‡ Other exceptions are gallium, bismuth, and certain alloys. These alloys are used in type metal that expands on solidifying, allowing the formation of raised type for printing. Another alloy, called gray iron, expands slightly on solidifying and is used to make metal castings against a mold, which reproduces its details. Many machine parts are made by casting iron (cast-iron parts).

SPECIAL FEATURE 14.3

The Pressure Cooker

If the pressure on a liquid is increased, its boiling point is raised. This is the principle of the pressure cooker, which is commonly used to cook foods (Fig. 14.13). In a closed container (with an appropriate safety valve), the heating of water or some other liquid causes increased pressure on the water due to the evaporated gas above the water. As a result, the boiling point is increased, and the water and food to be cooked can be heated to above 100°C. At a higher temperature, the food cooks in less time. For example, at a pressure of 1.4 atmospheres, the boiling point of water is 110°C; at 4.7 atmospheres, it is 150°C.

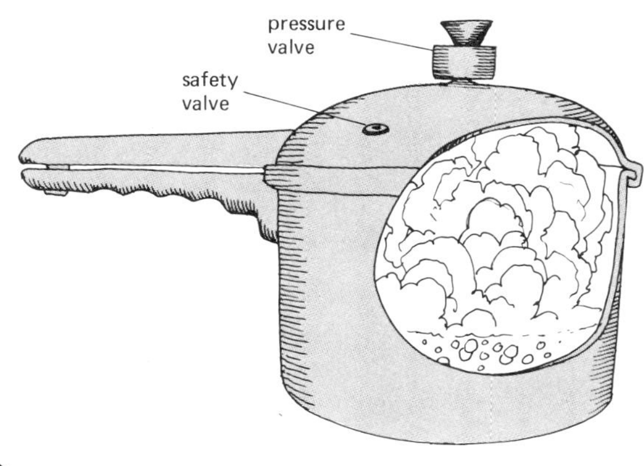

Figure 14.13 Pressure cooker. The boiling point of a liquid increases with increased pressure, and the liquid can be heated to higher-than-normal temperatures. This is the principle of the pressure cooker, which cooks foods more quickly as a result of higher temperatures.

the melting point of ice only 0.75°C. The primary cause of melting to produce the lubricating film is frictional heating between the skate blades and the ice.

Do you think it ever gets too cold to ice skate? It can, in the sense that you would not be able to glide on the skates. If it is really cold, the frictional heating between the skate blades and the ice may not be enough to lower the temperature below the melting point and provide a lubricating film of water. You would then "skate" as though you had left on your blade protectors—not in Olympic style.

The effect of pressure on the freezing point of ice can be observed by a demonstration, as shown in Figure 14.14. A fine wire with weights on each end is suspended over a block of ice. Because of the pressure of the wire, the freezing point is lowered, and the wire "cuts" slowly through the ice. But the block doesn't fall apart because the melted water refreezes. This melting and refreezing due to pressure differences is called regelation. It takes a while for the wire to cut through the block. The excitement is heightened if you have a game to guess when the cut will be complete, for example, in a Physics Olympics.§

§ If you aren't familiar with Physics Olympics, check the index of *The Physics Teacher* in your library. They are exciting contests that use physical principles that you are learning. You may want to have some Olympic events of your own.

As a final topic, let's consider the boiling and freezing points of aqueous (water) solutions. For example, when you dissolve table salt in water, how does this affect the boiling and freezing points of the water? When you put salt in a pot of water when cooking, will the salt water boil above or below 100°C? You will find that the boiling point of the solution is a bit greater than 100°C. The salt ions are attracted to the polar water molecules, and more energy is required to vaporize them.

How about on the freezing side? Here, the dissolved salt lowers the freezing point. The salt ions get in the way of the water molecules forming their hexagonal ice-crystal structure. Only when the water molecules are slowed down sufficiently (lower temperature) are the attractive forces large enough to cause freezing. In general, adding anything to water causes this effect.

QUESTION: Why is salt added to ice when making homemade ice cream?

ANSWER: Basically, to lower the temperature below 0°C so the ice cream mix will freeze. Heat is conducted from the mix through the inner metal "can" to the ice in the outer bucket. The metal can is rotated, and a stationary stirrer or "dasher" stirs the mix to keep it uniform and to promote

(a)

(b)

Figure 14.14 Regelation. (a) Weights suspended on a fine wire cause a large pressure on the ice under the wire, which lowers the freezing point. (b) The wire slowly "cuts" through the ice, which refreezes. (The soda was drunk while waiting.)

the heat transfer (mechanical convection). The ice cream mix, being largely water, freezes at slightly below 0°C. If pure ice is used, the temperature of the ice cream mix would be lowered to 0°C, and then no more heat would be conducted, since it would be at the same temperature as the ice-water mixture outside.

However, when salt is added to the ice, it mixes with the outer layers of the ice. The freezing or melting point of the ice-salt mixture is well below 0°C and it melts, taking heat from the ice and the surrounding water. This melting lowers the temperature of the salt, ice, and water mixture to below 0°C, and more heat can be conducted from the ice cream mix, causing it to freeze—what we all wait for impatiently.

We put salt on icy sidewalks and roads in the winter for the same freezing-point effect. The temperature is cold enough to freeze water and produce ice, but if it is not below the freezing point of the ice-salt mixture, the ice will melt. If it was really cold and the temperature was below the freezing point of the ice-salt, then the salt would not "melt" the ice.

SUMMARY OF KEY TERMS

Conduction heat transfer by molecular interaction with no net mass movement. Occurs chiefly in solids.

Thermal conductivity how effective a substance is in conducting heat.

Thermal conductor a material with good conductive heat transfer capability or thermal conductivity.

Thermal insulator a material with poor thermal conductivity.

Convection heat transfer by mass movement. Occurs in liquids and gases with movement of all or part of the fluid.

Convection cycle the cyclic motion in a fluid due to localized heating and convective heat transfer.

Radiation heat transfer by means of electromagnetic waves.

Change of state (phase) the transition from one state or phase of matter to another, for example, from solid to liquid.

Latent heat the heat energy associated with a phase change that is involved in the work of changing the state of the material without a change in temperature.

Latent heat of fusion the latent heat of a solid-liquid phase change. For water, this is 80 cal/g.

Latent heat of vaporization the latent heat of a liquid-gas phase change. For water, this is 540 cal/g.

Evaporization the vaporization of molecules from a solid or liquid.

Sublimation the change of phase from a solid directly to a gas.

Humidity the amount of water vapor in the air.

Absolute humidity the actual amount (mass) of water vapor in a given volume of air.

Relative humidity the ratio of how much water vapor a volume of air actually holds (absolute humidity) to what it could hold (saturated) at that temperature. This fraction is commonly expressed as a percent.

Dew point the temperature at which a volume of air is saturated or the relative humidity is 100 percent.

Boiling the condition in a liquid when its vapor pressure equals the pressure above the liquid.

Triple point the pressure and temperature at which a substance can coexist in all three phases.

Regelation melting and refreezing due to pressure differences.

EXERCISES

1. On a molecular level, why is a material a good thermal insulator?
2. Good silverware (knives, forks, and spoons) has an actual silver coating. Is thermal conductivity a consideration here?
3. Why do underground water pipes sometimes freeze only after it has been very cold for several days?
4. A medical emergency arises and you are told to boil some water quickly. In the kitchen you find two similar pots, one made of iron and the other of aluminum. Which one would you use?
5. (a) Is baked food more likely to "burn on the bottom" in an aluminum baking pan or a glass baking dish? (b) Counter tops in restaurants are sometimes made "burn proof" from cigarette burns by pressing a sheet of aluminum foil between the layers of varnished paper just below the plastic top during the manufacturing process. How does this help prevent burn marks when a lighted cigarette comes into contact with the counter?
6. How would the rate of thermal conduction (conductivity) of the bottom of a metal cooking pan vary with (a) the area and (b) the thickness? (*Hint:* Think in terms of heat "flow.")
7. Why do some metal pans (usually iron or steel) have a layer of copper on their bottoms? Is this only for looks?
8. Thermal underwear has a knitted structure with lots of holes (Fig. 14.15). Wouldn't a material without holes be a better insulator?
9. A bucket-brigade analogy is used to illustrate the molecular conductive process. Suppose a real bucket brigade passed buckets of hot water along the line. What type of actual heat transfer process is this?
10. Thermopane windows have double panes of glass separated by a small air space. Why are these windows better for insulation than single-pane windows? How do storm doors and windows help reduce your heating bill?
11. Foam insulation is sometimes blown between the outer wall and inner wall of a house. If air is a poor conductor, why bother with the insulation?
12. The outside coils of window air-conditioners have a fin network (Fig. 14.16). What is the purpose of these fins?
13. A big roaring fire in a fireplace is only about 10 percent efficient in heating a room. Why the low efficiency?
14. Discuss the energy balance and average temperature of the Earth if its only heat-loss mechanisms were conduction and convection.
15. (a) Why do we generally wear dark clothes in the winter and light-colored clothes in the summer? (b) On a hot,

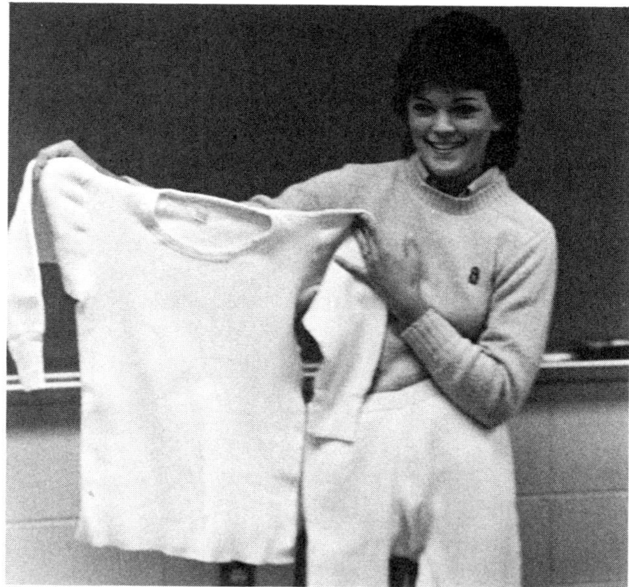

Figure 14.15 See Exercise 8.

Figure 14.16 See Exercise 12.

sunny summer day, it is possible to cook an egg on the hood of a car. Could this be done faster on the hood of a black car or on the hood of a white car? (c) Why does dirty snow melt faster than clean snow?

16. Homemakers often complain that their pie crusts do not brown on the bottom in shiny aluminum pie pans as they do in the older metal pie pans. Why is this?

17. A Thermos bottle is used to keep cold liquids cold and hot liquids hot. It consists of a double-walled, partially evacuated container with silvered walls (Fig. 14.17). Discuss how heat transfer is impaired in terms of conduction, convection, and radiation.

18. (a) When your skin is hot, the blood vessels in the skin dilate, or get larger in diameter. When the skin is cold (below 37°C), the blood vessels constrict. What is the purpose of this action? (b) Alcohol (taken internally) causes the blood vessels in the skin to dilate, and drinkers feel a warm "glow." Is this really a warming process for the body?

19. Why does it save energy to keep the thermostat set lower in the winter and higher in the summer?

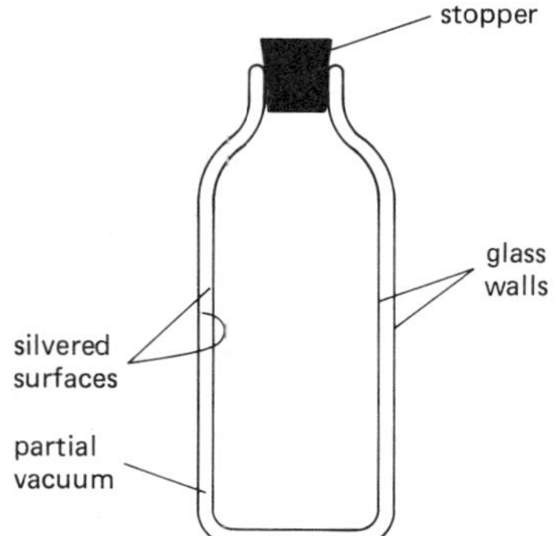

Figure 14.17
See Exercise 17.

20. Suppose you take cream in your coffee. If you have a hot cup of coffee you want to drink later, which method would keep the coffee hotter longer: to put the cream in it right away or wait until you are ready to drink it? [*Hint:* Think about temperature differences and cooling (heat transfer).]

21. The "radiator" of an automobile cooling system is where most of the heat is lost. This name implies that heat is lost mainly by radiation. Is this true?

22. When foods are broiled in an oven or a "toaster" oven, a heating element in the top of the oven is used. What's the purpose of this? Wouldn't a bottom element be more efficient in heating?

23. Some questions about microwave ovens:
(a) In heating frozen foods in sealed pouches, why do you first poke holes in the pouch?
(b) Why are microwave ovens built so they will not operate with the doors open?
(c) What is the purpose of the metal grating on the inside of the glass in the door?

24. Why do various substances have different boiling and freezing points? Would you expect the latent heats to be different?

25. Explain why the latent heat of vaporization of water is almost seven times greater than the latent heat of fusion.

26. Why do some liquids evaporate more readily than others?

27. People traveling in a hot region sometimes carry water in a porous canvas bag hung on the front bumper of a car or truck. What is the purpose of this?

28. (a) Why do we sometimes blow on the surface of a hot cup of coffee or a spoon full of hot soup?
(b) In some states, highway signs warn: "Bridge freezes before road" (Fig. 14.18). Why is this?

Figure 14.18 See Exercise 28(b).

29. To tell the wind direction, people sometimes wet a finger and hold it up. How does this help?
30. What is the wind chill factor if the air temperature is 20°F and the wind speed is 15 mi/h? (See Table 14.2.)
31. On a particular day there is a wind speed of 25 mi/h and a wind chill factor of 0°F. By how many degrees would the air temperature have to be lowered to have the same cooling effect in calm air?
32. Why do ice cubes in a refrigerator freezer get smaller with time, particularly in a frostfree freezer?
33. Fogs may be thought of as low-lying clouds. Why do fogs sometimes form in valleys overnight?
34. (a) Why does the mirror in the bathroom fog up when you take a shower?
    (b) Why does water condense on the outside of a glass containing an iced drink?
    (c) Why can you "see" your breath on a cold day?
    (d) When a teakettle boils, we say we can see the steam coming from the spout. Is this correct?
35. Freeze-dried coffee crystals are made from frozen coffee. How is this done?
36. A pan of water on a stove boils "faster" when the burner is on high heat, and boils "slower" with the burner on low heat. Is the temperature of the water greater for a fast boil? Explain.
37. Perfumes and colognes have an alcohol base. Why not use water, since it would be cheaper?
38. Why does covering a pot of water with a lid help the water boil more quickly?
39. (a) In a percolator coffee pot, water bubbles up through a central system and "perks" down through the coffee grounds (Fig. 14.19). What causes the water to rise in the stem? (b) Geysers also erupt for similar reasons. Water from underground streams fills long, vertical holes and is heated at the bottom from underground hot rocks. Explain why the boiling at the bottom starts at temperatures well over 100°C and why the spraying of water into the air at the surface is generally greater after the eruption gets underway. Why do geysers erupt periodically?

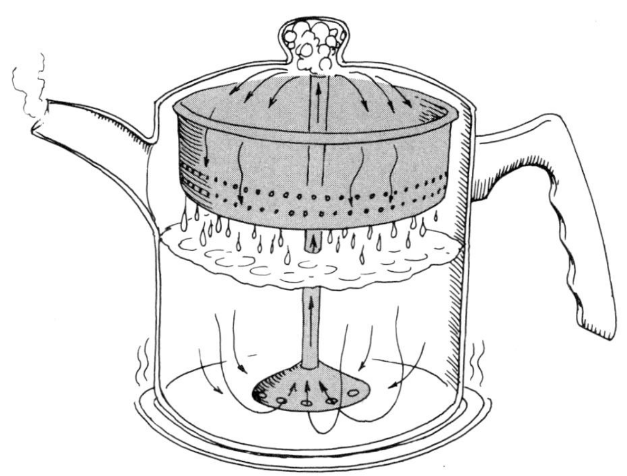

Figure 14.19 See Exercise 39.

40. Automobile cooling systems operate under pressure. (a) What is the purpose of this? (b) What would happen if you removed the radiator pressure cap immediately after turning off a hot engine, and why? (Don't try this—it's very dangerous.)
41. The antifreeze (ethylene glycol) used in automobile cooling systems has a freezing point of −11.5°C, which is lower than that of water. Why don't we replace the water completely with antifreeze?
42. Is regelation an important factor in making snowballs? If so, why can't balls be made from dry powdered snow?

15 Thermodynamics, Heat Engines, and Heat Pumps

Thermodynamics

As the name implies, thermodynamics deals with the transfer and actions (dynamics) of heat (Greek *therme,* meaning "heat"). In general, it is a broad and comprehensive branch of science that is concerned with all types of energy aspects, but chiefly the relationship between heat and mechanical energy. The formal development of thermodynamics began less than 200 years ago, primarily growing out of efforts to produce heat engines—devices for converting heat energy into mechanical work. These include steam engines, gasoline engines, diesel engines, jet engines, and any device that converts heat into work.

Thus far our discussion of heat has not involved the practical conversion or transformation of heat to mechanical work or energy. But this is an important consideration. Where would we be without the internal combustion engine of the automobile? Still using horses perhaps, which are, in a sense, complicated physiological heat engines.

Some of the general aspects of thermodynamics have been presented in previous chapters. However, there are other important principles that govern the utilization of heat and work. These laws of thermodynamics are basic in the operation and design of heat engines and heat pumps, which are main topics of this chapter. A heat pump is a device that uses mechanical energy or work to transfer heat from a lower-temperature source to a higher-temperature region. Can you think of a common heat pump? How about a refrigerator or an air conditioner? An air conditioner takes heat from a cool room and transfers it to the hot outdoors. It takes work or energy to do this. Check your electric bills in the summer.

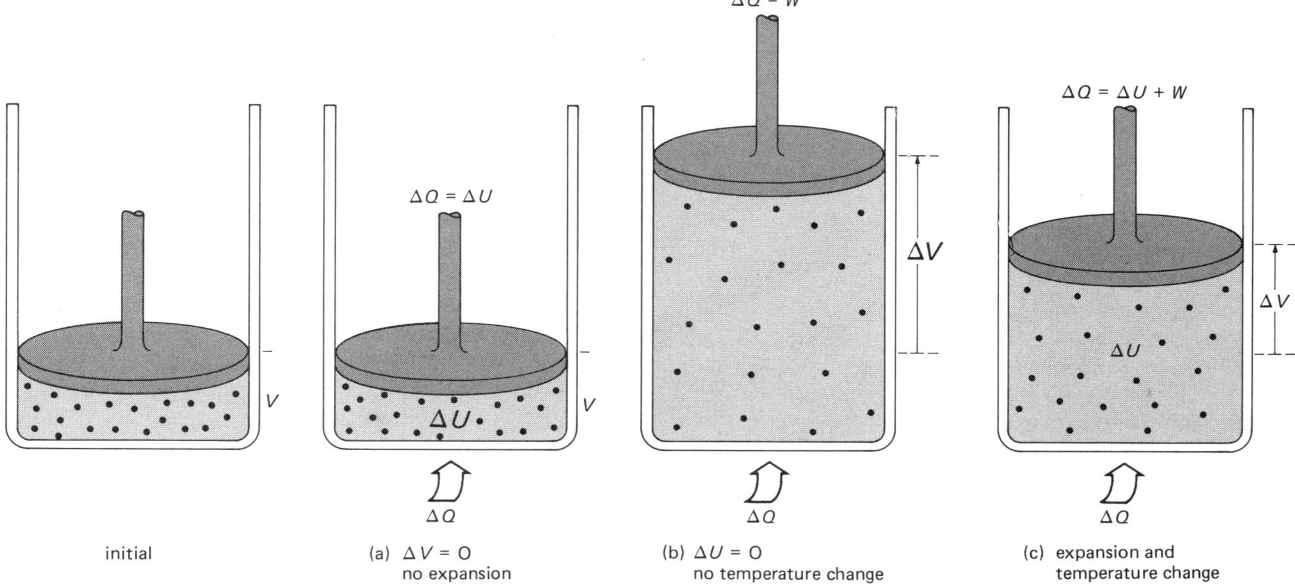

Figure 15.1 Processes and the first law of thermodynamics. Heat (ΔQ) added to a system can go totally into the internal energy ΔU with $\Delta V = 0$ (a) or into work W with $\Delta U = 0$ (b). In the general case of conservation of energy (c), $\Delta Q = \Delta U + W$.

The First Law of Thermodynamics

Since thermodynamics is concerned with energy transfer, it is important to keep track of or to account for the energy involved in a process. The first law of thermodynamics is simply the principle of the conservation of energy applied to a thermodynamic system. When heat is added to a system it doesn't disappear. Let's look at a quantity or system of perfect gas, as illustrated in Figure 15.1. When heat is added to the system, several things can happen or are possible: (a) If the volume is fixed (constant), the quantity of heat added (ΔQ) goes into increasing the internal energy (ΔU) of the gas, and $\Delta Q = \Delta U$. This is evidenced by an increase in the temperature (and pressure) of the gas. Recall that for an ideal gas $pV = NkT$ and $U \propto T$, in which T is the absolute temperature. No work is done in this case.

(b) Alternatively, with a movable piston the gas could expand and do work (W) in moving the piston, just as you would have to do work if you moved it manually. If all the added heat went into the work done *by* the gas as it expands, the internal energy of the gas would be the same afterward as before (same temperature), and $\Delta Q = W$.

Or (c), there could be a combination of these energy distributions, with heat going into increasing the internal energy *and* doing work, $\Delta Q = \Delta U + W$. This is the most general case, and the equation takes into account the other two cases with (a) $W = 0$ and (b) $\Delta U = 0$. This general conservation-of-energy principle is called the **first law of thermodynamics.**

Net heat transfer = change in internal energy + work

or

$$\Delta Q = \Delta U + W$$

Notice that this does not have to be a one-direction or a heat-added process. Suppose work was done on the system by an external force (Fig. 15.2). By the energy balance of the first law, the internal energy of the gas increases and/or heat is transferred from the system.*

QUESTION: What would happen when work is done on a system of gas by compressing it (a) if no

* To get the signs of the quantities in the equation of the first law to come out properly, work done *on* and heat removed *from* a system are taken to be negative ($-W$ and $-\Delta Q$).

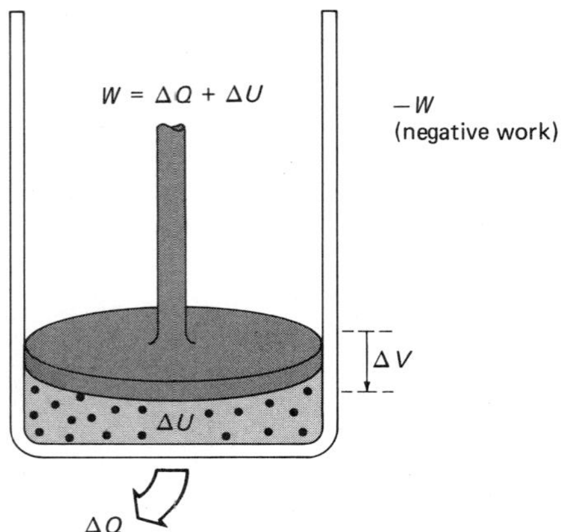

Figure 15.2 Work done *on* a system is taken to be negative. By the first law, the internal energy of the gas increases and/or heat is transferred from the system.

heat is removed from the system and (b) if the internal energy of the gas remains the same?

ANSWER: The first law tells you right away. (a) When work is done on the system and no heat is removed ($\Delta Q = 0$), the internal energy is increased by an amount equal to the work done.

(b) If the internal energy is unchanged ($\Delta U = 0$), then by the conservation of energy an amount of heat equal to the work done would have to be removed from the system.

In *any* case, the first law says in effect that you can't get more out of a system than you put in or that is already there. To see the last part of this statement, think about removing heat from a gas in a rigid container (no change in volume and no work). You could only remove a quantity of heat equal to the amount of internal energy the gas has. If you could do this (which you can't, in reality), the gas temperature would be at absolute zero. Why?

Heat Engines

A **heat engine** is a device that converts heat energy to work. There are many different types of heat engines available, and they come in many different sizes and shapes—gasoline engines on lawn mowers, diesel engines in trucks, and steam turbines used in electrical generation. Basically, they all operate on the same principle of adding heat to a fluid that uses some of the energy to do mechanical work.

In thermodynamic processes we are not generally concerned with the component parts of an engine, such as pistons, cylinders, and gears. For theoretical purposes any heat engine can be conveniently represented by a diagram, as shown in Figure 15.3. Heat is taken from a high-temperature (T_{hot}) source or reservoir, some of which is used to do useful work (work out), and the remainder is transferred to a low-temperature (T_{cold}) reservoir. Notice that the widths of the heat and work paths in the diagram are in keeping with the conservation of energy: heat in = work + heat out, or work = heat in − heat out.

In an actual engine, such as an internal-combustion gasoline engine, the high-temperature heat source is the exploding charge of vaporized gasoline and oxygen. Work is done by the expanding gases on a piston that is coupled to a crankshaft to give useful work output. The unused heat is "rejected" to the atmospheric surroundings (low temperature) through the engine exhaust system.

Most practical heat engines operate in a cycle or series of thermodyamic processes that brings the system back to its original condition. Heat could simply be added to a gas in a piston-cylinder arrangement, and we could have a one-shot work output in a single expansion process. However, in practical applications we want an engine to deliver work continuously. When an

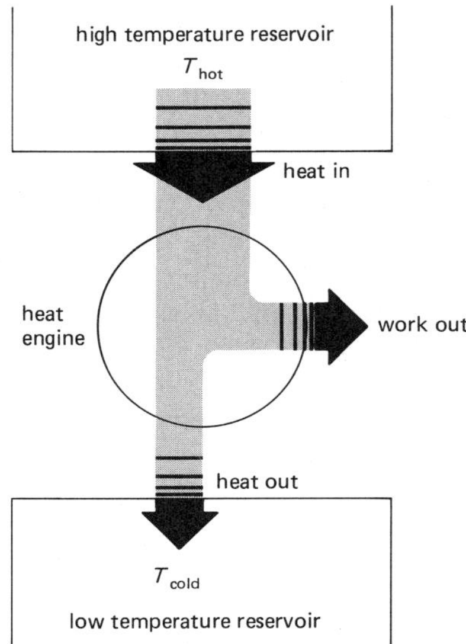

Figure 15.3 A heat-engine diagram. In general, a heat engine takes heat from a high-temperature reservoir, converts some of it to useful work output, and rejects the remainder.

SPECIAL FEATURE 15.1

The Drinking Bird Heat Engine

A novel example of a cyclic heat engine is the toy drinking bird (Fig. 15.4). It doesn't look much like a heat engine, but it falls under the definition of one. To start the "engine," you wet the absorbent flock material on the head and beak of the bird with water. The liquid inside the body is ether, which has a low boiling point and readily vaporizes at room temperature. The evaporation of ether in the lower part of the body creates pressure above the liquid. The ether in the tube does not evaporate as readily because the head is cooled by the evaporation of water from the flock material covering, and there is less vapor pressure in the head. The pressure difference causes the ether to be forced up the tube into the head.

The rising liquid raises the center of gravity of the bird above the pivot point, and the bird pitches forward for a "drink" (to rewet the flocking). In this position, the pressures in the head and body are equalized, and the ether drains back into the body. The bird pivots back and the cycle begins again.

The bird could be hooked up so its motion could do some useful work. This heat engine is cheap to run, but it is doubtful if "birdmobiles" will ever replace piston-engine automobiles.

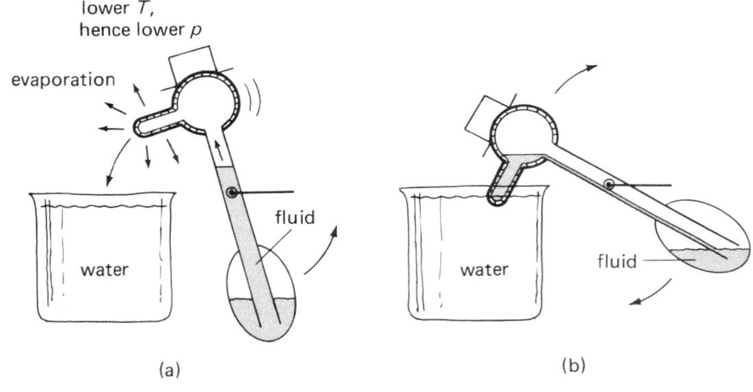

Figure 15.4 The "drinking" bird heat engine. A strange looking heat engine perhaps, but it does the job with work output.

engine operates in a cycle, we have work output during each cycle. The idea of cyclic operation is illustrated for a special case in Special Feature 15.1.

The cycle of a **two-stroke-cycle engine** is illustrated in Figure 15.5. Such internal combustion engines are used in lawn mowers, motorcycles, outboard motors, and chain saws. The two strokes refer to the up-and-down motions of the piston (each a stroke) during a cycle. The cycle of thermodynamic processes is also shown in the figure on a pressure-volume (p-V) diagram. This cyclic series of processes is called an Otto cycle, after the German engineer Nickolas Otto (1832–1891), who designed and built one of the first successful two-stroke gasoline engines. Go through the cycle comparing the figures to the diagram processes to see how the engine works.

Two-stroke-cycle engines have one power stroke per cycle; that is, power (work output) is produced each time the piston goes up and down. This is an advantage for lightweight engine applications. However, there is a disadvantage. Notice that there is a waste or loss of fuel with the venting of the exhaust gases (process 4-1, Fig. 15.5). This is avoided in a **four-stroke-cycle engine,** which has one more piston cycle (up and down) than the two-stroke-cycle engine for the complete thermodynamic cycle (Fig. 15.6). There are separate intake and exhaust strokes that prevent fuel loss as in the two-stroke-cycle engine. Follow the cycle through and check this out. Notice that the additional two strokes put a horizontal "leg" on the Otto cycle in the p-V diagram and that there is still only one power stroke per cycle (all four strokes make up one engine cycle, although this cycle requires two rotations of the crankshaft).

To keep engines running smoothly between power strokes, a flywheel attached to the crankshaft stores and

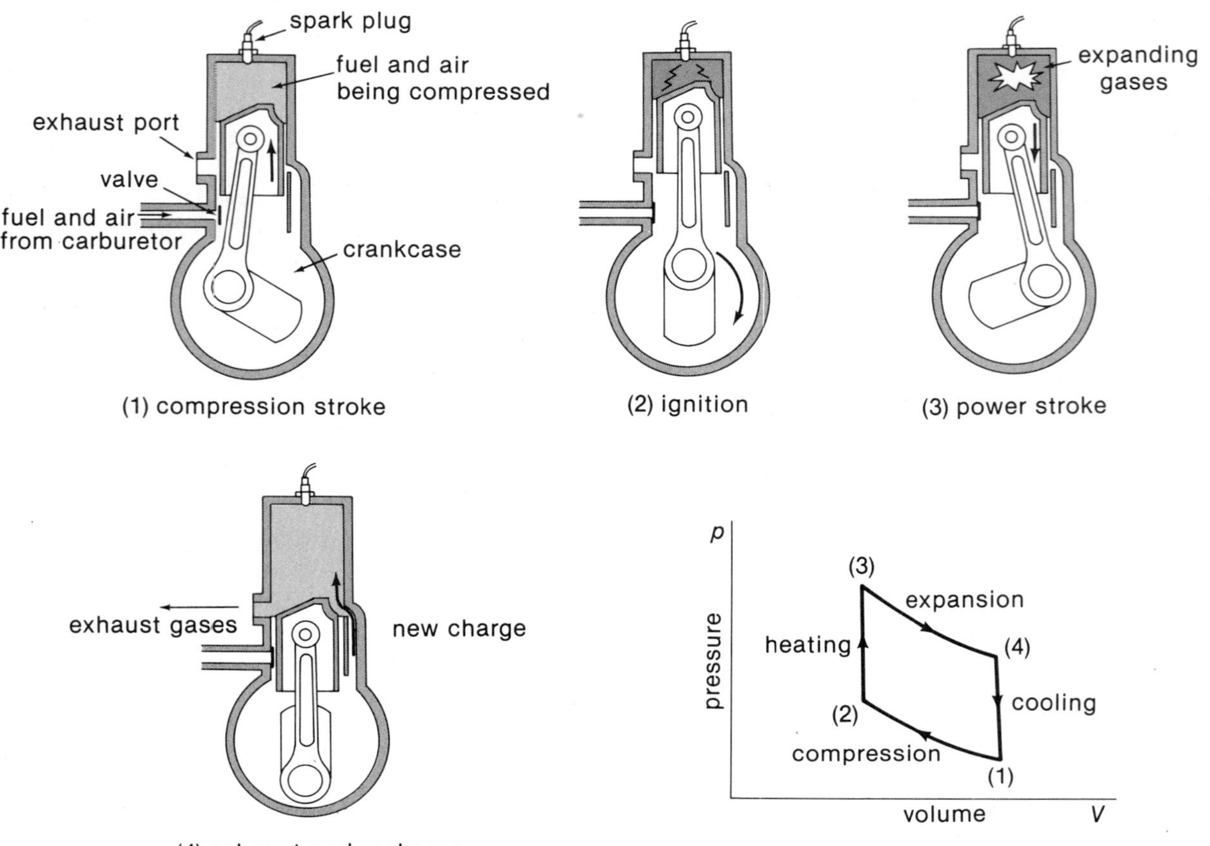

Figure 15.5 Two-stroke-cycle gasoline (heat) engine. The two strokes refer to the up-and-down motions of the piston (each a stroke), so there are two strokes per cycle. The (Otto) cycle of thermodynamic processes is shown in the p-V diagram.

supplies energy (see also Chapter 8, Fig. 8.10b). Most automobiles use four-stroke gasoline engines. To obtain greater work output, multicylinder engines—for example, those having four, six, or eight cylinders—are used. In each of these, all of the cylinders go through their four-stroke-cycle in two rotations of the crankshaft (to which the piston rods are attached). The cylinder firing or ignition occurs in a regular sequence for smoother power output. For example, in a four-cylinder engine, the timing is such that one of the cylinders fires every half-revolution of the crankshaft.

QUESTION: What is the difference between an automobile diesel engine and the common gasoline engine?

ANSWER: There are various differences, including fuel, stroke cycles, and ignition. The diesel engine uses a lower-grade fuel. Diesel "oil" is not as highly refined as gasoline, so it is usually cheaper. The diesel engine is a two-stroke-cycle engine, as compared to the common four-stroke-cycle gasoline auto engines.

Probably the most notable difference is the absence of spark plugs to ignite the fuel in a diesel engine (Fig. 15.7). Air in the cylinder during the compression stroke is compressed three to four times more than the pressure in the cylinder of a gasoline engine. This compression raises the temperature of the air high enough so the fuel ignites spontaneously when injected into the cylinder. The diesel engine is named after its inventor, Rudolf Diesel (1858–1913), a German engineer.

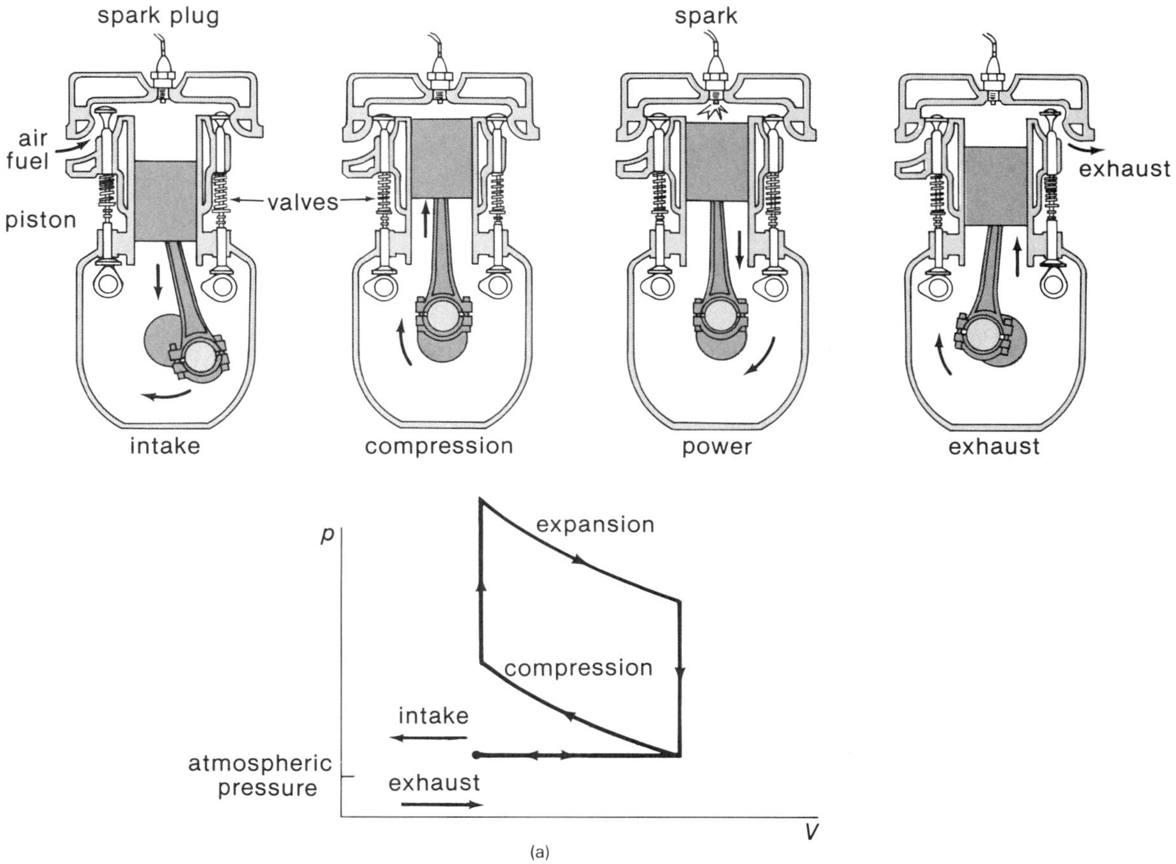

(b)

Figure 15.6 Four-stroke-cycle engine. Additional intake and exhaust strokes prevent fuel loss as in a two-stroke-cycle engine. The two strokes put a horizontal "leg" on the Otto cycle in a *p*-*V* diagram.

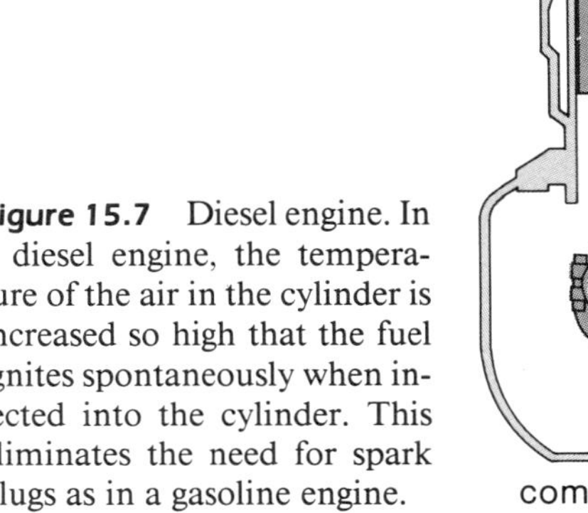

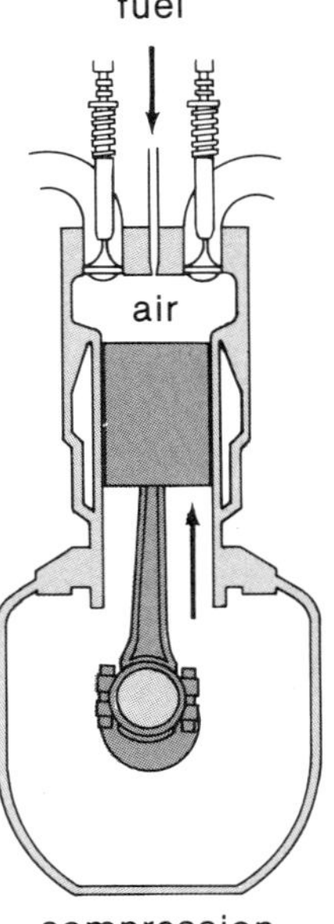

Figure 15.7 Diesel engine. In a diesel engine, the temperature of the air in the cylinder is increased so high that the fuel ignites spontaneously when injected into the cylinder. This eliminates the need for spark plugs as in a gasoline engine.

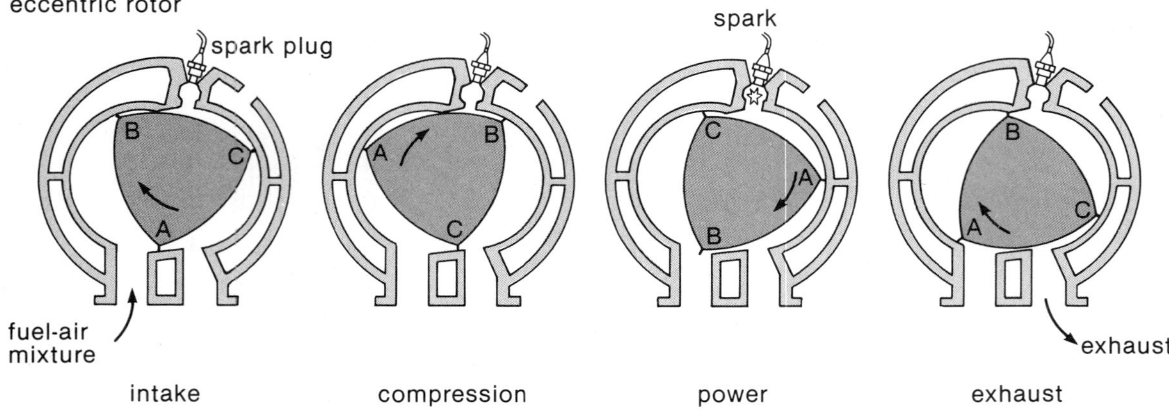

Figure 15.8 Wankel rotary engine cycle. Notice that there are four processes of a thermodynamic cycle going on at once, one at each rotor face in different stages of progress. This gives three power "strokes" per rotor revolution.

Efforts have been made continuously to develop engines of other designs, in particular, ones that would reduce the number of moving parts so as to minimize wear and the need for maintenance and adjustment. One of the serious contenders is the rotary engine. James Watt, the developer of the steam engine, tried to build a rotary steam engine in the late 1700's but could not overcome the problem of sealing between the rotary member and its casing.

A practical internal-combustion gasoline rotary engine was invented by Felix Wankel, a German engineer, in the early 1950's. The principle of the **Wankel rotary engine** is illustrated in Figure 15.8. An important feature of the engine is that it has only two moving parts, a triangular rotor with an internal gear and a geared output shaft.

In operation, the gases go through a four-step thermodynamic cycle similar to those in a piston engine. But in the rotary engine, these *separate* processes are taking place simultaneously—one at each rotor face in different stages of progress. (Think about this as you go through the diagrams in Fig. 15.8.) Since each four-process cycle takes place during one revolution of the rotor, there are three power impulses per revolution. In terms of power impulses per revolution of the output shaft, the rotary engine is equivalent to a six-cylinder piston engine.

Elimination of a crankshaft and piston rods makes the rotary engine relatively compact. They have been used in automobiles but do not give good fuel economy. It's a good idea, however, and with improved efficiency, rotary engines may someday give piston engines a run for the money.

Thermal Efficiency

Efficiency is what tells how "good" or how economical an engine is. It is common to measure cost efficiency in terms of fuel economy or how many miles per gallon (or kilometers per liter)—what we get out for what we put in. Recall from Chapter 6 that the mechanical efficiency of a machine is the ratio of the work output to the work input

$$\text{Efficiency} = \frac{\text{work out}}{\text{work in}}$$

which is the fraction (or percent) of useful work done by a machine.

In heat engines, a similar thermal efficiency is used. Instead of work input, there is heat input. The work output is equal to the heat in minus the heat out (see Fig. 15.3). Hence, the **thermal efficiency** is defined as follows:

$$\text{Thermal efficiency} = \frac{\text{work out}}{\text{heat in}} = \frac{\text{heat in} - \text{heat out}}{\text{heat in}}$$

For example, if a heat engine has a heat input of 1000 joules and rejects 600 joules while doing work in a cycle, its thermal efficiency is 40 percent:

$$[(1000\ \text{J} - 600\ \text{J})/1000\ \text{J} = 400\ \text{J}/1000\ \text{J} = 0.40\ (\times 100\%) = 40\%]$$

An automobile has an overall efficiency of less than 15 percent. Here a lot of heat is lost because of friction as well as being rejected to the surroundings through the exhaust and cooling systems. In practical terms, this means that more than 85 percent of the energy from the gasoline used in your car is "wasted" or does not go into doing the useful work of propelling the vehicle. (See Special Feature 6.1.

The Second Law of Thermodynamics

The first law of thermodynamics is concerned with energy balance or conservation. As long as the energy check sheet is balanced, the first law is satisfied. Suppose, however, that a heat engine operated so that *all* the heat input were converted into work (no heat out). This doesn't violate the first law, but something is wrong. In this case, the thermal efficiency of the engine would be 1 or 100 percent, and this just doesn't happen or has never been observed.

This fact is expressed in the **second law of thermodynamics,** which can be stated in several ways. One statement of the second law as it applies here is as follows:

> No heat engine operating in a cycle can convert heat energy completely into work.

Or, put another way,

> No heat engine operating in a cycle can have 100 percent efficiency.

If this were possible, the work output could be rechanneled back into heat input, and the engine would operate indefinitely as a perpetual-motion machine. It is possible for all the heat to go into work in a single expansion process, but for an engine operating in a cycle, heat must be rejected or lost somewhere.

Since a heat engine must lose some heat, what is the best we can do or what is the maximum possible efficiency? A French engineer, Sadi Carnot (1796–1832), studied this question and came up with the answer. He found that the thermal efficiency of a heat engine is limited by the operating temperatures (temperatures of the hot and cold reservoirs, see Fig. 15.3), and the **ideal** or **Carnot efficiency** is given by

$$\text{Ideal efficiency} = \frac{T_{\text{hot}} - T_{\text{cold}}}{T_{\text{hot}}} = 1 - \frac{T_{\text{cold}}}{T_{\text{hot}}}$$

where the temperatures are absolute temperatures.

This is the maximum possible theoretical efficiency under *ideal* conditions.† It sets an upper limit that can never be achieved. Actual thermodynamic conditions are never ideal, and friction is always present mechanically. But the ideal thermal efficiency does tell the engineer that a greater possible efficiency can be achieved by making T_{hot} higher and/or T_{cold} lower. For example, if a steam turbine took in superheated steam at 500 K (227°C) from a high-temperature (T_{hot}) reservoir and rejected it to a cold-temperature (T_{cold}) reservoir at 400 K (127°C), the ideal efficiency is $1 - (400/500) = 0.20$, or 20 percent. But if the cold-temperature reservoir were at 300 K (27°C), the ideal efficiency would be $1 - (300/500) = 0.40$, or 40 percent. (In actual practice, the steam is condensed on the low-temperature side, which relieves the "back pressure" on the turbine to make it more efficient.)

You could also increase the high temperature to get increased efficiency. Diesel automobiles typically get better "gas" mileage (are more efficient) than their similar gasoline-powered counterparts, even though diesel oil has a lower heat of combustion than gasoline (see Table 13.1). This is mainly because diesel engines operate at higher temperatures than gasoline engines, which makes for greater efficiency.

Heat Pumps

A **heat pump** is a device that transfers heat from a low-temperature reservoir to a high-temperature reservoir. This is the reverse function of a heat engine. A heat pump is represented by a diagram as shown in Figure 15.9. From experience you know that this heat transfer from a "colder" to a "hotter" reservoir will not happen spontaneously or by its own accord. This is expressed in another form of the second law:

> Heat will not flow spontaneously from a colder body to a hotter body.

Heat can flow spontaneously from a hotter body to a colder body. This is like heat flowing "down a temperature hill," as a ball would roll down a hill. The reverse, which would be like having a ball roll up a hill of its own accord, cannot happen. Obviously, heat transfer "up a

† These involve an ideal thermodynamic cycle called a Carnot cycle that consists of two isothermal (constant temperature) processes and two adiabatic (no heat transfer) processes.

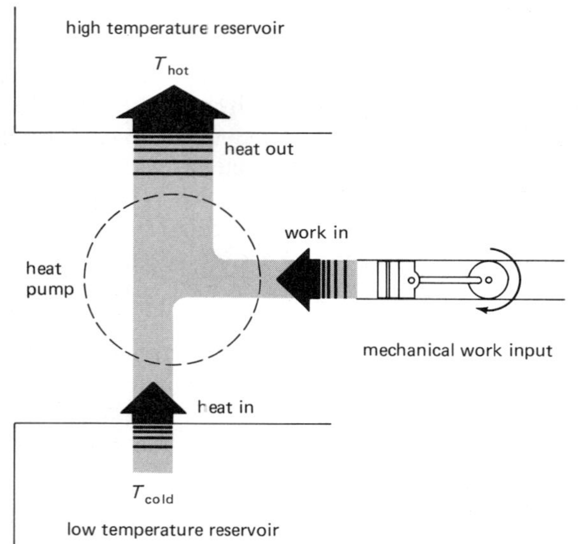

Figure 15.9 A heat pump diagram. In general, a heat pump takes heat from a low-temperature reservoir and transfers it to a high-temperature reservoir. This requires work input. A common practical heat pump is a refrigerator.

temperature hill" or a ball rolling up a hill requires something else. As you probably guessed or noticed from Figure 15.9, this is work (input).

One of the most common heat pumps is the kitchen refrigerator. Its function is to take heat from inside the refrigerator (low-temperature reservoir) and transfer it to the outside surroundings (high-temperature reservoir). The work is supplied by a compressor that uses electrical energy. A diagram of a refrigerator system is shown in Figure 15.10. A working substance, called the refrigerant, is used to absorb and transport heat. It is a fluid that readily undergoes a liquid-gas phase change at the operating temperature. The most common refrigerant nowadays is Freon, a fluorocarbon compound with a boiling point of about $-6°C$. (Sulfur dioxide was once used, and ammonia is a common industrial refrigerant, e.g., in ice plants.)

Let's see how your refrigerator works. It is convenient to divide the system into a high-pressure side and a low-pressure side (Fig. 15.10). Starting with the gaseous refrigerant in the compressor, which supplies the

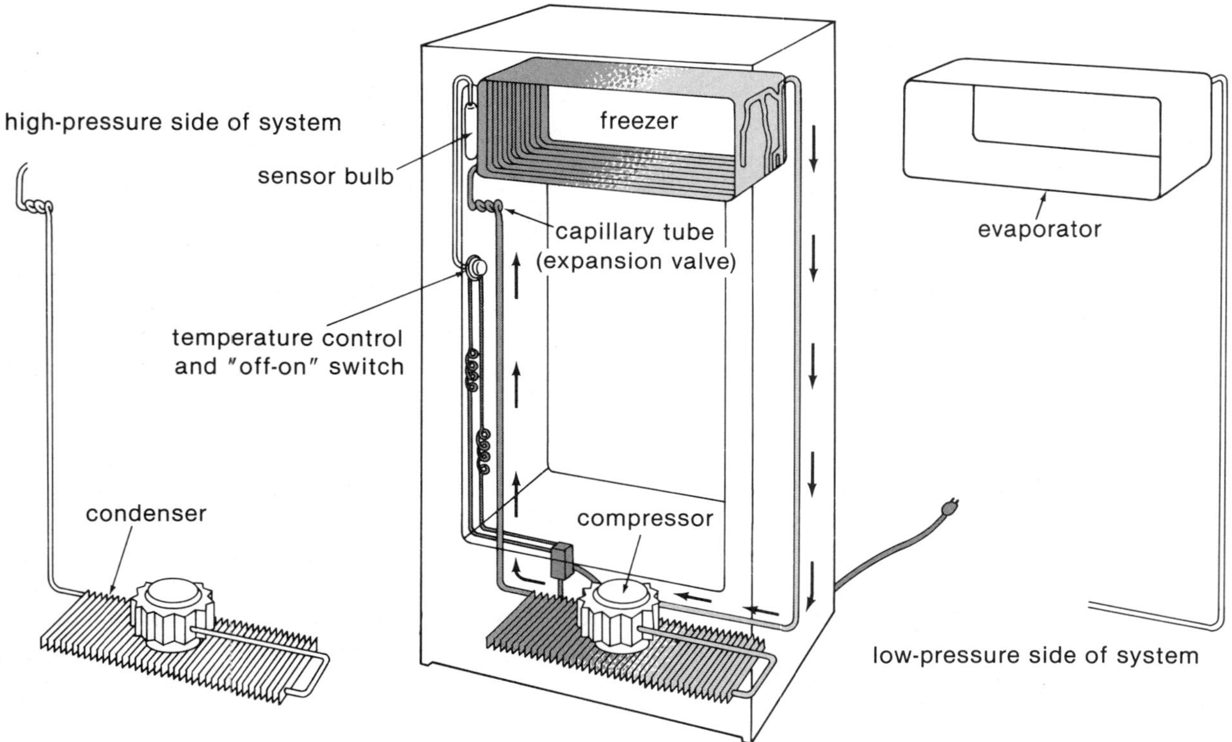

Figure 15.10 A refrigerator is a heat pump. A refrigeration process takes heat from inside the refrigerator (low temperature) and expels it to the outside environment (high temperature). Electrical energy provides the work to do this. See text for process description.

necessary work, the gas is compressed and emerges at a high temperature and pressure. It then passes into the condenser, where it is cooled and liquefies, and the heat from the refrigerant is rejected to the surroundings (heat out into the room).

The liquid refrigerant then passes through an expansion valve into the low-pressure side of the system. Work is done by the liquid in getting through the valve at the expense of internal energy. This lowers the temperature of the liquid. The cooled liquid refrigerant then flows through the evaporation coils (low-temperature reservoir), where it absorbs heat and vaporizes (boils). The gaseous refrigerant carrying heat goes into the compressor and the cycle begins again.

Air conditioners are also heat pumps that have cycles similar to refrigerators. In this case, the low-temperature reservoir is usually the inside of a car or a building, and heat is pumped to the outside environment.

The term heat pump is now applied to the year-round heating and cooling systems that are becoming more and more common (Fig. 15.11). When operating as a cooling system, the heat pump extracts heat from the air inside of the home and expels it to the outside, as an air conditioner does. In its other role as a heating system, the heat pump takes heat from the outside and delivers it to the inside of the home.

This is why heat pumps are more cost-efficient in the winter. No heating fuel is required. You may think that there would be little heat from the outside air on a cold day. But remember, the air has internal energy at any temperature, and there is a lot of air. When the outside temperatures goes very low, the heating efficiency of a heat pump goes down too. To keep the house warm enough on very cold days, heat pumps are equipped with auxiliary electrical heating systems. Also, water-exchange heat pumps are now available. Heat is exchanged with an in-ground reservoir, which has smaller temperature variations than the air.

Energy-saving solar heating and cooling systems are now important considerations. Solar heating is readily understood, but how does one cool with solar energy or heat? One method is discussed in Special Feature 15.2.

The Third Law of Thermodynamics

How cold can we make it by pumping heat? We know there is a lower limit to temperature—absolute zero. Can we get that low? The answer is no. This is the **third law of thermodynamics:**

> It is impossible to obtain a temperature of absolute zero.

One way of seeing this is as follows: Suppose we could obtain absolute zero and have a cold-temperature reservoir at that temperature. Then, with $T_{\text{cold}} = 0$ K, we could have a heat engine with an ideal efficiency of 100 percent with any high-temperature reservoir.

$$\text{Ideal efficiency} = 1 - \frac{T_{\text{cold}}}{T_{\text{hot}}} = 1 - \frac{0}{T_{\text{hot}}} = 1, \text{ or } 100\%$$

An ideal efficiency can never be practically attained, and moreover, the second law forbids a heat engine with 100 percent efficiency. So theoretically, absolute zero cannot be reached, as expressed in the third law.

The third law has never been violated experimentally, although scientists have come close to absolute zero—within 0.000001 (one-millionth) of a degree. Near absolute zero, each order of magnitude (factor of 10) reduction in temperature becomes more difficult with each step. For example, going from 0.1 K to 0.01 K is one step more difficult than in going from 1 K to 0.1 K. That is, more work is required. To obtain absolute zero, an infinite amount of work would be required.

Entropy

As a final topic of thermodynamics, we'll consider entropy.

You may have noticed that the various statements of the second law tell what *does not happen.* In so doing, it indicates the *direction* in which a thermodynamic process can take place. For example, heat flows from a hot body to a colder body. Knowing there is a temperature difference, you know the direction of the process. A more general property that gives the direction of a process was suggested by Rudolf Clausius (1822–1888), a German physicist. This property is called entropy, a name coined by Clausius.

Although defined formally in terms of temperature and heat, **entropy** in effect is a measure of disorder. It is found that a system naturally or spontaneously moves from a more orderly state to a more disorderly state. This is expressed by an increase in entropy, and in this context the second law can be stated as follows:

> The entropy of the universe increases in every *natural* process.

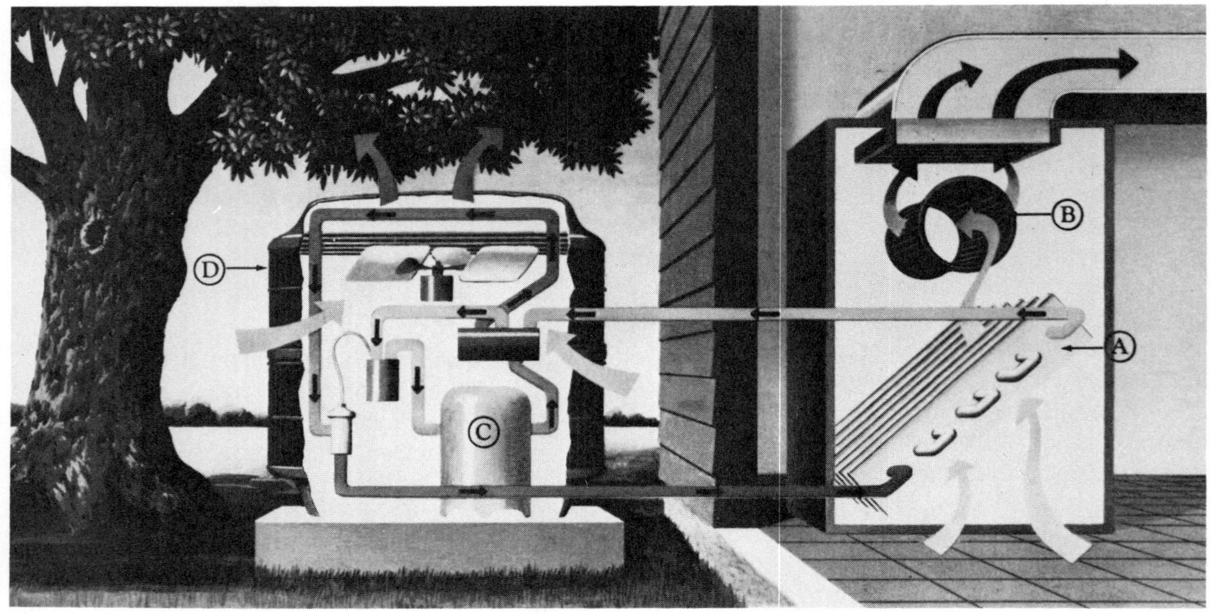

(a)

(b)

Figure 15.11 Heat pump operations. (a) Cooling cycle. Heat is absorbed from inside circulating air (A) and transferred and expelled to the outside (D). The compressor (C) does the required work. (b) Heating cycle. Heat is absorbed from the outside air by the refrigerant in the outdoor coils (A). The compressor (B) pressurizes the refrigerant and sends it to the indoor coils, where it gives up heat to the circulating air. In both cycles, heat is pumped from a "colder" to a "hotter" area.

SPECIAL FEATURE 15.2

Cooling from Heat: The Absorption Refrigerator

There are solar cooling systems, and there are natural-gas refrigerators that have burning flames. Gas refrigerators were once quite common. These refrigerators had freezer compartments at the top, yet if you looked at the bottom you could see a gas flame. How can you get "cold" from "hot" or cooling from heat? The point to keep in mind is that heat energy, like all forms of energy, is capable of doing work—the essential ingredient in pumping heat. The trick is how to do it.

There are several solar cooling cycles or systems. Let's look at one of these, absorption cooling. This is also the principle of the natural-gas refrigerator. A basic absorption refrigeration cycle is illustrated in Figure 15.12. It uses ammonia as a refrigerant and an ammonia-water solution as an *absorber*. To see how it works, start at the cooling unit or freezer [(1) in the figure]. Here, liquid ammonia absorbs heat from the cooling coils and evaporates (boils). The evaporation takes place in an atmosphere of hydrogen to speed up the process.

The ammonia–hydrogen gas mixture then goes to the absorber (2), where water absorbs the ammonia from the gas mixture. The hydrogen gas returns to the freezer, and the ammonia-water solution drains to the generator (3). Heat from a gas flame (or solar heat source) causes the liquid to bubble up through the tube, much as in a percolator coffee pot. Ammonia boils at a lower temperature (−2°C) than water, so the bubbles lifting the liquid to the rectifier (4) are practically pure ammonia vapor.

From the rectifier, the water drains back to the absorber, and the ammonia gas rises to the condenser (5). Here the ammonia gas is cooled and condensed back to liquid ammonia, which drains back to the freezer (1), and the cycle begins again.

There you have it—"cold" from "hot." The heat energy input merely supplies the necessary work needed for the cycle as electrical energy does in more common refrigerators.

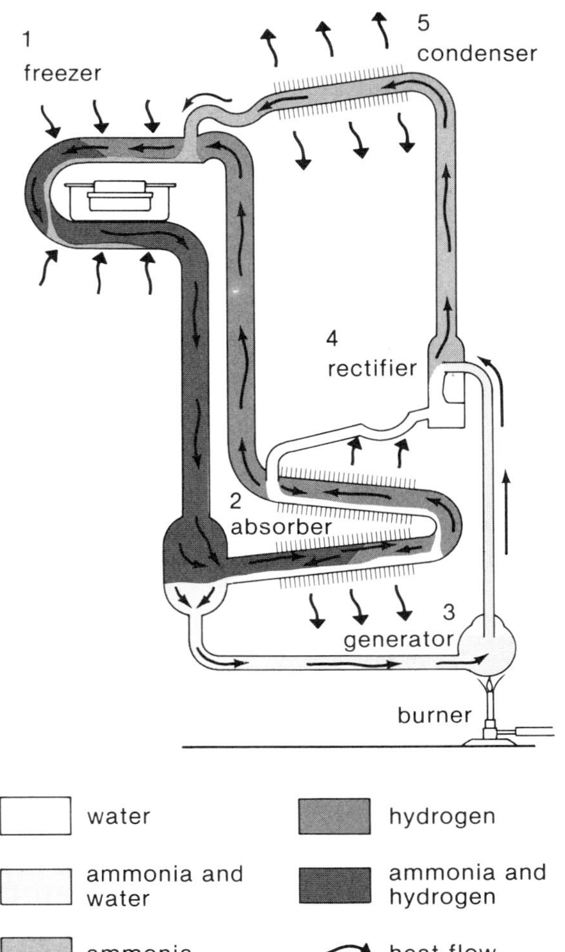

Figure 15.12 Absorption refrigeration cycle. In an absorption refrigerator, cooling is effected by heat, such as from a gas flame (or solar energy). Cold from hot? See text for description of operation.

For example, although not a thermodynamic example, the tendency of an orderly desk or room is toward a state of disorder or increased entropy (Fig. 15.13).

In the thermodynamic sense, entropy is a measure of the capability to do work or transfer heat. A system at a high temperature will naturally tend to do work on

(a)

(b)

Figure 15.13 Entropy on the increase. Entropy in effect is a measure of disorder, which increases in every natural process. Straightening up or making a messy desk orderly (decrease in entropy) requires work or an expenditure of energy, which creates entropy such that there is always a net entropy increase.

and/or transfer heat to its lower-temperature surroundings. In the process, the entropy of the system is increased, and the greater the entropy, the less *available* energy a system has. In terms of order, the heat energy is more "orderly" when it is more concentrated. When it is transferred or used (in a natural process), it is more "spread out" or "disorderly," and there is an increase in entropy.§

Since natural processes continually occur with heat transfer from "hotter" bodies to "colder" bodies, the entropy of the universe continually increases. In the limit, the universe should thermodynamically run down. The entropy would reach a maximum when everything is at the same temperature. This limit is called the **heat death of the universe.** The final temperature is estimated to be a few degrees above absolute zero. Not too warm, but this would occur billions of years from now, so don't worry.

Thermodynamics is an empirical science, and its "laws," like other physical laws, are not based on absolute impossibilities of things happening but on the experience that they have never been observed. Because of this, we can say with a great deal of certainty, "Entropy can be created but not destroyed" and "Energy can be neither created nor destroyed," which are the essences of the second and first laws of thermodynamics.

SUMMARY OF KEY TERMS

Thermodynamics the branch of science that deals with the relationship of heat and mechanical energy.

First law of thermodynamics the conservation of energy applied to thermodynamic systems: net heat transfer = change in internal energy + work ($\Delta Q = \Delta U + W$).

Heat engine a device that converts heat energy to work.

Thermal efficiency the ratio of work output to the heat input of a heat machine: work out/heat in = (heat in − heat out)/heat in.

Second law of thermodynamics no heat engine operating in a cycle can convert heat energy completely to work, or no heat engine can have 100 percent efficiency.

Ideal or Carnot efficiency the maximum possible theoretical efficiency of a heat engine under ideal conditions, and equal to $1 - \frac{T_{\text{cold}}}{T_{\text{hot}}}$.

Heat pump a device that transfers heat from a low-temperature reservoir to a high-temperature reservoir.

Third law of thermodynamics it is impossible to obtain a temperature of absolute zero.

Entropy a quantity that indicates the direction in which a thermodynamic process can take place. The entropy of the universe increases in every natural process. Entropy is a measure of disorder or the unavailability of energy of a system.

§ It should be noted that entropy can decrease in one part of a process, such as the transfer of heat by a heat pump. But this requires work or energy input from somewhere, which increases the entropy. Overall, there is a net entropy increase, and the entropy of the universe increases.

EXERCISES

1. Show that in terms of the first law, thermal efficiency is equal to $(\Delta Q - \Delta U)$/heat in. What is ΔQ? Is ΔU a consideration in a cyclic engine?
2. Draw a representative diagram of a heat engine with 100 percent efficiency.
3. In Figure 15.6, the power output (work/time) of the engine can be seen to be rated at 8 h.p. What will engines be rated in when we convert to the SI system?
4. Does a piston heat engine have to have more than one cylinder to be termed a heat engine? What is the purpose of multicylinder engines?
5. In an automobile engine, why does the engine run "rough" when its timing is off ("out of time") and the cylinders do not fire at exactly the right times? Does this have an effect on fuel efficiency or gas mileage?
6. What is the difference in the timings of a six-cylinder engine and an eight-cylinder engine? (See Exercise 5.)
7. An auto crankshaft is shown in Figure 15.14. Explain the purpose of its somewhat eccentric shape. What type of multicylinder engine is this crankshaft from? The camshaft is also shown. As it rotates, the cam lobs lift the valves. Why is there one lob for each valve? Compare with Figure 15.6.
8. During the compression stroke of a piston engine, how is the gas kept from leaking around the piston?
9. Some automobiles have tachometers. What does a "tach" tell you?
10. For a particular auto with a four-cylinder engine, the tachometer indicates that the engine is turning at a rate of 3000 rpm at a normal speed. (a) How many power strokes or outputs does the engine have each minute? (b) Would the tach rate be the same for a six- or eight-cylinder engine for the same power output? Explain. (c) If the car is driven at a faster speed without changing gears, what would the tach indicate? Why?
11. Identify the heat input and heat output for the drinking bird engine (Fig. 15.4).
12. Why doesn't the two-stroke-cycle diesel engine lose fuel like a two-stroke-cycle gasoline engine?
13. Diesel automobiles have starting heaters in their engines. What is the purpose of these?
14. Why is the number of power outputs per revolution of the output shaft of a single-rotor Wankel engine equivalent to that of a six-cylinder piston engine?
15. Why does a car run better after the engine is "warmed up" on a cold morning?
16. Give four statements of the second law of thermodynamics.
17. Rudolf Clausius stated the second law of thermodynamics in terms of heat pumps, similar to the one in terms of heat engines. What would Clausius' statement say about heat pumps?
18. It is possible to use the heat of the ocean to supply the mechanical energy to turn a generator and produce electricity. A diagram of this system is shown in Figure 15.15. In certain tropical areas, the surface layer of the ocean is about 25°C and the lower layers are about 5°C. Answer the following questions about the system:
    (a) Is the system a heat engine?
    (b) Why is ammonia used as a working fluid, and how does it produce mechanical work?
    (c) What would be the ideal efficiency of the system?
    (d) Do you think ocean thermal energy conversion will ever be economically feasible?
19. In electrical generation, the maximum practical temperature limit of superheated steam used in the turbine is

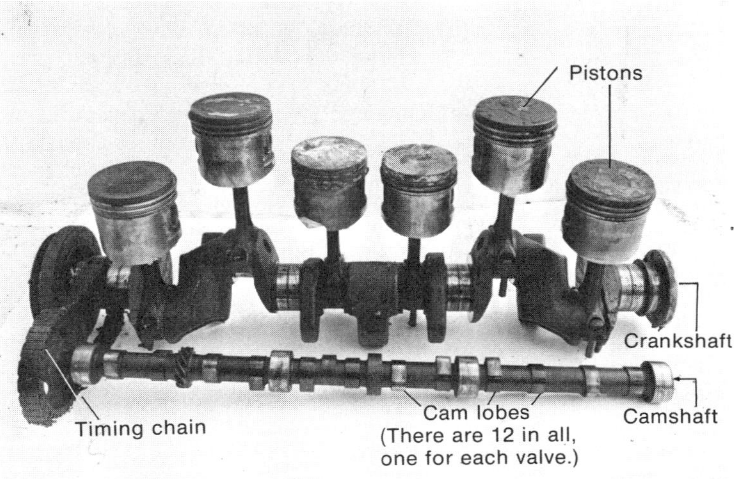

Figure 15.14 See Exercise 7.

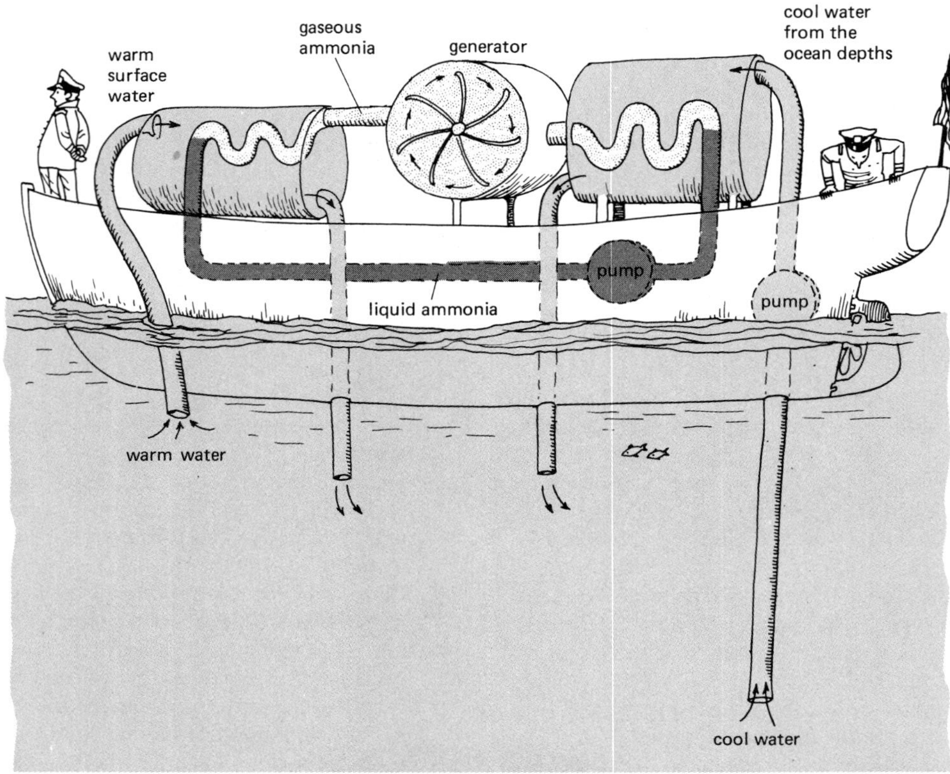

Figure 15.15 See Exercise 18.

about 540°C, owing to material limitations. (a) If the steam condenser or low-temperature reservoir operated at room temperature, what would be the ideal efficiency of the system? (b) The actual efficiency is about 35 to 40 percent. What does this tell you?

20. Given a choice between two heat engines with different operating temperatures, which would you choose? Why?

21. What would be the ideal efficiency of a heat engine if the temperature of the heat reservoirs were the same? How about the thermal efficiency?

22. Automobile engines can be water-cooled or air-cooled. Which type of engine would you expect to be more efficient, and why?

23. The heat out of a heat pump is greater than the heat in. Does this violate the first law? In atmospheric convection cycles (Chapter 14), air is transferred from higher, colder altitudes to lower, warmer levels. Does this violate the second law?

24. (a) Would leaving the refrigerator door open be a practical way to air-condition a kitchen? Explain. (b) Is it energy-efficient to put hot leftovers in a refrigerator? Explain.

25. Prior to refrigerators, "ice boxes" with blocks of ice in a top compartment were used to cool and keep foods from perishing. Is the old ice box a heat pump?

26. Rather than efficiency, heat pumps are rated in terms of coefficient of performance (c.o.p.). Thermal c.o.p. = heat removed (from low-temperature reservoir)/work. Ideal c.o.p. = $T_{\text{cold}}/(T_{\text{hot}} - T_{\text{cold}})$, where the T's are absolute temperatures.

(a) Show that the ideal c.o.p. for normal refrigerator operating temperatures is greater than one.

(b) The actual or thermal c.o.p. of a refrigerator is around 2.5. What does this tell you about the amount of heat pumped per work input?

(c) On moderately cold days, a heat pump used to heat a house has a thermal c.o.p. of about 3. If the work input was 4000 joules, how much heat would be delivered to the home? In terms of energy, how much of this would be cost-free?

(d) How does the c.o.p. of a heat pump used for home heating vary with outside temperature?

27. The heat reservoirs for the common heat pumps used in homes are both air (inside and outside). It has been suggested that a large tank of water be used for the outside reservoir. How would this help improve home heating in the winter? What could be a great disadvantage?

28. A puddle of water is often observed under a car after it has been driven with the air conditioner on. Where does this water come from?

29. On hot, humid days the coils on home and auto air-conditioners sometimes frost and "freeze up." The conditioned air is then less cool. Why? Why does the air compressor get hot when this happens? (Sometimes to the point that it is burned out.)
30. Suppose a temperature *below* absolute zero could be reached. What could the ideal efficiency of a heat engine be in this better-than-ideal (impossible) situation?
31. How does the entropy change in the process shown in Figure 15.16? Explain.
32. Does a quantity of water have more entropy when it is a liquid or when it is frozen? Explain.
33. How does the entropy of a gas in the cylinder of a heat engine change during the compression stroke of the piston?
34. Is heat energy generally becoming less available? Explain.
35. Some common sayings are, "You can't get something for nothing," "You can't even break even," and "I'll never sink that low." Are these expressions in any way descriptive of the laws of thermodynamics? Explain.

Figure 15.16 See Exercise 31.

PART FOUR
Sound

Don't make a sound! If you don't, there will be no disturbance, no vibrations, and hence, no sound. In general, we consider silence to be a condition in which no sound waves come to our ears.

A discussion of sound usually leads to an old question: If a tree falls in the forest and no one is there to hear it, is there sound? You may have been puzzled by this question. However, in this part of our study, you'll learn that the answer is simply a matter of definition. As has been learned, definitions and basic principles are very important in science. Physically, sound is the propagation of energy or a wave in some material medium, most commonly, air. On the other hand (ear?), sound is generally described as speech, music, noise, and other effects perceived by the human ear. So, if a tree falls in the forest, sound as a wave or disturbance propagation exists, but must it be heard to be sound? As you can see, it depends on how you look at or define sound.

In this short part on waves and sound, we'll look at the fascinating physical and physiological mixture of one of our major sensory links to the environment.

16

Vibrations and Waves

Vibrations and waves are quite common. We use the terms to mean various things. For example, some people talk about good vibrations ("vibes"), and we wave goodbye and wave flags. But these are different from the scientific meanings.

Examples of waves in scientific study are light waves and sound waves. Much of the information we receive about our world and communications are through our senses of sight and hearing. Waves generally involve vibrations, but the term *wave* is associated with a disturbance and energy transfer. A disturbance requires energy, and the energy of the disturbance moves or propagates as a wave. When a motorboat travels on a lake, people on the shore are soon aware of it, even if it is unseen or unheard, because the disturbance made by the boat is propagated to shore as water waves. So when some one tells you not to make waves, you are being told not to make a disturbance.

Vibrations

When something vibrates, it shakes or quivers. Automobiles usually have a few vibrations. The vibrational motion of a particle or a mass involves some type of back-and-forth motion or oscillation. Some examples of vibrating or oscillating objects are shown in Figure 16.1. They all repeat their motions periodically as a result of restoring forces. For the pendulum, the restoring force is not a mechanical one, but gravity.

To see how vibrational motion is described, let's consider a mass oscillating on a spring (Fig. 16.2). The restoring force of the spring is proportional to the dis-

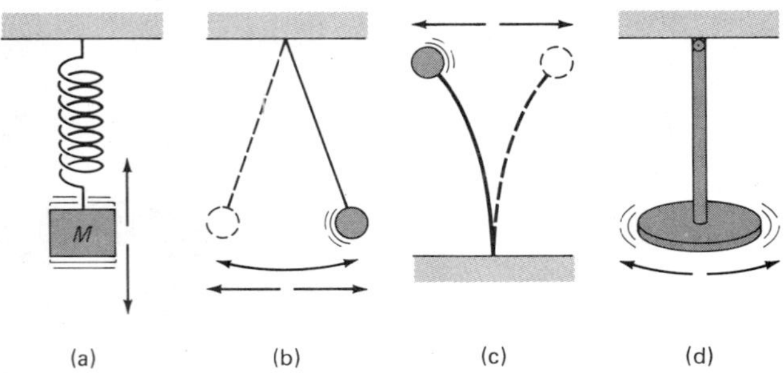

Figure 16.1 Vibrations and oscillations. (a) Mass on a spring. (b) Pendulum. (c) Mass on a flexible strip. (d) Disk on a rod (torsion oscillation).

Figure 16.2 Simple harmonic motion (SHM). (a) The SHM of an oscillator can be described by a sinusoidal curve, such as that made by a mass oscillating on a spring. The period T is the time for one oscillation, and the frequency is the number of oscillations or cycles per second ($f = 1/T$). (b) The stiffer the spring, the shorter the period and the greater the frequency of the vibration.

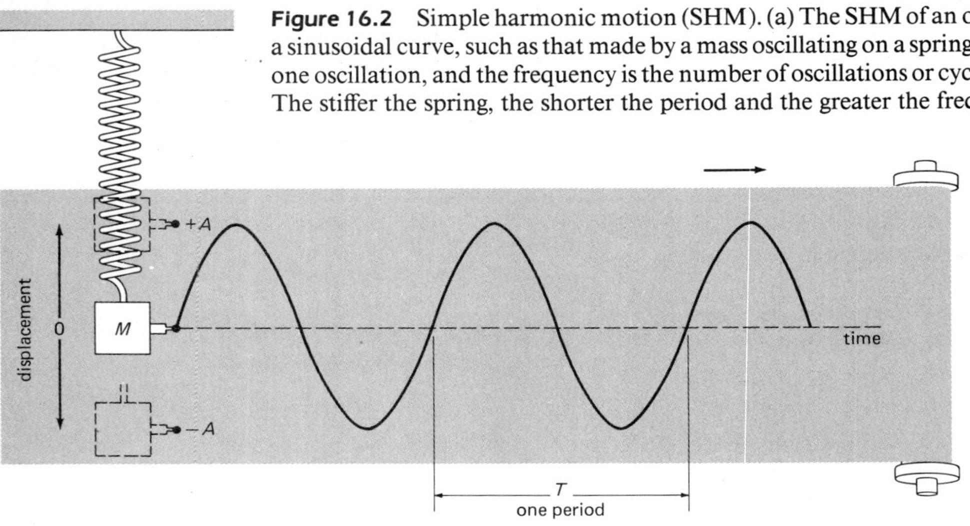

(a) generation of a sine curve

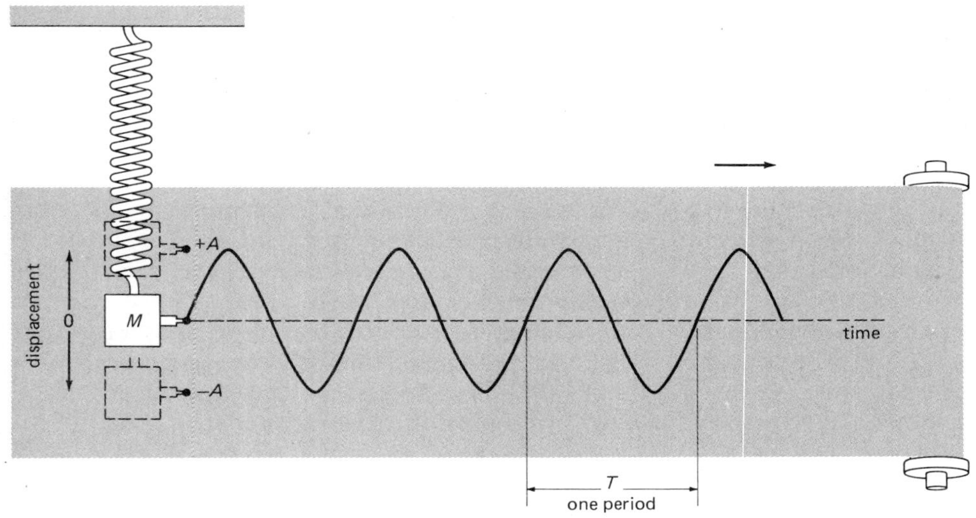

(b) shorter period, greater frequency

placement of the mass (Hooke's law, Chapter 10), and the mass is said to be in **simple harmonic motion.** A graphic picture of the motion is obtained by attaching a marker to the mass and moving a paper along behind the mass as it oscillates. This traces out a wiggly or wavy line called a sine curve.

The maximum displacement (distance up or down, $+A$ or $-A$, from the center or equilibrium position) of the mass is called the **amplitude** of the oscillation. This gives us a measure of the energy of the mass. If the mass had more energy, it would oscillate over a larger distance and have a greater amplitude. Since the motion is repeated over and over, we say it is cyclic or goes through a cycle. The time for the mass to go through one complete cycle is called the **period** (T) of the oscillation. On the sine curve, this is the time for tracing out one complete "wiggle." For example, if the tracing took 0.5 seconds, the oscillating mass has a period of 0.5 seconds ($T = 0.5$ s). That's pretty simple.

Another term used to describe the motion is **frequency** (f). This is the number of cycles the mass goes through each second, or the number of wiggles per second on the sine curve. The frequency and the period are related by

$$f = 1/T$$

So if the period of the mass is 0.5 s, then it has a frequency of $f = 1/T = 1/0.5 = 2$ cycles per second (cps). The SI unit of frequency is the hertz (Hz),* so we say the frequency is 2 Hz (1 Hz = 1 cps). This unit isn't as descriptive as cps, but that's what it is.

If a stiffer spring were used, there would be more hertz or oscillations per second, and the wiggles of the sine curve would be more bunched up on the same time scale (Fig. 16.2). Since the frequency of oscillation is greater, the period would be smaller.

A sine curve can be used to describe any type of simple harmonic motion, for example, a mass oscillating horizontally on a spring (on a frictionless surface) or a pendulum for small oscillations, as shown in Figure 16.3. Notice here that the oscillations start at the amplitude positions and that the curve begins at a "crest." We say that the oscillation or sine curve is *out of phase* with the one in Figure 16.2 by a quarter of a cycle. If the masses oscillated together and the forms of their sine curves coincided, then they would be *in phase.*

Of course, in actual vibrations or oscillations, energy is lost because of friction. To keep something oscillat-

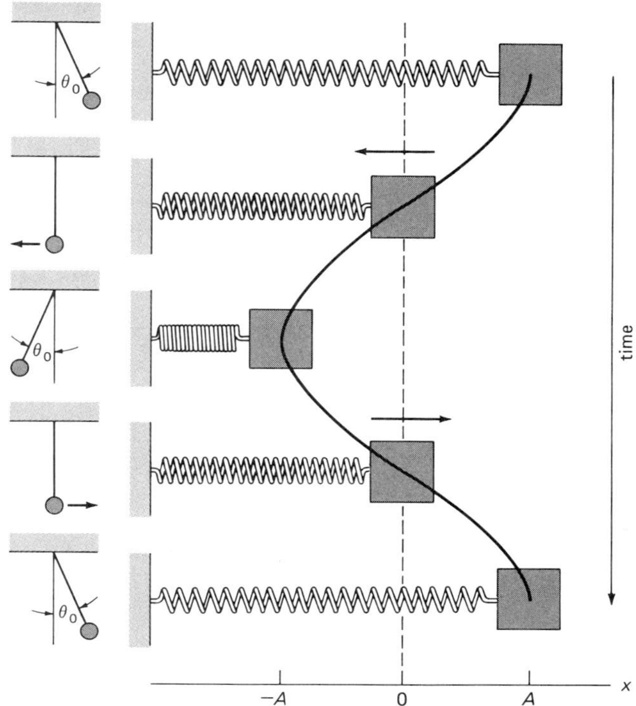

Figure 16.3 A sine curve also describes pendulum and horizontal-spring vibrations.

ing in a steady state with a constant amplitude, energy must be supplied by a driving force (Fig. 16.4). When this is removed, energy is lost, as evidenced by the decreasing amplitude, and the oscillation eventually dies out. This is called damped harmonic motion. Sometimes damping is desired and promoted by some means. For example, on dial bathroom scales the dial oscillates about your weight reading when you get on the scale. You want this oscillation to be "damped out" quickly so you don't have to wait for a long time to get an accurate reading.

So with these concepts under our belts, let's take a look at waves.

Waves

When you drop a pebble into a still pond, what happens? The pebble disturbs the water, and water waves propagate outward. What we actually see in the rearrangement of the water's surface is the motion of the disturbance, with energy being transferred outwardly in the water. The water wave (or disturbance) moves outwardly, but water is not carried with it. (See, for example, the Chapter introductory photo.) Hence, a

* After Heinrich Hertz (1857–1894), a German physicist and an early investigator of electromagnetic waves.

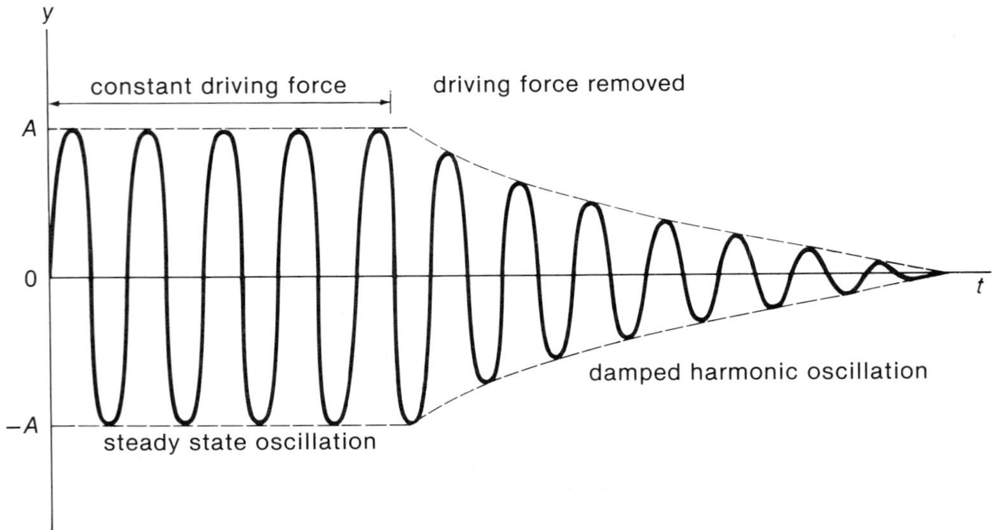

Figure 16.4 Steady-state and damped harmonic oscillations. Maintaining a steady state with a constant amplitude requires a constant driving force. If the driving force is removed, the amplitude "decays" or is damped as energy is lost.

wave is a disturbance that propagates through a medium or space.

Light and sound are both waves that propagate through space, but they are very different. Sound must propagate through a material medium—a gas (commonly air), a liquid, or a solid. If there is nothing to propagate the distrubance, there would be no sound waves. Light, on the other hand, is the propagation of energy in the form of electric and magnetic fields. Light waves can travel through a transparent medium, but none is required. These waves can travel through nothing or through empty space (vacuum).

But all waves originate from a disturbance. In the case of light, it is the motion of electrically charged particles. This kind of wave will be considered in more detail in a later chapter. For the time being, let's concentrate primarily on material waves.

CHARACTERISTICS AND TYPES OF WAVES

One of the simplest ways to "make waves" is to shake (disturb) the end of a stretched rope up and down (Fig. 16.5). The continual disturbance causes the rope "particles" to vibrate up and down, and the disturbances (waves) travel down the rope.

The wave form, which may be in the form of a sine curve, moves down the rope with a **wave velocity** (v). In general, this depends on the properties of the medium

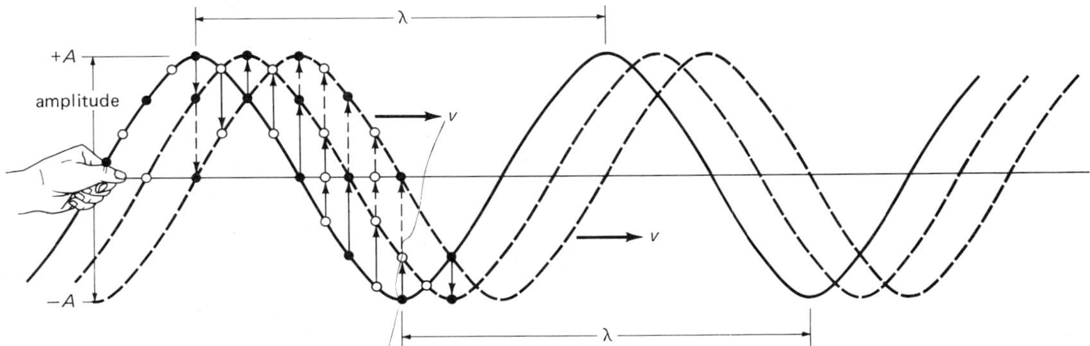

Figure 16.5 Wave characteristics. For a stretched-rope wave, the rope "particles" oscillate up and down with a maximum displacement (amplitude). The disturbance propagates down the rope with a wave velocity v. The distance between two crests, or two troughs, is called a wavelength (λ).

through which the disturbance travels. Wave velocity or speed is expressed in terms of how far the wave (disturbance) moves per time (length/time). A **wavelength** (λ) is the distance between two crests of a wave (or any two adjacent points that behave identically or oscillate in phase, e.g., two "trough" points, Fig. 16.5). Notice that a wave travels one wavelength during the time it takes for a particle to make one complete oscillation, which is the period (T) of oscillation. Hence, we may write

$$\text{Wave velocity} = \text{wavelength/period}$$

$$v = \frac{\lambda}{T}$$

This means that the wave disturbance moves with a speed such that one "wave" or wavelength passes by an observer in a time T.

Recall that the period is related to the frequency (f) of oscillation by $f = 1/T$. The frequency is the number of cycles per second, and in terms of the moving disturbance, it is the number of "waves" (cycles) passing by each second. So in terms of the frequency we have

$$\text{Wave velocity} = \text{wavelength} \times \text{frequency}$$

$$v = \lambda f$$

or

wave velocity = (length of a "wave") × (number of waves passing per second).

These equations give the distance a wave disturbance travels per time or its velocity (with direction added), and they hold for all kinds of periodic waves—in ropes and for light, sound, and water waves.

TRANSVERSE WAVES

The directions of the disturbance (particle) oscillations and wave velocities are used to make a distinction between different types of waves. In the case of a wave in a stretched rope (Fig. 16.5), the direction of the particle oscillations in the rope is perpendicular to the direction of the wave motion or velocity. Such a wave is called a **transverse wave.** This type of wave is sometimes called a shear wave because the disturbance gives rise to shear forces. (You shear a candy bar when you hold it and pull up with one hand and push down with the other and snap it or twist it in opposite directions with a torsional shear.)

Pure transverse material waves occur only in solids. We say that liquids and gases cannot support a shear, meaning that there is no restoring force to a shear disturbance. A shear cuts through liquids and gases, so pure transverse or shear disturbances cannot propagate in these media.

Light waves are transverse waves. The electric and magnetic fields oscillate perpendicularly to the direction in which the wave travels (Chapter 21).

LONGITUDINAL WAVES

Suppose you moved the end of a horizontal spring back and forth in the direction of the length of the spring. The disturbance would propagate down the spring in the form of alternate compressions and expansions (Fig. 16.6). How do the spring particles move here?

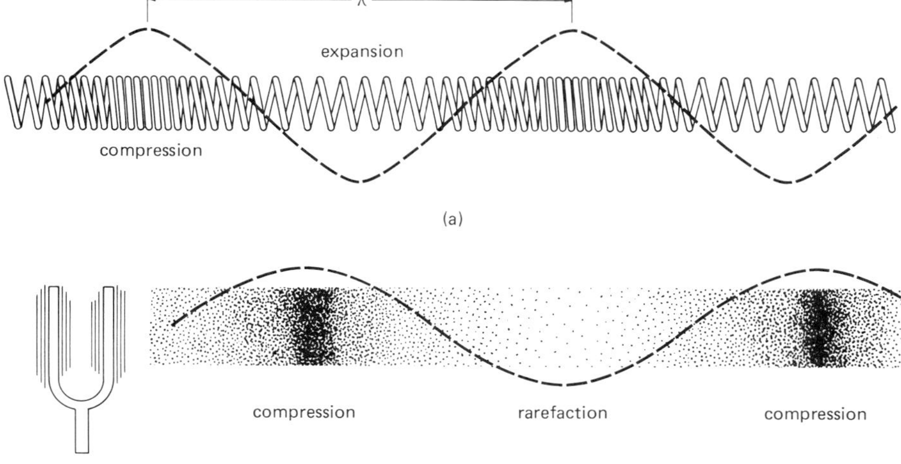

Figure 16.6 Longitudinal waves. In a longitudinal wave, the particles oscillate parallel to the direction of the wave propagation. Examples are waves in a long, horizontal spring (a) with compressions and expansions, and sound waves in air (b) with compressions and rarefactions.

Notice that they oscillate parallel to the direction of the wave motion. Such a wave is called a **longitudinal wave.** (Remember *long*itudinal—*along*.) Notice in the figure how a sine curve can be used to describe the wave form.

Longitudinal waves are sometimes called compressional waves because of the compressions involved. Compressional waves can propagate through all types of media—solid, liquid, and gas. Sound is a longitudinal or compressional wave (Fig. 16.6). In a gas, the compressions are regions of higher pressure and density. The intervening low-pressure and low-density regions are called rarefactions.

An application of the differences in the propagation properties of transverse and longitudinal waves is given in Special Feature 16.1.

SPECIAL FEATURE 16.1

The Earth's Liquid Outer Core

You may have heard that the Earth has a liquid center or, more correctly, a liquid outer core.* A good question to ask is, How do we know? We have never penetrated through the Earth's relatively thin crust. The deepest mine shafts and drillings of a few miles have only "scratched the surface." Hot molten material comes up from the upper mantle and erupts, forming volcanoes, but this occurs only in certain areas, and overall the mantle is solid.

Waves are used to probe and investigate the interior of the Earth. We can produce minor disturbances through explosions, but nature supplies some big ones free of charge through earthquakes. Earthquakes are caused by the sudden release of built-up stress along faults or cracks in the Earth, such as the San Andreas Fault in California. The energy from these disturbances propagates outwardly in the form of seismic waves. There are two general types of seismic waves—surface waves and body waves. The surface waves move along the Earth's surface and cause most of the earthquake damage.

Body waves travel through the Earth. There are P-waves, which are compressional (longitudinal) waves, and S-waves, which are shear (transverse) waves. The P and S stand for primary and secondary, which indicate their relative arrival at a location. Primary waves travel faster than secondary waves through a particular material and reach a seismic detection station first.

The longitudinal P-waves can travel through a solid or liquid, but the transverse S-waves cannot travel through a liquid. When an earthquake occurs on one side of the Earth, P-waves are detected on the opposite side and S-waves are not (Fig. 16.7). The absence of S-waves arriving in a "shadow zone" leads to the conclusion that the Earth must be liquid near the center. This is a highly viscous metallic liquid. When the P-waves enter and leave the liquid region, they are bent. This gives rise to a P-wave shadow zone, which tells us that only the outer part of the core is liquid.

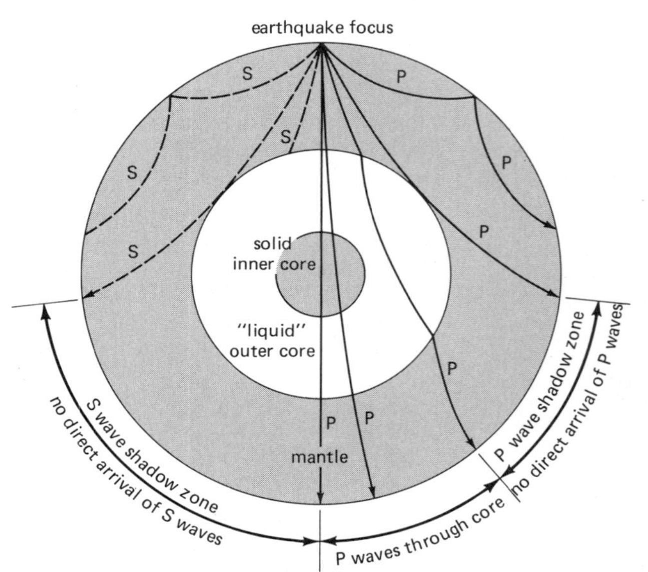

Figure 16.7 Evidence for the Earth's liquid outer core comes from waves generated by earthquakes. Transverse or shear waves cannot propagate through a liquid, but longitudinal or compressional waves can. Because of shadow zones where no waves are received, the Earth's outer core is believed to be a viscous liquid.

* The Earth is divided into three regions: an outer crust about 24–30 km (15–20 mi) thick; the mantle, 2900 km (1800 mi) thick; and a core with a radius of 3450 km (2150 mi). The solid inner core has a radius of about 1200 km (750 mi).

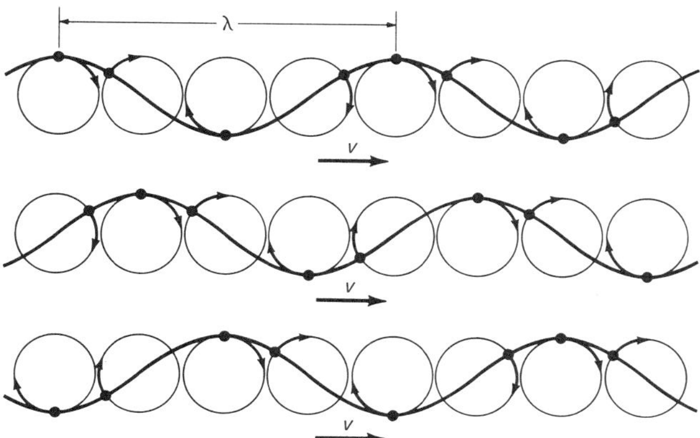

Figure 16.8 Water waves are combinations of transverse and longitudinal motions. The water "particles" move in more-or-less circular paths.

WATER WAVES

When we see waves on the surface of water, it is tempting to think of water waves as being transverse waves. Their profile looks like a transverse sine curve. However, these surface waves are combinations of transverse and longitudinal motions. The water "particles" move in more-or-less circular paths (Fig. 16.8). Notice that there is no net average motion of the water as a whole. You have probably noticed that a leaf or a fishing float or "bobber" just bobs up and down in place as water waves pass.

The diameter of the circular paths of the water particles decreases rapidly with depth (Fig. 16.9). In the open ocean a hundred meters or so below the surface, a submarine is undisturbed by the large waves due to a heavy storm on the ocean surface. However, as the wave approaches shallower water near a shore, the water particles have difficulty in completing their circular motion and are forced into more elliptical paths. The surface wave then grows higher and steeper. Finally, when the depth becomes too shallow, the water particles can no longer move through the bottom part of their paths, and the wave "breaks." The crest of the wave falls forward to form a surf (Fig. 16.9).

At some places with the aid of tidal influences, surfs can be very large, to a surfer's delight. Surfers make use of the high amplitude and energy of the breaking wave for long rides—with an occasional "wipe-out."

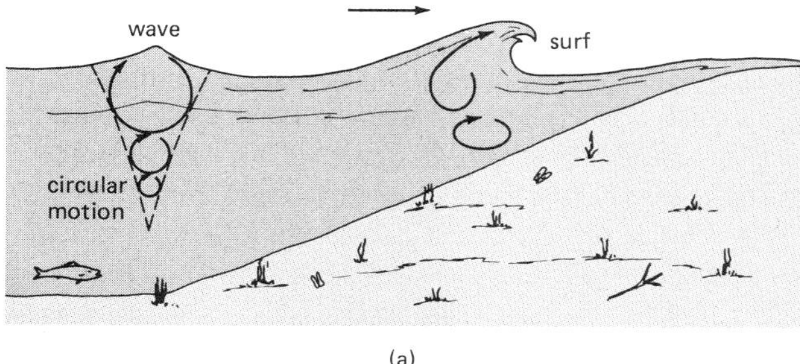

(a)

(b)

Figure 16.9 Surf. When water particles in a wave cannot complete the bottom parts of their paths, the wave "breaks," and the crest falls forward to form a surf.

Energy and Intensity

The energy of a wave is related to its amplitude (more accurately, it is proportional to the square of its amplitude). But we are often interested in the rate of energy transfer. When you strike a tuning fork, it is given energy and set into vibration. As the fork vibrates, it causes the air near it to vibrate, and energy is given to the air in the form of a sound wave (see Fig. 16.6). After a certain time the fork "runs down," and the production of sound waves ceases.

To improve the energy transfer to the air, suppose that after you strike the fork, you hold the base of its handle against a table top. The sound is louder, and the fork stops vibrating sooner than if it is not touching anything. The increased loudness of the sound occurs because the vibrating fork causes the table top to vibrate also. The large table surface is able to set more air in motion than the fork itself, and there is an area consideration in energy transfer.

To characterize the transfer of energy by a wave, we speak in terms of intensity. Intensity (I) is the rate of energy transfer through or to a cross-sectional area (A), or $I = \text{(energy/time)/area}$ (Fig. 16.10). Since energy per time is power, $I = \text{power/area} = P/A$, and the units of intensity are watts per square meter (W/m^2).

The intensity varies with the distance from the wave source. This can be seen by considering the intensities at various distances from a point source from which waves travel outwardly in all directions (Fig. 16.11). The area of a sphere around the sources is $A = 4\pi r^2$. So $I = P/A = P/4\pi r^2$, and the intensity from a small source "falls off" as $1/r^2$. For example, if you double the radius of the sphere or distance from the source, the intensity is only one fourth as great. At a farther distance from the source, the energy is "spread" thinner over a greater area, and the intensity is less. Imagine you have just so much paint to paint a house. Using all the paint, you'd have to spread it thinner on a larger house, so the paint "intensity" would be less.

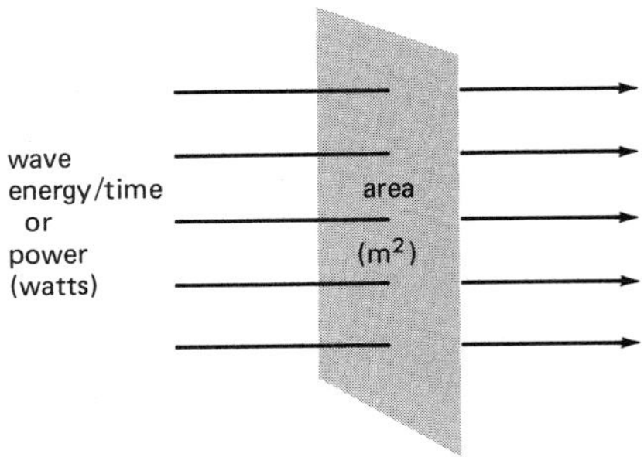

Figure 16.10 Intensity. Intensity is wave energy per time, or power passing through an area ($I = P/A$), and has units of watts/m² (W/m^2).

Interference

Suppose two waves traveling in different directions run into each other. What do you think happens? As they pass through each other and interfere, the disturbances combine and add up. This is called the **principle of superposition:** When two or more waves interfere, the height or displacement of the combined wave at any point is equal to the vector sum of the individual wave displacements.

To illustrate this, consider two similar wave pulses traveling in opposite directions in a stretched rope (Fig. 16.12a). As the waves interfere, the displacements add, and the wave form has a combined shape. Interference in which the adding of the displacements produces a taller combined wave form is called **constructive interference.** When exactly superimposed, the combined wave form is twice as tall as the individual waves ($A + A = 2A$, assuming they have the same amplitude).

However, remember that the displacement can be negative, such as in a downward pulse (Fig. 16.12b). Here, when the two pulses interfere and the (vector) displacements are added, the height of the combined wave form is smaller. This is called **destructive interference.** When exactly superimposed, the amplitude of the

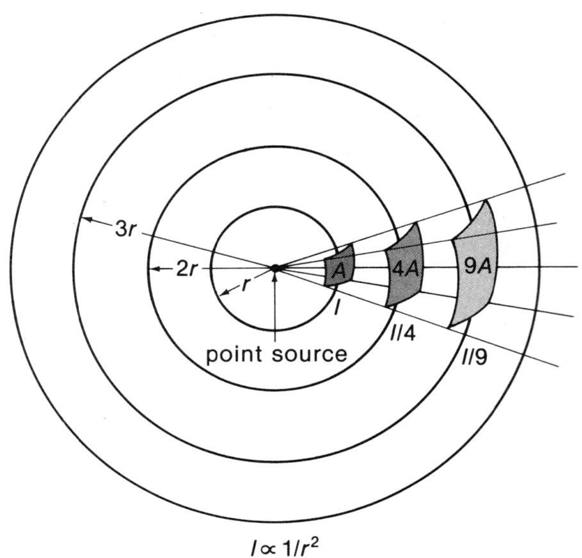

Figure 16.11 The intensity decreases or "falls off" as $1/r^2$ (distance from a point source). The energy propagating outward passes through increasingly greater spherical areas ($A \propto r^2$), so the intensity is less, $I = P/A \propto 1/r^2$.

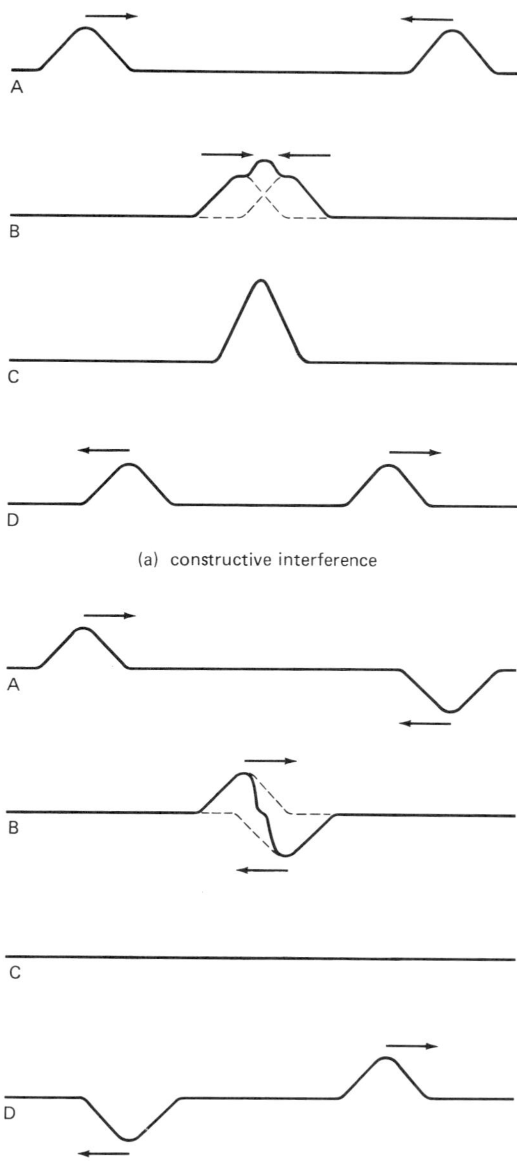

Figure 16.12 Wave interference. (a) Constructive interference occurs when the individual wave displacements add up or produce a greater combined wave form. (b) In destructive interference, the addition of displacements gives a smaller wave form.

combined wave form is zero [$A + (-A) = 0$]. The upward pulse "fills in" the downward pulse, so to speak, in what is called total destructive interference. The energy of the disturbance is not "destroyed." It is still there, stored in the rope.

In either case, after the pulses interfere they pass on down the rope as though nothing had happened. Check out the interfering waves in the Chapter introductory photo.

In some cases in which periodic waves continuously interfere, interference patterns are set up (Fig. 16.13). The waves spread out from each source and overlap or interfere. When a crest overlaps with a crest or a trough overlaps with a trough, constructive interference occurs. The waves are said to be in phase at these points. At other points where a crest overlaps a trough, destructive interference takes place. The waves are said to be completely out of phase here.

As the waves travel outward, these conditions occur at certain points and a pattern is formed, as shown in the Figure 16.13(b) for actual water waves. Along the dark lines radiating outward, crests overlap troughs,

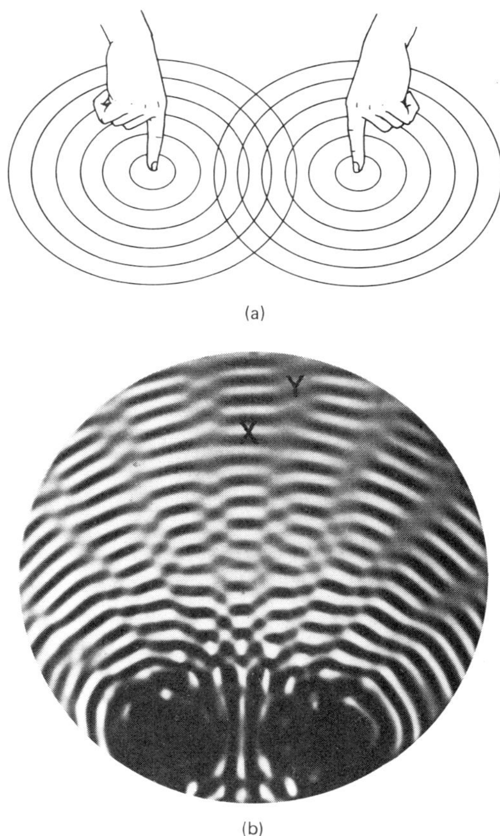

Figure 16.13 Interference. (a) Continuous waves give rise to interference patterns. (b) Dark radial lines (Y) are regions of destructive interference. The bright radial lines (X) are regions of constructive interference.

and destructive interference occurs. A small cork placed in one of these regions, say at Y, would be relatively motionless. The outward rays with alternate bright and dark lines are where constructive interference occurs (light is reflected from the tall crests). A cork placed in one of these regions, say at X, would bob up and down as the waves passed by.

Such interference patterns occur for all types of waves—water waves, sound waves, light waves, and so on—provided that the waves or periodic disturbances have the same frequency. When they don't, the waves still interfere, but the effect is different. We'll take a look at an example of this in the next chapter on sound.

Standing Waves

Interference also gives rise to standing waves. You have no doubt noticed that when you shake the end of a stretched rope up and down just right, the wave form appears to "stand" in the rope. This is due to the interference of the waves traveling down the rope and the waves reflected back from the other end.

When a wave form is reflected from a rigid support, the reflected wave form is inverted. This is illustrated for a wave pulse in Figure 16.14. The incident wave exerts an upward force on the support, and the support exerts a downward reaction force on the rope, causing the reflected pulse to be inverted. For a continuous wave, the reflected wave is continuously inverted. (If the support can move freely, such as a light ring that can move up and down on a rod, the reflected wave is not inverted.)

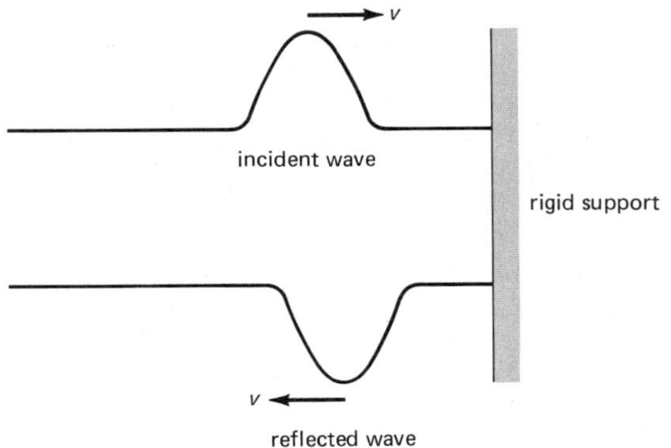

Figure 16.14 When a wave is reflected from a rigid support, the wave form is inverted.

When two waves of equal amplitude and wavelength traveling in opposite directions interfere, they are continuously in phase and out of phase. Suppose the traveling waves at a particular instant were as shown in Figure 16.15(a). Their combined wave form would be a straight line (destructive interference). When each wave moves one fourth of a wavelength (the distance of half a loop), they are in phase (Fig. 16.15b). In another one-fourth wavelength movement the waves are completely out of phase, and in another one-fourth wavelength they are back in phase, and so on.

This action gives rise to a **standing wave** with the wave envelopes clearly visible, as shown in the photograph of a vibrating string. Notice that there are points along the string that do not move. These are called nodal points or **nodes.** The amplitude (maximum displacement) points of the envelopes are called **antinodes.**

CHARACTERISTIC FREQUENCIES

If you try to form standing waves in a rope, you'll find that they will only occur when you shake the rope just right—at the right frequencies. Otherwise, the waves will not interfere to form a standing wave. This is because of what we call boundary conditions. Notice in Figure 16.16 that a node is required at each end of the rope (a *near* node at the shaking end). As a result, only a certain number of wave loops can "fit in" the rope. The first possibility is one loop, or half a wavelength (two loops per standing wavelength), then two loops (one wavelength), three loops ($1\frac{1}{2}$ wavelengths), and so on.

The speed of the waves in the rope is fixed by the tension in the rope and the mass along its length. So for particular wavelengths there can be only particular or **characteristic frequencies,** by the equation $\lambda f = v$ or $f = v/\lambda$. The wave forms of these characteristic frequencies are sometimes called normal modes of oscillation. The lowest frequency is called the *fundamental frequency* or first harmonic. The next frequency, which is twice the fundamental frequency, is called the second harmonic (or sometimes the first overtone), and so on. Each harmonic is a multiple of the fundamental frequency.

Get a rope and try this for yourself. It is difficult to get more than the lowest three or four harmonics in the rope. Standing waves occur in stringed and wind musical instruments. The latter involve standing waves in air, as will be learned in the next chapter when we look at how musical instruments make sounds or tones.

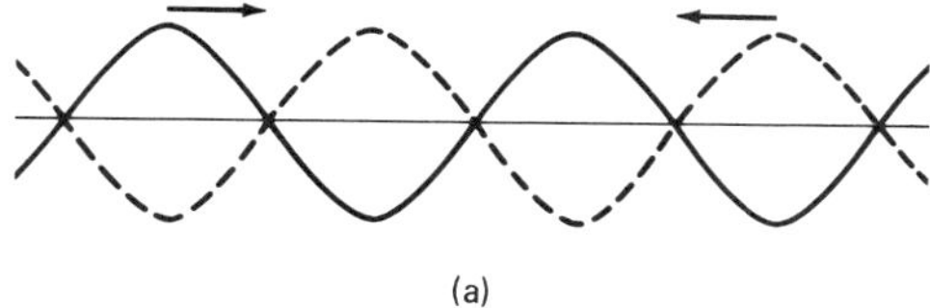
(a)

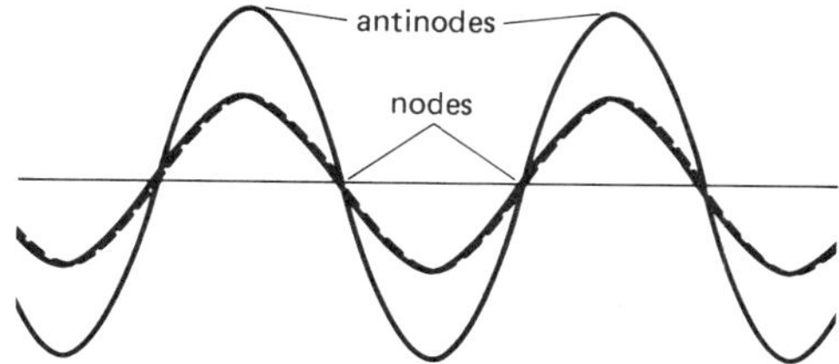

(b) waves move 1/4 λ

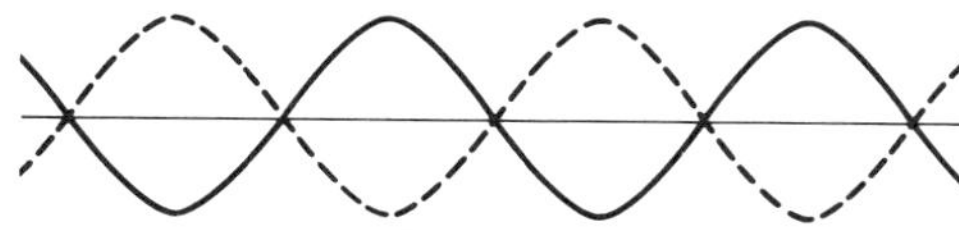
(c) another 1/4 λ

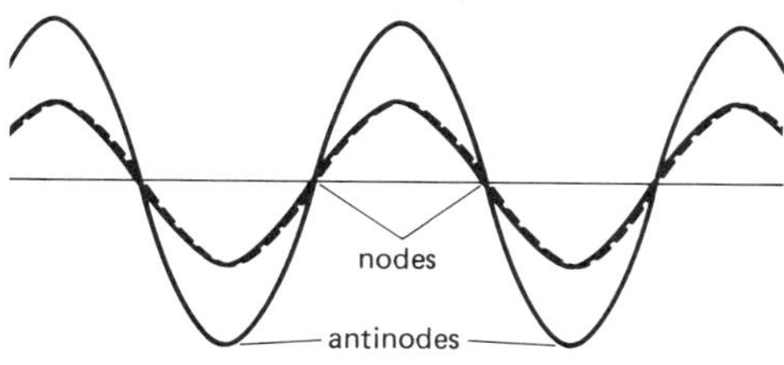

(d) another 1/4 λ

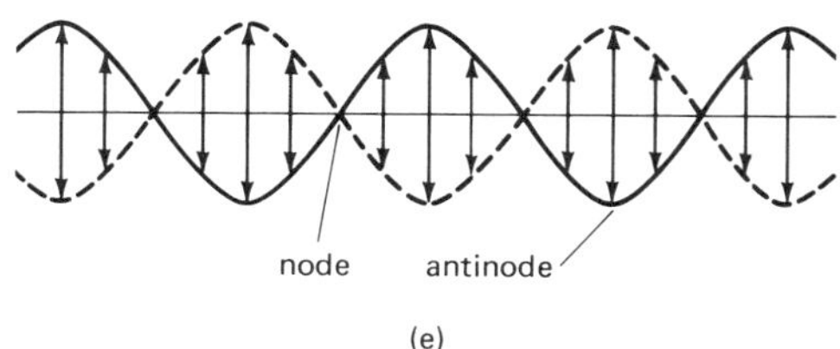

(e)

Resonance

All things vibrate or oscillate at a characteristic frequency or frequencies. The characteristic frequencies of an object or material are *natural* frequencies of oscillation and depend on such factors as mass, elasticity or restoring force, and geometry (boundary conditions). In a sense, these are the frequencies at which an object or material "wants" to vibrate.

When an oscillator is driven at a frequency corresponding to a particular characteristic frequency, we say that the oscillator is driven in **resonance.** At a resonant frequency, there is maximum energy transfer from the driving source to the oscillator. This is not to say that energy will not be transferred at other driving frequencies. However, the oscillations may be erratic, as when you shake a rope at a frequency other than one of its characteristic frequencies.

A rope of string has many possible characteristic frequencies, as we have seen. But in some cases there is only one characteristic frequency, for example, in a child's swing or pendulum. A swing is basically a simple pendulum, which has only one characteristic frequency or just a fundamental frequency. This is determined by the length of the pendulum. When a swing is pushed with this (resonant) frequency, it swings higher and higher (greater amplitude). You don't have to push hard, but you need the right rhythm or frequency so energy is transferred to the swing smoothly. Most of us have experienced what happens if we change the driving frequency slightly and push the swing before it reaches the return amplitude of its swing. Another example is bouncing a ball (Fig. 16.17). To have the ball

Figure 16.15 Traveling wave interference. Waves traveling in opposite directions in a string are continuously in phase and out of phase (a to d). Interfering waves of equal amplitude and wavelength give rise to standing waves (e, f).

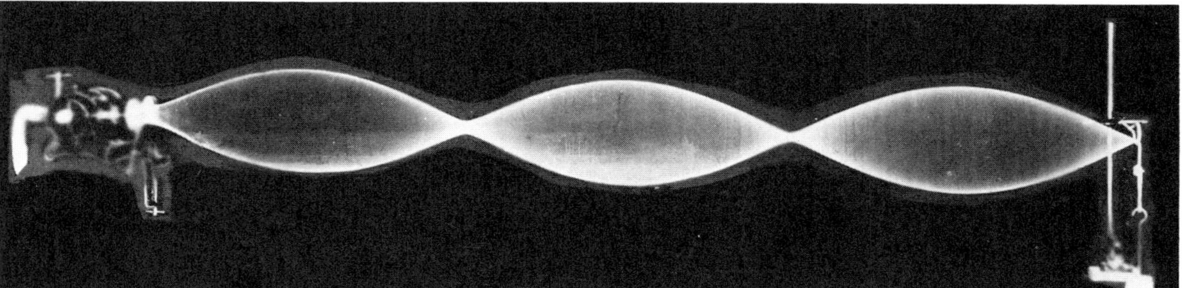
(f)

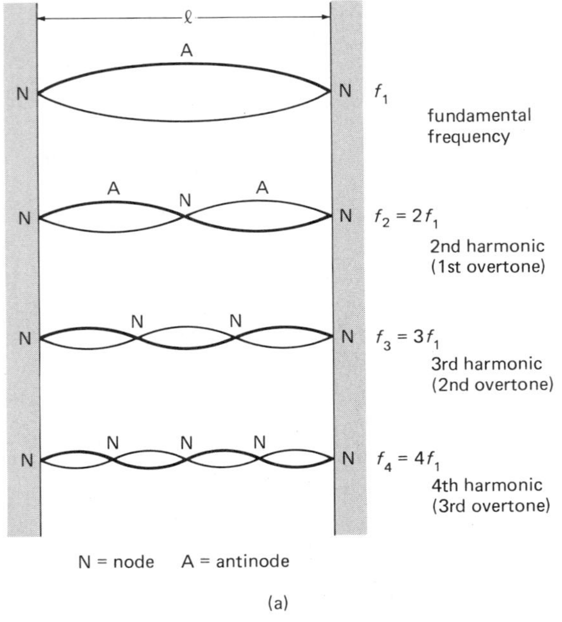

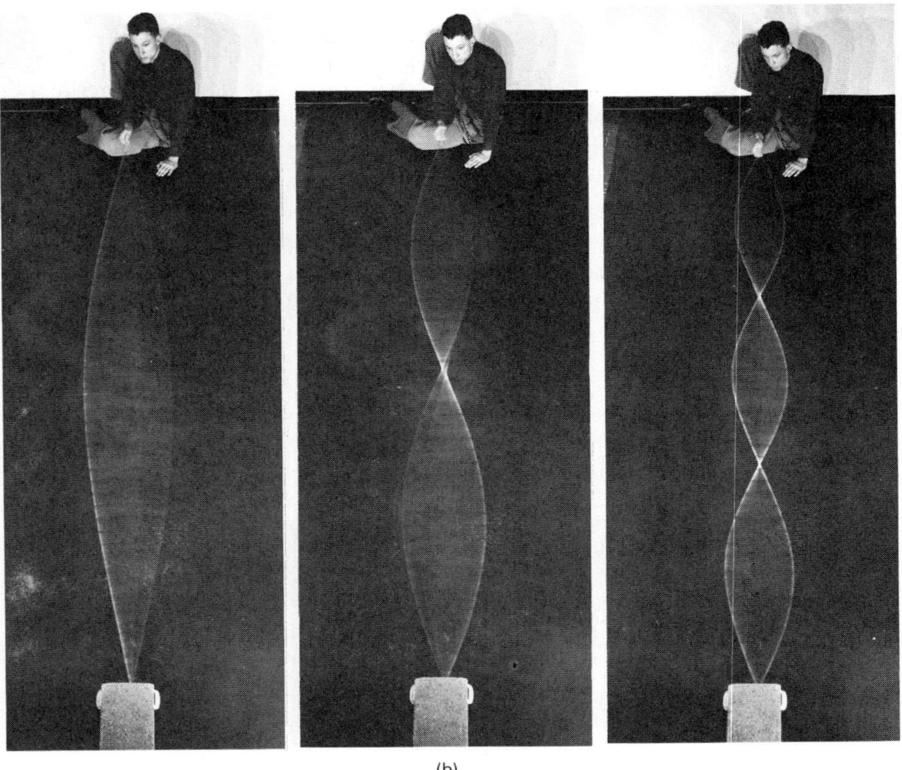

Figure 16.16 Characteristic frequencies. Boundary conditions permit only certain combinations of loops or half-wavelengths ($\lambda/2$). This allows only certain characteristic frequencies of vibration.

Figure 16.17 Resonance. Similar to pushing a swing, you must push or drive a bouncing basketball at its "bounce" frequency for maximum energy transfer. If this big fellow stooped over and shortened the bounce distance, he'd have to drive the ball at a greater frequency. Why?

bounce smoothly, you must drive it at its natural "bounce" frequency or in resonance.

A mass oscillating on a spring also has only one translational characteristic frequency. This depends on the spring constant, or how "stiff" the spring is, and the mass. However, there is another possibility. The mass may oscillate in a rotational "twisting" or torsional mode (see Fig. 16.1), and there is a characteristic frequency for this mode. Looking at the atoms of molecules as being held together by "springs," you might expect molecules to have certain natural oscillations, and they do. As was learned in Chapter 14, water molecules absorb infrared and microwave radiations (heat transfer). This is because the water molecule has a natural vibrational frequency in the infrared frequency region. The infrared radiation drives the molecule in resonance, and there is energy (heat) transfer. The water molecule also has a natural rotational oscillation at a lower characteristic frequency. This is in the microwave frequency region, and microwave resonance in water molecules is the principle of microwave ovens (see Chapter 14).

The maximum energy transfer of the resonance condition is also important in electrical circuits, as will be learned in a later chapter. For example, when you tune in a certain station frequency on your radio, you are adjusting a circuit so it will resonate electrically at that frequency. This gives you maximum energy transfer to the circuit by radio waves at that frequency, and you can pick up a particular station at its assigned frequency.

An example of an unwanted resonance is given in Special Feature 16.2.

SUMMARY OF KEY TERMS

Vibration an oscillation or back-and-forth motion.

Simple harmonic motion periodic motion in which the restoring force is proportional to the displacement (Hooke's law).

Sine curve a graphical curve used to describe simple harmonic motion and wave forms.

Amplitude the maximum displacement of an oscillation from the center or equilibrium position.

Period the time (T) to complete one cycle in periodic motion.

Frequency the number of cycles per second in periodic motion: $f = 1/T$. The unit of frequency is the hertz (Hz).

Damped harmonic motion periodic motion in which energy is lost and the amplitude of the oscillation decreases with time.

Wave a disturbance that is propagated through a medium or space.

Wavelength the distance (λ) between two crests of a periodic wave (or any two adjacent points that behave identically, e.g., two troughs).

Wave velocity the velocity of a wave, where the magnitude is given by $v = \lambda f$.

Transverse wave a wave in which the particle or oscillation motion is perpendicular to the direction of the wave motion or velocity. This is sometimes called a shear wave.

Longitudinal wave a wave in which the particle motion is parallel to the direction of the wave motion or velocity. This is sometimes called a compressional wave.

Intensity the rate of energy transfer through an area: $I =$ (energy/time)/area = power/area. The units of intensity are watts/m^2 (W/m^2).

Principle of superposition when two or more waves interfere, the height or displacement of the combined wave at any point is equal to the vector sum of the individual wave displacements.

Constructive interference interference in which the adding of the displacement produces a taller combined wave form.

Destructive interference interference in which the addition of the displacements produces a smaller combined wave form.

Standing wave a steady-state oscillating wave form due to the interference of two waves of equal amplitudes and wavelengths traveling in opposite directions. The motionless points of a standing wave are called nodes, and the amplitude points of the envelopes are called antinodes.

Characteristic frequencies the particular frequency or frequencies at which an object or material naturally vibrates.

Resonance driving an oscillator at a characteristic or natural frequency that gives maximum energy transfer from the driving source to the oscillator.

EXERCISES

1. In Figure 16.2 the sine curve motion of mass oscillating on a spring is demonstrated. How could you show that an oscillating pendulum traces out a sine curve with time?
2. For a pendulum swinging in simple harmonic motion (small arcs), what positions of the swing correspond to the amplitude of the sine curve that describes the motion? What is the total energy at these points?
3. If a mass oscillating on a spring is given more energy by a

SPECIAL FEATURE 16.2

Galloping Gertie: The Tacoma Narrows Bridge Collapse

The Tacoma Narrows Bridge near Tacoma, Washington, was opened to traffic on July 1, 1940. It was 2800 ft (855 m) long and 39 ft (12 m) wide. The steel girders used in its construction were 8 ft (2.4 m) tall. During its first several months of use, many transverse modes of vibration were observed. On the morning of November 7, 1940, winds gusting up to 40–45 mi/h started the main span vibrating, with nodes, of course, at the main towers and at several places in between. The vibrational frequency was 36 Hz with an amplitude of 1.5 ft (Fig. 16.18).

At 10 AM the main span of the bridge began to vibrate in a torsional (twisting) mode in two segments with a frequency of 14 Hz. This was apparently due to the loosening of a cable by which the roadway was suspended. Twisting in the wind, the bridge was nicknamed "Galloping Gertie."

The wind drove the structure near the critical velocity for the torsional mode, and this vibration built up by resonance. Shortly after 11 AM, the main span broke and collapsed. The bridge was rebuilt using the original anchorages and tower foundations, but with a new design that stiffened the structure and increased the resonance frequency so that high winds would not set it into resonance vibrations—and it worked.

Figure 16.18 Tacoma Narrows bridge collapse (1940). Gale winds set the bridge into resonance (torsional) vibration. The bridge, nicknamed "Galloping Gertie," oscillated for almost an hour, then collapsed.

driving force, how does this affect the (a) amplitude, (b) frequency, and (c) period of the oscillation? Explain.

4. If an object oscillates with a frequency of 10 Hz, how long does it take to make one complete cycle?

5. Suppose it takes you 20 minutes to walk around the block twice. (a) What would be the period of your motion? (b) What would be your frequency in hertz? What does this mean?

6. A mass oscillates with a frequency of 5 Hz. How many cycles does it go through in (a) 2 s and (b) 0.6 s?

7. Two masses oscillate up and down on springs with the same amplitude and frequency. Compare the motions when they are (a) in phase, (b) one-fourth cycle out of phase, and (c) completely out of phase. (Draw some diagrams and show where the masses are relative to each other at a particular instant.)

8. Radio station frequencies (radio waves) are in the kHz and MHz (kilohertz and megahertz) ranges. What are the corresponding ranges of the periods of these radio waves? (The speed of electromagnetic radio waves is 3×10^8 m/s.)

9. To maintain a steady-state oscillation, how much work must be done by the driving force? What happens if more or less than this amount of work is done?

10. Give some examples in which damped harmonic motion is undesirable and some in which such motion is desirable.

11. An automobile has springs and shock absorbers to provide a smooth ride. Explain how they smooth the ride.

12. How does a bottle or twig bobbing up and down in a lake demonstrate that waves carry energy?

13. How do electromagnetic waves carry energy? What disturbances cause EM waves?

14. After a motorboat passes by on a lake, an observer on the shore notices that the waves hit the shore every 2 seconds and that the distance between the crests of the waves is about 2 m. What is the speed of the water waves? Does this depend on the speed of the boat? Explain.

15. When a motor boat moves across a lake, water waves continually lap the shore for some time after it has passed. But few or no waves reach the shore from a row boat. Why is this?

16. In general, the wave speed in a medium is fixed by the elasticity and particle mass of the medium. With the wave speed constant in water, how is the distance between the crests of the surface waves affected when the disturbance frequency is increased?

17. Light waves in vacuum travel at a speed of 300,000 km/s. The frequency of visible light is on the order of 10^{14} Hz (one followed by 14 zeroes). What is the order of magnitude of the wavelength of visible light?

18. Sine curves representing two waves are shown in Figure 16.19. Which wave has the greater (a) amplitude, (b) frequency, (c) wavelength, and (d) period?

19. The wave velocities of some transverse and longitudinal waves are in the upward direction. What are the oscillation directions of the waves?

20. If earthquakes produced only S-waves, would we be able to tell whether the inner core was solid? Explain.

21. Neglecting tidal influences, why are surfs along some coastal areas greater than others? Do tides influence surfs? Explain.

22. How would the intensity you receive from a point source vary if you (a) moved four times as far from the source? (b) If you moved two thirds of the distance closer to the source?

23. Two waves of equal amplitude interfere. What can you say about the amplitude of the combined wave form when the waves are (a) in phase, (b) out of phase, and (c) completely out of phase?

24. Is energy destroyed in destructive interference? Explain. What is "destroyed"?

25. Explain why interference patterns are formed; that is, what causes a pattern?

26. If a transverse wave and a longitudinal wave interfered in a medium, what would be the general wave form of the combined wave?

27. When a stretched rope is oscillated at its fourth harmonic frequency, how many standing wavelengths are formed in the rope?

28. A long, limber rod is clamped at one end, leaving the other end free. The boundary conditions for a standing wave in the rod are a node and an antinode at the respective ends. In this case, only the odd harmonics are possible, e.g., f_o, $3f_o$, $5f_o$, and so on. Draw a few possible standing wave forms and demonstrate this in terms of wavelength.

29. When you drop a wrench or a metal bowl, it makes a

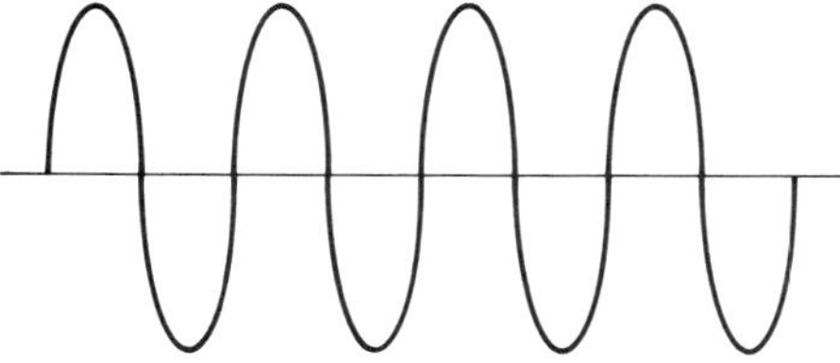

(a)

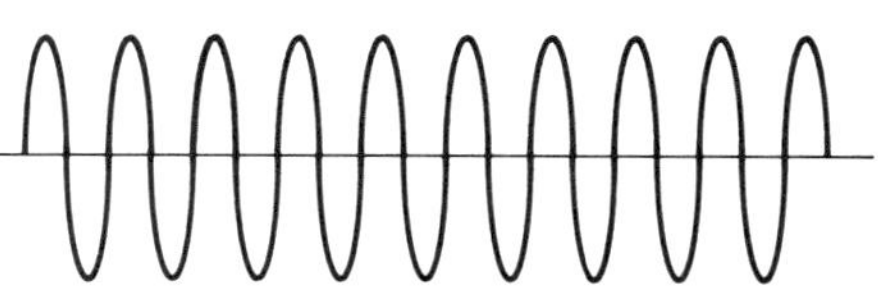

(b)

Figure 16.19 See Exercise 18.

certain sound if it hits a hard surface. Why is this? Why does a baseball bat make a different sound?

30. It is common for the "front end" of an automobile to shimmy or vibrate when it is out of alignment or out of balance. However, this usually occurs only at a certain speed, and speeding up or slowing down causes the vibration to cease. Why is this?

31. When getting on a swing that you have agreed to push, a fellow student informs you that the resonant frequency for the swing is 0.33 Hz. How often should you push the swing to give a smooth, high ride?

32. A swing has only one characteristic frequency f_o. However, it can be pushed smoothly at frequencies of $f_o/2$, $f_o/3, f_o/4$, and so on. Explain why, and describe the effect of the different driving frequencies.

33. When marching across small bridges, troops are ordered to break step or cadence. Why might marching in step be undesirable?

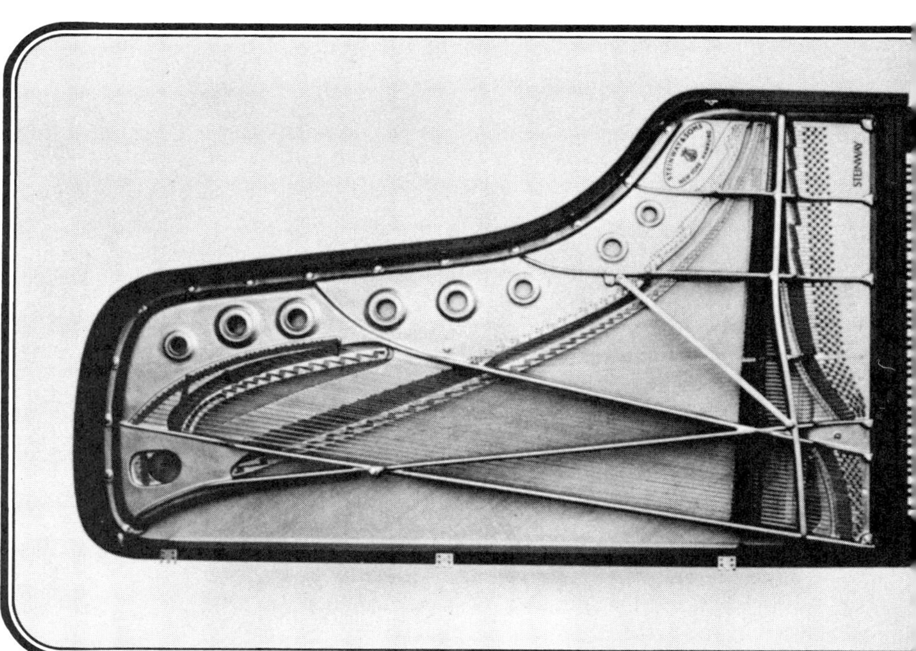

17

Sound and Music

The Nature of Sound

In the technical sense, sound waves are longitudinal waves that are propagated in solids, liquids, and gases. Without matter there are no sound waves. Most sound comes to us through the air. If we could live in a vacuum, there would be no normal sounds.

Before there can be sound waves in air, there must be vibrations of some material body. When the tires of an automobile "squeal" in a fast start, we hear the sound because something (tires) vibrates and disturbs the air. A sound pulse results from a single disturbance, such as a hand clap or explosion, and a continuous wave results from a continuous disturbance. An example of the latter is a tuning fork (Fig. 17.1). The fork is essentially a bar bent in a "U" shape that vibrates at its fundamental frequency (antinodes at each end). Higher harmonics are negligible, and a single tone or frequency is heard.

In the ear, the eardrum (a thin membrane) is set into vibration by the pressure compressions and rarefactions of the sound wave. These vibrations are transmitted by bone structures to the inner ear, where they are picked up by the auditory nerve.* When these vibrations have a frequency between 20 Hz and 20 kHz (20,000 Hz), nerve impulses are initiated that are interpreted by the brain as sound.

The vibrational frequency range or spectrum is divided into three general regions (Fig. 17.2). Below the audible frequency region of human hearing is the infrasonic region (0 to 20 Hz). The longitudinal P-waves generated by earthquakes lie in this region (Chapter 16). Above the audible region is the ultrasonic region,

* The inner ear contains a liquid, which is set into motion. This motion is picked up by hair cells, which initiate signals to the nervous system. The fluid and hair cells of the inner ear are also important in maintaining our balance, or equilibrium.

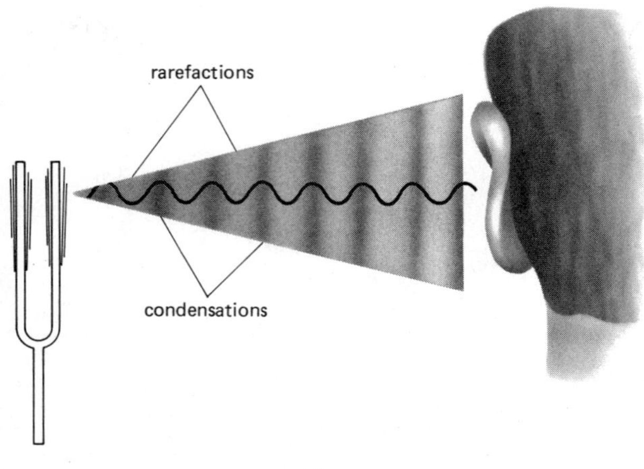

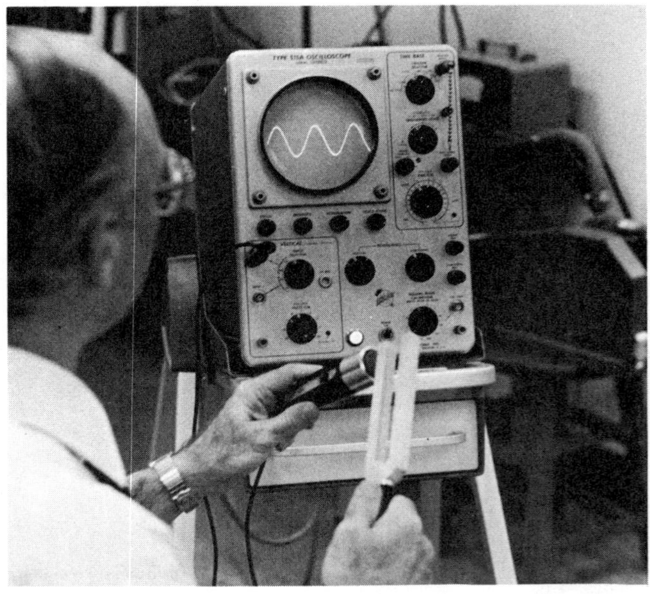

Figure 17.1 Tuning fork. A tuning fork is essentially a bar bent in a "U" shape that vibrates at its fundamental frequency, giving rise to sound disturbances in air (a). The form of the wave from a tuning fork is displayed electronically on an oscilloscope (b).

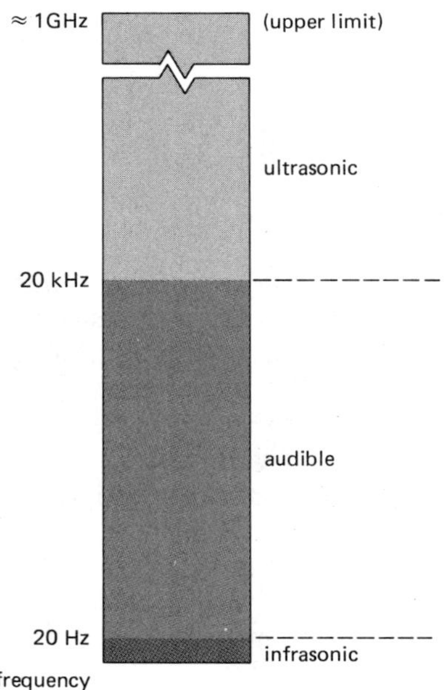

Figure 17.2 The sound frequency spectrum. We hear only sound disturbances with frequencies between 20 Hz and 20 kHz (audible region). Below this is the infrasonic region. Above this is the ultrasonic region, which has an upper limit of about a billion hertz that is set by the limit of material elasticity.

which includes the elastic vibrations of crystals, such as quartz. The sound frequency spectrum does not continue indefinitely. There is a general upper limit of about one billion hertz, which is the upper limit set by material elasticity.

Humans cannot detect sound in the ultrasonic region, but other animals can. For example, the hearing frequency response of dogs extends beyond that of humans. Whistles with ultrasonic frequencies, which do not disturb people, are sometimes used to call dogs. Some practical applications of ultrasound are given in Special Feature 17.1.

One of the most common audible sounds is the human voice. The energy for sounds associated with the voice originates from the muscle action of the diaphragm, which forces air from the lungs (Chapter 12). To produce variable sounds, this steady stream of air must be disturbed or "modulated." The basic modulating organ is the larynx (the voice box), across which are stretched two membrane-like bands called vocal cords. The vocal cords form a slit-like opening. As air is forced from the lungs, the flexible vocal cords allow the voice box slit to widen and narrow. Their vibrating and breaking up of the stream of air produces a variation in the pressure, and hence sound waves. The effect of the vocal cords is similar to that of the reed in some musical instruments, for example, the clarinet. The vibrating reed converts a steady air flow into a modulated one.

SPECIAL FEATURE 17.1
Ultrasonics

Ultrasonic sound is in the frequency region above the audible region (>20,000 Hz). This high-frequency sound finds many uses and applications. For example, audible-frequency sound waves have a limited propagation range in water. However, ultrasonic waves can travel for kilometers and are highly directional. As a result, ultrasound is used in marine applications. For example, the depth of the ocean is determined by depth "sounding" techniques using a fathometer. A beam of ultrasound is directed downward from a ship and is reflected from the ocean floor. The depth is computed by knowing the ultrasound speed and elapsed time. This detection and ranging technique is called sonar. Ultrasonic sonar is used not only to detect and determine the range of ships and submarines, but also that of schools of fish. In addition, combining or modulating audible sound waves with ultrasound makes underwater radio communication possible.

Ultrasound is used in many technical applications. One of the best known in ultrasonic cleaning (Fig. 17.3a). Ultrasound in liquid baths is used to clean metal parts. Ultrasonic vibrations (small wavelengths) can loosen traces of foreign matter from otherwise inaccessible places. Jewelers make use of ultrasonic baths to clean rings and other jewelry.

Another important industrial application of ultrasound is ultrasonic drilling in the machining of very hard materials (Fig. 17.3b). By means of an abrasive paste and an ultrasonic vibrator, the material is rapidly worn away. Since the ultrasonic drill does not rotate, the vibrator tip can be orientated to produce holes or surfaces of any shape. Ultrasonic soldering irons are also available. These are particularly useful in the soldering of aluminum. The ultrasound removes the aluminum oxide coating on the surface and eliminates the need for fluxes.

An ultrasonic beam or pulse traveling in metal can be used to detect flaws (Fig. 17.4). When the ultrasound strikes a flaw, which has different properties than the surrounding medium, reflection and refraction occur. The echo pattern is monitored, and an irregularity indicates the presence of a flaw in the metal. Such techniques provide a means of nondestructive testing of metal castings and other metal objects, such as airplane parts.

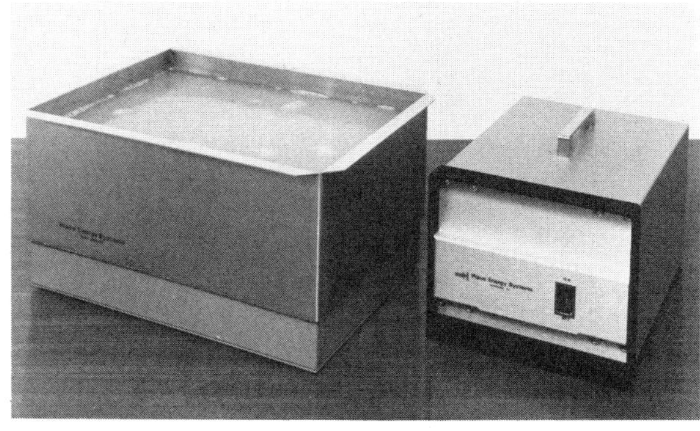

(a)

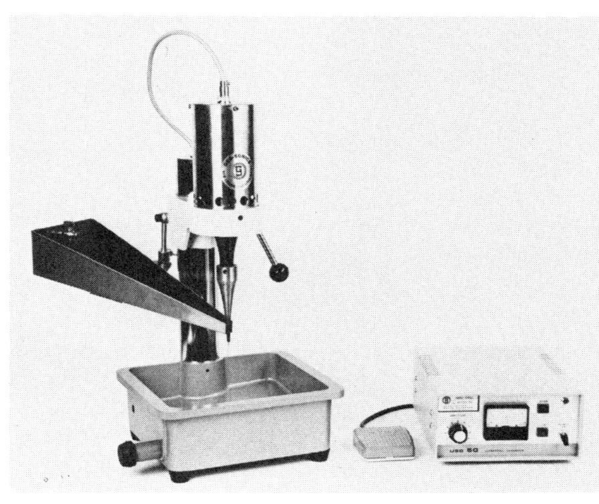

(b)

Figure 17.3 Ultrasonic applications. (a) An ultrasonic cleaning tank. (b) An ultrasonic drill makes holes in glass, ceramics, and gem stones by ultrasonic vibrations rather than by rotary motion, as in ordinary drills.

In the medical field, ultrasonic "sonar" can be used to view soft internal tissues and organs, such as the liver or spleen, which are nearly invisible to X-rays. Ultrasound can also be used to "view" a fetus at different stages of development without the dangerous effects presented by X-rays. The different degrees of reflection of scanned areas are monitored and stored in a computer. The computer then reconstructs an "echogram" of the region (Fig. 17.5). Ultrasonic brain scans are

(Continued)

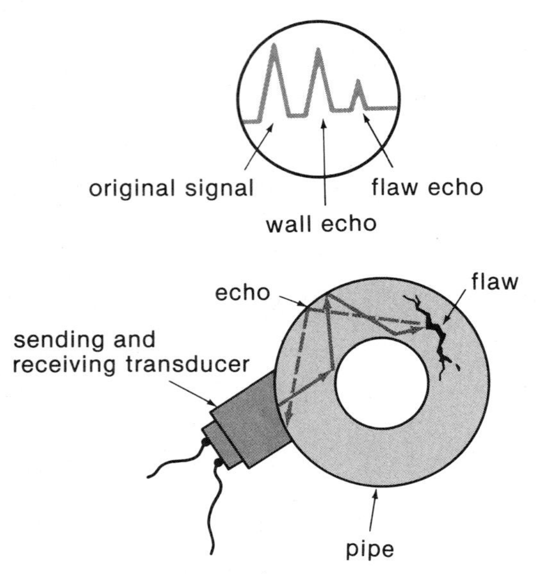

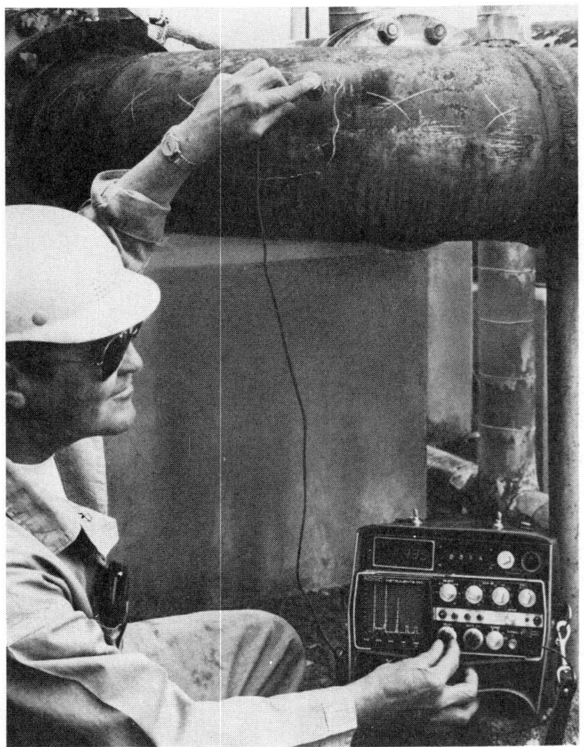

Figure 17.4 Ultrasonic nondestructive testing. Echoes reflected from flaws cause irregularities in the observed reflection patterns.

used to detect tumors and cerebral hemorrhage. By applying the Doppler effect to reflected ultrasound, doctors can detect and monitor movements such as the actions of heart valves, the flow of blood, and the beating of fetal hearts.

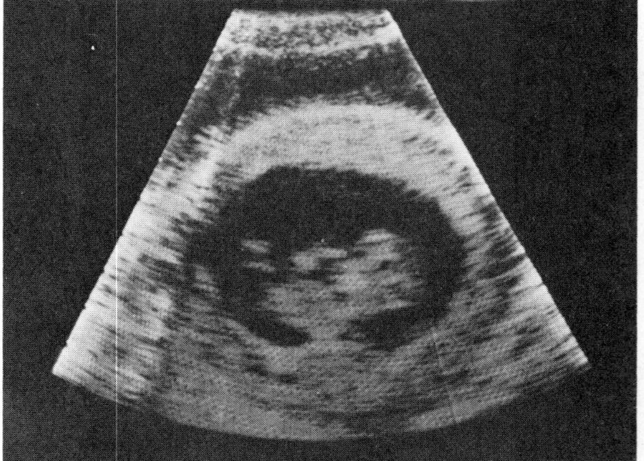

Figure 17.5 Medical ultrasonics. An ultrasonic scan of a human fetus.

The control of the vocal cords determines the frequency range of a person's singing voice.

The modulated wave from the larynx is further modulated in the numerous (resonance) cavities of the throat, mouth, and nose. Some vocal cavities have controllable parameters, such as the tongue and lips, that can be used to produce a wide variety of sounds. Sound out the vowel letters *a*, *e*, *i*, *o*, *u* (and sometimes *w* and *y*)

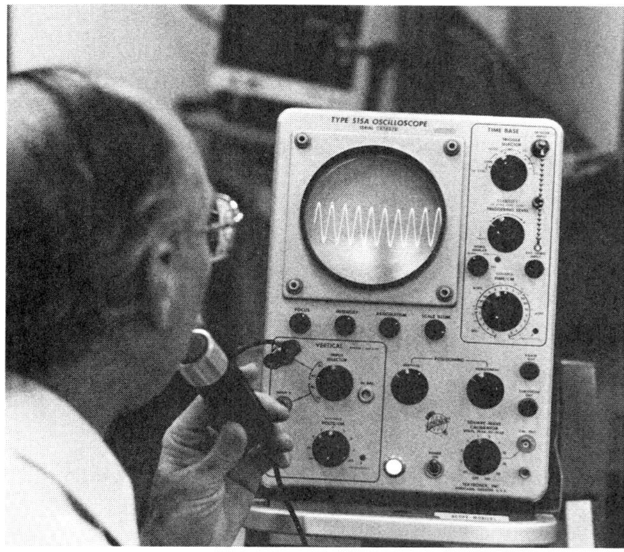

(a)

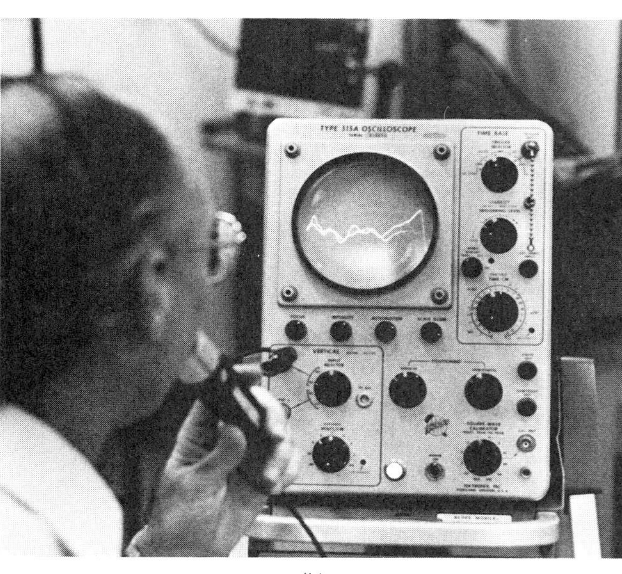

(b)

Figure 17.6 Human sounds displayed electronically. (a) Whistling a note gives a simple wave form. Compare with Figure 17.1. (b) Voice sounds have complex wave forms as a result of various overtones from the vocal cavities.

and notice the changes in the position of your lips and tongue.

The wave forms of human sounds can be complex, owing to various overtones from the vocal cavities (Fig. 17.6). Voice sounds are quite specific for individuals and provide "voice prints," similar to finger prints. However, the use of voice prints for legal indentification is still controversial.

Speed of Sound

If you put your ear to a railroad rail and someone far down the track hits the rail with a hammer, you'll hear the disturbance in the rail before you hear it in the air. This is because sound travels faster in a solid than in a gas. The speed of sound in a liquid is intermediate.

This should come as no surprise, since there are differences in the stiffness or elasticity of different materials. In a highly elastic material, the restoring forces between the molecules cause a disturbance to propagate faster. Solids are more elastic than liquids, which in turn are more elastic than gases. Hence, the speed of sound is different in each. As a general comparison, sound travels about 15 times faster through a solid than through air and about four times faster through water than through air.

In air, the speed of sound is 331 m/s at a temperature of 0°C. The temperature affects the speed of sound in a medium because as the temperature increases, so does the speed of its molecules. As a result, the molecules collide more often, and a disturbance is transmitted more quickly to neighboring molecules or through the substance. In air, the speed of sound increases by about 0.6 m/s for each degree Celsius increase above 0°C. At room temperature (20°C) this would be an increase of $0.6 \text{ m/s} \times 20 = 12 \text{ m/s}$, and the speed of sound is $331 + 12 = 343 \text{ m/s}$. So sound travels faster on a warmer day, and you hear things sooner.

Notice that the speed of sound in air is quite different from the speed of light in air or vacuum (300,000,000 m/s). You may notice this difference sometimes when watching a ball game, say from the center-field stands. It is common to see a batter hit the ball, then hear the "crack" of the bat later. You see the ball being hit almost instantaneously because the speed of light is so fast, but if you are 150 meters from home plate, the sound disturbance takes about half a second to travel to you.

This effect is also used to estimate the distance of a thunderstorm. A bolt of lightning is seen almost instantaneously. The resulting thunder comes rumbling along behind at about $\frac{1}{3}$ km/s (33 m/s or $\frac{1}{5}$ mi/s). By timing the interval between seeing the lightning and hearing the thunder, for example, by counting seconds

—"one-thousand-one, one-thousand-two," and so on —you can compute the distance the thunder traveled or how far away the storm is. Say you counted 6 seconds. Then, $\frac{1}{3}$ km/s × 6 s = 2 km, and the storm is approximately 2 kilometers (1.2 mi) away.

Reflection of Sound

When a sound wave strikes the surface of a material different from the one in which it is traveling, some of the wave disturbance is reflected and some is transmitted. If the surface is very rigid and smooth, a large portion of the wave is reflected.

When you mention the reflection of sound, one usually thinks of echoes. **Echoes** are produced when the reflecting surface is far enough away that we can distinguish between the direct sound and the reflected sound. If you were in the mountains and let out a shout, you might hear an echo reflected off a mountain surface. Suppose you heard the echo 2 seconds after you shouted. Then you would know that the reflecting surface was about $\frac{1}{3}$ km away, since sound travels at about $\frac{1}{3}$ km/s (1 second out and 1 second back).

This principle is used in the practical applications of sonar (Special Feature 17.1) and radar (Special Feature 17.2). Sonar makes use of ultrasonic waves to detect and determine the distance of a reflecting object in water, while radar uses radio waves in air. You can also tell if an object is moving, the direction, and how fast it's going using other wave principles we'll discuss shortly.

Sound reflections can cause problems in concert halls and auditoriums by giving rise to an "echo chamber" effect, caused by receiving multiple reflections at different times (Fig. 17.7). After a short time (depending on the size of the room), the multiple reflections are received so close together that they form a diffuse, continuous sound called **reverberant sound.** If this persists too long, the reverberant sound interferes with subsequent sounds from the source—for example, a singer or an orchestra—and there is no clarity. We find this to be displeasing, and damping of the reflections is promoted by furnishing the hall with pad-

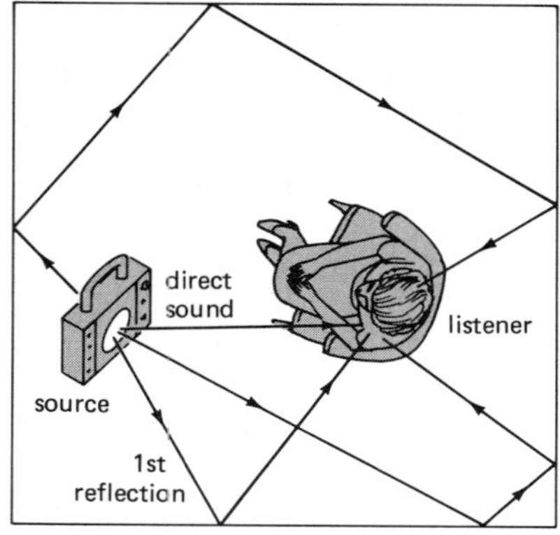

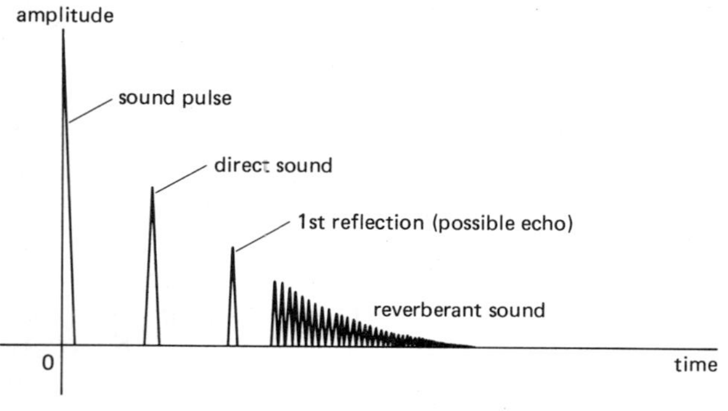

(a)

(b)

Figure 17.7 Acoustic reflections. (a) Reflections can give rise to an echo-chamber effect. Problems can also arise from reverberant sound that persists too long and interferes with subsequent sounds. (b) Concert halls are designed to prevent this.

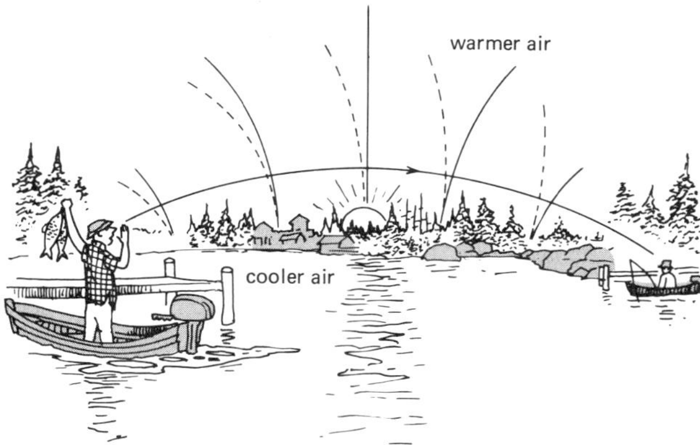

Figure 17.8 Refraction of sound. Because of different densities of air layers at different temperatures, sounds are refracted or bent and can be heard. Under uniform conditions the sound waves (dashed lines) may be too faint to be heard.

ded seats, drapes, and so on, that absorb a great deal of the sound rather than reflect it. On the other hand, if the reverberant sound is dampened too much, the fullness of the musical tones is lost.

Refraction of Sound

On a calm summer evening or night, it is possible to hear distant voices or sounds quite distinctly that ordinarily would not be heard. This is due to the refraction or "bending" of sound that occurs when a wave passes from one medium into another or into a different density region of the same medium.

In order for this to occur in air, the atmospheric conditions must be such that a cool air layer is near the Earth with a warm air layer higher up. These conditions often occur over bodies of water at or after sunset (Fig. 17.8). Notice in the figure that the dashed lines represent the waves under regular conditions. They spread outward from the source and ordinarily would be too faint to be heard by someone on the opposite side of the lake. However, as the waves move through the warmer air they travel faster. As a result, the upper part of the wave is "bent" or refracted downward.† This increases the sound intensity received by the person on the opposite side of the lake, so the sound can be heard. Refraction will be considered in more detail in Chapter 24 for light waves.

† This is somewhat analogous to the bending of a bimetallic strip due to thermal expansion (Chapter 13). The metal on one side of the strip travels (expands) farther than the other, and the strip is bent.

Resonance

Resonance can be easily demonstrated with sound, in particular, with two tuning forks of the same frequency (Fig. 17.9). The "natural" frequencies of the tuning forks are usually stamped right on them. Suppose both tuning forks have the same frequency. If one is set into vibration with the other close by and then the vibrating fork is stopped after a short time, sound can still be heard. On investigation, you will find that the other tuning fork is vibrating and producing sound as a result of resonance energy transfer. (The wooden boxes amplify and direct the sound.)

If tuning forks of different frequencies are used, there is no observable sound energy transfer. Just a small piece of putty or gum stuck on a receiving fork with the same frequency will change its natural frequency enough so that resonance is not observed.

QUESTION: Opera singers are said to be able to shatter crystal glasses with their voices. How is this possible?

ANSWER: Opera singers with powerful voices have been able to do this (Fig. 17.10). If you wet your finger and move it around the rim of a crystal glass, you can get it to "sing" or vibrate at its resonance frequency. The wavelength of the sound is the same as the distance around the rim of the glass.

If a singer is able to sustain a tone with the proper frequency, the resonance energy transfer will increase the amplitude of the vibrations of the glass to the point that it shatters.

The Doppler Effect

When you drop a pebble into a still pond, the water waves travel out in all directions on the surface, generally with the same wavelength and frequency. The same is true for a stationary sound source. But what if the source is moving? You have probably heard a truck horn blowing as the truck approaches and passes by and noticed a distinctive change in the sound pitch or frequency. (The perceived sound frequency is commonly referred to as pitch.) The sound gets shriller (higher frequency or pitch) as the source approaches

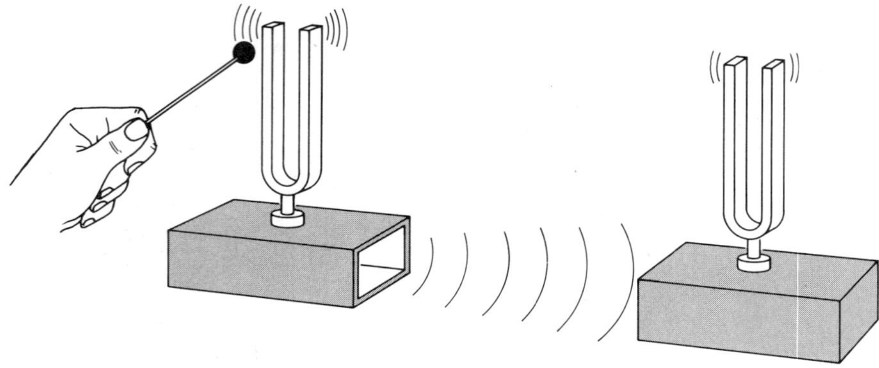

Figure 17.9 Sound resonance. Sound waves from one tuning fork set another tuning fork of the same frequency into resonance vibration.

Figure 17.10 Resonance in action. See Question and Answer on previous page. (This glass wasn't shattered by a singer, but the effect is the same.)

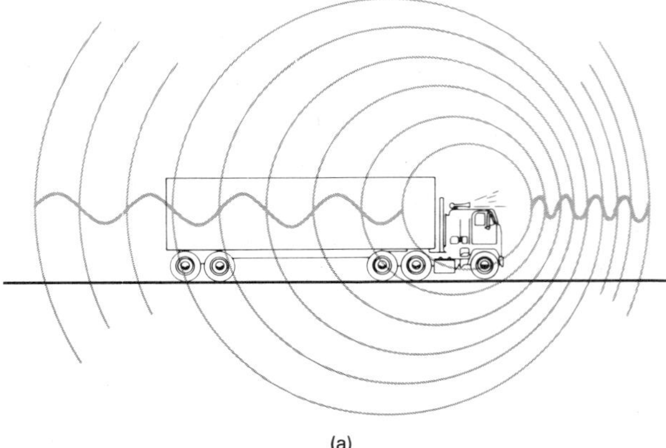

(a)

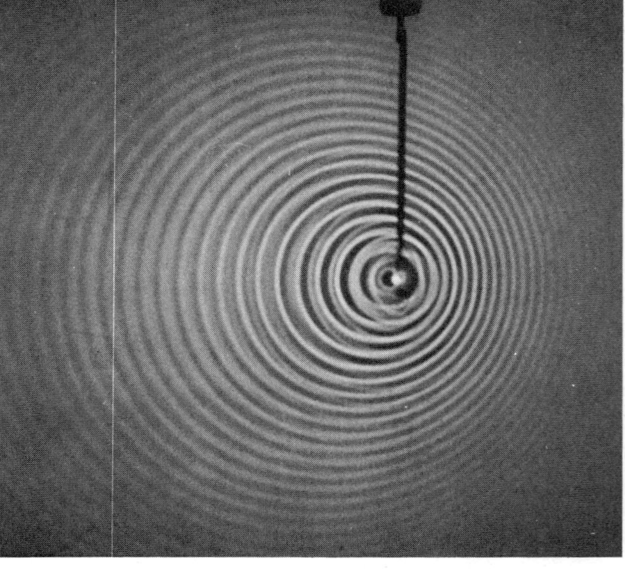

(b)

Figure 17.11 The Doppler effect. The waves in front of a moving source are "bunched up" and have a higher frequency, while the waves behind are spread out and have a lower frequency.

and becomes lower in pitch (lower frequency) as it recedes. This change in frequency due to the relative motion between the source and the observer is called the **Doppler effect.**‡ (There is a similar effect if the source is stationary and the observer is moving.)

‡ After Christian Doppler (1803–1853), the Austrian scientist who first described the effect.

The effect of the moving source is to "bunch up" the waves in front of the source, and hence to increase the frequency (Fig. 17.11). Also, the waves behind the moving source are "spread out," which decreases the frequency. Notice that the Doppler "shift" or change in the frequency or wavelength (from when the sound is stationary) will depend on the velocity of the source.

The Doppler effect also applies to light waves. If we examine light coming to us from distant galaxies, we find that the frequencies of the light components (colors) have been shifted compared to stationary light sources here on Earth. The observed shift is toward the red end of the visible spectrum — the long-wavelength or low-frequency end. This Doppler "red shift" implies that the galaxies are moving away from us and supports the idea of an expanding universe. One theory proposes that this resulted from a "big bang," which occurred when all the matter in the universe was in one big mass. Since we know the speed of the galaxies or the expansion rate, we have calculated that the big bang or universe expansion began about 15 billion years ago (see Chapter 30).

A more "down-to-Earth" or terrestrial application of the Doppler effect is discussed in Special Feature 17.2.

SPECIAL FEATURE 17.2

Radar and the Doppler Effect

The Doppler effect is used in the determination of the velocity of a moving object by sonar (ultrasonic waves in water) and radar (radio waves in air). Let's discuss radar here, since it is more familiar. Radar stands for *ra*dio *d*etecting *a*nd *r*anging. The detecting and ranging (distance) comes from reflection and the time delay and wave speed, as discussed previously. Add the Doppler effect and there's more.

Suppose a "Smokey Bear taking a picture" (CB radio jargon for a state trooper using radar) sitting along the side of the highway directs a beam of radio waves toward a car, and the waves are reflected back (Fig. 17.12). If the car being observed is parked or not moving, the reflected waves have the same frequency. But if the car is moving toward the police car, the reflected waves have a higher frequency, or are Doppler-shifted. In effect, the moving car acts like a moving source, and there is a double Doppler shift — coming and going. The magnitude of the shift depends on the speed of the car. A computer quickly calculates this and displays it for the officer to see. If this is above the speed limit, the next thing the driver might hear is the Doppler shift of a siren!

If the car is moving away from the patrol car, the reflected waves would have a lower frequency. The speed is still calculated from the magnitude of the frequency shift. It's a bit more complicated, but the speed of the car can be determined even when the patrol car is moving. The computer does this easily and can get you coming and going.

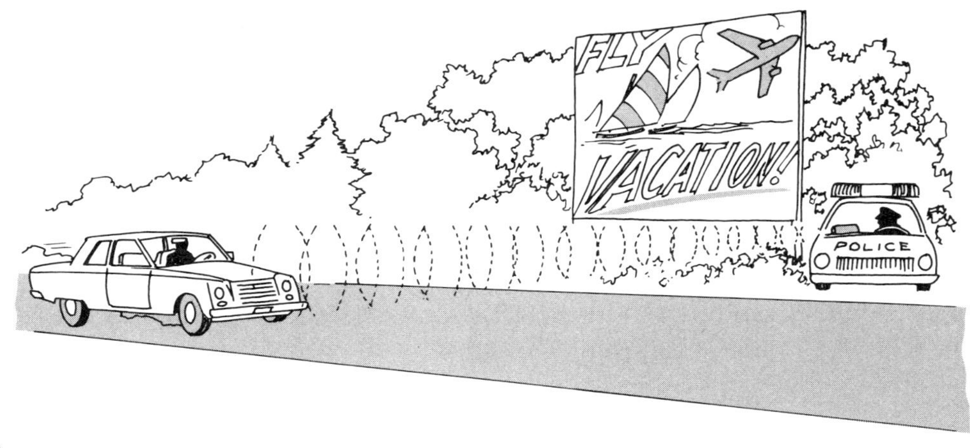

Figure 17.12 The Doppler effect in action. The Doppler shift is used in radar to catch speeders.

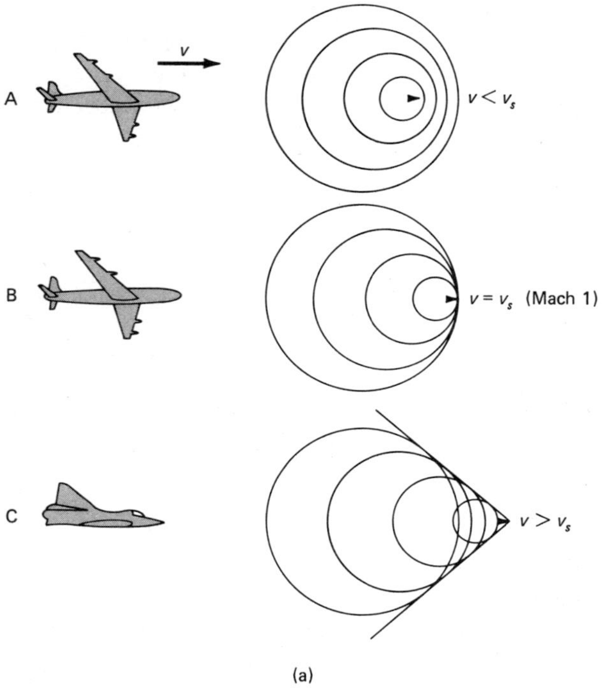

(b)

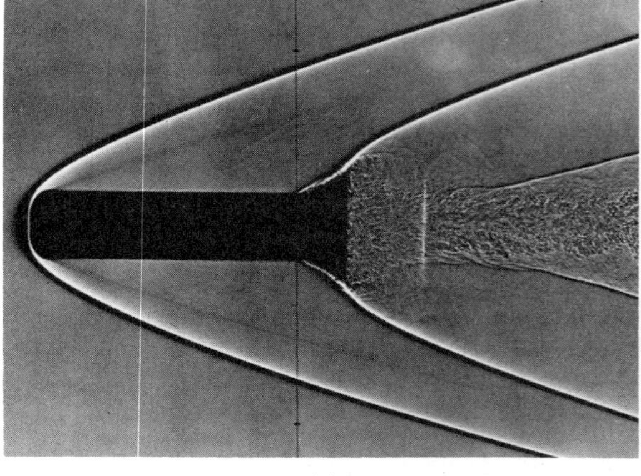
(c)

Figure 17.13 Bow waves are formed by airplanes in air (a) similar to those formed in water by boats or ducks (b). The shock wave boundaries are clearly seen for a gun-launched projectile traveling at Mach 7 in a wind tunnel (c).

SONIC BOOMS

A sonic boom results when a sound source travels faster than the speed of sound, and interference comes into effect. You have no doubt experienced sonic booms from high-speed jet aircraft that rattle your windows and shake your doors. In fact, the jet aircraft are flying faster than the speed of sound (≈ 335 m/s or 750 mi/h).

As a sound source approaches the speed of sound, the waves come closer together in the forward direction (Fig. 17.13). When a source travels at the speed of sound or faster, the waves interfere and build up in a V-shaped pressure ridge or shock wave, commonly called a bow wave. A similar bow wave is formed by the bow of a speed boat moving through water at a speed greater than the wave speed in water. The speeds of fast-flying aircraft or objects are measured in Mach numbers. A jet aircraft flying at the speed of sound is traveling at Mach 1. At one and one-half times the speed of sound, the jet is traveling at Mach 1.5; at Mach 2, twice the speed of sound, and so on.

The bow wave from a supersonic aircraft trails out and downward behind the aircraft. When this shock-wave region passes over, a sharp "sonic boom" is heard.

Beats

Have you ever heard musicians tuning their instruments? A particular note (frequency) is sounded on one instrument, for example, a piano, and the other instruments are adjusted until they have the same tone. As this is done, you can hear pulsating fluctuations that vanish when things are in tune. These fluctuations are called **beats** and result from an interference effect.

When two waves interfere in space, a wave pattern is formed when the waves have the same amplitude and frequency (Chapter 16). Recall that this gives regions in which a tone would be heard and others in which there would be no sound (destructive interference). But when two sounds of equal intensity (amplitude) with slightly different frequencies interfere, a pulsed disturbance is formed. The amplitude of the combined traveling wave varies, and a pulsating tone is heard, being louder when the regions of constructive interference are received (Fig. 17.14).

The number of sound pulsations per second is called the **beat frequency,** f_b. This is given by the difference in the two tone frequencies, $f_b = f_2 - f_1$. For example, if the tones are from two tuning forks with frequencies of

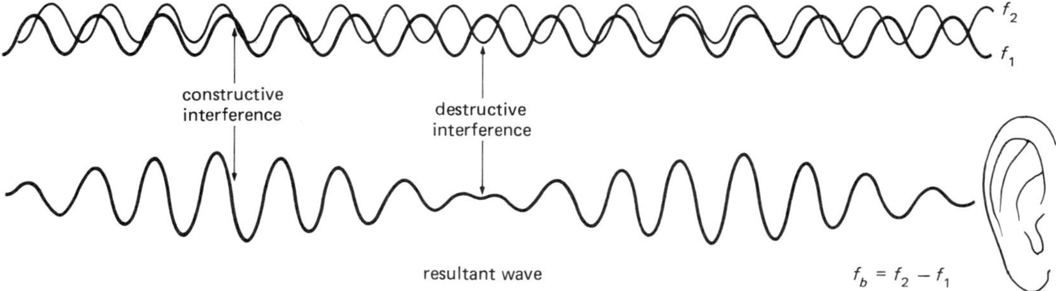

Figure 17.14 Beats. Traveling waves with equal amplitudes and slightly different frequencies give rise to pulsating fluctuations, or beats.

516 Hz and 512 Hz, the beat frequency is $f_b = 516 - 512 = 4$ Hz, or four pulsations or beats each second. Musicians try to tune their instruments together until no beats are heard, which means $f_2 = f_1$, and they are in tune. The human ear can detect up to about 7 beats per second.

Sound and Hearing

Sound waves are physical disturbances, and the physical properties that describe these waves are quantities that can be measured without reference to the ear or hearing. Basically, these properties are intensity, frequency, and wave form. The response of the human ear varies from person to person. The physiological terms used to describe the sensations of the ear are loudness, pitch, and quality (or timbre). These sensory effects and the physical wave properties have the following general correlations:

Sensory effect	↔	*Physical wave property*
Loudness	↔	Intensity
Pitch	↔	Frequency
Quality	↔	Wave form (overtones)

However, there is not a one-to-one correlation. The physical properties are objective and directly measurable. The sensory effects are subjective and vary with individuals. This is similar to our subjective temperature sense of touch and objective physical measurements with a thermometer.

Let's take a look at these similar, yet different, objective wave and subjective sensory effects.

LOUDNESS AND INTENSITY

How loud does a sound have to be for us to hear it? Another way of saying this is, What is the intensity threshold of hearing? We can just hear a 1000-Hz tone with an intensity of 10^{-12} W/m^2. Thus, for us to hear a sound, it must not only have a frequency in the audible range, but also a sufficient intensity. As the intensity is increased, the perceived sound gets louder. At an intensity of about 1.0 W/m^2, the sound is uncomfortably loud and may be painful to the ear. This intensity is sometimes referred to as the threshold of feeling or pain.

Think a moment about this intensity range of hearing for our ears. The intensity at the threshold of feeling is a trillion times (10^{12} = 1,000,000,000,000) greater than the intensity at the threshold of hearing!

Because the ear is sensitive to such an enormous range of intensities, the unit of watt/m^2 gives some pretty big numbers when we compare one intensity to another. Therefore, we commonly use another comparative measure of intensity called the **sound-intensity level,** which has units of **decibels** (dB).§ Here, sound intensities are compared to a standard level—the threshold of hearing. Using a logarithmic function, the large range of intensity of human hearing is compressed into a more manageable scale (Fig. 17.15). The threshold of hearing intensity has a sound level of 0 dB, and the threshold of feeling has a sound level of 120 dB.

The decibel scale is not linear. That is, if the intensity of a sound is doubled, the dB level is not doubled. In fact, if the sound intensity is doubled, the dB reading

§ The bel unit is in honor of Alexander Graham Bell, the inventor of the telephone. A decibel is one-tenth of a bel.

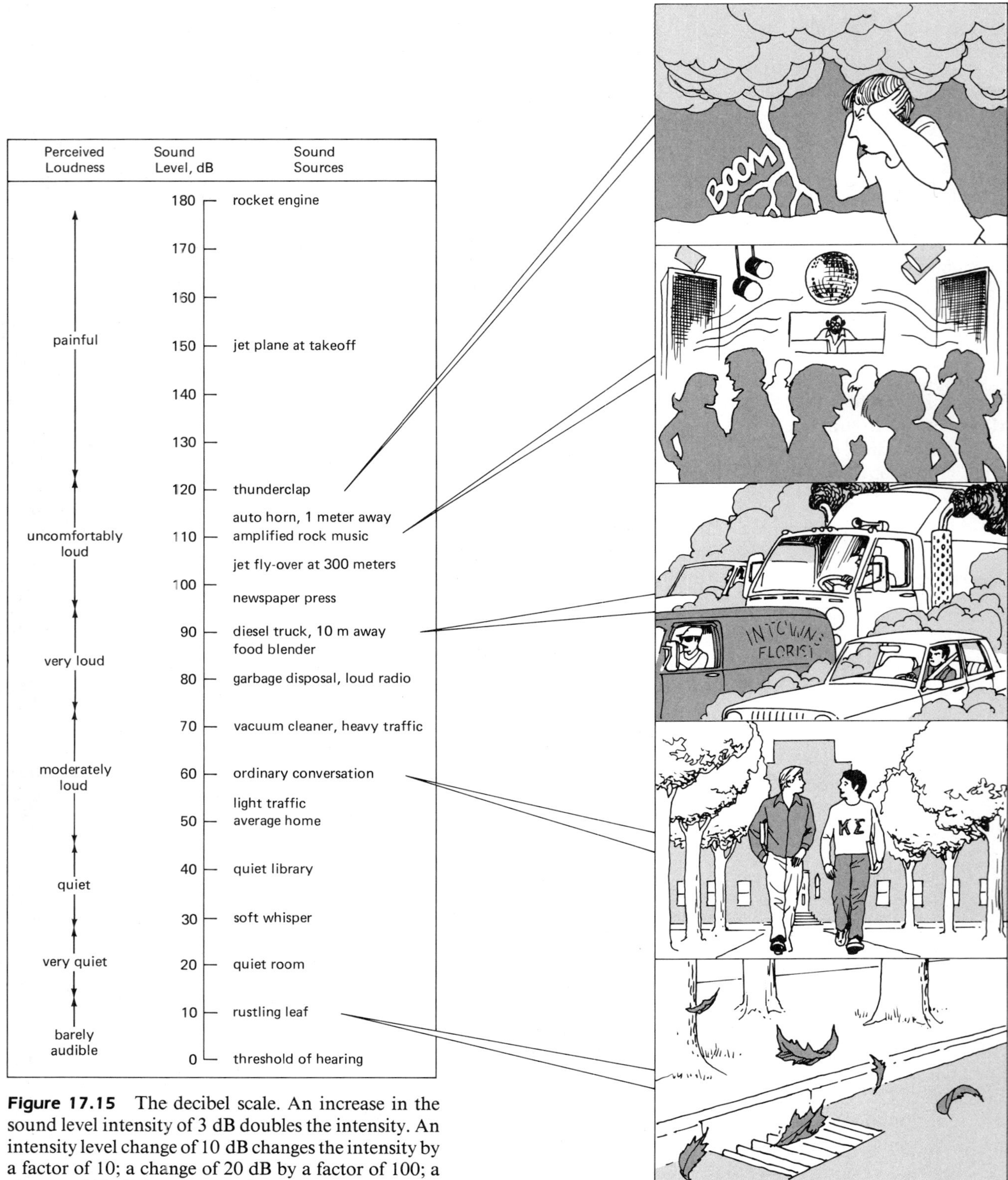

Figure 17.15 The decibel scale. An increase in the sound level intensity of 3 dB doubles the intensity. An intensity level change of 10 dB changes the intensity by a factor of 10; a change of 20 dB by a factor of 100; a change of 30 dB by a factor of 1000; and so on.

increases by three. For example, if a sound has an intensity of 50 dB, a sound with twice the intensity has a sound level of 53 dB.

Comparisons are conveniently made on the dB scale in factors of tens of decibel differences:

An increase of 10 dB increases the sound intensity by a factor of 10.
An increase of 20 dB increases the sound intensity by a factor of 100.
An increase of 30 dB increases the sound intensity by a factor of 1000, and so on.

Thus, a sound with an intensity of 80 dB, for example, has an intensity 100 times greater than that of a sound with an intensity level of 60 dB. Too much sound intensity can be bad for the ear. See Special Feature 17.3.

Loudness is related to the sound intensity, but loudness is very subjective and ear-dependent. Here's the problem. The ear does not respond equally to all frequencies. For example, two tones with different frequencies, but with the *same* intensity levels (W/m^2), are judged by the ear to have *different* loudnesses.

PITCH AND FREQUENCY

Pitch and frequency are often used interchangeably, but frequency is objective and pitch is subjective. For example, if a pure frequency tone is sounded at two

SPECIAL FEATURE 17.3

Noise-Exposure Limits

Sounds with very high intensities can be dangerous. Above the threshold of pain (about 120 dB), sound is painfully loud to the ear. Brief exposures to levels of 140 to 150 dB can rupture eardrums and cause permanent hearing loss. Consequently, ear protectors or ear valves must be worn in some occupations, and noise levels must be monitored (Fig. 17.16). Longer exposure to lower sound (noise) levels can also damage hearing. For example, there may be a hearing loss for a certain frequency range. (Have you ever noticed a temporary hearing loss after listening to the music of a loud band for a long time?) Federal standards now set permissible noise-exposure limits, as listed in Table 17.1. For example, 6 hours is the maximum time that a worker can be exposed to sound-intensity levels of 92 dB. For a sound level of 95 dB, 4 hours is the maximum exposure time. Notice that this is a doubling of intensity.

Figure 17.16 Occupational loudness. The sound-level intensity of a machine is monitored with a sound-level meter. Federal regulations set limits for sound-level exposure times.

Table 17.1 Permissible Noise Exposure Limits*

Maximum duration per day (hours)	*Sound level (dB)*
8	90
6	92
4	95
3	97
2	100
$1\frac{1}{2}$	102
1	105
$\frac{1}{2}$	110
$\frac{1}{4}$ or less	115

*When the daily noise exposure is composed of two or more periods of exposure of different intensity levels, their combined effect should be considered.

intensity levels, nearly all listeners will agree that the more intense sound has a lower pitch (perceived "frequency"). This effect is most pronounced for low frequencies around 100 Hz. The ear is most sensitive over the frequency range of about 500 to 5000 Hz. Here, the pitch of a tone is relatively independent of intensity.

Pitch discrimination is the ability to distinguish between two tones of different frequencies as being different in pitch. Individuals who can name the pitch of a tone on a musical (frequency) scale without comparison to an external standard are said to have "perfect pitch." There is disagreement as to whether this is a special ability or comes with training. It does seem odd that most of us are able to remember visual properties, such as color or shapes, exactly but not note pitches. Why do you suppose this is the case?

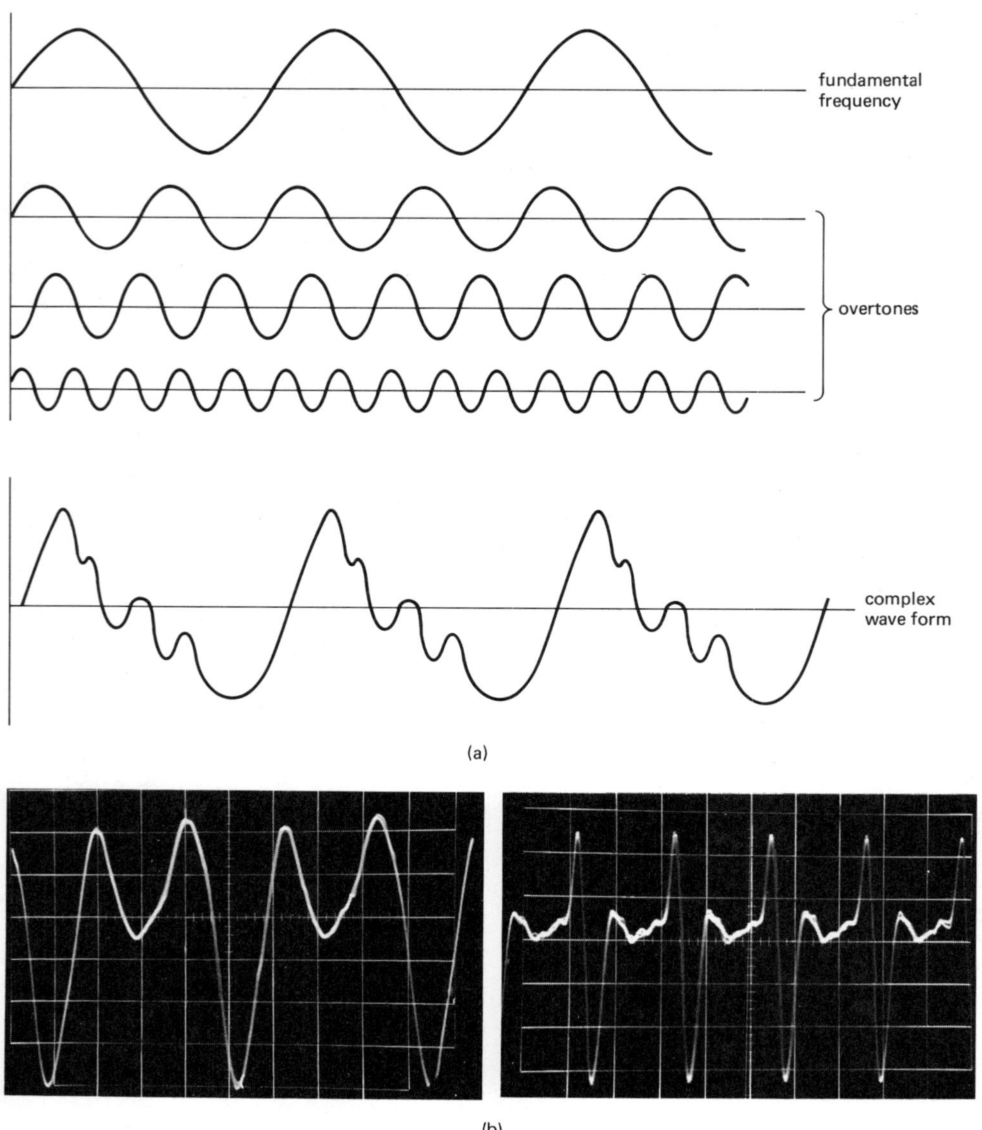

Figure 17.17 Quality. (a) The quality of a sound depends on the number and combination of overtones. (b) Oscillograph displays of sound for a tone from a French horn (left) and trumpet (right). The instruments have different qualities.

QUALITY (TIMBRE) AND WAVE FORM

The quality of a tone is the characteristic that enables us to distinguish it from another with the same intensity and frequency. Tone quality depends on the wave form or number of harmonics or overtones present (Fig. 17.17). You can objectively produce a tone with the same intensity and frequency as your favorite singer, but a different combination of overtones gives the singer's voice a better quality or a more pleasing sound, along with a certain "richness."

The quality of a tone is highly subjective. Some sound qualities are pleasing to some people and cultures and not to others. Japanese music and instruments may sound displeasing to some Americans, and vice versa. Some combinations of tones or chords generally sound pleasing, but other combinations are termed discordant or as having a lack of quality. Such sounds may be classified as noise instead of music by some. Have you ever been told or thought yourself that hard rock music is noise? Different strokes (or quality) for different folks.

Music

Music soothes (the savage breast), inspires, and delights. A Beethoven symphony can evoke a mood of sadness, whereas a lively march can get your foot tapping to the music. Music is very personalized. We all have our favorite songs. Such a wide variety of reactions and feelings makes the psychological interpretation of music very complex.

When tones played together are judged by a listener to be pleasant, the combination is referred to as a consonance or consonant sound. On the other hand, if the combination is judged to be harsh or unpleasant, it is called dissonance or a dissonant sound.

Dissonance is important to music. A bit of dissonance breaks up the blandness of continual consonance, adding excitement or restlessness. But music has become increasingly dissonant. Modern music is much more dissonant than older classical music. Why do you think this is? Is it actual subjective preference or a desire to be different?

MUSICAL SCALES

We designate particular musical notes or the fundamental frequency of a tone by constructing musical scales. As with temperature scales, there are various ways to do this. Musical scale intervals are based on pleasing or consonant sounds when the frequencies of two notes form ratios of whole numbers. The simplest scale in Western-culture music is the *just diatonic scale.* The international standard frequency of 440 Hz for the note A_4 is the starting point for constructing the consonant intervals defined by whole-number ratios, as shown in Figure 17.18.

Starting with a C note on a piano and playing successive white keys to the right from a C note to the higher C note, this gives the familiar "do–re–me–fa–sol–la–ti–do," where each note is represented by a letter (Fig. 17.18). This is the C-major scale. Notice that the frequency doubles from C_4 to C_5 (from 264 Hz to 528 Hz). This interval is called an octave (Latin,

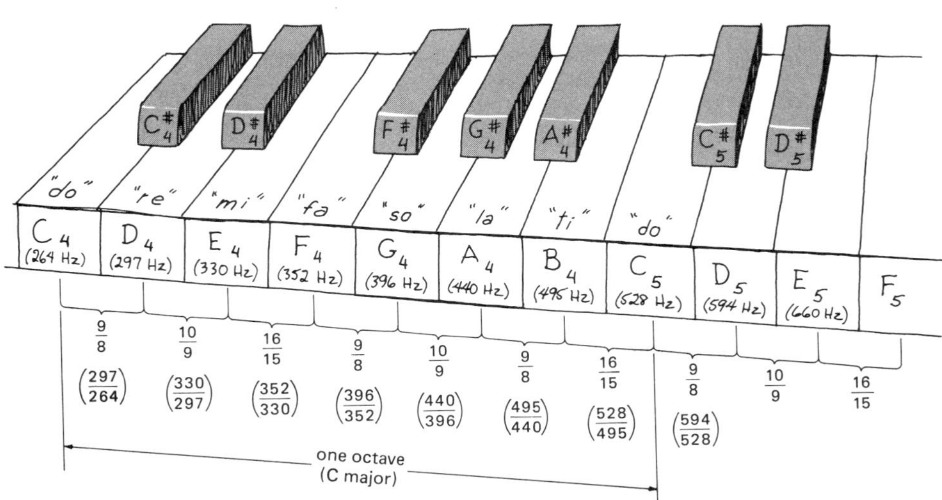

Figure 17.18 Notes and frequencies of the just diatonic music scale.

meaning "eight"), with eight steps or notes. A piano keyboard covers seven octaves, and the C_4 of the middle octave is called "middle C." The frequency of each note, D, E, and so on, doubles with each octave interval; for example, the frequency of D_5 is twice that of D_4.

If you look at the intervals or frequency ratios in Figure 17.18, you will quickly notice an oddity. There are relatively larger intervals ($\frac{9}{8}$ and $\frac{10}{9}$), while others are smaller ($\frac{16}{15}$). The larger intervals are called whole intervals, or tones, and the smaller intervals are called half intervals, or semitones. On the piano, this means that some adjacent white keys differ by a tone and others by a semitone. The scale would be more uniform if we had a complete range of semitones. The black keys of the piano provide this and divide an octave into 12 intervals or steps.

The just diatonic scale has many consonances, but there are serious problems because the tone and semitone intervals are not uniform along the scale, even with 12 steps. When played in another major scale or key, say D-major (do–re–me sequence starts with D), certain notes are not the same because of the nonuniform semitone interval; for example, an E note is different in these different major scales.

The problem could be alleviated if a scale had 12 equal semitone intervals in an octave. Such a scale is called the *equally tempered scale.* The standard frequency for this scale is still $A_4 = 440$ Hz (Table 17.2). The advantage of this scale is that all intervals are the same in any key. Also, the sharp (#) of one note and the flat ($\flat$) of the next higher note are the same frequency, for example $C^{\#}$ and $D^{\flat}$. This is not generally the case on the just diatonic scale. However, the equal semitone interval ratios are no longer whole numbers, and some consonance is lost.

Table 17.2 The Equally Tempered Scale for the Central Octave of a Piano

Note	*Frequency (Hz)*
C_4	261.6
$C_4^{\#}$ (or $D_4^{\flat}$)	277.2
D_4	293.7
$D_4^{\#}$ (or $E_4^{\flat}$)	311.1
E_4	329.6
F_4	349.2
$F_4^{\#}$ (or $G_4^{\flat}$)	370.0
G_4	392.0
$G_4^{\#}$ (or $A_4^{\flat}$)	415.3
A_4	440.0
$A_4^{\#}$ (or $B_4^{\flat}$)	466.2
B_4	493.9
C_5	523.2

Pianos are usually tuned to the equally tempered scale, since they can be used to play music in many keys. Some quality may be lost, but if pianos were tuned to the just diatonic scale, compositions in some keys would sound fine while others would be particularly dissonant.

MUSICAL INSTRUMENTS

Do you play a musical instrument? If so, it falls into one of the three general instrument categories: string, wind, and percussion. All involve tones resulting from standing waves.

String instruments include the guitar, violin or fiddle, bass, and piano (see Chapter introductory photo for the latter). Here, standing waves are formed in strings fixed at each end. When a string instrument such as a guitar or violin is being tuned, what does the musician do? He or she tightens or loosens the strings, that is, adjusts the string tension. This changes the pitch or frequency of the string sound until no beats are heard. Actually, the wave speed in a string is being adjusted. This increases with the tension or tightening. Adjusting the wave speed changes the frequency, since the length between the ends (between the bridge and neck) or the fundamental wavelength of the string is fixed, and $\lambda f = v$.

The wave speed also depends on the linear mass density of the string. You have probably noticed that the high-frequency strings on a string instrument are thin and light as compared with the low-frequency strings, which are thick and heavy.

Once the strings on a violin or guitar are tuned to certain notes, other notes of the musical scale are obtained by changing the effective lengths of the strings by fingering (Fig. 17.19). This shortens the fundamental wavelength of a vibrating string and increases the frequency. A piano, on the other hand, has a string for each note or frequency, and tuning is a major operation.

The quality or overtone combination of a string instrument's sound depends on how the strings are excited. This may be done by plucking, bowing, or striking (as with a hammer in a piano). Where, as well as how, a string is excited affects the quality of the sound.

Figure 17.19 On stringed instruments, different notes on the musical scale are obtained by changing the effective lengths of the strings.

The bodies and cavities of string instruments are important to the quality and amplification of sound. Vibrating strings themselves disturb little air and would produce little sound intensity, but they are coupled with the instrument body, for example, through the bridge on a guitar and violin. The body acts as a resonance cavity for amplification and the production of overtones that gives the instrument a particular quality. Recall how our different vocal cavities give rise to different voice qualities.

The making of the bodies of string instruments is more of an art than a science. For example, for a fine violin, maple wood for the bottom plate is better when cut in December or January and taken from the south side of the tree. The spruce wood for the top plate is better when taken from a tree exposed to the sun. For the piano, a cast-iron frame introduced in 1855 by Henry Steinway gives this instrument its brilliance over its ancestor, the clavichord.

In wind instruments, standing waves are set up in air columns. For instruments such as trumpets and trombones, the vibrations of the player's lips are the driving force. For woodwinds such as the clarinet and saxophone, a vibrating reed does the job. For the flute and piccolo, the fluttering air stream resulting from blowing against a hole sets the air column into vibration.

To visualize how standing waves in air columns are formed, consider open and closed pipes, as illustrated in Figure 17.20. A closed pipe is really open at one end (the antinode end) so that sound can get out. An open pipe has antinodes at each end. An antinode position is where the interference gives the standing wave in air the maximum vibrations and the combined amplitudes of the traveling waves give an amplification of the sound.

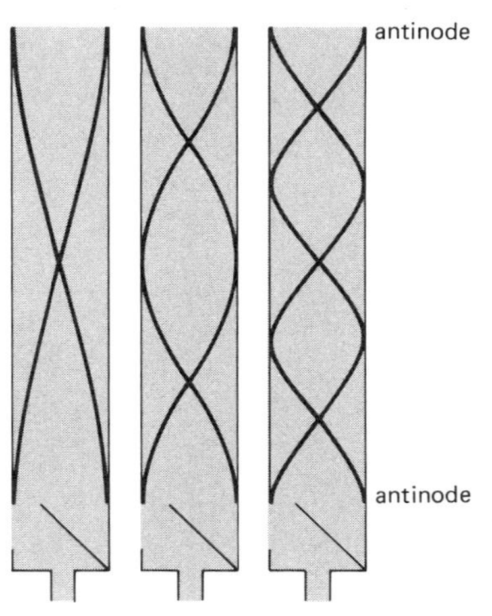

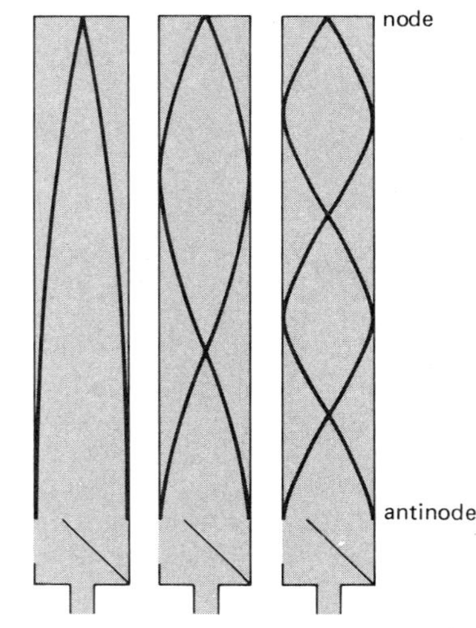

Figure 17.20 In an organ, standing waves are set up in pipes, giving fundamental frequencies and overtones.

Different harmonics can fit in the pipes and give the sounded note quality. A pipe organ has a pipe for each note, similar to a piano having a string for each note. For instruments such as the clarinet, trumpet, or saxaphone, the length of the "pipe" is varied, and different notes are obtained by means of holes or "stops" and valves (on a trumpet). For a trombone, the length of the "pipe" is physically varied by moving the slide. Standing waves can be formed in cone-shaped and flared pipes as well as in cylindrical pipes. In percussion instruments, such as drums and cymbals, standing waves are set up in two-dimensional membranes or surfaces. This analysis is more complicated.

In summary, music is in the ear of the beholder, just as beauty is in the eye of the beholder. The introduction of dissonant musical sounds by Igor Stravinsky in his works in the early 1900's made audiences furious. Today, we have musical tones produced by synthesizers, which produce sounds electronically. These look more like computers than musical instruments. Here the emphasis is more on pure tones than on quality, and some people find "Moog music" (named after Robert Moog, a pioneer in electronic sound) somewhat displeasing. Indeed, music is a very personal thing, and you have to "make your own kind of music."

SUMMARY OF KEY TERMS

Audible frequency region region of human hearing frequency response, 20 Hz to 20 kHz. Lower than 20 Hz is the infrasonic region, and above 20 kHz is the ultrasonic region.

Speed of sound (temperature dependence) the speed of sound v_s increases with temperature. In air, $v_s = 331 + 0.6\ T_c$ m/s, where T_c is the temperature in degrees Celsius.

Echo reflected sound that can be distinguished from the direct sound because of the distance of the reflecting surface.

Refraction the bending of waves as they pass from one medium into another or into a different-density region of the same medium.

Doppler effect a change in the frequency of a wave due to relative motion between the source and observer.

Sonic boom the shock wave associated with aircraft traveling at supersonic speeds.

Beats the pulsating fluctuations due to the interference of waves with equal amplitudes but slightly different frequencies (f_1 and f_2). The beat frequency is $f_b = f_2 - f_1$.

Loudness the subjective sensory effect related to the intensity of sound.

Decibel a unit of sound-intensity level that gives a comparative measure of intensity.

Pitch the subjective sensory effect related to wave frequency.

Quality the subjective sensory effect related to the wave form or number of overtones in a wave.

Consonance a combination of sounds judged pleasing to the ear.

Dissonance a combination of sounds judged displeasing to the ear.

Musical scale a scale of note frequencies with constant intervals defined by whole-number ratios of frequencies.

Octave a musical scale interval defined by a doubling of frequency or a ratio of 2 : 1.

EXERCISES

1. Why do flying insects produce a high-pitched buzzing sound?
2. Some people say there are two things necessary to have sound, and others say there are three things. Why the difference?
3. An alarm clock is hung in a closed glass jar. When the alarm goes off, it can be heard. Why? What would happen if the air were pumped out of the jar? Why can you still hear the alarm clock?
4. Children often make a tin can–string "telephone" (Fig. 17.21). Explain the sound principles involved.
5. What happens to the energy when a sound "dies out"?
6. The speed of sound in air is given for still air. How would wind affect the propagation of sound? What happens when you shout into the wind?
7. Sound travels faster in moist or humid air. Why is this? (*Hint:* The major components of air are N_2 and O_2. In a volume of moist air, some of these molecules are replaced by H_2O molecules. Check out their masses, using the periodic table.) Could this be the basis of the weather saying, "Sound traveling far and wide, a stormy day will betide"?
8. In a popular lecture demonstration, the instructor breathes in helium (He) gas. When the instructor talks,

Figure 17.21 See Exercise 4.

his or her voice has a high-pitched, "Donald Duck" sound. What causes this? (*Hint:* The vocal cavities and wavelengths are still the same.)

9. How many times faster is the speed of light than the speed of sound in air?
10. How much faster does sound travel on a warm day when the temperature is 25°C than on a cold day, when the temperature is 0°C?
11. What happens to the wavelength of a sound wave if the frequency is doubled?
12. The distance between two condensations of a sound wave is 2 meters. Could this sound be heard by a human ear if it were intense enough? (Take $v_s = 340$ m/s and recall that $\lambda f = v$.)
13. What are the wavelength limits of the audible range for human hearing? (Use $v_s = 340$ m/s.)
14. While backpacking, you notice that there is a 4-second interval between seeing the lightning and hearing the thunder from an approaching storm. Approximately how far away is the storm?
15. Why is an echo not as loud as the direct sound?
16. Why do sounds in an empty building or room sound "hollow"?
17. After a snowfall, it "sounds" particularly quiet. Why is this?
18. When you hear a jet aircraft flying overhead and look in the direction of the sound to see the plane, you may have trouble finding it there. Explain. Why can you see a high-flying jet with contrails (condensation or vapor trails) and not hear it?
19. Singing in the shower is a popular pastime (Fig. 17.22). Why does your voice sound fuller in the shower?
20. Is there a resonance effect in bouncing a ball (for example, when dribbling a basketball)? Explain.
21. It is common to be able to hear music, particularly the bass part, through walls or ceilings (floors if you are on top). What does this tell you about the natural frequency of the walls?
22. Is there a significant Doppler effect when a source moves at right angles to or by a stationary observer at some distance?
23. If a sound source and an observer are both moving with the same velocity, is there a Doppler effect?
24. Binary stars revolve about each other. How is this detected using the Doppler effect? We also know that the Sun rotates. How is this known?
25. Does a sonic boom occur only when a jet plane "breaks the sound barrier"?
26. How fast would a "jet" fish have to swim to create an "aquatic boom"? (Give a numerical answer.)
27. Associate some Mach numbers with the adjectives *subsonic, sonic,* and *supersonic.*
28. A rocket travels at Mach 1.5 in the lower atmosphere, where the speed of sound is 330 m/s. What is the speed of the rocket in m/s?

Figure 17.22 See Exercise 19.

29. Do beats have anything to do with the beat of the music?
30. When you hear a sound 1000 times the intensity of the faintest sound you can hear, what is the intensity level of the sound in decibels?
31. Is the effect the same when the sound level is increased from 10 to 20 dB as from 40 to 80 dB? Explain.
32. Can there be a negative dB level, for example −10 dB?
33. Estimate the dB levels that might be heard from the sound sources shown in Figure 17.23.
34. On the low-frequency strings of instruments, a second wire is wrapped around the main wire. What is the purpose of this?
35. What is the effect when the base of the handle of a vibrating tuning fork is placed against a table top? Is a similar effect used in musical instruments? Explain.
36. Suppose you break a string on a guitar and the only spare available is lighter than the broken string. Using this string in a pinch, how would you compensate so the string produces the proper tone?
37. What is the relationship between octaves and overtones on a string instrument?
38. Guitar keyboards have raised ridges or frets to show where the fingers should be placed, but violins do not. Why do you suppose this is the case?
39. Give some examples of subjective sensory effects or differences for the senses other than hearing.
40. What is the difference between a pure tone and a musical note?
41. Normal voice sounds range from 500 to 5000 Hz in frequency. High-fidelity music requires a much greater frequency range. Why?
42. Stereophonic music is recorded with two microphones and sound tracks and is played back through two speakers. How does this provide better quality or listening? Quadraphonic systems use four microphones, two forward and two back. How does this enhance the sound?

Figure 17.23 See Exercise 33.

PART FIVE
Electricity and Magnetism

Gravity (Chapter 5) is one of the four fundamental forces that we experience daily. The other fundamental force that can be experienced outside of the nucleus is the electromagnetic force. We usually separate the electromagnetic force into an electric force and a magnetic force for convenience of study, but these forces are closely interrelated. Basically, you can't have a magnetic force without an electric force, as will be learned in this section.

In an overall sense, gravity holds the world together on a macrocosmic scale — the solar system and the universe (although the latter is expanding). As was learned in Part Two (Properties of Matter), the electric force is analogously important in holding matter together on a microcosmic scale — atoms and molecules. Even so, gravity and electromagnetism are used in many practical applications.

In this part of our study, we will concentrate on the aspects and applications of electricity and magnetism. Consider this: How would your life be without the use of electricity and magnetism? We get glimpse experiences of this when our domestic or business electrical service goes off for short periods of times. Indeed, we have become an electric and magnetic society. For example, our automobiles would not run, or even start, without these effects. Many people might consider this part of our study the most important part of practical physics. See if you do.

18

Electrostatics— Charges at Rest

To kick off the study of electricity, we will start with the simplest case: charges at rest, or electrostatics. In this chapter you will find that many of the principles studied previously apply to electricity, for example, force, field, work, energy, and so on. Electricity and gravity (Chapter 5) are the most common fundamental interactions we experience. There are many similarities in the principles used to describe them, but there are also differences, since they result from different fundamental properties. Let's take a look.

Electric Charge and Force

What is electricity? Suppose you were asked this question. What would you say? We use the term electricity frequently, but many people would have trouble telling you what it is. A typical answer is that it is a current. But what's a current? Finally, they get around to saying that it has something to do with electric charge. If you ask what is electric charge, an appropriate answer would be that electric charge is a fundamental property.

Basically, we don't know what electric charge is, only what it does. (The term fundamental hides our ignorance.) Electric charge is involved in a fundamental interaction called the electric force, just as mass is involved in the fundamental gravitational interaction or force. But on a comparative scale, the electric force is billions upon billions, even trillions upon trillions, of times stronger than gravity.

Electric charge is associated with the atomic particles, the electron and proton. As was learned in previous chapters, it is the electric forces between these particles that hold solids and liquids together. The electron and the proton have the same magnitude of electric charge, which is measured in the standard unit of

Table 18.1 Particles and Electric Charge

Particle	*Electric charge*	*(Mass of particle)*
Electron	$e^- = -1.6 \times 10^{-19}$ C	$(m_e = 9.11 \times 10^{-31}$ kg)
Proton	$p^+ = +1.6 \times 10^{-19}$ C	$(m_p = 1.67 \times 10^{-27}$ kg)

the coulomb (C) (see Table 18.1). Even though the particles have the same size or magnitude of charge, there is a difference, as evidenced by their force interactions. Unlike the gravitational force, which is always attractive or seeks to bring masses together, the electric force can be either attractive or repulsive. This is expressed in the **law of charges:**

Like charges repel, and unlike charges attract.

To designate this difference, we speak of positive (+) and negative (−) charge (Fig. 18.1). Negative charge is arbitrarily associated with that of the electron and positive charge with that of the proton. The charge of the electron (or proton) is the smallest unit of free charge we have ever found in nature. All multiple charges, commonly designated by the letter q, are whole-number or integer multiples of the charge of the electron or fundamental charge.*

Figure 18.1 The law of charges.

Coulomb's Law

The magnitude of the electric force between two charges is given by an expression discovered by the French physicist Charles A. deCoulomb (1736–1806), after whom the standard unit of charge is named:

$$F = \frac{kq_1q_2}{r^2}$$

This expression is known as **Coulomb's law.** The q's are the charges (in coulombs), and r is the separation distance of the charges (Fig. 18.2).

Notice how similar this formula is to that of Newton's law of gravitation, $F = Gm_1m_2/r^2$. (Is Nature trying to tell us something?) There is a big difference in the constants G and k. Recall that the universal gravitational constant G is 6.67×10^{-11} N-m²/kg². The Coulomb constant k is 9.0×10^9 N-m²/C². This is a much larger number and tends to indicate that the electric force is much stronger than the gravitational force. However, the comparative magnitudes of the masses (m's) and charges (q's) must also be considered. Another obvious difference between these forces is that gravity only attracts, while electrical forces may either attract or repel. The vector direction of the Coulomb force is given by the law of charges.

The gravitational and electrical forces may perform similar functions (Fig. 18.3), but each has its own place

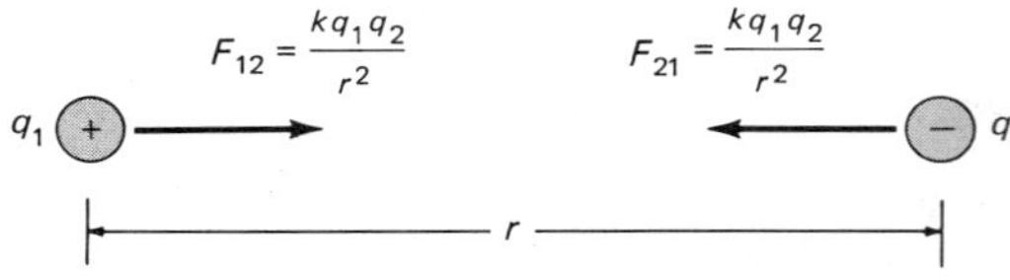

Figure 18.2 Coulomb's law gives the magnitude of the equal and opposite forces between two electric charges.

* Current theory predicts the existence of charges $\frac{1}{3}$ and $\frac{2}{3}$ that of the electronic charge. These are associated with subatomic particles called quarks. Experimental evidence supports the quark theory, but it is believed that they do not exist as free particles and are permanently bound inside the atomic nucleus. We'll learn more about quarks in Chapter 28.

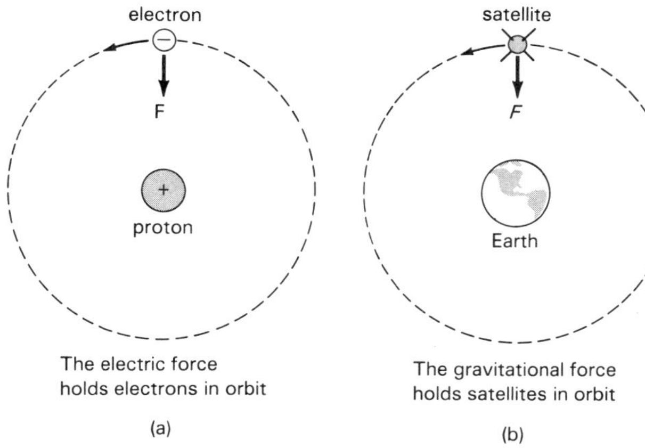

Figure 18.3 The electrical and gravitational forces may perform similar functions, such as supplying the centripetal acceleration for (a) an orbiting electron in an atom and (b) an orbiting satellite around the Earth.

in the scheme of things. The gravitational force provides the centripetal force to hold a satellite in orbit, while the electrical force provides the centripetal force to hold an electron in orbit in an atom. There is also a gravitational force between the electron and proton (mass particles), but this is very small compared with the electrical force between the particles and is usually neglected. We often refer simply to charges rather than to particles or masses with charges.

QUESTION: Just how much stronger is the electrical force than the gravitational force?

ANSWER: To get a comparison, consider the situation of the simple (hydrogen) atom, as illustrated in Figure 18.3. Then, the electrical and gravitational forces are:

$$F_e = \frac{kp^+e^-}{r^2} \quad \text{and} \quad F_g = \frac{Gm_p m_e}{r^2}$$

where the m's are the masses of the particles and r is their separation distance. When we form a ratio, the r's cancel, and

$$\frac{F_e}{F_g} = \frac{kp^+e^-}{Gm_p m_e}$$

Putting in the values of k and G and the magnitudes of the charges and masses from Table 18.1, we find that

$$\frac{F_e}{F_g} = 10^{40}$$

or

$$F_e = 10^{40}\, F_g$$

This means that the electrical force (F_e) is 10,000 trillion, trillion, trillion times the gravitational force (F_g)! That's even bigger than the national debt, which is currently between one and two trillion (10^{12}) dollars.

Conductors, Insulators, and Semiconductors

Just as there are thermal conductors and insulators, there are also electrical conductors and insulators. Recall from Chapter 14 that metals are good thermal conductors because some of the electrons in a metal are "free" to move around (not permanently bound to a particular molecule or atom). The electrical conductivity of a material also depends on the electron bonding or how tightly the atoms of a substance hold on to their electrons. In some materials, such as metals, some of the outer electrons in the atoms are relatively free to move. Such materials are called **conductors.**

In other materials, such as wood or glass, there is little electron mobility. The electrons are firmly attached to their atoms. These materials are called **insulators** (poor conductors). Ionic solids, such as NaCl, are poor conductors because the transferred electrons are tightly bound to the ions, and the ions are "fixed" in their lattice positions (Chapter 10). However, they do become good conductors when melted or dissolved in water. In these cases, the ions (charged particles) are free to move in the liquid.

In some materials, the charge mobility is in between that of good conductors and poor conductors. Such materials are called **semiconductors.** In some instances we "dope" or add impurity atoms to poor crystalline conductors, such as pure germanium and silicon, so they become semiconducting. These solid-state semiconductors have given rise to a whole new field of solid-state electronics with component names like transistors and silicon "chips." You probably own a "transistor" radio, and solid-state circuitry is now a common term. For example, automobiles now have solid-state ignitions. Solid-state semiconductors will be considered in more detail in Chapter 22, after we learn more about the fundamentals of electricity. For the time being, notice where they "fit in" in terms of the relative conductivity in Figure 18.4.

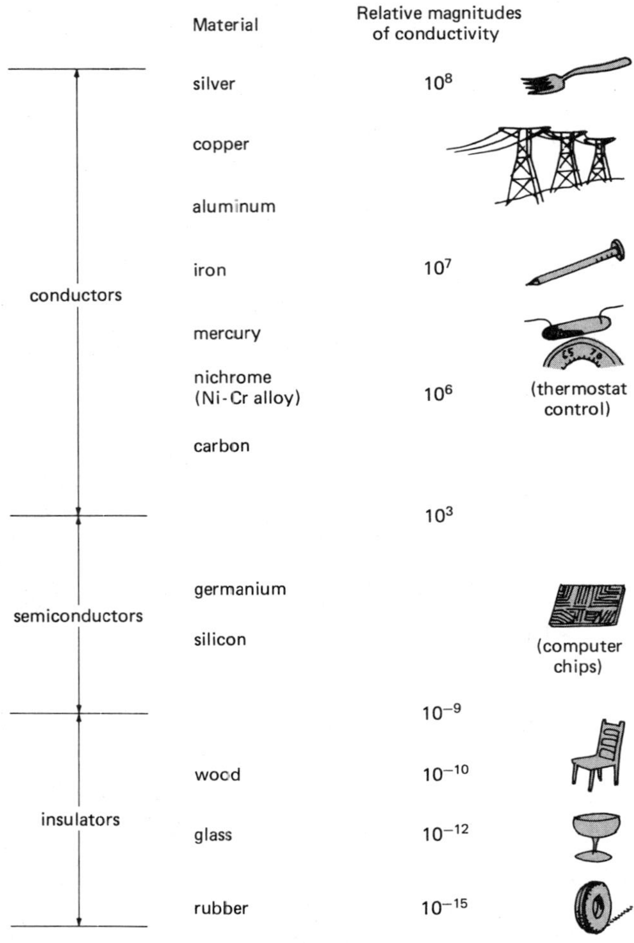

Figure 18.4 The relative conductivity of some materials.

Electrostatic Charging

What does it mean when something is electrically charged, and how do we do it? Being made up of atoms yourself, you contain a bunch of electric charges, but ordinarily you are not "charged." An atom is electrically *neutral.* That is, it has an equal number of protons and electrons, or just as many plus charges as negative charges. However, if we transfer a charge (an electron) from one atom to another, as in ionic bonding, we have a *charged* particle or ion.

Hence, an electrically charged object involves an imbalance of electrons and protons. If an object has more electrons than protons, it is negatively charged. If it has more protons than electrons, it is positively charged. The difference or imbalance of charge is called the *net charge.* For example, a sodium ion (Na^+), having lost an electron, has a net charge of $+1$ (one more proton than electrons). If the electron is transferred to a chlorine atom, the chlorine ion (Cl^-) has a net charge of -1.

The total net charge of the neutral sodium and chlorine atoms together (or separately) is zero. The total net charge of the ions together is still zero ($+1 - 1 = 0$). The charge was simply transferred or moved. The fact that the net charge of a system is constant is called the *conservation of charge.* (The net charge may be other than zero, but it remains constant.)

The conservation of charge means that charge can neither be created nor destroyed *or* that charge must be created or destroyed in pairs. It was once thought that only the first part of the statement was true, but the latter part also applies. Charge pairs can be created or destroyed. Oppositely charged particle-antiparticle pairs can be created from energy. When a charged particle and its oppositely charged antiparticle (Chapter 9) interact, they are both annihilated or destroyed, with the mass going into radiant energy or other particle pairs (conservation of mass-energy). So, on a big scale, the net charge of the universe is constant (but no one knows for sure what this net charge is).

CHARGING BY FRICTION

Have you ever shuffled across a carpeted floor and been zapped by a spark when you reached for the door knob? (Then maybe you tried it by bringing your finger close to the ear or nose of a friend or your little brother or sister.) This happens because you are electrically charged and there is enough electric force to cause the air molecules to be ionized (electrons freed from molecules). There is then a flow of charge, giving rise to a spark. Or, you may have stroked a cat and heard the crackle (mini-thunder?) or seen the sparks in the dark. In both cases, electrons were transferred by friction from one object to another, giving each a net electric charge.

The outer electrons of the atoms of some materials are loosely bound and can be freed and transferred to another object. For example, when a hard rubber rod is rubbed with a piece of fur, electrons are transferred to the rod from the fur. With an excess of electrons, the rod is negatively charged. (How about the fur?) Similarly, when a glass rod is rubbed with silk, the glass gives up electrons to the silk and becomes positively charged (deficiency of electrons). Since rubber and glass are poor conductors, the charge stays on the rods for some time before leaking off, if the air is dry.

Figure 18.5 A common type of electroscope.

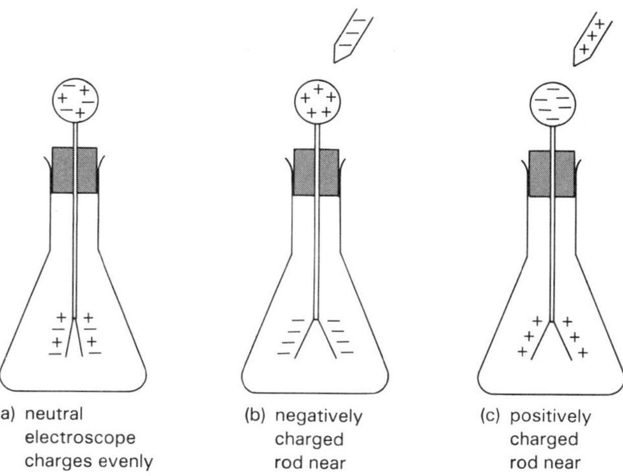

Figure 18.6 An electroscope can be used to tell if an object is electrically charged.

QUESTION: Why do static electricity effects, like sparks jumping to door knobs, occur best on dry days?

ANSWER: Charging by friction occurs best in dry air. In moist, humid air an invisible film of water condenses on objects, which makes the surfaces conductive. Charges quickly leak off, and there is no build-up of charge.

You can demonstrate whether there is a net charge on the rods with an electroscope (Fig. 18.5). For example, if you bring a negatively charged rod near the metal bulb of the electroscope, by the law of charges electrons in the bulb are repelled (Fig. 18.6). Since the metal is a conductor, the electrons move as far away as possible — to the metal foil leaves. The leaves, which now have excess negative charges, are repelled apart or diverge. When the rod is removed, the electrons redistribute over the conductor, and the leaves come together. If a positively charged rod is brought near the bulb, electrons are attracted to the bulb, leaving the foil leaves positively charged, and they diverge in this case too.

The electroscope tells you the rods are charged, but not how they are charged (positive or negative). However, this can be determined by charging the electroscope with a known type of charge.

CHARGING BY CONTACT

Electrons can be transferred from one object to another simply by touching or contact. For example, if a negatively charged rod is touched to an electroscope bulb, electrons are transferred to the bulb. The electrons spread to all parts of the metallic surfaces, and with an excess of electrons the leaves diverge (Fig. 18.7). The electroscope is then negatively charged.

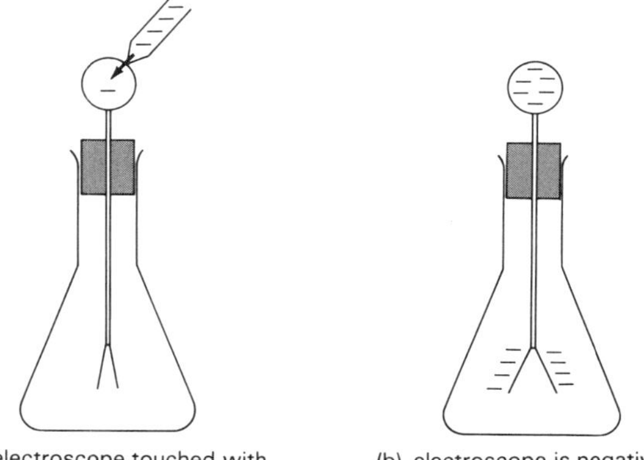

Figure 18.7 Charging by contact.

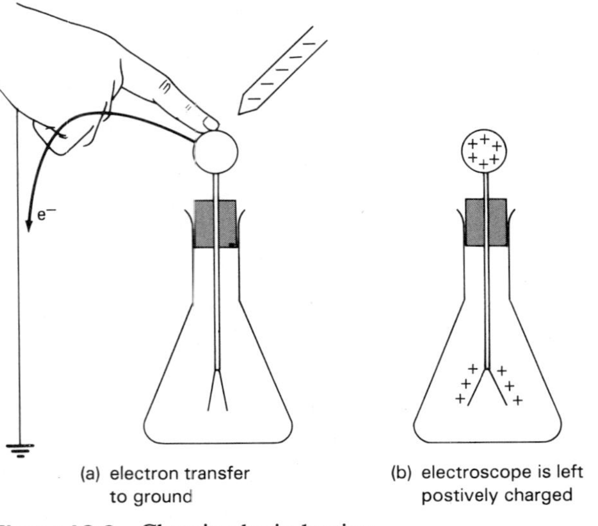

Figure 18.8 Charging by induction.

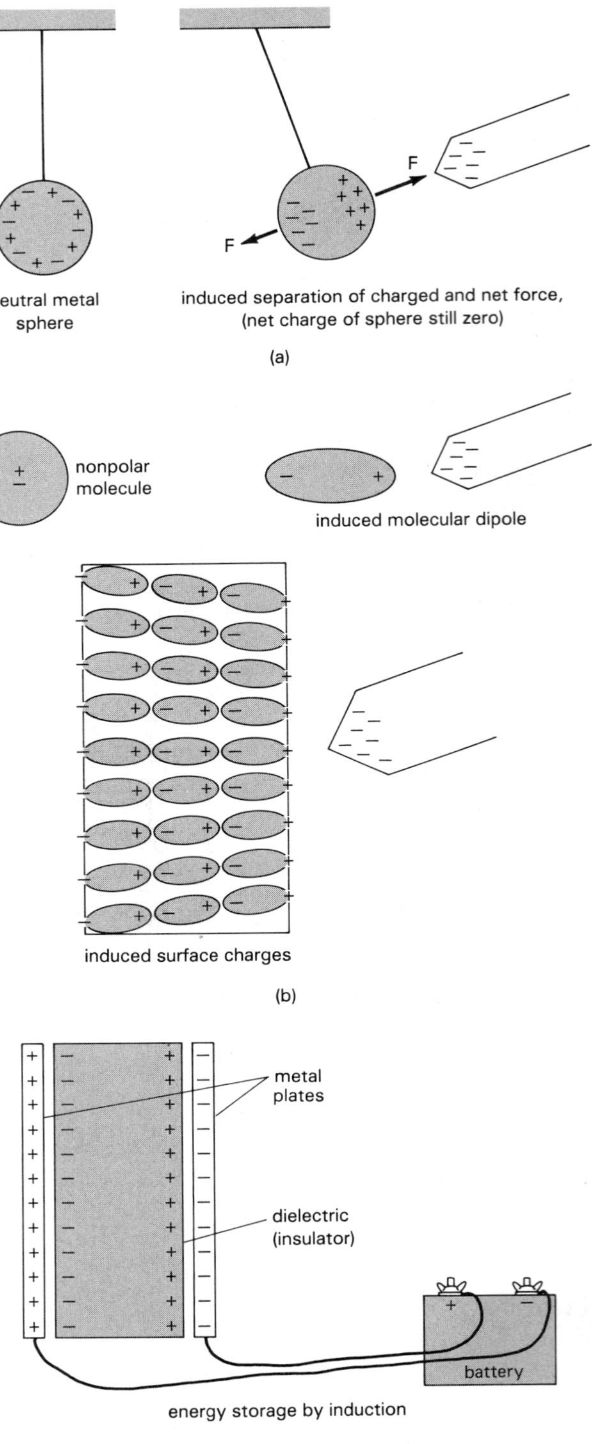

Figure 18.9 Polarization or separation of charge by induction. (a) In a metal ball. (b) In molecular dipoles. (c) Energy storage by induction in a capacitor with a dielectric.

With a charged electroscope you can now tell whether a rod is negatively or positively charged. For example, with a negatively charged electroscope, if a negatively charged rod is brought near the bulb, the leaves will diverge further. If a positively charged rod is brought near the bulb, the leaves will come together. This demonstrates that there are two types of electric charges and repulsive and attractive forces.

CHARGING BY INDUCTION

An object can be charged without friction or contact of a charged rod. Charging can be done simply by bringing a charge close by. For example, suppose a charged rod is brought near an electroscope bulb, as shown in Figure 18.8. Touching the bulb with the finger "grounds" the electroscope.† This provides a path for the electrons to get farther away from the charged rod, and some of them are conducted to ground.

When the finger is removed, the electroscope is left with a deficiency of electrons, and it is positively charged. How could you test to see if it were positively charged using the negatively charged rod?

Charging by induction need not involve a removal of charge. An object can have regions of charges on different surfaces. For example, if a charged rod is

† Electrical ground refers to earth (ground) or some other large conductor that can receive or supply electrons without significantly affecting its own electrical condition. By analogy, this is like adding a cup of water to or taking a cup of water from the ocean.

brought near an insulated metal ball, as shown in Figure 18.9(a), there is a separation of charge and a net electrical force, even though the net charge of the ball as a whole is zero. The induction results in a *polarization* or separation of charge.

All materials, including insulators, can be polarized by induction to some degree. Some materials have permanent dipoles (Chapter 9, a molecule with different regions of charge), and molecular dipoles can be *induced* in others. Induction results in the orientation of the dipoles and opposite surface charges of equal magnitude on opposite sides of the material, as illustrated in Figure 18.9(b).

This principle is used in a capacitor, which can be used to store electrical energy. When two metal plates are connected to a battery, the battery does work in charging the plates, and we say energy is "stored" in the capacitor (Fig. 18.9c). However, when an insulating material, called a dielectric, is placed between the metal plates and the capacitor is charged, work also goes into inducing dipoles (separating charge) or orienting permanent dipoles. When the charged capacitor is disconnected from the battery, the "work" is stored in the capacitor as electrical energy. Depending on the type of dielectric material, the amount of energy stored because of work done in inducing or orienting dipoles can be many times that of charging a capacitor without a dielectric.

Some common examples of combinations of charging by friction and induction are shown in Figure 18.10. When you rub an inflated balloon on your hair (or sweater), it will stick to a wall or ceiling. The balloon is charged by friction, and the charged balloon induces an opposite charge on the wall, which provides the attractive electrical force. Similarly, a hard rubber comb after being run through the hair attracts bits of paper.

(a)

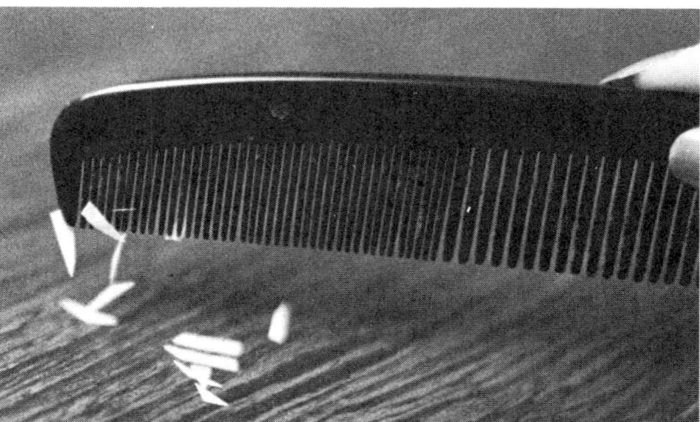

(b)

Figure 18.10 Examples of electrostatic forces. (a) Balloons stick to walls. (b) Bits of paper are picked up by a charged rubber comb.

"Static electricity" can be undesirable as well as used practically. You may have noticed a chain or conductive strip dangling to the ground from a gasoline truck. This is intended to prevent charge build-up and sparking, which would be highly undesirable.‡ Similarly, medical personnel in operating rooms wear shoe coverings with conducting strips to prevent electrostatic sparking in the presence of flammable gases. You can now buy products to prevent "static cling" in clothes (Fig. 18.11). The electrostatic precipitator discussed later in Special Feature 18.2 is an example of a practical use of electrostatics. You'll find another in Chapter 22 in the electrostatic copier.

‡ Tires in contact with the road develop a negative charge. By induction, parts of the metal body of the vehicle near the tires are positively charged, and sparking could occur between the truck and a nearby grounded object or oppositely charged body. A chain to ground would conduct some electrons from the truck's body, but this would leave it positively charged and still prone to sparking.

Lightning and Lightning Rods

During the development of a storm cloud, there is a separation of charge, presumably from the frictional separation of water droplets. This gives rise to regions of different charge in the cloud (Fig. 18.12). The bot-

(a)

(b)

Figure 18.11 Products to prevent static cling.

tom of the cloud is generally negatively charged, and an opposite charge is induced on the surface of the earth.

Air is a poor electrical conductor, but when the charge build-up is great enough, the electric force ionizes the air, and lightning occurs. This can be from cloud to cloud or from cloud to earth, primarily a tree or tall building — the shortest distance. The downward ionized path is nearly invisible. When this makes contact with positive charges near the earth, large numbers of electrons flow to the ground. The bright lightning flash or return stroke we see actually propagates *upward* from the ground and results from electron flow to the cloud along the original ionized path. We visually perceive the stroke to be heading downward.

Ben Franklin is often said to have been the first to demonstrate the electrical nature of lightning. In 1750 he suggested an experiment using a metal rod on a tall building. But a French physicist named d'Alibard set up the experiment first and drew sparks from a rod during a thunderstorm. Franklin later performed a

Figure 18.12 Lightning results from a separation of charge in a storm cloud. The ionization of the air allows the conduction of electric charges between a cloud and earth or from cloud to cloud. Ben Franklin demonstrated the electrical nature of lightning.

similar experiment using a kite. He drew sparks with his knuckles from a key hanging at the end of the kite cord (Fig. 18.12).

Ben was extremely lucky that he wasn't electrocuted. Under no circumstances should you try to duplicate this experiment. Others have and were killed. An insulator, such as a tree, is usually shattered when struck by lightning, whereas a conductor may melt and fuse. People near a lightning strike are often knocked unconscious by the electrical shock. If you come upon someone who has been "struck by lightning" and the person is still alive, give mouth-to-mouth resuscitation if the person is not breathing and keep him or her warm to treat physical shock.

QUESTION: Is there any truth in the common sayings, "Lightning never strikes twice in the same place" and "Rubber tires on cars make them safe from lightning"?

ANSWER: These are general misconceptions. Lightning can and does strike in the same place. For example, the Empire State Building is struck, on the average, about 23 times a year. As you can imagine, this building has an elaborate lightning-rod system (Fig. 18.13).

The National Weather Service tells us that there are *no* absolutely safe places from lightning, although enclosed automobiles and buildings are among the safest. Perhaps rubber tires, being insulators, make people feel safer (like wearing rubber-soled shoes). But lightning, which travels kilometers through insulating air, would not be halted by a few inches of rubber. The lightning would travel through the metal frame, then jump to ground through the air, or along a wet tire surface, or through the tire itself. The latter is a greater possibility now with steel-belted radial tires.

SPECIAL FEATURE 18.1

The Lightning Rod

One of the practical outcomes of Franklin's work with lightning was the lightning rod, which was described in *Poor Richard's Almanac* in 1753. It consists simply of a pointed metal rod that is attached to the top of a building; the rod is connected by a wire to a metal rod driven into the earth (grounded). Franklin wrote that the rod "either prevents the stroke from the cloud or, if the stroke is made, conducts the stroke to earth with the safety of the building" (see Fig. 18.13).

The idea that a lightning rod may prevent a lightning stroke comes from the fact that charges tend to accumulate at points or sharp edges, and charges would leak off or be attracted here. This could conceivably leak off the cloud-induced charge and prevent lightning from occurring, at least in the rod area. However, the general view today is that the main function of a lightning-rod system depends on the principle of establishing contact with the downward ionized "leader" from a cloud before it reaches the structure to be protected and discharging the charge flow harmlessly to ground.

Figure 18.13 Lightning strikes the Empire State Building, which has an elaborate lightning "arrestor" system.

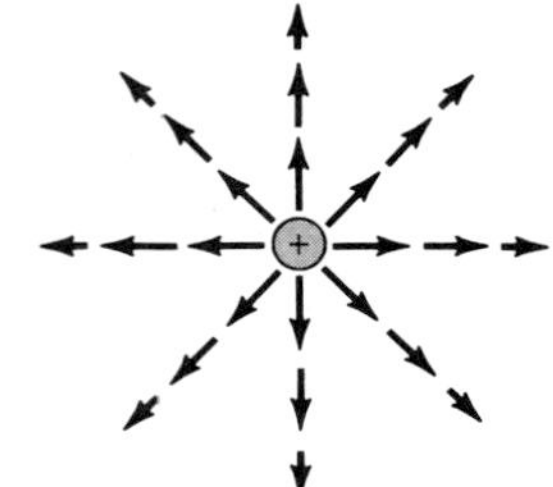

electric field vectors near postive charge

(a)

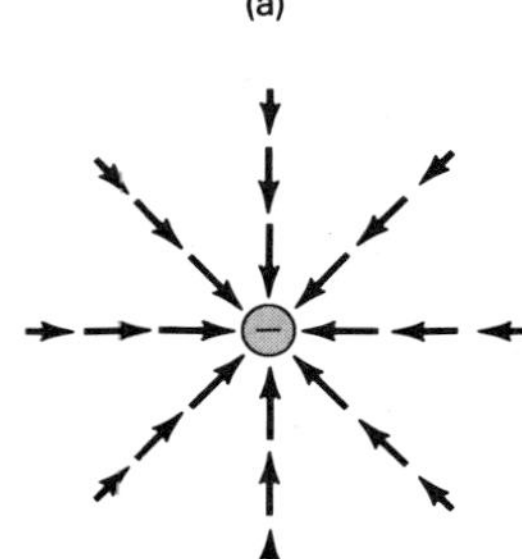

electric field vectors near negative charge

(b)

Figure 18.14 Electric-field vectors. The electric field points away from a positive charge (a) and toward a negative charge (b).

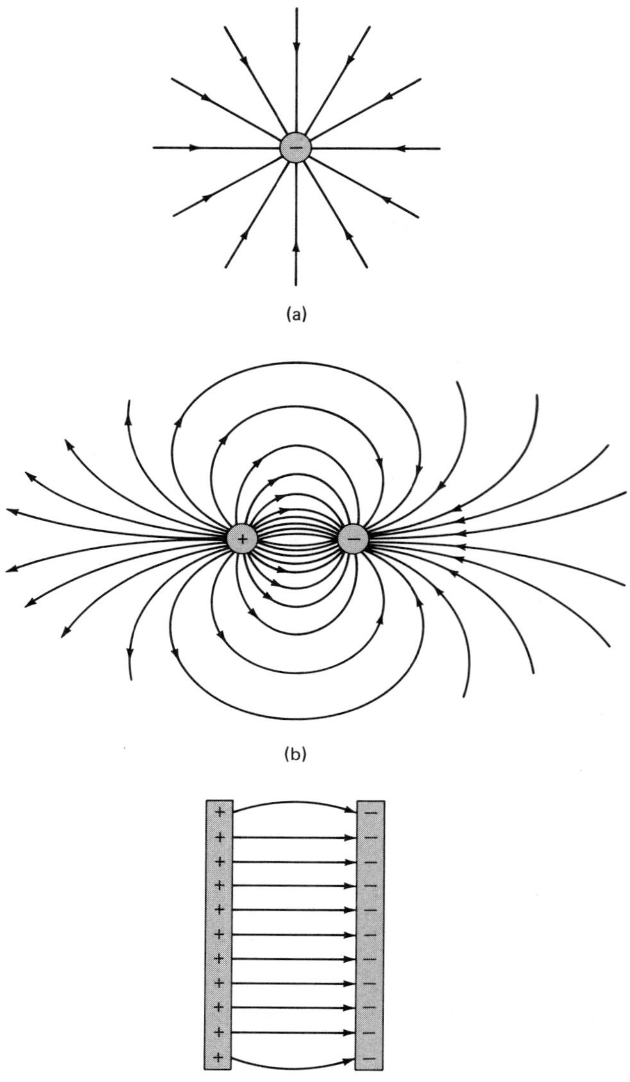

Figure 18.15 Lines of force. When the electric-field vectors are joined, lines of force are formed for (a) a point charge, (b) an electric dipole, and (c) charged parallel plates.

Electric Field

Recall from Chapter 5 that another way of looking at the gravitational interaction is to consider the possible effect on a mass rather than the cause. This was done in terms of gravitational field in space, which described the force properties around a mass or mass distribution. We can think of any other mass as interacting with the field rather than with the mass producing it. For example, the gravitational effects on the Space Shuttle can be described by the Earth's gravitational field without considering the Earth itself.

Similarly, an electric charge produces an electric field around it that interacts with any other charges in the field. Like a gravitational field, the electric field can be represented by vector arrows (Fig. 18.14). The electric-field arrows represent the electric force per unit charge, or

$$E = \frac{F}{q}$$

Electric field = force/charge

This is the force per charge brought into the field, *not* on the charge producing the field. The units of electric field are newton/coulomb (N/C).

Since the electric force can be either attractive or repulsive, the direction of the electric field vector arrows are designated to be in the direction of the force that a positive charge would experience at that location.

We say that an electric field is mapped out by placing a small, positive test charge at various positions. In Figure 18.14 the arrows are away from the positive charge, as prescribed by the law of charges for a positive test charge. If the charge producing the field is negative, the electric field arrows are in the opposite direction, or toward the charge.

The field arrows can be joined together to form "lines of force" (Fig. 18.15). The closer together the lines of force are, the stronger the electric field in that region. Thus, a positive charge in an electric field would experience an acceleration in the direction indicated by the arrowheads. A negative charge in an electric field would experience force in the opposite direction. Why?

The electric fields for an electric dipole and charged parallel plates are shown in Figure 18.15. Notice that the electric field between charged parallel plates is uniform (the same everywhere, except in the fringe areas near the edges). This is a capacitor without a dielectric. We say that the energy resulting from the work done in charging the plates is stored in the electric field. Remember that we talked about electromagnetic radiation propagating and carrying energy through space. This is a wave combination of an electric field and a magnetic field (discussed in a later chapter) and is another example of energy being stored in a field.

An important environmental application of electric fields is given in Special Feature 18.2.

SPECIAL FEATURE 18.2

The Electrostatic Precipitator

A practical application that uses an electric field and electric force is the Cottrell or electrostatic precipitator, which is used to remove particulate matter from flue gases. The developmental work on this type of precipitator was done by F. G. Cottrell, an American physical chemist, in the early 1900's and was used in the smelting industry in 1912.

The basic principle of an electrostatic precipitator is illustrated in Figure 18.16. The positively charged plates are called discharge plates, and the negatively

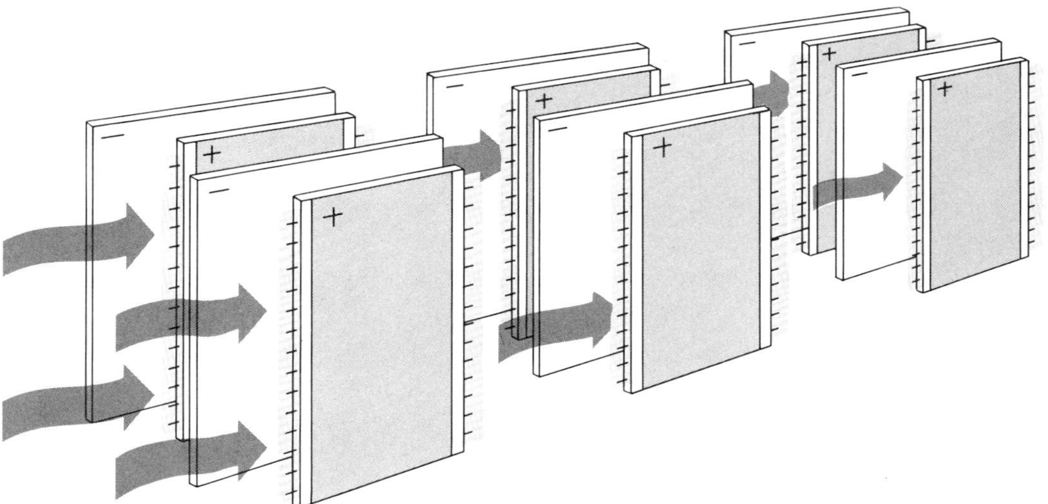

Figure 18.16 Electrostatic precipitation. The large electric field between the plates gives rise to corona discharge at the needle projections on the positive plates. Particles in the flue gases passing between the plates are ionized, attracted toward, and collected on the negative plates. Other designs use wires instead of positive plates.

(Continued)

charged plates are called collection plates. As a result of a large electric field between the plates, a corona discharge occurs around the needle projections on the positive plates. (Some precipitator designs use wires as positive discharge electrodes.)

The particulate matter, such as soot, in flue gases passing through the precipitator chamber is ionized. The positively charged particles then move toward and collect on the negative collection plates. The particulate matter accumulates and falls to the bottom of the chamber, where it is removed.

Electrostatic precipitators are capable of removing more than 90 percent of the particulate matter from flue gases (Fig. 18.17). They are commonly used in fossil-fuel electrical generating plants and industrial operations in which there is a large amount of particulate matter in the flue gases resulting from incomplete combustion and impurities.

Figure 18.17 Electrostatic precipitation is extremely efficient. The two smokestacks on the left are equipped with precipitators; the two on the right are not.

Electric Potential

We use electrical energy every day to do work. This energy is available when we need it, so it must be stored as potential energy. To visualize how this energy is stored by charges, consider the situation in Figure 18.18. Remember, potential energy is the energy of position.

When a charge is moved toward a like charge or away from an unlike charge, work is required against the electric forces. This is like doing work in compressing and stretching electrical "springs." As a result, the electrical charges have potential energy. If a charge is released, it moves or accelerates as its potential energy is converted into kinetic energy.

Rather than dealing with total potential energy in practical electrical cases in which there are numerous charges involved, it is more convenient to consider the electric potential energy per charge, or the **electric potential,**

$$\text{Electric potential} = \frac{\text{potential energy}}{\text{charge}}$$

The unit of electric potential is the **volt** (V), and we commonly speak of electric potentials in terms of **volt-**

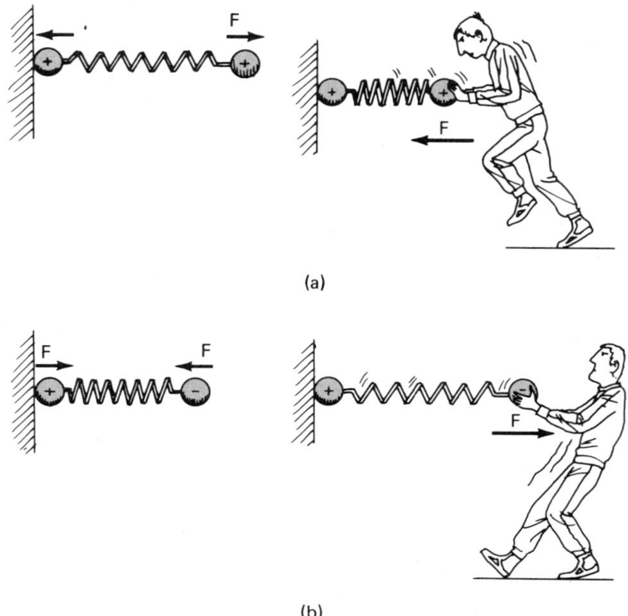

Figure 18.18 Electric potential energy. When like charges are brought closer together (a) or unlike charges are separated (b), work is done against the electric force, which is "stored" as electric potential energy. This is analogous to doing work in compressing and stretching electrical "springs."

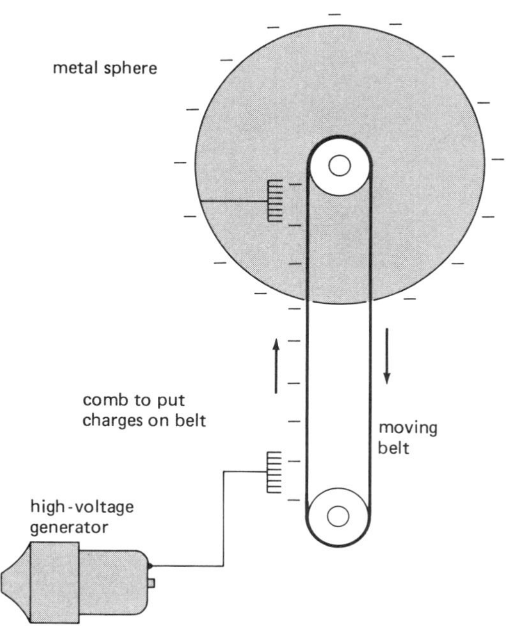

Figure 18.19 A diagram of a Van de Graaff generator. The accumulated charge on the sphere results in an electric potential of millions of volts.

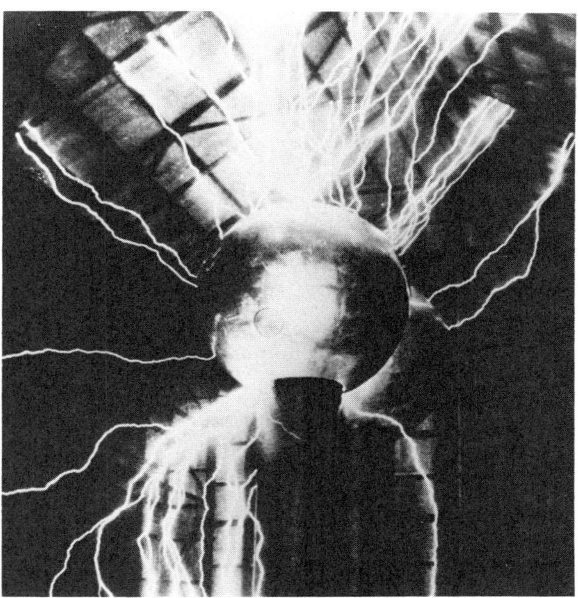

Figure 18.20 Big volts: Ionization discharge in air from a Van de Graaff generator. The ionization of air requires an electric field of about 3 million volts per meter.

ages.§ You are no doubt familiar with these common terms. We have 12-volt batteries in our cars, and the lamps and other appliances in our homes operate on 110 V. More about this later. Here, we're getting the basic fundamentals and electrical jargon.

Incidentally, it is interesting to note that the volt and other common electrical units are the same in the SI metric and British systems. So, electrically, we have already converted to SI units.

The Van de Graaff Generator

An application that uses the principles discussed in this chapter is the Van de Graaff electrostatic generator, which was invented by the American physicist Robert Van de Graaff around 1930. It generates electrostatic potentials of *millions* of volts. The high voltages are used to accelerate charged particles that can be used to probe the nucleus and to produce X-rays for medical and industrial applications.

The basic design of the Van de Graaff generator is shown in Figure 18.19. Electric charge is "sprayed" on a motor-driven insulating belt by a corona or spark discharge from a row of metal points. The charges are deposited on the inside of a metal sphere and, being repelled from each other, are distributed over the outside of the sphere. This leaves the inside uncharged and able to receive more charge from the belt. The operation results in the build-up of a huge charge on the sphere and the "stored" electric potential of millions of volts.

When the sphere is charged to a high enough potential, corona discharge can occur through the ionization of air (Fig. 18.20). At STP (standard temperature and pressure, 0°C and 1 atm), the electric field required for the ionization of air is about 3 million volts per meter (V/m).‖ Although the corona discharge is rather spectacular, it is usually undesirable, since it limits the voltage or amount of charge that can be accumulated on the sphere. The discharge can be suppressed and higher voltages achieved by surrounding the charged sphere with a gas of high ionization potential under high pressure. Electric fields greater than 10 million volts/m can then be achieved.

Some other effects of the high voltage of a Van de Graaff generator are shown in Figure 18.21. Notice the human electroscope with the hair taking the place of the leaves.

§ The volt is named in honor of Alessandro Volta (1745–1827), an Italian scientist who invented one of the first batteries.

‖ In addition to N/C, volt/meter (V/m) is a unit of electric field. This comes from the relation $E = V/d$ (volt/meter).

(a)

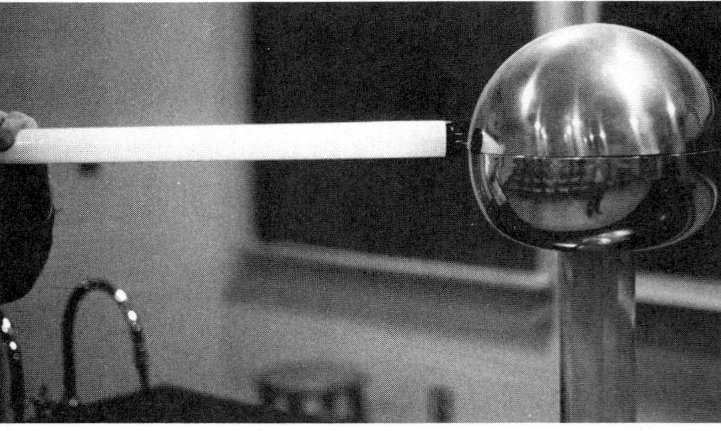
(b)

Figure 18.21 Effects of high Van de Graaff voltages. (a) A human electroscope. This young fellow is insulated, and when touching the Van de Graaff sphere he is charged to the same potential. (b) An electric field strong enough to light a fluorescent bulb. The lamp only lights out to the hand. The person can "turn down the lights" simply by moving the hand toward the generator along the tube.

SUMMARY OF KEY TERMS

Electric charge the fundamental property associated with electric force.

Law of charges like charges repel and unlike charges attract.

Coulomb's law the relationship that gives the electrostatic force between two charges, $F = kq_1q_2/r^2$.

Conductor a material in which electrons are free to move.

Insulator (nonconductor) a material in which there is little electron mobility.

Semiconductor a material with charge mobility in between that of a good conductor and a poor conductor (insulator).

Conservation of charge the net charge of a system is constant.

Electrical induction the redistribution of charge in an object due to the influence of a charged body nearby but not in contact.

Electrical ground the earth (ground) or some other large conductor that can receive or supply electrons without significantly affecting its own electrical condition.

Electric field the force field surrounding a charged body. Electric field = force/charge, in the direction a positive charge would experience a force.

Electric potential the electric potential energy per unit charge. This is commonly called voltage and measured in volts.

EXERCISES

1. About how many electrons would it take to make up a charge of 1 coulomb? How many protons would it take?
2. Two negative charges of -1 C each are placed at opposite ends of a meterstick. (a) Could a free electron be placed somewhere on the meterstick so it would be in static equilibrium (zero net force)? How about a proton? (b) What would happen if the electron or proton were placed on the meterstick other than at the 50-cm position?
3. A charge of $+1$ C is placed at one end of a meterstick and a charge of -1 C at the other end. Could a free electron be placed somewhere on the meterstick so it would be in static equilibrium (zero net force)? How about a proton?
4. The gravitational force is weaker than the electric force. It is easy to feel or experience the gravitational force, for example, when you pick up a heavy object. There are electrons and protons all around. Why don't we generally feel the electric force?
5. A large charge of $+Q$ and a small charge of $-q$ are a short distance apart. How do the electric forces on each charge

(a)

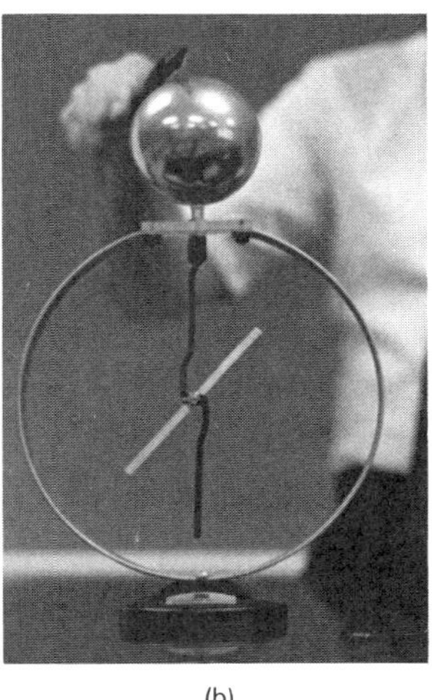
(b)

Figure 18.22 (a) Electroscope uncharged. (b) Electroscope charged. See Exercise 10.

compare? Is this comparison described by another non-electrical physical law?

6. Is the Coulomb constant k a universal constant like G? Explain.
7. An electron is a certain distance from a positive charge. If the electron is moved twice as far away from the other charge, how is the electric force affected?
8. An electrically neutral object (like yourself) can be given a net charge by several means. Does this violate the conservation of charge? Explain.
9. Given an object and an electroscope, how could you tell if the object was charged? If it was, how could you tell what type of net charge it had?
10. Another type of electroscope, called a Braun electroscope, is shown in Figure 18.22. Explain its principle of operation. Does this type of electroscope have any practical advantages over the foil-leaf electroscope?
11. When thin plastic food wrap is pulled from its roll in a box and cut off, it often sticks together. Explain why.
12. An electroscope is negatively charged, and its leaves diverge. What would happen to the leaves (and why) if you touched the bulb with (a) your finger and (b) a glass rod that had been rubbed with silk?
13. Will a charged electroscope remain charged indefinitely? Explain.
14. How could you charge an electroscope negatively by using induction? How could you prove it was negatively charged?
15. What causes static cling in clothing? Why is this more of a problem on a dry day?
16. Why is dust so difficult to get off a phonograph record?
17. We commonly rub balloons on our hair to charge them electrostatically. How could a bald-headed person charge a balloon so it would stick to the wall?
18. An inflated balloon is charged and placed against a smooth ceiling. What determines if the balloon will stick to the ceiling?
19. (a) Getting zapped by a spark when you are about to touch a door knob involves charging by friction and induction. Explain why. (b) Is the chain sometimes seen dangling from a gasoline truck really a safety device? Explain.
20. What would happen in Figure 18.9(a) if the rod is touched to the metal sphere?
21. Why must work be done in charging a pair of capacitor plates even without a dielectric between the plates? Why is more work required when there is a dielectric between the plates?
22. Why must a capacitor dielectric be an insulating material?
23. Two metal spheres on insulating rods are in contact, as shown in Figure 18.23. How could you charge both spheres by induction without directly touching the metal spheres? How would the spheres be charged (positively or negatively)?
24. Why is it not a good idea to seek shelter under a tree in a thunderstorm?
25. When it starts to rain, and sometimes before, life guards make everyone get out of a swimming pool. Why is this a wise precaution?
26. It is said that there is no absolutely safe place from lightning. Does this apply to astronauts in an orbiting space shuttle or on the moon?
27. How is the direction of the force acting on a charged particle due to an electric field determined?
28. Sketch the electric field in the vicinity of two isolated (a) positive charges and (b) negative charges. (Don't forget the directional arrowheads.)

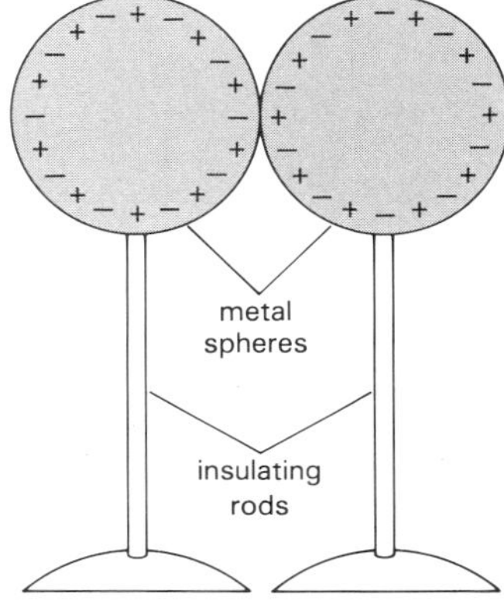

Figure 18.23 See Exercise 23.

29. A net charge on a metal conductor resides on the outside of the conductor. Why is this the case?
30. Why do electric field lines always begin on positive charges and end on negative charges?
31. When a closed, hollow metal object, such as a metal box, is charged by induction by outside charges or is given a net charge by contact, the electric force or electric field inside the object is zero. Why? (See Exercises 29 and 30.)
32. Exercise 31 is an example of electrostatic shielding. A closed metal object acts as an electrostatic shield. If there are charges inside the shield, these will set up their own electric field. But the field inside will be unaffected by charges outside and will experience no force because of them. Would it be possible to construct a gravitational shield? Explain. (Electrostatic shielding is common in electronic equipment, for example a radio set, where it is desired to keep the electric fields originating in one part of the set from interfering with components in another part.)
33. Would it be safe to get inside a Van de Graaff generator sphere and then have it charged?
34. Explain why the boy's hair in Figure 18.21 stands on end. Why is he insulated from ground?
35. We say that energy is stored in an electric field. How might this be described by a three-dimensional field representation? (*Hint:* See Chapter 5.)
36. One charged object has twice the electric potential energy as another. Does it necessarily have twice the electric potential? Can it have?
37. Knowing the charge and voltage of an object, how could you determine its electric potential energy?

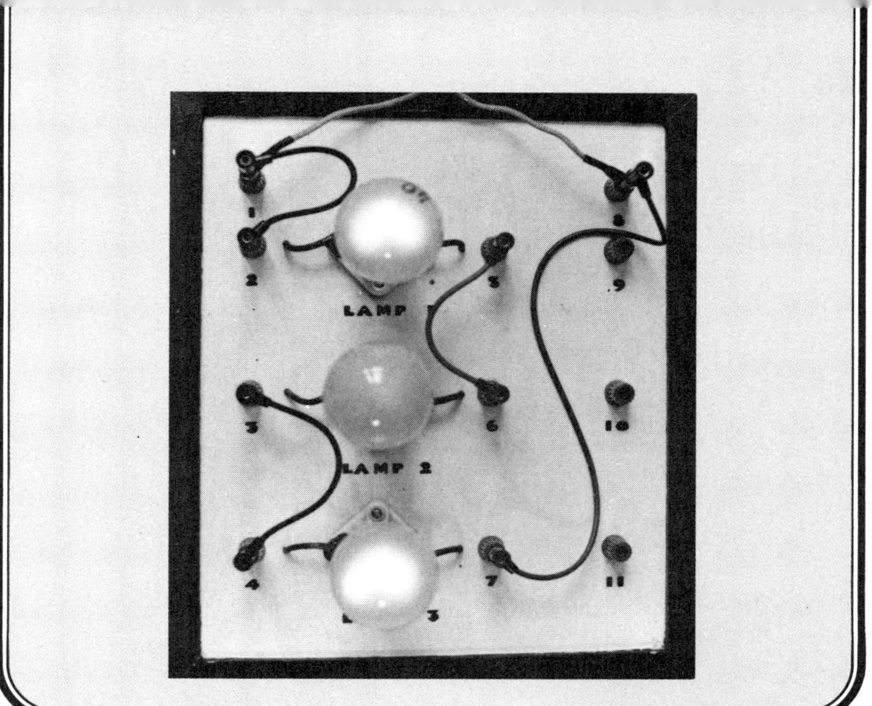

19 Electric Current—Charges in Motion

Electrostatic electricity is important, but the vast majority of electrical applications involves electric currents or charges in motion. When you turn on a flashlight or a car's headlights, there is a flow of electrons in a wire circuit. The lamp filament (a wire too) is part of the circuit, and here electrical energy is turned into heat and light (radiant energy). Similarly, when you switch on a light in the home there is a flow of electrons in a circuit, but in this case the electrons move alternately back and forth. More about this later. First, a little history.

Early investigators thought electricity was the result of different fluids—positive and negative. Recall that the early theory of heat also considered it to be a fluid. This is probably why we speak about a "flow" of charges and about heat flow. (A fluid is a substance that can flow.) Ben Franklin advanced a single-fluid theory of electricity. He postulated the existence of a tenuous, imponderable fluid. All bodies normally contained a certain quantity of the fluid. A surplus or deficit gave rise to electrical properties. With an excess, a body was positively excited, and with a deficit it was negatively excited. This theory gave rise to a later idea that it was the positive electric charges that flowed or moved. Today, we know it's the electrons that move in a conductor.

Electric Current

In solid conductors, particularly metals, it is the outer atomic electrons that are free or relatively free to move. As we know, the atomic nuclei with the positively charged protons are fixed in the lattice structure. (In ionic liquid conductors, both positive and negative *ions* can move.)

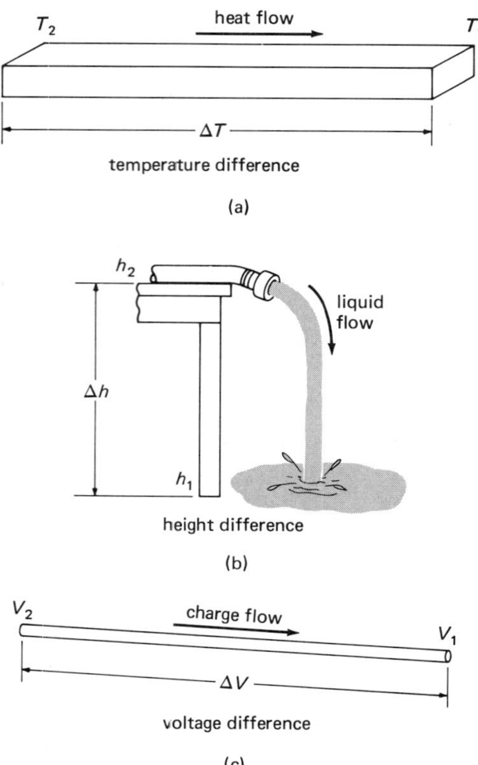

Figure 19.1 "Flows" generally require a difference in conditions.

But what causes electrons to move or an electric current to flow in a conductor, such as a common metal wire? Stop and think for a minute (maybe for a few seconds). What causes heat to flow or water to flow downhill? A temperature *difference* or a height (potential energy) *difference* (Fig. 19.1). In the electrical case, it is an electric potential or voltage (energy/charge) *difference.* In a sense, the electron flow or current is down an electrical "hill," from a higher to a lower potential. This could be due to an electrostatic separation of charge.

Of course, the "flows" in Figure 19.1 would only be temporary. When thermal equilibrium (equal temperatures), equal heights, and equal potentials are reached, or when there are no differences, the flows cease. To have a sustained heat flow would require a heat pump and a system of thermal conductors or a circuit to maintain the temperature difference. The same is true for water flow and electric current (Fig. 19.2). In the water "circuit" the path for the water flow is complete, and the pump supplies the work to maintain the potential difference for a continuous flow. The water flow can do work, such as turning a paddle wheel. Similarly,

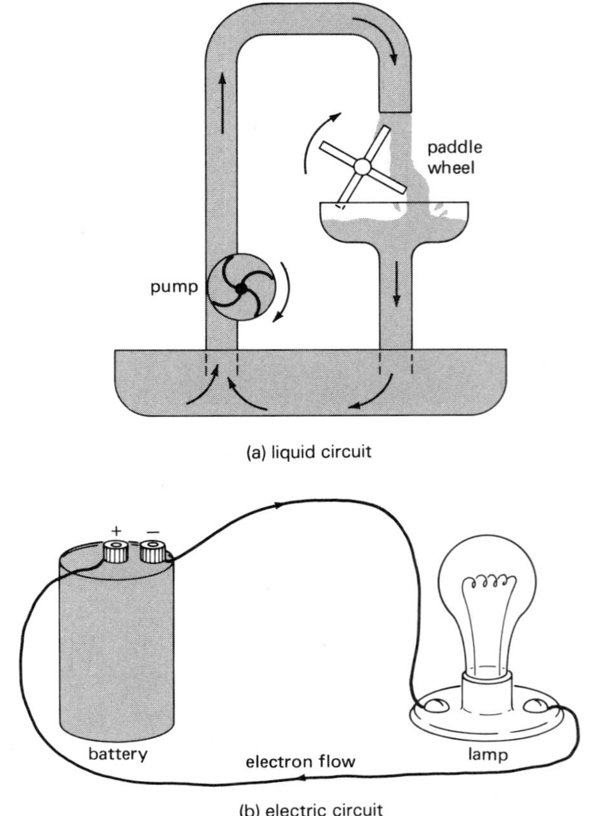

(b) electric circuit

Figure 19.2 Circuit analogies. For sustained flow, a completed circuit and a method to maintain a potential difference are required for both liquid and electrical circuits.

> A sustained electric current requires a complete circuit and a voltage source.

Conducting wires usually complete an electric circuit, and the voltage source, which does work to maintain the electric potential difference, may be a battery. The electric current can light a bulb in the circuit.

In terms of charge movement, the **electric current** is the net amount of charge q that passes through an area of a conductor per time:

$$\text{Electric current} = \frac{\text{charge}}{\text{time}}$$

or

$$I = \frac{q}{t}$$

The units of electric current are then coulomb/second. This is given the name of **ampere** (A), in honor of the French physicist Andre Ampere (1775–1836), an early investigator of electricity. It is common to speak of current in "amps," for short. Small currents are expressed in milliamps (mA, 10^{-3} A) and microamps (μA, 10^{-6} A).

VOLTAGE SOURCES

A voltage source or electrical "pump" is required to maintain a potential difference for the flow of current. There are many types of voltage sources, which are often called "seats" of electromotive force (emf). This is an unfortunate name because electromotive force is not a force, but the energy per charge (voltage) given by the source. The most common types of voltage sources are batteries and generators.

A battery produces a voltage by chemical means. There is a certain "terminal" voltage between or *across* the battery terminals. For example, most automobile batteries are 12-volt sources. When you attach the battery cables to the positive (+) and negative (−) terminals, the circuit has a "pumping potential" of 12 V. (The positive terminal has a potential of + 12 V and the negative terminal is grounded or at 0 V. The voltage difference between the terminals is then $\Delta V = V_2 - V_1 = 12 - 0 = 12$ V.) The electron flow in the circuit is away from the negative terminal toward the positive terminal. We call this **direct current (dc)** because the flow of charge is direct, or in *one direction.*

As long as the chemical action of the battery is maintained, the voltage is constant with time (Fig. 19.3). We often say that a car battery is a 12-VDC (12 volts–direct current or 12 volts-dc) source.

The voltage of a generator, on the other hand, fluctuates with time (see Chapter 21). It can alternate between positive and negative values (Fig. 19.3). As the polarity of the voltage alternates back and forth (positive and negative), so does the direction of the electron current. We call this **alternating current (ac).** Common household voltage is 120 VAC (volts–alternating current or volts-ac).* The voltage frequency is 60 Hz. This means the electrons in the conductor move back and forth 60 times per second.

* You may hear household voltage given as 110 VAC, 115 VAC, or 120 VAC. This is because the voltage varies between 110 V and 120 V.

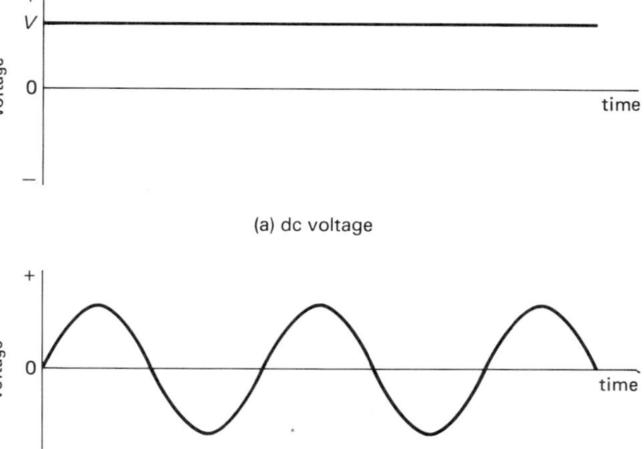

Figure 19.3 Voltages. (a) A dc voltage, such as that maintained by a battery, is relatively constant with time. The magnitude of a dc voltage may vary, but the polarity (+ or −) does not change, so the current is always direct or in one direction. (b) The polarity of an ac voltage, such as common household voltage, alternates with time, so the direction of the current alternates (ac).

Alternating voltage has advantages, and its production will be considered in a later chapter. For the time being, we'll take a closer look at direct-current battery sources. Before doing that, however, let's consider current flow in a bit more detail.

CHARGE FLOW

The idea of electron flow is convenient in describing electrical current, but don't get the idea that the electrons flow around the circuit like water through pipes. Water circuits are helpful analogies, but they can be misleading, as is the term "current flow."

In the absence of a voltage difference, the electrons in a metal wire move around randomly and chaotically at very high speeds—on the order of 1000 km/s (about two million miles per hour). With the electrons moving around in all directions, there is no net flow one way or the other. With a potential difference from a battery across the wire, the electrons still collide and move chaotically, but there is a general charge movement along the wire toward the lower potential. This electron movement has an average or **drift velocity** on the order of 0.1 cm/s—not much of a flow. In the case of alternating current, the electrons move periodically back and forth and there is *no* net current flow.

Yet when you switch on a light, it comes on instantaneously. Telephone conversations in the form of electric signals travel hundreds of kilometers through telephone lines. If this depended on actual electron flow, telephone conversations would be pretty slow. What then travels through a circuit? It is the electric field. The circuit wires act as guides for the electric force field that drives the conducting electrons. A battery or a generator connected to a circuit causes an electric field to travel through the circuit with a speed near the speed of light. The energy in the field does the work in the circuit. For example, the alternating electric field associated with an alternating voltage source causes the electrons in a lamp filament to vibrate back and forth, which causes the filament to get hot and emit light.

So it's really energy that flows in a circuit, producing only a slight motion of the electrons. But we still talk in terms of current and electron flow because they are convenient terms for describing the overall electrical effects.

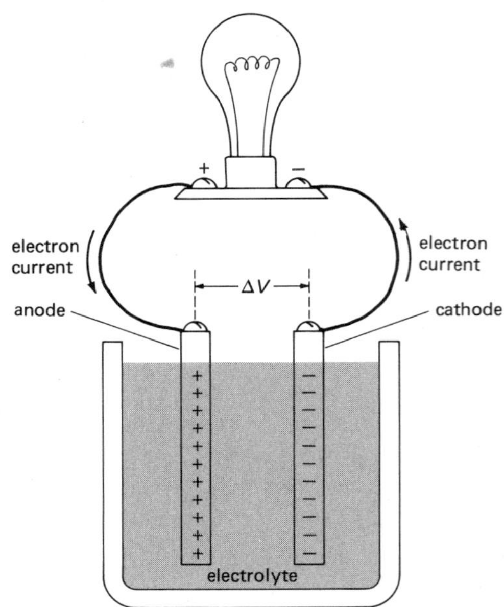

Figure 19.4 A simple battery consists of two electrodes and an electrolyte. Chemical action gives a potential difference across the terminals, and in a completed circuit a current is maintained by the chemical reactions and ionic conduction in the battery.

Batteries

A battery converts chemical energy into electrical energy in a chemical cell. The term battery refers to a "battery" of cells, although we often call a single cell a battery.

The Italian scientist Alessandro Volta, after whom the volt unit is named, is credited with constructing one of the first practical batteries. Basically, a battery consists of two electrodes in an electrolyte (Fig. 19.4). With the appropriate materials (Volta used zinc and copper electrodes in a dilute electrolyte solution of sulfuric acid) one electrode, called the **cathode,** becomes negatively charged, and the other electrode, the **anode,** becomes positively charged as a result of chemical processes. That is, a potential difference develops across the battery electrodes or terminals.

When a battery is connected to a circuit, the potential difference causes a flow of electrons that is maintained by chemical reactions and ionic conduction in the battery. We commonly say that the battery "delivers" current to the circuit, or that the circuit "draws" current from the battery.

There is a wide variety of batteries available for all the battery-operated devices in existence today. Probably the most common batteries are the single-cell 1.5-volt D-cell flashlight battery and the multi-cell 12-volt storage battery used in automobiles (Fig. 19.5). The flashlight battery is called a "dry" cell. It is not actually dry, only relatively so compared to an automobile battery. The electrolyte (ammonium chloride and other additives) is in the form of a paste. A dry cell becomes "dead" when its active materials have been depleted. Most dry cells cannot be recharged.

A storage battery, on the other hand, can be recharged. In the common lead storage battery, there are many plate electrodes. The positive electrode is lead oxide (PbO_2), and the negative electrode is spongy (porous) metallic lead. The electrolyte is a water solution of sulfuric acid (H_2SO_4). When current is drawn from the battery, both electrodes change chemically to lead sulfate ($PbSO_4$), while the electrolyte is converted to water and the acid solution becomes more dilute. The "strength" of a battery may be tested by determining the specific gravity of the electrolyte solution with a hydrometer (Chapter 11). If a current is passed through a storage battery in the opposite direction to that of the normal current flow, the battery is recharged, with the electrodes and electrolyte restored to their original conditions (reverse chemical action).

The 12-V auto battery actually consists of six 2-V cells connected in series—the positive terminal of each

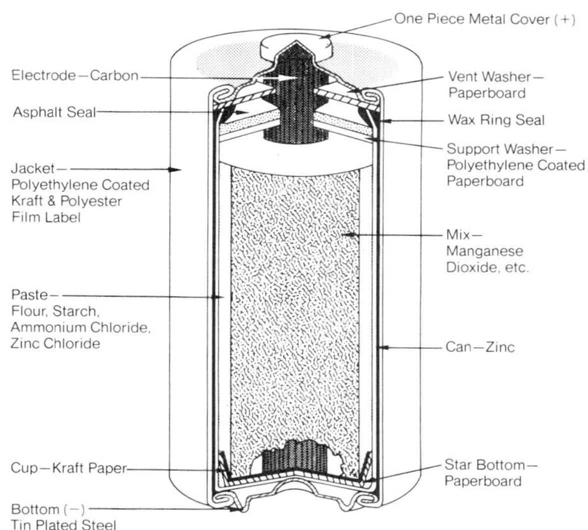

(a) dry cell

(b) storage battery

Figure 19.5 Common batteries. (a) A dry cell, such as the D-cell battery, is commonly used in flashlights. (b) An automobile storage battery.

cell connected to the negative terminal of the next cell. When connected in this fashion, the voltages of the cells add. This is analogous to a series of pumps raising water to successively higher gravitational potentials (Fig. 19.6).

Batteries may also be connected in parallel. In this case, all the positive terminals have a common connection, as do the negative terminals (Fig. 19.7). When batteries with the same voltage are connected in parallel, the potential difference across the combination is the same as the voltage of the individual batteries. However, each battery supplies an equal fraction of the current delivered to the circuit, similar to the pumps in the parallel water analogy. The lamp in Figure 19.7 could be lighted using only one battery. An advantage of connecting three batteries in parallel would be that

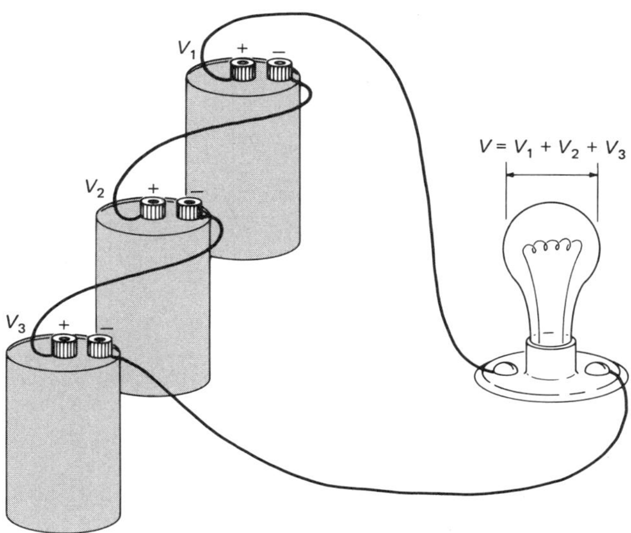

(a) batteries in series

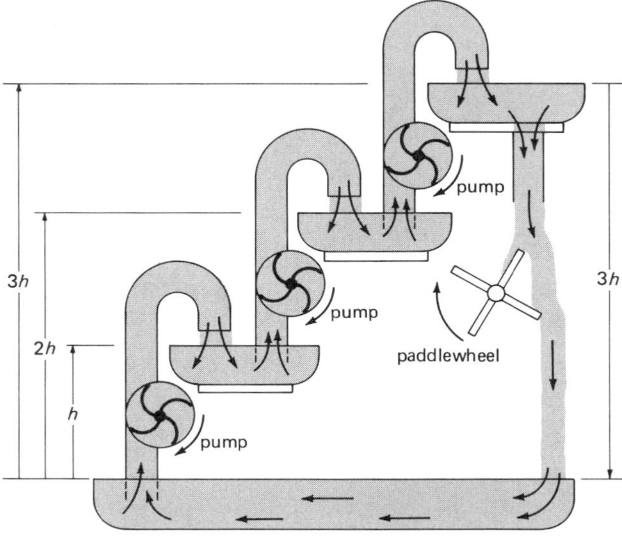

(b) pumps in series

Figure 19.6 Batteries and pumps in series. Each battery or pump adds to the potential difference in the respective circuits.

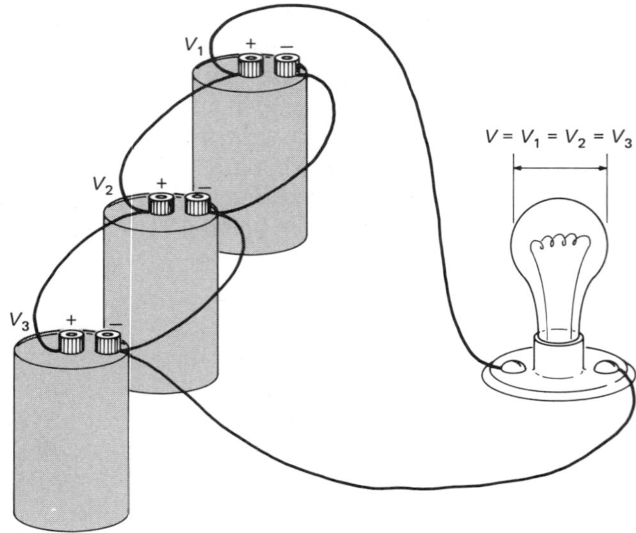

(a) batteries in parallel

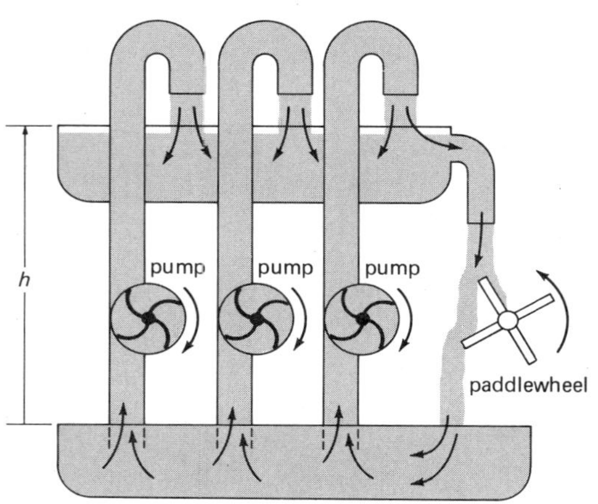

(b) pumps in parallel

Figure 19.7 Batteries and pumps in parallel. Each battery or pump contributes a fraction of the "flow" in the respective circuits. With similar batteries or pumps, the potential difference in a circuit is the same as for an individual battery or pump.

you could go three times as long before you had to change batteries, since each battery would supply one third of the current.

QUESTION: Are the batteries in an ordinary flashlight connected in series or parallel?

ANSWER: They are connected in series by contact (Fig. 19.8). The rounded, raised piece of metal on the top of the battery is the positive terminal, and the metal bottom is the negative terminal (see Fig. 19.5). The batteries are put in the flashlight with plus (+) to minus (−) terminal contact, so they are in series. The spring on the base end of the flashlight (which is part of the circuit) provides pressure so there is good contact.

When you switch on a flashlight, you are making or "closing" (completing) the circuit. When you switch it off, you break or "open" the circuit. This is the same for a light switch in a house.

Electrical Resistance

Motion is generally accompanied by resistance, for example, frictional resistance between objects and internal resistance or viscosity in fluid flow. Similarly, when there is charge flow, there is electrical resistance. On the atomic level, this opposition to charge flow arises from collisions between electrons and lattice atoms or ions of a material. Hence, **electrical resistance** is a material property.

But resistance also depends on such things as the size of a conductor and its temperature. For example, the resistance of a wire conductor depends on its length and cross-sectional area. Think about a pipe with fluid flow. The longer the pipe, the more resistance there is. However, the greater the diameter or cross-sectional area of the pipe, the more fluid flow there would be, indicating

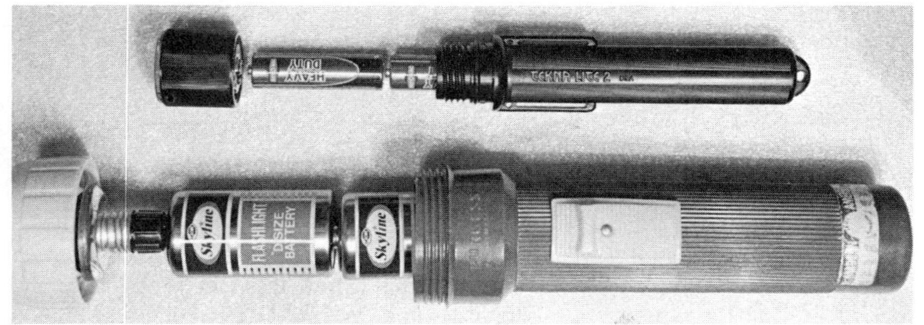

Figure 19.8 Series or parallel battery connections in flashlights? See Question and Answer.

less fluid resistance. Similarly for an electrical wire conductor, the longer the wire, the greater the resistance, and the greater its diameter or cross-sectional area, the smaller the resistance.

Temperature also affects electrical resistance. As a general rule, the resistance of most metallic conductors increases as the temperature increases, while the resistance of most nonmetallic conductors decreases as the temperature increases. The increased atomic or ion motion in metals generally hinders the flow of charge through more collisions, whereas in nonmetals (e.g., carbon and semiconductors) there is a greater flow of charge or decreased resistance. Hence, cooling a metal conductor decreases its resistance. The extreme case of superconductivity is discussed in Special Feature 19.1.

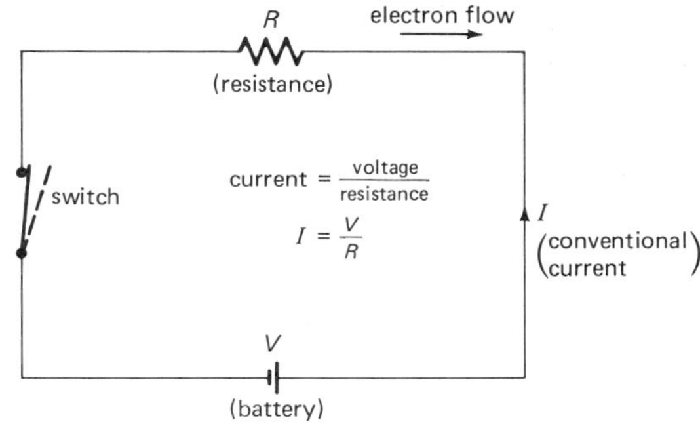

(a) electrical circuit

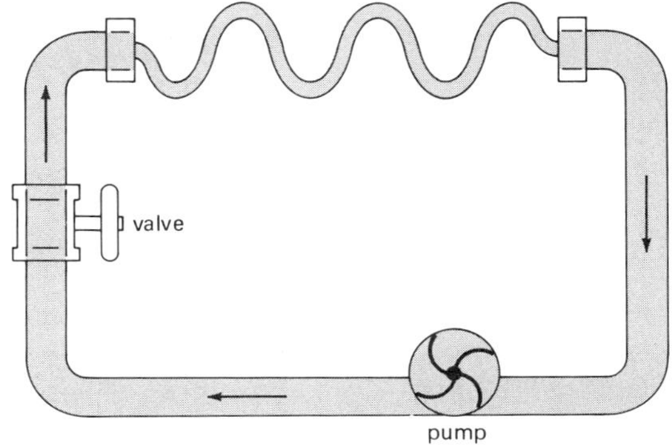

(b) water analogy

Figure 19.9 Ohm's law. (a) For a give voltage *(V)*, the current *(I)* is determined by the resistance *(R)* in the circuit, $I = V/R$. (b) A liquid-circuit analogy for the electrical circuit. The bent tubing offers resistance to the liquid flow.

Ohm's Law

How much current flows in a circuit depends not only on how much resistance there is in the circuit, but also on the voltage source. The greater the electric potential, the greater energy per charge there is. The relationship of the current (I), the voltage (V), and the resistance (R) for many conductors is expressed by **Ohm's law,** which was formulated by Georg Ohm (1789–1854), a German physicist:

$$\text{Current} = \frac{\text{voltage}}{\text{resistance}}$$

or

$$I = \frac{V}{R}$$

Ohm's law shows that the greater the voltage, the greater the current flow, and the greater the resistance, the smaller the current flow.

An electrical circuit and diagram, along with a water analogy, illustrating the elements of Ohm's law are shown in Figure 19.9. Notice how the circuit elements

SPECIAL FEATURE 19.1

Superconductivity

The electrical resistance of a metal conductor can be decreased by cooling. You may wonder how far one can go with this decrease. Strangely enough, for certain materials called superconductors, you can go all the way. At very low temperatures (a few kelvin) materials such as lead and mercury exhibit *superconductivity,* and the electrical resistance drops to zero! In this state, an electrical current established in a superconducting loop would persist indefinitely, with no resistive losses. Currents of several hundred amps induced in a superconducting lead ring have been observed to persist constantly for several years. In a sense, the conducting electrons of a superconductor never collide with the lattice. However, at normal operating temperatures we are always faced with resistance and energy loss.

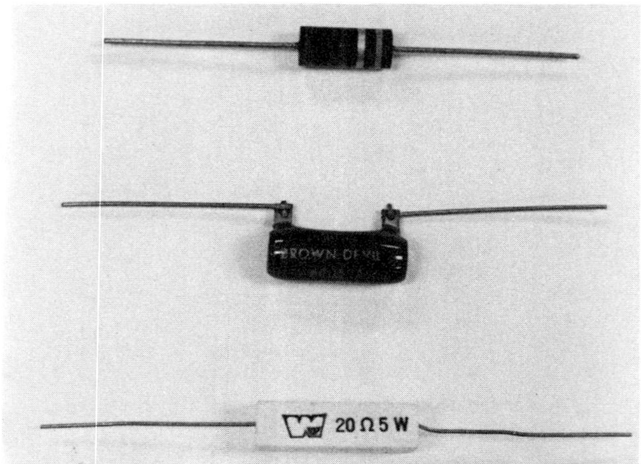

Figure 19.10 Commercially available resistors. The value of the top resistor is given by color-coded bands.

(a)

(b)

Figure 19.11 Power ratings. (a) The light bulb uses 60 watts (60 J/s) at 120 V. (b) The power rating is expressed in terms of current and voltage *($P = IV$)*.

are represented by symbols in the circuit diagram. It is customary to designate the *conventional current I* in the direction that positive charges would flow, even though the actual electron net flow is in the opposite direction—perhaps a leftover from Franklin's single-fluid theory.

The unit of resistance is the ohm (Ω, a capital Greek omega, used to avoid confusion with a capital O or zero). If the battery in Figure 19.9 were 12 volts and the lamp resistance were 6Ω, then by Ohm's law the current in the circuit would be $I = V/R = 12\text{V}/6\Omega = 2$ amps. (The connecting wires are considered to have negligible resistance.) Commercially available resistors (Fig. 19.10) are often put in circuits to affect the current flow.

It should be kept in mind that not all conductors are "ohmic" or follow Ohm's law. In general, metals are ohmic conductors, but some other materials, such as carbon, are not.

Electric Power

When an electric current flows, the electric charge moves from a higher to a lower potential, and work is done. This results in the conversion of energy from one form (electrical energy) to another. Examples of these other forms include heat energy in an electric toaster, radiant energy in a light bulb, and mechanical energy in turning a motor.

Recall that the rate at which work is done (or electrical energy is dissipated) is power—work/time (or energy/time). **Electric power** is given by the product of the current and the voltage:

$$\text{Electric power} = \text{current} \times \text{voltage}$$

or†

$$P = IV$$

Electric power is expressed in watts (W) or kilowatts (kW). The power requirements or wattage ratings are usually indicated on electrical components (Fig. 19.11). For example, a 60-W light bulb uses or dissipates 60 joules of electrical energy per second. Unfortunately, all of this current does not go into light (radiant) energy. About 95 percent is dissipated as heat and only 5 percent as light. The filament temperature of a common light bulb is on the order of 2500 to 3000°C. The electrical heating effect is sometimes called "joule heat" or I^2R losses.‡ The heat-energy conversion

† $IV = (\text{charge/time}) \times (\text{work or energy/charge})$
$= \text{work or energy/time} = \text{power} = P$.

‡ Ohm's law is $I = V/R$ or $V = IR$, and since $P = IV$ we can write the equation for power as (a) $P = IV = I(IR) = I^2R$ or (b) $P = IV = (V/R)V = V^2/R$.

comes about from electrons colliding with the lattice atoms or ions of a conductor and the transfer of energy, which increases the internal (heat) energy of the conductor (Chapter 13).

In applications in which heating effects are wanted, as in hair dryers and electric heaters, joule heating is promoted. This is done by having a heating element of relatively low resistance (but much higher than that of the connecting wires) so there will be a large current to take advantage of the square of the current in the I^2R losses. Another way of seeing this is through the equation for power in the form of $P = V^2/R$ (see footnote). For a fixed voltage, when the resistance is decreased, the power increases. A typical hair dryer may have a wattage or power rating of 1200 W. This is the energy consumption (per time) of the hair dryer. Most of this energy goes into joule heat in the dryer heating element. Some is used to turn the motor of the hair-dryer blower. The connecting wires in the hair-dryer cord also dissipate joule heat, but the resistance of the copper wire is only a very small fraction of an ohm. Even with a large current flowing in the hair-dryer circuit, the I^2R loss of the cord is a relatively small percentage.

Typical power requirements for some common household appliances are given in Table 19.1. Simply dividing the wattage rating by 120 V gives the current an appliance draws or uses ($I = P/V$).

Table 19.1 Typical Power and Current Requirements of Some Common Household Appliances

Appliance	*Power*	*Current ($I = P/V$) $V = 120$ V*
Air conditioner		
room	1500 W	12.5 A
central	4500 W	37.5 A
Blanket, electric	180 W	1.5 A
Blender	800 W	6.7 A
Coffee maker	1625 W	13.5 A
Dishwasher	1200 W	10.0 A
Food processor	330 W	2.75 A
Heater		
portable	1400 W	11.7 A
water	4500 W	37.5 A
Iron	1100 W	9.2 A
Microwave oven	625 W	5.2 A
Refrigerator		
regular	400 W	3.3 A
frost-free	500 W	4.2 A
Stove		
range top	6000 W	50.0 A
oven	4500 W	37.5 A
Television		
black-and-white	50 W	0.42 A
color	100 W	0.83 A
Toaster	950 W	7.9 A

Electric Circuits

As we know, a sustained flow of current requires a complete path or circuit (along with a voltage source). When there is a break in the path, we say that it's an open circuit. Switches are convenient means of opening and closing circuits and are commonly used to turn circuit components, like appliances, on and off.

It is also important to know how components are connected in a circuit. There are two basic ways of doing this, in *series* and in *parallel*. Each has different voltage and current characteristics and different applications. For example, are the lights in a circuit in your home wired in series or in parallel? Let's take a look and see.

SERIES CIRCUITS

As the name implies, the components in a series circuit are in an in-line series so the same current flows through all the components. This is shown for a light bulb circuit in Figure 19.12. If you trace around the circuit with your finger, you will note that it goes through all the bulbs, just as the current flow does.

Suppose there was a 12-V battery in the circuit "pumping" the current or "raising" the charge to a higher potential difference or voltage. The individual voltage "drops" across the lamp resistances would not be 12 V, but they would add up to 12 V: $V = V_1 + V_2 + V_3 = 12$ V. We express this by saying that the voltage drops around a circuit are equal to the voltage rise (of the battery). This is really just a statement of the conservation of energy. If all the lamps were the same, there would be a 4-V voltage drop across each. This is analogous to the potential energy or height drop shown in the water circuit in Figure 19.13. The total resistance of a series circuit is the sum of the individual resistances. For example, if R_1, R_2 and R_3 were 10Ω, 15Ω, and 30Ω, respectively, the total (equivalent) resistance would be 55Ω.§ This means that the three resistances could be replaced with a single 55Ω resistance—the battery wouldn't know the difference.

§ In a series circuit the total resistance is the sum of the individual resistances, $R_s = R_1 + R_2 + R_3$ in this case.

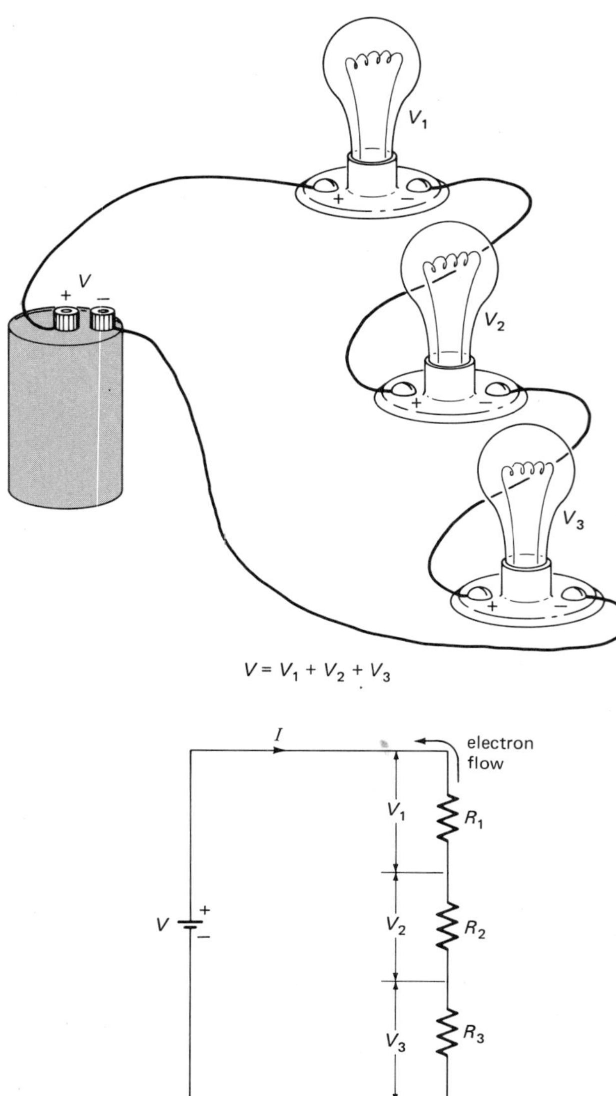

Figure 19.12 Series circuit. The current flow through each light bulb (resistance) is the same, and the voltage "drops" across the light bulbs add up to the voltage "rise" of the battery.

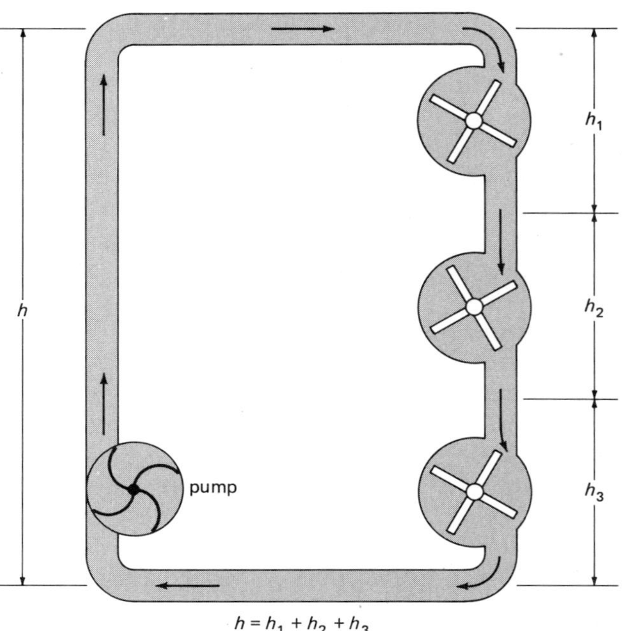

Figure 19.13 A liquid circuit with paddle wheels (resistances) in series. Compare with the electrical circuit in Figure 19.12.

Now suppose you hooked light bulbs in series in a household circuit with the 120-V source provided by the electric company. You'd have a problem. The light bulbs are rated for a certain wattage at 120 V, but there would only be 40 V across each, so they would glow very dimly. What would happen if one of the light bulbs blew out? The circuit would be opened and no current would flow, so all of the lights would go out. This wouldn't be very practical in a house circuit, even if the voltage were adequate. How would you like to have a light bulb in a lamp in a room blow out, and all the other lights on that circuit go dead, along with a radio you may have plugged into the circuit? Such problems are avoided by wiring circuits in parallel.‖

PARALLEL CIRCUITS

In a parallel circuit, all the components have a common connection to the high-potential side of the voltage source and a common connection to the low-potential side (Fig. 19.14). In this manner, the light bulbs are connected in "parallel" branches instead of being in a series. The voltage drop across each bulb is the same and is equal to the voltage rise of the battery, $V = V_1 = V_2 = V_3$.

However, the current from the voltage source divides at the common junction, and a fraction of the current flows through each bulb. If the bulbs are the same and have equal resistances, each conducts one third of the total current (conservation of charge). If the bulbs are different, the greatest part of the current flows

‖ Do not get the idea that series circuits are totally impractical. There are many electrical applications other than household circuits in which series-circuit characteristics are wanted.

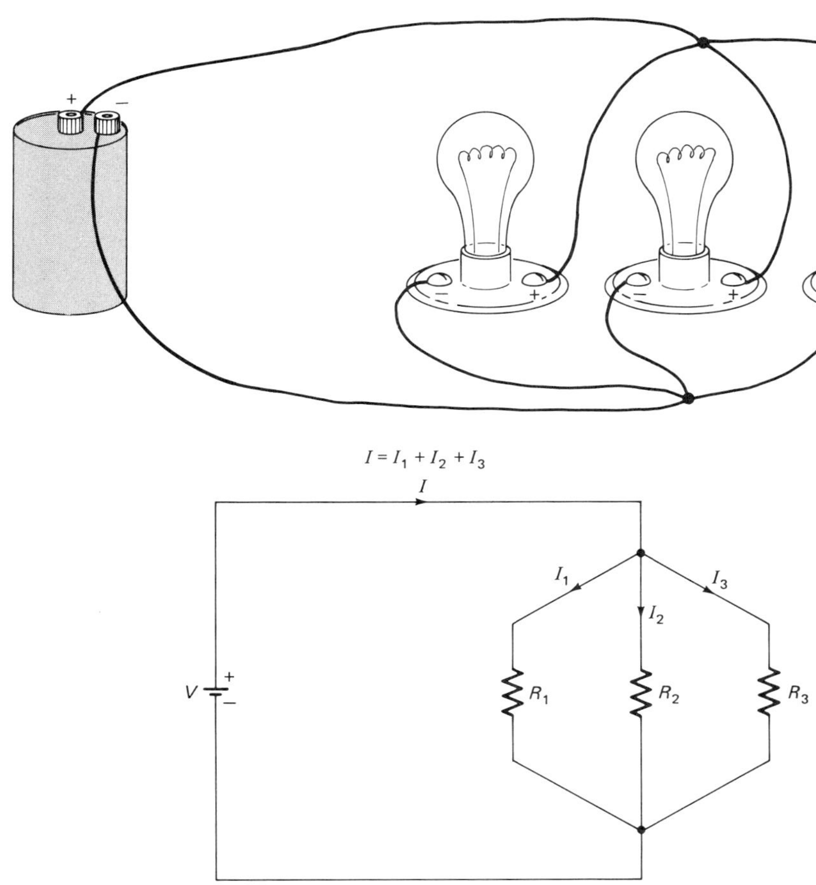

Figure 19.14 Parallel circuit. The voltage drop across each light bulb (resistance) is the same, and the current from the battery divides among the bulbs according to their resistances.

through the bulb with the least resistance—following the path of least resistance. Not all the current follows the path of least resistance; it divides proportionately. A parallel water circuit is shown in Figure 19.15. Think of what happens to the water flow from the pump when it reaches the common parallel pipes. It divides proportionately among the pipes, but the potential drop for each is the same.

In a parallel circuit with a 120-V source, we don't have the previous problem of a (household) series circuit. The voltage across all the bulbs in parallel is 120 V, which is what they are rated for. If one bulb blows out, the circuit paths through the other bulbs are not disrupted, and the bulbs remain lighted. Because of these characteristics, household circuits are wired in parallel. When you plug an appliance into a wall socket, you are connecting it in parallel into a circuit. A household circuit is illustrated in Figure 19.16. Most appliances operate on 120 V. The two power lines coming into a house are at voltages of +120 V and −120 V.

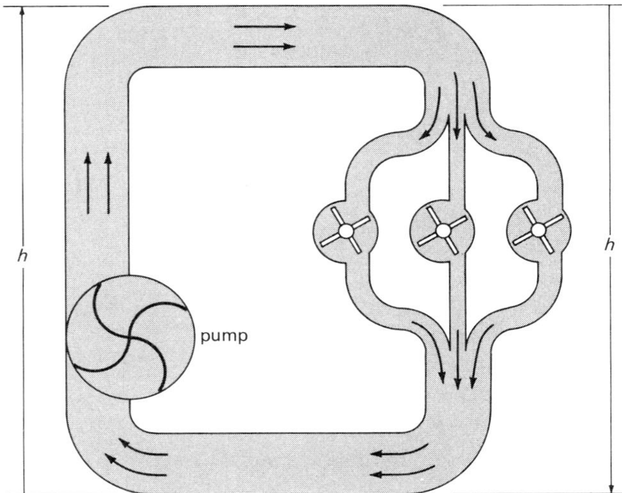

Figure 19.15 A liquid circuit with paddle wheels (resistances) in parallel. Compare with the electrical circuit in Figure 19.14.

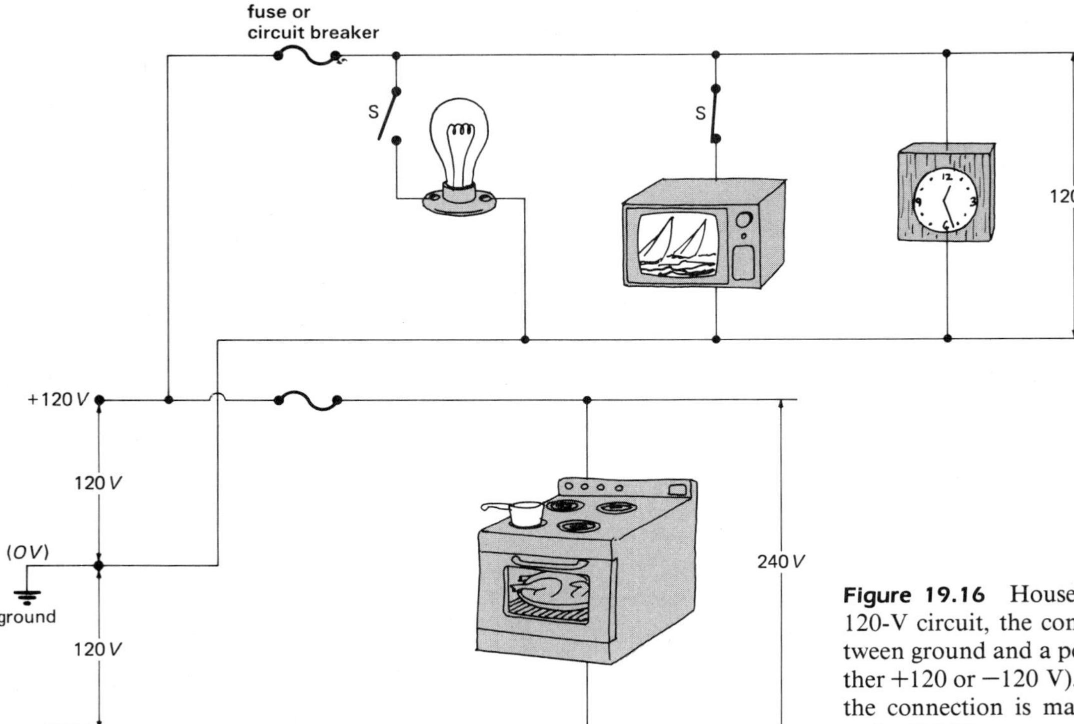

Figure 19.16 Household circuits. For a 120-V circuit, the connection is made between ground and a potential of 120 V (either +120 or −120 V). For a 240-V circuit, the connection is made between +120 V and −120 V.

Connecting between one of these "hot" lines and ground (0 V) gives a potential or voltage difference of 120 V. Connecting across both lines gives a voltage difference of 240 V, which is needed for the power requirements of some air conditioners and electric stoves.¶

Notice the fuse in the circuit in Figure 19.16. This is an important safety feature, as will be learned in the next section. (A circuit breaker serves the same purpose.) You might think that the more appliances or resistances you put in a parallel circuit, the lower the current flow would be. But, oddly enough, the more individual resistances added to the circuit, the less the total resistance—sort of "opposite" to a series circuit. This can cause problems with large current flows and joule heating, as discussed in the Electrical Safety section that follows.

In fact, the total resistance of the circuit is always less than the smallest appliance resistance. For example, suppose the parallel resistances R_1, R_2, and R_3 in Figure 19.14 were 10Ω, 15Ω, and 30Ω, respectively. The total resistance of the arrangement is 5Ω.** That is, the three resistances could be replaced with a 5Ω resistance without the battery knowing the difference. Recall that these three resistances connected in series had a total resistance of 55Ω, so more current flows in the parallel circuit ($I = V/R$).

QUESTION: Are Christmas-tree lights wired in series or in parallel?

ANSWER: Most of the newer strings of Christmas-tree lights have a shunt resistance in the bulb wired in parallel with the bulb filament. Each of these parallel-bulb combinations is wired in series along the string. In this manner, when one bulb blows out, the others remain lit because the shunt resistance still completes the circuit at that point.

At one time, Christmas-tree lights without shunt resistors were wired directly in series. When

¶ When regular household voltage is considered to be 110 V, this difference is 220 V.

** In a parallel circuit, the total resistance R_p is given by $1/R_p = 1/R_1 + 1/R_2 + 1/R_3 + \ldots$. In this case, $1/R_p = \frac{1}{10} + \frac{1}{15} + \frac{1}{30} = \frac{3}{30} + \frac{2}{30} + \frac{1}{30} = \frac{6}{30}$, and $R_p = \frac{30}{6} = 5\Omega$.

one bulb blew out on these older strings of bulbs, they all went out. Then you had to hunt for the bad bulb to replace it. There was big trouble when more than one bulb blew at once.

Electrical Safety

The importance of electrical safety cannot be overstressed. Electrical accidents can result in property damage, personal injury, and death. This does not mean that you should fear electricity, just that you should use it properly and safely.

The common methods of protecting against property damage through circuit overloads (too much current) and overheating are fuses and circuit breakers. All electrical circuits in the home are required to be protected by these means. When too much current flows in a circuit, it becomes hot (joule heat) and perhaps melts the insulation and starts a fire. An overload may also burn out and damage appliances. Electronic equipment commonly has fuses to protect the component parts from overloads. Even your automobile circuits are fused for overload protection.

A **fuse** is essentially a short piece or strip of metal with a low melting point. When the current in a fused circuit exceeds the fuse rating, for example 15 or 20 amps, the joule heat melts or vaporizes the fuse strip. The fuse "blows" and the circuit is opened.

Another common problem is that the insulation on wires may become worn, for example, on an extension or appliance cord. If the bare wires touch each other or if a high-voltage or "hot" wire touches ground, this is called a short circuit, since the path of the circuit is effectively shortened. This provides a low-resistance path, and a large current flows, which also blows the protecting fuse.

There are various types of fuses. The common type in older homes is the Edison-base fuse (Fig. 19.17). Its base is the same as that of an ordinary light bulb. Edison-base fuses are interchangeable in the fuse socket; for example, a 30-amp fuse can be placed in a 15-amp circuit. Because of this, Edison-base fuses are no longer permitted in new installations. To prevent interchanging, a nontamperable fuse is used. This fuse, called a "Type-S" fuse (trade name "Fustat") will not fit or screw into an Edison-base socket. An unremovable adapter is installed in the socket. Each differently rated Type-S fuse has an adapter with different-sized threads. As a result, a 30-amp Type-S fuse will not screw into a 15-amp adapter (Fig. 19.17).

It is more common now to use **circuit breakers** in place of fuses (Fig. 19.18). If the current in a circuit exceeds a certain value, the breaker is activated, and a magnetic relay (switch) breaks or opens the circuit. The circuit breaker can be "reset" or closed by a manual switch (see Chapter 20, Fig. 20.11).

In either case—whether a circuit is opened when a fuse blows or when a circuit breaker "trips"—steps

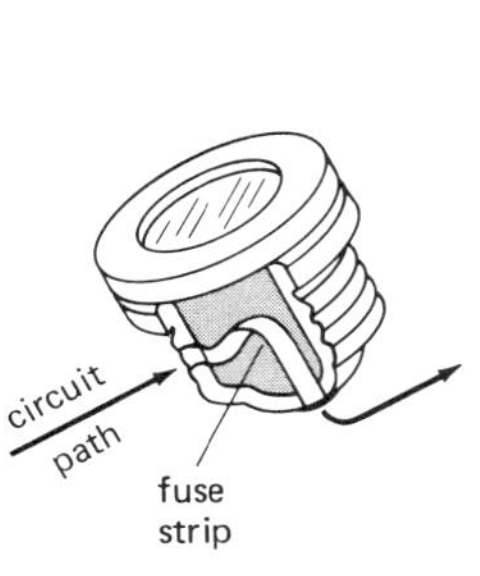

Figure 19.17 Fuses. A fuse "blows" or opens the circuit when the joule heat from a current larger than the rated value melts the fuse strip. Edison-base fuses (one shown in lower left of photo) are interchangeable, and any value of fuse can be put into a socket. Type-S fuses (top of photo) have an adapter installed in the socket specific for a rated fuse so they cannot be interchanged. Notice the different threads on the fuses.

Figure 19.18 A circuit-breaker panel. If the current in a circuit exceeds a certain value, the breaker is activated, and a magnetic relay (switch) "breaks" or opens the circuit. The circuit breaker can be "reset" or closed by a manual switch.

should be taken to find out why and to remedy the problem. Remember, these are safety devices, and when they open a circuit they are telling you something—namely, that the circuit is overloaded or shorted.

Switches, fuses, and circuit breakers are always placed in the hot (high-voltage) side of the line, so as to interrupt power flow to the circuit element. However, fuses and circuit breakers may not always protect you from electrical shock. An example is illustrated in Figure 19.19. A hot wire inside a tool (e.g., an electric drill) or appliance—for example, from the winding of a motor—may come into contact with the metal frame or housing. If you touch the frame, watch out. See Special Feature 19.2 for some possible effects.

To prevent this, a grounding wire is used. The circuit is then completed (shorted) to ground, and the fuse in the circuit is blown. This is why many electrical tools and appliances have three-prong plugs (Fig. 19.20). The grounding prong is usually U-shaped. In the wall receptacle, this connection runs to ground. The use of a three-prong plug in an older-type "two-hole" socket requires an adapter. The grounding lug, or in some cases a wire, of the adapter should be connected to the

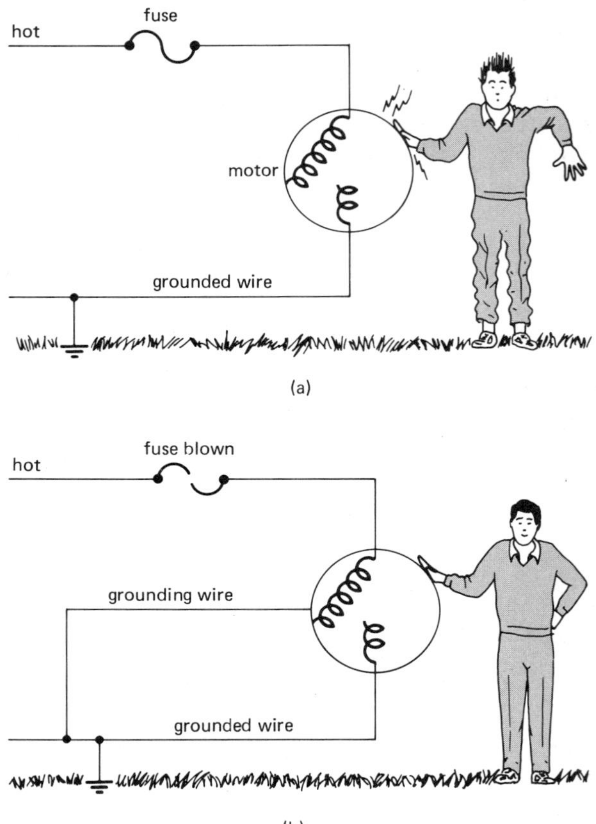

Figure 19.19 Grounding. (a) A hot wire in contact with the metal casing of a motor or tool, which is not connected to ground, gives a potentially dangerous situation. (b) If the casing is grounded, the circuit would be completed to ground and the fuse blown. This is the purpose of the grounding prong of a three-prong plug.

SPECIAL FEATURE 19.2

Personal Safety and Electrical Effects

Personal safety in working with electricity involves common sense and a fundamental knowledge of electricity. The main thing to remember is not to become part of a circuit yourself. As with any other circuit component, the current drawn by the human body depends on its resistance and the voltage source, $I = V/R_{body}$. If the skin is dry, there is usually a high resistance, and an electrical shock may only be an uncomfortable and surprising tingle. However, if the skin is wet—for example, with perspiration—the resistance may be low enough to allow injurious and fatal current flows. Table 19.2 shows that only milliamps are needed.

If you come into contact with a "hot" wire, say at 120 V, current could flow through parts of your body. If the circuit is completed through your hand (finger-to-finger), a shock and perhaps a burn can result. However, if the circuit is completed through the body (hand-to-hand or hand-to-foot), there can be serious results, depending on the amount of current flow. One may lose muscle control and not be able to let go.

Table 19.2 The Effects of Electric Currents on Humans*

Current (mA)	*Effect*
1	Barely perceptible
5–10	Mild shock
10–15	Difficult to let go
15–25	"Muscular freeze"; cannot let go
25–50	Breathing difficulty
50–100	Breathing may stop; ventricular fibrillation
>100	Death

* Varies with individuals

(Muscles are controlled by nerves, which are activated by electrical impulses.) Larger amounts of current can cause breathing difficulties, and the heart may have uncontrolled contractions (ventricular fibrillation). If the current is greater than 100 mA (0.1 A), death can result. This isn't much current on a relative basis. A 60-watt light bulb operating on 120 V draws $I = P/V = \frac{60}{120} = 0.5$ amp of current. Keep in mind that it is the current that kills, not the high voltage itself.

Figure 19.20 A three-prong plug and adapter. The adapter has a grounding lug or wire that should be connected to the grounded receptacle by means of the plate-fastening screw. Otherwise, the whole purpose of the grounding system is defeated.

grounded receptacle by means of the plate-fastening screw. Many times people do not ground the grounding lug, which defeats the whole purpose of the safety grounding system.††

Have you ever tried to plug something in and the plug wouldn't go, but when you turned it over, it fit? If you take a look at a wall receptacle, you'll find that one of the slits is bigger than the other. If you look at the plug, you will find that one of the prongs is bigger than the other (Fig. 19.21). This is called a polarized plug. *Polarizing* in the electrical sense refers to a method or identification by which proper connections can be made. The original intent of this type of plug was as a safety feature. The small slit in the receptacle is the hot side and the large slit the neutral or ground side, if properly connected. The housing of an appliance could then be connected to the ground side all the time by

†† When connecting the grounding lug to the plate-fastening screw, it is assumed that the receptacle box is grounded by means of a third dedicated grounding wire (not the zero potential *ground* wire which is a current-carrying wire in the circuit). If this is not done, the three-prong plug does not serve its intended purpose.

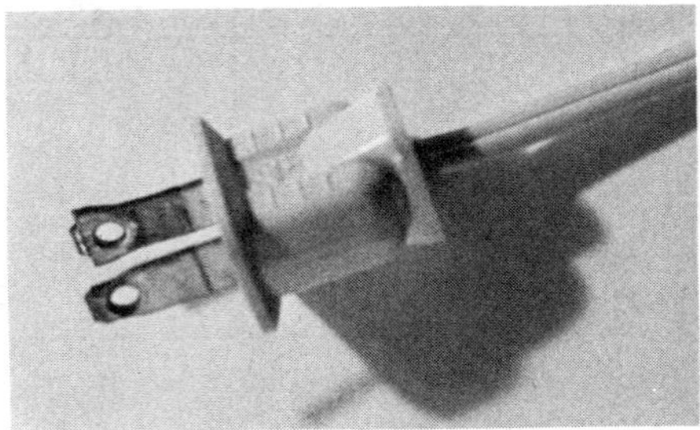

Figure 19.21 A "polarized" plug—an early attempt at electrical safety that is still with us. It's a good backup system *if* the receptacle and appliance are wired properly.

means of the polarized plug. The effect would be similar to that of a three-prong plug. Then, if a condition such as in Figure 19.19 occurs, the hot wire is shorted to ground, which blows a fuse or trips a circuit breaker.

However, this system left too much to human error. If the receptacle or the appliance is not wired (polarized) properly, a dangerous situation can exist. The polarization is insured with a "dedicated" third grounding wire as in a three-prong plug system, which is the accepted safety system. The original two-prong polarized plug system remains as a general backup safety system, *provided* it is wired properly.

SUMMARY OF KEY TERMS

Electric current a movement or flow of electric charges that does work through energy transport. Units of electric current are coulomb/second or ampere (A, commonly stated as "amps" for short).

Voltage source an electrical "pump" that supplies a potential or voltage difference for the flow of current, commonly a battery or generator. The unit of voltage (difference) is the volt.

Direct current (dc) electric current that flows only in one direction because of the fixed polarity of the voltage source, e.g., a battery.

Alternating current (ac) electric current in which the charge motion alternates back and forth owing to a changing polarity of the voltage source.

Battery a single cell or series of cells that converts chemical energy into electrical energy.

Electrical resistance the opposition of a material to electric current flow, which is measured in ohms (Ω).

Ohm's law the relationship of the current, potential difference (voltage), and resistance in many conductors.

$$\text{Current} = \frac{\text{voltage}}{\text{resistance}}$$

or

$$I = \frac{V}{R}$$

Electric power the time rate of electrical energy transfer or rate of doing work, which is given by the product of the current and voltage. The unit of power is the watt (W).

$$\text{Electric power} = \text{current} \times \text{voltage}$$

or

$$P = IV$$

Series circuit a circuit in which the components (such as resistances) are arranged in a series such that the same current flows through each.

Parallel circuit a circuit in which the components (such as resistances) are arranged in parallel branches with common junctions such that the current divides among the branches.

Electrical safety important procedures, practices, and devices that prevent electrical accidents that could result in property damage and personal injury.

EXERCISES

1. People often say that they left the lights in the house "burning." Is this a correct statement? How do you think it might have originated?
2. If a wire is connected to two objects and current flows in the wire, what do you know about the electric potentials of the objects? Which way will the current flow?
3. If a net charge of 10 coulombs passes through the cross-sectional area of a conductor in 2 seconds, how much current flows in the conductor?
4. Voltage sources "pump" energy into a circuit. Where does the energy go?
5. We are told not to waste electricity. How is electricity "wasted"?
6. What is electromotive force? What is a "seat" of electromotive force?
7. If the drift velocity of electrons in a conductor is so small, why does an auto battery influence the starter as soon as you turn on the ignition switch?
8. Since the charge moves back and forth in alternating current, there is no net current flow. What "flows" in an ac circuit, and what does this do to give electric-current effects, say, in a light bulb?

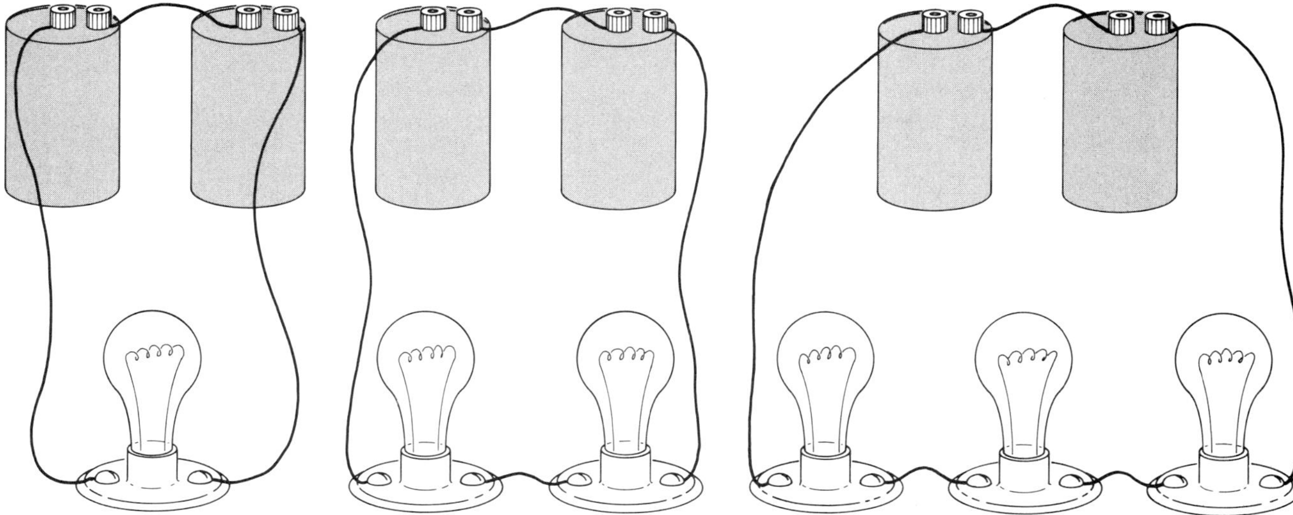

Figure 19.22 See Exercise 13.

9. How does a battery complete a circuit with regard to current flow?
10. Suppose that two batteries in series both are placed backward in a flashlight so the base of one of the batteries makes contact with the bulb. Would the flashlight work? Explain.
11. On a very cold day, a car battery is very sluggish in starting the engine because the current output of the battery is reduced. Why is this?
12. If you had two 1.5-V D-cells and another 9-V dry cell, what would be the maximum voltage you could get by connecting the batteries together, and how would you connect them?
13. For the three circuits shown in Figure 19.22, in which case would the bulb(s) be (a) the brightest and (b) the dimmest, and why? (All the batteries and all the bulbs are the same.)
14. How are the circuit contacts made to an ordinary light bulb?
15. If you switched the lead wires to a circuit rapidly back and forth between the terminals of a battery, would there be an "alternating" current in the circuit? Explain.
16. When you plug in a lamp in the home and turn it on, do electrons flow through the plug to the lamp? Explain.
17. A closed switch in a circuit has very little resistance. What is the resistance of an open switch?
18. Dim lights set certain moods. One way of dimming lights is by using a variable-resistance dimmer switch, which usually has a rotating knob. What causes the lights to dim when the knob is rotated? (*Hint:* Think in terms of Ohm's law.)
19. Which has the greater resistance, a 75-W or a 100-W light bulb? If the tungsten filament in each bulb were the same length, how would they otherwise differ?
20. A 75-W light bulb and a 100-W light bulb are connected in a circuit, as shown in Figure 19.23. Which bulb has the greater current through it? Which bulb has the greater voltage drop across it?
21. Sketch a battery circuit that has a 75-W light bulb and a 100-W light bulb in series. Which bulb has the greater

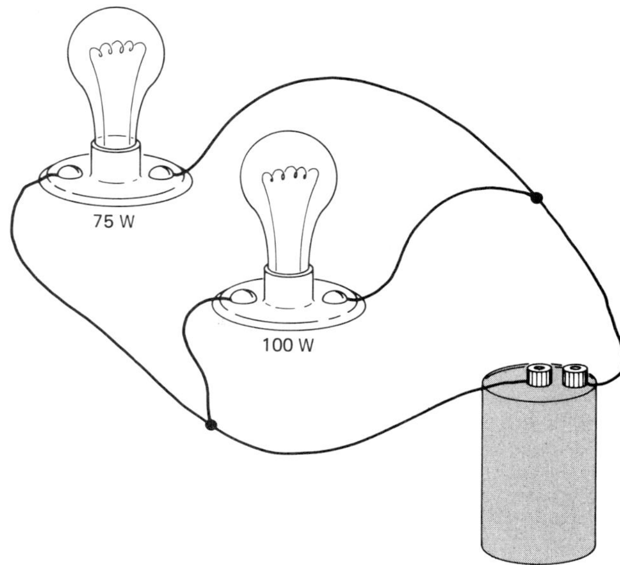

Figure 19.23 See Exercise 20.

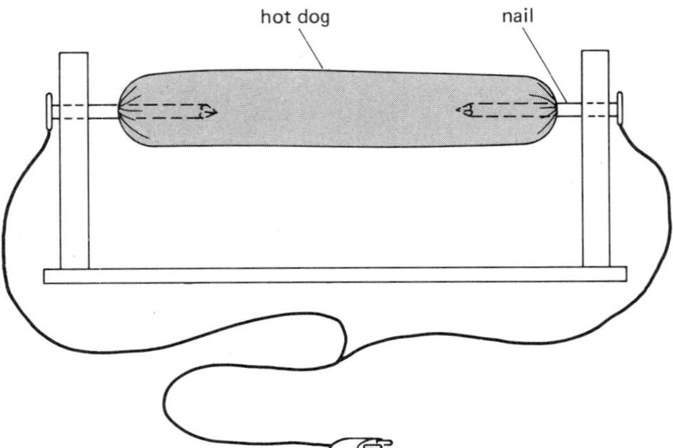

Figure 19.24 See Exercise 25.

current through it? Which bulb has the greater voltage drop across it?

22. A hair dryer rated at 1200 W operating on 120 V draws 10 amps of current ($I = P/V$). The dryer cord has a resistance of 0.02 ohms. How many watts are dissipated in the cord as joule heat? What fraction is this of the total wattage?
23. Although the connecting wires of a circuit have small resistances, they still have I^2R losses and heat up. How does this affect the resistance of a wire? How does the change in resistance affect the joule heat?
24. How much more power does an electric can opener use (rating shown in Fig. 19.11) than a 100-W light bulb?
25. A hot dog can be cooked by "plugging it in," so to speak, see Figure 19.24. (This is a lecture demonstration to be done by the teacher and should *not* be attempted at home. It can be dangerous.) It takes only a minute or so to be heated. There are now commercially available hot-dog cookers that operate on the same principle and cook several hot dogs at one time. What causes a hot dog to cook this way? Would the hot dogs be "wired" in series or in parallel in the cooker?
26. Are auto headlights wired in series or in parallel? How could you find out if you weren't sure?
27. Some light bulb questions (several billion bulbs are used in the United States each year):
    (a) The gas in a light bulb is not air, but a mixture of argon and nitrogen at low pressure. Why isn't air used? Wouldn't it be cheaper?
    (b) After long periods of use, a gray spot develops on the inside of the bulb. (Check this out on a used bulb.) What is the gray spot? (*Hint:* It's metallic.)
    (c) Why does too much current cause a bulb to blow out?
28. Are the ceiling lights in a room all that is in that circuit, or are there usually some wall plugs in the circuit, too?
29. What do you think would happen if a 120-V appliance were plugged into a 240-V outlet? (The plugs are different, so this can't normally be done.) How about a 240-V appliance plugged into a 120-V outlet?
30. An electric wall receptacle is sometimes called an outlet. What does it let out?
31. If two 6-V batteries connected in series are in a circuit with a resistance of 4 ohms, (a) how much current flows in the circuit? (b) If the batteries were connected in parallel, how much current would flow in the circuit? (c) If another 2-ohm resistance were placed in the circuit in series with the other resistance, how much current would flow in the circuit with the batteries in parallel and series connections? (d) What is the voltage drop across the resistances in each case?
32. A 100-W color television is used on the average of 2 hours each day. If the cost of electricity is 6 cents per kilowatt-hour, how much of your monthly (30 days) electric bill is for television use?
33. What six items are the biggest users of electrical energy in a typical home? (See Table 19.1, but take hours of use into consideration.)
34. Why is a switch placed in the high-voltage side of a circuit?
35. Why are Edison-base fuse boxes no longer used in new home installations?
36. A foolhardy and dangerous thing to do when no spare Edison-base fuses are available is to put a penny in the socket behind the blown fuse. Why is this dangerous?
37. Why can birds sit on high-voltage power lines (which are bare and have no insulation) and not be electrocuted?
38. Warning signs commonly state, "Danger—High Voltage" (Fig. 19.25). Wouldn't it be more appropriate to say, "Danger—High Current"? Explain.
39. If a power line happens to fall on a parked automobile in which you are sitting, you should sit still and not touch

Figure 19.25 See Exercise 38.

any metal parts of the car. Why? Wouldn't the power line be shorted to ground and trip some protective device in the line?

40. Some people cut off the grounding prong of a three-prong plug so it can be used in a two-prong receptacle. Is this a wise thing to do? Explain.
41. What's wrong with the wiring arrangement shown in Figure 19.26, and how should it be properly wired?
42. A warning on the handle of a hair dryer reads, "Danger. Electrocution possible if used or dropped in tub. Unplug after each use." Explain the reason for the warning.
43. An old saying used by people who work around electrical voltages is, "You should keep one hand in your pocket." What is the reason for this saying?

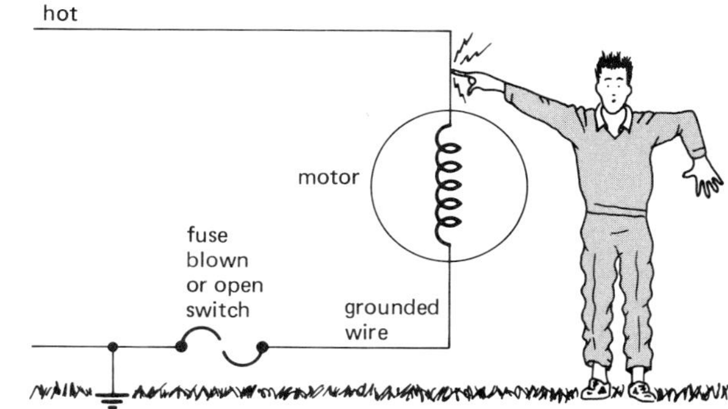

Figure 19.26 See Exercise 41.

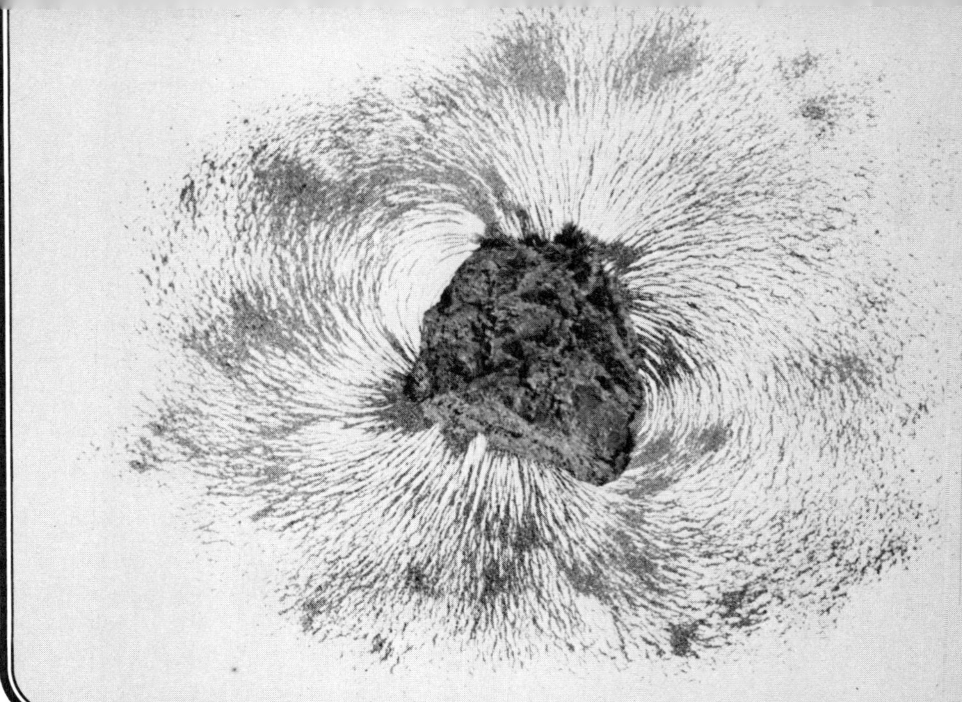

20

Magnetism

Electricity and magnetism are closely related. In fact, we will find they are inseparable, although we talk about electric forces and magnetic forces. Electric forces arise from stationary or moving electric charges, while magnetic forces involve moving electric charges. Since both forces ultimately arise from electric charges, the general term *electromagnetic force* is sometimes applied to both.

Electromagnetic forces are basic in the operations of motors, generators, and electric meters and in many other practical applications. The basic principles and applications of electromagnetism are the subjects of this and the next chapter.

Magnets and Magnetic Fields

Like most people, you have probably been fascinated by the attractive and repulsive forces between magnets —a hands-on example of force-at-a-distance. Magnets are readily available today because we know how to make them. But they were once quite scarce and existed only as natural magnets or rocks that were found in nature.

Natural magnets, called lodestones, were found as early as the sixth century B.C. in the province of Magnesia in ancient Greece, from which magnetism derives its name. Lodestones could attract pieces of iron and other lodestones. For centuries, the attractive properties of natural magnets were attributed to supernatural forces. Early Greek philosophers believed that a magnet had a "soul" that caused it to attract pieces of iron (see Chapter introductory photo).

Sometime around the first century A.D., the Chinese learned to make artificial magnets by stroking pieces of iron with a natural magnet. This led to one of the first practical applications of magnets, the compass, which implied that the Earth itself has magnetic properties. (The Chinese are said to have developed the compass, but several other peoples claim this invention, too.)

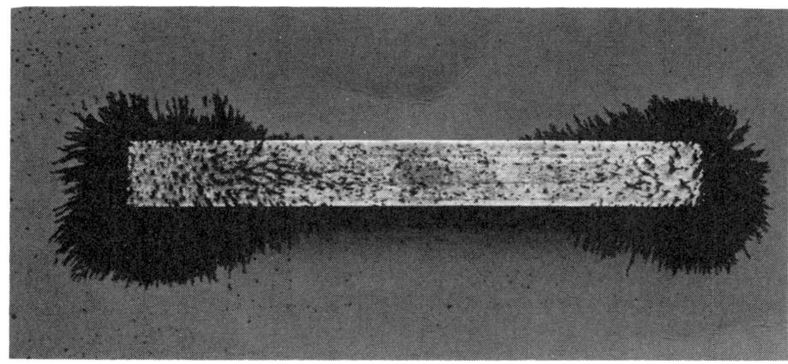

Figure 20.1 Magnetic poles. Regions of concentrated magnetic strength are called magnetic poles, as shown by attracted iron filings.

The most familiar magnets are the common bar magnet and the horseshoe magnet (a bar magnet bent in the form of a horseshoe). If a bar magnet is dipped into iron filings, the filings cling to the magnet in concentrations near the ends of the bar (Fig. 20.1). These regions of apparently concentrated magnetic strengths are called **magnetic poles.** The poles are distinguished as the north (seeking) pole and the south (seeking) pole. This comes from the properties of a compass. The north magnetic pole "seeks" and points north, and the south magnetic pole points south.

We have never been able to isolate a single magnetic pole, as we have done with electrical charges. That is, we have never found a magnetic "monopole" to occur by itself—the poles always occur together. If you break a magnet in two, you get two smaller dipole magnets. The reason for this is discussed later in the chapter.

When magnets are brought close together, it is quickly observed that the magnets attract each other in some cases and repel in others. This action is described by the **law of poles,** which is similar to the law of charges:

> Like magnetic poles repel, and unlike magnetic poles attract.

That is, N-N and S-S repel, and N-S (or S-N) attract, as illustrated in Figure 20.2.

Also as in the electric case, it is convenient to describe the magnetic force in terms of a magnetic field. The **magnetic field,** commonly called a B field, is the magnetic force per *pole.** (Recall that the electric field is the electric force per *charge.*) Its direction is determined by the direction of the force experienced by a north magnetic pole. For example, if a compass is placed in the vicinity of a magnet, the magnetic field can be mapped out and is in the direction of the north pole of the compass (Fig. 20.3).

The magnetic field can be "seen" by iron filing patterns. When iron filings are sprinkled on a piece of paper over a magnet, they become induced magnets and line up with the field, as shown in Figure 20.3.

QUESTION: According to the law of poles, the north pole of a compass is attracted towards the Earth's south magnetic pole. Why does it point north?

ANSWER: This is because the Earth's south

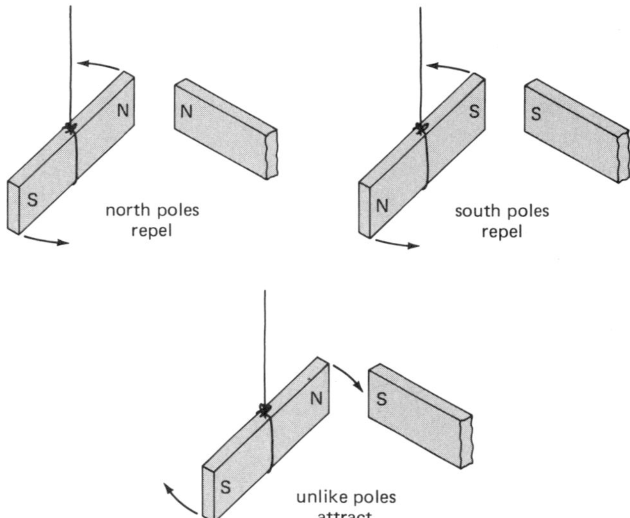

Figure 20.2 The law of poles. Like magnetic poles repel, and unlike poles attract.

* This is an older designation used here for illustration. Technically, the magnetic field is defind as the magnetic force "per moving charge," since magnetic fields are produced by moving electrical charges, as will be learned. The unit of the magnetic field in the SI system is the tesla (T), in honor of Nikola Tesla, a Yugoslav scientist who worked in the United States.

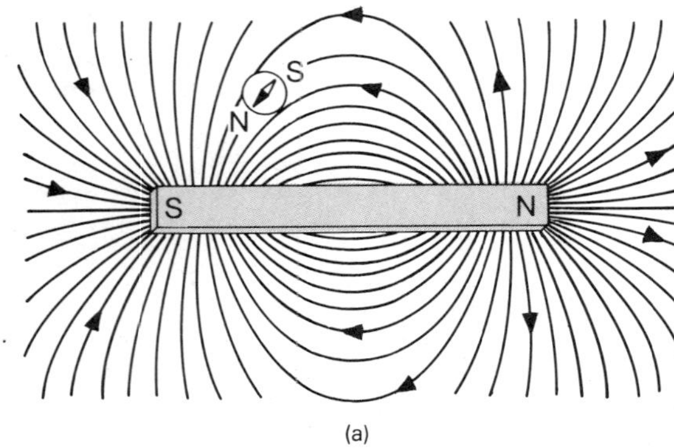

(a)

Figure 20.3 Magnetic fields. The magnetic field at a particular point is the magnetic force per pole in the direction of the force experienced by a north magnetic pole. (a) Iron filings line up with the magnetic field and show its pattern (b and c).

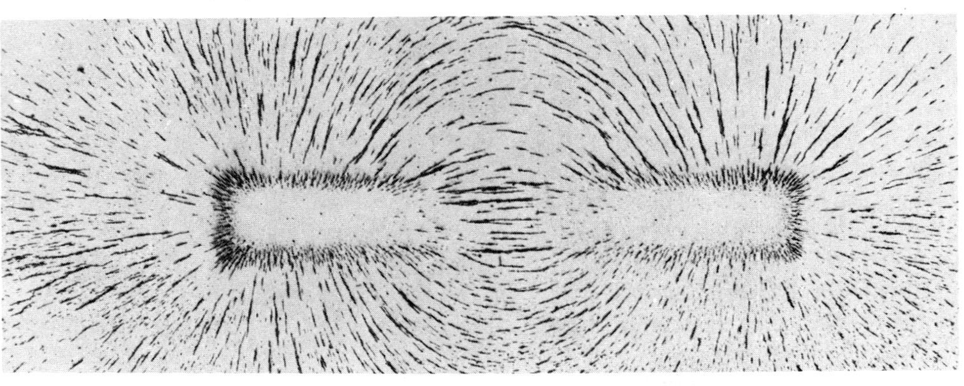

(b)

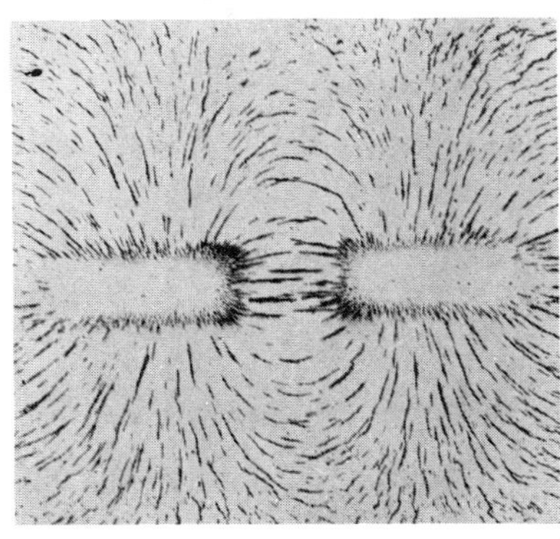

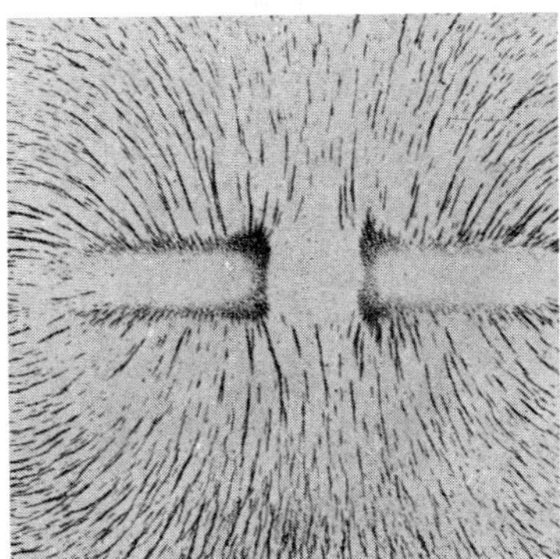

(c)

magnetic pole is near its *geographic* north pole. The Earth's magnetic field is similar to that of a bar magnet located in the Earth (Fig. 20.4). When a compass needle lines up with the Earth's magnetic field, its north pole points north (magnetic north), but toward a south magnetic pole. The Earth's magnetic poles do not coincide with the rotational geographic poles. The south magnetic pole is located north of Prince Wales Island, Canada, at about 75°N, 101°W. This is some 1600 km (1000 mi) from the geographic north pole (true north). The other magnetic pole is even farther removed from the south geographic pole, being located in Wilkes Land, Antarctica (69°S, 14°E).

The Earth does not have a big magnetized chunk of iron or a bar magnet as used in Figure 20.4 for illustration. It is believed that the Earth's magnetic field is associated with its liquid outer core and rotation. The magnetic poles also "wander" slowly and are not always in the same place. There is also evidence that the

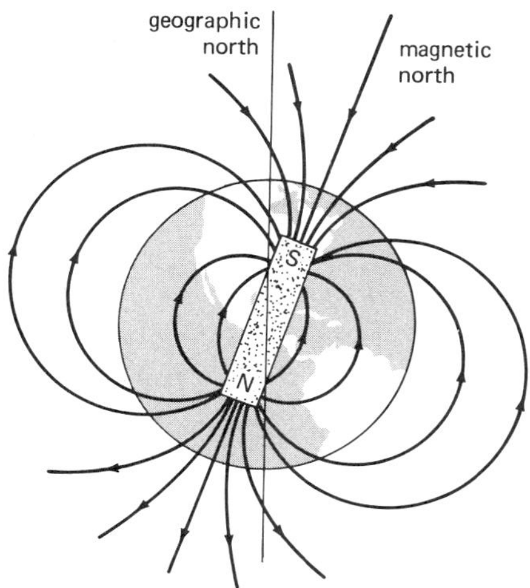

Figure 20.4 Earth's magnetic field. The magnetic field of the Earth is similar to that of an imaginary bar magnet inside the Earth, with the south magnetic pole at magnetic north of the compass.

Earth's magnetic poles have switched polarity at various times in the past, most recently about 700,000 years ago. During a period of pole reversal, the south magnetic pole is near the south geographic pole. We are not certain why these changes occurred, or if it will happen again. More about this later in Special Feature 20.1.

Electromagnetism

Now we can produce magnetic fields at will—even turn them on and off. In electrical experiments carried out in 1820, the Danish physicist Hans Christian Oersted discovered a relationship between electricity and magnetism. It was found that

> An electric current produces a magnetic field.

This can be shown by an arrangement illustrated in Figure 20.5. When the circuit is open and there is no current flow, the compass needle points in its normal north-south direction. However, when the switch is closed and there is an electric current in the circuit, the compass needle is deflected, indicating the presence of a magnetic field other than that of the Earth.

Using a compass to map out the magnetic field around a straight wire segment, we find that the field lines are in concentric circles around the wire (Fig. 20.6). The circular sense of the magnetic field lines is given by a right-hand rule. If the thumb of the right hand is in the direction of the *conventional* current flow (direction *positive* charges would flow), the fingers curl around the wire in the circular sense of the field. At any point on a field-line circle, the field direction is tangential to the circle.

The magnetic field around a current-carrying loop of wire is somewhat like that of a bar magnet (Fig. 20.7). If a number of loops of wire are wound in a tight coil, called a solenoid, the fields of the wire loops add, such that the field outside a current-carrying solenoid is very much like that of a bar magnet, as well as having a relatively uniform field inside the coil. There are important and practical differences, however. The strength of the magnetic field of a solenoid can be varied by varying the current, as well as by being turned on and off. You can reverse the magnetic field by reversing the direction of the current in the coil.

Electromagnetism gives us some clues to the Earth's magnetic field. See Special Feature 20.1.

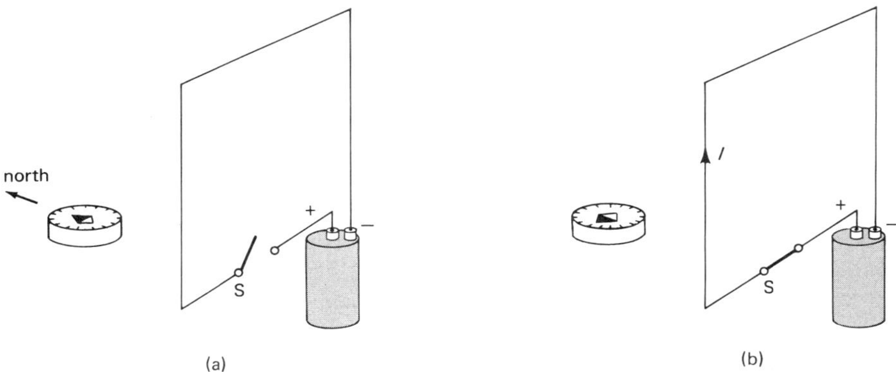

Figure 20.5 An electric current produces a magnetic field. When the switch is closed and current flows in the circuit, the compass needle is deflected, indicating the presence of a magnetic field.

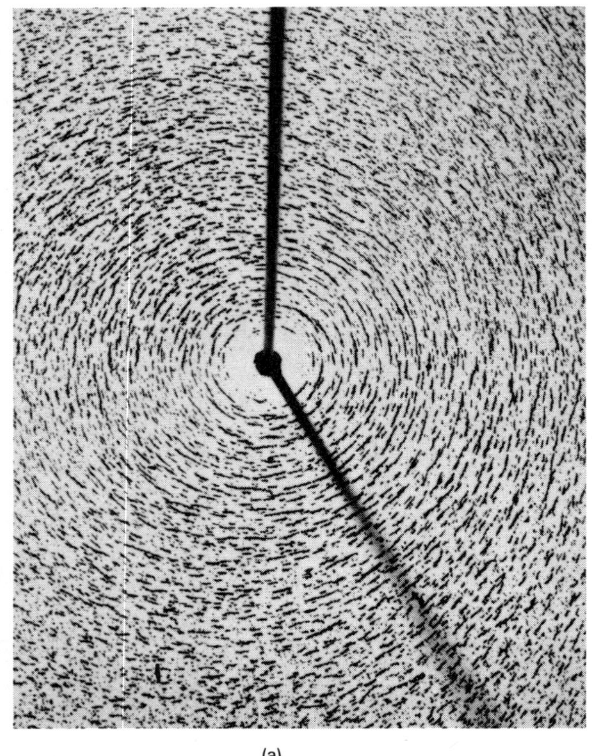

(a)

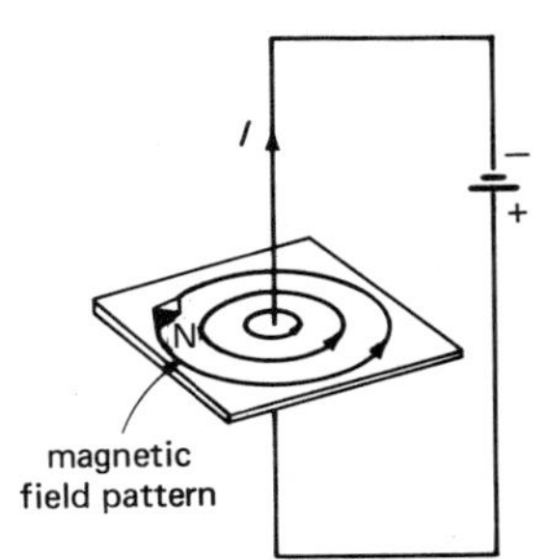

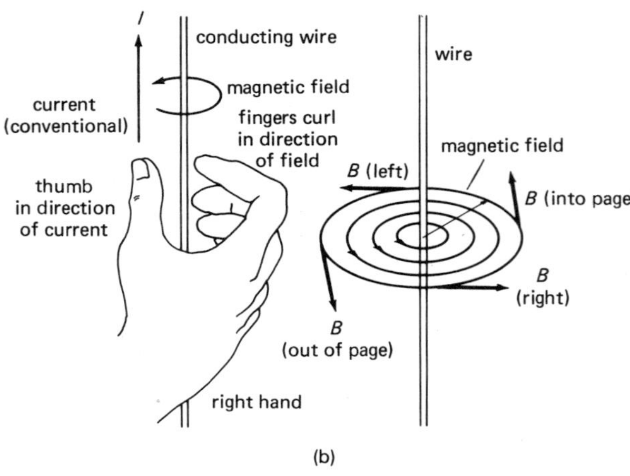

(b)

Figure 20.6 (a) The magnetic field is in circles around a current-carrying straight wire. (b) The circular sense of the magnetic field is given by a right-hand rule. The magnetic field vector is tangential to a circle at any point.

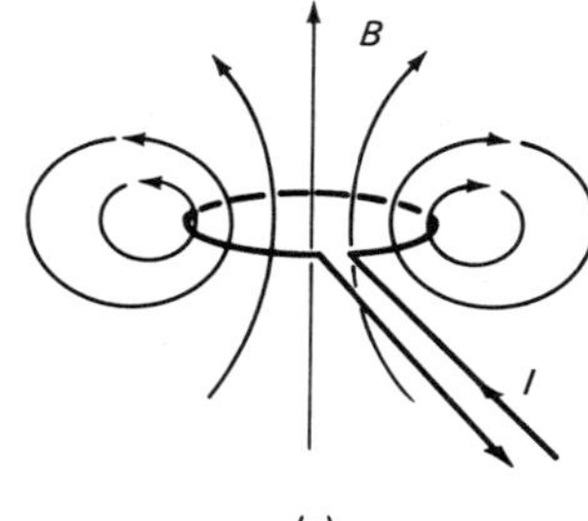

(a)

Figure 20.7 Magnetic fields around (a) a current-carrying loop and (b) a current-carrying solenoid. The effects of the loops of wire in a solenoid with a large number of loops give a relatively uniform magnetic field inside the coil. Notice that the fields of the loop and the solenoid are similar to that of a bar magnet.

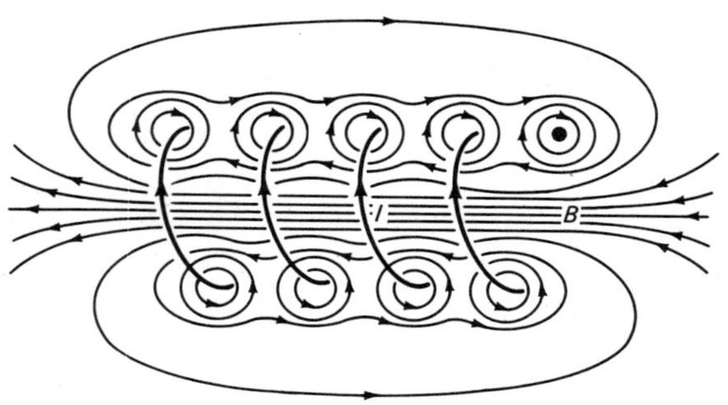

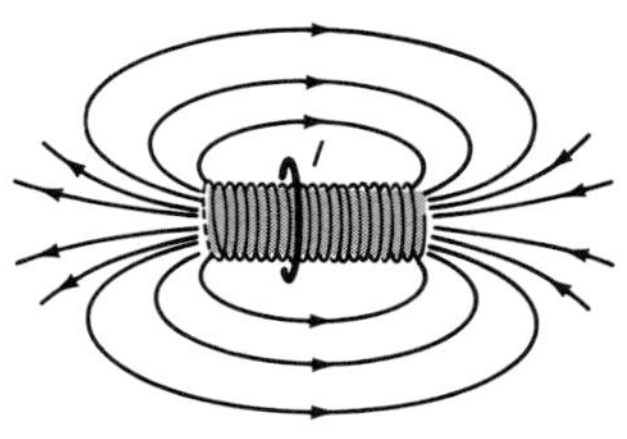

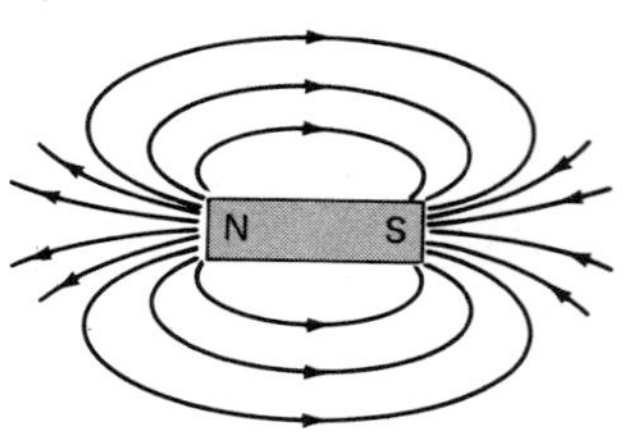

(b)

SPECIAL FEATURE 20.1

The Earth's Magnetic Field

We don't really know what causes the Earth's magnetic field. But the fact that a current produces a magnetic field leads scientists to speculate that there may be a flow of charge in the Earth's liquid outer core, which is chiefly iron, that gives rise to the magnetic field. This is thought to be associated with the Earth's rotation. Some of the other planets have magnetic fields, whereas others have none or very weak magnetic fields. For example, the magnetic fields of Saturn and Jupiter are about 1000 and 10 times that of the Earth's, respectively. These planets are largely gaseous and have fast periods of rotations of 10 to 11 hours. Venus and Mercury, on the other hand, have very weak magnetic fields. These planets are about as dense as Earth but rotate relatively slowly (243 and 58 days, respectively).

Another theory is that the Earth's magnetic field arises from charge in motion in thermal convection cycles in the liquid outer core due to heat coming up from the central core. If the heat flow subsided and the directions of the convection cycles were reversed with subsequent heat flow, this could account for the Earth's magnetic pole reversals.

QUESTION: If magnetism is produced by moving electric charges, what produces the magnetic field of a bar magnet?

ANSWER: Recall that the electrons in an atom or ion revolve in orbits about the nucleus. This orbital motion of an electronic charge is a current and gives rise to a magnetic field. Also, electrons have intrisic "spin" contributions to the magnetic field. This cannot be accounted for by classical mechanics, such as for a particle in orbit, and its origin is not known. We assume it derives from the spinning of the electron on its own axis, hence the term "spin."

All materials contain atomic electrons. Whether a material is magnetic or can be made into a permanent magnet depends on whether the electron magnetic fields add or cancel. This is the topic of the next section. Read on.

Magnetic Materials

A bar magnet easily picks up nails, paper clips, and iron filings, and we say that these are magnetic materials. On the other hand, a magnet has no observable effect on nonmagnetic materials such as wood or aluminum. The magnetic properties of a material depend on the magnetic fields of its electrons, as discussed in the previous Question and Answer. The magnetic properties of the most common magnetic materials depend on the electron spin, so we'll concentrate on this. Because of the electron spin, each electron acts as a tiny magnet. If two electron spins are in the same direction, their magnetic fields add. However, if a pair of electron spins are in opposite directions, their magnetic fields cancel. In most atoms, the electron magnetic fields cancel each other, and most materials are nonmagnetic. Common magnetic materials, called ferromagnetic materials, are iron, nickel, and cobalt. In these materials there is not a complete field cancellation; for example, an iron atom has four uncancelled electron spin fields. As a result, each atom acts like a little magnet, and we say the atom has a magnetic moment (similar to an electric dipole moment). There is an interaction among adjacent atoms, and large groups of them line up with each other. These groups of aligned atoms are called **magnetic domains.**

However, an ordinary piece of iron by itself is not a magnet. This is because the magnetic domains are randomly oriented, and their effects cancel (Fig. 20.8). In the presence of a magnetic field, the domains are induced into alignment, and the iron becomes magnetized. The degree of magnetism depends on the degree of alignment. When the magnetic field is removed, thermal motion generally causes the domains to go back into a random orientation, and the magnetism is lost. However, after being in a strong magnetic field, the domains may retain some alignment and magnetism. You may have noticed this in a paper clip that has been around a permanent magnet.

"Permanent" magnets are made by placing a piece of iron or some other ferromagnetic material or alloy in

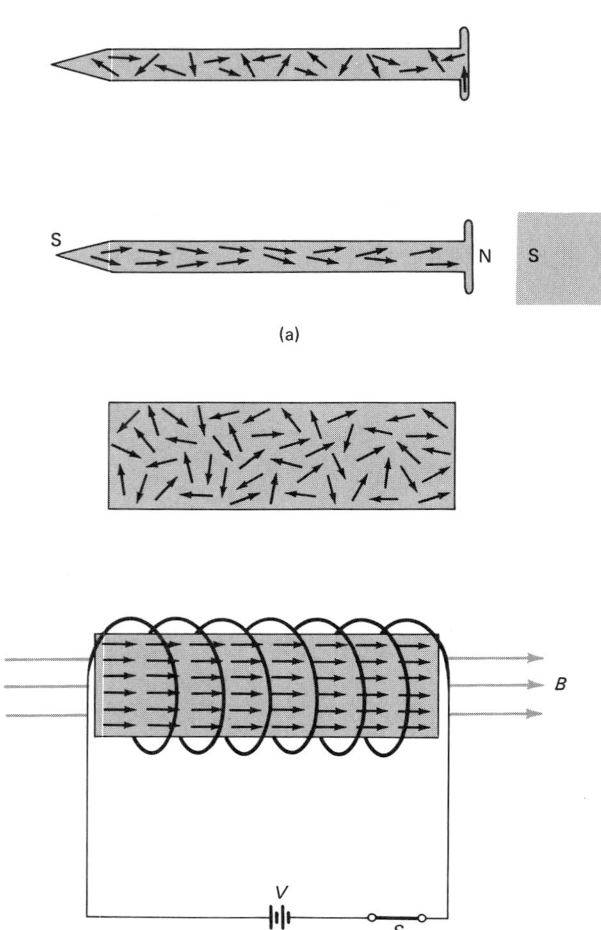

Figure 20.8 Magnetic domains. Ordinarily, the magnetic domains in a magnetic material such as iron are randomly oriented, and their effects cancel. A magnetic field from a magnet (a) or from a current-carrying wire (b) causes the domains to be aligned, and the iron becomes magnetized.

a solenoid in an oven. The iron is heated up and the magnetic field of the solenoid is turned on. This magnetic field is maintained as the iron cools and the domains are "locked in" their alignment. Relatively light-weight magnets are made of Alnico, an aluminum-nickel-cobalt alloy. The magnetism of a permanent magnet can be destroyed by heating it or striking it on a hard surface. Why? Also, you can now understand why breaking a magnet in two gives two smaller magnets.

The magnetism of natural magnets arises from iron minerals in the rocks. When molten material containing these minerals is extruded upward, as in volcanic eruptions, and solidifies in the Earth's magnetic field, the rock becomes magnetized. The orientation of the magnetization is in the direction of the Earth's magnetic field. These solidified rocks, called igneous rocks, are worn down by erosion. The rock fragments are carried away by water, and they eventually settle in bodies of water where they become the layers of future sedimentary rocks.

In the settling process, the magnetic fragments become generally aligned with the Earth's magnetic field. By studying this remanent magnetization in geologic rock formations, scientists have learned about changes that have taken place in the Earth's magnetic field, such as pole reversals.

QUESTION: Are coins (money) magnetic?

ANSWER: Some are and some are not. It depends on their metal content. In general, the U.S. "silver" coins aren't magnetic, but Canadian coins are. This is because Canadian coins still contain nickel (a ferromagnetic material). Get a magnet and a few coins and see for yourself.

Have you ever put a Canadian coin in a coin-operated machine and not have it work or "go down"? This is because the machines are equipped with magnets to stop metal slugs or washers (usually iron) from entering the coin mechanism so as to prevent damage and economic losses (Fig. 20.9). Coins containing ferromagnetic nickel are likewise attracted to the magnets.

Electromagnets

A solenoid with an iron core is the principle of an electromagnet used to pick up magnetic metal objects, for example, in a junkyard or in the application shown in Figure 20.10. When the solenoid current is turned on, the iron core becomes magnetized and attracts magnetic materials. By turning off the current, they can be dropped or placed where wanted.

Electromagnets can be used to produce very strong magnetic fields—the greater the current, the greater the magnetic field. But there are limitations. With very large currents there are also large I^2R losses in the magnet coils. Large electromagnets used in scientific work are often water-cooled to carry away the joule heat.

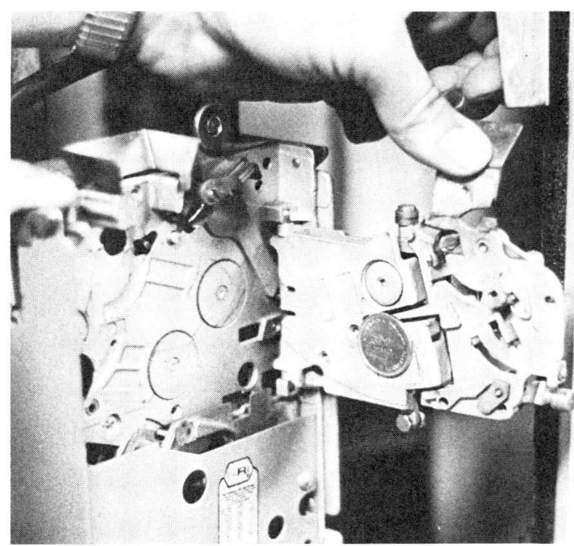

Figure 20.9 A Canadian coin sticks to a magnet in the mechanism of a coin-operated machine.

There are also superconducting magnets. If a superconducting material is used for the coils and they are cooled below a certain critical temperature (a few kelvins), the resistance and I^2R losses drop to zero. (See Special Feature 19.1.)

You have probably used electromagnets without knowing it. A common application is the relay, which is an electromagnetic switch. One application is shown in Figure 20.11(a). When the coil is energized, the electromagnet attracts the contact arm and opens the high-voltage circuit. Notice that relay switching allows low-voltage control of high-voltage circuits. High-voltage contact switches tend to become pitted and burned, which causes poor contact. The low-voltage control of lights in homes is coming into increasing use. Also, large-current motors and machines can be controlled from a distance without having to run heavy wires to the controlling switches. This provides economy and safety.

A simplified diagram of an electromagnetic circuit breaker is also shown in the figure. With a safe amount of current in the circuit, the electromagnet is not strong enough to attract the contact arm and break the circuit. But when too much current flows in the circuit, the breaker switch is tripped. See if you can use your knowledge of circuits and electromagnets to explain why a doorbell rings continuously when the button is pushed [Fig. 20.11(c)]. Another common use of a magnetic device is described in Special Feature 20.2.

Figure 20.10 An electromagnet. Pellets of iron oxide used in a steel-making process are lifted by an electromagnet.

Magnetic Forces on Moving Electric Charges

As we have learned, a current produces a magnetic field. This can cause a magnet (compass) to move as a result of a force interaction, as was illustrated in Figure 20.5—a case of a stationary current-carrying wire giving rise to a force on a movable magnet. Nature tends to show symmetry, and we can have a stationary magnet giving rise to a force on a movable current-carrying wire (Fig. 20.12). Another way of saying this is that

> A moving electric charge or current in a magnetic field experiences a force.

Notice in the figure how the force is reversed when the direction of the current is reversed. This could also be done by turning the magnet over, thereby reversing the direction of the magnetic field.

This force is experienced by a single moving charged particle or a beam of charged particles, as well as a current in a wire, in which case there is a force on the wire. Positive and negative charges experience forces in opposite directions. In any case, there must be charge in motion. The force is greatest when the charge moves perpendicular to the magnetic field. It becomes less as the charge motion becomes nearly parallel to the field and is zero when the charges move parallel to the field. The size of the force also depends on the magnitudes of the magnetic field and the electric charge and on the

Figure 20.11 Electromagnet applications. (a) A magnetic relay is an electromagnetic switch. (b) Diagram of a circuit breaker used for electrical safety (Chapter 19). (c) A doorbell circuit.

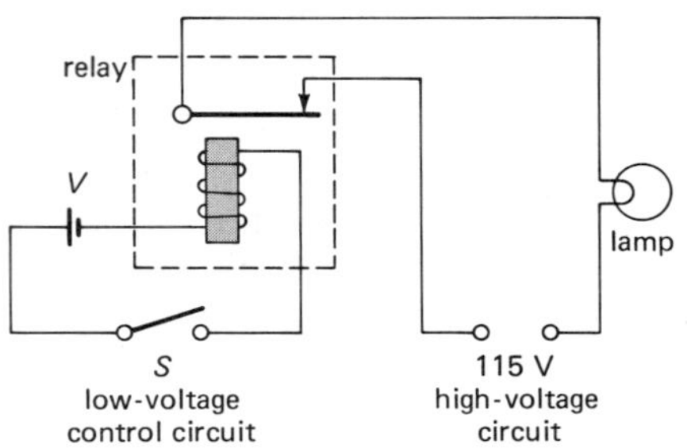

(a) relay

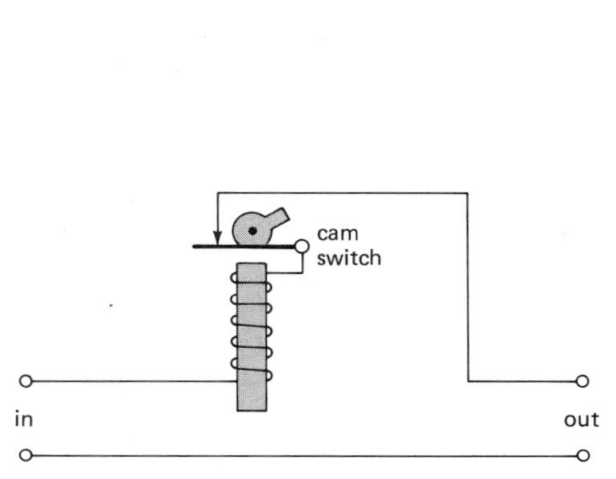

(b) circuit breaker

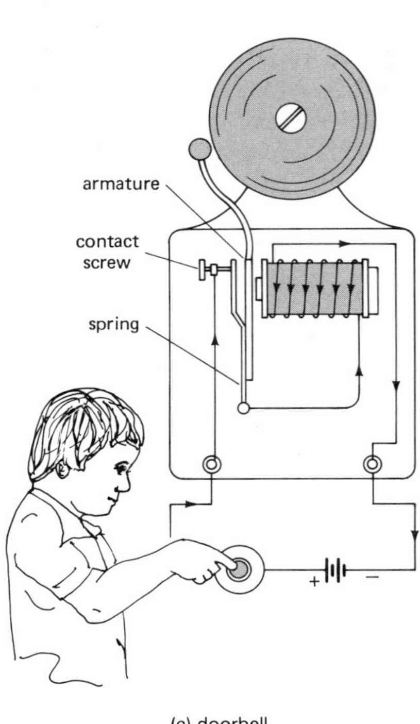

(c) doorbell

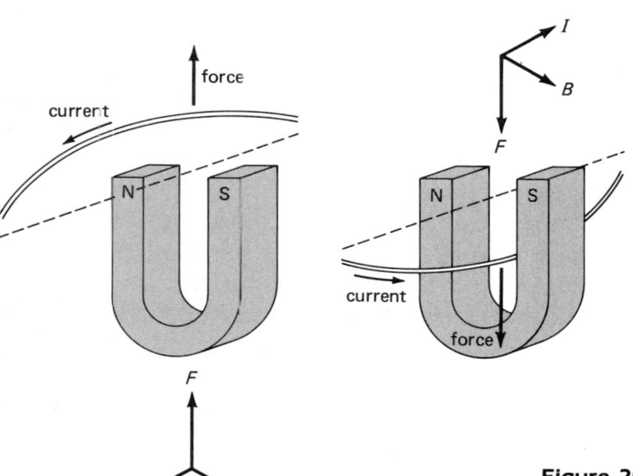

Figure 20.12 A moving electric charge or current in a magnetic field experiences a force. The current shown here is the conventional or positive-charge current.

SPECIAL FEATURE 20.2

The Auto Starting System

You may have heard of (or had trouble with) the "solenoid" associated with the starter of your car. This solenoid is similar to a relay, except that the electromagnet contains a movable core (Fig. 20.13). A solenoid is an electromagnetic device that converts electrical energy into linear motion (of the core). When the electromagnet is energized, the core is drawn into the coil, providing mechanical action.* A spring may be used to return the core when the coil circuit is opened (not shown in the figure).

In the starting system of an automobile, when the starter switch is closed, a low-current starter solenoid is activated. The movement of the core or plunger mechanically engages the starter gear and closes a switch that connects the cranking motor to the automobile (high-current) battery. When the starter switch is released, the solenoid is de-energized, and a spring within the solenoid assembly pulls the gear out of mesh and interrupts the current flow to the cranking motor.

Solenoids are also used in home door chimes. Here you usually just get a "ding-dong" sound rather than continuous ringing, as with a door bell (see Fig. 20.21 and Exercise 27 at the end of the chaper).

* The core becomes a magnet with the opposite polarity to that of the coil, and the core is "drawn" (attracted) into the coil.

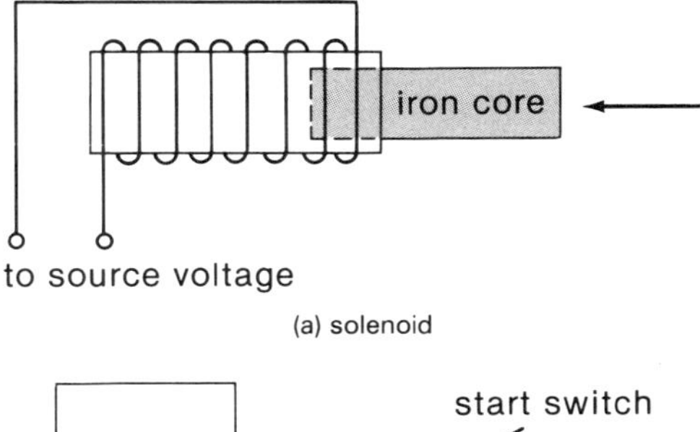

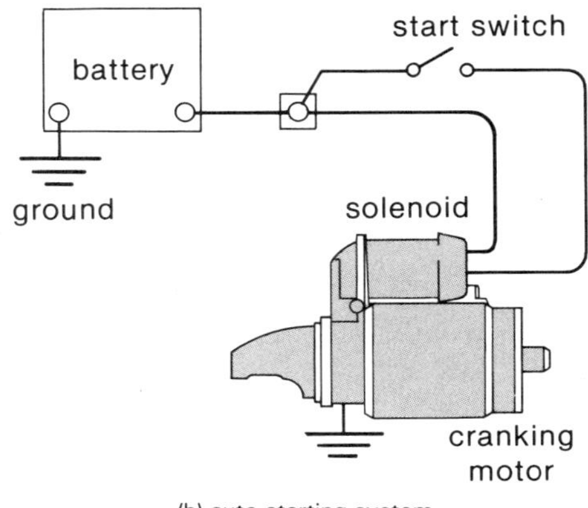

Figure 20.13 A solenoid. (a) A magnetic solenoid is similar to a relay, except the electromagnet contains a movable core. (b) Solenoids are used in the starting systems of automobiles to engage the starter gear.

speed of the moving charge. The direction of the force is always perpendicular to both the direction of the charge motion and the magnetic field direction.

It is interesting to note that the ampere is defined by the mutual force interaction of two parallel, current-carrying wires (see Table 1.2). One wire produces a magnetic field, and the other wire experiences a force (and vice versa).

Motors

A force being available, we can usually do something useful with it. In the electromagnetic case with a force on a wire, we can build an electric motor, which converts electrical energy into mechanical energy. Think of how life would be without electric motors. We'd still be winding up clocks and phonographs, carrying water, and doing a lot of other manual work.

The basic principle of a motor is illustrated in Figure 20.14. Practical motors have many windings or loops of wire on the rotating armature. As can be seen in the figure, the electromagnetic forces on the current-carrying wire loop in the magnetic field of a permanent magnet are perpendicular to the wire and give rise to a rotating torque. The split-ring commutator is essential to the operation of this dc motor. In starting from rest, if the commutator were not split or in two pieces, the loop would go through only one half-cycle, since it would be in static equilibrium and stop when vertical (see figure), when the torque would be zero.

But just at that moment, the end of the loop that was on the negative part of the commutator makes contact with the positive part (and vice versa for the other end

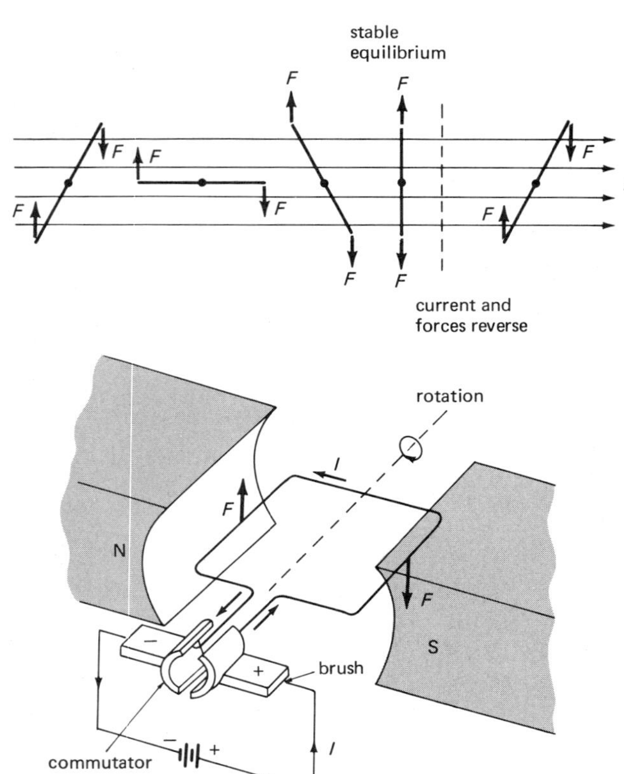

Figure 20.14 Motors. (a) The principle of a motor. A current-carrying loop in a magnetic field experiences a force or a torque and rotates. A dc motor has a split-ring commutator to change the current direction in the loop. This is done automatically in the windings of an ac motor. (b) A universal motor used in vacuum cleaners. Note the many windings.

of the loop), and the current and forces are reversed. The rotational inertia of the armature carries it through this position, and the loop rotates through another half-cycle. This process repeats, and the motor armature rotates continuously—pretty clever. Ac motors have two separate rings that aren't split for continuous contact. Since the current alternates or changes direction itself in this case, there is no need to change contacts.

Small dc motors are made with permanent magnets, but large dc and ac motors usually have electromagnets that give larger magnetic fields and larger forces on the armature.

The Galvanometer

The electromagnetic force on a current-carrying wire is the principle of the galvanometer, which is used in electrical meters that measure current and voltage (ammeters and voltmeters). A galvanometer movement is shown in Figure 20.15. A small current in the pivoted coil between the pole faces of a permanent magnet produces a force on the coil, causing it to rotate, and the attached needle is deflected. The greater the current, the greater is the force and the needle deflection. A restoring spring returns the needle (and coil) to its zero position when there is no current in the coil.

A large current through the fine-wire coil would burn it out, so in an ammeter a low "shunt" resistance is connected in parallel with the coil (Fig. 20.16). The shunt resistance then carries most of the current. The meter is calibrated to read the total current. By having a selection of shunt resistances, an ammeter can read various ranges of current. A voltmeter also has a galvanometer, but it has a large resistance in series with it to limit the current when placed across a high voltage. The meter is calibrated in volts instead of amps by using Ohm's law ($V = IR$).

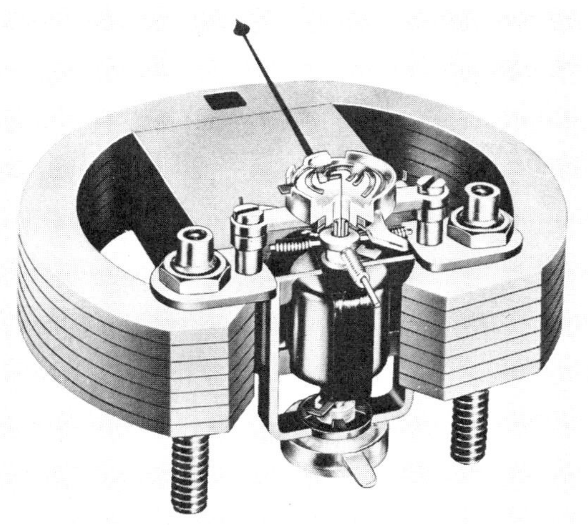

Figure 20.15 A galvanometer movement. When current passes through the pivoted coil between the pole faces of the magnet, a torque causes the coil to rotate. The deflection, as indicated by a pointer, is proportional to the current.

The Earth's Magnetosphere and Ionosphere

The forces on charged particles in a magnetic field give rise to many interesting geophysical effects associated with the Earth's magnetic field. The Earth is continually being bombarded by charged particles and radiation from the Sun and by cosmic rays. Cosmic rays (mostly protons) come from outside the solar system, perhaps from distant stars, or maybe they are vagabonds left over from the big bang. We're not sure. The Sun continually hurls off clouds of high-velocity ionized particles (electrons and protons) and radiation. We call this stream of emissions the solar wind.

The Earth's magnetic field is deformed slightly, owing to the "pressure" of the solar wind (Fig. 20.17). On the daylight side of the Earth, the field lines are compressed slightly, while on the night side they are stretched out in a long tail. This confines the geomagnetic field to a region scientists call the **magnetosphere.** Notice how the Earth and its magnetosphere look like an object in a wind tunnel with a bow shock wave (Chapter 17).

The magnetosphere acts as an obstacle to the solar wind and deflects a great deal of it. But some charged

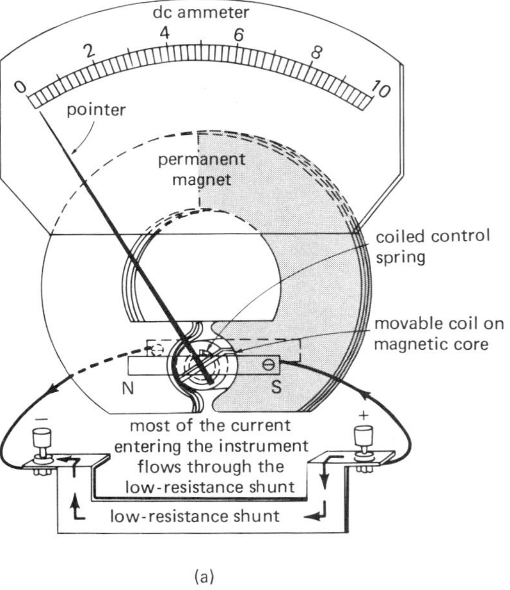

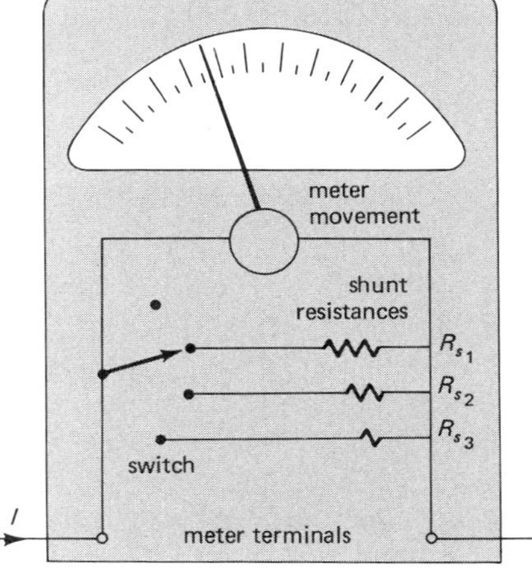

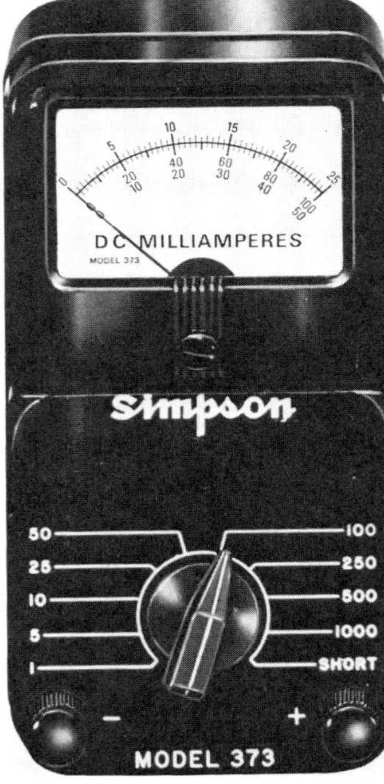

Figure 20.16 The ammeter. (a) Exposed view. The parallel shunt resistance carries most of the current and prevents the galvanometer coil from being burned out. (b) With a selection of shunt resistances, an ammeter can read various ranges of currents.

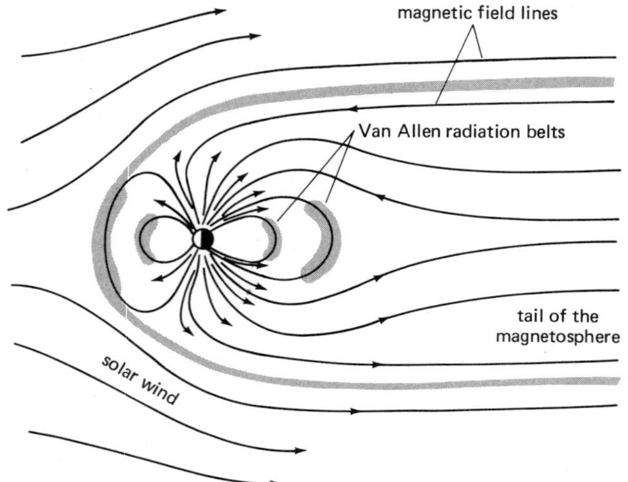

Figure 20.17 Earth's magnetosphere. The region of the geomagnetic field is called the magnetosphere. This is shaped by solar winds. Two large, donut-shaped regions of magnetically confined charged particles are called Van Allen radiation belts.

particles enter the geomagnetic field and experience forces perpendicular to the direction of the field lines. As for any charged particle in a magnetic field, it moves in a corkscrew path around a field line (either north or south from the equator). Thus, the magnetic dipole field of the Earth acts as a trap (or "magnetic bottle") and confines many particles at certain altitudes. The particles bounce back and forth and drift around the Earth. Two large, donut-shaped regions at altitudes of several thousand kilometers are called the Van Allen radiation belts (Fig. 20.17).† Several other radiation belts were made in the lower Van Allen belt region in the late 1950's and early 1960's as a result of high-altitude nuclear explosions. These involved experiments to determine the behavior of charged particles in the Earth's magnetic field.

At lower altitudes (100–400 km) there are noticeable concentrations or layers of ions. Hence, this region is called the ionosphere. Here, where the air is more dense, nitrogen and oxygen molecules are ionized by energetic particles and radiation from the Sun. The layers vary in ion density and are called the D, E, and F layers (Fig. 20.18).

Since the production of ions requires direct solar radiation, the concentration of charged particles varies from day to night, particularly in the D and E layers, which weaken at night (the D layer virtually disappears). The upper F layer, however, is present both night and day because of the low density of the atmosphere in this region, and the ions and electrons do not recombine as rapidly as they do at lower altitudes.

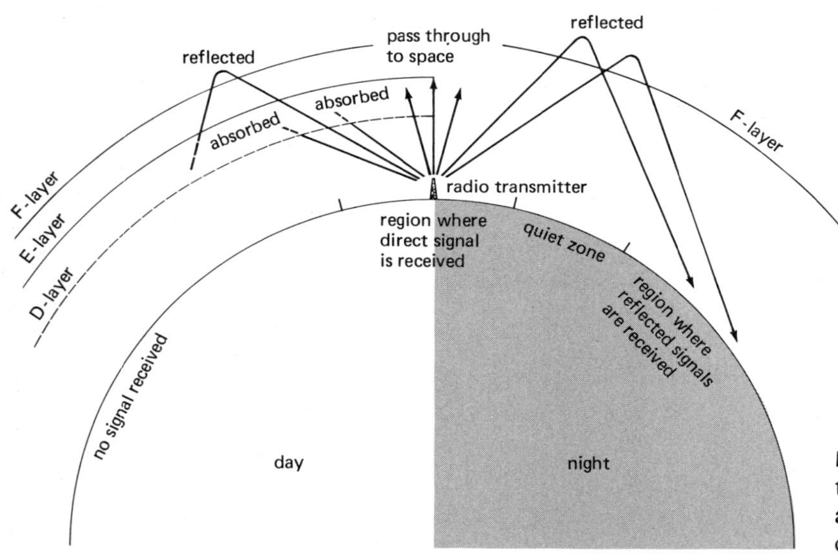

Figure 20.18 The ion layers and radio transmission. Reflections from the F-layer allow radio transmission around the Earth's curved surface.

You may have had some contact with the ion layers through AM radio reception. Have you ever turned on a radio at night and been surprised to receive a station

† After James A. Van Allen, an American scientist who identified the radiation belts in 1958 from data from instruments he had on board the first U.S. satellite, Explorer I.

hundreds of kilometers away in another part of the country? This is because the F layer reflects medium- and high-frequency radio waves (Fig. 20.18). Ham-radio operators make good use of this in reflecting the straight-line radio waves around the curvature of the earth for long-distance transmissions. During the day, when it is present, the D layer absorbs most radio waves as they pass through and on the return reflection from the higher E and F layers. However, at night the absorbing D layer all but disappears, and the E layer is very weak. Of course, you may be in a "quiet zone" and not receive the station. In this zone, you are too far from the station to receive the direct signals and not in position to receive the reflected radio waves.

Radio and other communications are also disturbed during periods of increased solar activity, such as periods of maximum sunspots and solar flares. Solar flares are violent magnetic storms on the Sun that spew out enormous quantities of charged particles and radiation. One spectacular sight that apparently results from these solar storms is aurora—aurora borealis (northern lights) in the northern hemisphere and aurora australis (southern lights) in the southern hemisphere (Fig. 20.19). These eerie, flickering lights generally occur in the Earth's polar regions and follow a timetable of being a day or so after a solar disturbance.

Figure 20.19 The aurora borealis or northern lights, as seen in Alaska.

It is believed that the stream of charged solar particles is trapped in the Earth's magnetic field, and the particles are guided toward the poles. They excite or ionize oxygen and nitrogen molecules, which emit light on de-excitation or recombination—the glow of the aurora.

SUMMARY OF KEY TERMS

Magnetic poles regions of concentrated magnetic strength, designated as the north pole and south pole.

Law of poles like magnetic poles repel, and unlike magnetic poles attract.

Magnetic field the magnetic force field surrounding a magnet or electric charge in motion. At a particular location, the magnetic force or pole in the direction of the force experienced by a north pole.

Magnetic domain a group of aligned magnetic atoms whose interaction gives the domain region magnetic properties as a whole.

Ferromagnetic materials strong magnetic materials, such as iron, nickel, and cobalt.

Magnetic force a force experienced by an electric charge (or current) when moving in a magnetic field in a direction other than parallel to the field lines.

EXERCISES

1. Which way would a compass located at the geographic North Pole point? How about when located at the nearby south magnetic pole?
2. Are nails attracted to either pole of a bar magnet? Why or why not?
3. Nails hanging on a magnet are induced magnets (Fig. 20.20). Identify the poles on each of the nails.
4. Why do airplane pilots and ship navigators following a compass course have to make corrections to stay on a course charted on a map?
5. If the Earth's magnetic field underwent another pole reversal, what effects would this have on such things as navigation and the Van Allen belts? Think of the magnetic field weakening, going to zero, and building up again with reverse polarity.
6. Why do iron filings show magnetic field patterns?
7. A long, straight horizontal wire carries a direct current to the north. In what general directions would a compass needle point if the compass were placed (a) above the wire, (b) below the wire, and (c) on either side of the wire?

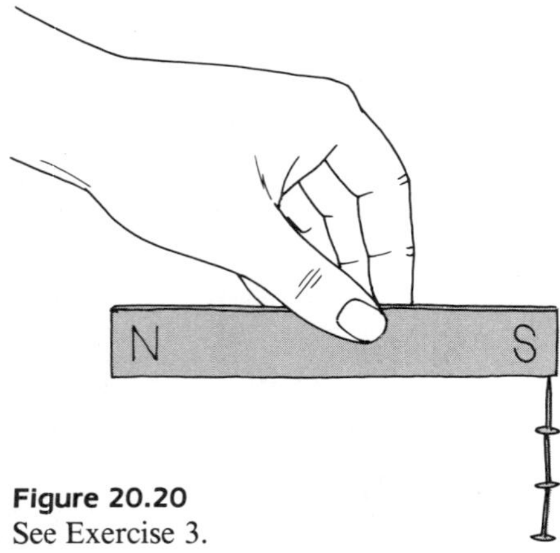

Figure 20.20
See Exercise 3.

What would happen in each case if the wire carried alternating current?

8. Why does stroking a needle over a permanent magnet magnetize the needle? Should the needle be stroked back and forth? Explain.
9. Why does striking a magnet with something or on a hard surface weaken its magnetism? How about heating the magnet?
10. A fellow student (who hasn't yet taken physics) thinks that a bar magnet results from a flow and separation of north and south magnetic poles, similar to electrostatic charging by induction. How could you disprove this theory?
11. Why doesn't breaking a magnet in two give single magnetic poles? How about if you kept breaking it into smaller pieces?
12. Do the poles of a magnet have definite points of location? Discuss the concept of poles in terms of electromagnetism.
13. Could you make a compass out of an electrically charged needle?
14. You are given four items—a piece of string, a needle, and two identical iron bars, one of which is a permanent magnet. Using the string, needle, and unmagnetized bar only one time each, give three ways you could identify which bar is the magnet. (*Hint:* Two ways are pretty easy. If you get stuck on the third, keep in mind that the unmagnetized bar becomes an induced magnet when the magnetic field lines are generally along the length of the bar.) What would happen if you put the end of one bar at the mid-length point of the other?
15. If a current-carrying solenoid were suspended on a string so it could rotate freely, what would happen?
16. Could an electromagnet be made without a metal core? What are the advantages of an iron core? Does it add to the field?
17. A junkyard electromagnet is used to lift scrap iron. Does it make any difference if the magnet is operated on ac or dc?
18. There are both dc and ac motors, but inside a motor there must always be an alternating current. Why is this, and how is this done in a dc motor?
19. A charged non-metallic ball is suspended on a string. Would it be possible to get the charged ball swinging by using a stationary magnet (without touching it, of course)? What would happen if you set the ball swinging with the stationary magnet nearby?
20. What would happen in Figure 20.13(a) if the current were reversed and the magnet were rotated around the wire so the pole ends pointed downward?
21. An electric field can be used to distinguish between positively and negatively charged particles. Could you do this with a magnetic field? Explain.
22. An electron and a proton are traveling along the same straight line, say up the middle of this book. (a) What would happen if a uniform magnetic field were turned on in a direction perpendicular to and out of the page? (b) Parallel to the page in a top-bottom direction? (c) Parallel to the page in a side-to-side direction from right to left?
23. An ammeter is a low-resistance instrument and is connected in-line or in series in a circuit. A voltmeter is a high-resistance instrument and is connected across or in parallel with a circuit component to determine the voltage drop across it. Why are high and low resistances critical for these instruments? (*Hint:* Think in terms of voltage drops and distribution of current.)
24. What is the difference between an electromagnetic relay and a solenoid?
25. Distinguish between electromagnetic and electromechanical devices.
26. Explain how the doorbell in Figure 20.11 works.
27. Explain how the door-chime mechanism shown in Figure 20.21 works.
28. Why do most of the energetic charged particles in cosmic rays and the solar wind not reach the Earth's surface?
29. Why do ham-radio operators generally make long-distance transmissions at night or early in the morning?

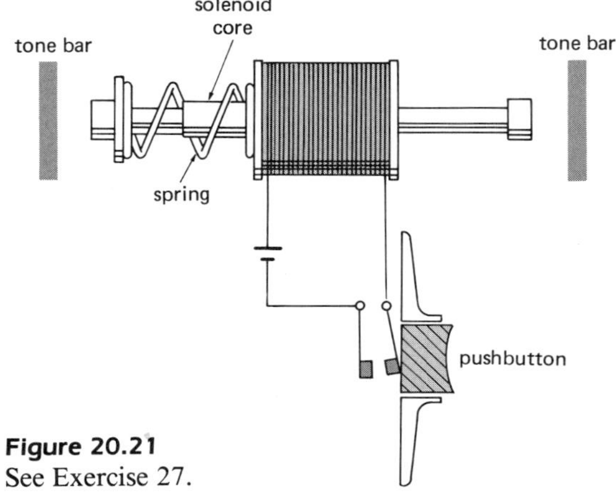

Figure 20.21
See Exercise 27.

21

Electromagnetic Induction

As we learned in the last chapter, a current produces a magnetic field. With the similarities and symmetries in electromagnetic interactions, you may have thought about the reverse situation: Can a magnetic field produce a current? The answer is yes, under certain conditions.

After it was found in 1820 that a current produces a magnetic field, several scientists of the day asked themselves the same question. In about a decade, electromagnetic induction was discovered, and voltages and currents were induced in circuits using magnetic fields. This important discovery forms the basis for the production of electricity we use in our homes and at work. Have you ever wondered how electricity is generated? You'll find out in this chapter.

Faraday's Law of Induction

The trick in inducing a voltage or current in a circuit, say a wire loop, with a magnetic field is that the number of field lines through the loop must vary with time. There are several ways of doing this. For example, a magnet could be brought toward a loop of wire, as illustrated in Figure 21.1. When the magnet is stationary, there is no current in the loop. If the magnet is brought quickly toward the loop, the number of field lines through the loop increases, and a current is induced in the wire, as indicated by the deflection of the galvanometer.

If the magnet is pulled away quickly, the number of field lines through the loop decreases. The galvanometer then shows another deflection, but in the *opposite* direction, indicating that the polarity of the induced voltage and the direction of the current have changed. The magnitude of the induced voltage depends on how fast the change in the number of field lines through the loop takes place. If you moved the magnet very slowly, the induced voltage would be too small to detect. With an appreciable induced voltage, the current in the loop depends on the voltage and resistance of the loop (Ohm's law, $I = V/R$).

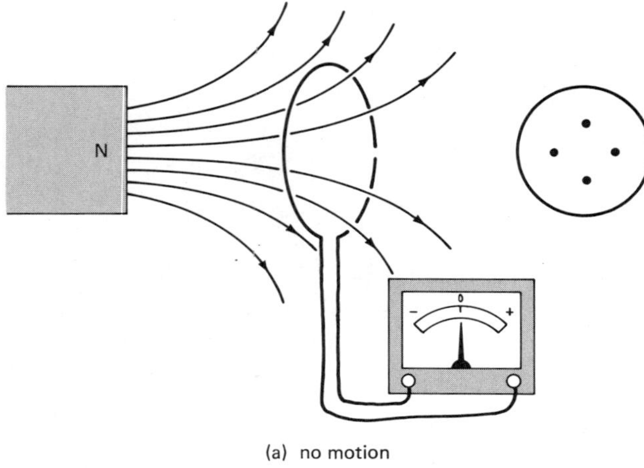

(a) no motion

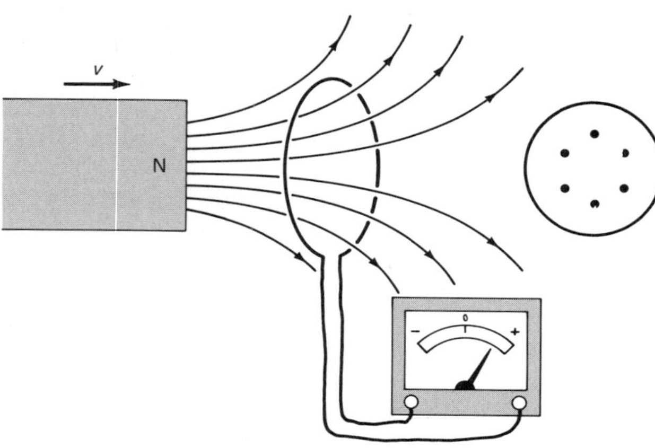

(b) magnet moved toward loop

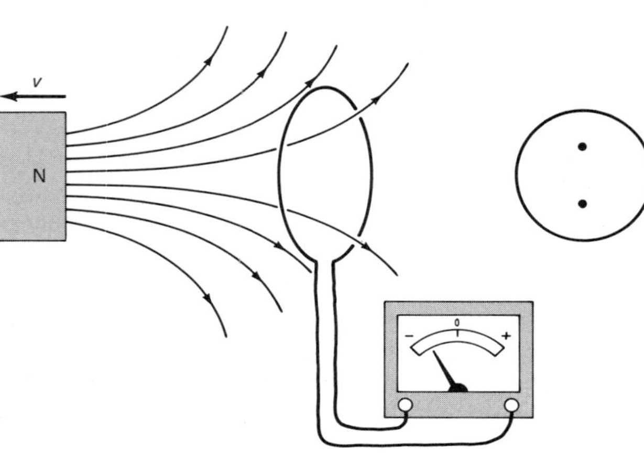

(c) magnet moved away from loop

Figure 21.1 Magnetic induction. A change in the number of magnetic field lines through a wire loop by a moving magnet induces a current in the wire.

In a practical demonstration of electromagnetic induction in this manner, a coil with several loops of wire would be used to increase the magnitude of the induced voltage. The change in the field lines through each loop contributes to the voltage. For example, a circuit with a coil of ten loops would have ten times the induced voltage of a single loop.

The same effects would be observed if the magnet were held stationary and the loop moved (Fig. 21.2). The change in the number of the field lines depends on the relative motion of the magnet and loop, so either (or both) could be moved. In one case, the field lines cut through the wire, and in the other, the wire cuts through the field lines.

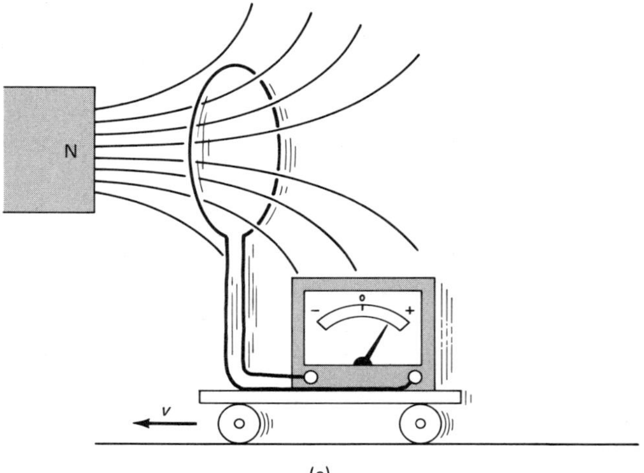

(a)

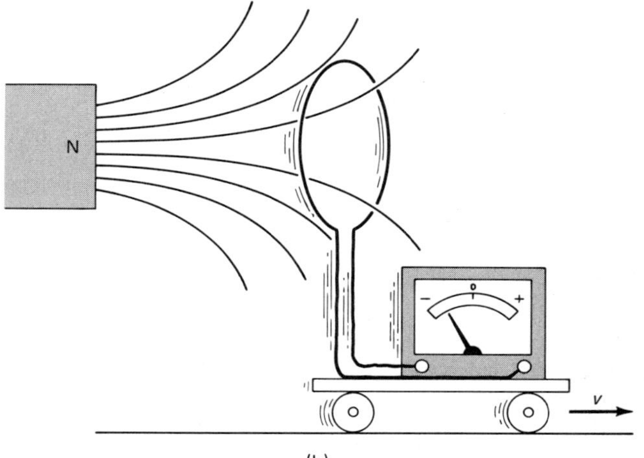

(b)

Figure 21.2 Magnetic induction with a stationary magnet and moving loop. The change in the number of field lines depends on the relative motion of a magnet or loop, so either or both could induce a current when moving.

Such experiments were done independently by Michael Faraday in England in 1831 and by Joseph Henry in the United States at about the same time. The result was what we call **Faraday's law of induction,** which is generally stated:

> A voltage is induced in a conducting loop when there is a change in the magnetic field through the loop.

QUESTION: Would there be an induced voltage or current in a loop when the plane of the loop is moved perpendicularly to a uniform magnetic field (up and down).

ANSWER: No. Although the wire of the loop cuts through field lines, the number of field lines through the loop does not change with vertical motions, since the field is uniform. For every field line coming into the loop, another one leaves it, so the number of field lines through the loop stays constant.

You may have already thought of how a voltage could be induced in a loop in a uniform magnetic field. If a loop is rotated in the field, the number of field lines through the loop changes (Fig. 21.3). The field lines through the loop go from a maximum number when the plane of the loop is perpendicular to the field to zero when it is parallel to the field. On the next half-cycle or rotation, this goes from zero to a maximum again. This ideal will be important later, when we discuss generators.

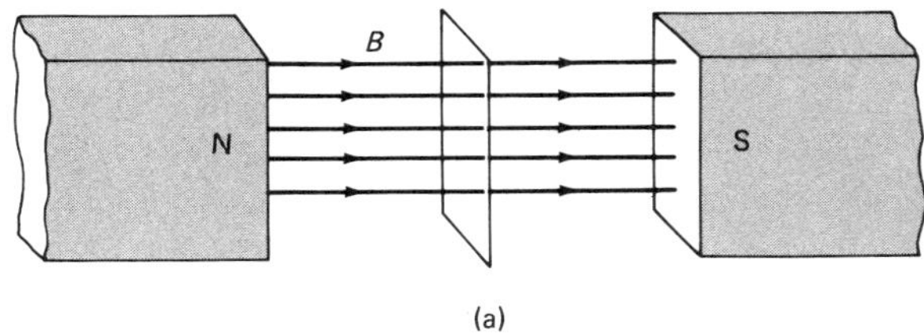

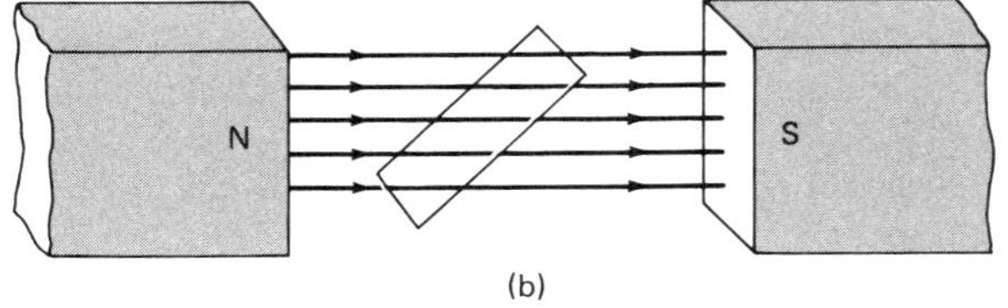

Figure 21.3 Magnetic induction by rotating a loop in a magnetic field. The number of field lines goes from some maximum value (a) to zero when the plane of the loop is parallel to the field.

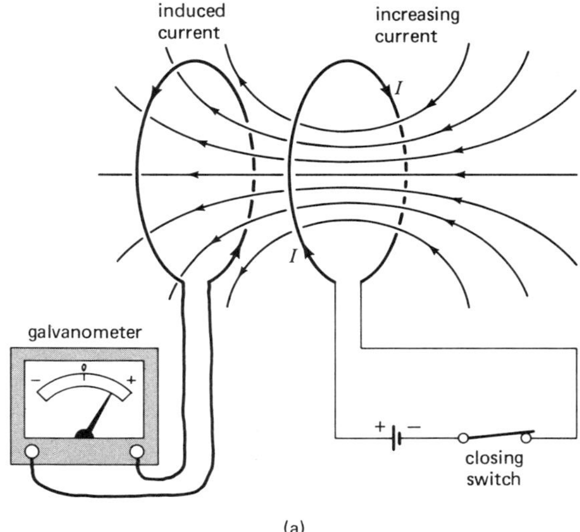

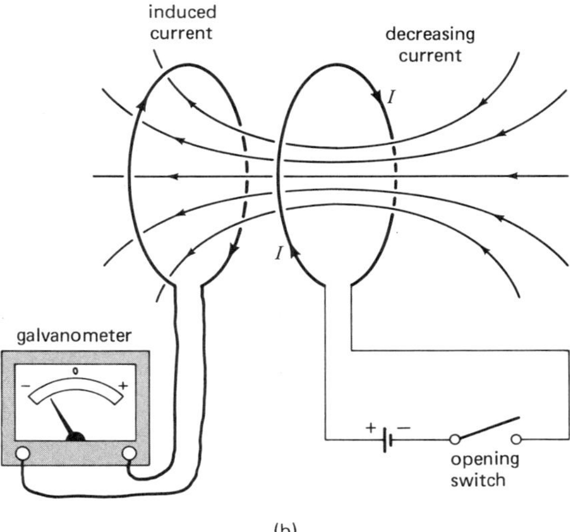

Figure 21.4 Magnetic induction by a time-varying magnetic field. (a) When the switch is closed, a current builds up in the right loop and a magnetic field builds up (changes) through the other loop, inducing a current in the loop. (b) The current and the magnetic field decrease when the switch is opened, and the direction of the induced current is reversed.

Another way to induce a voltage without having any moving parts is to have the magnetic field change with time. Consider the circuits with parallel loops shown in Figure 21.4. When the switch in the battery circuit is closed, the current goes from zero to some constant value. This is done in a very short time, and the increasing current produces an increasing magnetic field around the loop. This field goes through the second loop, the field through this loop changes (increases, since originally it was zero), and a voltage is induced in the loop. This process is called **mutual induction** because it depends on the interaction of two circuits.

When the current in the battery circuit is steady, so is the magnetic field, and there is no induced voltage in the second loop. (Why?) However, when the switch is opened, the current goes to zero, and the magnetic field decreases or "collapses." The decreasing change of the field through the second loop gives rise to an induced voltage with a reversed polarity and the current in the opposite direction.

You might think that there is a "moving part," since the switch has to be opened and closed. But suppose the first loop were hooked to an ac source. Would the effect not be the same? We'll return to this idea later in the discussion of transformers.

Several effects of electromagnetic induction are given in Special Feature 21.1.

SPECIAL FEATURE 21.1

Demonstrations of Electromagnetic Induction

An apparatus used to demonstrate some effects of electromagnetic induction is shown in Figure 21.5. Here, the coil is connected to an ac source, and the iron core concentrates the alternating magnetic field. A current is induced in a conductor placed over the core, as can be seen from the lighted bulb and loops of wire. Since a current is induced, there are I^2R (joule heat) losses in the conductor. By means of this heat, water can easily be boiled in a special hollow cylinder. This is known as induction heating and is used in industrial processes. A current-carrying loop in the presence of a magnetic field also experiences a force (Chapter 20). As a result of the induced current in a metal ring, there is a net upward force, and the ring "levitates" on the core. If pushed down and released, the ring jumps off into the air.

(a)

(b)

Figure 21.5 Electromagnetic induction effects. Alternating current in the coil produces an alternating magnetic field that is concentrated by the vertical iron rod (a). This produces a current in the loops of wire, as shown by the glowing bulb. A current gives I^2R losses, and water can quickly be boiled in the hollow metal container by induction heating. (The slight variation on the dark rod is condensed water vapor.) A current in a magnetic field experiences a force. This is upward in this case, and a metal loop "levitates" on the rod. (The hollow metal container is fastened down behind the rod.) (b) If a ring is pushed down over the rod and quickly released, the upward force causes it to jump into the air.

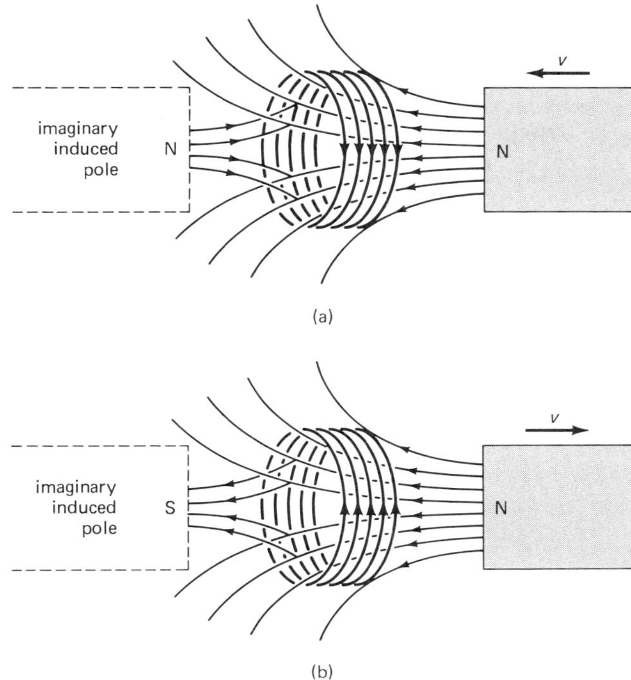

Figure 21.6 Lenz's law. The effect of Lenz's law in opposing the change of a moving magnet is as though there were an imaginary induced magnetic pole to oppose the change.

Lenz's Law

Knowing the direction of the induced voltage and current is often important. This is given by **Lenz's Law:***

> An induced current is in such a direction that its effects oppose the change that produces it.

For example, suppose a magnet is brought toward a coil (Fig. 21.6). The change producing the induced current is an increase of field through the coil. By Lenz's law, the induced current is in the direction that produces a magnetic field in the opposite direction to that of the magnet, so as to oppose the change or keep the field from increasing (field lines in opposite directions cancel). Use the right-hand rule to show that this is the case for the current in the coil in the figure [thumb in the direction of the conventional (positive change) current, fingers in the direction of the field].

In effect, the induced magnetic field in the coil is like that of a magnet with a like pole toward the pole of the permanent magnet so as to oppose it. Similarly, if the magnet is moved away from the coil, the induced current is in the opposite direction, so the induced magnetic field adds to the decreasing number of field lines. The coil behaves as an attractive magnet in this case.

Lenz's law is really a rather subtle statement of the conservation of energy. To see this, consider a dc motor (Chapter 20, Fig. 20.14) and suppose that an induced current did not oppose the change producing it. In a dc motor with a current from an external source in its armature loops, the loops experience a torque and turn in a magnetic field. This turning changes the number of field lines through the loops (Fig. 21.3), so there is also an *induced* voltage and current in the loops. If the field of the induced current added to that of the permanent magnet, then there would be a greater change in the number of field lines, more induced current, more field lines, more induced current, and so on—a runaway condition. This would be a violation of the conservation of energy, which doesn't happen because the effects of the induced current oppose the change that produces it.

So, in a dc motor the induced voltage and current are in the opposite direction and oppose the voltage and current from the external source This gives rise to what is called a counter or back emf.† The net voltage to drive current through the resistance of the motor coils is less than the source voltage by the amount of the back emf (voltage). The back emf depends on the rotational speed of the armature—the faster the rotational speed, the greater the change of field lines and the greater the back emf. But the back emf tends to reduce the armature current, and hence the torque on the armature, which may allow the rotation speed to decrease. Thus, the back emf controls the rotational speed of the armature to some extent.

A back emf can be induced in a stationary coil. For example, if a coil is in a battery circuit and the circuit switch is closed, the current and magnetic field build up in the coil. The field lines of any one of the coil loops cut through the other loops, and a back emf opposes the current build-up. This is called **self-induction.** When the switch is opened, the back emf is in the opposite direction, so the induced current opposes the current decrease. In an ac circuit, there is a continuing back emf or opposition to current flow. A coil or *inductor* in an ac

* After Heinrich Lenz, the German physicist who deduced this law in 1834.

† emf = electromotive force. See Chapter 19.

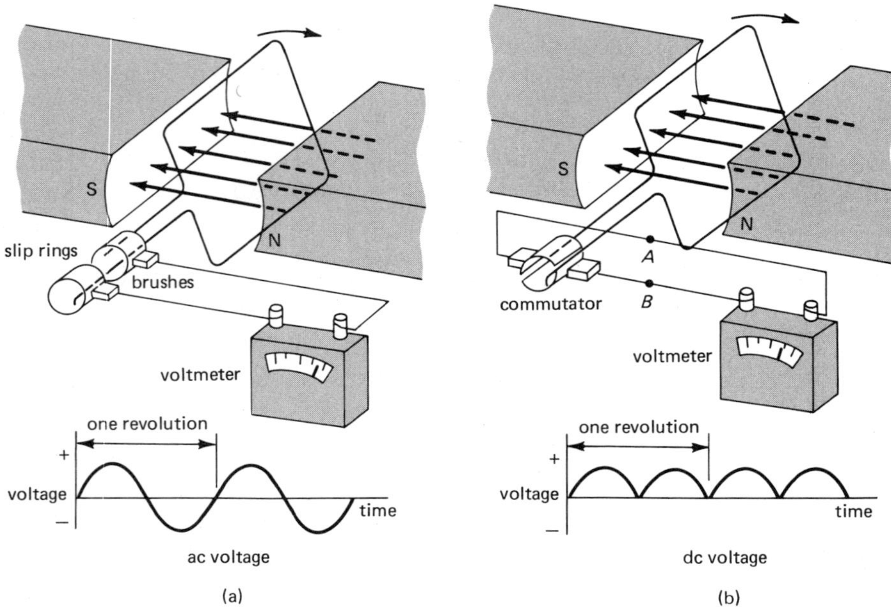

Figure 21.7 Simple generators. (a) In an ac generator, the output voltage alternates in polarity, and the current alternates. (b) In a dc generator, the voltage output varies but does not change polarity, so the current is direct.

circuit acts somewhat like a resistance, except that the inductor opposition to current flow depends on the frequency of the driving voltage source. Why?

Generators

You have probably suspected that electromagnetic induction is used to generate electricity. Recall that a motor is a device that converts electrical energy into mechanical energy. A generator has the opposite function.

> A generator is a device that converts mechanical energy into electrical energy.

Diagrams of simple ac and dc generators are shown in Figure 21.7. In the ac generator, the polarity of the induced voltage changes each half-cycle (by Lenz's law), and there is an ac voltage output. An ac generator is sometimes called an "alternator." The ac electricity we commonly use has an alternating frequency of 60 Hz, and the 110–120 V voltage is a special time-averaged value, not the peak or amplitude voltage.

A dc generator can be made by using a split-ring commutator as in a dc motor. Here, the voltage polarity in the loop changes, but the change of commutator contacts each half-cycle maintains the same output polarity. The voltage output varies or pulsates.

Direct-current generators are not used much anymore. It is easier to "rectify" or convert ac to dc. (How this is done will be discussed in the next chapter.) For example, "alternators" are now used in automobiles, and the alternating current is rectified to direct current. (Why?) Alternators are more efficient than the old-style dc generators because of greater outputs at lower rotational speeds.

Of course, there is a great deal more to the generation of electricity than we have mentioned here (Fig. 21.8). Steam from the heat of combustion of fuels (or nuclear energy) or hydropower is used to turn turbines (mechanical energy) that drive the electrical generators. A generator has several sets of armature coils (not one loop). When several electromagnets are used, the magnetic field can be made to rotate, and the armature remains stationary. The induction effect is the same—relative motion.

Transformers

Another important application of electromagnetic induction is the transformer. There's one outside your house, and you've probably seen them in other electrical equipment (Fig. 21.9). As shown in the figure, a **transformer** is simply two coils (primary and secondary) wrapped around an iron core. There are no moving parts.

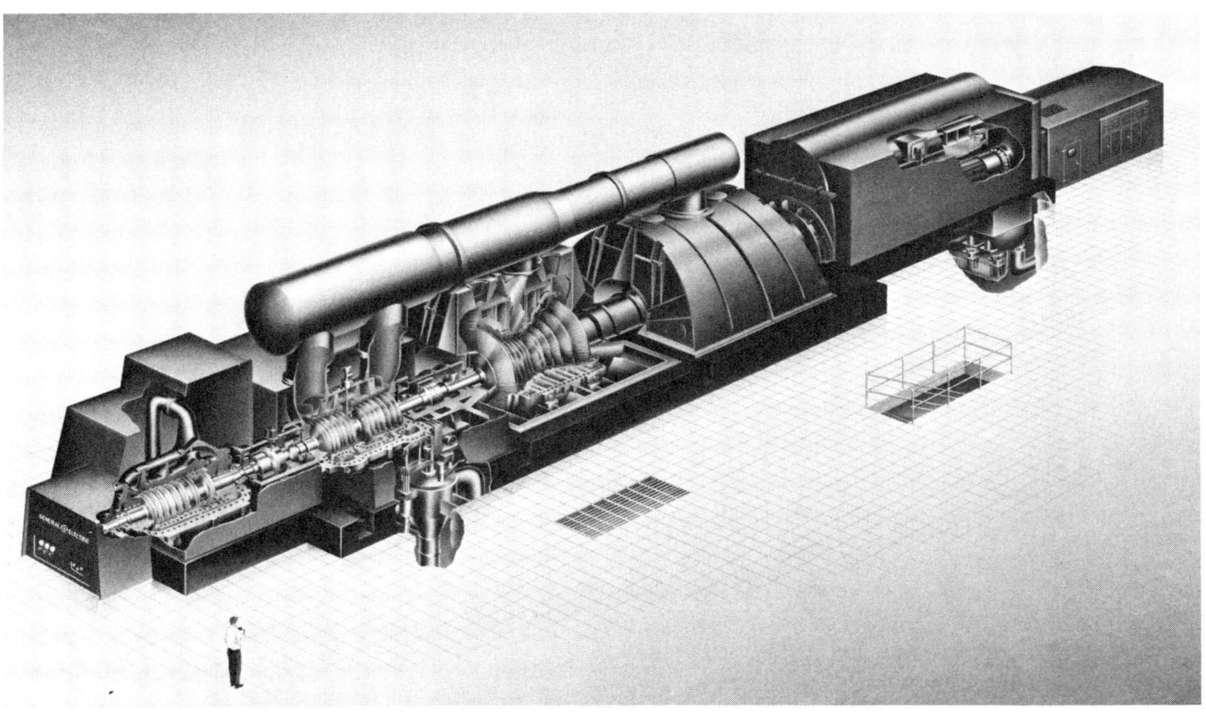

(a)

(b)

Figure 21.8 Electrical generators. (a) A cutaway view of a steam-turbine generator unit. The unit is rated at more than 500 megawatts and is capable of producing enough electricity for a city of more than half a million people. (b) A turbine, or hydroelectric generator. Falling water turns the turbine.

Suppose you put a battery and a switch in the primary-coil circuit and closed the switch; what would happen? Similar to the case of the parallel loops in Figure 21.4, while the current is building up from zero, the magnetic field increases. In the transformer, this field also goes through the secondary coil and induces a voltage. If the secondary has more windings than the primary, then the voltage pulse in the secondary is increased by this factor, for example, twice the voltage if there are twice as many windings. A similar situation occurs when the switch is opened.

But for continuous operation, why not use an ac source? Here, the current turns on and off automatically 120 times a second for 60-Hz alternating current. Then if the primary input voltage were 120 V and there were twice as many windings on the secondary, the output voltage of the secondary would be 240 V! For obvious reasons, we call this a **step-up transformer.** By

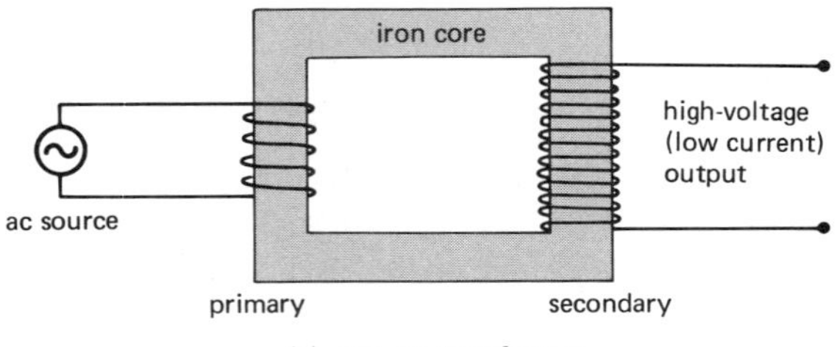

(a) step-up transformer

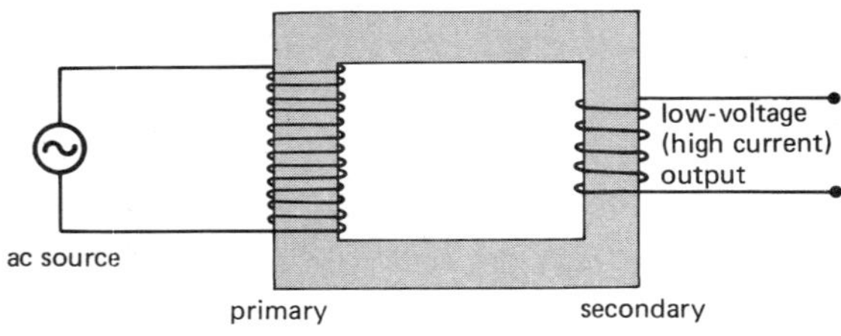

(b) step-down transformer

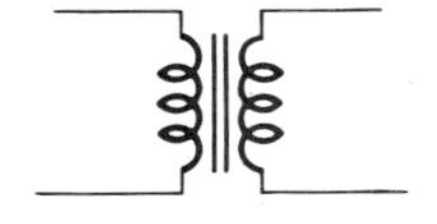
(c) transformer circuit symbol

(d)

Figure 21.9 Transformers. Transformers can be used to step up or step down ac voltage, depending on the relative number of turns on the transformer primary and secondary.

the same reasoning, if the secondary has fewer windings than the primary, the output voltage will be less than the input voltage, and we have a **step-down transformer.** For example, for a 120-V input and a secondary with half the number of windings as the primary, the output voltage would be 60 V. This can be expressed by the ratios:

$$\frac{\text{Secondary voltage}}{\text{Number of secondary windings}} = \frac{\text{Primary voltage}}{\text{Number of primary windings}}$$

or

$$\frac{V_s}{N_s} = \frac{V_p}{N_p} \quad \text{or} \quad V_s = \frac{N_s}{N_p} V_p$$

For example, for the case of twice as many windings on the secondary ($N_s = 2N_p$, or $N_s/N_p = 2$), the voltage is stepped up by a factor of two, $V_s = 2V_p$. Very large voltages can be obtained using step-up transformers. See Special Feature 21.2 for a practical application of this.

In stepping up the voltage, it might seem that you are getting something for nothing, but not so. Nature doesn't work that way, as you've no doubt learned by now. Remember that the voltage is energy/charge, and you can't double the voltage without giving something up. To see what this is, let's look at the conservation of energy in terms of electrical power (energy/time and $P = IV$). Assuming that the power losses (joule heating) in the transformer are negligible, we have

$$\text{Power to primary} = \text{power out of secondary}$$

or

$$I_p V_p = I_s V_s$$

Hence, if you step up the voltage, say from 120 V to 240 V, then the current is *stepped down* by the same factor. For example, if the primary current is 10 amps, then the secondary current is 5 amps [(10 amps)(120 V) = (5 amps)(240 V)]. So a step-up transformer steps up the voltage but *steps down* the current. Can you guess what a step-down transformer does? You've got it—it steps down the voltage but *steps up* the current.

Electric Power Transmission

The first central electric generating and distributing facility in the world went into operation in 1882. It was built by Thomas Edison on Pearl Street in New York City and supplied direct current at 110 V through underground mains to an area roughly 3.2 km (2 mi) in diameter. For some years, dc electricity was practically

SPECIAL FEATURE 21.2

The Auto Spark Ignition System

The common ignition system that "fires" spark plugs in an automobile engine has an ignition "coil," which is actually a transformer (Fig. 21.10). The 12 volts from the battery would not cause the spark discharge in the spark-plug gap to ignite the air-fuel mixture. When a spark plug is supposed to fire, a switch (the "breaker points" or "points") breaks the circuit between the battery and the primary of the coil. (The points are located in the distributor, which distributes the secondary voltage to the spark plugs in turn.) Voltages of 12,000 V or more are generated in this manner.

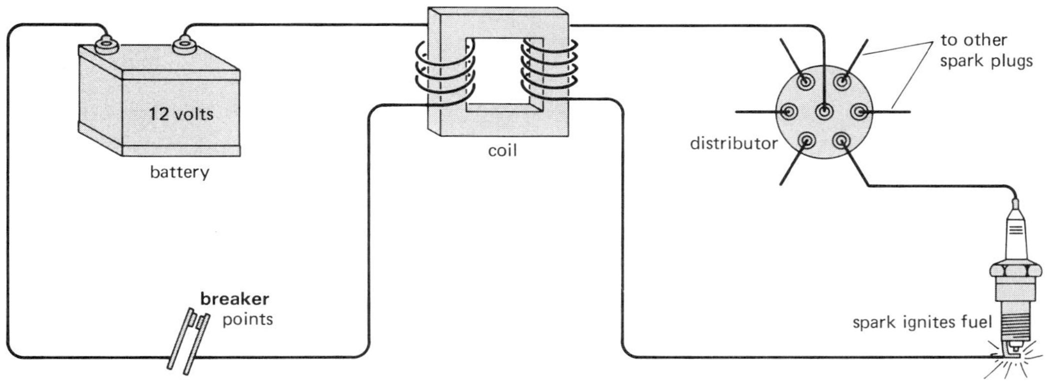

Figure 21.10 A diagram of an auto ignition system showing the ignition coil or step-up transformer.

the only type in use, but today almost all the electricity generated is ac. When ac systems first appeared, there were some definite opinions against alternating current. Advocates of dc branded ac as dangerous because of the high voltage used.‡

The development of the transformer and of the induction motor helped lead to ac electric power systems. The transformer provided a convenient way to vary voltages and made high-voltage power transmission possible. The ac induction motor is simple and economical. In your home, the motors on the washing machine, furnace fan, and refrigerator are induction motors. In such motors, the current in the armature or rotor is induced by the changing magnetic field of alternating current in the stationary (stator) field coils. With a current in the rotor in a magnetic field, the rotor experiences a force and rotates.

A diagram of a typical ac power transmission system is shown in Figure 21.11. Notice the high voltages used in the transmission lines. Line losses are due to the I^2R (joule heat) losses of the power lines themselves. Although metal wires are used, there is still appreciable resistance, particularly over long distances. For example, heavy aluminum wire has a resistance of about 0.3Ω per kilometer. Aluminum has almost, if not completely, replaced copper in long-distance transmission lines because of its lightness, strength, and cost. An important point to notice is that the I^2R losses depend on the *square* of the current.

Let's do a little arithmetic to show why electrical power is transmitted at such high voltages. Suppose a generator has an output of 10 amps at 240 V, and we want to transmit this electric power over a 50-km line with a 15Ω line resistance (50 km × 0.3Ω = 15Ω). If

‡ "My personal desire would be to prohibit entirely the use of alternating currents. They are unnecessary as they are dangerous. . . . I can therefore see no justification for the introduction of a system which has no element of permanency and every element of danger to life and property.

I have always consistently opposed high-tension and alternating systems of electric lighting, not only on account of danger, but because of their general unreliability and unsuitability for any general system of distribution." Thomas A. Edison, 1889. (From *A Random Walk in Science*, compiled by R. L. Weber, The Institute of Physics, London, 1973.)

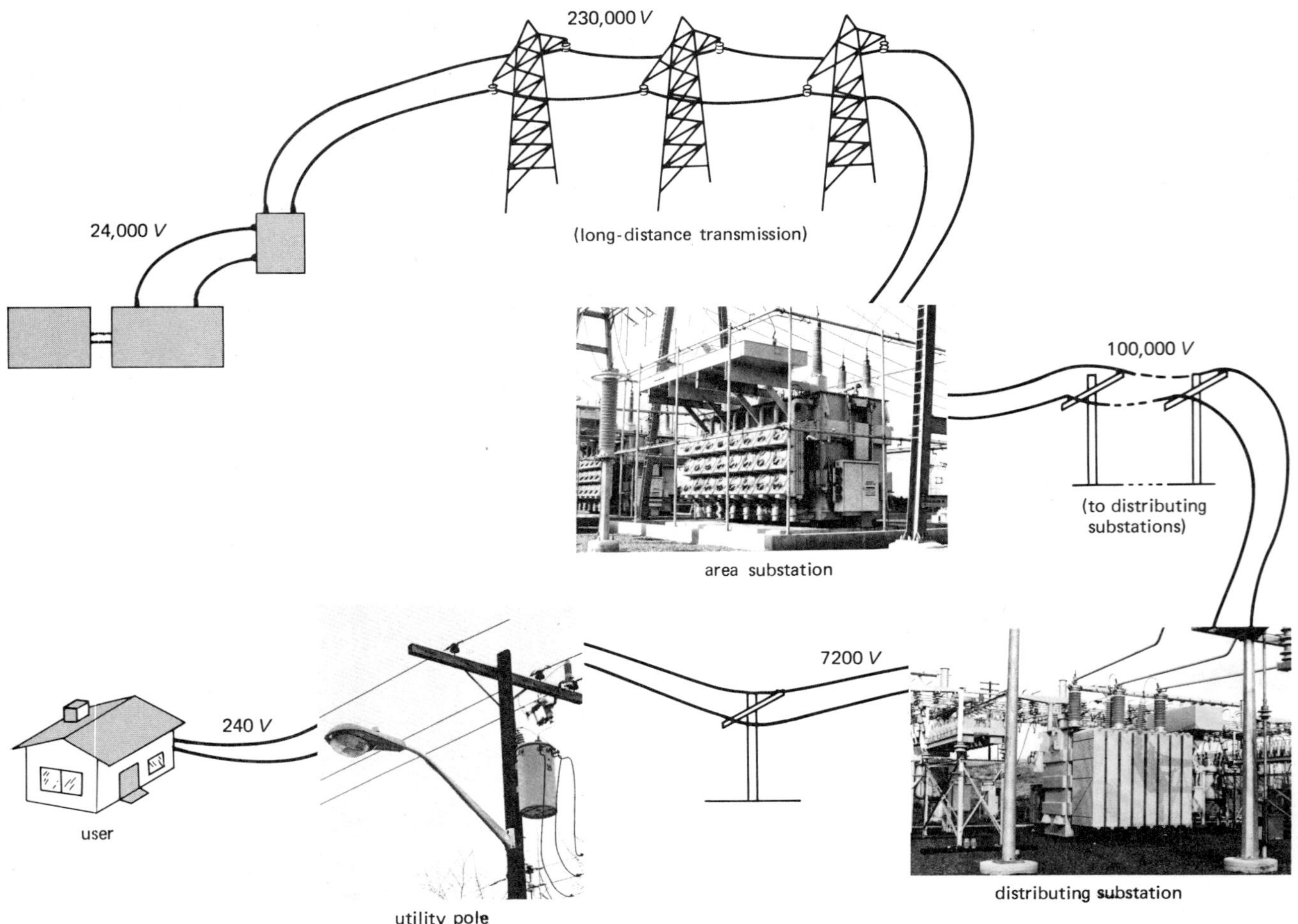

Figure 21.11 A diagram of a typical ac power transmission system.

we just sent this power over the line as generated, the line I^2R losses would be

$$P_{\text{loss}} = I^2R = (10\ \text{A})^2(15\Omega) = 1500\ \text{W}$$

Hence, the line losses would be more than 60 percent of the input power, $P = IV = (10\ \text{A})(240\ \text{V}) = 2400\ \text{W}$.

But suppose we use a step-up transformer and step the voltage up by a factor of 10 to 2400 V before transmission. The current would be stepped down by a factor of 10 to 1. The line losses in this case would be

$$P_{\text{loss}} = I^2R = (1\ \text{A})^2(15\Omega) = 15\ \text{W}$$

The line losses are reduced by a factor of 100 (as a result of the I^2 term) and would be less than 1 percent of the power input!

After high-voltage transmission, the power voltage is stepped down in a series of steps (Fig. 21.11) until the output voltage is the normal 240 V at your home.

Another source of power loss is through "leakage." At high transmission voltages, the air surrounding the lines may be ionized. This provides a conduction path to nearby trees, buildings, or the ground and gives rise to leakage losses. Long insulators are used on power-line towers to prevent the wires from coming too close to the tower (Fig. 21.12).

Leakage losses are generally greater during wet weather because moist air is more easily ionized than dry air. Under certain atmospheric conditions, the ionizing of air near power lines with very large voltages gives rise to corona discharge. The corona can sometimes be seen and the accompanying crackling noise heard.

Figure 21.12 Extra-high-voltage (500,000-volt) transmission line. Notice the long insulators at the towers and the insulators holding the suspended lines apart.

Electromagnetic Waves

We have considered electromagnetic waves or radiation generally in the study of heat and energy transfer. However, we are now in a position to understand the electromagnetic nature of these waves that are all around us. James Clerk Maxwell (1831 – 1879), an English scientist, collectively summarized the relationships between electricity and magnetism. The relationships between electric and magnetic fields are expressed in what are known as **Maxwell's equations.** A qualitative description of the general results is as follows:

> A changing magnetic field produces a changing electric field.
> A changing electric field produces a changing magnetic field.

Here, changing means varying with time.

The first statement reflects the fact that a changing magnetic field gives rise to an induced voltage, or *an electric field in space.* The second statement implies that a changing electric field in space gives rise to a changing magnetic field in space.

Electromagnetic waves are radiated by accelerating electric charges. Suppose electrons oscillate back and forth in simple harmonic motion. This could occur in a radio transmitter, where the oscillations have frequencies on the order of one million hertz. During the oscillations, the electrons are accelerated and radiate electromagnetic waves (Fig. 21.13). The continual oscillations of the charges produce oscillating electric and magnetic fields. These propagate outward, with the wave energy contained in the fields. The oscillating electric and magnetic fields are perpendicular to each other and to the direction of propagation.

Recall that electromagnetic waves need no medium of propagation and that all travel in vacuum at the

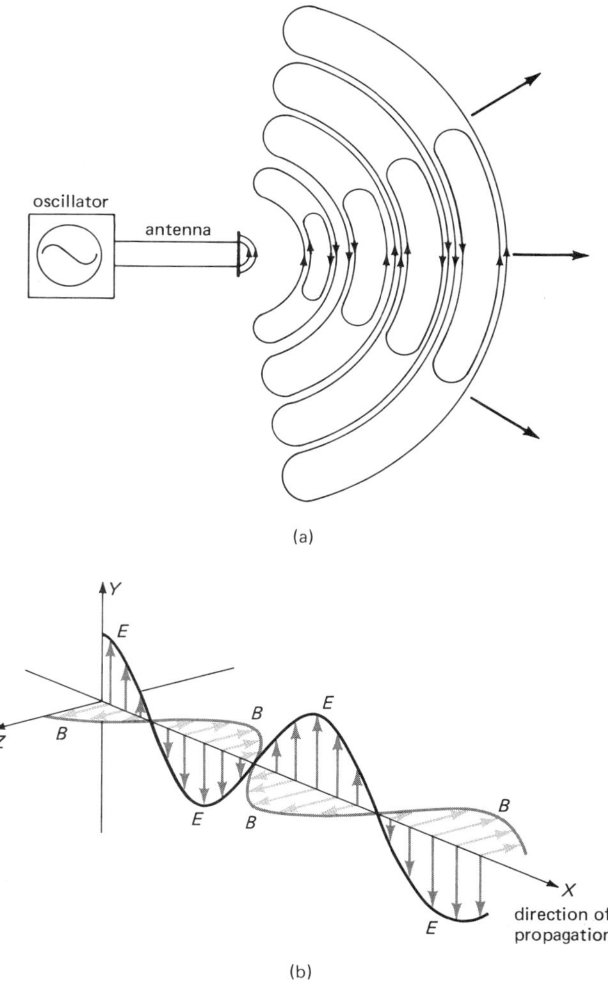

Figure 21.13 Electromagnetic waves are generated from electron oscillations in antennas (a). The oscillating electric and magnetic fields radiating outward are perpendicular to each other and to the direction of propagation (b).

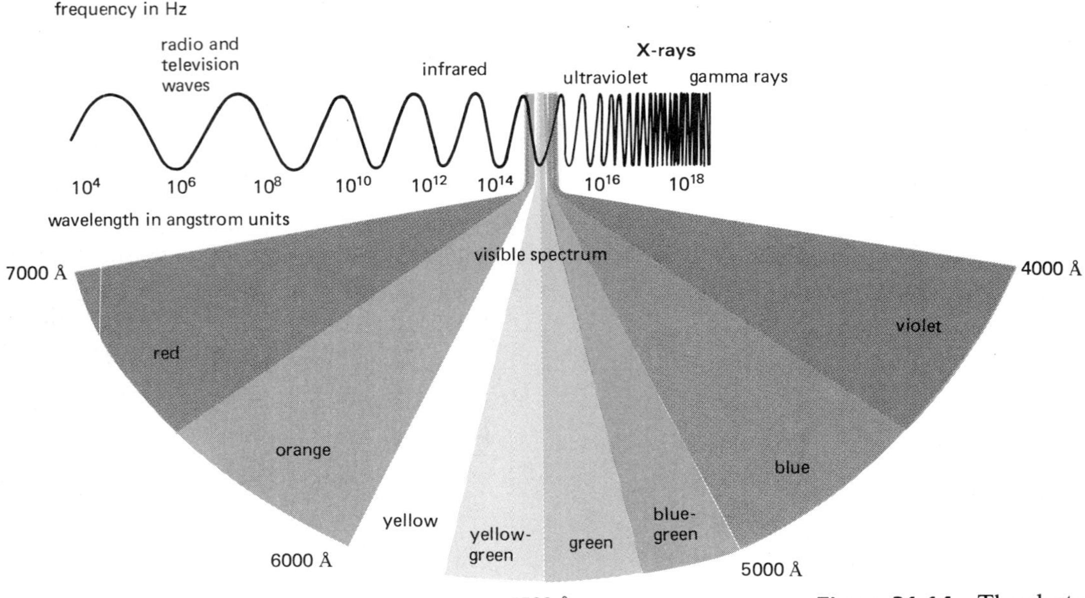

Figure 21.14 The electromagnetic spectrum.

same speed—the speed of light ($c = 3 \times 10^8$ m/s = 30,000 km/s ≈ 186,000 mi/s). What distinguishes one wave from another is the frequency or wavelength ($\lambda f = c$), which forms an electromagnetic (EM) spectrum (Fig. 21.14). Because of the short wavelengths involved with electromagnetic waves, the small length unit of the angstrom (Å) is used, where 1 Å = 10^{-10} m = 10^{-8} cm. Notice that the greater the frequency of a wave, the shorter its wavelength, and vice versa ($\lambda = c/f$).

Similar to the sound-frequency spectrum, different regions of the EM spectrum are given different names. Let's take a look at these types of waves in these frequency regions.

POWER FREQUENCY WAVES

Electromagnetic waves of the 60-Hz power frequency result from ac currents moving back and forth in electrical power lines and circuits. With wavelenghs of about 5000 km (3100 mi), these waves are of little practical importance. Occasionally you may pick them up as a 60-Hz ac hum in a stereo.

RADIO AND TV WAVES

AM radio waves are generally in the kHz range and FM radio waves in the MHz range. Higher frequencies are used in "short-wave" radio. Television channels are also in the MHz range. More will be said about these waves and devices in the next chapter.

MICROWAVES

Microwaves, with a frequency range of 10^9 to 10^{11} Hz, are used in communication and radar applications. Probably the most common use today is in microwave ovens (Chapter 14).

INFRARED RADIATION

Just below the visible region is the invisible infrared region. This type of radiation was considered earlier in the greenhouse effect (Chapter 7) and in radiation heat transfer (Chapter 14).

VISIBLE LIGHT

The visible spectrum is a very small part of the total EM spectrum, occupying a region between about 4×10^{14} Hz and 7×10^{14} Hz. As the activator of one of our major senses, visible light will be treated in greater detail in a later chapter.

ULTRAVIOLET RADIATION

The nonvisible ultraviolet (uv) radiation has been mentioned previously in a discussion of atmospheric ozone layer absorption (Chapter 7). Most of the uv radiation from the Sun is absorbed in the atmosphere, but we go out looking for the small transmitted portion when we go sunbathing to get a tan. The protective mechanism of the skin against uv radiation is pigmentation. The degree of penetration of uv radiation depends on the amount of melanin pigment in our skin, which all people have in varying amounts, except albinos. When skin is exposed to sunlight, the ultraviolet radiation induces the production of more pigment, causing the skin to "tan." People with little pigment may initially redden and burn. We now have chemicals or "sunscreens" with molecules that absorb part of the uv radiation before it reaches the skin and prevents severe burning. Incidently, ultraviolet radiation is also absorbed by ordinary glass, so you cannot get a tan through glass windows. You feel the "heat" coming through the window as a result of transmitted infrared radiation.

Ultraviolet radiation can be harmful to the eyes. Welders wear goggles or face masks with special uv-absorbing glass to protect their eyes from the large amounts of uv radiation produced by welding arcs. If you're a (snow) skier, you may also wear protective glasses, particularly in going cross-country. Sunlight reflected from snowy surfaces can produce "snow-blindness" due to ultraviolet radiation.

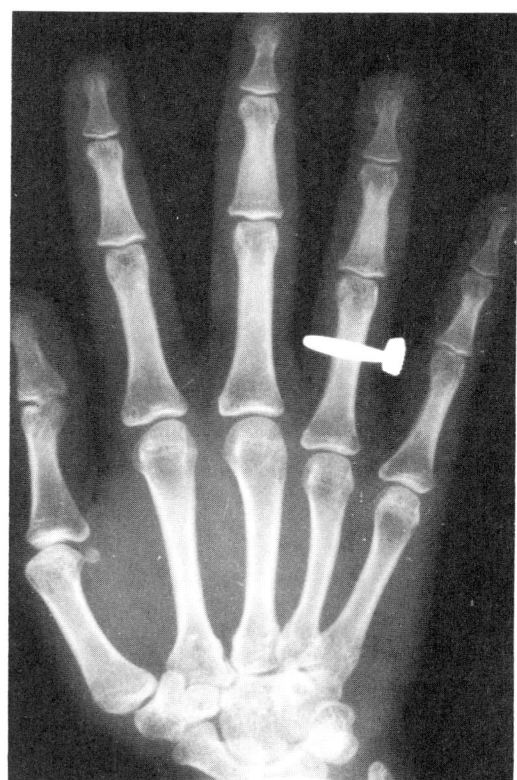

Figure 21.15 X-rays. Variations in X-ray absorption give a contrasted picture of internal structures (and external structures—the ring).

X-RAYS

Most of us have had an "X-ray" to check for a broken bone or some other medical condition. X-rays are highly penetrating electromagnetic radiations that can pass through many solid bodies that are opaque to other types of radiations. More dense, solid objects absorb more of the X-rays than less dense materials. In the human body, the bones absorb more X-rays than the surrounding tissue. As a result, a contrasted picture of internal body structures is obtained (Fig. 21.15). X-rays are used to check metal pieces for internal cracks, for example, stress cracks in airplane wings. At the airport, your bags get a shot of X-rays as they go through the security check so the security officer can check out any metal objects in your bag. (When you walk through the check port and your keys set off a buzzer, this is a magnetic effect; the metal in the keys takes energy from a magnetic field, and this sets off an alarm.)

The basic principle of an X-ray tube is illustrated in Figure 21.16. Electrons emitted from the heated cathode are accelerated toward a metal anode "target" by

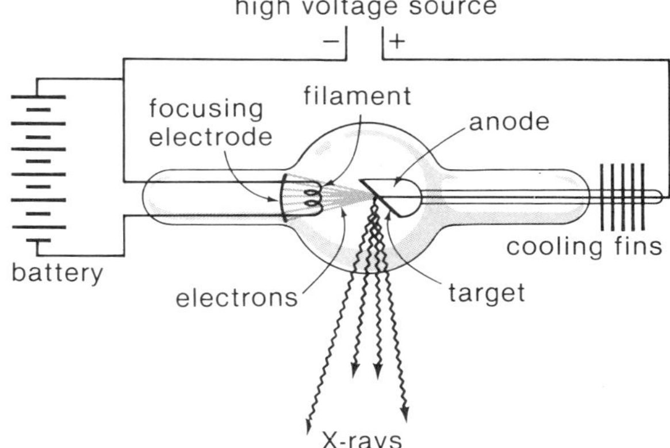

Figure 21.16 An X-ray tube. High-energy electrons from the cathode are decelerated by the electrons in the target anode. Radiation (energy) in the form of X-rays is emitted as a result of this "braking" action.

an applied high voltage across the tube (typically 50 to 150 kV). As the "cathode rays" collide with the target, they interact electrically with the electrons of the target material. The repulsive electric forces cause the cathode electrons to slow down. In doing so, they lose energy in the form of electromagnetic radiation, X-rays. In German the name for X-rays is *bremsstrahlung,* or "braking rays."

GAMMA RAYS

The gamma rays (γ-rays) at the upper end of the electromagnetic spectrum are produced in nuclear reactions and other processes in particle accelerators. This type of radiation will be considered in later chapters, where these processes are discussed.

SUMMARY OF KEY TERMS

Electromagnetic induction the inducing of a voltage or electric field as a result of a changing (time-varying) magnetic field.

Faraday's law of induction a voltage is induced in a conducting loop when there is a change in the number of field lines through the loop.

Lenz's law an induced current is in such a direction that its effects oppose the change that produces it.

Generator a device that converts mechanical energy into electrical energy.

Transformer a device that increases or decreases the magnitude of a voltage or current through electromagnetic induction. The change depends on the number of primary and secondary windings:

$$\frac{\text{Secondary voltage } (V_s)}{\text{No. of secondary windings } (N_s)} = \frac{\text{Primary voltage } (V_p)}{\text{No. of primary windings } (N_p)}$$

and

$$\text{Power to primary } (I_p V_p) = \text{power out of secondary } (I_s V_s)$$

if there are no losses.

Electromagnetic waves propagating electric and magnetic fields radiated from accelerated charged particles.

EXERCISES

1. Give three ways you could induce a voltage in a wire loop. Can you make it four or five ways?
2. If a thin coil with a large number of loops is rotated in the Earth's magnetic field, a voltage is induced in the coil. Since the Earth's magnetic field is "free," why isn't this method used to generate electricity?
3. Does the orientation of a loop with respect to a magnet make any difference in inducing a voltage through relative motion? Explain.
4. A voltage is induced in a stationary loop when a magnet is brought toward it or when the loop is brought toward a stationary magnet. What would be the effect if (a) both the magnet and loop were brought toward each other? (b) if both the magnet and loop had the same velocity (both magnitude and direction)?
5. Suppose the loop in Figure 21.3 were moved parallel to the field. Would there be an induced voltage in the loop? Explain.
6. A metal frame with a movable bar is in a uniform magnetic field (Fig. 21.17). (a) If the bar is moved to the right as shown in the figure, would there be an induced voltage in the frame? Explain. If so, what is the direction of the (conventional) current? (b) What would be the case if the bar is moved in the opposite direction?
7. Referring to Figure 21.17, consider the effect on an electron in the bar being in motion in a magnetic field. In what general direction is the magnetic force on the electron? How does this relate to the induced current?
8. Where does the induced voltage in a loop come from? Is this a case of getting something for nothing? (*Hint:* Remember that voltage is energy per charge.)

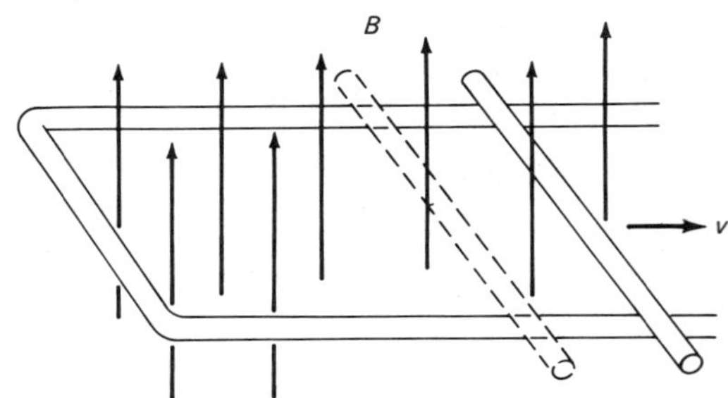

Figure 21.17 See Exercise 6.

Figure 21.18 See Exercise 11.

9. In inducing a current in a loop by bringing a magnet toward the loop, would it make any difference in either the magnitude or direction of the induced current if the north or south pole end were used? Explain. (*Hint:* Think in terms of the induced magnetic field opposing the change.)
10. When a magnet is brought toward a loop or coil, the magnetic field from the induced current is like that of a repelling magnet (see Fig. 21.6). As a result, work must be done in moving the magnet into a coil. Why is this?
11. A bar magnet is dropped through a coil (Fig. 21.18). (a) What will be observed on the galvanometer? (b) As the magnet enters, will the acceleration be greater or less than *g*? How about as it leaves? Explain.
12. Distinguish between self-induction and mutual induction.
13. Suppose the rotational frequency of a generator is increased from 60 Hz to 120 Hz. What effect would this have on the output voltage?
14. A motor and a generator perform opposite functions. Yet someone makes the statement that a motor really acts as a motor and a generator at the same time. Is this true?
15. An inventor proposes to revolutionize electrical generation with a new type of generator, as shown in Figure 21.19. (a) What type of voltage does the generator produce? (b) The inventor states that the internal frictional losses of the spring are practically negligible, so the magnet will oscillate indefinitely and therefore provide a cheap source of electrical power. Would you financially back this invention? Explain.

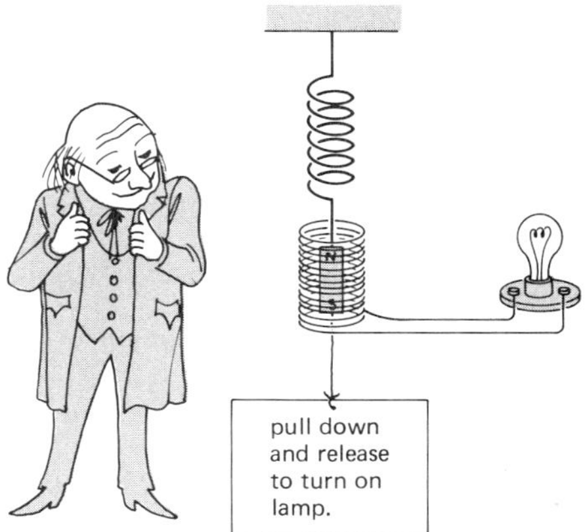

Figure 21.19 See Exercise 15.

16. Give three ways the voltage output of a generator could be increased.
17. What is actually transformed by a transformer?
18. Would a transformer operate on (a) a battery? (b) a dc generator?
19. Would two coils act as a transformer without an iron core? If so, why not omit the core and save money?
20. What would be the effect(s) if the primary and secondary of a transformer had the same number of windings? What kind of "step" transformer would this be?
21. Some transformers have various terminals or "taps" on the secondary so that connecting to different taps puts different fractions of the total number of secondary windings into a circuit. What is the advantage of this?
22. A fluorescent lamp needs an initial voltage of 12,000 V to ionize the internal gases. This is supplied by a transformer with a 120-V input. How many times more windings are on the transformer secondary than on the primary?
23. An ac source has a 10-amp output. A particular circuit requires a 2-amp input. How could you accomplish this? Explain.
24. A person has a single transformer with 50 windings on one part of the core and 500 windings on the other. Is this a step-up or a step-down transformer? Explain.
25. Does a transformer operate on self-induction or mutual induction? Does it have both? If so, what are the effects?
26. In the ignition system shown in Figure 21.10, if a voltage of 12,000 V is required to fire a spark plug, what can you say about the "coil" windings?
27. If the line losses in electric power transmission are reduced by stepping up the voltage, why not transmit electric power at even higher voltages than those normally used?
28. Are all electromagnetic waves the same? Explain.
29. How do radio waves differ from microwaves and X-rays?
30. Why do people usually get darker or tanner when continually exposed to sunlight?

31. When you are sunbathing and the sun "goes behind a cloud," it feels cooler. It is also common to get a sunburn on a cloudy, cool day. Explain these experiences in terms of types of electromagnetic waves. (*Hint:* Consider absorption and transmission in the "greenhouse" effect.)
32. How do "sunscreens" in suntan oils and lotions prevent sunburns?
33. In movies and in actual practice, money is sometimes "marked" and messages are written with substances that can be seen only under ultraviolet light. Why is ultraviolet light so special?
34. Why is an X-ray tube sometimes called a cathode-ray tube?
35. Electromagnetic waves are generally radiated by accelerating electric charges. Yet X-rays are called "braking rays" instead of "accelerating rays." Explain why.

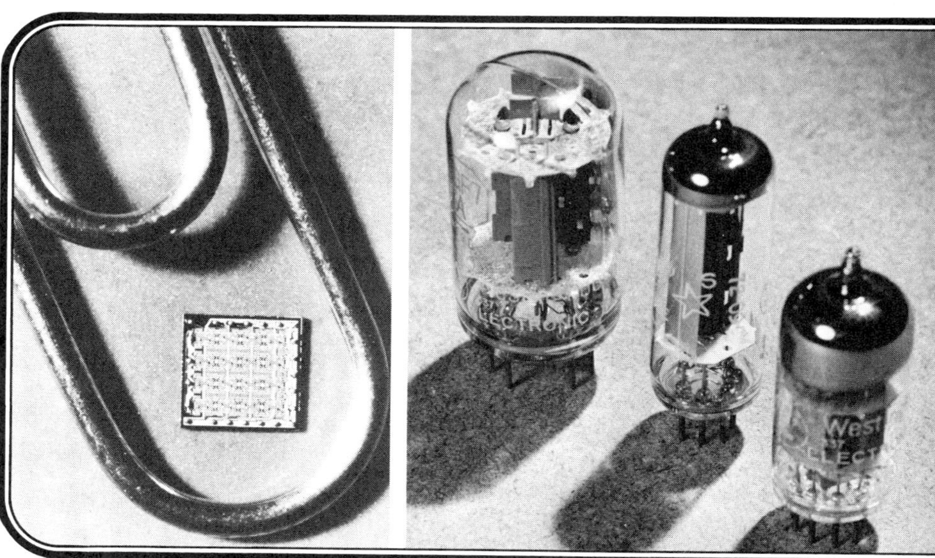

22 Electronics and Communications

Electronics is everywhere — radio, television, calculators, digital clocks, video games, and more — we live in an electronic age. Without electronics, there would not be many of the things that have become so commonplace in our lives. Today when someone mentions electronics, we tend to think of the spectacular advances made in the last 40 years that have given us new terms such as semiconductor, transistor, and silicon "chip." However, electronics was around long before this, in the heyday of the vacuum tube.

Basically, electronics has to do with the actions of electrons that we manipulate and control. The discovery of the electron vacuum tube in the late 1800's gave electronics its start. Early radios, TVs and even computers contained "tubes." You may still have a tube radio or TV around the house. Beginning in the 1940's, other discoveries were made that revolutionized electronics. Work in solid-state physics showed that the functions of vacuum tubes could be done by solid-state semiconductors. This allowed a great reduction in size and in power consumption. For example, we now have small "transistorized" radios that operate on small batteries.

Since the key to electronics is the control of electrons, how do we do it? The answer is, with electric and magnetic fields as studied in the previous chapters. Now we'll further apply those principles you studied so hard to learn so you can more fully appreciate and understand the electronic age.

The Electron Tube

The electron tube has been replaced in large part by solid-state devices. However, it is instructive to consider electron tubes to see how their operations are related to semiconductor diodes and transistors. Thomas Edison made the discovery that forms the

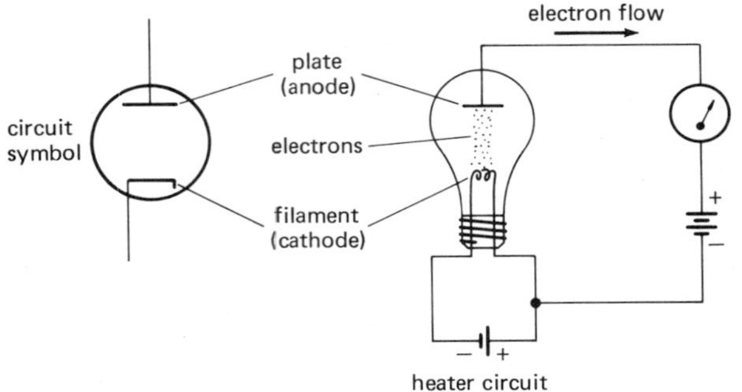

Figure 22.1 A diagram of an electron tube circuit and the circuit symbol for a diode tube.

basis of the electron tube. In experimenting with his lamp, Edison placed a metal plate in a bulb and connected the plate to a battery (Fig. 22.1). He observed that current would flow in the plate circuit when the positive terminal of the battery was connected to the plate. However, when the negative terminal of the battery was connected to the plate, no current would flow.

Edison recorded and even patented this so-called **"Edison effect,"** but he did not apply it. Without the benefit of modern theory, the Edison effect was also unexplained. You can do this right away. Electrons are "boiled off" the hot lamp (cathode) filament and are attracted to the positive (anode) plate. When the polarity is reversed, the electrons are repelled from the negatively charged plate, and no current flows. Simple enough. When the plate is positive and the plate voltage is increased, the electron plate current also increases, up to a *saturation* point at which all the emitted electrons are attracted to the plate.

With two electrodes, an anode and a cathode, such tubes are called **diodes** (*di*, meaning "two"). There is usually a partial vacuum in an electron tube, since air molecules would hinder the electron flow and cause the filament to oxidize. As a result, electron tubes are commonly referred to as *vacuum tubes.*

Rectification: A One-Way Street

One of the great advantages of the vacuum tube is that it can be used for **rectification,** that is, to change ac to dc. As was noted, the plate current can be turned on and off like a "valve" by switching the battery leads or polarity. (In Great Britain, electron tubes are called valves.) If an ac source is used instead of a battery, the polarity of the plate changes automatically each half-cycle. This effect is shown in Figure 22.2, where an ac input is applied through a transformer. The output voltage across the load resistance has only positive pulses and so is dc. This is called **half-wave rectification.** Vacuum tubes were used to do this in the early 1900's in converting radio signals from ac to dc. **Full-wave rectification** can be obtained by using two diodes, as shown in Figure 22.3.

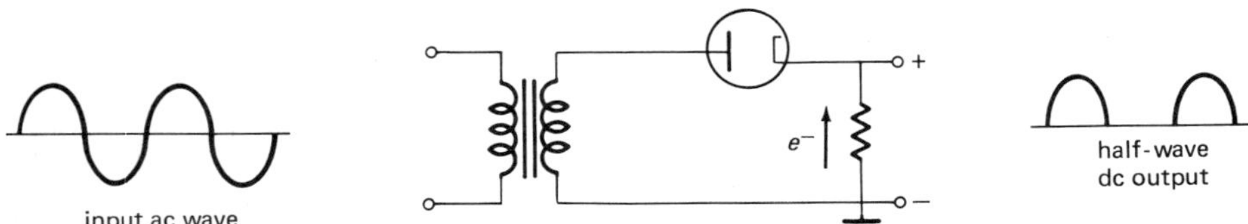

Figure 22.2 Half-wave rectification with a diode. The tube will conduct only during the input half-cycle with the proper polarity, which gives a half-wave dc output.

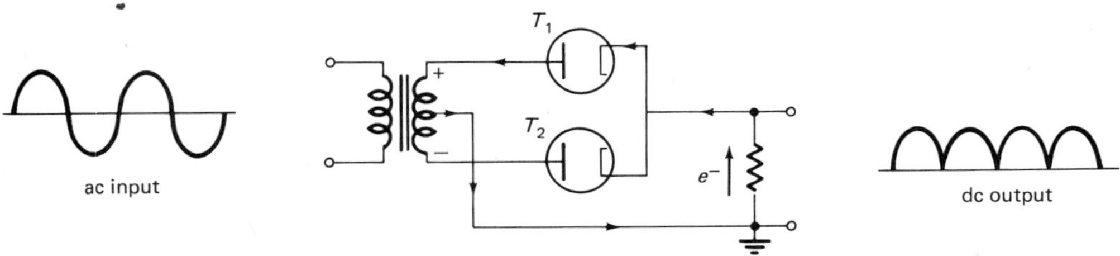

Figure 22.3 Full-wave rectification with two diodes. The tubes conduct during alternate half-cycles to give a full-wave dc output.

The Triode and Amplification

In 1907 Lee DeForest, an American scientist, made an important improvement to the vacuum tube by adding a third element called a **grid.** Its effects had such a tremendous impact on radio electronics that DeForest is often called the "father of modern radio." The grid consists of a fine-wire coil or mesh located between the plate and the cathode (Fig. 22.4). Electrons from the cathode can easily pass through the grid.

The grid can be used to control the plate current, hence its common name, *control grid.* If a positive "bias" voltage is put on the grid, the tube is quickly saturated, since this voltage adds to the plate voltage in drawing electrons from the cathode. However, if the plate has a negative bias (voltage), the electrons will be impeded, and less current will flow. The more negative the grid voltage is made, the smaller the plate current becomes (Fig. 22.5). Finally, a voltage is reached at which the plate current goes to zero. This is called the **cut-off voltage,** and the tube is "cut-off."

These characteristics are the basis for signal amplification. When something is amplified, it is increased or made bigger by adding energy. Amplifiers are important parts of radio, TV, and stereo sets. Here, electronic circuits amplify either voltage or current, or both (power, $P = IV$).

Let's see how a triode can amplify a signal. Electronically, a signal means a small voltage, usually alternating, such as a radio signal or wave. Suppose for the circuit in Figure 22.6 that an input signal to the grid of the triode has a peak voltage of 1 V. Let the grid bias be -4 V and the tube characteristics be as in Figure 22.5. The signal will cause the grid voltage to increase (and decrease) by one volt. From Figure 22.5 we see that a 1-volt change in the grid voltage produces a 1-mA (0.001 A) change in the plate current.

Now, the plate current must flow through the load resistance in the plate circuit. A change in the plate current will therefore produce a change in the voltage across the resistance. With a 25 kΩ (25,000Ω) resistance, for a 0.001-A current change this would be

$$V = IR = (0.001\ \text{A})(25{,}000\Omega) = 25\ \text{V}$$

Hence, a 1-volt change in the grid voltage produces a 25-volt change across the resistance! How about that? This is a voltage amplification or gain with an amplification factor of 25.

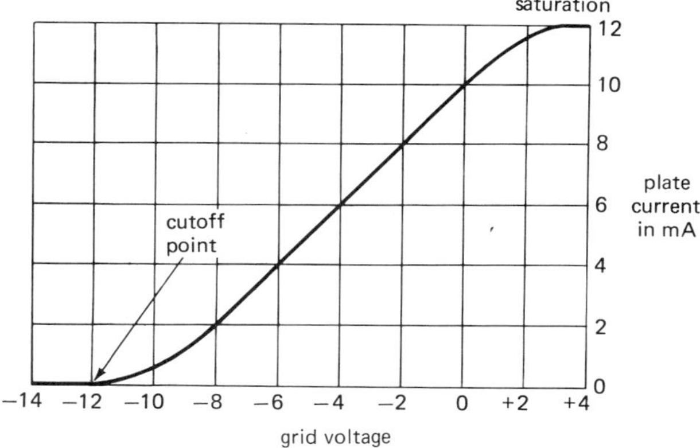

Figure 22.5 Grid voltage versus plate current. With a positive grid voltage or "bias," the tube is quickly saturated. With a negative bias, a cut-off voltage is reached, at which the plate current is zero.

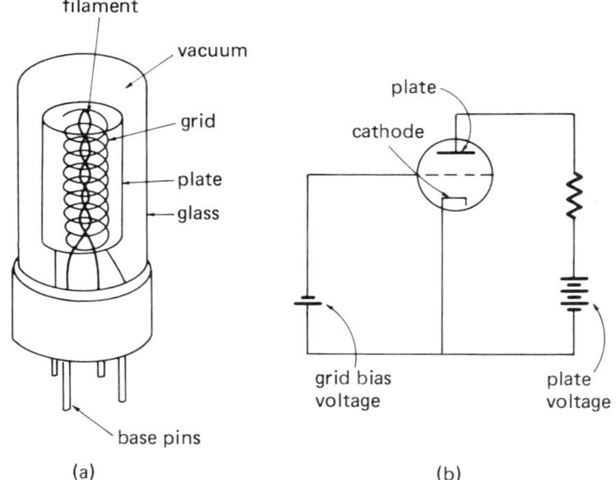

Figure 22.4 Triode tube. (a) The third element, a grid, consists of a fine-wire coil or mesh between the plate and cathode. (b) A triode circuit showing the grid bias voltage that can be used to control and amplify a signal.

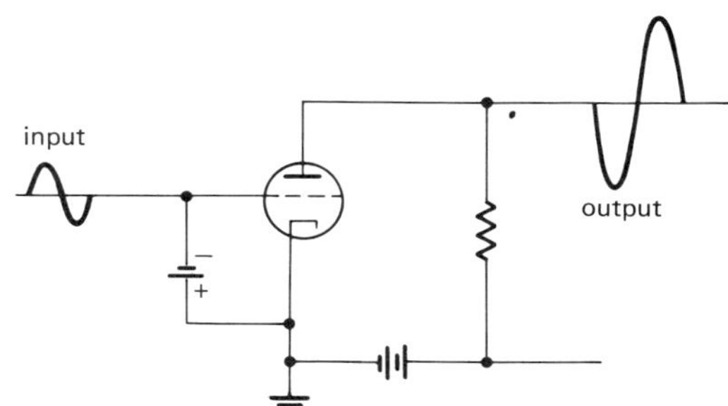

Figure 22.6 Simple amplifier. A 1-volt change in the grid voltage produces a 25-volt change across the resistance (25 kΩ), or the output is amplified. See text for description.

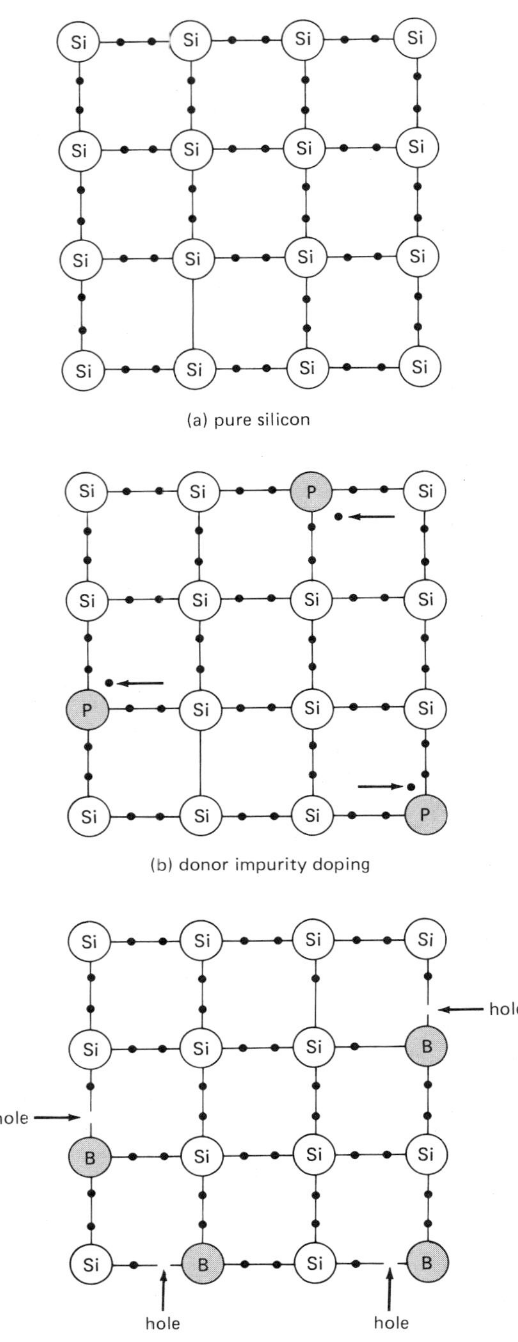

Figure 22.7 Doped semiconductors. (a) Pure silicon is a good insulator. (b) Impurity atoms that donate extra electrons add to the conduction and give an N-type semiconductor. (c) Impurity atoms lacking electrons create "holes" that add to the conduction and give a P-type semiconductor.

Solid-State Semiconductors

Solid-state physics is the study of all kinds of solids. Of particular interest in electronic applications are semiconductors. Recall that **semiconductors** are a group of materials with intermediate electrical conductivity.

Two of the most important semiconductor materials are silicon and germanium. The atoms of these elements have four outermost electrons, which are shared with other atoms in a crystalline lattice. This is illustrated for silicon in Figure 22.7. In pure crystals of silicon, all the electrons are bound, which leaves none free to conduct electricity. Hence, pure silicon (and germanium) is a good insulator.

However, there are usually some impurity atoms available that make it a semiconductor. This is done on purpose by "doping" or adding impurity atoms to the pure material. For example, phosphorus atoms might be added to silicon (Fig. 22.7). Phosphorus atoms have five outermost electrons, so there is one extra electron per impurity atom in the doped lattice. If a voltage is applied to the material, the extra electrons can move in the crystal, and it is semiconducting. An impurity that contributes extra electrons is called a **donor impurity,** and a semiconductor with negative (electron) charge carriers is called an **N-type semiconductor.**

Similarly, pure silicon may be doped with boron atoms (Fig. 22.7). Here we have three outermost electrons, and so there is a "hole" or vacancy in the lattice. So now what happens when a voltage is applied? In this case, electrons can transfer to fill the holes, thereby leaving other holes behind. Hence, a flow of electrons in one direction is equivalent to a flow of holes in the opposite direction. In effect, the holes behave like positive charges, and we refer to positive holes. An impurity, that creates a "hole" or vacancy is called an **acceptor impurity,** and a semiconductor with positive "holes" as charge carriers is called a **P-type semiconductor.**

Some materials become semiconductors when exposed to light. A common application of this is discussed in Special Feature 22.1.

The Junction Diode

Using N-type and P-type semiconductors, solid-state diodes and transistors can be made that take the place of vacuum tubes. When a thin piece of P-type semiconductor is placed in contact with a piece of N-type semiconductor, a **junction diode** is formed. At the junction between the two crystals, the charge carriers of each tend to diffuse into the other crystal (Fig. 22.8). This charge-separating action sets up a potential barrier across the junction. That is, a small voltage or potential exists between the regions close to the junction, which prevents further diffusion.

The potential barrier gives the joined N-P crystals the unidirectional current flow characteristic of a diode. If an external voltage is applied to the crystals with the P-type positive and the N-type negative (Fig. 22.8), the potential barrier is reduced, and current flows in the circuit. With this polarity, the diode is said to be biased in the forward direction or to have forward bias.

However, if the polarity is reversed so the diode has a "reverse bias," the potential barrier is increased, and little or no current flows across the junction. Hence, a junction diode has the same directional current-flow characteristics as the vacuum-tube diode. For example, diode circuits are used for rectification. The basic circuits are the same as those shown in Figures 22.2 and 22.3 with the vacuum tubes replaced by diodes. Another application of diodes is discussed in Special Feature 22.2.

The Transistor

The next step is adding some control by making a transistor, which is the solid-state analog of the triode vacuum tube. This is done by sandwiching a thin piece of

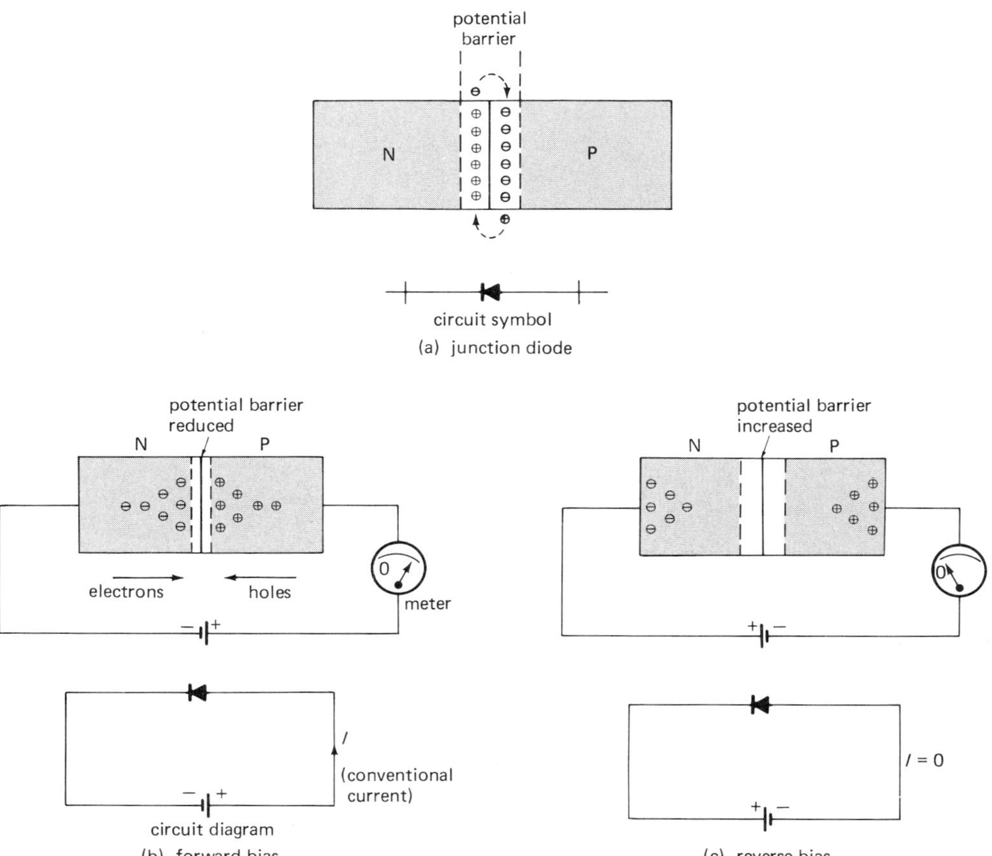

Figure 22.8 (a) A junction diode is formed by the contact of two semiconducting materials (N-P). The solid-state diode behaves like a tube diode or "valve" and conducts current when biased in the forward direction (b) but not when reverse-biased (c).

SPECIAL FEATURE 22.1

Xerography and Electrostatic Copiers

Xerography is a dry process by which anything written, typed, printed, or photographed can be copied. The word is coined from the Greek *xeros,* meaning "dry," and *graphein,* meaning "to write."

The heart of xerography is the photoconductor. Some semiconductors are light-sensitive, or photoconductors. In the dark, a photoconductor such as selenium is a good insulator and hence can be electrostatically charged. However, when light strikes a photoconductor, it becomes conductive and permits the electrical charge to leak away from the part hit by the light.

In transfer electrostatic copying, a photoconductor-coated plate, drum, or belt is electrostatically charged and then illuminated with a projected image of the writing or object to be copied (Fig. 22.9). The illuminated portions of the plate become conducting and discharge.

The plate then comes into contact with a negatively charged black powder called toner or "dry ink." The charge-retaining portions of the plate attract the toner, and it adheres to these regions. Paper is then placed over the plate and a positive charge is delivered to the back of the paper. This attracts the toner from the plate surface to the paper. The image is then fused into the paper permanently by heat, and a copy of the original image is produced.

one type of crystal between two pieces of the other type of crystal, sort of like making a ham sandwich (Fig. 22.10). Either type of crystal can be the "ham," and we refer to NPN and PNP transistors. In either case, one of the outer crystals is called the *emitter,* the center crystal is called the *base,* and the other outer crystal is the *collector.* These correspond to the cathode, grid, and plate, respectively, in a triode vacuum tube. In effect, a **transistor** is two diodes back-to-back.

To get an idea of how a transistor works, consider the circuit, called a common-base circuit, for the NPN transistor shown in Figure 22.11. Notice that the NP junction of the emitter-base circuit is forward-biased. The electrons from the negative terminal of the battery enter the emitter and flow across the forward-biased junction. The base is a very thin section, and most of the electrons stream right on through the base into the collector, despite the reverse bias of the base-collector

SPECIAL FEATURE 22.2

LEDs — Light-Emitting Diodes

The LED displays used for hand calculators, cash registers, digital clocks, gas pumps, and so on are applications of a special type of semiconductor diode. When a semiconductor diode is biased in the forward direction, electrons and holes flow in opposite directions, as we have seen. Some of the electrons and holes combine in the process, and when they do, energy is released in the form of light. If one of the semiconducting materials is transparent, such as gallium phosphide, light is emitted from the diode (LED). LEDs can be shaped in the form of numbers and letters for use in readout displays. (LEDs have been replaced in large part by LCDs or liquid-crystal displays. This type of display will be discussed in Chapter 23.)

Solar cells are made by using the reverse process. If a diode has a thin, transparent P-type layer, incoming light will pass through the crystal to the junction and strike the N-type (electron carrier) material. This can give electrons enough energy to be kicked across the junction into the P-type material. As a result, a voltage develops across the diode, similar to what happens in a battery, and a "solar current" would flow in a circuit connected to the diode. (See Chapter 7 for more on solar-cell uses.)

How Xerography Works

1 Surface of a plate coated with a photo-conductive metal is electrically charged as it passes under wires.

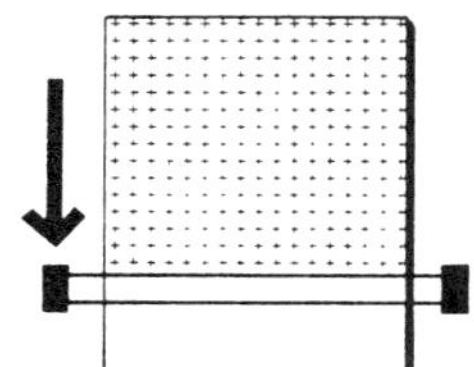

2 Plus marks represent positively charged plate.

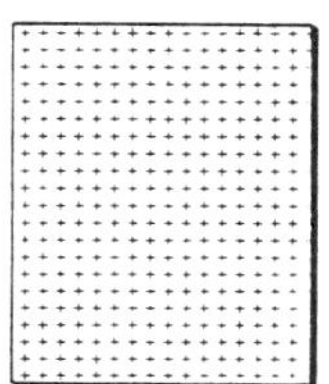

3 Original document is projected through a lens. Plus marks represent latent image retaining positive charge. Charge is drained in areas exposed to light.

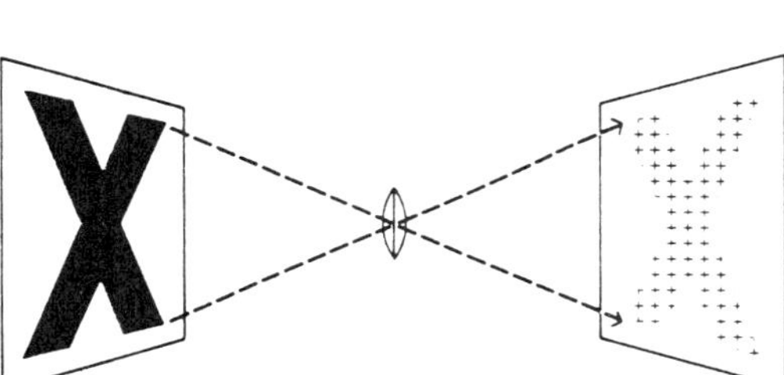

4 Negatively charged powder (toner or "dry ink") is applied to the latent image, which now becomes visible.

5 Paper (or other material) is placed over plate and given positive charge.

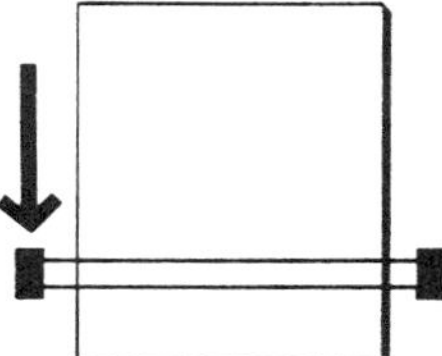

6 Positively charged paper attracts dry ink from plate, forming direct positive image.

7 Image is fused into the surface of the paper or other material by heat for permanency.

Figure 22.9 Electrostatic copying. Xerography process steps and electrostatic copier. This Xerox 8200 copier/duplicator uses a photoconductor-coated belt rather than a drum, as commonly used in older models.

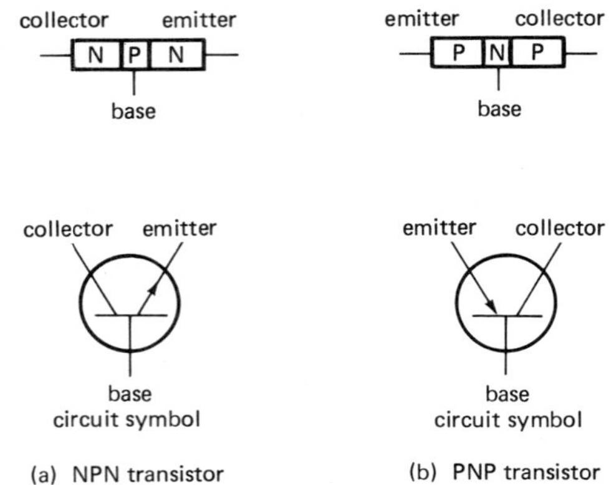

Figure 22.10 The transistor is a solid-state triode. (a) An NPN transistor and (b) a PNP transistor.

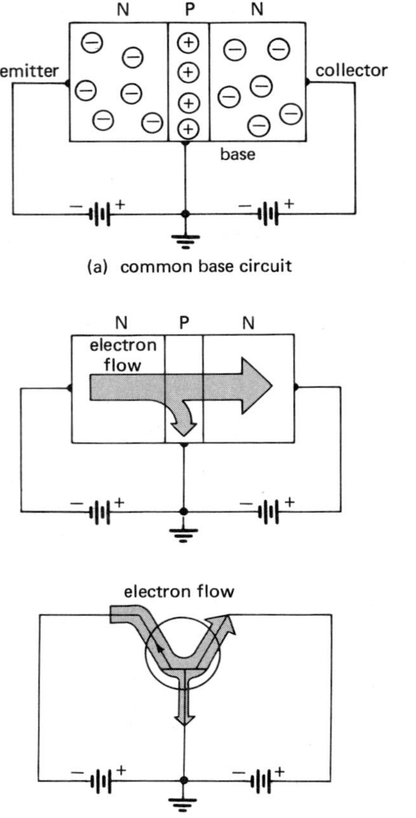

Figure 22.11 Transistor circuit. (a) A common-base circuit. (b) Charge flow in the circuit. There is no current amplification, but there is a voltage amplification. See text.

junction. The reverse bias does give a high resistance direction, but once the electrons make the junction, it's a "downhill slide" for them into the collector. Only about 5 percent or less of the electrons remain in the base-emitter circuit.

The output current in this case is slightly less than the input current, so there is certainly no current amplification or gain. But what about the voltage gain? That's a different story. The resistance of the forward-biased NP (emitter-base) junction is low, and the resistance of the reverse-biased PN (base-collector) junction is high. Since about the same current flows through both junctions, the output voltage is much greater than the input voltage ($V = IR$), and there is voltage amplification. This can be very high, depending on the circuit components.

This is how the transistor is said to have gotten its name. "Trans" stands for *trans*fer and "istor" for re*sist*ance. The gain of a transistor essentially comes about by a "transfer of resistance," or "transistor."

A transistor can be connected in other ways to achieve a current gain, and there are various types of transistors with different characteristics. The solid-state electronics we use so frequently gets complicated very quickly. This discussion here was meant to serve only as an introduction and to give some idea of how diodes and transistors work. These devices have all but replaced vacuum tubes in many applications. Solid-state diodes and transistors offer great advantages in size, power consumption, and material economy (Fig. 22.12). No heating element (filament) is required for

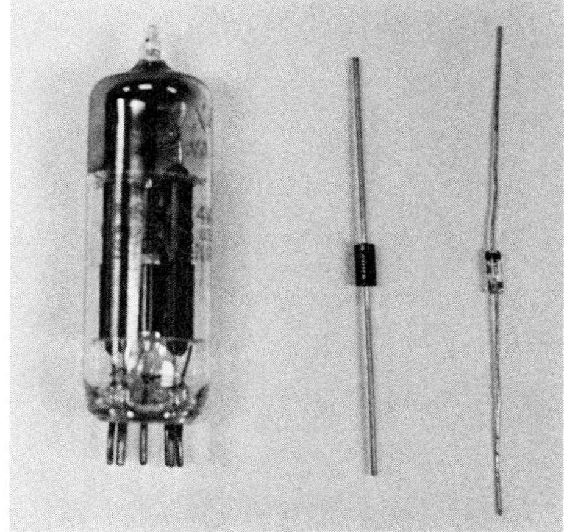

Figure 22.12 Tube and solid state. Note the relative sizes of the solid-state components. On a chip, these can be minuscule. See Chapter introductory photo.

diodes and transistors, as it is in the case of vacuum tubes. This gives a difference of milliwatts versus watts in power consumption.

Printed and Integrated Circuits

With the development of solid-state components, there have been rapid strides toward circuit miniaturization. This has been particularly true and necessary in aerospace applications, where size and weight are of prime importance. However, it is also noticeable in everyday applications, for example, pocket calculators, radios, and mini TVs.

One method of assembling the miniature components in a convenient package is on a printed circuit (Fig. 22.13). The "printing" of the circuit takes the place of wires and is done on a copper-clad fused glass cloth board. The circuit is printed on the board by hand or applied by a silkscreen process using an acid-resistant paint. When placed in an acid bath, the unpainted portions of the copper are etched away, leaving the desired circuit connections. A photographic method is also used. The copper board is coated with a photosensitive material. A mask of the desired circuit is placed on the board, which is exposed to light. In the developing process, the light-exposed surfaces are etched away.

After the circuit has been etched, small holes are drilled at the proper locations for the mounting of the leads of the circuit elements, for example, resistors, capacitors, diodes, and tube sockets, which are soldered in place. Printed circuit plug-in modules make the assembly and repair of electronic circuits quite easy. A television may have its "works in a drawer," and repairs are made simply by replacing a circuit board.

An even further miniaturization has come about through the use of integrated circuits (ICs). An integrated circuit may consist of many diodes, transistors, and resistors fabricated on a silicon "chip," usually only a few millimeters in size (Fig. 22.14, and Chapter introductory photo). Single crystals of lightly doped silicon are drawn from molten material (about 1100°C). These are cleaved into thin wafers. Then hundreds of identical IC chips are fabricated on the surface of a wafer in a series of steps. The chips are then separated and packaged for use.

You may have wondered how components are fabricated on the chips. This is done by several techniques. For example, diodes and transistors can be made by vapor deposition. Successive deposits of N- and P-type

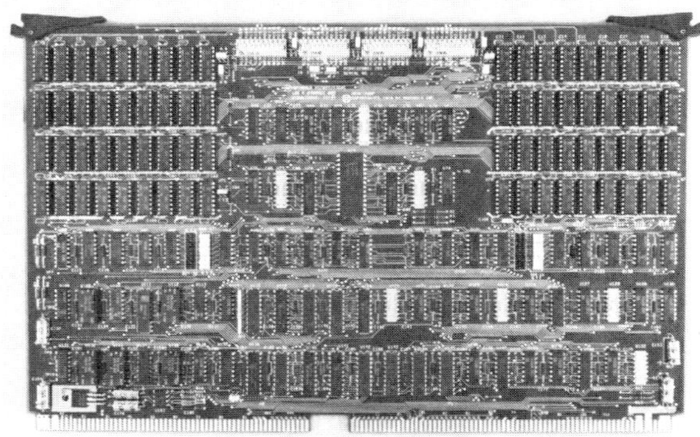

(a)

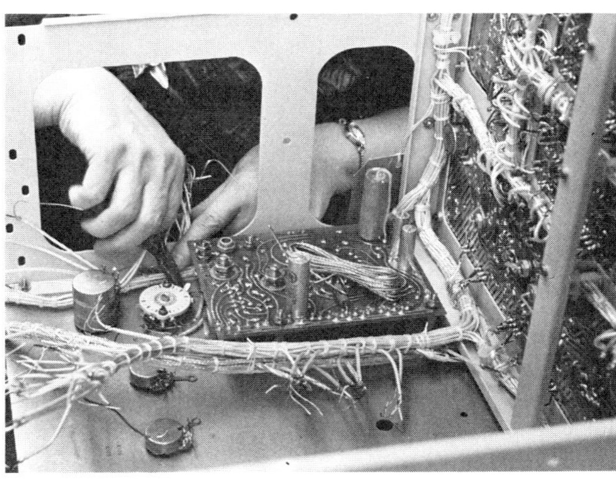

(b)

Figure 22.13 Printed circuits. (a) A printed-circuit memory module. (b) Printed-circuit boards being used in macro circuits.

impurities diffuse into the wafer to create diodes or the PNP or NPN structures needed for transistors. A focused beam of ion impurities can also be used in a process called ion implant.

Resistors can be made from N- or P-type regions. However, these have wide tolerances or ranges, which makes for poor precision. For precision applications, a small amount of material such as nichrome (nickel-chromium alloy) is deposited on the wafer, and a laser is used to burn away the nichrome until the desired resistance is obtained.

Sounds good, but how can these components be connected together in circuits? The basic means of obtaining interconnecting circuits is by a photosensitive technique. The silicon wafer is covered with a thin layer of photosensitive plastic. A photographic negative with

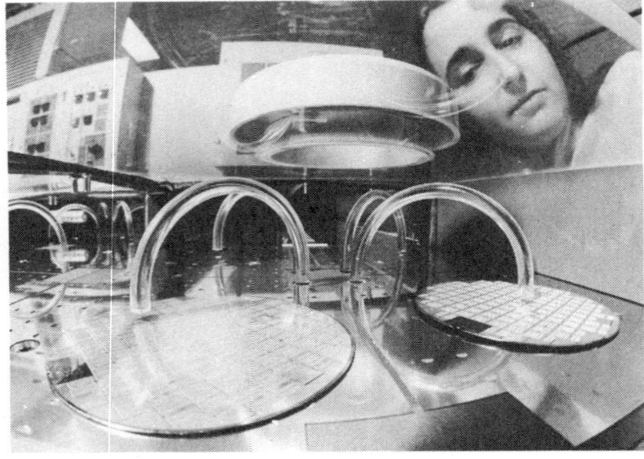

(a)

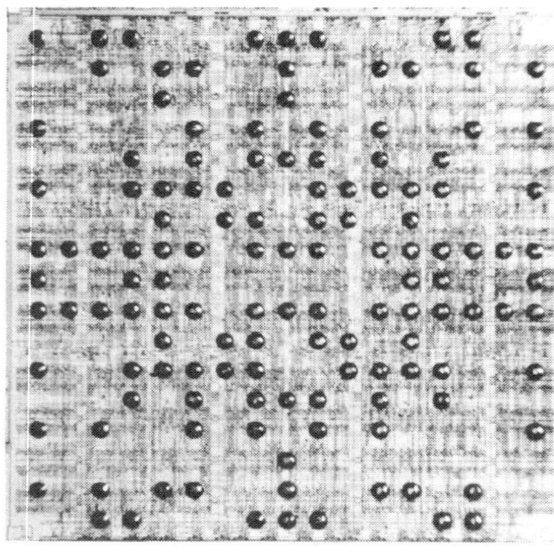

(b)

Figure 22.14 Integrated circuits. (a) The processing of razor-thin slices of silicon. The thin wafers float on a cushion of air, and the tube-like optical sensor loops control their movement. (b) The finished product—a high-performance logic chip for a computer. This chip contains more than 700 logic circuits.

the desired circuit pattern for the chips is placed on the wafer, which is then exposed to ultraviolet light. Where the light strikes the plastic coating, it hardens and remains in place when the unexposed portion is washed away. A thin layer of metal is then deposited over the wafer. The hardened plastic is dissolved, and the metal in these regions is removed with it. This leaves a metal pattern on the surfaces where the plastic was originally removed that interconnects the components. It is amazing that a single chip can contain hundreds of circuits.

Communications

Communicating is the process of conveying or transferring information. A variety of ways have been developed to do this. We can talk directly to each other. Out of hearing range, we could use smoke signals or signal lights. Today, our long-distance communications are primarily by electronic means, for example, telephone, radio, and television.

One of the first electronic means of communication was the telegraph. This was basically a long-distance electromagnet or relay. The manual opening and closing of a switch or key at one end of a line, opened and closed the electromagnetic relay at the other. This was hooked to a "sounder" that made audible sounds. Messages were transmitted in code, usually Morse code, which is combinations of dots and dashes (named after Samuel F. B. Morse, who is credited with inventing the telegraph).

MICROPHONES AND SPEAKERS

In the telegraph, information is carried by means of electrical impulses or signals, and there is no voice communication involved. To send and receive verbal communications, we must first change the sound to electrical signals for sending and then later change the electrical signals back to sound for receiving. As we all know from experience, this is done by microphones and speakers (sometimes called loudspeakers and sometimes very loud). So, before looking at radio communications, let's see how sound energy is converted to electromagnetic energy, and vice versa.

Diagrams of two common types of microphones or "mikes" are shown in Figure 22.15. The carbon microphone has a "button" of carbon granules. When sound waves striking the diaphragm compress the carbon, there is better contact between the carbon granules, and the resistance is reduced. More current then flows in the circuit. Similarly, when the diaphragm bounces back, the carbon resistance increases, and the current is less. Hence, the sound waves produce an electrical signal pattern in the circuit. You use a carbon microphone all the time in using a telephone.

Another common type of microphone is the "dynamic" microphone. Here, sound waves striking the diaphragm cause a coil to move back and forth on a magnet. With the coil moving in a magnetic field, there is an induced voltage and current in the coil with the same pattern as the sound wave that caused the motion.

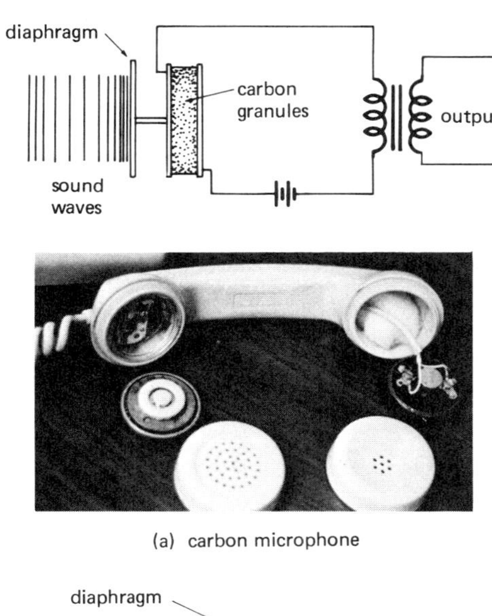

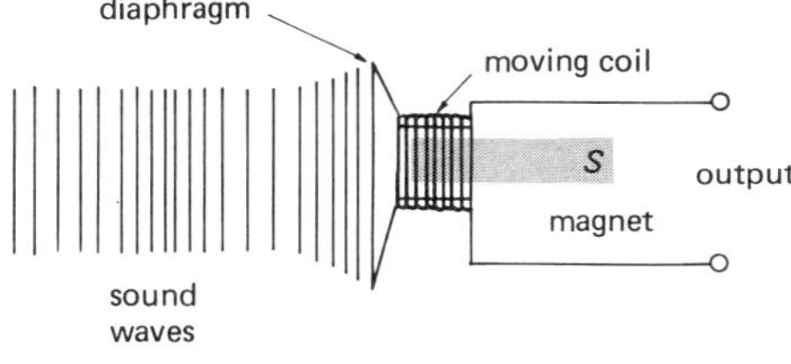

(a) carbon microphone

(b) dynamic microphone

Figure 22.15 Microphones. (a) In a carbon microphone, the sound waves striking the diaphragm change the packing and electrical resistance of the carbon granules. These microphones are commonly used in telephones (photo). (b) In a dynamic microphone, a voltage is induced by a moving coil on a magnet.

Sound reproduction through speakers is also usually done by electromagnetic means. As illustrated in Figure 22.16, a voice coil is attached to a speaker cone. When the electrical current of a "sound pattern" is in the coil, the interaction with the magnetic field of the magnet causes the coil to move back and forth. (Do you remember why?) The coil motion causes the speaker cone also to move back and forth, causing disturbances in the air, or sound waves. The quality and frequency response of the speaker depends on its construction and design.

A similar type of receiver used in telephones (Fig. 22.16) and headsets uses a stationary coil with many turns of wire on the poles of a permanent magnet. The diaphragm is this case is soft iron, which is held in place and under tension by the magnet. Current signals in the coils strengthen and weaken the pull of the magnet on

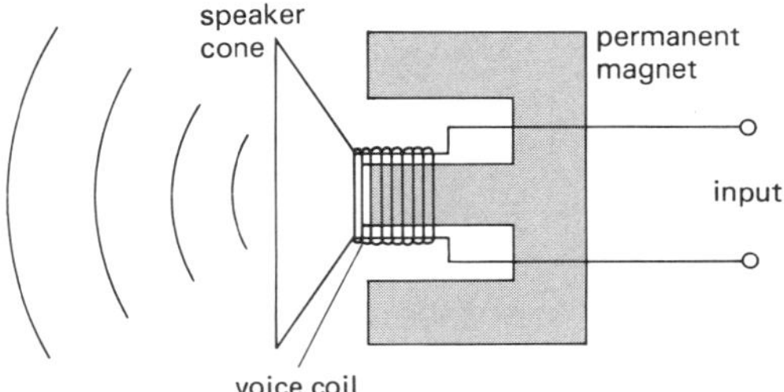

(a) loud speaker

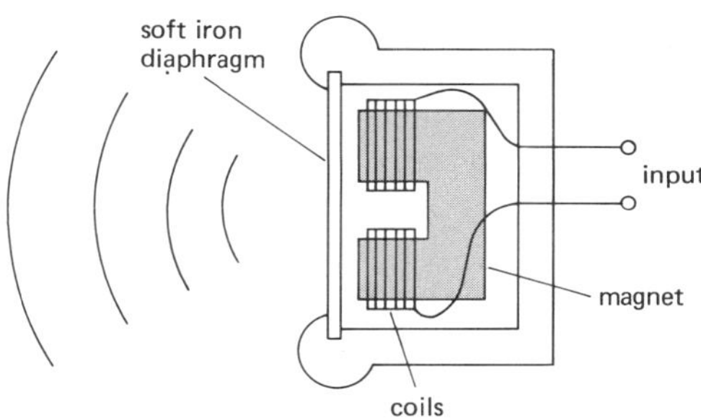

(b) telephone or headset receiver

Figure 22.16 Loudspeakers. The variations in the electrical input cause the coils to move in a magnetic field. This drives a speaker cone (a), as shown in the photo, or a metal diaphragm (b) in the most common telephone receivers (see Fig. 22.15).

SPECIAL FEATURE 22.3

Woofers and Tweeters

You high-fidelity enthusiasts might be wondering about your woofers and tweeters. To provide optimum response for different sound frequencies, three different speakers are generally used (see Fig. 22.16). Low-frequency speakers are called woofers, and high-frequency speakers are called tweeters. There are intermediate speakers for intermediate frequencies. All three can be obtained in one speaker unit, called a three-way speaker.

the flexible iron disk, causing it to vibrate and produce sound.

TAPE RECORDINGS

A common application of a magnetic material is in tape recordings. In making a recording, a plastic tape coated with a thin film of iron oxide is run past a recording head, where a magnetic field is set up in the coil by the electric signals from a microphone. The magnetic field is strong enough to magnetize the iron oxide film, and the magnetic pattern corresponds to the original sound patterns. When played back, the magnetized tape passes over a pickup head, and a voltage is induced in a coil in the head. This is amplified, and the signal is converted to sound in a speaker. This recording may be erased by running the tape through a uniform magnetic field to remove the magnetic variations in the iron oxide film.

RADIO

Early radio was called "wireless" because no wires were needed for transmission, as is the case with the telegraph and the telephone. Radio waves are generated in transmitter antennas, as discussed in the last chapter. When you listen to radio stations, the radio dial is set in the kHz and MHz radio frequency range. How then do radio waves carry sound, since radio frequencies are well above the audible range of hearing? Recall that in a telephone line the electrical signals have the same frequencies as the audible sounds that produce them.

Radio waves carry the electrical reproductions of the sound waves in a sort of piggy-back fashion. This is accomplished by a process called **modulation.** The radio wave acts as a "carrier" wave for sound waves. The assigned frequency of a radio station is the frequency of its carrier wave. The sound waves are electrically impressed on the radio waves so they modulate the carrier (Fig. 22.17). The carrier wave may be amplitude modulated (AM) or frequency modulated (FM).

The AM frequency spectrum or "band" for radio is from 550 to 1600 kHz, and the FM frequency band is

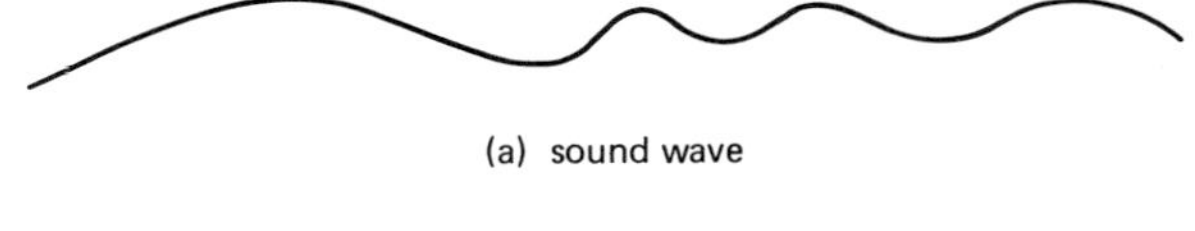

(a) sound wave

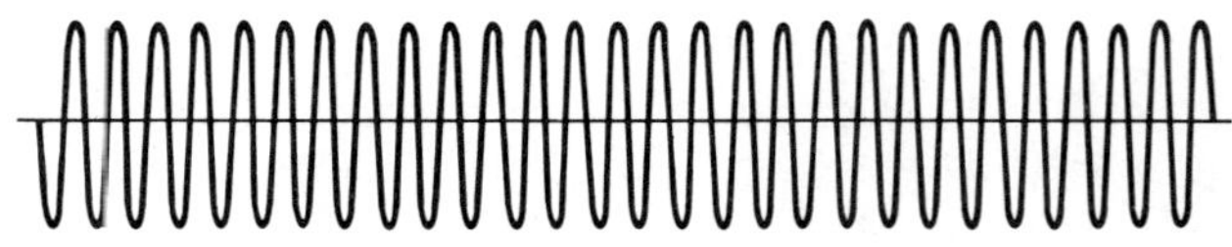

(b) unmodulated carrier wave

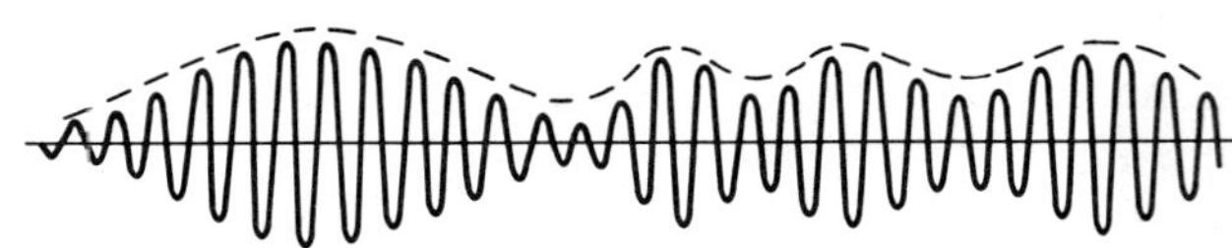

(c) amplitude modulated (AM) wave

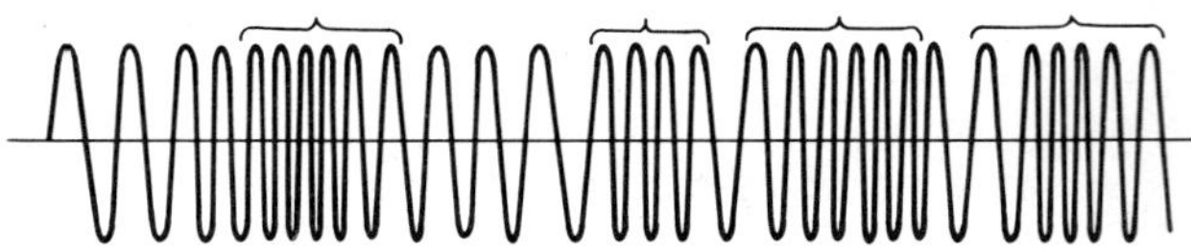

(d) frequency modulated (FM) wave

Figure 22.17 Modulation. A sound wave (a) can be impressed on an unmodulated carrier wave (b) so that it is (c) amplitude modulated (AM) or (d) frequency modulated (FM).

from 88 to 108 MHz. (Check your radio dial.) Each station is assigned a particular band width or channel. For AM stations, this band width is 10 kHz wide, with 5 kHz on each side of the station's assigned carrier frequency. A band width of 10 kHz can carry all of the necessary frequencies for audible communications. For FM stations, the channel width is 200 kHz wide. With wider band widths, a greater range of frequencies can be transmitted by FM radio. This adds to the fidelity of the sound, particularly music with higher frequencies, and allows for stereophonic sound. Static is also less of a problem. In AM applications, static from various sources also modulates the wave and is carried along. This unwanted amplitude modulation is ignored in FM reception.

Tuning

With all of the radio stations available, how do you select a particular one for listening? Another way of putting this is, What are you really doing when you "tune in" a radio station? Actually, you are adjusting a circuit so that it resonates at a particular frequency.

Consider a circuit with a coil (inductor) and a capacitor, as shown in Figure 22.18. With the capacitor charged, when the switch is closed it discharges, and current flows in the circuit. A magnetic field builds up in the coil, and energy is transferred to the inductor. As the capacitor discharges to zero, the magnetic field begins to collapse, and a current is induced to prevent this (Lenz's law). The current recharges the capacitor, and energy oscillates back and forth between the capacitor and inductor in this so-called tank circuit. The frequency of the energy oscillation depends on the values of the capacitance and inductance.

This is a case of **electrical resonance.** Similar to the mechanical resonance of a pendulum or swing (Chapter 16), the circuit has a characteristic resonance frequency at which it oscillates. When driven at this resonance frequency, there is maximum energy transfer to the circuit. The resonance frequency can be changed by using a variable capacitor. This is what you are changing when you turn the tuning knob of a radio. When the resonant frequency of the circuit is the same as the carrier frequency of a radio station, it is driven or receives energy from that station and not from others at different frequencies.

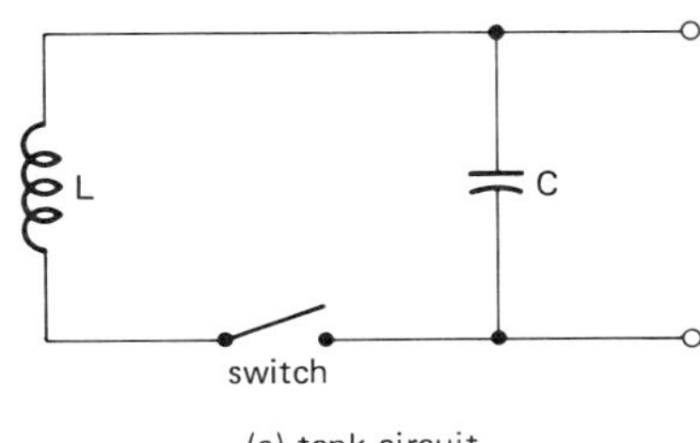

(a) tank circuit

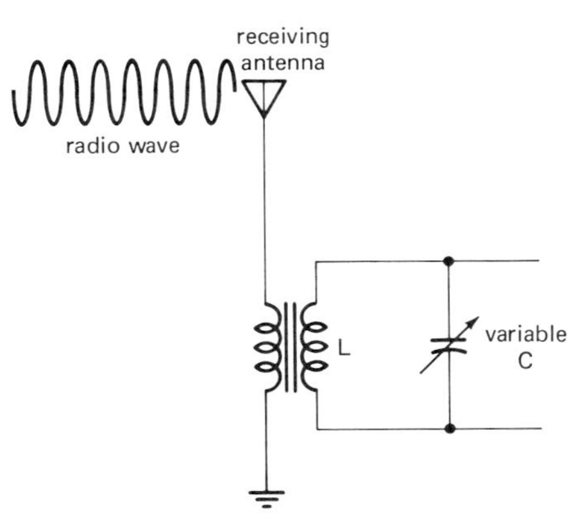

(b) tuning circuit

Figure 22.18 Tuning. (a) A "tank" circuit, which has a certain electrical resonance frequency, can be used as a tuning circuit. (b) Varying the capacitance changes the resonant frequency and radio stations transmitting on assigned frequencies are selected or tuned in.

Detection

Having received a modulated radio wave, if it were fed directly into a speaker, you would hear nothing. Looking at an amplitude-modulated wave again (Fig. 22.19), with equal positive and negative values, the average value of the wave is zero. In order to detect the variation in the amplitude of the wave, which carries the music or sound, the wave must be rectified. This is where a diode comes in. Passing only current in one direction, there is half-wave rectification. (A transistor could also be used to get amplification as well as rectification.) This is called **detection** (or demodulation).

Putting this all together, we have a simple AM radio receiver (Fig. 22.19). Even with half-wave rectification, there are thousands or millions of vibrations per second fed into the headphones or speaker. But the speaker diaphragm cannot vibrate this fast. Instead, it responds to an average value that has the same form as the sound wave modulation.

Of course, ordinary radios have many more components, for example, amplifiers, filters, and so on, but you could build a simple radio, as shown in Figure

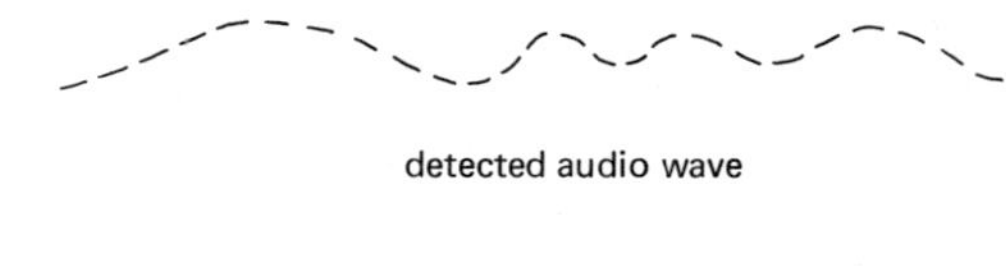

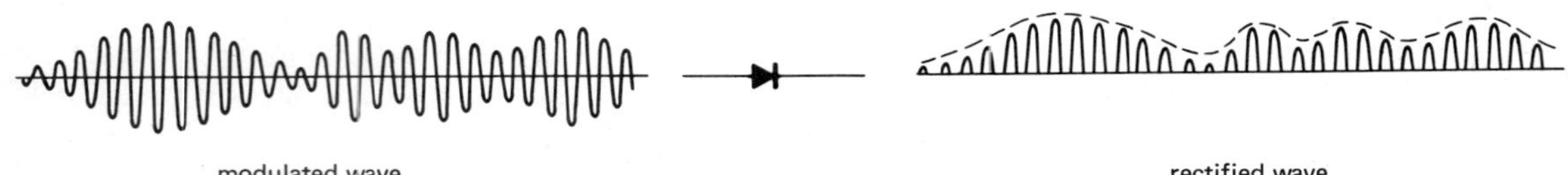

Figure 22.19 Detection or demodulation. An AM wave is a half-wave rectified by a diode. A speaker can then respond to the detected audio portion of the wave.

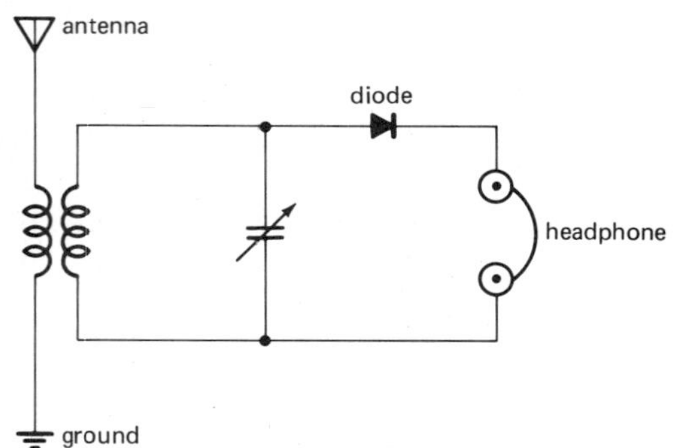

Figure 22.20 A simple radio. The diagram shows the basic components of a simple radio that works. Of course, ordinary radios have many more components.

22.20. Notice that it doesn't have a battery. Unless you are very close to a powerful radio transmitter, a long antenna would be needed to collect enough energy to operate the headphones. You don't really need the tuning circuit for the radio to operate, but without it you'd pick up several stations at once.

TELEVISION

What did people do before TV? Usually they listened to the radio and played checkers. Now a great deal of time is spent watching (and listening) to TV. Television combines both sight and sound, using an AM picture signal and an FM audio signal (Fig. 22.21). With communication satellites, we now receive live television broadcasts from around the world. Many people, particularly those who can remember pre-TV days, find the medium of television amazing. For others, it is just a source of "soaps." Some speculate about the sociological implications of television on our society, from educational to detrimental.

In any case, TV is an electronic advance of the last 40 years. The detailed analysis of the various TV circuits is quite complicated. We consider here only some of the

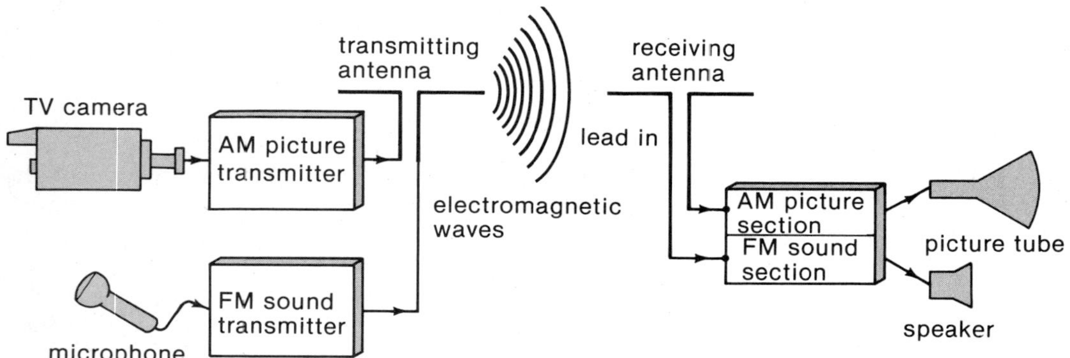

Figure 22.21 Schematic diagram of television transmission and reception. TV uses an AM picture signal and an FM audio signal.

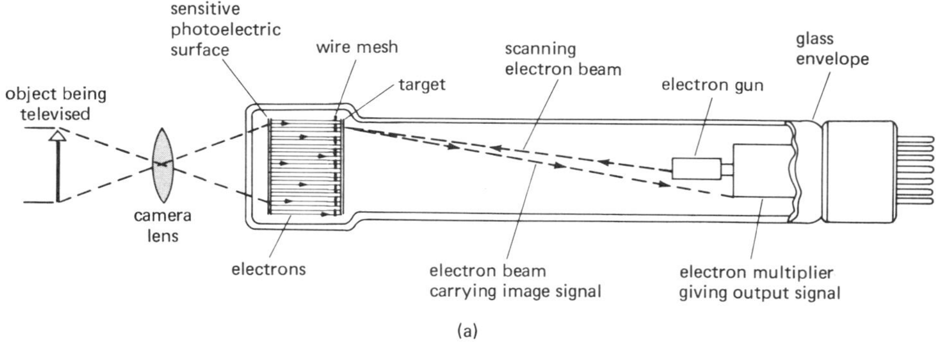

(b)

Figure 22.22 Television camera. (a) The picture signal in a TV camera is formed by an image-orthicon tube. (b) An actual TV camera.

basic components. The picture signal in the TV camera is formed by an image-orthicon tube (Fig. 22.22). Light from objects activates an array of photosensitive cells, which release electrons. These are attracted to a target screen. As a result, there is an "electron image" of the objects on the screen. This is scanned by an electron beam, which converts the image information into electrical signals.

The picture signals are transmitted, along with the accompanying FM sound, to the receiver. The picture tube is a cathode ray tube (CRT), as illustrated in Figure 22.23. The electron beam is accelerated toward the screen and deflected by means of magnetic coils so that the beam scans a fluorescent screen in a 525-line pattern in a fraction of a second. Odd-numbered lines, called a line field, are scanned, then the even-numbered line field, so as to give a better picture. Both field scannings together are called a frame, and the beam scans 30 frames a second—pretty fast! The signals from the television camera reproduce the camera image on the screen in a mosaic picture of light and dark spots, and a black-and-white picture is formed.

Color television is somewhat more involved. The color camera separates the light into three beams by filters, and most picture tubes have three electron beams, one for each of the primary colors that can be mixed together to form different colors (see Chapter 25). In the three-beam picture tube, the beams are directed through a shadow mask onto the screen (Fig. 22.24). On the screen are thousands of phosphor dots arranged in triads. Each dot will glow with its specific color when hit by an electron beam. The emissions

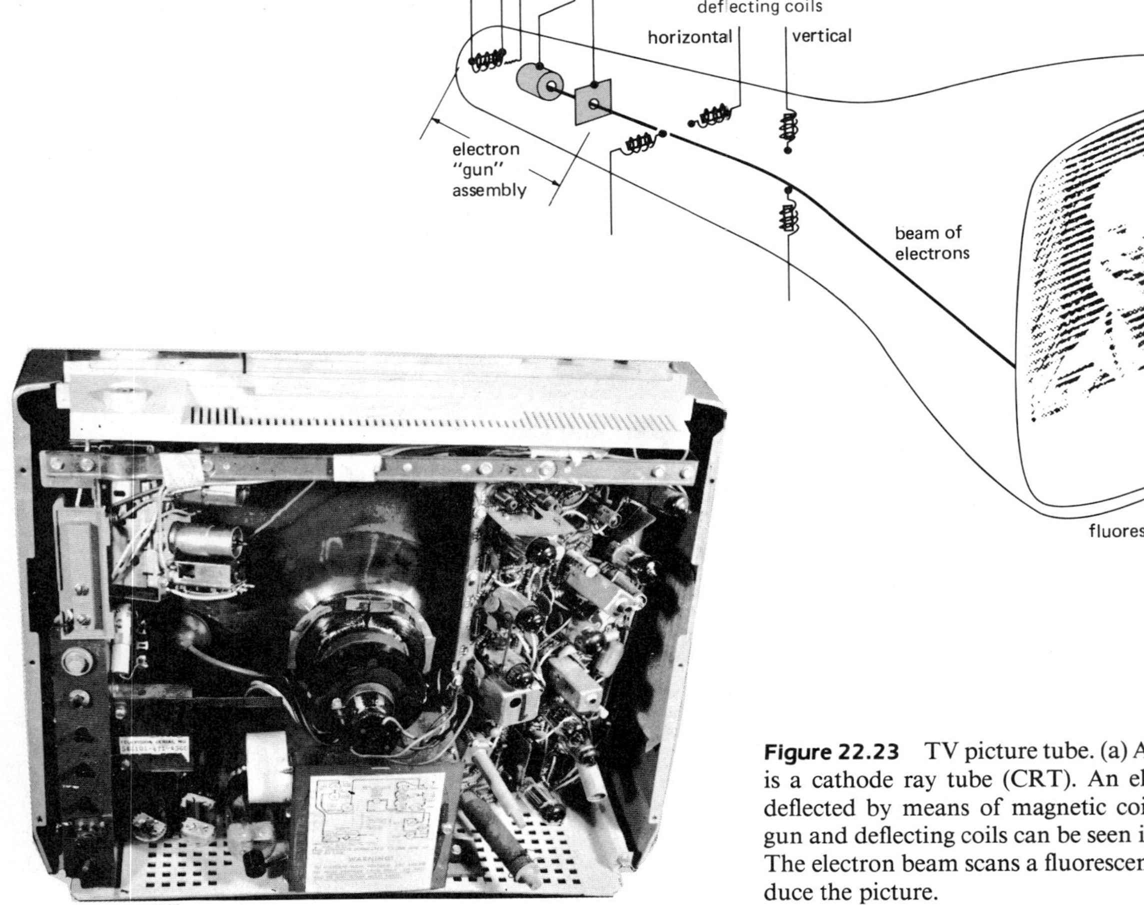

Figure 22.23 TV picture tube. (a) A television tube is a cathode ray tube (CRT). An electron beam is deflected by means of magnetic coils. An electron gun and deflecting coils can be seen in the photo. (b) The electron beam scans a fluorescent screen to produce the picture.

start of 1st downward scan, called 1st field

start of 2nd downward scan, called 2nd field

at end of 2nd field, entire picture (called frame) is completed

at end of 1st field, spot returns to start 2nd field

(b)

from the three dots of the triad cannot be distinguished separately at normal viewing distances. The eye sees the light from these primary color sources as a single color.

Some color sets use a tube with a single electron beam. This is known as a chromatron tube. The chromatron tube uses the principle of sequential dot display. Sets of primary-color phosphor dots are excited sequentially in rapid succession by the single beam. The beam is modulated so that each color dot contributes to producing the right color.

For either black-and-white or color picture tubes, high voltages—on the order of thousands of volts (up to 20,000 volts)—are used on the picture screen anode of the tube. The deceleration of electrons at the tube screen can produce X-rays, particularly in high-voltage color tubes, and therefore must be properly shielded. This is why you are warned not to sit too close to the TV—there may be some X-ray leakage. Even when

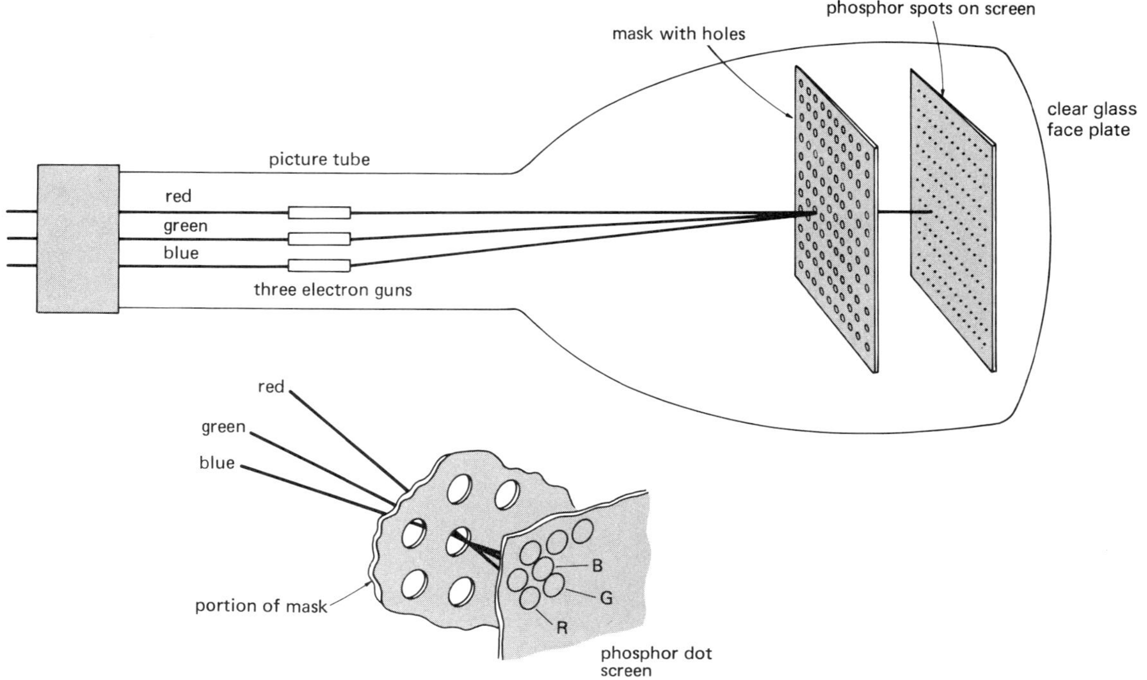

Figure 22.24 Color production. In many color-TV receivers, the color beams are directed through a screen mask and excite phosphor-dot triads on the screen.

the set is shut off, a dangerous potential can remain because of capacitive effects. Extreme caution should be used when working on a TV set. In most cases, this should be left to an experienced technician.

Television channels are divided into two groups: VHF (very high frequency), with Channels 2 to 6 between 54 and 88 MHz (low band) and Channels 7 to 13 between 174 and 216 MHz (high band), and UHF (ultra-high frequency) with Channels 14 to 83 between 470 and 890 MHz. TV channels have relatively wide bands of 6 MHz and greater, since they must carry more information than radio.

SUMMARY OF KEY TERMS

Electronics the branch of physics that deals with the actions and effects of electrons.

Edison effect the flow of electrons in a vacuum tube with proper polarity.

Rectification the conversion of alternating current to direct current.

Cathode the filament or negative terminal of a vacuum tube.

Anode the plate or positive terminal of a vacuum tube.

Grid the third element of a triode used to control the plate current.

Doping the adding of impurity atoms to a material to make it semiconducting.

Donor impurity an impurity that contributes extra electrons.

N-type semiconductor a material with negative (electron) charge carriers.

Acceptor impurity an impurity that creates holes or vacancies.

P-type semiconductor a material with positive "holes" as charge carriers.

Junction diode a solid-state diode formed by contacting P- and N-type semiconducting materials.

Transistor the solid-state analog of a vacuum tube triode.

Modulation the impressing of sound waves on a radio carrier wave through modifications in amplitude (AM) or frequency (FM).

EXERCISES

1. What would happen if the diode in Figure 22.2 were turned around?
2. Draw circuits for (a) half-wave rectification and (b) full-wave rectification using solid-state diodes.
3. A bridge circuit for full-wave rectification is shown in Figure 22.25. (Notice that the transformer does not have to be center-tapped in this case.) Prove that the current through the load resistance R_L is in the same direction for each half-cycle of the input voltage, or that there is full-wave rectification.
4. If a piece of either N-type or P-type semiconductor were put into a battery circuit, would there be conduction in each case? Explain. What if the polarity were reversed?
5. When a solid-state diode is biased in the forward direction, is the P-type material positive or negative?
6. When a solid-state diode is conducting, which way do the holes flow across the junction?
7. Do solar cells supply ac or dc? Explain.

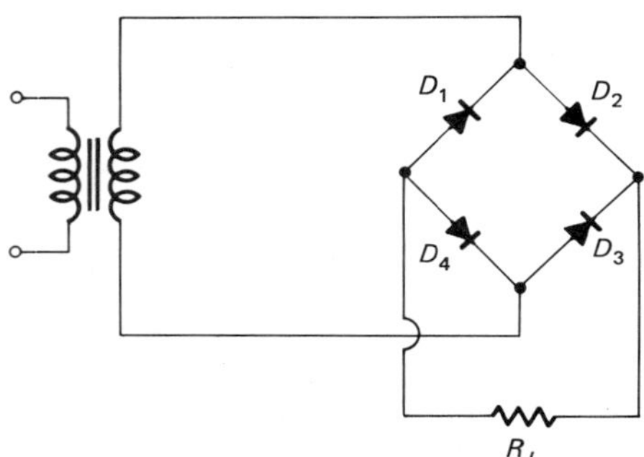

Figure 22.25 See Exercise 3.

8. What would be the effect if a bunch of solar cells were connected in (a) series and (b) parallel?
9. Suppose the NPN transistor in Fig. 22.11 were replaced by a PNP transistor. Would there be any difference? Explain.
10. Why is a transistor like two diodes "back-to-back"? Would "head-to-head" be a better description?
11. What is the energy conversion associated with (a) microphones and (b) speakers?
12. A dog is a woofer and a bird is a tweeter. Is there any correlation between the sounds of these animals and woofer and tweeter speakers?
13. What would be the effect if single intermediate speakers were used in a stereo system instead of three-way speakers?
14. Explain how radio waves in the kHz and MHz ranges can carry sound. How is this sound detected?
15. There can be several AM radio stations in the same town. Theoretically, how many different 10-kHz-wide channels or radio stations can the AM frequency band accommodate in one town?
16. FM radio channels are 200 kHz wide. How many FM stations could the FM frequency band theoretically accommodate in one town? (*Helpful hint:* 200 kHz = 0.2 MHz.)
17. Suppose you were building a simple radio. Could you get by without (a) a battery? (b) a tuning circuit? (c) a detector?
18. Why does dust cling to the front of a TV picture tube?
19. How long does it take for the electron beam in a TV tube to scan one complete picture or frame?
20. What would be the result if all of the dots in all the color triads in a color TV picture tube were equally excited?

PART SIX
Light

The Nature of Light

The light entering our eyes brings to us the visual information about our world — the beauty of sunsets; the features of objects, including our own and other persons' "looks"; the existence of stars; and so on. It is a small wonder that the nature of light has occupied scientific thought from the earliest times. Have you ever asked yourself the question, "What is light?"

One theory, dating back to the ancient Greeks, considered light to be composed of "corpuscles" or particles, and this theory survived relatively unquestioned for many centuries. But, evidence began to be noted that supported a wave nature of light. Leonardo da Vinci (circa 1500), noting the similarity between sound echoes and the reflection of light, speculated that light might have a wave nature. By the seventeenth century a definite controversy existed over the nature of light. Isaac Newton favored a theory based on light being a stream of particles.

However, by the late nineteenth century there appeared to be overwhelming evidence in support of the wave theory. Experiments showed that light exhibits wave properties such as interference and the Doppler effect, so it must be a wave. Then, studies in electromagnetism showed that light is a special kind of wave — an electromagnetic wave (Chapter 21).

Interesting enough, we have been forced to return to a "particle" description of light in certain cases. In some instances light behaves like a wave, while in other instances it must be described as a particle or packet of energy in order to explain what is observed. This has given rise to what is called the *dual nature of light*. Have you ever heard of this? For our present discussion, we will focus on the wave nature of light and look at some interesting practical applications of light. The particle nature of light will be discussed later in the text (Part Seven: Modern Physics).

23

Light Waves

Light Emission

Electromagnetic radiation spans a wide range of frequencies, but light is generally taken to mean electromagnetic radiation in or near the visible region. For example, radio waves are electromagnetic waves but are not generally thought of as light. To produce radio waves, electrons are oscillated back and forth in metal antennas hundreds of thousands of times a second. The oscillation frequencies are controlled by crystals, for example, quartz, which vibrate naturally at radio frequencies. These resonant frequencies depend on the elasticity of the crystal material and on the size and shape of the crystal. In general, the higher the frequency, the smaller the crystal, in much the same way as a high sound frequency or pitch generally comes from smaller bells.

In order for light to radiate from an antenna, the electrons would have to oscillate many billions or trillions of times each second. This is beyond the range of crystal vibrations in both size and elasticity. The "antennas" of visible light emissions are atoms themselves. In our simplified model of the atom (Chapter 9) we speak of transitions rather than vibrations. When atoms become excited, electrons are raised to higher orbits or shells (Fig. 23.1). These are states of higher energies. The electrons generally remain in the higher energy levels for only fractions of seconds and then spontaneously return to their normal orbits or states. When they do so, energy is emitted as light or electromagnetic radiation, with different frequencies of light corresponding to transitions between different energy levels.

The excitation in most materials is short-lived, but the electrons in some materials remain in excited states for long periods of time—minutes and hours. As a result, light is emitted from these materials long after they have been excited, and they can glow in the dark.

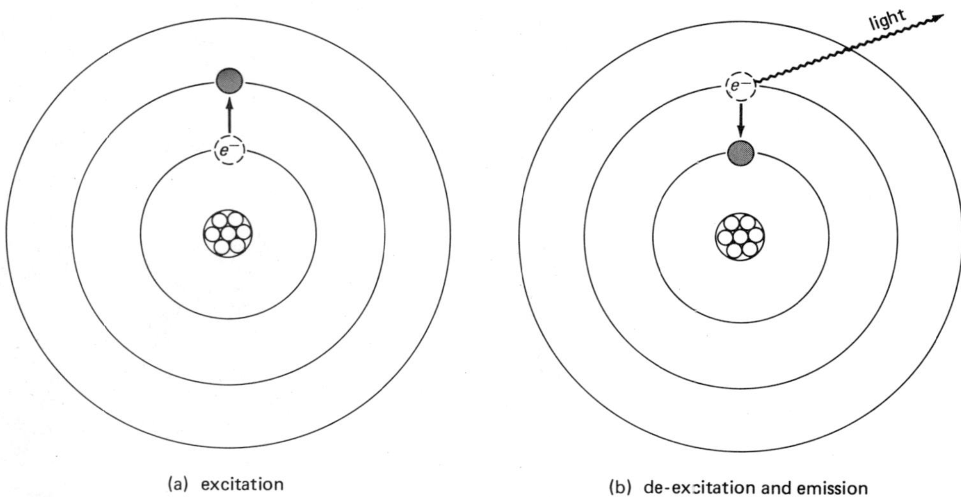

Figure 23.1 Light radiation. (a) An atomic electron is excited to a higher orbit. (b) In the de-excitation, the electron returns to a lower orbit, with energy being emitted as light or electromagnetic radiation.

You have probably seen novelty items and toys that have done this, as well as some luminous clock dials. This process of electrons being "hung up" in excited states is called **phosphorescence.** As you might expect, phosphorus is a good phosphorescent material.

Atoms may be excited by various means, all of which involve a transfer of energy. Phosphorescent materials are excited by exposing them to light. (Remember that waves carry energy.) In the case of a gas discharge tube, electrons boiled off the cathode excite the gas molecules in the tube through collisions. Depending on the type of gas in the tube, light of certain frequencies or colors is emitted. For example, neon tubes used in "neon signs" emit light primarily in the red end of the visible spectrum. Other gases produce different colors, since different atoms and molecules have different orbit spacings and emit different, but particular, light frequencies which determine the color of the emitted light. Light from a sodium lamp, for example, is yellow, and the mercury vapor in mercury lamps commonly used for street lights has a large number of transitions, emitting light in the blue end of the visible spectrum.

The situation is different if we look at light from an incandescent light bulb. In the common light bulb, a solid tungsten filament glows as a result of electrical joule heating and emits visible light of all wavelengths or colors. This is called white light, which is a mixture of all visible wavelengths or frequencies. (Light with a single frequency or wavelength is called **monochromatic**—"one color".) In solids, the atoms are closely

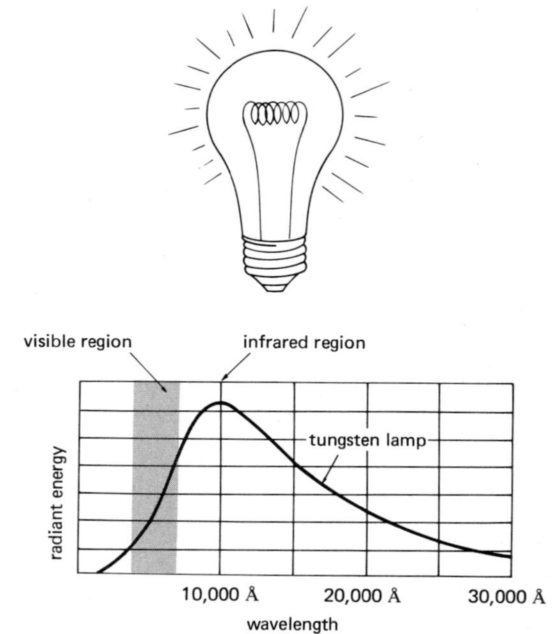

Figure 23.2 The radiant energy spectrum for an incandescent lamp. Note that only a small portion of the energy is emitted in the visible region, making the lamp very inefficient as a light source.

packed together. This influences the spacings of the atomic electron levels to the point that an electron from one atom can jump to the level of another atom. There are so many possibilities that transitions for all the frequencies in the visible spectrum exist and occur. The incandescent lamp is very inefficient. Most of the radiation is in the infrared portion of the spectrum (Fig. 23.2), and 95 percent or more of the energy is radiated as heat rather than visible light. (Recall that infrared radiation is called heat waves.)

Fluorescence

Another common light source is the fluorescent lamp. This is basically a gas-discharge tube containing mercury vapor. When the lamp is in operation, electrons from the electrode excite the mercury atoms, which emit ultraviolet radiation in the de-excitation transitions (Fig. 23.3). The ultraviolet light is absorbed by a thin coating of phosphor material on the inner surface of the glass tube, and the phosphor emits visible light in a process called **fluorescence.** A fluorescent material absorbs ultraviolet light and emits light at a lower frequency, either because part of the energy goes into the kinetic energy of an entire atom or because an excited electron jumps from orbit to orbit, or energy level to energy level, down the "energy ladder," perhaps skipping a rung now and then instead of going directly from the initial excited state to its normal state (Fig. 23.4).

In either case, many frequencies of light are emitted that combine to produce white light. Because of the fluorescent conversion and lower operating temperatures, fluorescent lamps are considerably more efficient than incandescent lamps. Notice in the graph in Figure 23.3 that most of the light is emitted in the visible region. Different types of phosphors emit light of different colors. The phosphors in the so-called "cool white" fluorescent lamps emit nearly all visible colors and give very white light. "Warm white" lamps have phosphors that emit more red light, giving a warm glow.

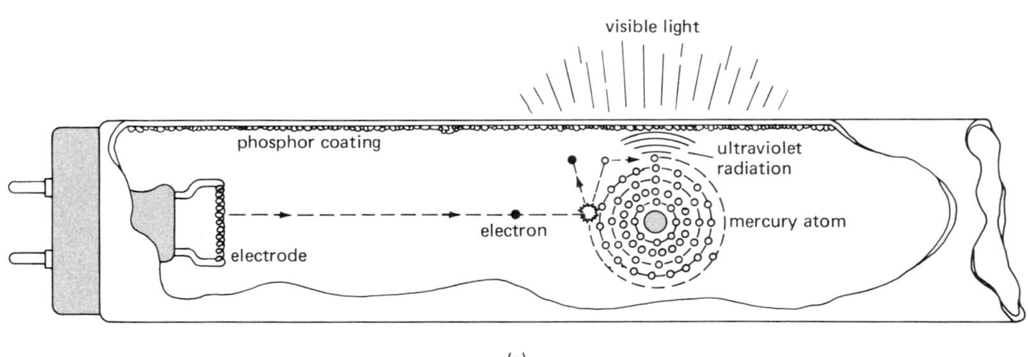

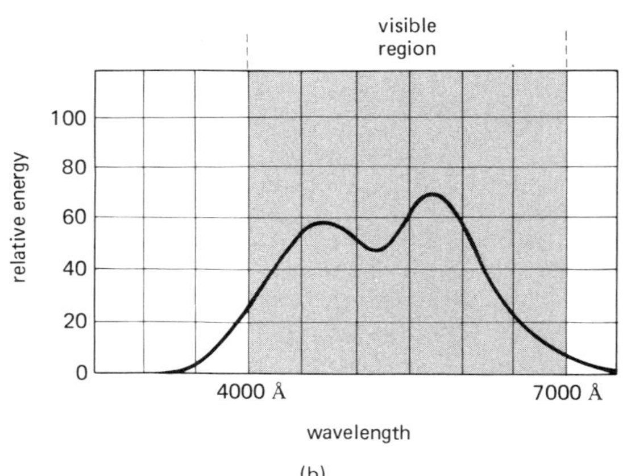

Figure 23.3 The fluorescent lamp. (a) Ultraviolet light emitted by mercury atoms is absorbed by the phosphor material on the inner surface of the bulb and is re-emitted as visible light. (b) As a result, most of the emitted light is in the visible region of the spectrum. (Compare with Figure 23.2.)

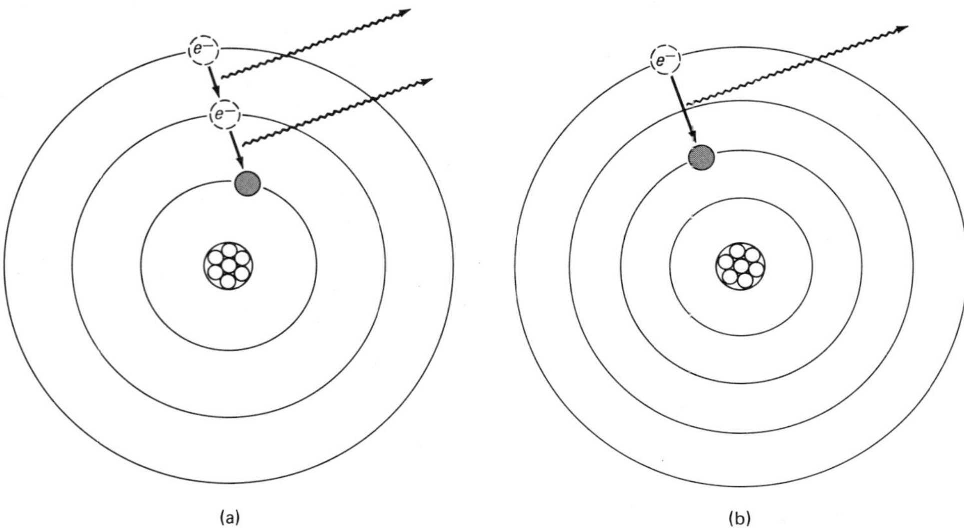

Figure 23.4 In the de-excitation process, an electron may jump from orbit to orbit (or energy level to energy level) down the "energy ladder" (a). It may also skip a "rung" or two in going back to its normal unexcited state (b).

The fluorescent excitation by ultraviolet (uv) radiation has another common use. "Black lights" used at discos emit radiation in the violet and near-ultraviolet regions. The uv black lights cause fluorescent paints and dyes on signs, posters, and performers' clothes to fluoresce with brilliant colors. Many products, such as laundry detergents, are packaged in boxes that have bright colors containing phosphors in the ink. These fluoresce somewhat under the store's fluorescent lamps and appear brighter to get your attention. In the dark, when illuminated with black light, the boxes glow brilliantly. Some natural minerals are also fluorescent. Illumination with black lights (uv) is a method of identifying these materials.

Interference

As a wave, light shows interference phenomena. Recall from Chapter 16 that waves superimpose and interfere constructively and destructively. One of the earliest demonstrations of interference of light was performed in 1801 by Thomas Young, an English scientist. This experiment helped establish the wave nature of light, as well as giving a means to measure the wavelength of light. It is often called Young's double-slit experiment because he used light from a single source passing through two small slits (Fig. 23.5). The slits act as two similar "sources" that emit the same light together.

Because of the path difference from the slits to the screen, there are bright and dark regions of constructive and destructive interference. At positions where the waves arrive in phase—for example, wave crests arriving at the same time—there is constructive interference, and a bright "fringe" is seen on the screen. At intervening positions, the waves arrive out of phase, and dark fringes occur due to destructive interference. By taking a photograph of the interference pattern and measuring the distance of one of the bright fringes from the central maximum, which is on the order of millimeters, the wavelength of a monochromatic light source can be computed from the interference conditions and geometry of the experiment. This is a neat way of measuring something that is less than one-millionth of a meter.

An interference effect involving another electromagnetic wave is sometimes observed on TV. You have probably noticed how a TV picture flutters when an airplane flies over. This arises when TV waves are reflected from the plane. The TV set then receives two signals, one reflected from the plane and the direct signal from the television transmitter. The waves travel different distances and may arrive at the TV set either in or out of phase. But as the plane moves at a high speed, the path length changes, and the interference at the set shifts rapidly back and forth between constructive and destructive interference, giving a fluttering in the picture.

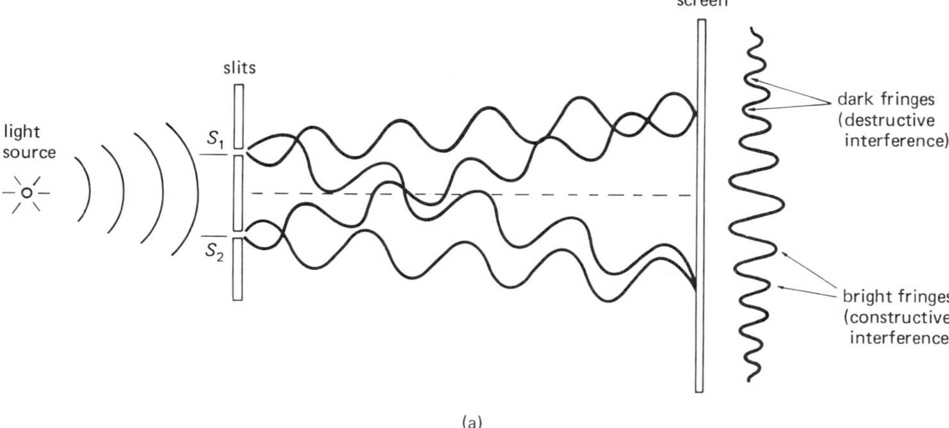

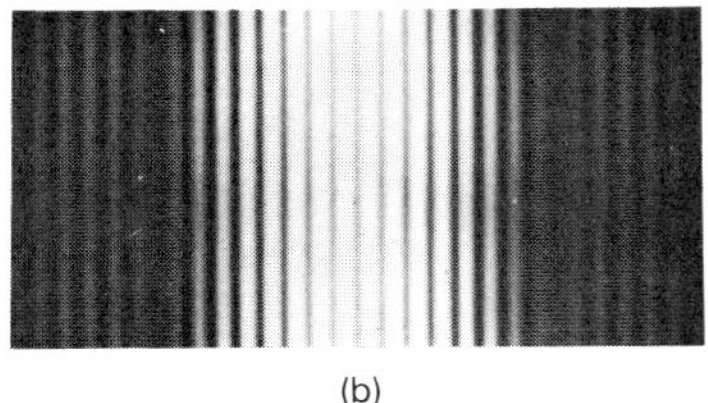

Figure 23.5 Young's double-slit experiment. (a) Light from the slits arriving in phase at some points on a screen interferes constructively and gives bright fringes. Destructive interference gives dark fringes. This produces an interference pattern (b).

Thin Film Interference

Have you ever noticed the multicolored patterns on oil or gasoline slicks on a wet pavement or in soap bubbles and wondered what caused them? These are common examples of interference—thin film interference. Even though the films are transparent, there is some reflection at the surfaces, and interference occurs for light reflected from the top and bottom surfaces of a film (Fig. 23.6). When the film thickness is so that the reflected waves are in phase, constructive interference occurs, and the light is reflected from the film. When the film thickness is such that the reflected waves are completely out of phase, destructive interference occurs for the reflected waves, and the light is transmitted (not destroyed).

The path difference of the reflected waves also depends on the angle at which the light strikes the film. In any case, when white light falls on a thin film, some wavelengths are reflected and others are transmitted. For a film of relatively uniform thickness, the reflection may be selective for a particular wavelength region of the visible spectrum, and the film is seen to be of one particular color. However, oil and soap films generally vary in thickness, and a vivid display of different colors corresponding to different thicknesses of film is seen.

For the oil film in Figure 23.6, for normal (direct) incidence, destructive interference or transmission occurs for light when the film thickness is one fourth of a wavelength. The path difference is $\frac{1}{4}\lambda$ down and $\frac{1}{4}\lambda$

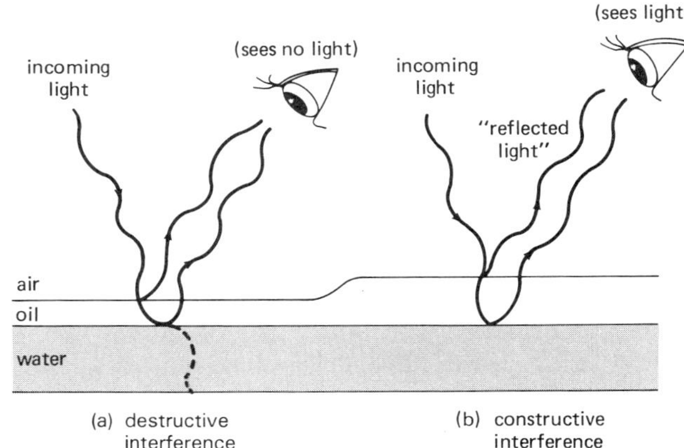

Figure 23.6 Thin film interference. Reflections from the top and bottom surfaces give rise to interference. (a) If the film thickness is such that the reflected waves are completely out of phase, destructive interference occurs, and no light is seen. (The light is transmitted, not destroyed.) (b) With film thicknesses for which in-phase constructive interference occurs, light is seen (reflected).

SPECIAL FEATURE 23.1

Nonreflecting Lenses

Thin films are used to make nonreflective glass lenses, such as those used on cameras and binoculars. Interference properties allow for the reduction of reflected light from the lenses or the greater transmission of light, which is needed for exposing camera film or for binocular viewing. A nonreflective lens has a thin film coating with a thickness of $\frac{1}{4}\lambda$ for a particular light component. This is usually chosen in the yellow-green portion of the visible spectrum (about 5500 Å wavelength) for which the eye is most sensitive. Other wavelengths are still reflected and give the glass a somewhat bluish appearance. You have probably noticed this bluish hue for coated lenses in cameras and other optical instruments.

The film coating serves a double purpose. Not only does it promote nonreflection from the front of the lens, but it also cuts down on back reflection (Fig. 23.7). Some of the light transmitted through the lens is reflected from the back surface. This could be reflected again from the front surface of an uncoated lens and give rise to a poor image on the photographic film in a camera. However, the reflections from the two film surfaces of a coated lens interfere destructively, and there is no reflection.

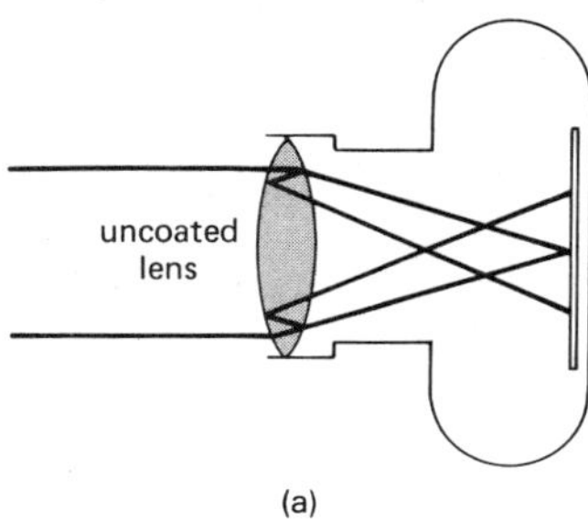

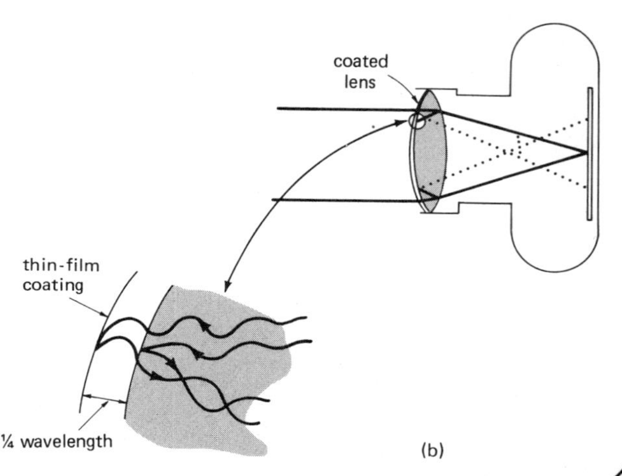

Figure 23.7 Nonreflective coatings for lenses. (a) Reflections from the interior surface of a lens can give a poor focus or image on the camera film. (b) With a thin film coating, the reflections from the film surfaces interfere destructively, and there is no reflection.

back, so the two reflected waves are out of phase by half a wavelength, or a crest meets a trough when the reflected waves interfere.* An application of this type of destructive thin film interference is given in Special Feature 23.1.

Thin film interference is used to check the smoothness and uniformity of optical components such as mirrors and lenses. Glass plates, called optical flats, are ground and polished so they are as flat or as smooth as possible. The smoothness of the flats can be checked by putting the plates together so there is a uniform air wedge between the flats (Fig. 23.8). If they are smooth, regular interference patterns occur as a result of the uniformly increasing path differences between the plates. If a surface (or surfaces) is not smooth, there would be an irregularity in the interference pattern in that region. Once you have an optical flat, it can be used to determine the flatness of a reflecting surface such as a glass mirror or metal plate.

A similar technique is used to check the smoothness and symmetry of lenses (Fig. 23.9). When a curved lens is placed on an optical flat, there is a curved air wedge between the lens and the flat. The regular interference pattern in this case is a set of concentric bright and dark circular fringes or rings. They are called **Newton's rings,** after Sir Isaac, who described this interference

* When light is reflected from a more "optically dense" medium, the wave is inverted or undergoes a 180° phase shift. This is analogous to a wave in a rope being reflected from a rigid support (see Chapter 6). For the case in Figure 23.6, the reflections at both surfaces undergo phase shifts, so there is no change in the phase difference of waves. On the other hand, for a soap film the top surface reflection is phase-shifted, but the bottom surface reflection is not (as in a rope with a movable support). As a result, destructive interference occurs for film thicknesses of $\lambda/2$ for normal (direct) incident light.

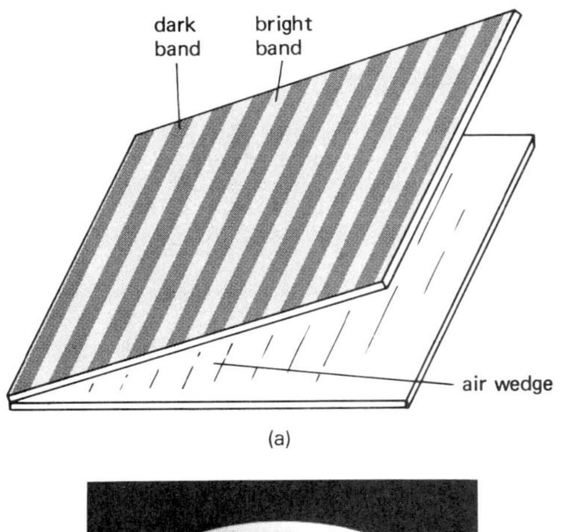

(a)

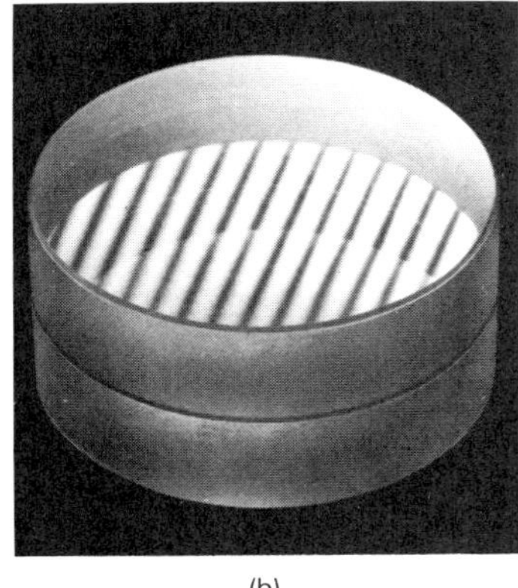

(b)

Figure 23.8 An optical flat and air wedge can be used to check the smoothness of a reflecting surface. If the surface is smooth, a regular interference pattern occurs.

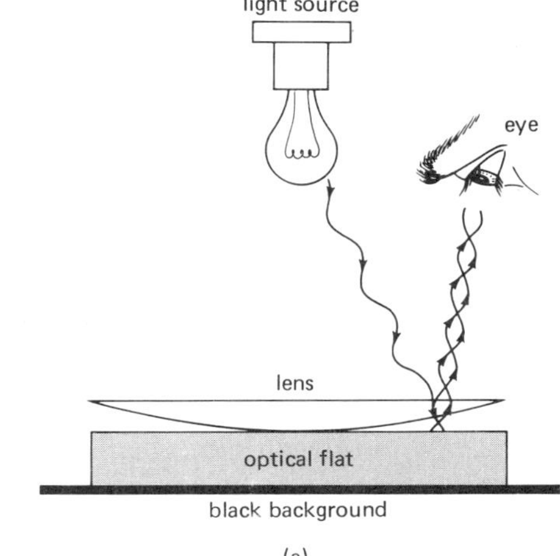

(a)

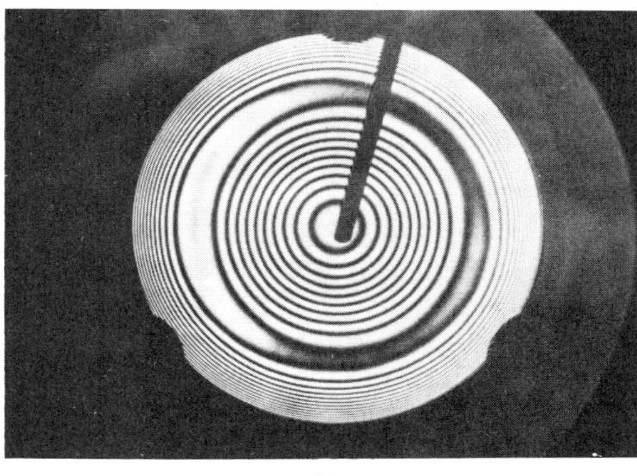

(b)

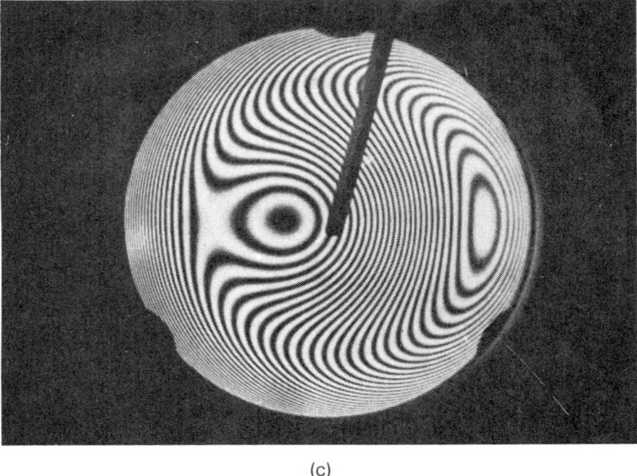

(c)

Figure 23.9 Newton's rings. (a) Lens inspection using an optical flat. (b) Circular interference fringes or rings indicate that the lens is of high quality and smoothly ground. (c) A distorted interference pattern indicates irregularities.

effect. If there are irregularities in the lens surface, a distorted ring pattern is observed.

Diffraction

We think of light waves and waves in general as traveling in straight lines, unless they are deviated from their straight-line paths by reflection or refraction (Chapter 17). However, waves may be deflected or "bent" by another means called **diffraction,** which is the bending of waves around corners. The diffraction of sound waves is a common experience. For example, even though you cannot see someone around the corner of an open doorway, you can hear them speaking. If the sound waves traveled in straight lines, you would not receive the sound. But since you do, the sound waves must be diffracted or bent around the corner of the doorway.

You might wonder why light waves are not similarly bent around the corner of the doorway so you could see someone as well as hear them. This is because diffraction depends on the wavelength of the wave and the size of the diffracting object. Recall from Chapter 9 that for you to be able to "see" or detect an object with a wave, the object must be the right size to influence the wave. For example, long-wavelength water waves may pass by a thin reed without any indication of its presence being given by the wave. But if the wavelength of the waves approaches the size of the reed, the waves are affected.

Sound waves have wavelengths with a range of a few centimeters to a few meters, so they are diffracted or bent around common-sized objects such as trees and people. An "object" may also be an opening, such as a door or a window. Here, too, sound is diffracted or bent around the corners of the opening. The effect of the relative size of an object or opening and wavelength on diffraction is illustrated in Figure 23.10. Diffraction effects are greater when the object or opening size is of the same order as or smaller than the wavelength. Consider radio waves: The AM band (550 to 1600 kHz) ranges in wavelength from 550 m to 190 m, while the FM band (88 to 108 MHz) ranges from 3.4 m to 2.7 m. Hence, FM radio waves may be blocked by large objects such as buildings or hills, while AM radio waves are diffracted around them and are received in a "shadow" area.

With light waves you have to have some mighty small objects or openings to produce diffraction. Recall that the wavelength of visible light is of the order of 10^{-5} cm. Hence, diffraction is not observed for ordinary-sized objects or openings, and their corners cast sharp shadows when illuminated with light. However, the shadow around a small pinhole may be fuzzy as a result of diffraction effects.

Light diffraction is used practically through diffraction gratings. A diffraction grating consists of a large number of equally spaced parallel lines — thousands of lines per centimeter. A transmission grating has lines cut on a piece of glass, typically 6000 lines per cm (Fig. 23.11). The transparent spaces between the grooved lines act as a series of parallel slits. A reflection grating can be made by cutting lines on a piece of metal instead of glass. Light diffracted by the grating slits interferes, and an interference pattern similar to that of Young's double-slit experiment is formed. In effect, the grating extends the interference from a two-slit situation to a large number of slits.

When illuminated with white light, the bright fringes of the interference pattern are spectra of wavelengths or colors. Each wavelength is diffracted at a slightly different angle, which spreads out or separates the component wavelengths into a spectrum. You have probably noticed this colorful diffraction-grating effect on a phonograph record when it is viewed at a glancing angle. The spacings between the record grooves act as individual reflectors, and in effect the record is a reflection grating that separates white light into a spectrum of

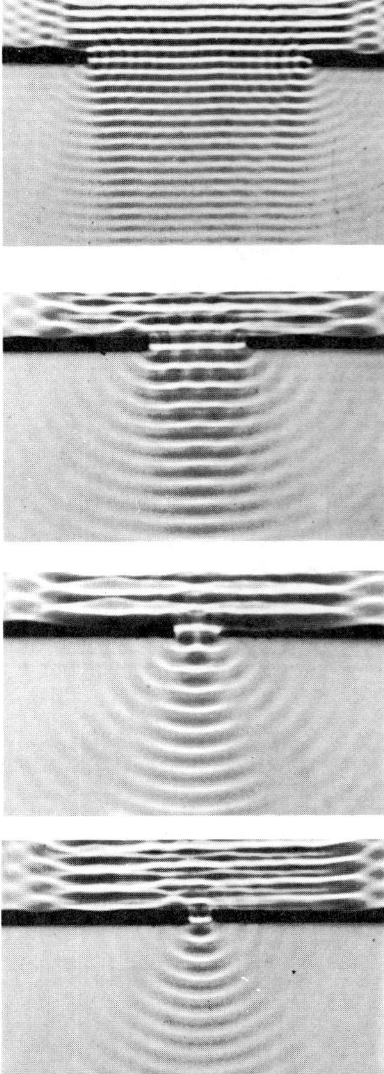

Figure 23.10 The effect of the size of an opening and wavelength on diffraction. Diffraction effects are greater when the opening (or object) size is of the same order as or smaller than the wavelength, as shown here for water waves.

Figure 23.11 Diffraction grating. A master diffraction grating being ruled by a diamond-point scribe. Typically, there are about 6000 lines per centimeter.

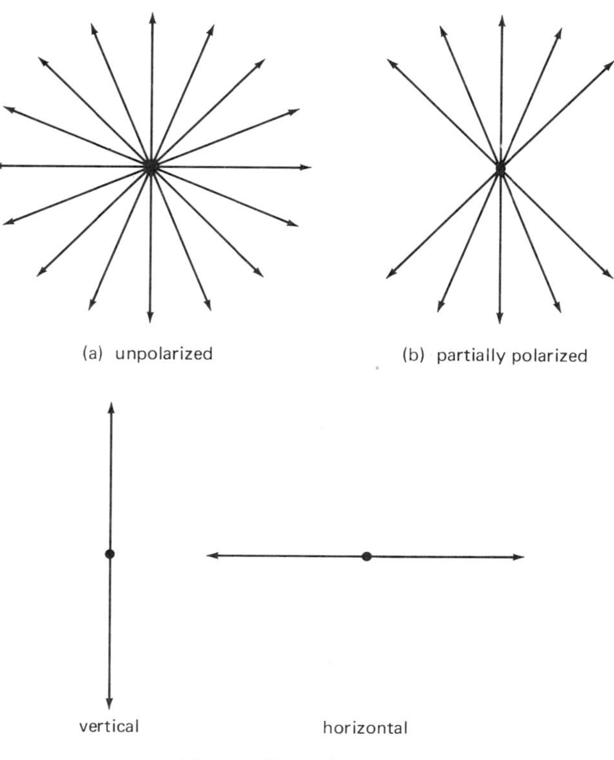

Figure 23.12 Polarization is graphically represented by field vectors.

colors or wavelengths. Because of this property, diffraction gratings are used in instruments called spectrometers to select specific wavelengths of light. When the grating is slightly rotated, the various wavelengths "parade" by a given point and particular wavelengths may be selected. These wavelengths are used to optically analyze substances.

Polarization

Another interesting wave property of transverse light waves is polarization. This is the principle of the "polarized" sunglasses which many of us use. Recall how light is an electromagnetic wave with vibrating electric (E) and magnetic (B) field vectors perpendicular to the direction of propagation (Chapter 21). Light from an ordinary light source consists of a large number of waves emitted by the atoms or molecules of the source. Each atom produces a wave with its own orientation of the E (and B) vibration corresponding to the direction of the atomic vibration. However, with numerous atoms, all directions are possible. This is schematically represented in Fig. 23.12 in terms of the electric field or E vectors. When the field vectors are randomly oriented, we say the light is unpolarized. If there is some partial preferential orientation of the field vectors, the light is partially polarized. If the field vectors are in a plane, the light is linearly polarized or plane polarized (or simply polarized).

Light may be polarized by several means. One common method is by absorption. Certain crystals have the ability to selectively absorb field vector components with certain orientations, and light passing through such crystals is polarized. Around 1930, Edwin Land, an American scientist, found a way to align tiny needle-shaped crystals in sheets of transparent celluloid, thereby producing a thin, polarizing material now known commercially as Polaroid.

Improved Polaroid films have been developed using polymer materials. During the manufacturing process, the film is stretched so as to align the long chain molecules of the polymer. With proper treatment, the outer (valence) molecular electrons can move along the oriented chains. As a result, the molecules readily absorb light with E vectors parallel to the oriented chains and transmit light perpendicular to the chains. The direction perpendicular to the oriented chain molecules is commonly called the *transmission axis* or *polarization direction.* Hence, when unpolarized light falls on a po-

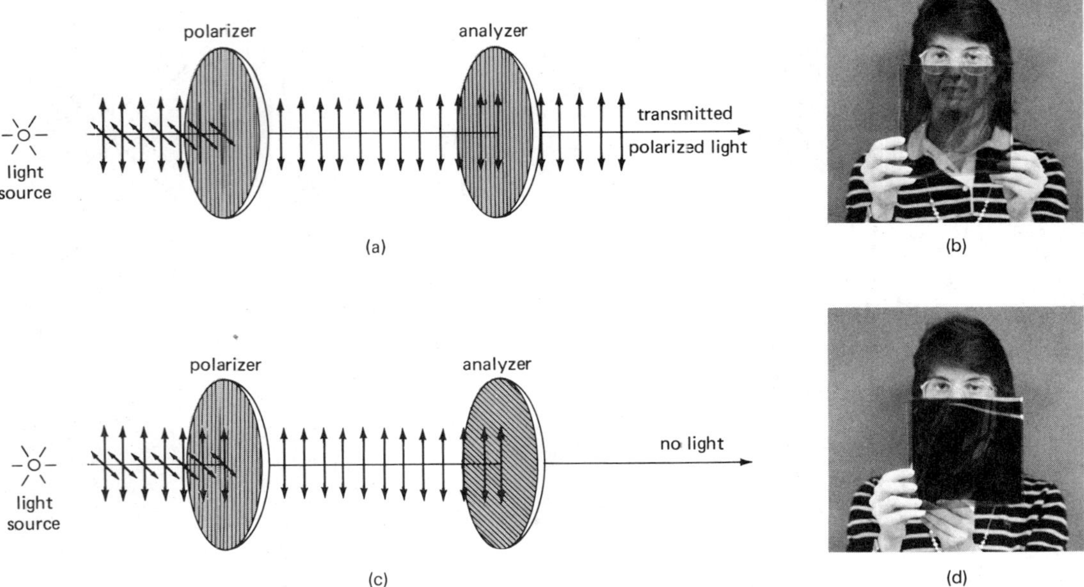

Figure 23.13 Polaroid films. When the film sheets are oriented with the same polarization direction, the transmitted light is polarized (a and b). When one of the sheets is rotated 90° ("crossed Polaroids") little or, ideally, no light is transmitted (c and d).

larizing sheet, the sheet acts as a polarizer and polarized light is transmitted (Fig. 23.13).†

The human eye cannot distingish between polarized and unpolarized light. Hence, we need an analyzer to detect polarized light, which can be another sheet of polarizing film. When the analyzer has the same orientation as the polarizer, light is transmitted. But, when rotated 90°, or when the polarizing sheets are "crossed," the analyzer absorbs the polarized light, and ideally, no light is transmitted (Fig. 23.13). You can demonstrate this yourself with a couple pairs of polarizing sunglasses. Try it next time you are at the drug store or somewhere where they sell polarizing sunglasses (if you don't personally have two pairs of polarizing sunglasses). Incidently, polarization proves that light is a transverse wave instead of a longitudinal wave because longitudinal waves cannot be polarized. Why not?

Light is also partially polarized by reflection. The direction of polarization is parallel to the reflecting surface, and the degree of polarization depends on the material and the angle at which the light strikes the surface. This is why we use polarizing sunglasses (Fig. 23.14). Light reflected from roads, water, and other surfaces gives a large intensity and glare. By properly aligning the polarizing lenses of the glasses (transmission axis vertically), some of the partially polarized reflected light is absorbed, which reduces the glare.

There is also polarization by scattering. In the scattering process, when light strikes a suspension of particles, such as the molecules of the air, electrons are set into vibration by the light, and light waves are reradiated. The oscillating electrons do not radiate in the direction of their oscillation, so the scattered light is partially polarized. As a result, sunlight scattered by the atmosphere and received at the Earth's surface is partially polarized. You can check this yourself by looking at a blue sky through polarizing sunglasses and then rotating the glasses. Light from different areas of the sky will show different degrees of transmission depending on the degree of polarization, which depends on the relative direction of the Sun. It is believed that some insects, such as bees, use the polarized sky light for navigation. Scientists have put bees in a box with a Polaroid film over the top. When the film is rotated, the flying bees change direction.

In closing out this chapter, let's take a look at a

† A common analogy is to view a polarizer as a picket fence through which a random wave in a rope will pass only parallel to the pickets and be polarized in this direction. However, keep in mind that in a polarizing sheet, the blocking "pickets" (absorbing molecular chains) are perpendicular to the transmission axis.

(a)

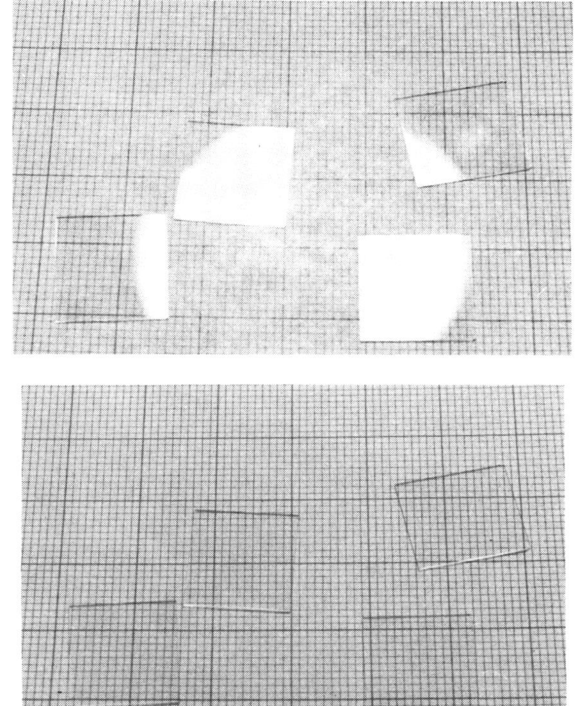

(b)

Figure 23.14 Glare reduction. Light reflected from surfaces is partially polarized (a). Polarizing sunglasses reduce the intensity or glare by blocking out the polarized light, as shown in (b) for glass surfaces.

couple of practical applications of polarization. One is discussed in Special Feature 23.2. Another involves movie entertainment. Recently, there has been a comeback of 3-D (three-dimensional) movies. This is where objects seemingly jump out of the screen at you when you view a movie with special glasses, which give a third dimension or depth perception to a two-dimensional picture on the screen. We have visual depth perception in part because we see simultaneously with both eyes, each having a slightly different view or perception. The brain interprets the differences of the views in terms of depth perception.

Three-D motion pictures are projected on the screen by two projectors, each with a slightly different image (Fig. 23.15). One image is for one eye and the other for the other eye so that the brain will interpret this as depth or a third dimension. So that you get a different image for each eye, the light from each projector is polarized, but in perpendicular directions to each other. The 3-D glasses are really polarizing glasses with perpendicular polarization directions for each eye. Hence, one eye sees one image and the other eye sees the slightly different image. Viewed without the polarizing glasses, the combined pictures on the screen appear blurred.

Other methods of using two images and color to fool the brain are now being explored for TV applications. We may soon be watching 3-D television programs.

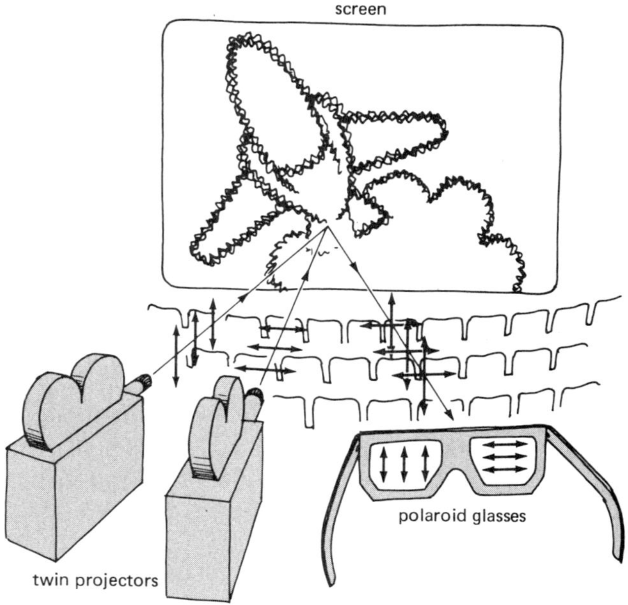

Figure 23.15 Three-D movies use twin projectors and polarizing glasses so that slightly different images are seen by each eye, giving the perception of depth or a third dimension.

SPECIAL FEATURE 23.2

LCDs—Liquid Crystal Displays

A now very common application of polarization is in LCDs or liquid crystal displays. These displays are found on calculators, wrist watches, and even gas pumps. One type of LCD makes use of the light-polarizing properties of some liquid crystals. (Recall that liquid crystals are liquids that show some degree of molecular order. See Chapter 10.) Such crystals have the ability to rotate or "twist" the polarization direction of polarized light by 90° (Fig. 23.16). In a so-called twisted nematic display, a liquid crystal is sandwiched between crossed polarizer sheets and backed by a mirror. Light falling on the surface of the LCD is polarized, twisted, reflected, and twisted again, and then it leaves the LCD. Hence, the display appears light when illuminated.

But when a voltage is applied to the crystal, it loses its twisting property, and the crystal appears dark when illuminated because the light is blocked by the polarizing sheet and not reflected. Such voltage-induced dark regions of the crystal are used to form numbers and letters of the display. You can show that the light coming from such an LCD is polarized by a Polaroid sheet (or polarizing sunglasses), as shown in Figure 23.16.

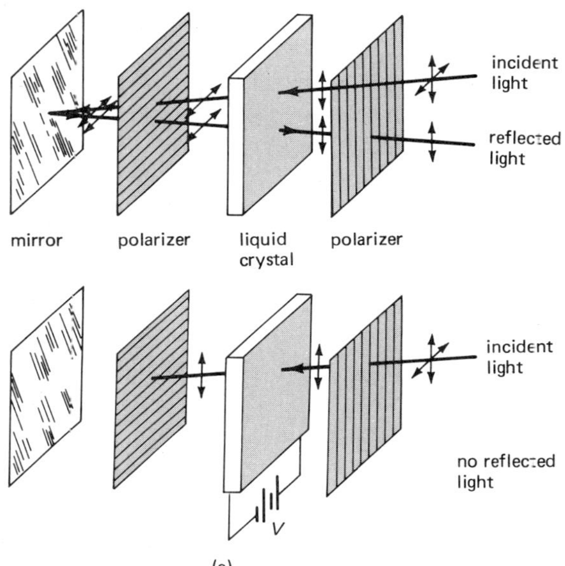

(b)

Figure 23.16 Twisted nematic LCD. (a) When no voltage is applied to the liquid crystal, it "twists" the light polarization through 90°. When voltage is applied, there is no twisting, and light does not pass through the second polarizer and is not reflected. The crystal then appears dark. (b) The bright areas of an LCD are due to reflected polarized light, as can be shown with a polarizing sheet (analyzer).

SUMMARY OF KEY TERMS

Phosphorescence the process whereby the electrons in some materials remain in excited states for long periods of time and emit light long after being excited, or "glow in the dark."

Fluorescence the process whereby certain materials absorb ultraviolet light and emit light in the lower-frequency visible region.

Thin film interference the interference of light reflected from the top and bottom surfaces of thin films.

Newton's rings the interference ring pattern seen in a lens on an optical flat as a result of thin film interference.

Diffraction the bending of waves around the corners of objects or openings that occurs when the wavelength is on the order of the size of the object or opening or smaller.

Polarization a preferential orientation of the field vectors of light waves.

EXERCISES

1. Cold is the absence of heat. Is dark the absence of light? Explain.
2. Quartz crystals are now used in watches. How can a quartz crystal keep time?
3. What determines how long a phosphorescent material will glow in the dark? What determines how brightly a material will glow?
4. During the summer, "brownouts" or voltage reductions are experienced in some large cities due to electrical demand exceeding the supply. How would this affect the light emitted from an incandescent lamp? (*Hint:* See Special Feature 13.3.)
5. Mercury-vapor street lights have a bluish hue, while the newer sodium arc lamps have a yellow appearance. Explain this difference.
6. At a disco, the black light over the band blows out and there are no more bulbs available. The proprietor decides to use an infrared lamp instead to get the fluorescent effects. Is this a good idea? Explain.
7. Why can we still see things when only black lights are used?
8. Some soap manufacturers add a blue dye to their detergents that fluoresces with a slight blue color. What is the purpose of this? (*Hint:* "Whiter whites" have a cleaner appearance when they are slightly bluish in color.)
9. When washed dishes aren't rinsed well, they often show a display of colors after they dry. What causes this?
10. What would be the effect if a thin film were illuminated with monochromatic light instead of white light? Consider films of uniform and nonuniform thicknesses.
11. Will a thin film always be nonreflecting (or will destructive interference occur) for normal incidence when the film thickness is one fourth of a wavelength of the light? Explain.
12. Explain why the film thickness on nonreflecting glass is one fourth of the wavelength of yellow-green light. What would happen if the film thickness were half a wavelength?
13. The film on most nonreflecting lenses is just the thickness to produce destructive interference for yellow-green light. If you wanted to have destructive interference primarily for blue light, would you make the film thinner or thicker? Explain.
14. For destructive interference with thin films, is the light really "destroyed"? Explain. How about the light for the dark fringes or destructive-interference regions in the interference pattern of Young's experiment?
15. Why are camera lenses coated to make them nonreflecting?
16. Is there a phase shift or shifts for light during the back reflections from a coated camera lens? Explain. How does this affect the interference?
17. Suppose you want to use thin-film interference to make "reflecting" glass or a mirror. How could you do this? (Discuss in terms of film thickness *and* type of material, e.g., what if the film material were more optically dense than glass?)
18. Explain why a uniform air wedge between two optical flats produces an interference pattern of alternate bright and dark lines when illuminated with monochromatic light. What would be seen if white light is used?
19. Television waves can be diffracted. Which would be expected to show the greater diffraction for common-sized objects that might be encountered: (a) VHF stations or UHF stations? (b) Low-band VHF stations or high-band VHF stations? (*Hint:* See Chapter 22 and recall that $\lambda = c/f$, where $c = 3 \times 10^8$ m/s.)
20. What size objects or openings would be good to diffract TV waves? (*Hint:* See Chapter 22 and recall that $\lambda = c/f$, where $c = 3 \times 10^8$ m/s.)
21. Some of the vivid colors of the fanned tail of a peacock are due to diffraction. Why is this? (*Hint:* Think in terms of feather diffraction gratings.)
22. How do polarizing sunglasses reduce glare?
23. Does it make any difference how the plane or direction of polarization of the lenses of sunglasses is oriented? Explain.
24. Given two pairs of sunglasses, how could you tell if they were nonpolarizing or polarizing, or could you?
25. Suppose you had two sheets of Polaroid film and you

held one in front of the other while looking through them (see Fig. 23.13). (a) If you rotated one of the sheets through one complete rotation, how many times would the Polaroids be crossed or darkened? (b) Think about this one—rotating both Polaroids through one complete rotation at the same time, but in opposite directions. How many times would the Polaroids cross or darken in this case? (c) What would be the case if the Polaroids were rotated in the same direction and at the same speed?

26. Rather than wear the flimsy glasses provided at a 3-D movie, Joe Cool decides to wear his designer polarizing sunglasses instead. Is this a cool idea?

24 Reflection and Refraction

Were it not for the reflection of light, our world would appear dark, except in the vicinity of light sources. Look around you. You see things because of light reflected from them. Without reflection, we would not see the moon. (How about the stars?) Many optical phenomena involve refraction. Without refraction, there would be no rainbow, and eyeglasses, which many of us wear, would not help correct vision problems. Reflection and refraction as they relate to sound were touched upon briefly in Chapter 17. In this chapter we'll take a look at these effects as they involve light.

Light Rays

The wave nature of light was quite important in considering interference phenomena in Chapter 23. The wiggly wave forms allow us to visualize and analyze constructive and destructive interference. However, for reflection and refraction it is convenient to think of light *rays*. This is the representation of light in what is called geometrical or ray optics. In effect, we ignore the wave nature of light and represent it as rays or straight lines.

To see how this is done, consider the diagrams in Figure 24.1. As waves spread out circularly (or spherically) from a point source, the adjacent portions of a wave that are in phase are said to form a **wave front.** For example, if a line were drawn through adjacent crests, it would define a circular wave front. The wave front propagates outward with the same velocity as the wave and in a direction perpendicular to the wave front. A line drawn perpendicular to the wave fronts that indicates the direction of propagation is called a **ray.** There are also plane wave fronts, for example, at distances far from a point source, and the rays are parallel. We often

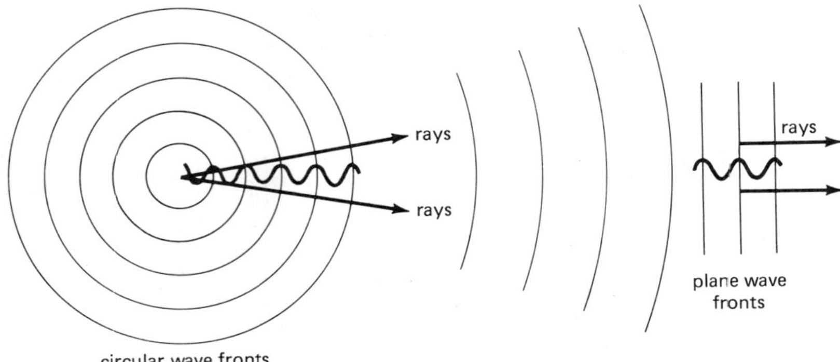

Figure 24.1 Rays. From a point disturbance, waves spread out with circular wave fronts. Lines perpendicular to the wave fronts in the direction of the propagation are called rays. At great distances the rays are parallel, and a group of parallel rays forms a beam.

speak of a beam of light, by which we mean a group of parallel rays. Unless reflected, refracted, or diffracted, light travels in a straight line, as represented by a light ray.

Reflection

Reflection might be thought of as light "bouncing" off a surface. Actually, it is much more complicated than this. It involves absorption and emission associated with complex electron vibrations in the atoms of the reflecting medium. But with ray optics we don't have to worry about this. We can describe what happens very simply in terms of rays. It is found that light rays are reflected from a smooth surface in a particular way (Fig. 24.2). Measuring the angles of the incident ray and reflected ray from the "normal," that is, a line perpendicular to the reflecting surface, gives the **law of reflection:**

The angle of incidence is equal to the angle of reflection, or $\theta_i = \theta_r$.

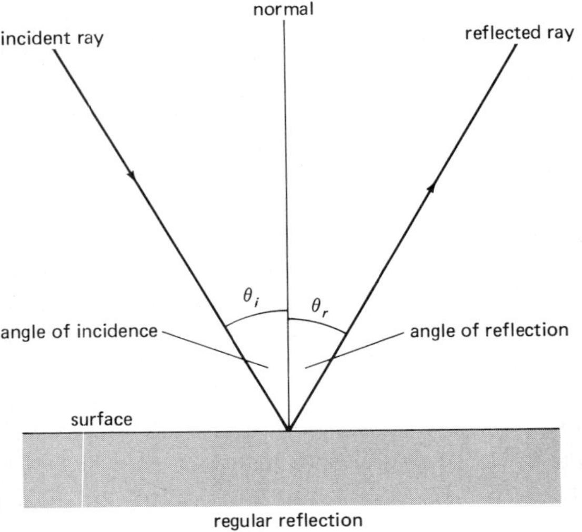

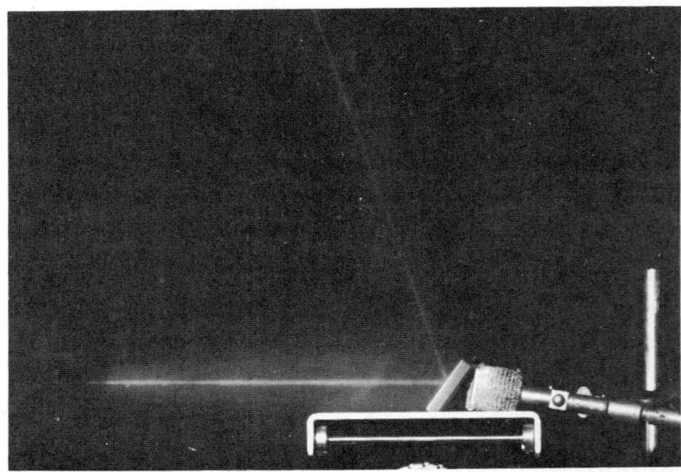

Figure 24.2 Law of reflection. The angle of incidence θ_i is equal to the angle of reflection θ_r, where the angles are measured from a normal (line) perpendicular to the reflecting surface.

QUESTION: If you have a mirror positioned in a doorway so you can see someone coming down the hall, can they see you?

ANSWER: Yes. As children we probably all played this game, thinking that the person could not see us around the corner. But the law of reflection works both ways or in both directions; the rays are just reversed. You see the person because light rays are reflected to your eye. Light rays from your face at the same angle are reflected back along the same path. This is called reverse-ray tracing. If you can see their eyes, they can see you.

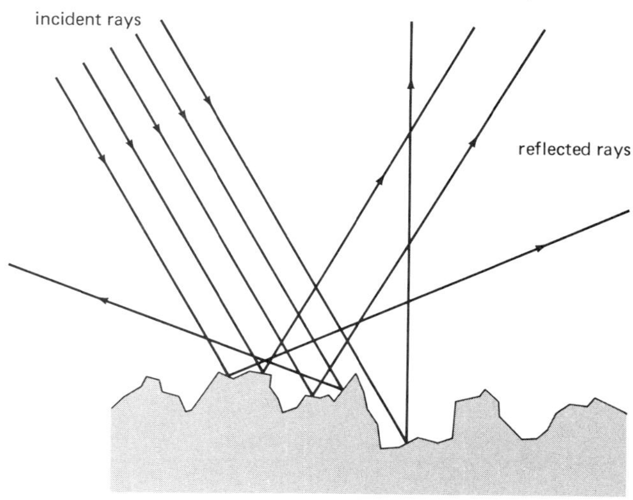

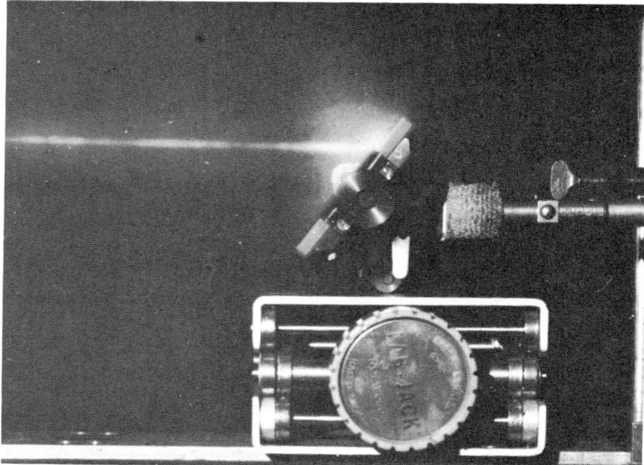

Figure 24.3 Diffuse or irregular reflection. The law of reflection still applies, but the surface irregularities cause the reflections to be diffuse.

The reflection from very smooth or mirror surfaces such as in Figure 24.2 is called **regular** or **specular reflection.** Of course, not all surfaces are smooth enough to act as mirrors in which we can see images. The pages of this book are reflecting light to your eye; otherwise, you could not see them. Other than light sources, in general we see objects because of the light they reflect.* A page of this book is relatively rough or irregular, and the reflection from such surfaces is called **irregular** or **diffuse reflection** (Fig. 24.3).

The law of reflection still applies in diffuse reflection. It's just that the small reflecting surface irregularities are at different angles to each other, and so are the reflected rays. It is diffuse reflection from particles in the air that allows us to see light beams such as those from a flashlight or searchlight. Light is reflected from the beam by dust particles and so on; otherwise, we wouldn't see it (no light to our eye). This is why we can see "shafts" of light when sunlight shines through the leaves of trees in a forest or through clouds. The latter is referred to as "the Sun drawing water" in weather folklore and is said to be indicative of rain. (See Exercise 4 at the end of this chapter.)

MIRRORS

When you look at your "reflections" in a mirror, these are regular reflections. In general, a **mirror** is any smooth surface that regularly reflects light. This may be a smooth water surface (see Chapter introductory photo). When the water surface is rough, the reflection is diffuse, and the mirror quality is lost. A surface is "mirror"-smooth when the surface irregularities are on the same order as or smaller than the wavelength of light (10^{-5} cm for visible light).

Commercially, flat or **plane mirrors** are available as polished metal surfaces, but most commonly, mirrors are made of glass with surfaces that have been coated or "silvered" with compounds of tin and mercury. (Silver compounds can be used, but they are expensive.) A glass mirror may be front-coated or back-coated, depending on the application. Common household mir-

* The black print on this page absorbs most of the incident light, and we "see" it by contrast with the reflected white light from the white portion of the page. More about this in the next chapter in the discussion on color.

rors are usually back-coated, but mirrors for telescopes are front-coated (Chapter 25).

QUESTION: At night, a glass windowpane acts as a mirror. Why isn't it a mirror during the day?

ANSWER: When light strikes a transparent medium, most of the light is transmitted, but some is reflected. During the day, the light reflected from the inside of the windowpane is overwhelmed by the light coming through the window from the outside. However, in the evening or at night when the light transmitted from the outside is reduced, the inside reflections can be discerned, and the windowpane acts like a mirror.

Because of regular reflections, mirrors produce images. We look at our images in mirrors everyday, usually starting off with the bathroom mirror in the morning. Flat mirrors, such as bathroom and dressing-table mirrors, are called plane mirrors. Light reflected by a plane mirror causes us to see images "inside" the mirror. An image appears to be located the same distance behind the mirror as the object is in front of the mirror (Fig. 24.4).

There is also a right-left reversal of the object and image in a plane mirror. As we have all observed when looking into a mirror, if you raise your right hand, your image will raise the left hand. Your image's hair will appear to be parted on the opposite side as yours (unless you part your hair in the middle, have no part, or are bald). A practical example of the right-left reversal of a plane mirror is shown in Figure 24.5.

Another type of mirror is a spherical mirror, which is just a reflecting surface that has a spherical contour. There are two kinds of spherical mirrors, depending on which surface of the spherical section is reflecting (Fig. 24.6). If the concave (inside) surface is reflecting, this is a **converging spherical mirror.** Notice in the figure how rays parallel to the mirror axis converge or come to a focus at a point, which is called the **focal point.** The converging properties of these spherical mirrors allow light to be converged to form an upside-down image on a screen (when object is beyond focal point). The image may also be magnified. Converging spherical mirrors are used as cosmetic mirrors. When you bring

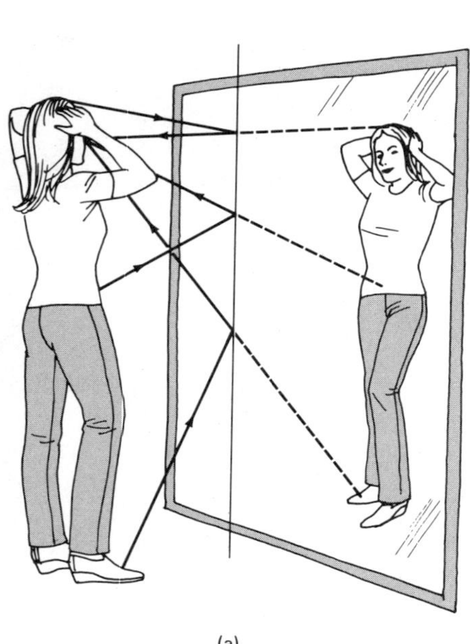

(a)

(b)

Figure 24.4 Plane (flat) mirrors. (a) The image in a plane mirror appears to be located the same distance behind the mirror as the object is in front of the mirror. (b) It's not hot. The finger is being held over the image of a candle behind a glass window in a laboratory cabinet that acts as a mirror.

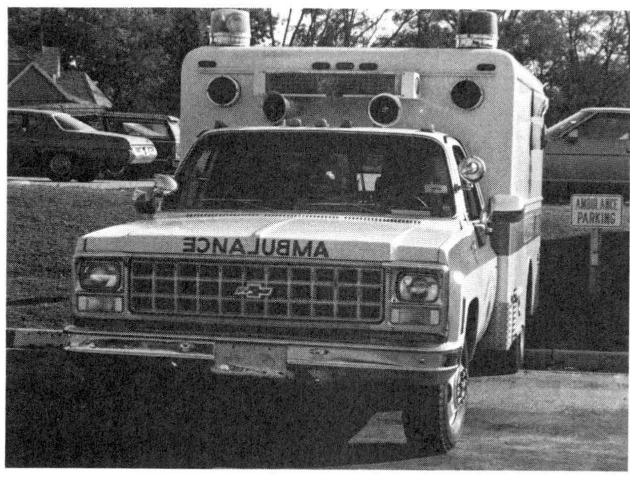

(a)

(b)

Figure 24.5 Right-left reversal. (a) The spelling of *AMBULANCE* is reversed so that it will be seen spelled properly in a car's rear-view mirror (b). Notice that the sign painter had a problem with reversed lettering with the *N*.

your face close to such a curved mirror, you see a right-side up magnified image of yourself (when you, the object, are closer to the mirror than the focal point).

If the convex (outside) surface is reflecting, this is a **diverging spherical mirror** (Fig. 24.6b). Here, light rays parallel to the mirror axis are reflected as though they come from an inside focal point. A reflecting, spherical

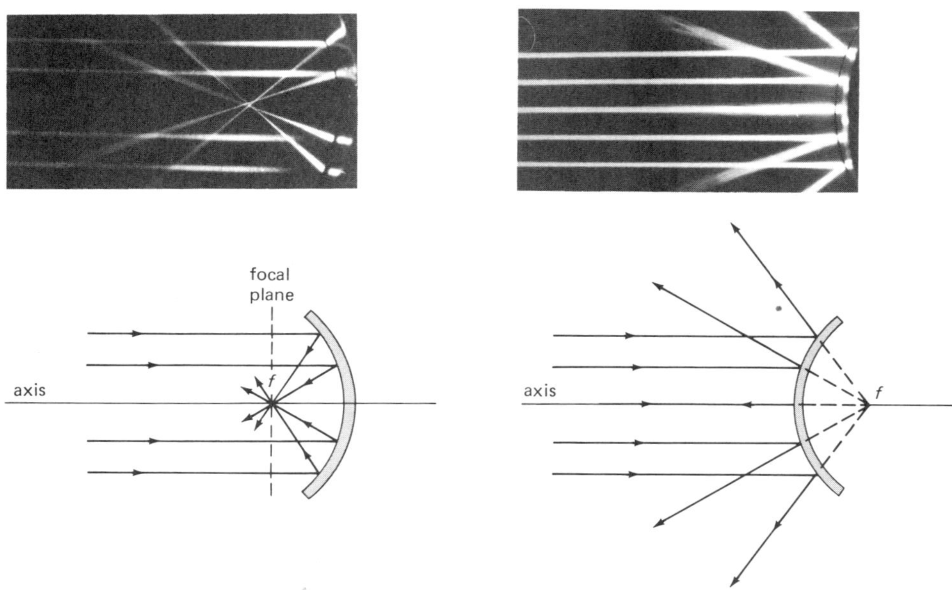

(a) concave converging mirror

(b) convex diverging mirror

Figure 24.6 Spherical mirrors. (a) Concave. Rays parallel to the mirror axis converge at the focal point *f*. (b) Convex. Rays parallel to the mirror axis diverge on reflection, as though they come from a focal point on the opposite side of the mirror.

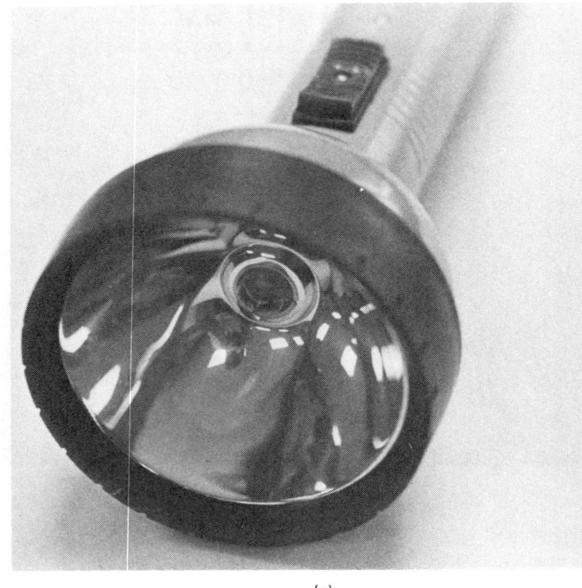

(a)

(b)

(c)

Figure 24.7 Spherical mirror applications. (a) A bulb at the focal point of a spherical flashlight reflector gives a parallel beam. (b) Convex spherical mirrors provide a wide-angle view for store monitoring. (c) A spherical "mirror" for converging radio waves from satellites. The detector is located at the focal point of the TV "dish" antenna.

Christmas-tree ornament or a spherical car hubcap would act as a diverging mirror.

Some practical applications of spherical mirrors are shown in Figure 24.7. The first two are similar to the diagrams in Figure 24.6 but with reverse-ray tracing. If a light source is at the focal point of a converging mirror, the rays are reflected in a parallel beam. This is the purpose of the spherical reflector in a flashlight or searchlight. Large diverging mirrors are commonly used in stores for monitoring. By reverse-ray tracing, light coming in at wide angles is reflected into a parallel beam by a diverging mirror. As a result, we have a wide field of view when looking into a diverging mirror.

Another example of a converging "mirror" is the satellite TV "dish" antenna. This is not an optical mirror. Instead of light, radio waves from distant satellites, which form nearly parallel rays, are focused or converged at the focal point, where a sensor or detector is located.

Refraction

Back in Chapter 17, the refraction of sound waves was discussed briefly. Recall that refraction is the change or deviation of a wave's direction of propagation when it passes obliquely from one medium into another or into a different-density region of the same medium.† Similarly, light is bent when it passes from one medium into

† *Obliquely* means slanting or inclined at an angle, as opposed to direct or normal incidence (incident angle of 0°).

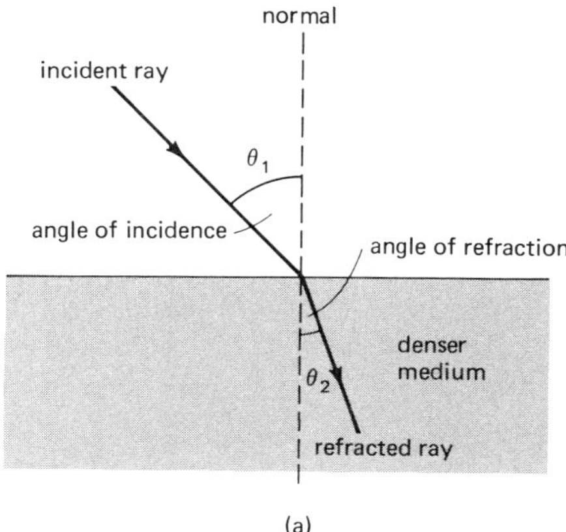

(a)

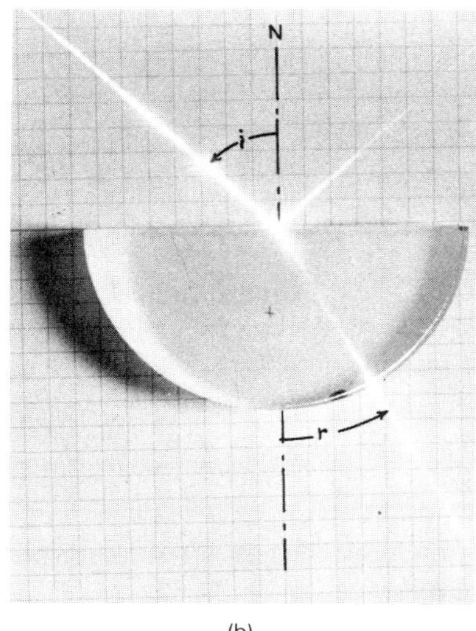

(b)

Figure 24.8 Refraction. A light ray is deviated or "bent" when it enters another medium. When entering an optically denser medium, the ray is bent toward the normal ($\theta_2 < \theta_1$). Notice in the photo that some of the incident light is reflected as well as refracted.

another (Fig. 24.8). The directions of the incident and refracted rays are expressed in terms of incident and refracted angles (θ_1 and θ_2, respectively), which are measured from a normal or perpendicular line to the surface boundary of the medium. When light passes obliquely into a denser medium, for example, from air into glass or water, light rays are bent toward the normal ($\theta_2 < \theta_1$).

It is the slowing down of the light as it enters the denser medium that gives rise to refraction. This involves complex processes, but in a simple manner, one might expect the passage of light by atomic absorption and emission through the denser medium to take longer. For example, the speed of light in water is about 75 percent of that in air or vacuum, and in glass it is about 67 percent or less.

To help you understand how light is bent or refracted when it passes into a medium, consider a group such as your school band marching across a field. The marching column then obliquely (at an angle) enters a muddy, wet region (Fig. 24.9). As the marchers in the rows hit the wet, slippery region, they keep marching at the same cadence (frequency). But slipping in the muddy earth, the marchers in that portion of the row are slowed down. The marchers in the other part of the same row on solid ground continue on with the same stride, and as a result, the direction of the marching column is changed when it is in the muddy region. (This change in direction with changes in marching speeds is also seen when a marching column turns a corner.) We might think of the marching rows as wave

Figure 24.9 Marching "refraction". When the rows of the marching band enter a slippery (muddy) region, the marchers in a particular row in the mud slip and change speed (while keeping the same cadence or frequency). As a result, the direction of the marching column is changed.

(a)

(b)

Figure 24.10 Refraction effects. (a) When refraction is not taken into account, the fish appears to be at a different location than it actually is, which makes catching difficult. (b) The bending of light rays makes the pencil appear to be almost severed.

fronts. In this case, the wave frequency (cadence) remains the same, but the wavelength and wave velocity change in the medium. Notice in Figures 24.8 and 24.9 that by reverse-ray tracing, when light enters a less dense medium, it is refracted or bent away from the normal.

We have all observed refraction effects. Have you ever tried to catch a fish in a bowl or tank, and the fish is seemingly never where it appears to be? This is because it isn't where it appears to be when we think of light traveling in straight lines, as we usually do (Fig. 24.10). Owing to refraction, the fish is actually at a lower depth than we think it to be. Our minds don't usually take refraction into account. An object in a glass of water appears to be bent or almost severed. Refraction in the atmosphere also produces some interesting effects. See Special Feature 24.1.

TOTAL INTERNAL REFLECTION

As we have learned, light entering a less-dense medium is bent away from the normal. How far can the light be bent? The answer is "all the way," so to speak. If the incident angle is increased, the refracted light rays will just graze the surface at some critical angle of incidence (Fig. 24.12). Beyond this critical angle, the incident light is totally reflected back into the medium.‡

This total internal reflection allows specially cut pieces of glass (prisms) to be used as mirrors that reflect light at angles of 90° or even 180° (backward), as illustrated in Figure 24.13. The critical angle of glass is about 42°, so light internally incident on a surface at a greater angle, for example, 45°, is totally internally reflected. By the law of reflection, the angle of reflection is equal to the angle of incidence, so the light ray is changed in direction by 90°. Two 90° internal reflections give a 180° change in direction.

Internal reflection also enhances the brillance of cut diamonds. A diamond is cut with many faces or facets —58 in the so-called brilliant cut. Light entering the diamond is internally reflected so it emerges from the upper facets, giving rise to the diamond's sparkling brilliance. Check this out on a diamond ring.

Because of total internal reflection, light can be "piped" from one location to another in glass or plastic rods (Fig. 24.14). On entering the "light pipe" at an angle greater than the critical angle of the pipe material, the light undergoes repeated internal reflections and follows the contour of the pipe.

Flexible light pipes can be made by fusing bundles of

‡ Although called *total* internal reflection, a small part of the light is always transmitted into the second medium.

SPECIAL FEATURE 24.1

Atmospheric Refractions: Mirages, Hot Air, and Twinkling Stars

Atmospheric refraction effects are common. One of these effects is a mirage. The term *mirage* generally brings to mind a thirsty person in a desert "seeing" a pool of water that really isn't there. This is not the mind playing tricks but is an optical illusion. A more common mirage is the "wet spot" we often see on a hot highway in the summer that we can never seem to reach. What we are actually seeing is a reflection of the sky (Fig. 24.11a). The warm air layer near the road is less dense than the cooler air above it. This density difference causes sky light to be refracted or bent to the eye. In the desert, a water mirage may be due to the refraction of light from the leaves of a tall, distant palm tree.

It is also refraction that allows us to "see hot air" rising from hot surfaces such as a road or a hot plate. If you stop and think for a moment, you'll know that you can't see air. What you are seeing is the refraction or bending of light waves passing through the different-density regions of the rising convection currents. Similarly, atmospheric motions and density changes refract incoming starlight, causing stars to twinkle.

When watching a sunset (or moon set), you may have noticed that the Sun near the horizon appears to be flattened (Fig. 24.11b). This is also due to atmospheric refraction. The density of the atmosphere decreases with altitude, so the rays from the top and bottom portions of the Sun on the horizon are refracted by different degrees. This produces the observed flattening. Rays from the sides of the Sun on a horizontal plane are generally refracted by the same amount, so the Sun still appears circular along its sides. Also, because of refraction the sun can be seen for several minutes after it actually sinks below the horizon.

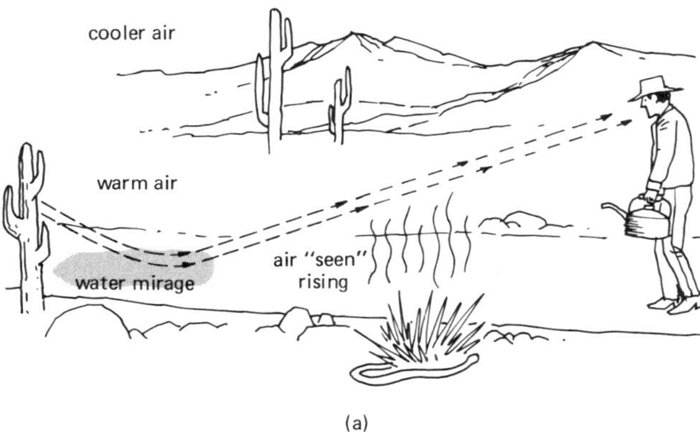

Figure 24.11 Atmospheric refraction. (a) Layers of air at different temperatures and densities act like different media and refract or bend skylight so we see a water mirage on hot road surfaces. Refraction also accounts for being able to "see" hot air rising. (b) The sun appears flattened because of different refraction of light rays from the top and bottom portions.

transparent fibers together. This has given rise to the relatively new field of **fiber optics** (Fig. 24.15). Images can be transmitted through the fiber bundles. With two sets of fibers in a well-randomized fiber bundle, light can be transmitted through one set and returned by the other. This arrangement is used in a fiberscope, which allows visual inspection of places that cannot be viewed directly. Such techniques have made it possible for physicians to view internal parts of the human body, such as the stomach and heart valves. In addition, fiber-optics light pipes or "wires" are now being used to optically transmit telephone conversations.

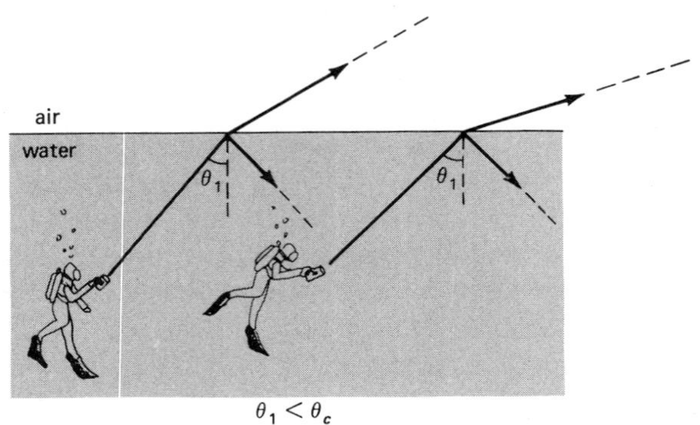

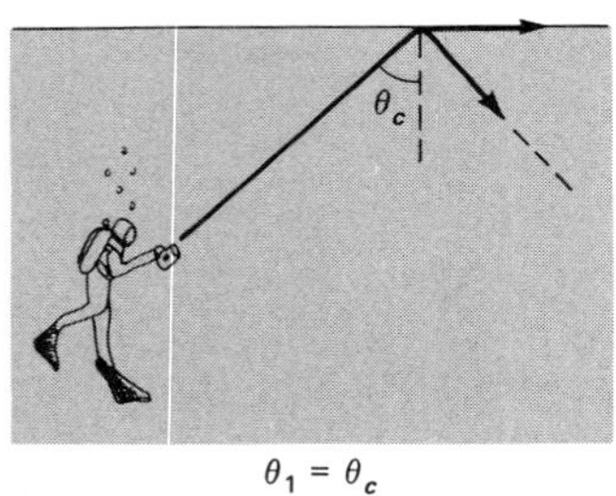

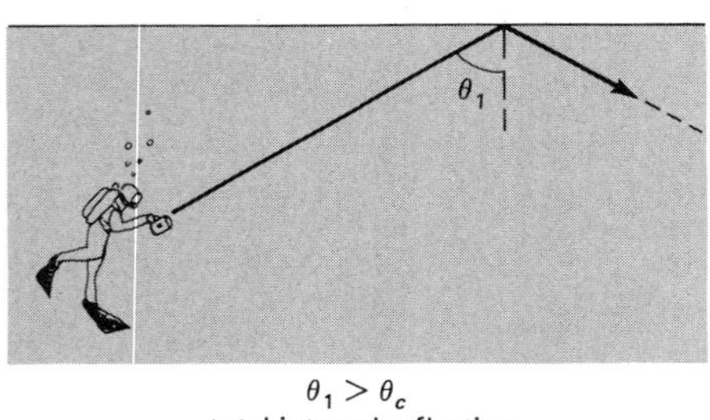

Figure 24.12 Total internal reflection. When light is directed at a less-dense medium with an incident angle greater than a certain critical angle (θ_c), the light is totally reflected, and the surface acts as a mirror.

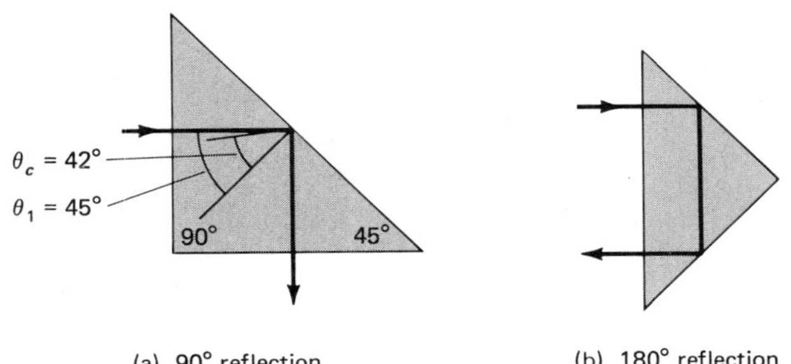

Figure 24.13 Glass prisms become mirrors, using internal reflection.

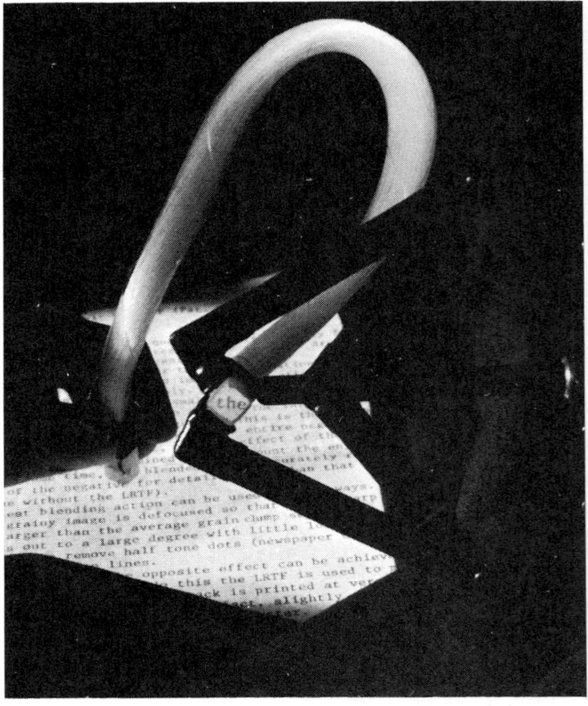

(a)

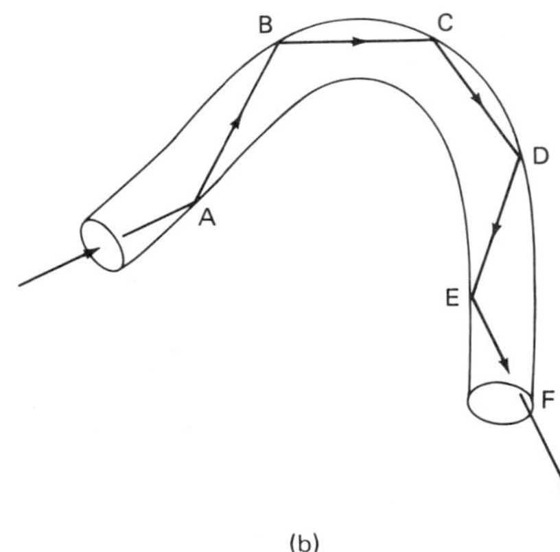

(b)

Figure 24.14 Internal reflection is the principle of light "piping" in fiber optics. (a) The angle of incidence is greater than the critical angle of the material for each reflection, and the light is internally reflected along the pipe (b).

Dispersion

When white light is passed through a prism, a spectrum of colors is observed to emerge (Fig. 24.16). This separation of light into its component frequencies or colors is called **dispersion.** To get an idea of how this occurs, consider the electron oscillators of the atoms making

(a)

(b)

Figure 24.15 Fiber optics applications. Light and images can be transmitted through light pipes and fibers, as in decorative lamps (a), making it possible to view otherwise inaccessible places (b).

up a transparent medium. When light is transmitted through the medium, it is transferred by atoms through absorption and emission by the electron oscillators. Since an oscillator has a resonant frequency, there is greater absorption of the components of the light with frequencies closer to electron-oscillator resonant frequency. The natural or resonant frequency of atoms in most transparent materials is in the ultraviolet region. Hence, light in the higher-frequency blue end of the visible spectrum is absorbed more by the electron oscillators and takes a longer time to get through the material or travels more slowly than light at the lower-frequency red end of the visible spectrum. Blue light is then bent or refracted more than red light. As a result of the different velocities for light with different frequencies, the components of white light are refracted at different angles and spread out into a spectrum of colors or frequencies.

Because of dispersion, prisms are used in spectrometers to analyze the components of light. In addition to a diamond's brilliance due to internal reflection, it is also said to have "fire" as a result of colorful dispersion. Dispersion can cause problems with lenses, as will be learned later in this chapter, but if not for dispersion, we would not have one of the beautiful sights of nature — the rainbow. See Special Feature 24.2.

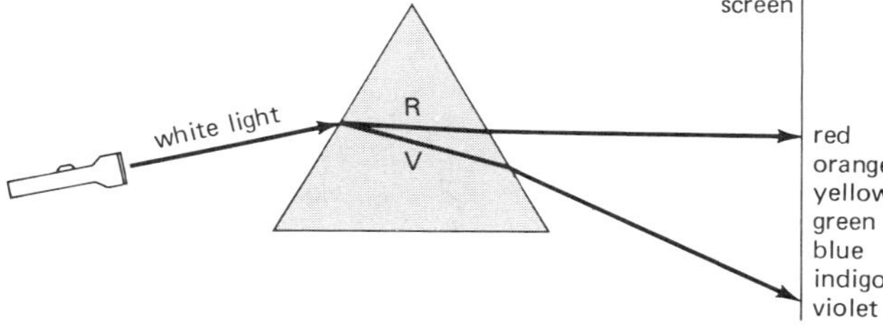

Figure 24.16 Dispersion. The dispersive properties of glass cause a prism to separate white light into a spectrum of colors.

Lenses

Refraction is to lenses as reflection is to mirrors. Many of us wear lenses in frames on our noses or in contact with our corneas to correct vision problems. The term *lens* comes from the Latin word for lentil, a plant with a seed that resembles a common lens shape. An optical lens is made from a transparent material, most commonly glass but also plastic and crystal, and usually has a spherical contour to one or both surfaces. There are a

SPECIAL FEATURE 24.2
The Rainbow

What causes a rainbow? Rain, of course, or more correctly, water droplets in the air after a rain, and the Sun must be there too. But we do not see a rainbow after every rain, so there must be some special conditions. Basically, whether or not we see a rainbow depends on the relative positions of the Sun and the observer. You may have noticed that the Sun is behind you when a rainbow is seen.

The rainbow is produced by refraction, dispersion, and internal reflection in water droplets. Occasionally more than one rainbow can be seen. The main or primary rainbow is sometimes accompanied by a fainter and higher secondary rainbow arc. Have you ever seen a secondary rainbow? For the primary rainbow, light is reflected once inside a water droplet (Fig. 24.17). Being refracted and dispersed, the sunlight is spread out into a spectrum of colors.

Because of the conditions on the refraction and internal reflection, the angles between the incoming and outgoing visible-light colors lie in the narrow range of 40° for violet to 42° for red. This means that you see a rainbow only when the Sun is positioned so that the dispersed light is reflected to you through these angles. With millions of water droplets in the air, a colorful arc is seen, running vertically in color from violet to red. (Below the rainbow arc the light from the droplets combine to form a bright, illuminated region.)

The less frequently seen secondary rainbow is caused by a double reflection in the water droplets. This gives rise to an inversion of the sequence of colors in the secondary rainbow from that in the primary rainbow.

As the Sun rises, less of the rainbow arc is seen. In fact, an observer on the ground cannot see a primary rainbow when the Sun's altitude (angle above the horizon) is greater than 42°. The primary rainbow is then below the horizon and will not be seen unless viewed from a height. As an observer is elevated, more of a rainbow arc is seen. It is common to see a completely circular rainbow from an airplane (no rainbow's end, no pot of gold). You have probably seen a circular "rainbow" in the fine spray of a garden hose.

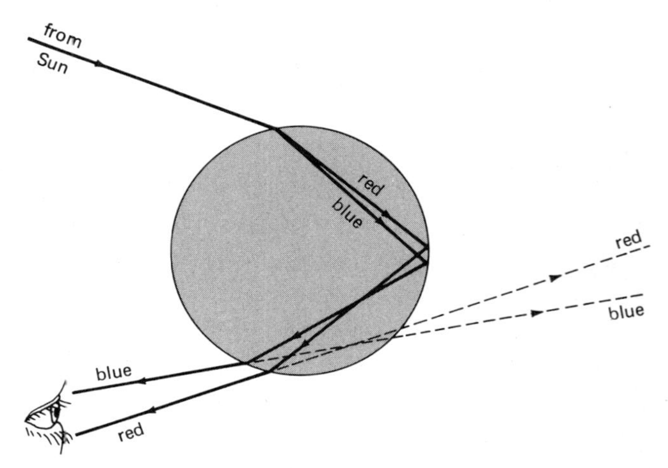

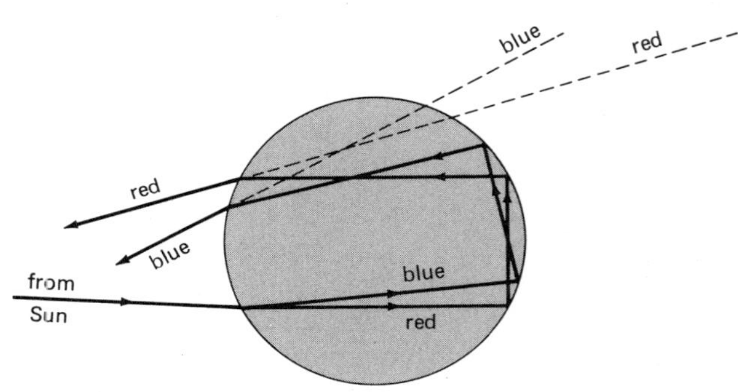

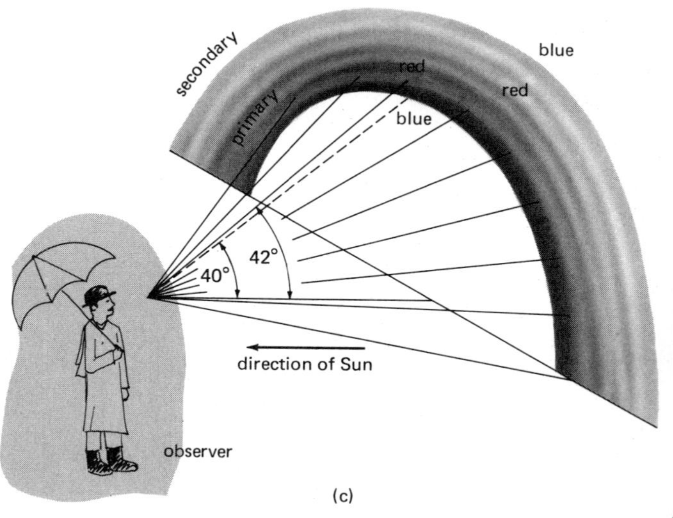

Figure 24.17 Dispersion, refraction, internal reflection, and rainbows. (a) Single and (b) double internal reflections can occur in water droplets, giving rise to primary and secondary rainbows. (c) Dispersion separates the component colors of sunlight at different angles, and an observer sees different colors coming from different altitudes, or a rainbow of colors.

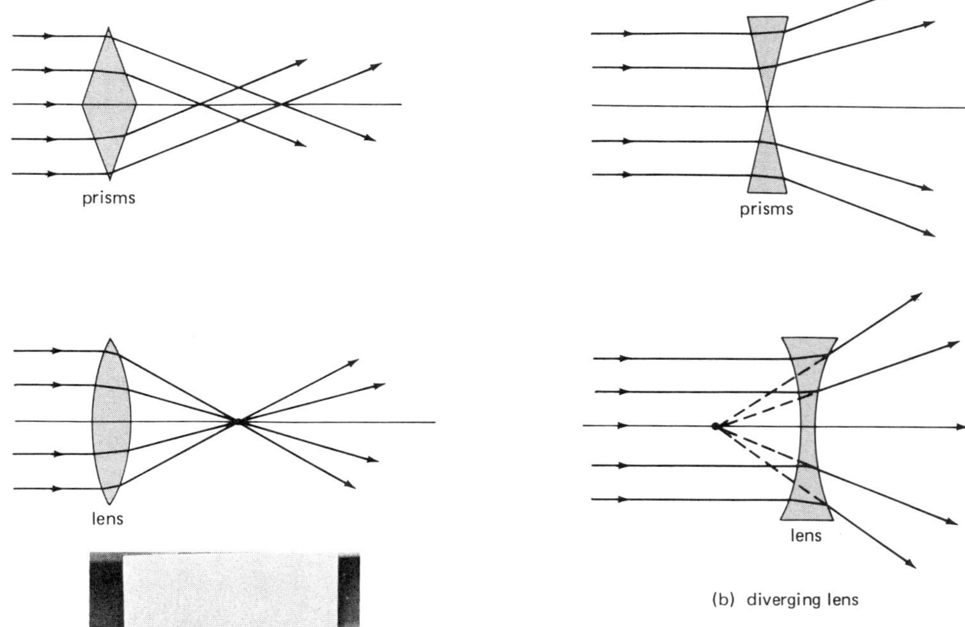

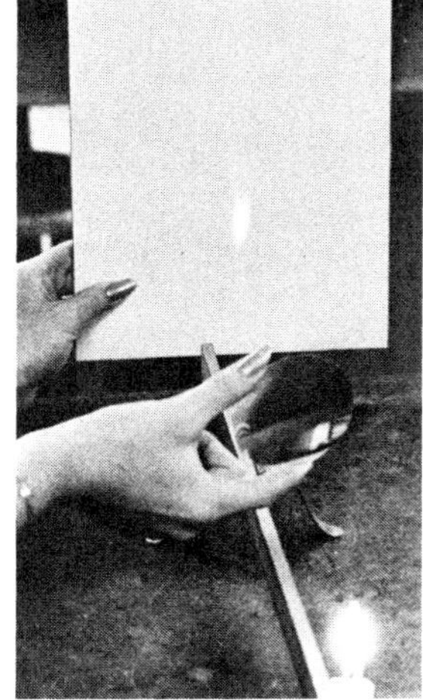

(a) converging lens

Figure 24.18 Lenses. (a) A converging lens is similar to two prisms placed base-to-base and can be used to form an image on a screen. (b) A diverging lens is similar to two prisms placed tip-to-tip, and the light is diverged so that no image can be formed on a screen.

variety of lens shapes, but we will limit our discussion to two simple types.

If two prisms are placed base to base, they tend to refractively converge incident light rays parallel to their common base line (Fig. 24.18). As such, the prisms approximate a spherical converging lens. The converging light energy will form an image of an object on a screen. We use lenses in projectors to project images on screens. A single lens inverts the image, but another lens can be used to invert it again so that it is right side up on the screen.

Similarly, if two prisms are placed point to point, they approximate a spherical diverging lens. The light energy diverges, so an image cannot be formed on a screen with a diverging lens. However, diverging lenses have various uses, as we will discuss in the next chapter.

ABERRATIONS

Lenses are used to form images, magnify, and so on in many optical instruments. Examples of some of these will be considered in the next chapter. Lenses are introduced here to give you an idea of how they incorporate refraction. Spherical lenses may not form perfect images, owing to material defects and to various natural and inherent effects called **aberrations.** Aberrations cause blurred or out-of-focus images. Let's consider a couple of the more common ones.

Spherical aberration results from rays nearer the edges of a converging spherical lens being brought to focus in front of the focal point (Fig. 24.19). This can be remedied by covering the outer part of the lens, as with

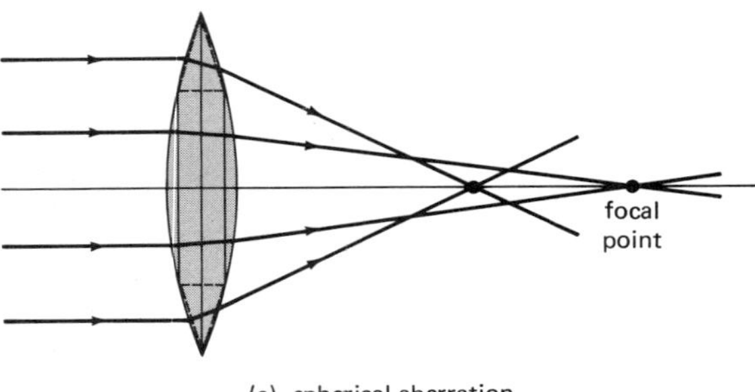

(a) spherical aberration

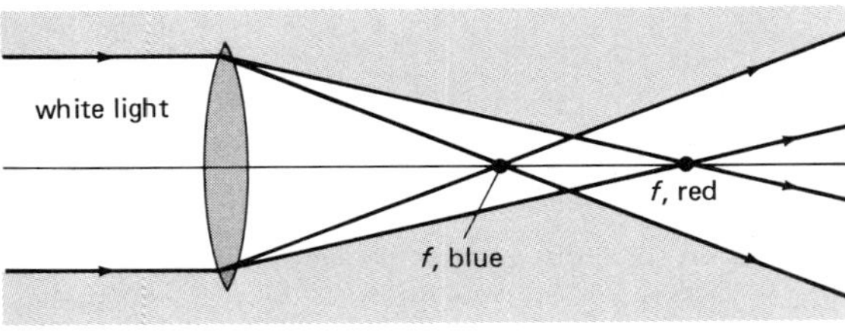

(b) chromatic aberration

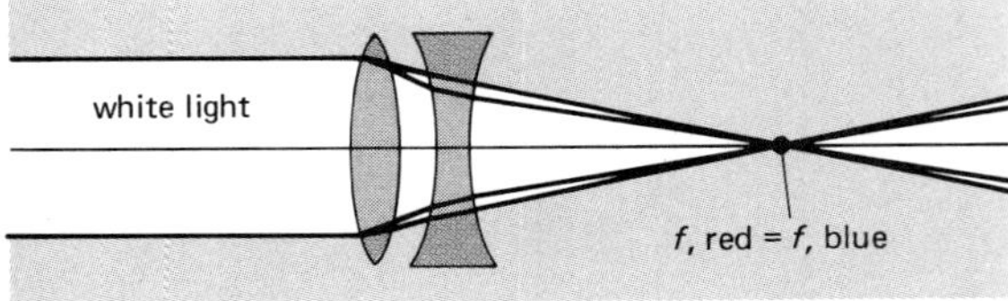

(c) achromatic doublet

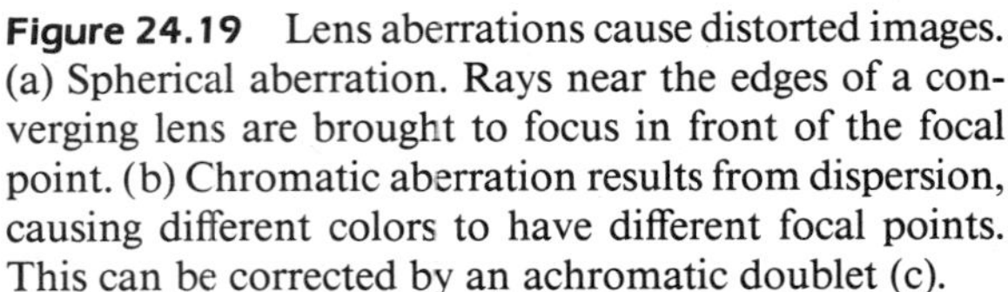

Figure 24.19 Lens aberrations cause distorted images. (a) Spherical aberration. Rays near the edges of a converging lens are brought to focus in front of the focal point. (b) Chromatic aberration results from dispersion, causing different colors to have different focal points. This can be corrected by an achromatic doublet (c).

a diaphragm in a camera. Spherical aberration also results for spherical converging mirrors. This can be eliminated by making parabolic-shaped converging mirrors. All the reflected rays from a parabolic mirror converge at a focal point or in the focal plane.

Chromatic aberration arises from dispersion. The dispersion of light passing through a lens results in rays of different colors coming together at different points. This aberration effect can be eliminated by using a lens combination called an achromatic doublet (achromatic means "without color"). The converging and diverging lenses are made of two different materials, so one lens refractively counteracts the dispersion of the other.

SUMMARY OF KEY TERMS

Wave front the front or form defined by adjacent portions of a wave that are in phase, for example, a line drawn through adjacent crests, which may form a circular, spherical, or plane wave front.

Ray a line drawn perpendicular to wave fronts that indicates the direction of wave propagation and represents the wave.

Reflection the return of incident light from a surface such that the angle of incidence is equal to the angle of reflection (law of reflection). Reflection from a smooth, mirror surface in which images can be seen is called regular reflection. Reflection from a rough surface is called irregular reflection and is in all directions.

Converging spherical mirror a mirror surface on the inside of a spherical section that, on reflection, converges or focuses rays parallel to the mirror axis at a focal point.

Diverging spherical mirror a mirror surface on the outside of a spherical section that, on reflection, diverges rays parallel to the mirror axis so that they appear to come from a focal point inside the mirror.

Refraction the bending or deviation of a wave's direction of propagation when it passes at an angle from one medium into another or into a different-density region of the same medium. When light passes into a denser medium, it is bent or refracted toward the normal.

Total internal reflection the reflection of light at a transparent-medium surface when the angle of incidence is greater than some critical angle. At incident angles above the critical angle, the light is totally internally reflected instead of being transmitted.

Dispersion the refractive separation of light into component colors as a result of light of different frequencies having different velocities in a medium.

Lenses glass, plastic, or crystal forms that use the refractive properties of light to form images, magnify, and so on.

Converging lens a lens that converges or focuses rays parallel to the lens axis at a focal point.

Diverging lens a lens that diverges rays parallel to the lens axis as though the rays come from a focal point on the incident side of the lens.

Aberration the distortion of an image due to the failure of the light to converge at a point or in a plane. Aberration can result from light rays passing through the edge of a lens or reflected from the edge of a spherical mirror that do not converge with the rays transmitted near the axis (spherical aberration) and from dispersion in a lens (chromatic aberration).

EXERCISES

1. A supplementary statement to the law of reflection that is often omitted is that the incident and reflected rays lie in a plane. Discuss what this means.
2. For regular reflection, what is the case with an angle of incidence of (a) 0° and (b) 90°?
3. If the moon had a very smooth surface, how would a full moon appear?
4. Why are only parts of the laser beam visible in Figure 24.20? What kind of reflection is this? (Blackboard erasers are being clapped together in the background.)
5. When you look at a window-glass "mirror" at night, two similar images, one behind the other, are often seen. Why are there two images?
6. If you walk toward a plane, full-length mirror, what does your image do? How fast does the image move? Is the image in step with you (like marching in a band)?
7. In detective movies and also in the study of children's activities, one-way mirrors are used for secret observations. (A one-way mirror is seen as a mirror from one side but can be seen through from the other side.) Reflecting sunglasses are another example. How do one-way mirrors work? (*Hint:* At night a glass windowpane is a one-way mirror.)
8. A good reflecting surface such as a plane mirror has a reflectivity of about 95 percent, or 5 percent of the incident light is absorbed on reflection. If four mirrors are set up so that an incident beam is reflected from one to the other, what percent of the incident light energy will be reflected from the fourth mirror? (*Hint:* The answer is *not* 80 percent.)
9. Explain the purpose of dual truck mirrors, as shown in Figure 24.21.
10. Operating room and dentist's lamps have large spherical reflectors. What is the purpose of these?
11. A dentist sometimes uses a small converging spherical mirror to examine cavities when filling them. What is the advantage of a converging mirror over a plane mirror?
12. When light enters a denser medium, how is the wavelength of the light affected? (*Hint:* Think about the marching analogy given in the text.)
13. Is there refraction for incident angles of (a) 0° and (b) 90°? Explain.
14. Explain how refraction causes the pencil in Figure 24.10 to appear almost severed.
15. A design or picture on the bottom of a swimming pool has varying distortions when viewed through the water in a filled pool. What causes these distortions?
16. While sitting by a swimming pool on a fixed bench in the spring before the pool is filled, a person notices that he can see only a limited portion of the design on the bottom of the pool (Fig. 24.22). However, when the pool is filled,

Figure 24.20 See Exercise 4.

Figure 24.21 See Exercise 9.

(a)

(b)

Figure 24.22 See Exercise 16.

he notices that he can see more of the design previously hidden by the pool wall. How is this possible?

17. Does atmospheric refraction have any affect on the length of the day? That is, would the daylight hours be longer or shorter if we had no atmosphere?
18. A thick beer or root-beer mug looks as though it holds more than it actually does because the liquid appears to be closer to the sides than it actually is (Fig. 24.23). Explain this "optical illusion" that makes you think that you're getting more for your money.
19. Is it possible to have total internal reflection with light in air incident on water? Explain.
20. Consider light in air incident on a glass surface. What is the maximum angle of refraction that can be achieved? (*Hint:* Think of reverse-ray tracing.)
21. We sometimes say that total internal reflection can be

Figure 24.23 See Exercise 18.

used to make an internal "mirror." Would it be possible to see an image in such an internal mirror? Explain.

22. Design a periscope to be used to see over and around objects using (a) plane mirrors and (b) prisms.
23. (a) How much of the underwater world could you see looking at various angles of incidence? (b) How would a fish see the "above-water" world when looking up at various angles?
24. Why is a spectrum of colors not seen when light passes through a glass windowpane?
25. A prism can be used to separate white light into a spectrum of colors. Could another prism be used to recombine the spectrum colors into white light? Explain. (*Hint:* Think of ray tracing.)
26. Is there an absence of rainbows after some rains, or is

Figure 24.24 See Exercise 31.

there always a rainbow and an observer may not be in the proper location to see it? Explain. (Rains after dark are excluded.)

27. Why are rainbows seen in the form of "bows" or circular arcs? Why is the secondary rainbow fainter than the primary rainbow?

28. Do two observers at different locations see the exact same rainbow? Explain your answer.

29. With dispersion and a single internal reflection in a water droplet, it might appear that the red arc of the primary rainbow would be below the violet or blue arc (see Fig. 24.18). However, the colors of the primary rainbow are seen to run vertically from violet to red. Explain why this is the case.

30. A halo or ring is sometimes seen around the moon or Sun as a result of high thin cirrus clouds that are composed of ice crystals. What causes these halos? (*Hint:* Keep in mind that the clouds are between the light source and the observer.)

31. A magnifying glass is a converging lens. Sunlight can be focused to a small spot using such a lens (Fig. 24.24.) What is the small spot an image of? Why are holes burnt in paper or leaves when the small bright spot is focused on them? (If you've ever focused the spot on your skin, you know it gets hot.)

32. What would be the effect if a small light source were placed at the focal point of a converging lens?

33. Discuss the types of aberration that may affect the images formed by (a) front-coated spherical converging mirrors and (b) back-coated spherical converging mirrors.

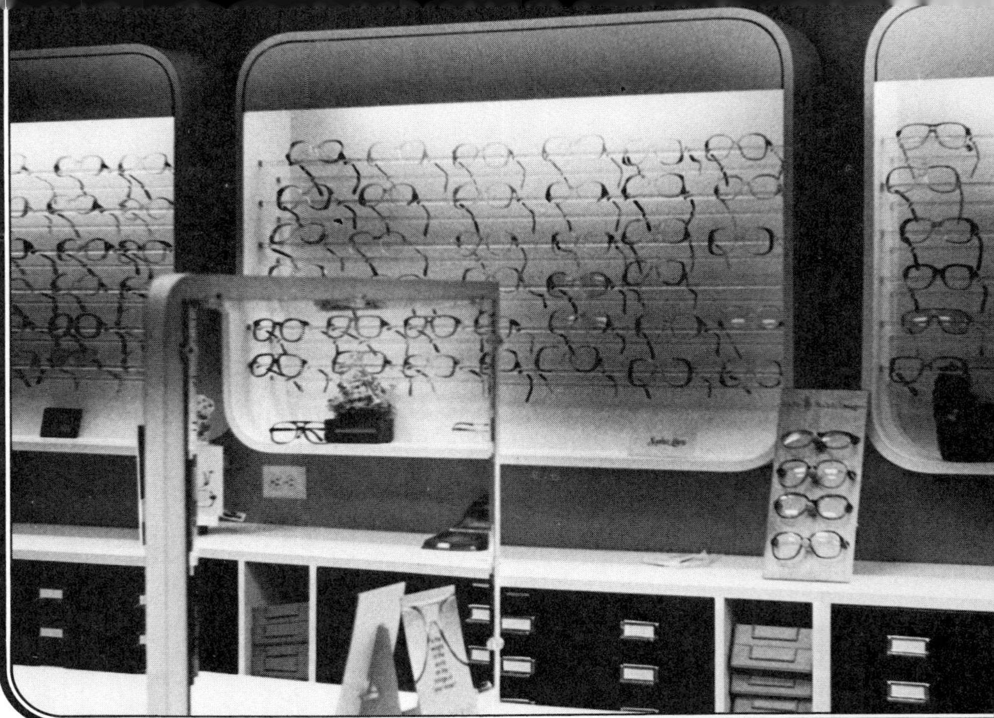

25

Vision and Optical Instruments

Vision or sight is perhaps our most important sense. We tend to take vision for granted — until things start to appear blurred. Then we can often correct the problem with lenses (glasses). Even when we don't have vision defects, we use optical instruments containing mirrors and lenses to improve or extend our visual observations. For example, with microscopes and telescopes we are able to see very small and very distant objects that are not visible to the unaided eye.

In this chapter we'll take a look at these things so you can understand how you "see" the world.

The Human Eye

The eye is a unique optical instrument and is one of our most valuable possessions. Without it, we would not see the beauty of the sunset or be able to read this and other books to gain knowledge. The anatomy of the eye is shown in Figure 25.1. The eyeball is nearly spherical, with a white outer covering called the sclera (the "white of the eye"). Light enters the eye through a curved, transparent tissue called the cornea. Behind the cornea is a circular diaphragm, the iris, which has a central hole called the pupil. Our knowledge of everything we see depends on the information conveyed through the tiny pupils of our eyes. The size of the pupil aperture or opening is adjusted by muscle action and controls the amount of light entering the eye. For example, in very bright light the iris diaphragm closes, and the pupil becomes very small. How about in dim light?

A converging crystalline lens composed of glassy fibers is situated behind the iris. The shape or curvature of the crystalline lens is controlled by the ciliary muscles. By adjustments and changes in the curvature of the lens, which is called accommodation, the images of

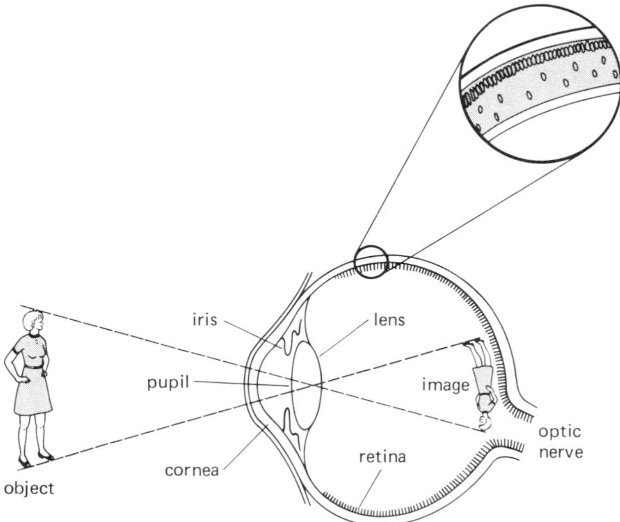

Figure 25.1 The anatomy of the human eye.

objects at different distances from the eye can be focused on the **retina** on the back wall of the eyeball. The eyeball contains a fluid in front of the lens and a gelatinous material in the space behind the lens.

The operation of the eye is similar in several ways to that of a simple camera (Fig. 25.2). Both have a lens, but the curvature of the camera lens cannot be changed. Instead, the camera lens is moved back and forth so as to focus images on the film. Both have variable diaphragms and both have a shutter. The shutter of the eye is the eyelid, which, unlike the shutter of a camera, is open for continuous exposure. Because of the refractive properties of the converging lenses, both form inverted or upside-down images on a light-sensitive surface (the film of the camera and the retina of the eye).

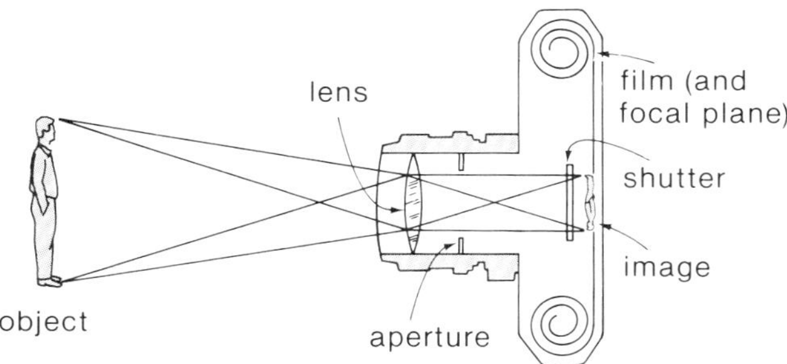

Figure 25.2 The operation of a camera is similar in several ways to that of the eye. Compare with Figure 25.1.

Figure 25.3 Demonstration of the blind spot. Hold the book at arm's length and, with your right eye closed, look intently at the black cross. Slowly bring the book toward your face. What happens to the square and dot?

QUESTION: If the images in our eyes are upside down, why don't we see the world that way?

ANSWER: By some means, the brain learns early in life to interpret the inverted image of the world right side up. Experiments have been done with persons wearing special glasses that give them an inverted view of the world. After some initial run-ins, they become accustomed to and function quite well in their "upside-down" world.

The retina "film" of the eye is composed of two types of photosensitive cells called **rods** and **cones** (names descriptive of shapes). The more numerous rods have a greater sensitivity to light and can distinguish between low light intensities for twilight (black-and-white) vision. The cones respond selectively to certain colors of light, some to one color and others to other colors. Cones are considerably less sensitive to light than are rods. This is why we cannot see color in very dim light. The rods and cones of the retina are connected to optic nerve fibers, which relay the light-stimulated signals to the brain.

In the region where the optic nerve enters the eyeball (Fig. 25.1), there are no rods or cones. As a result, your eye has a "blind spot" for which there is no optical response. Figure 25.3 can be used to demonstrate this blind spot. Hold the book at arm's length and, with your right eye closed, look intently at the black cross. Then slowly bring the book toward your face. At certain points, you will see the square and dot alternately disappear and reappear. Where did they go? Nowhere. Their images were just crossing your eye's blind spot. The blind spot is not noticed in ordinary vision because of eye and object movements and binocular vision.

Binocular (two-eyed) vision also accounts for some of our depth perception, which depends on slight differences between the shapes and positions of the images on the retinae of the two eyes. This occurs because the eyes are set several centimeters apart and get slightly different views of objects. Some other visual effects are discussed in Special Feature 25.1.

SPECIAL FEATURE 25.1

Optical Illusions

"Seeing is believing," goes the old saying. But this is not always the case, or at least you shouldn't believe everything you see. We can be visually fooled by several means. For example, fooling our depth perception causes objects to seem to "jump" out of the screen in 3-D movies (Chapter 23), and atmospheric refraction produces mirages (Chapter 24).

Some optical illusions have to do with the eye itself and interpretations by the brain. For example, in Figure 25.4 you may see fleeting patches of gray on the white areas between the black diamonds (spots before your eyes!). This is because the stimulation of one area of the retina can affect the sensations in an adjacent region, producing an illusion.

Also in Figure 25.4 is a design that can be used to illustrate an "after-image." When a particular area of the retina is stimulated, it may "remember" the stimulation after you have shifted your gaze to another object. Look intently at the white dot on the black Y for 15 to 20 seconds (quite a long time), then transfer your gaze to the white dot between the E's. You'll see a ghostly white after-image Y.

Other optical illusions are shown in Figure 25.5.

(a) (b)

Figure 25.4 Optical illusions. (a) Fleeting gray patches are usually seen on the white areas between the black diamonds (spots before your eyes!). (b) This one calls for concentration. See text for description.

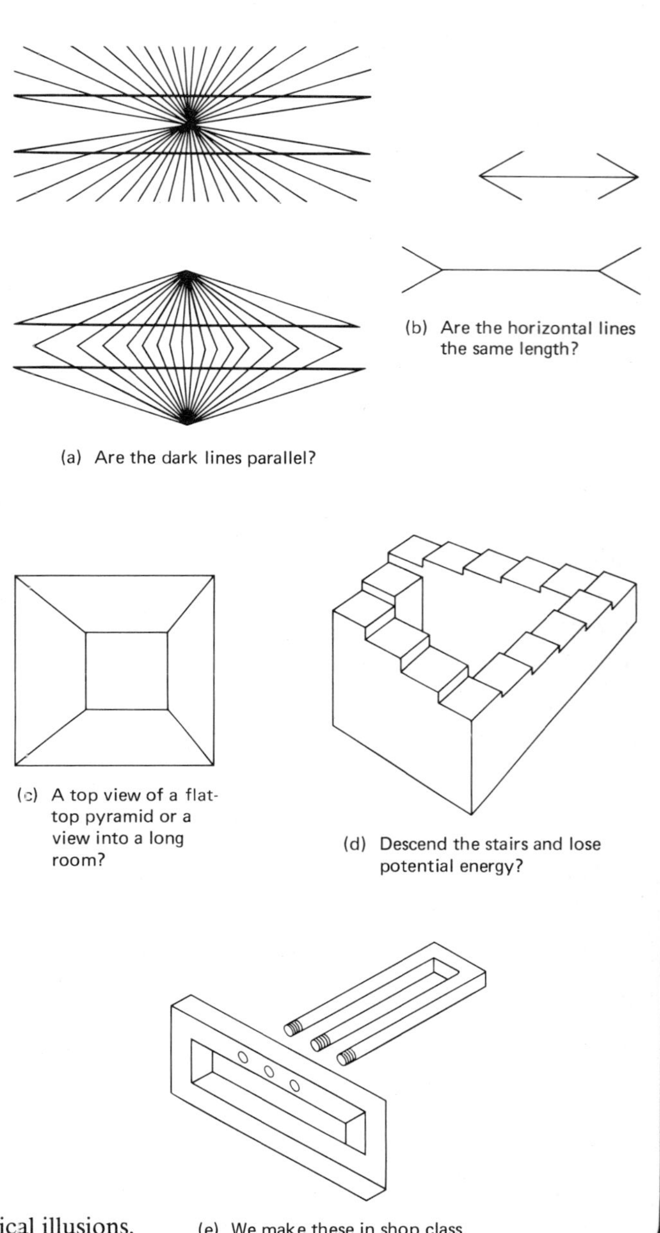

(a) Are the dark lines parallel?

(b) Are the horizontal lines the same length?

(c) A top view of a flat-top pyramid or a view into a long room?

(d) Descend the stairs and lose potential energy?

(e) We make these in shop class.

Figure 25.5 Some geometrical illusions.

Visual Defects

Optical illusions as discussed in Special Feature 25.1 aren't visual defects. Defects generally occur because the eye is abnormal for some reason. This can be due to a variety of reasons, including disease and injury. Let's take a look at several common vision defects that can be corrected with glasses.

The points between which the eye can see distinctly are called the far point and the near point. The far point

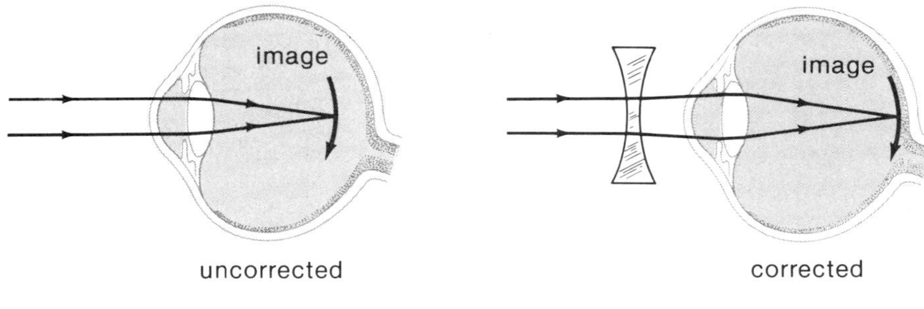

(a) nearsightedness

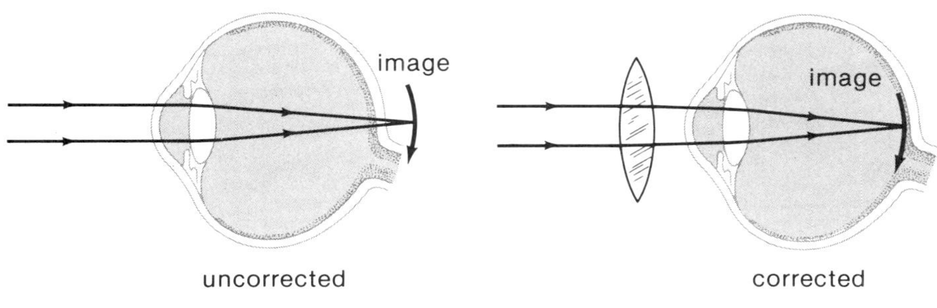

(b) farsightedness

Figure 25.6 Common visual defects. When the image is not formed on the retina, corrective lenses bring things back into focus for (a) nearsightedness and (b) farsightedness.

is normally without limit (infinity), and the near point depends on the accommodation of the crystalline lens. Bring your finger slowly toward your nose. At some point (your near point), you will see the finger blur or see two images. The normal eye produces sharp images on the retina for objects between the near and far points. However, a person may be nearsighted or farsighted and see blurred images (Fig. 25.6). This results from images not being focused on the retina because the size and shape of the eyeball and the crystalline lens are not properly matched.

Nearsightedness, or myopia, arises when the image is formed in front of the retina. A nearsighted person can see close or near objects clearly, but not distant objects. This defect is corrected by wearing glasses with a diverging lens for the nearsighted eye. The lens diverges the incoming rays so that the image formed by the crystalline lens is moved backward to the retina (Fig. 25.6a).

Farsighted persons can see far objects clearly, but near objects are blurred or out of focus. **Farsightedness,** or hyperopia, is due to the image being formed behind the retina. A converging lens will correct this by converging the incoming rays so that the image is moved forward to the retina (Fig. 25.6b).

Farsightedness (not being able to see close objects clearly) occurs naturally with age. You may have noticed an older person without glasses holding reading material away from him- or herself, even at arm's length. Children can see objects clearly as close as 10 cm from the eye (Table 25.1). However, as a person grows older the ciliary muscles weaken, and the crystalline lens loses its elasticity or hardens, limiting the eye's accommodation. As a result, the near point recedes with age. When a person's "arms get too short," he or

Table 25.1 Approximate Near Points of the Normal Eye

Age (yr)	*Near point (cm)*
10	10
20	12
30	15
40	25
50	40
60	100

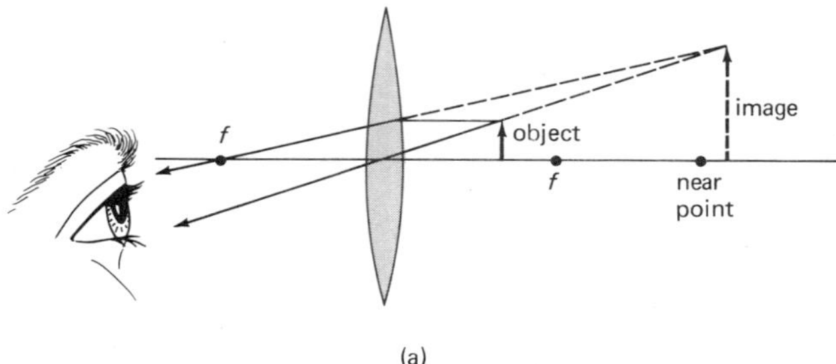

(a)

(b)

Figure 25.7 Near-point correction. The near point of the eye recedes with age, and closer objects appear blurred. (a) A converging lens "projects" the image of a close object beyond the near point, and it can be seen clearly. (b) People with this normal farsightedness commonly wear glasses for reading.

she has to get "reading glasses." The converging lens forms an image of a close object outside the near point, which the eye sees clearly (Fig. 25.7). Notice that the image is larger than the object. Reading glasses magnify things slightly.

This type of farsightedness may be called a *normal* defect, since it occurs naturally with age. So the next time you see someone holding something at arm's length to read, don't laugh. You'll be there someday.

Another common defect, **astigmatism,** occurs when the cornea and/or the crystalline lens of the eye are not perfectly spherical. As a result, the light rays have different focuses in different planes or directions. That is, a viewed object may be distinct in one direction and blurred in another. Astigmatism may be corrected with a lens having a greater curvature in the plane in which the cornea or crystalline lens has deficient curvature.

Bifocal glasses are sometimes used to correct a combination of defects. The bifocal lens was invented by Ben Franklin and consists of two lenses on the same piece of glass. (Ben glued two lenses together.) For example, a small lower lens can correct for farsightedness (for reading), and the upper lens can be used to correct for nearsightedness or astigmatism. In some cases, "trifocals" are used to correct three conditions.

Color Vision

Color is actually a physiological sensation of the brain in response to the light excitation of cone receptors in the retina. Many animals have no cone cells and are color-blind or live in a black-and-white world. For example, a color TV would be no better than a black-and-white set in a dog's world. In the human eye, the cones are sensitive to light with frequencies between about 7.5×10^{14} and 4.3×10^{14} Hz (4000 Å to 7000 Å wavelength). Different frequencies of light are perceived by the brain as having different colors. The association of colors with particular light frequencies is subjective. Monochromatic light has a particular physical frequency, but as pitch is to sound, color is to vision—it may vary from person to person.

The concept of color vision is not well understood. One of the most popular theories is that three different types of cones are contained in the retina, each of which responds to light from different parts of the visible spectrum (Fig. 25.8). The "blue" cones have maximum response for light with a wavelength around 4300 Å, the "green" cones for wavelengths around 5500 Å, and the "red" cones for wavelengths around 5800 Å.*

Combinations and different degrees of cone stimulation give rise to intermediate colors. For example, when red and green cones are equally stimulated by light of a particular frequency, the brain interprets this as yellow in color. But when the red cones are stimulated more strongly than the green cones, the brain "sees" orange.

Color-blindness results when one type of the primary cones is lacking. Occasionally this occurs because of failure to inherit the appropriate gene for the cone formation. The color genes are found in the female sex chromosome, so almost all color-blind people are male (about 4 percent of the male population). For example, if a person completely lacks red cones, he can see green

* The cone "color" indicates only a general response region of the spectrum. For example, light with a wavelength of 5800 Å is orange in color, but the major cone response in this end of the spectrum is from a "red" cone.

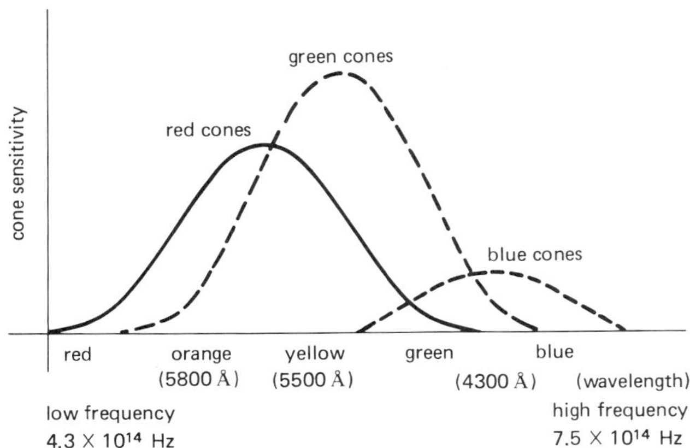

Figure 25.8 Cone sensitivity. Different cones of the eye are believed to respond to different frequencies of light to give three basic color responses.

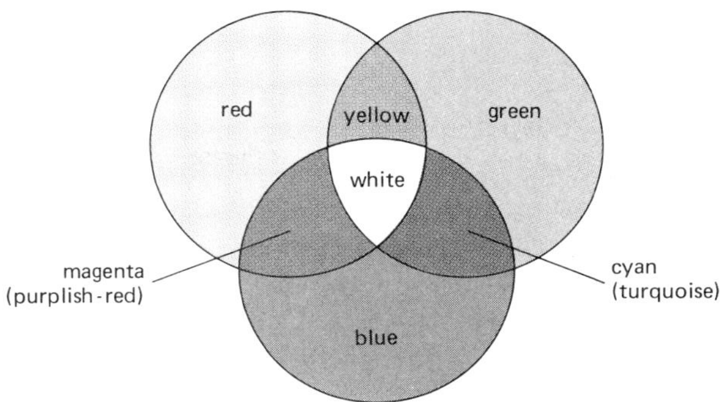

Figure 25.9 Additive method of color production. Light mixing of the primary colors (additive primaries) produces the colors as illustrated in the regions of overlapping light beams.

through orange-red colors by use of his green cones. However, he is not able to distinguish among these colors satisfactorily because he has no red cones to contrast with the green ones. A similar condition exists if the green cones are missing. In either case, it is difficult or impossible to distinguish colors of the larger wavelengths, and the condition is called red-green color blindness.

Color Mixing

White light is a mixture of all visible frequencies or colors, although perhaps not of equal intensities. For example, sunlight has a predominant yellow-green component. Even so, the composite colors of white light can be easily demonstrated by dispersion with a prism.

When colors of light are mixed together, it is found that one doesn't need all colors to make white light—just red, green, and blue. When light beams of these colors are projected on a screen so the beams overlap, additive mixtures of colors and white are produced (Fig. 25.9). Evidently the excitation of all types of cones in the eye causes the combination of signals to be interpreted as white and other combinations as different colors. This is referred to as the **additive method of color production.**

By adding varying amounts of red, green, and blue light, we find that any color of the visible spectrum can be generated. Recall that the triad dots on a color TV screen have red, green, and blue phosphors (Chapter 22). As a result, **red, green, and blue** are called the **additive primaries or primary colors.** Not only does a mixture of the additive primaries appear white to the eye, but many pairs of color combinations do also. The colors of such a pair are said to be complementary colors. For example, the complement of blue is yellow, of red is cyan (turquoise), and of green is magenta (purplish-red). This is not surprising, since from Figure 25.9 we can see that a combination of red and green is interpreted by the brain as yellow. Presumably, yellow light stimulates the red and green cones, which along with blue cone stimulation is a "white" combination.

Thus we see that an object has color because of the light coming from it. Other than light sources, objects have color when illuminated with white light because they reflect the wavelengths of the color they appear to be. The light coming from an object can be looked at in terms of selective reflection or selective absorption. When white light strikes a colored surface, certain light frequencies cause resonance electronic oscillations in the surface atoms, somewhat as sound causes a tuning fork to resonantly vibrate. The electrons of a particular atom have a narrow range of vibrational frequencies. The light components in this range are absorbed and re-emitted (reflected). The other light frequencies are absorbed by the material but go into heat (internal energy) rather than being re-emitted.

The paper of this page has atoms with a great range of resonance frequencies. When white light strikes the page, enough frequencies are reflected to your eye so that it appears white. The black print, on the other hand, has very few atoms with vibrations in the visible range of frequencies, so the light is almost totally absorbed, and the print appears black. (The light goes into the internal energy of the print ink. Recall how a thermometer with its bulb painted black registers a higher temperature than one with an unpainted bulb.) When white light strikes transparent red glass or a red rose, only red light is transmitted through the glass or reflected from the rose. All of the other colors are absorbed.

Selective absorption is important in the mixing of pigments for color production, such as in making paints and dyes for clothing. The additive method discussed previously for light does not apply here. You probably know or can guess, or any artist will tell you, that if you mix red, green, and blue pigments or paints you won't get white as in the case of light. You'd end up with paint with some sort of dark brown color.

But how about mixing blue and yellow paints? You might know that this produces green. The same effect is obtained with light by passing white light through blue and yellow filters (Fig. 25.10). This is because the blue pigment in the paint or filter absorbs the wavelengths or colors, except those in the blue region of the spectrum. Likewise, the yellow pigment in the paint or filter selectively absorbs the wavelengths of light, except those in the yellow region. The wavelengths in the intermediate green region are not strongly absorbed by either pigment, and hence, green light is transmitted through the filters or is reflected from the paint mixture.

This is an example of the **subtractive method of color production or mixing.** A mixture of absorbing pigments results in the subtraction of colors, and the eye sees the color that is not subtracted or absorbed. Three particu-

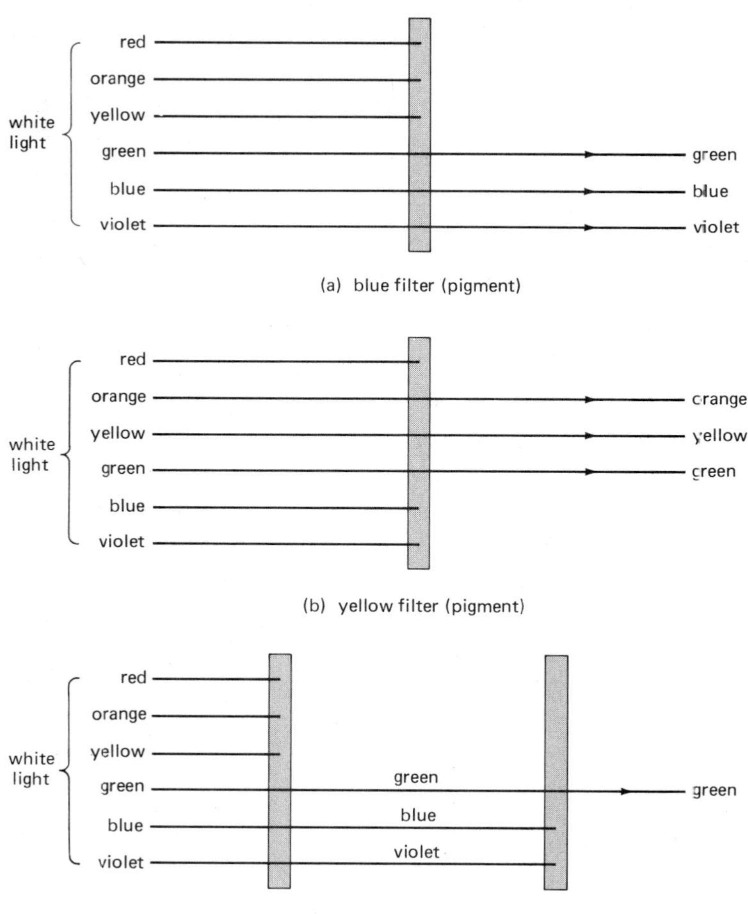

Figure 25.10 Selective absorption. A particular filter (or paint pigment) absorbs wavelengths or colors in a particular region. By subtracting (or absorbing) component colors of white light, a particular color is produced.

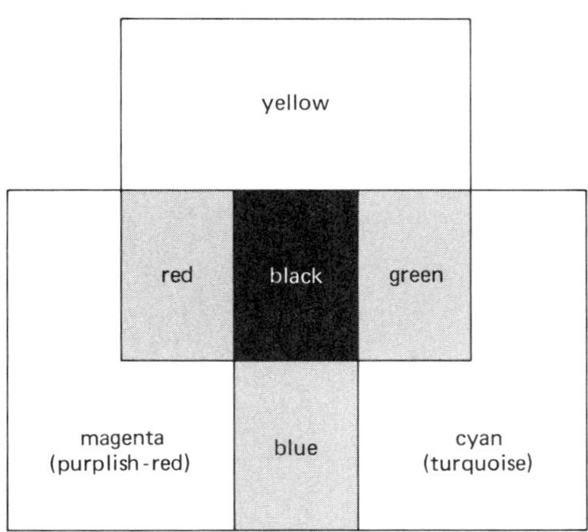

Figure 25.11 Subtractive primaries. When the subtractive primaries or pigments—cyan, magenta, and yellow—are mixed, the colors (or lack of color) are obtained, as illustrated in the overlapping regions.

lar pigments—**cyan, magenta, and yellow**—are called the **subtractive primaries or primary pigments** (colors). Various combinations of two of the subtractive primaries produce the three additive primary colors (red, blue, and green), as illustrated in Figure 25.11. Notice that the magenta pigment "subtracts" the green color from the cyan-yellow overlap. As a result, magenta is sometimes referred to as "minus green." (Think about adding a magenta filter to Figure 25.10.) Similarly, cyan is "minus red" and yellow is "minus blue."

When all of the primary pigments are mixed in the proper proportions, almost all of the wavelengths of the visible spectrum are absorbed or subtracted, and the mixture appears black. People in the paint and dye businesses commonly refer to the subtractive primaries as being red, yellow, and blue, because magenta is a shade of red and cyan is a shade of blue. In paint stores, paint of any color can be produced by mixing proper combinations of subtractive primaries.

QUESTION: Suppose a yellow banana and a red apple are illuminated with green light. How would they look?

ANSWER: The banana would appear green. Since green is close to yellow, some green light would be reflected by the "yellow" oscillators in the banana peel. The apple would appear dark because most of the green light would be absorbed. If there were some leaves on the apple stem, you'd see green leaves and a dark apple. See Special Feature 25.2 for another color phenomenon.

SPECIAL FEATURE 25.2

Why the Sky Is Blue (and Sunsets are Red)

If we lived on the moon, which has no atmosphere, the sky would appear black, except in the vicinity of the Sun. This is what astronauts on the moon see. On Earth, however, our sky is blue as a result of the scattering of sunlight in the atmosphere. As sunlight passes through the atmosphere, the nitrogen and oxygen molecules of the air absorb some of the light and re-emit it. The light is scattered in all directions from the free gas molecules.

This scattering is selective, with the resonant frequencies of the small molecular oscillators in the ultraviolet region. The frequencies of light in the visible region are below the resonant frequencies but are close enough to be absorbed and scattered somewhat, particularly in the blue end of the spectrum. Thus, the light in the visible spectrum is preferentially scattered, with the light at the blue end of the spectrum being scattered about ten times more than light at the red end. This is called Rayleigh scattering after Lord Rayleigh, the British scientist who explained it. Some of the scattered light reaches the Earth, which we see as blue sky light (Fig. 25.12).

It should be kept in mind that *all* colors are present in sky light, but the dominant wavelength or color lies in the blue. You may have noticed that the sky light is more blue directly overhead or high in the sky and less blue toward the horizon, becoming white just above the horizon. This is because there are fewer scatters along a path through the atmosphere directly overhead (your zenith position) than toward the horizon, and multiple scattering along the horizon path gives rise to the white appearance. By analogy, if you add a drop of milk to a glass of water and illuminate the suspension with intense white light, the scattered light has a bluish hue. And yet, a glass of milk is white (due to multiple scatterings). Atmospheric pollution may enhance the milky white appearance of the sky.

The scattering of sunlight by the atmospheric gases

(Continued)

Figure 25.12 Atmospheric scattering accounts for (a) the blueness of the sky and (b) red sunsets. See Special Feature 25.2 for details.

and small particles gives rise to red sunsets. Generally it might be thought that since the distance sunlight travels through the atmosphere is greater to an observer at sunset, then most of the higher-frequency colors of the visible spectrum are scattered from the sunlight and only light in the red end of the spectrum reaches the observer. However, it has been shown that the dominant color of this light, due solely to molecular scattering, is orange.* Hence, there must be scattering by small particles in the atmosphere that shifts the light from the setting (or rising) sun toward the red. Foreign particles in the atmosphere are not necessary to give a blue sky; they even detract from it. Yet, such particles are necessary for deep red sunsets and sunrises, which occur most often when there is a high-pressure air mass to the west (for sunsets) and to the east (for sunrises), since the particle concentration is generally higher in a high-pressure air mass than in a low-pressure air mass.

The beauty of red sunrises and sunsets is often made more spectacular by layers of pink-colored clouds. The cloud color is due to the reflection of red light. Since the water droplets of clouds scatter visible light of all wavelengths about equally, the clouds do not affect the color of the light, but merely diffusely reflect the incident red light.

* Bohren, C. F. and Fraser, A. B., Colors of the Sky, *The Physics Teacher*, Vol. 23, No. 5, p. 267.

QUESTION: If the scattering by atmospheric gases is greater for visible light of greater frequency, why isn't the sky violet? It should be scattered more than blue light.

ANSWER: Violet light is scattered more than blue light, but the sky is blue for a couple of reasons. First, and more important, the eye is more sensitive to blue light than to violet light. See Figure 25.8. Second, sunlight contains more blue light than violet light. The greatest color component is yellow-green, and the distribution generally decreases toward the ends of the spectrum.

Optical Instruments

There is a wide variety of optical instruments that use mirrors, lenses, prisms, diffraction gratings, fiber optics—any optical application to achieve the desired purpose of the instrument. Before leaving the subject of "optics," let's consider some of the optical instruments that you use or are probably familiar with.

PROJECTORS

The projection of the images of slides and films on a screen is commonplace. We all like to show slides of our vacation or of the party last week or to show a home movie. This is done by means of a projector. The basic components of a slide projector are shown in Figure 25.13, along with some common types of projectors. A pair of "condensing" lenses concentrates light from a source on a slide. The slide is placed just beyond the focal point of the converging lens, and a magnified image of the slide is formed on the screen. Because of the refractive properties of the lens, which are indicated by the light rays in the diagram, the image of an upright object is inverted. Therefore, slides are placed in the projector upside down so the image on the screen is seen right side up.

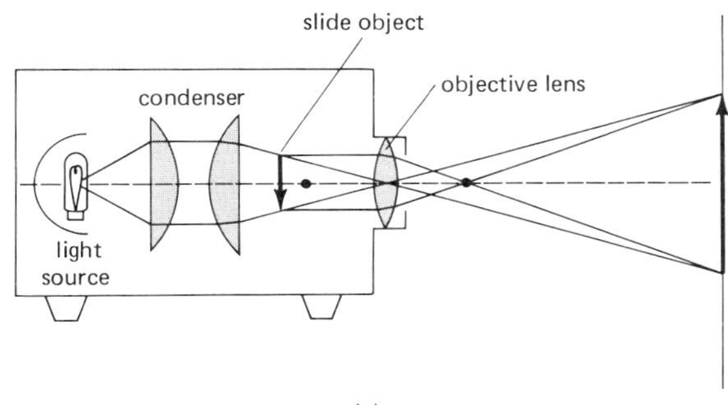

(a)

(b)

Figure 25.13 Optical instruments. (a) Diagram of a simple projector and (b) various kinds of projectors.

Motion picture projectors have basically the same components, along with the machinery to advance the film at a given speed. By quick advancement of one slightly different picture frame after another, the projected image is given the illusion of motion.

A sound motion picture also involves optical methods. When a motion picture is filmed, a recording amplifier drives a special lamp that exposes a narrow "sound track" along the side of the film. This track is a variation in optical density (light and dark regions) that follows the variations in the lamp voltage induced by sounds. To reproduce the sound when the film runs through the projector, an optical system sends a beam of light through the sound track to a photocell or phototube. The variations in the light and dark portions of the track cause fluctuations in the light beam, which are monitored by the phototube, and the electrical impulses are fed into an amplifier-speaker system that reproduces the original sounds.

MICROSCOPES

When we want to see small objects better, we use a magnifying glass or a microscope. The single, converging-lens magnifying glass is called a simple microscope. The common microscope is a compound microscope because it uses more than one lens. To see how a lens magnifies, think of how we judge the size of an object. A car in the distance looks quite small, as compared with when it is close by (Fig. 25.14). The difference is the angle of view. When the car is close by, it looks much larger because it is viewed through a much wider angle.

A lens can make an object appear larger than it is because the refraction or bending of light widens the angle of view (Fig. 25.15). But the eye sees the light as if it had been traveling in a straight line, and the object appears to be much larger than it really is, or magnified.

To obtain greater magnification than that given by a single lens or simple microscope, a compound microscope may be used. Here, more than one lens is used, as

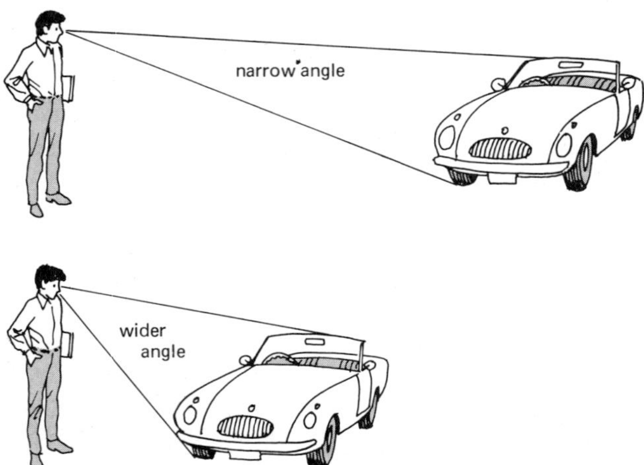

Figure 25.14 The angle of view determines how large an object appears.

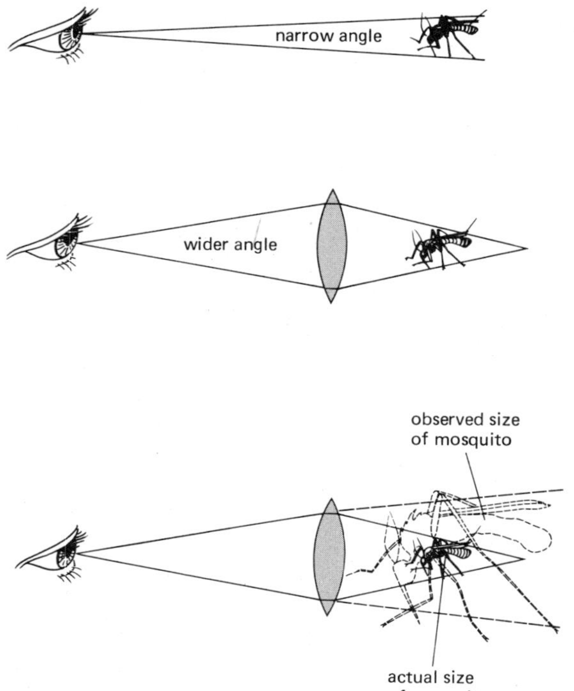

Figure 25.15 Magnification. A single lens or simple microscope expands the angle of view so that an object appears larger.

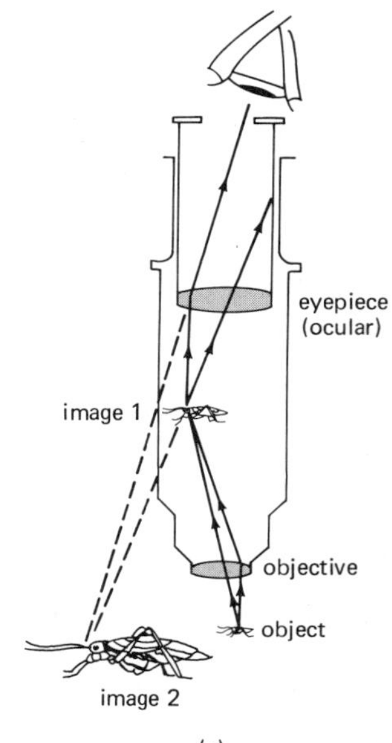

(a)

(b)

Figure 25.16 A basic compound microscope uses two lenses, an objective and an eyepiece, for greater magnification.

shown for a basic compound microscope in Figure 25.16. The objective (lens) forms a magnified image of the object, in much the same way as the lens of a projector does. The eyepiece or ocular (lens) acts as a magnifying glass and further magnifies the image as seen by an observer looking into the microscope. Actual microscopes are much more involved; only the basic principles are presented here.

TELESCOPES

Telescopes are used to see distant objects or to make them appear closer. Telescopes collect and concentrate light energy to form images. This is particularly the case for astronomical telescopes used to view distant stars and galaxies. Smaller telescopes are used to view things on earth, for example, a transit telescope used by a surveyor. There are two general types of telescopes—refracting telescopes, based on lens refraction, and reflecting telescopes, based on mirror reflection.

The Refracting Telescope

The optics of an astronomical refracting telescope are similar to those of a microscope, but the objective lens is made very large so as to collect a large quantity of light from a distant object so an image may be seen (Fig. 25.17). The final image of an astronomical telescope is inverted or upside down, but this poses no problem in astronomical work.

However, it is not convenient to have an inverted image when viewing objects on earth with a telescope. In a *terrestrial* telescope, the image of an object is viewed right side up which is important, for example, in a surveyor's telescope (Fig. 25.18). This is accomplished by using a diverging lens as an eyepiece, which is called a Galilean telescope, after Galileo, who built one in 1609. A third, "erecting" converging lens may be placed between the objective and the eyepiece to invert the image so that the final image is right side up. This is the principle of the "spyglass" telescope, which must be extended to a great length (like the type used in pirate movies). The inconvenient length increase can be avoided by using internally reflecting prisms. This is the principle of the common prism binoculars (Fig. 25.18c).

The Reflecting Telescope

To form images of distant stars and galaxies, enough light energy must be collected or received by the telescope. The amount of light entering a refracting telescope can be increased by increasing the size of the objective lens. However, there are physical limitations to this approach in grinding the lens, as well as material defects and aberration effects. The largest refracting telescope is the 40-inch (102-cm; objective diameter) telescope at the Yerkes Observatory in Williams Bay, Wisconsin, shown in Figure 25.17.

Another approach is to use a reflecting telescope.

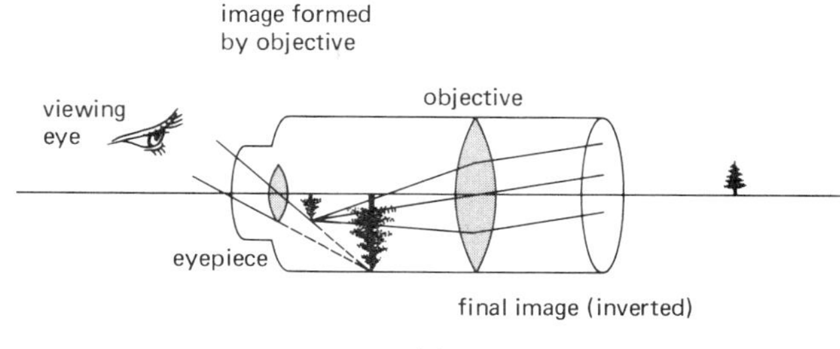

(a)

(b)

Figure 25.17 Refractive astronomical telescope. (a) The image is inverted, but this poses no problem in astronomical work. (b) The 40-inch (102-cm; objective diameter) refracting telescope at Yerkes Observatory.

The basic components of this type of telescope are shown in Figure 25.19. In this case, the objective is a concave mirror, which collects and focuses the light. To make the image conveniently accessible to the eye or a camera, the rays may be deflected by a plane mirror to the side of the tube and observed, with the image magnified by the eyepiece. However, if the telescope is large enough, the observations may be made inside the telescope tube. This is the case for the 200-inch (5.1-m; mirror diameter) Hale Observatories telescope on Palomar Mountain in California, which is currently the world's largest reflecting telescope (Fig. 25.19). There are plans calling for even larger reflecting telescopes.

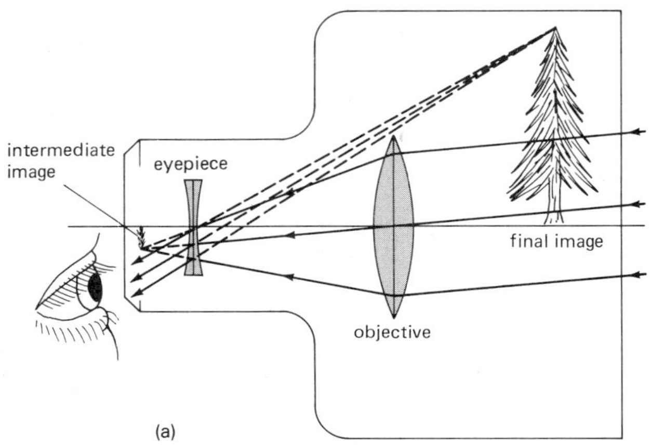

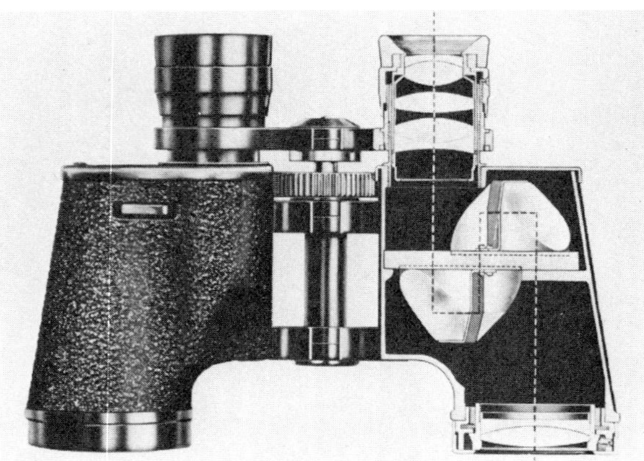

Figure 25.18 Terrestrial telescopes. (a) Using a diverging lens for an eyepiece, the final image is upright. (b) A surveyor's transit is a type of terrestrial telescope. (c) Binoculars shorten the length of a telescope by using prisms.

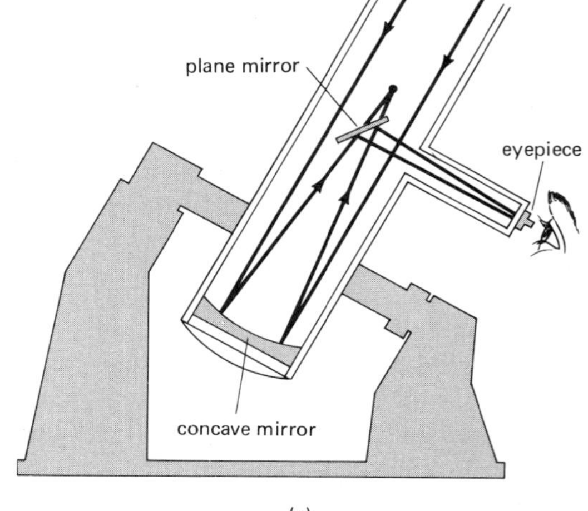

Figure 25.19 Reflecting telescope. (a) A concave mirror collects and focuses the light. A mirror may be used to bring the rays to a side eyepiece. (b) The Hale 200-inch (5.1-m; mirror diameter) telescope. The glass mirror weighs 15 tons and is 24 inches thick at the edge and 20 inches thick at the center. A thin coating of aluminum provides the reflecting surface. (c) Observations are made at the prime focus position at the center of the telescope. See also Figure 30.1.

Large mirrors can be constructed more easily than large lenses, since only one surface need be ground and silvered. Reflecting telescopes are also free of chromatic aberration. Parabolic mirrors are usually used, which makes them free of spherical aberrations as well.

Plans are also being made for orbiting space telescopes that will be free of atmospheric effects and background light from cities that hamper Earth-bound observation. An infrared (IR) telescope has been put into orbit about the Earth as part of the Space Shuttle program. It surveyed the "infrared sky" and sent back a great deal of data before it stopped transmitting. Among the initial findings are what is believed to be very small solid particles orbiting the star Vega, the first direct evidence of solid material around a star other than the Sun, and the discovery of five new comets in the solar system. Another type of telescope, the radio telescope, which gives another "view" of the sky, is considered in Special Feature 25.3.

SPECIAL FEATURE 25.3

Radio Telescopes

Although not an optical instrument, the radio telescope is an important astronomical tool. Stars and galaxies emit radio waves as well as light waves. This fact was discovered accidentally by an electrical engineer named Carl Jansky in 1931 while working on a static problem in intercontinental radio communications. He found that an annoying static hiss was coming from a fixed direction in space, which later was found out to be the center of our galaxy (the Milky Way).

Thus, Jansky discovered radio waves coming from space. This was quickly recognized as another source of astronomical information, and radio telescopes were built (Fig. 25.20). A radio telescope operates similarly to reflecting light telescopes, inasmuch as radiation is collected by a large-area reflector and focused to form an "image." However, the parabolic collector of a radio telescope does not look like a mirror surface, since it is covered with wire mesh. To avoid confusion, the collector of a radio telescope is referred to as a "dish" instead of a mirror. You now commonly see satellite TV antenna dishes.

The radio telescope dish is not a mirror for light waves, but it is for radio waves. This is because electromagnetic waves cannot detect any hole or surface irregularity that is smaller than the wavelength of the radiation. Since radio waves range from about 1 centimeter to several meters in wavelength, the wire-mesh surface on the metal dish acts as a good reflecting surface for such waves.

Another noticeable difference between optical and radio telescopes is that there is no film or eyepiece for a radio telescope. Instead, the radio waves are detected by an antenna positioned at the focal point of the dish. The signals are amplified, and the information received by the telescope is displayed on a recorder so it can be "seen."

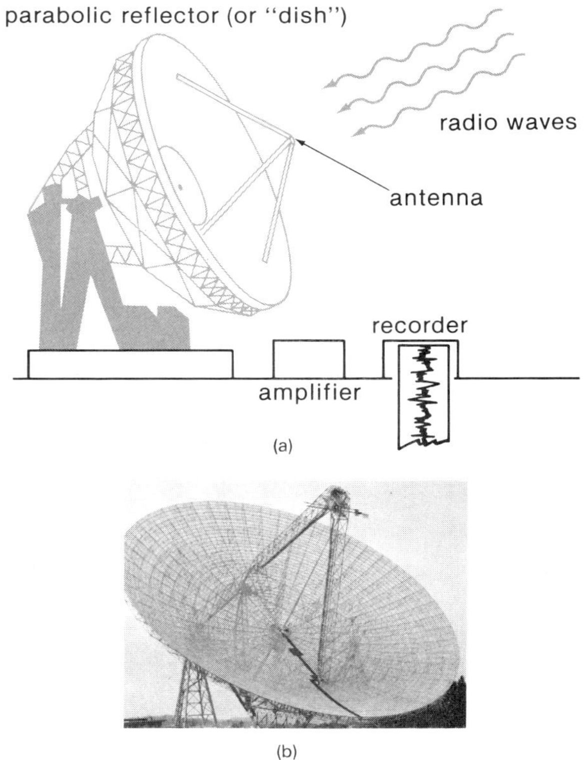

Figure 25.20 Radio telescope. (a) The telescope "dish" reflects radio waves to a focus at the antenna or detector, which feeds the signal into an amplifier and recorder. (b) The radio telescope at Green Banks, West Virginia.

(Continued)

Radio telescopes have added a new dimension to astronomy. They supplement optical telescopes and offer some definite advantages. Radio waves pass freely through the huge clouds of dust that exist in our galaxy and hide a large part of it from optical view. Radio waves are easily able to penetrate the Earth's atmosphere, whereas a large part of the incoming light is reflected and scattered.

Radio astronomy has extended the dimensions of the known universe almost twofold. With radio telescopes astronomers can detect galaxies that are two times farther away than those detected by optical telescopes. Although the emission of radio waves may be less intense, the penetration of the radio waves and construction of large dishes permit their detection (Fig. 25.21). Visual light from distant galaxies is blanketed by background illumination of the night sky. Radio telescopes may be operated around the clock, and observations can be made day or night.

Figure 25.21 The 305-meter (1000-ft) diameter radio telescope at Arecibo, Puerto Rico. The dish was made by placing wire mesh over a natural bowl in the mountains. The antenna is mounted on a trolley suspended by cables 500 feet above the valley floor. Although the dish is fixed, signals are received from many directions as a result of the Earth's rotation and revolution.

SUMMARY OF KEY TERMS

Crystalline lens the converging lens of the eye that focuses incoming light.

Retina the "film" on the back surface of the eyeball on which images are formed and transmitted to the brain.

Accommodation the changing of the curvature of the crystalline lens so that the images of objects at various distances are formed on the retina.

Rods photosensitive cells of the retina that are responsible for twilight (black-and-white) vision.

Cones photosensitive cells of the retina that are responsible for color vision.

Nearsightedness a vision defect in which a person can see near objects clearly, but no distant objects.

Farsightedness a vision defect in which a person can see far objects clearly, but not near or close objects.

Astigmatism a vision defect caused by the cornea and/or the crystalline lens not being spherical, so the image on the retina is out of focus in particular planes or directions.

Additive primaries or primary colors the colors red, green, and blue. Light of these colors can be mixed or "added" to give any color of the visible spectrum or white.

Complementary colors pairs of colored light combinations that appear white to the eye, for example, blue and yellow.

Subtractive primaries or primary pigments cyan, magenta, and yellow pigments, which can be mixed to give paints or dyes that appear as any color or black through selective absorption of light.

Rayleigh scattering the preferential scattering by molecules of the air of high-frequency visible light that gives rise to the blueness of the sky.

Refracting telescope a telescope that uses lenses to view distant objects.

Reflecting telescope a telescope that uses a converging mirror to view distant objects.

EXERCISES

1. Many people wear "contacts." On what part of the eye are contact lenses worn?
2. How are the operations of a camera shutter and of the eyelid generally different? Can you think of an instance when a camera shutter is operated like an eyelid?
3. If upside-down images are formed on the retina, how do we know that the world isn't really upside down when we see it?
4. Does the eye "camera" have black-and-white or color film? Is there an analogy with the type of film used in a camera?
5. What is meant by "twilight vision"?
6. Moonlight is reflected sunlight. Why don't we see color when viewing objects in moonlight?
7. During eye examinations, the ophthalmologist usually dilates your eyes. This is done by adding fluid to your eyes that causes the pupils to widen and remain open so the inside of the eyeball can be examined. The fluid also temporarily paralyzes the ciliary muscles. How does this affect the person whose eyes are dilated?
8. Answer the questions in Figure 25.22.
9. A person wears bifocals to correct for nearsightedness and astigmatism. How would the bifocal lens be ground for these vision defects?
10. Why are "reading glasses" that are used to correct normal farsightedness also "magnifying glasses"? (See Fig. 25.7.)
11. In very rare instances of color-blindness, a person lacks blue cones (called "blue weakness"). What colors would this person have difficulty in distinguishing?
12. Is white a color? Is black? Explain.
13. Suppose that the beams of red and blue spotlights overlapped each other on a wall. (a) What color would the wall appear to be if it were white? (b) How about a blue wall? (c) A green wall?
14. What do light filters filter out?
15. In a department-store window display, white light passing through a red filter falls on cyan- and yellow-colored objects. What colors are seen by someone looking at the display through the store window?
16. How do different-colored spotlights influence the appearance of a performer's clothes, for example, a red spotlight on a performer in a dark blue suit? What color of spotlight would you use if you wanted to make the performer really stand out?
17. What color would a green chalkboard appear to be when viewed through yellow sunglasses? How about when viewed through rose-colored (red) glasses? If the chalk-

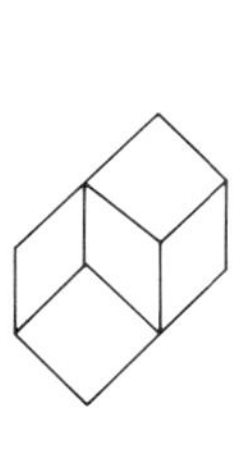

(a) Which cube does not have all of its sides?

(b) Are the diagonal lines parallel?

(c) Are all the men the same size?

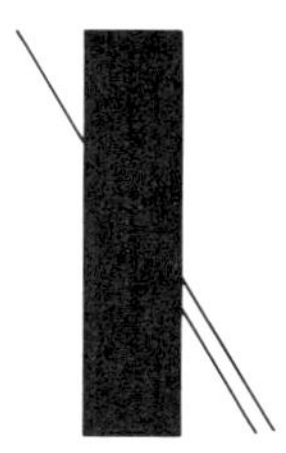

(d) Is the lower line on the right an extension of the line on the left?

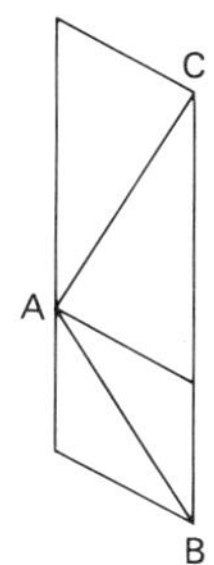

(e) Is line AB equal in length to line AC?

Figure 25.22 See Exercise 8.

board had writing on it in white chalk, what color would the writing appear to be in these cases?

18. How would the American flag appear when illuminated with (a) red light, (b) blue light, and (c) green light?
19. Could you ignite a piece of paper with a magnifying glass and sunlight more quickly if the paper were white or black? Explain.
20. Why is the inside of a camera black?
21. Why do fluorescent lamps tend to enhance blue clothing more than incandescent lamps?
22. Two complementary colors produce white, but if a colored light from a spotlight falls on an object with its complementary color, it appears black. Why?
23. Why does a white piece of paper appear white, red, blue, or the color of whatever type of light illuminates it?
24. By which process of color production could red and cyan (purplish-red) colors be combined to get white?
25. Can white be obtained by the subtractive method of color production? Explain.
26. Is color TV based on the additive or subtractive method of color production? Explain.
27. A painter mixes paints with the primary pigments of (a) yellow and blue, and (b) blue and red. What are the colors of the resulting paint mixtures?
28. Clouds generally appear white, but nimbus clouds that threaten or produce rain appear dark. Why is this?
29. Stars are in the sky during the daytime as well as at night. Why can't we see them? Do astronauts on the moon see stars during the day?
30. Why does the sun generally appear yellowish-orange at sunset? Why are there not red sunsets every evening?
31. There is an old weather saying, "Red sky at night, sailors' delight. Red sky at morning, sailors take warning."‡ Explain the possible validity of this prediction, considering that fair weather is generally associated with high-pressure air masses and that the conterminous (being adjacent or in contact) United States lies in the Westerlies wind zone (air mass movement generally from west to east).
32. Could slides be put in a projector right side up and have a right-side-up image on a screen if other lenses were added to the system? Explain.
33. How are animated cartoons made?
34. Why are upside-down images no problem in astronomical telescopes?
35. A detective uses a small astronomical telescope to view a crime committed in a distant apartment building. How would he describe what he saw? If you were a defense lawyer at the accused person's trial, how would you try to discredit the detective's testimony? (Would it have helped if the detective had stood on his head when observing the crime?)
36. What determines how far telescopes can be used to see objects? Do reflecting telescopes or refracting telescopes offer any greater advantage in this respect?
37. Why do astronomers want to put telescopes in high-altitude orbits above the Earth? Would being closer to stars and galaxies make a difference?
38. The Palomar Mountain Hale Observatories reflecting telescope's mirror is 200 inches in diameter. What is the diameter of the mirror in meters? How does the circular area of the mirror surface compare to the floor area of your room? (The area of a circle is $\pi d^2/4$, where d is the diameter.)
39. Spherical and parabolic mirrors can be used for solar heating, for example, to cook foods and even to light cigarettes (see also Fig. 7.14). Discuss the advantages and disadvantages of this source of heating.

‡ A similar quote is made in the Bible. Check out Matthew 16:1–4.

PART SEVEN
Modern Physics

The term *modern physics* generally implies the developments in physics since around 1900. But there's more to it than that. Modern physics also implies a revolution in scientific thought. There were many developments in "classical" physics during the 1700's and 1800's. The principles of mechanics, gravitation, heat and thermodynamics, electricity and magnetism, optics—all the things we have studied—were reasonably well understood. Scientists felt secure in their descriptions of nature. The theories perhaps needed a little "polishing" to make them just right.

Then, in the late 1800's, things began to happen. The prevailing classical theories failed to explain various experimental results—sometimes rather badly. There were problems with explaining electromagnetic wave propagation, line spectra, and photoelectricity (photocells). In the 1900's new areas of physics opened up, for example, nuclear physics. Here, too, there were some problems with classical theories. By the scientific method, something had to change—theories needed to be modified or scrapped.

It would have been foolish to throw out the classical theories that explained so many phenomena, particularly when there were no replacements. What evolved was the modification of classical theories and principles with new ideas *and,* in some instances, completely new theories to explain the modern physics observations. The remaining chapters of this text are about the "new" or modern physics. We'll kick it off with a "classic" example of how new thoughts and ideas had to be added to physics in the case of relativity.

26

Relativity

The Background

The background of the theory of relativity is full of discovery and debate. For most people, the term *relativity* or the theory of relativity immediately brings to mind Albert Einstein (and vice versa). Indeed, Einstein did formulate a theory of relativity, which consists of two main parts: the special theory (circa 1905) and the general theory (circa 1915). But before looking at this theory, let's set the stage for its development.

The theory of relativity involves light, in particular, the speed of light. From common experience we see that light appears to travel instantaneously from one place to another. Measuring its speed is no easy task. Galileo tried to do this by measuring the time it took light signals to travel several kilometers. Galileo's signals came from a lantern. When a covering bucket was removed, a light beam was sent to an assistant, who removed a bucket from another lantern and sent a signal back to Galileo. As you can imagine, the experimental results were not very good (in fact, they were useless), since the assistant's and Galileo's reaction times were involved. (For typical people, this reaction time would allow light to go all the way around the Earth.) The speed of light was subsequently measured using astronomical methods.

The first successful terrestrial method of measuring the speed of light was carried out in 1849 by the French physicist H. L. Fizeau. The idea of Fizeau's cogwheel method is illustrated in Figure 26.1. If the cogwheel is not moving, then the light from a candle will go through the opening between cogs 1 and 2, travel to the mirror, be reflected back, and pass again through the opening between the cogs to an observer behind the candle. However, if the wheel is rotated, the beam is "chopped" up. When the wheel is rotating slowly, a segment of the light beam — for example, that chopped off between cogs 1 and 2 — will be reflected back and

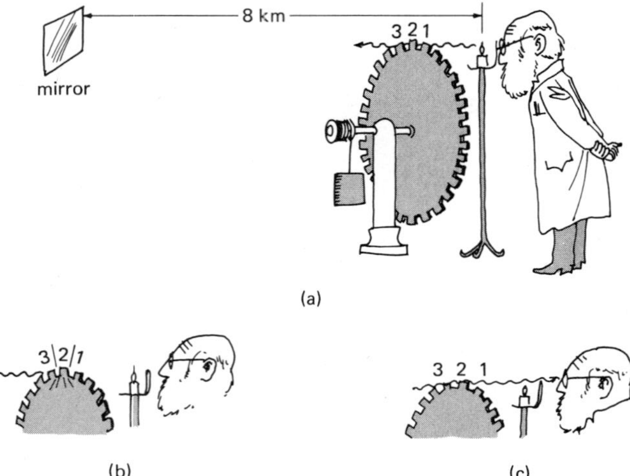

Figure 26.1 Fizeau's cogwheel method for measuring the speed of light. See text for description.

will arrive while cog 2 is in front of the observer. But at a faster rotational speed, the cog will be out of the way and the light-beam segment will pass between cogs 2 and 3 and will be seen by the observer. The time it takes for the next cog space to rotate in front of the observer can be calculated from the rotational speed, and if we know the distance traveled by the light, the speed of light can also be calculated.

Fizeau's results were not extremely accurate, but they were pretty good considering the equipment. His experimental result for the speed of light was about 5 percent off the present-day accepted value:

(speed of light)* $c = 3 \times 10^8$ m/s ($\simeq$ 186,000 mi/s)

At this rate, it takes about 8 minutes for light to reach us from the Sun. The closest star to our solar system, Alpha Centauri, is about 4.3 light years away. A light year is the *distance* light can travel in a year (about 10 trillion kilometers or 5.8 trillion miles). Hence, the light we see from Alpha Centauri is over four years "old," since it takes 4.3 years for the light to reach us. How about the light from a star 1000 light years away?

The Ether

With the wave nature of light being demonstrated through interference experiments (Chapter 23) and the speed of light being reasonably well known by astronomical methods prior to the 1800's, scientists turned their attention to the consideration of the medium that carried or propagated light waves. From general experience, it was thought that a medium was necessary for wave propagation. For example, sound waves propagate in air (and other media). It could be shown that sound waves could not travel through a vacuum—some material substance was needed for their propagation. Similarly, you couldn't have water waves without water. By such reasoning, it was also believed that light waves had to have a medium of propagation.

How then did light propagate through the void of space, for example, from the Sun to Earth? Not being able to see how light could travel through nothing, scientists created a hypothetical medium which was called lumeniferous **ether,** or just plain ether. This substance was presumed to occupy interstellar space and to be present in all materials through which light travels. The idea of the existence of the ether seemed so logical that it quickly gained widespread acceptance.

It is easy to *postulate* the existence of something, but by the scientific method, a theory must be substantiated by experiment. So scientists set out to detect the ether experimentally. This set the stage for Einstein's theory of relativity.

The Michelson-Morley Experiment

Since the ether permeated all space, it was reasoned that it was the one thing that remained fixed in the universe. This belief led to an idea about how to detect the ether and prove its existence. The Earth's average speed through the stationary ether in its revolution about the Sun is about 30 km/s. Moving through the ether would cause an ether "wind" to blow over the Earth (Fig. 26.2). This is similar to the "wind" you experience when traveling in a car or on a motorcycle through still air.

From experience with real winds and sound waves, it was known that the measured speed of sound in air depends on the wind speed and direction. For example,

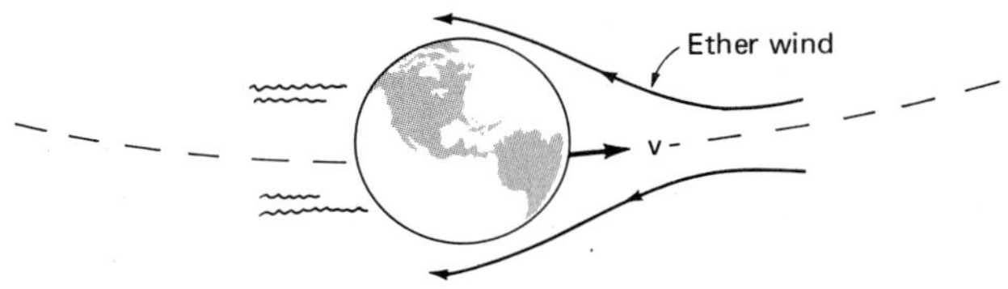

Figure 26.2 The orbital motion of the Earth through the ether should give rise to an "ether wind."

* This is the speed of light in vacuum (or approximately in air). Recall from Chapter 24 that the speed of light in a transparent material is less than c.

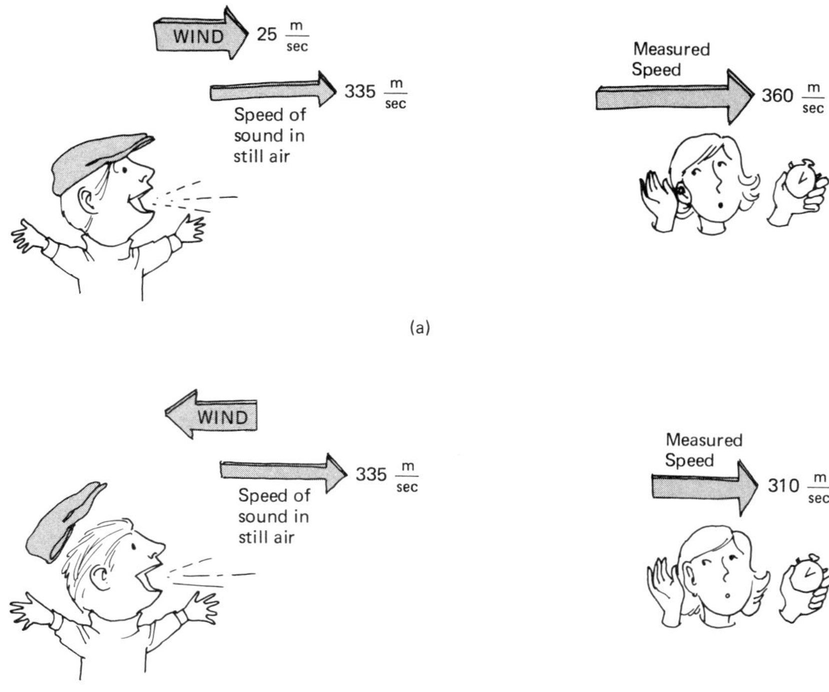

Figure 26.3 Speed measurements with velocity vector addition. (a) The speed of sound in still air is enhanced if the air is moving (wind), so an observer measures the sum of the speeds. (b) If the wind is blowing in a direction opposite that of the sound propagation, the speed is measured to be less.

suppose the speed of sound in still air were 335 m/s. If there were a 25-m/s wind blowing in the direction in which the sound wave was propagating, then an observer would measure the sound speed to be 335 + 25 = 360 m/s (Fig. 26.3). Similarly, if the wind were blowing in the opposite direction, the observer would measure the sound speed to be 335 − 25 = 310 m/s. Thus, the motion of the air, or wind, changes the observed speed of sound. It was reasoned that the ether wind should do the same with the speed of light and thereby would prove that the ether existed.

This was the purpose of the famous Michelson and Morley experiment, which was performed in 1887 by two American scientists, A. A. Michelson and E. W. Morley.† An atmospheric wind analogy of the experiment is shown in Figure 26.4. Two equally powerful planes fly the same distance in a race. The plane flying perpendicular to the wind must steer slightly into the wind on both legs of the race, which slows the plane somewhat. The other plane, flying directly into the wind on the first leg of the race, is slowed up much more, but the wind adds to the speed on the return leg. Even so, it can be shown by analysis or by an actual race that this plane always loses the race or takes a longer time to cover the distance.

If the atmospheric wind were replaced by the ether wind and the air planes by light beams, the light beams would then arrive back at the finish line at different times, as the airplanes do. Michelson used an ingenious method to develop an instrument to measure the expected time difference (Fig. 26.5). The method was based on the wave interference of light. The light beams interfering at a detector (finish line) produce a pattern of alternate bright and dark fringes. This pattern is noted for one position of the apparatus, which is called an interferometer, and then the apparatus is turned through an angle, say 90°. If this change in direction changes the travel times of the light beams, as it should in an ether wind, then there would be a shift in the interference pattern.

The interferometer was sensitive enough to detect a time difference resulting from adding or subtracting the Earth's orbital speed of 30 km/s to or from the 300,000 km/s speed of light, but nothing was observed! Measurements were made at different times (day and night, different seasons) and at different locations (America and Europe), but always with a null result. There was no fringe shift!

† Because of his work in the development of optical instruments and light investigations, Albert A. Michelson was the first American to receive the Nobel Prize in physics (1907). Edward W. Morley, Michelson's associate, was a chemist by training.

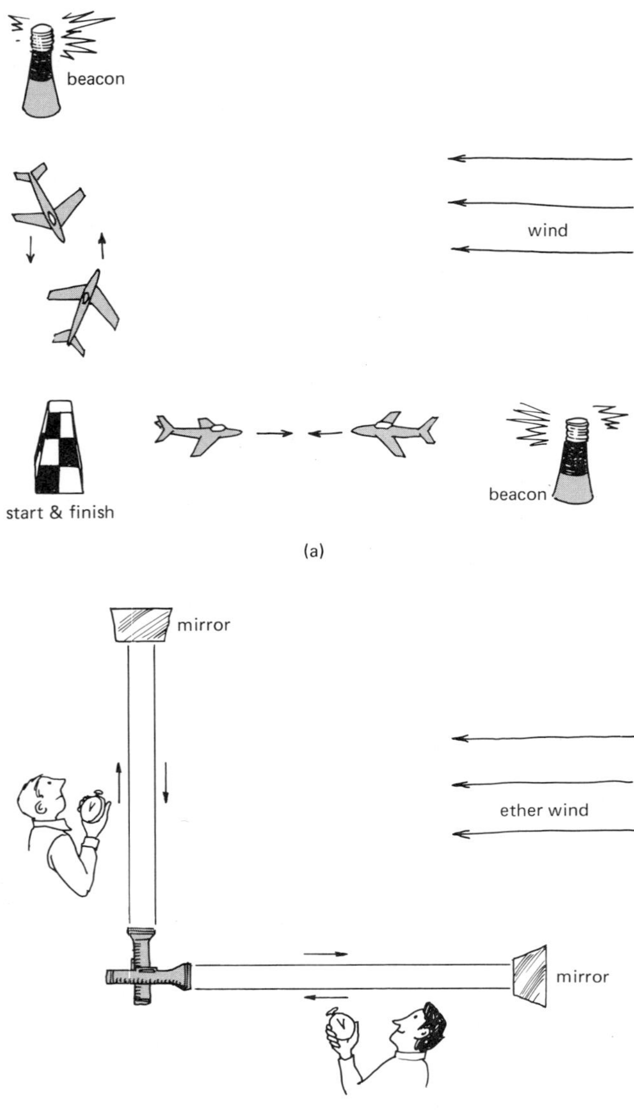

Figure 26.4 Principle of the Michelson-Morley experiment. (a) Airplane analogy. When equal round-trip distances are flown as shown, the trip parallel to the wind always takes longer. (b) Similarly, light beams would classically be expected to arrive back at the starting point at different times.

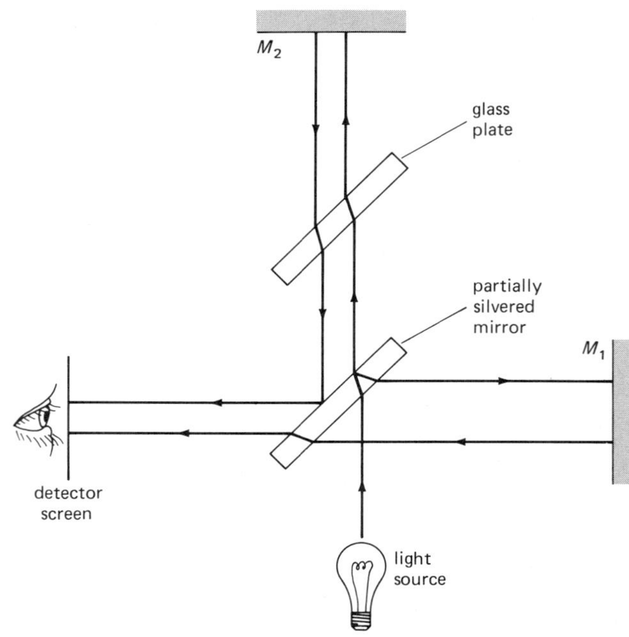

Figure 26.5 The Michelson interferometer. Light beams arriving at the detector show an interference pattern. If the apparatus is rotated 90°, any differences in the travel times of the light beams would cause a shift in the interference pattern.

Where was that ether? Several explanations were suggested to make the null result of the Michelson-Morley experiment consistent with classical wave theory. Perhaps the ether in the immediate vicinity of the Earth was "dragged" along with it so the interferometer was at rest with respect to the ether. This would explain the lack of an interference fringe shift, since there would be no wind. But if this were the case, light waves coming to the Earth would also be dragged along, since they travel in the ether. As a result, we would always see the light from a distant star coming from the same direction—which is not the case.

Another popular explanation at the time was offered by G. F. Fitzgerald, an Irish physicist. He suggested that all objects contracted or shrank in the direction of their motion through the ether. This so-called Fitzgerald contraction would adequately explain the results of the Michelson-Morley experiment if the arm of the interferometer moving in the direction of the ether wind was shortened just enough so that no shift in the interference pattern would occur. It was a clever idea, but not without problems, since similar contractions weren't observed in other situations.

This was the unsettling state of affairs at the turn of the century that paved the way for a major reassessment of physical theories. In 1905 Albert Einstein published his special theory of relativity, which set forth the currently accepted explanation for the Michelson-Morley experiment. (See Special Feature 26.1 and Fig. 26.6.) At the time, Einstein's theory was rejected by many scientists. It required some thinking that went against "classical common sense."

SPECIAL FEATURE 26.1

Albert Einstein

Albert Einstein (1879–1955) was one of the greatest figures in physics. Born on March 14, 1879, in the city of Ulm, Germany, he received his early education in Germany. School instruction seemed to bore him, and he showed no particular intellectual promise. In 1896 at the age of 15, he entered the Swiss Federal Polytechnic school in Zurich to be trained as a teacher of physics and mathematics. In 1900 he received his diploma and acquired Swiss citizenship. After graduation, he found work in the Swiss Patent Office in Berne, where his main duty was the preliminary examination of patent applications. This job provided a livelihood and left him with ample time to work on fundamental problems in physics.

In 1905 he received a Ph.D. degree from the University of Zurich, and in the same year he published three papers of immense importance. Each contained a great discovery in physics. One dealt with quantum theory and an explanation of the photoelectric effect (Chapter 27). This paper formed the basis for his receipt of the 1921 Nobel Prize in physics. Another addressed molecular motions and sizes and an analysis of Brownian motion (Chapter 13). The third paper provided the special theory of relativity, which revolutionized modern ideas about space and time.

Following this, Einstein had no difficulty in becoming a professor at various universities. In 1915, while at the University of Berlin, he published his paper on the general theory of relativity, which provided a new theory of gravitation that included Newton's theory as a special case. During the late 1920's the political situation in Germany deteriorated severely. In 1933, when Hitler and the Nazis came to power, Einstein emigrated to the United States and joined the Institute for Advanced Study in Princeton, New Jersey, where he settled permanently. He became a U.S. citizen in 1940. The rest of his life was spent working on a unified theory that would include both gravitation and electromagnetism.

Near the beginning of World War II and following the discovery of nuclear fission, Einstein was asked to write a letter to President Roosevelt by other emigrant scientists, who along with him recognized the tremendous military potential of nuclear fission, particularly if Germany should develop it first. Einstein's famous letter was instrumental in starting the Manhattan Project and in the development of the atomic bomb. After World War II, Einstein devoted much of his time to organizations advocating agreements to end the threat of nuclear war.

Figure 26.6 Albert Einstein.

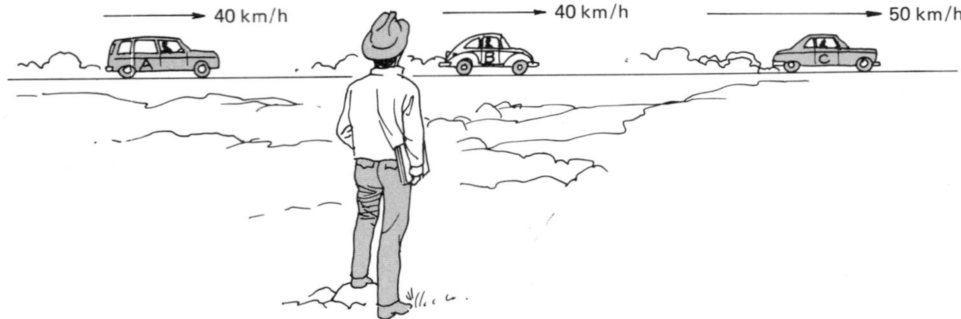

Figure 26.7 Relative motion. A "stationary" observer sees the cars moving as shown. But an observer in car A sees car B not moving and car C moving with a speed of 10 km/h.

The Special Theory of Relativity

The special theory is based on two postulates:

I. The Principle of Relativity. All the laws of physics are the same for all observers moving at a constant velocity with respect to one another.

II. The speed of light in free space is the same ($c = 3 \times 10^8$ m/s) for all observers regardless of the motion of the source or the motion of the observer.

Let's take a look at each of these postulates to get a grasp of their meanings. The first postulate, The Principle of Relativity, has subtle implications. It is really an extension of Newtonian relativity, which is similarly stated for the laws of mechanics, to *all* the laws of physics. The fact that physical laws or experimental results of these must be the same for all observers moving uniformly with respect to each other was recognized by Newton, as well as Galileo.

Another way of stating the first postulate is that the laws of nature are the same in a laboratory at rest as they are in any *uniformly* moving laboratory. Everything would appear the same. For example, if you were in a ship or an airplane moving with a constant velocity, the making and pouring of a cup of coffee or throwing something up and catching it would be exactly the same as when the ship or plane were at rest. Or, looking at it another way, you cannot tell, by any experiment, whether you are at rest or moving uniformly.

Suppose a person is watching uniformly moving automobiles, as illustrated in Figure 26.7. The velocities of the cars are shown referenced to the ground or the "stationary" observer. But if car A is taken as a *reference system,* car B is not moving and car C is moving with a speed of 10 km/h. As a matter of fact, with respect to car A, the "stationary" observer is moving with a speed of 40 km/h in the direction opposite to that of car C. Hence, the motion is *relative.*

But who is at rest? No one and everyone, in a sense. Anyone moving uniformly with respect to someone at "rest" is entitled to consider himself or herself to be at rest and the other person to be moving uniformly. "A state of rest" then is one for which Newton's first law of motion (law of inertia) holds. We call this an inertial system—one at rest or in uniform motion.‡

What all this means is that there is no "absolute" reference frame with the unique property of being at rest with respect to everything else—like the ether. So, the ether is rejected in Einstein's theory.

The Constancy of the Velocity of Light

The second postulate is a bit more difficult to understand, if not believe. It says that the speed of light (in vacuum) has the same value c with respect to any inertial observer, or that everyone measures the speed of light to be the same no matter how fast they are moving. This postulate is a revolutionary statement, which seems to defy common sense and experience.

Classically, we expect and observe velocities to add (as vectors). For example, in Figure 26.8 the two inertial observers measure the ball to have different horizontal speeds. Relative to the observer on the train, which travels at 30 m/s, if the speed of the ball is

‡ The special theory of relativity is limited to inertial systems by the first postulate. Accelerated systems are treated in the general theory of relativity, discussed later in the chapter. In accelerating systems, we must either modify the laws of physics or introduce (pseudo) forces not found in inertial systems (see Chapter 4).

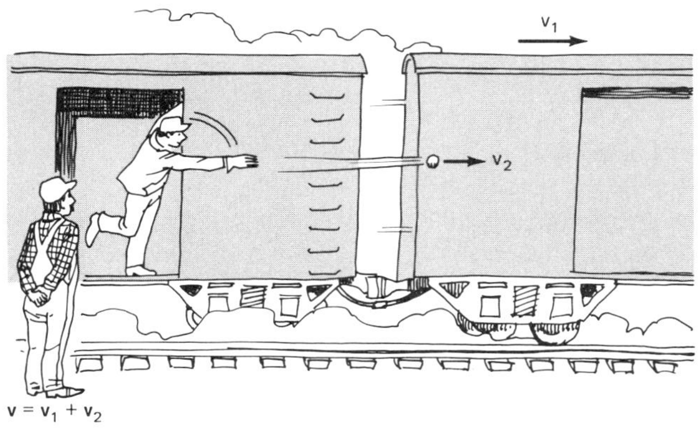

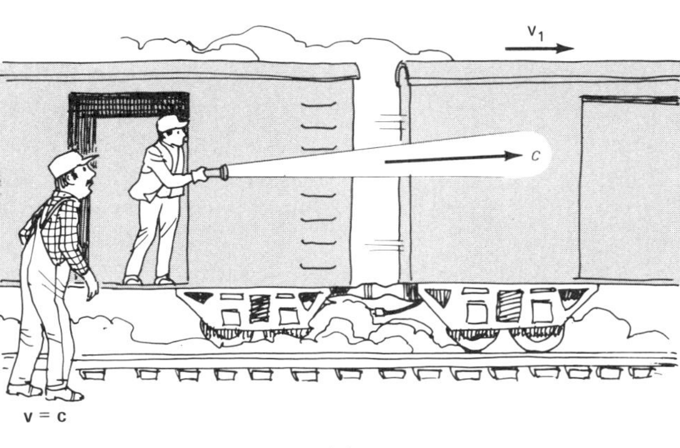

Figure 26.8 Velocity addition. (a) Classically, the thrown ball has a velocity v_2 relative to the moving thrower, and the velocity of the ball is different for an observer in another system. (b) Relativistically, light has the same speed for all observers.

10 m/s, then relative to the observer on the ground, it has a speed of 30 + 10 = 40 m/s. Suppose that the ball is replaced with a beam or pulse of light. Would the speed of light then be 30 m/s + 300,000,000 m/s for the stationary observer? Not according to the second postulate. Both observers would measure the speed of light to be $c = 3 \times 10^8$ m/s. This would be the case even if both the light source and the observer were moving toward (or away) from each other.

The constancy of the velocity of light explains the null result of the Michelson-Morley experiment, but as you can imagine, the idea didn't receive much acceptance at the time. The postulates of the special theory go on to predict other rather "strange" things, which are described in the following sections. However, the test of a theory is the scientific method, and Einstein's theory has passed the test in every instance that has been tried.

Time Dilation and Length Contraction

Two of the "strange" predictions of the special theory involve the measurements of time and length. The strangeness comes about because we do not commonly observe the predictions at our slow pace. Noticeable relativistic effects require speeds that are appreciable fractions of the speed of light. When this occurs, with one system moving at a constant velocity with respect to another, the special theory predicts that an observer will measure *different* times and lengths in the *different* systems. That is, when an observer compares times and lengths that *he* measures in the "moving" system to similar measurements made in his own system, he finds they are different. There is a time dilation—that is, a clock in the moving system appears to run more slowly—and a length contraction—i.e., a length, such as that of a meterstick, in the direction of the motion is measured to be shorter.

These ideas are illustrated in Figure 26.9. When both observers are at rest relative to each other, both of their clocks run together, and the lengths of objects are the same. But when things get moving, an observer finds (measures) that the moving clock runs more slowly and that lengths in the moving system are shorter. Of course, the situation is *relative.* An observer in the spaceship considers himself to be at rest and the "stationary" observer to be moving relative to him. The spaceship observer would measure the same time dilation and length contraction when studying (measuring) a clock and lengths in the other system.

The magnitudes of the measured time dilation and length contraction are given by the following equations:

$$t = \frac{t_o}{\sqrt{1 - \dfrac{v^2}{c^2}}} = \gamma t_o$$

$$L = L_o \sqrt{1 - \frac{v^2}{c^2}} = \frac{L_o}{\gamma}$$

where v is the relative speed of the systems, c is the speed of light, and t_o and L_o are the *proper* time and length intervals in the moving system, or the intervals in the system in which the clock and the length, say a meter-

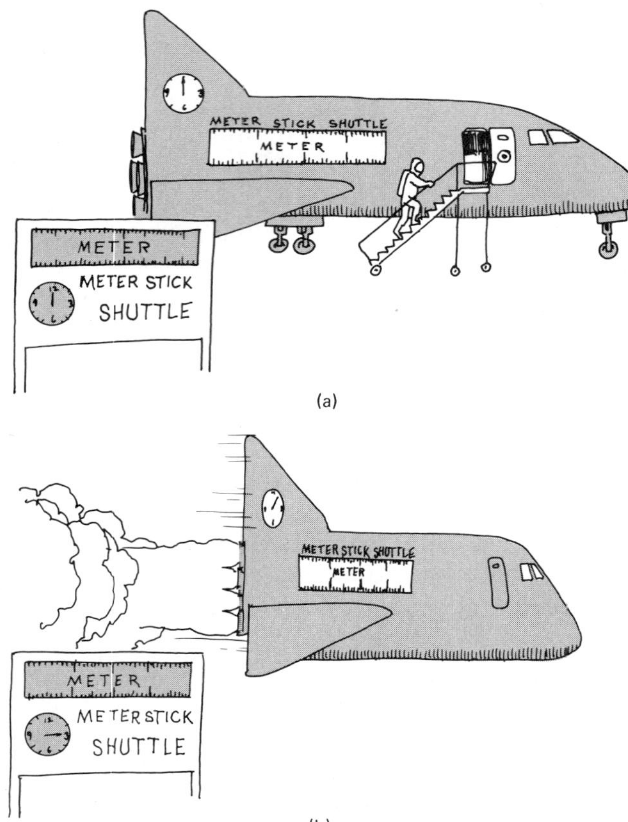

Figure 26.9 Time dilation and length contraction. When one system moves at a constant velocity relative to another, the special theory predicts that an observer will measure different times and lengths in different systems. The moving clock will appear to run more slowly (time dilation), and lengths in the moving system will appear to be shorter (length contraction).

stick, are actually located. The t and L are the time and length intervals that are *measured* by a "stationary" observer in the other system. For convenience, the expression $1/\sqrt{1 - v^2/c^2}$ is often written as γ. Since v is always less than c, γ is always greater than one. Some values of γ for speeds that are appreciable fractions of c are given in Table 26.1.

For example, suppose a spaceship moved with a constant velocity relative to an observer with a speed of $0.6c$ (six-tenths the speed of light). Then, with $\gamma = 1.25$, this gives $t = 1.25t_o$ and $L = L_o/1.25 = 0.8L_o$. This means that when the hour hand on the clock in the moving system goes around one time ($t_o = 1$ hour), the observer in the "stationary" system observes this to take $t = 1.25t_o = 1.25(1) = 1.25$ hours, or 1 hour and

Table 26.1 Some Values of $\gamma = 1/\sqrt{1 - (v/c)^2}$

v	γ
$0.1c$	1.01
$0.2c$	1.02
$0.3c$	1.05
$0.4c$	1.09
$0.5c$	1.15
$0.6c$	1.25
$0.7c$	1.40
$0.8c$	1.67
$0.9c$	2.29
$0.99c$	7.09
$0.995c$	10.0
$0.999c$	22.4

15 minutes, according to his clock. Or when the observer sees 0.8 hour (48 minutes) pass on the moving clock (proper time), one hour has elapsed on his own clock: $t = 1.25t_o = 1.25(0.8) = 1$ hour. Hence, he observes things to be happening at a slower rate in the moving system.

Also, according to the above length equation, a meterstick on the spaceship ($L_o = 1$ m) would be measured to be contracted or shorter in the direction of motion. The meterstick would appear to be 80 cm long as compared with the observer's own meterstick ($L = L_o/\gamma = 1/1.25 = 0.8$ m $= 80$ cm). Notice that if the spaceship were moving with a speed of $0.999c$, events on the spaceship would appear to be in really slow motion and "crunched up." For every hour that elapsed on the observer's clock, less than 3 minutes would be measured to pass on the moving clock, and the moving meterstick would be measured to be less than 5 cm long!

QUESTION: All of this sounds nice, but is there any experimental evidence for time dilation or length contraction?

ANSWER: Yes. To observe these effects, we need to have a situation in which something is moving at a very high speed. This is provided by subatomic particles. One such particle, called a muon, has the same charge as an electron but is about 200 times more massive. Muons are created in the atmosphere as a result of cosmic ray collisions with the nuclei of the gas molecules of the air. The muons then approach the Earth with speeds near that of light ($\sim 0.998c$).

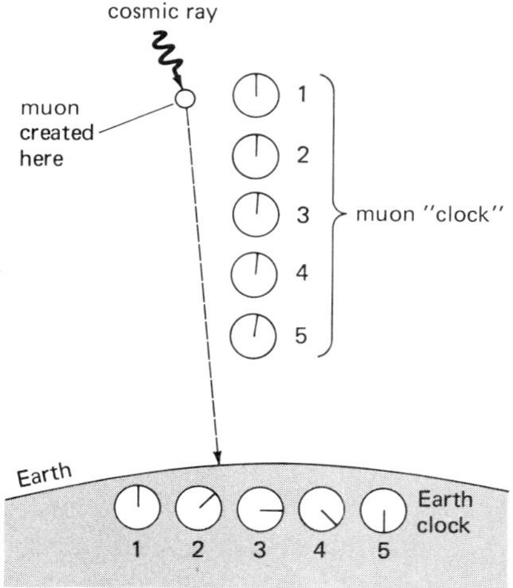

Figure 26.10 Experimental evidence of a time dilation in muon decay. See Question and Answer.

However, the muon is unstable and quickly decays into other particles. The average lifetime of a muon at rest in the laboratory is about 2 μs (2×10^{-6} s). During this time, even traveling at the speed of light, a muon would travel only 600 m ($d = ct$). But muons are created at altitudes of several kilometers, so one would expect the vast majority of the muons to be decayed away before reaching the Earth. Yet an appreciable number of muons are detected at the Earth's surface.

The disagreement arises because time dilation was not taken into account. The muon decays by its own clock. To an observer on Earth, the fast-moving muon "clock" runs more slowly (Fig. 26.10). A speed on the order of $0.998c$ gives a $\gamma = 15$, and for a $t_o = 2$ μs, the time on an Earth clock is

$$t = \gamma t_o = (15)(2\ \mu\text{s}) = 30\ \mu\text{s}$$

During this time, according to an Earth observer, the muon would travel a distance

$$d = vt = (0.998c)\gamma t_o$$
$$= (0.998c)(30\ \mu\text{s}) = 8900\ \text{m}$$

and many muons would reach the Earth before decaying.

You can also look at this from a muon's point of view. Suppose it were created at an altitude of 8000 m (Earth system). Because of length contraction, the muon traveling at $0.998c$ measures this distance to be

$$L = L_o/\gamma = 8000\ \text{m}/15 = 530\ \text{m}$$

which can be traveled in its 2-μs lifetime.

The Twin Paradox

Time dilation gives rise to another popular relativistic topic—the twin paradox. According to the result of the special theory, a clock in a moving system runs more slowly than a clock in an observer's system. For example, with a $\gamma = 4$, for every hour that elapses on an observer's clock, only one fourth of an hour ticks off on a clock in a moving system by his observations. Similarly, observing 1 year of events in the moving system takes 4 years in the observer's time frame. Keeping in mind that we "age by the clock" (heart beat and age measured by proper time), a good question quickly arises: Does an observer in one system age more quickly than another in a relatively moving system?

The "twin paradox" states the problem in terms of a set of twins (Fig. 26.11). Suppose one of the twins takes a high-speed space journey that takes 40 years according to the twin who stays on Earth. Would the Earth twin be older than the space twin on return? With $\gamma = 4$, the Earth twin would spend 40 years observing the 10 years (t_o) that elapsed on the spaceship clock. If the twins were 25 years old at blast-off, then does the 35-year-old returning space traveler find his twin getting ready for retirement at age 65? The answer is yes.

But what happens from the point of view of the space traveler? From the *relative* part of relativity, you know that the space twin looking back at the Earth "sees" his twin's clock running slowly. So from this point of view, the Earth twin ages more slowly, and he is the one who is younger.

The difficulty with the analysis is that the space traveler does not remain in an inertial system. What happens when he is accelerating in starting and stopping? To investigate the paradox in detail, the general theory, which treats accelerating reference frames, should be applied. However, by special applications of the special theory, it too predicts that the twin who has been accelerated will be younger than the one who stayed at home. You may be wondering how we know the "correct" answer, since this requires experimental testing.

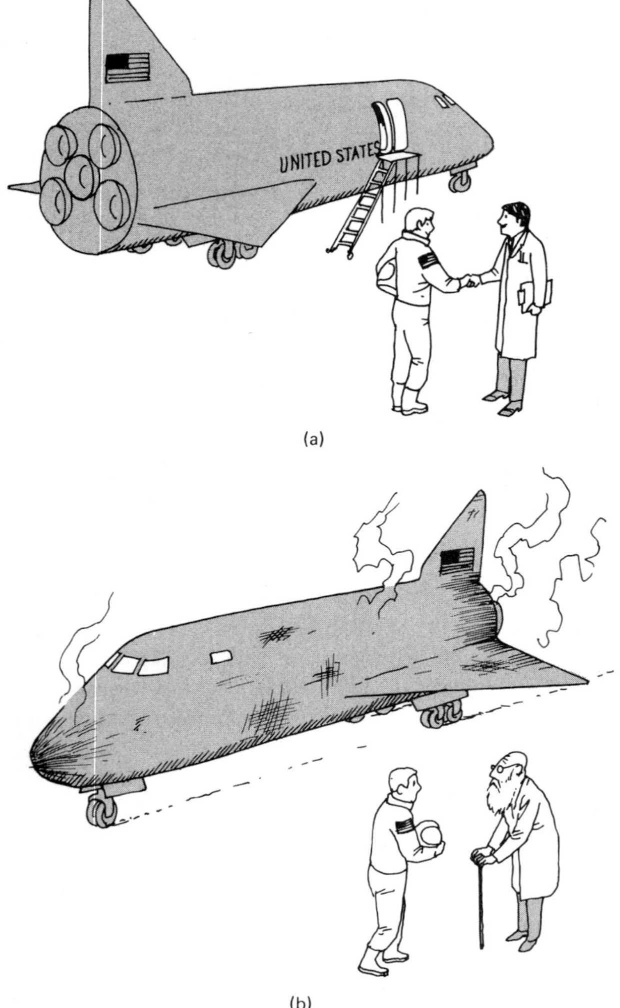

(a)

(b)

Figure 26.11 The twin paradox. The theory of relativity predicts that identical twins (a) would age differently if one went on a space trip (b).

The twin paradox *has* been tested. Not with real twins, of course, but with atomic-clock "twins." Cesium atomic clocks (see Table 1.2)—four of them—were flown around the world in *opposite* directions on commercial aircraft.§ The clocks had been previously synchronized with stationary cesium-clock "twins" on Earth. Afterward, the moving clocks were found to be "out of sync" (showed a different time) in accord with the relativistic predictions. Ultra-accurate cesium clocks had to be used instead of common wrist watches because the time differences were on the order of billionths of a second. Even so, the flying clocks came back "younger."

§ Hafele and Keating, *Science,* 117, 4044, July 14, 1972, pp. 166–170.

In a more recently reported experiment (1985), rather than physically moving clocks, researchers used Earth-based clocks located in different countries. Pairs of these Earth stations simultaneously viewed signals from global positioning satellites, which, depending on the sequence observed, gives an east-west effect similar to the flying clocks. Time differences were noted as predicted by the theory of relativity. This research will help in synchronizing clocks around the world to subnanosecond accuracy.

Experiments with unstable particles accelerated to high speeds in particle accelerators also provide experimental testing that involves the twin paradox. These particles are stable for only a certain time, and it is observed that accelerated particles live longer, on the average, than those at rest or unacclerated.

The twin paradox gives rise to such limericks as

> A precocious student quite bright
> Could travel much faster than light,
> He departed one day
> In an Einsteinian way
> And arrived on the previous night.

Mass in Motion

If relative motion affects the measurements of two fundamental properties, length and time, can mass be far behind? The answer is no. The theory of relativity predicts that an object's mass increases as its speed increases. Similar to time dilation or increase, there is a mass increase or "dilation" according to the equation

$$m = \frac{m_o}{\sqrt{1 - \frac{v^2}{c^2}}} = \gamma m_o$$

where m_o is the rest mass of the object, or the mass of the object when measured in its rest frame. The observed mass m is commonly called the apparent relativistic mass.

There are experimental observations of the relativistic mass increase. Particle accelerators accelerate beams of charged particles, usually electrons or pro-

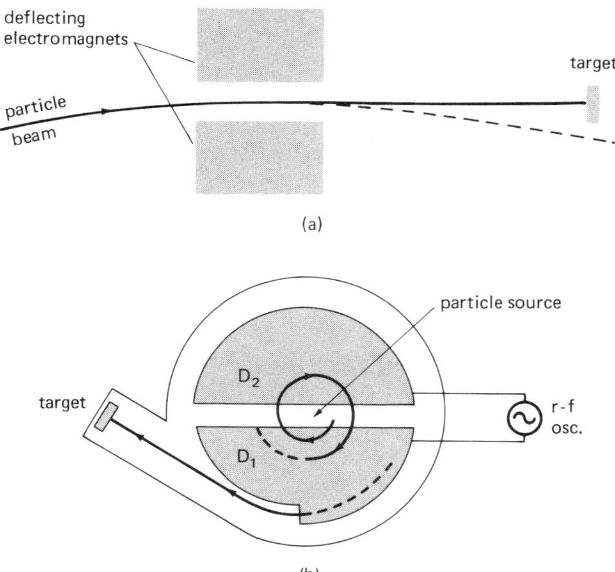

Figure 26.12 Accelerators and relativity. When a particle is accelerated to high speeds, a greater magnetic field is needed to deflect it than would be predicted using the particle's rest mass (a). In a cyclotron (b) the particle travels more slowly than expected. Both indicate a mass increase.

tons, to very high speeds using electric fields. The directions of the particle beams are controlled by magnetic fields. In linear accelerators, beams are deflected or turned by "deflecting" magnets (Fig. 26.12). It is found that a greater magnetic field is required to deflect the particles than would be needed as predicted by using their rest mass, indicating a mass increase. In circular accelerators such as cyclotrons, the particles are given a timed "kick" of energy in certain parts of their orbits. As the particles gain speed, it is found that they arrive increasingly late. The "lateness" is explained by a relativistic increase in the masses of the particles, which follows the previous equation.

The mass equation can be used to illustrate the basis of the phrase, "Nothing can travel faster than the speed of light." From the equation, we see that as v approaches c, the term v^2/c^2 approaches a value of one $[v \sim c,\ v^2/c^2 \sim c^2/c^2 = 1]$. This makes the term $\sqrt{1 - v^2/c^2}$ approach zero, and the mass m is infinitely large. Since we don't have enough energy to accelerate a mass that grows infinitely large (which is difficult to imagine), the theory of relativity sets an upper limit for the speeds of objects. We could modify the foregoing phrase somewhat: "Nothing (no material object) can travel *as fast as the speed of light.*"

Classical or Relativistic?

The ideas of length contractions, time dilations, and mass increases may seem "strange," inasmuch as we do not ordinarily observe examples of these effects. This is because our everyday observations are "classical" observations. To observe relativistic effects, we need speeds that are appreciable fractions of the speed of light. Our fastest race cars, and even our fastest rockets, do not even come close.

But notice that the theory of relativity takes our relatively slow-paced world into account. With v very much less than c, the ratio v/c is very small or practically zero. Since $v^2/c^2 = (v/c)^2$, the relativistic equations become

$$t = \frac{t_o}{\sqrt{1-(v/c)^2}} \approx \frac{t_o}{\sqrt{1-0}} = t_o$$

$$L = L_o\sqrt{1-(v/c)^2} \approx L_o\sqrt{1-0} = L_o$$

$$m = \frac{m_o}{\sqrt{1-(v/c)^2}} \approx \frac{m_o}{\sqrt{1-0}} = m_o$$

Hence, the relativistic equations predict that at relatively slow (classical or nonrelativistic) speeds, we will observe classical results.

Mass and Energy

One of the most remarkable results of Einstein's special theory of relativity is the relationship between mass and energy. It had previously been believed that both energy and mass could be neither created nor destroyed. Energy could be changed in form, but the total energy was constant, just as was the total mass. What Einstein showed was that mass is a form of energy, and the relationship for this is given by the famous equation

$$E = mc^2$$

where c is a constant, the speed of light.‖

The validity of this relationship has been shown in nuclear processes such as fission and fusion (Chapter 7 and Chapter 29). Here, the total mass of the products after a reaction is less than that before the reaction. The

‖ Technically we should write $E_o = m_oc^2$, where E_o is the rest energy and m_o is the rest mass, but it is common to write and say $E = mc^2$. The total mass referred to in this discussion is the total rest mass. The total relativistic energy of a moving particle (with kinetic energy K) is $E = K + E_o$ or $mc^2 = K + m_oc^2$, so for a particle at rest, $E = E_o = mc^2$.

mass loss is just equal to the energy gained in the reaction according to $E = mc^2$. In other reactions, energetic particles can cause nuclear reactions in which there is more total mass after the reaction but less energy, by an amount $E = mc^2$. Hence, we now talk about the conservation of mass-energy instead of separate conservation laws.

The mass-energy equivalence does not mean that we can convert any amount of mass into energy at will. If we could, our energy problems would be solved. For example, if only 1 gram of mass could be completely converted into energy, the amount would be equivalent to the electrical energy used on the average in 20,000 homes in a month. The c^2 in $E = mc^2$ is a big number, so a little mass goes a long way in creating energy.

Mass-energy conversion takes place on a limited scale in nuclear reactions, as will be considered in more detail in Chapter 29. It also takes place in ordinary chemical reactions, such as in the burning of a match. Mass is changed into energy that is lost as heat and light. However, the mass loss is so slight that it is difficult to detect.

On the other hand, the amount of mass lost by the Sun in mass-energy conversion is enormous. If the energy radiated by the Sun is converted into units of mass using Einstein's $E = mc^2$ equation, it is found that the Sun loses about 4.5 billion kilograms of mass per second (about 5 million tons per second in customary weight units). You might wonder if the Sun will waste or radiate away to nothing. Don't worry. At this rate, in a billion (10^9) years the Sun will lose only about 1/100 of one percent of its mass. This is still a good bit of matter—equal to about 25 Earth masses.

(a)

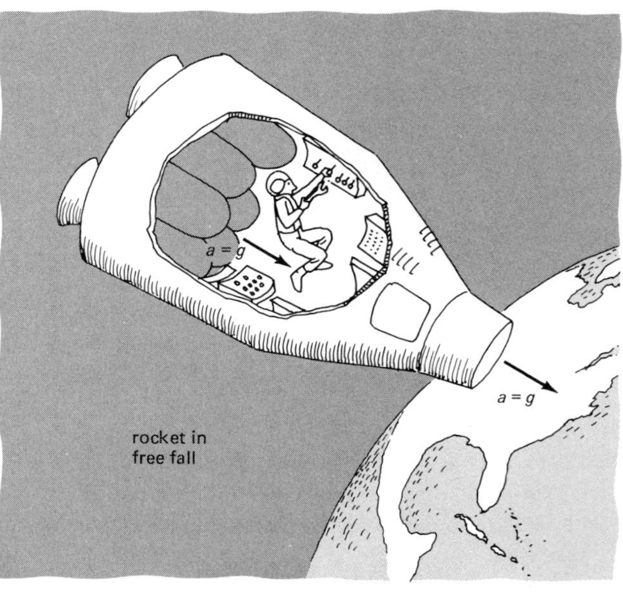

(b)

Figure 26.13 Equivalence. The conditions and laws of physics are the same for an astronaut in (a) an isolated rocket in free space and (b) an accelerated system in free fall.

The General Theory of Relativity

The special theory, after being put forth by Einstein in 1905, gave scientists a great deal of food for thought. Then, ten years later in 1915, he did it again and brought forth his general theory of relativity. This complex mathematical theory extends to relativistic applications to accelerated systems. As you might imagine, the general theory was also met with a bit of skepticism. We have already seen some of its predictions—that a twin in an accelerated system grows older (twin paradox) and that gravity is a "warping" of a space-time continuum (Chapter 5).

The general theory is a result of Einstein's belief that the laws of physics should be the same in *all* reference frames, both nonaccelerated (inertial) *and* accelerated. It is essentially a gravitational theory directed toward this end. To illustrate the equivalence in accelerated and nonaccelerated systems, consider the astronaut in the two situations shown in Figure 26.13. If the spaceship were at rest in free space (negligible gravity), then objects, including the astronaut, would float around in the ship in a state of "zero g." Then suppose the spaceship were in free fall near the Earth. As discussed in Chapter 5, the astronaut would observe the same conditions in a state of *apparent* weightlessness in the accelerated system. In fact, no experiment could be per-

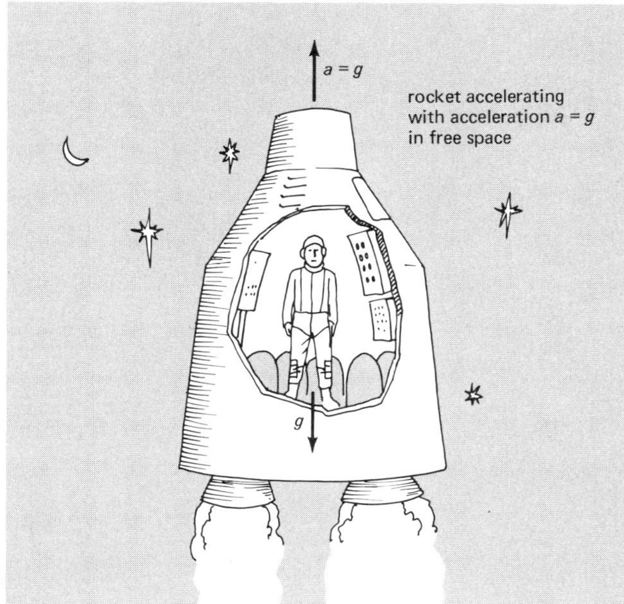

(a)

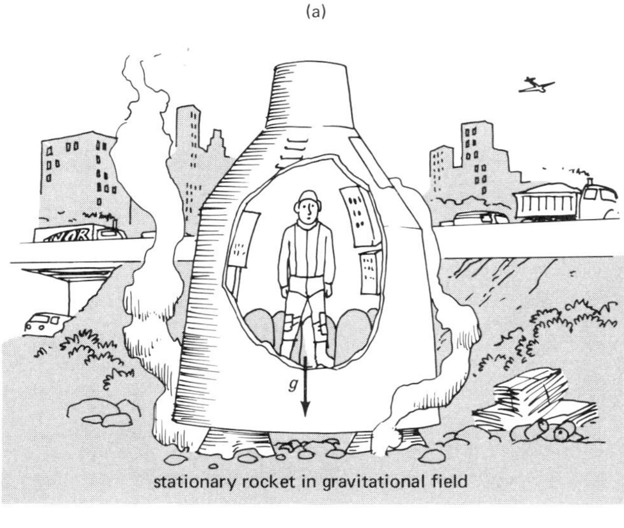

(b)

Figure 26.14 Equivalence. The conditions and laws of physics are the same for an astronaut in (a) an accelerating rocket or (b) a gravitational field.

formed in the closed spaceship system that could distinguish between the nonaccelerated and free-fall conditions.

Further, let the astronaut in free space turn on the rocket engines and accelerate with an acceleration equal to g (Fig. 26.14). The astronaut would then feel the reaction force of the floor or "weight," and objects would fall "down" to the floor. Everything would be exactly the same as in a stationary ship in a gravitational field with a force of 1 g. These results are summed up in the **principle of equivalence:**

> No experiment performed inside an accelerated, closed system can distinguish between the effects of a gravitational field and the effects of an acceleration.

In other words, observations made in an accelerated system are indistinguishable from observations made in a gravitational field.

According to Einstein's general theory, the principle of equivalence holds not only for mechanical phenomena, such as in dropping an object, but for *all* phenomena, including electromagnetic phenomena. This gives rise to some new ideas and predictions. For example, suppose a ball is thrown parallel to the floor of a stationary spaceship in a gravity-free region (Fig. 26.15). The ball would be observed to follow a straight-line path according to Newton's first law. But if the spaceship were accelerating, say with $a = g$, the astronaut would observe the ball to follow a curved (parabolic) path to the floor. (An outside observer would see the ball moving in a straight line, and the floor of the spaceship accelerates up to the ball.)

By the principle of equivalence, the astronaut could not distinguish between an accelerated system and a gravitational field. The ball follows a path just as if it

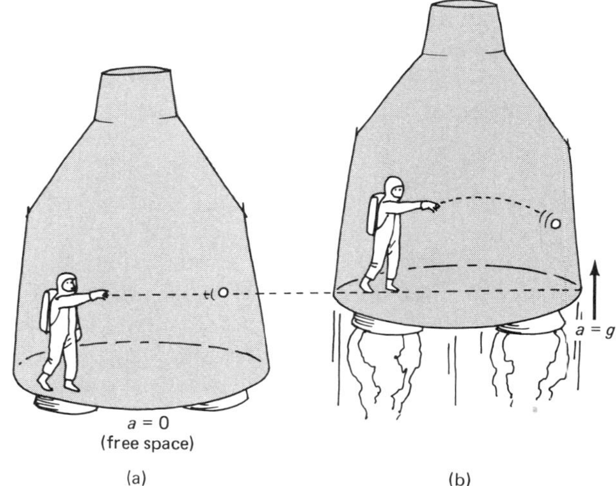

Figure 26.15 Effects of acceleration. In an accelerated system in free space (a), a thrown ball follows a straight path. In an accelerated system (b), the astronaut sees the thrown ball deflected downward as though in a gravitational field.

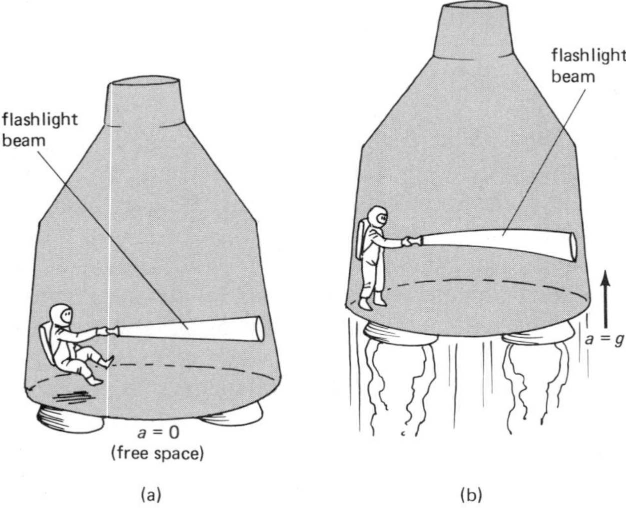

Figure 26.16 Effects of acceleration. In an accelerated system, a light beam appears to be deflected downward. By the principle of equivalence, the same should occur in a gravitational field.

were thrown in a stationary spaceship sitting on a launching pad on Earth. Since the principle of equivalence holds for *all* phenomena, let's replace the ball with a light beam (Fig. 26.16). In an accelerated system, *or equivalently in a gravitational field,* it is predicted that the light beam would be bent or curved. That is, the general theory predicts that light should be bent in a gravitational field!

No such bending of light is commonly observed in the Earth's gravitational field. This is because the effect is too small. To test the prediction experimentally, large distances and a very strong gravitational field are needed, such as that of the Sun—and this is what has been used. Distant stars appear to be motionless and are measured to be a constant angular distance apart (at night, Fig. 26.17). The same stars are "out in the daytime" at other times of the year (as a result of the Earth's revolution about the Sun). Light from one of the stars may pass near the Sun, but any bending would not be observed because the starlight is masked by sunlight scattered in the atmosphere (which is why we can't see stars in the daytime). But when a total eclipse occurs, the moon coming between the Earth and the Sun blocks the sunlight, and an observer in the moon's shadow can see stars during the darkened midday.

During a total solar eclipse, a distant star appearing near the surface of the Sun will have an apparent location that is different from its actual location, owing to the bending of light by the Sun's gravitational field (Fig. 26.17). The apparent angle between the two stars on opposite sides of the Sun should be slightly larger during the eclipse. Such measurements have been made during total eclipses, and the stars were measured to be slightly farther apart, as predicted by Einstein's general theory. Other experimental evidence is also available, as discussed in Special Feature 26.2.

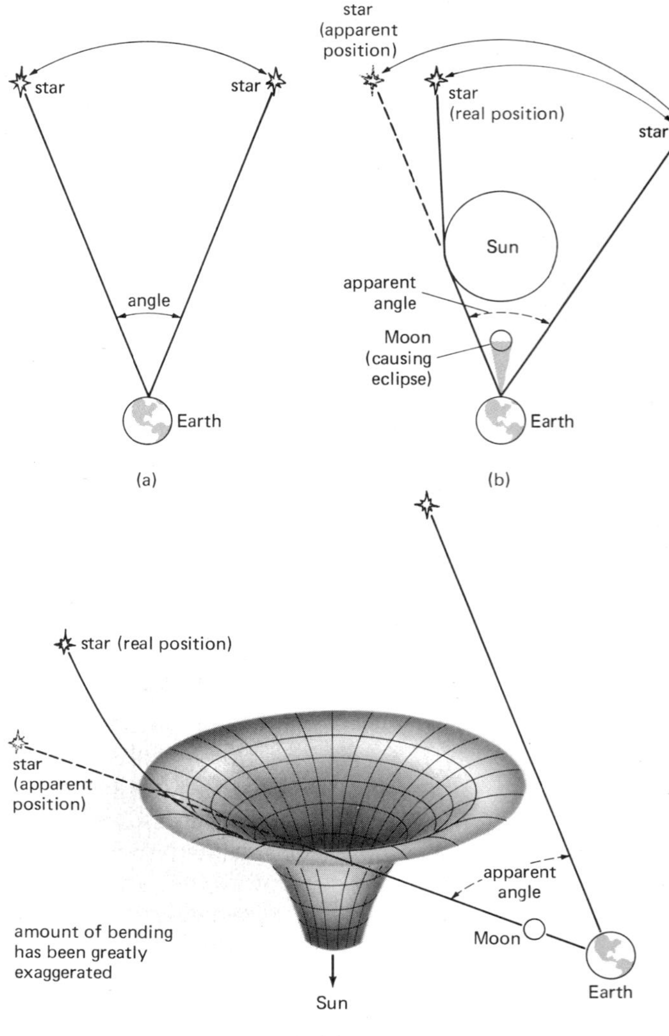

Figure 26.17 Experimental evidence of gravitational attraction of light. (a) Normally, two stars are viewed to be a certain angular distance apart. (b) During a solar eclipse, with one star behind the Sun, the star can still be seen because of solar gravitational effects on light, as evidenced by a larger apparent angular separation. (c) This effect illustrated in Einstein's space-time gravity. Light is deflected in the vicinity of the Sun, as a mass would be.

SPECIAL FEATURE 26.2

Space-Age General Relativity

In addition to light sources appearing to be displaced slightly when seen near the Sun, the radiation from them is also delayed slightly in reaching the Earth. We have no way of measuring the delay in receiving light from stars, but we can detect it in radio broadcasts from space probes because we know where they are and when the signals should arrive at Earth. The experiment has been performed with several space probes, but most precisely with the Viking landers on Mars. When Mars is on the far side of the Sun, signals from a Viking lander must pass through a gravitational field or a space-time warp or "sink" caused by the Sun (Fig. 26.18). Signals from Viking have been observed to be delayed by about 100 microseconds. This is as though Mars had jumped about 30 km out of its orbit, which is very close to what is predicted by the theory of relativity.

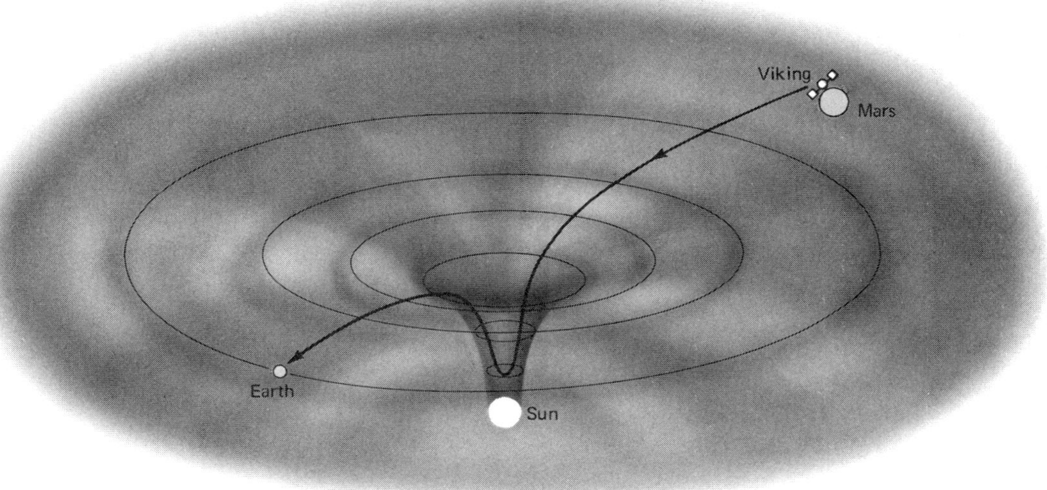

Figure 26.18 More evidence of the general theory. Signals from the Viking lander on Mars were delayed when Mars was on the far side of the Sun because they had to pass through the gravitational field (space-time warp) of the Sun.

Actually, this result of general relativity was presented way back in Chapter 5. Recall that a black hole is a burned-out, gravitationally collapsed star that is so incredibly dense that light coming very close to it would be "bent" into the black hole and could not escape. Of course, a distant star and a less-distant black hole would still be relatively motionless, and we would never see the star if its light were captured by the black hole. If and when the existence and location of a black hole are established by binary-star light and X-ray sources (see Chapters 5 and 30), then we might test the general theory with some *big* light bending, if we could get a spaceship in an orbit around the black hole (but not *too* close an orbit!).

Another result of the general theory is that gravitation affects time—gravity causes time to slow down, and the greater the gravitational field, the greater the slowing of time. There is evidence of this effect in the gravitational "red"-shift of light from the Sun. The gravitational field of the Sun is quite large, and it would be expected to cause the electronic vibrations in atoms to slow down. As a result, light emitted from these atoms would have a lower frequency or would be shifted toward the lower-frequency red end of the visible spectrum. When light from the Sun is compared with similar light from a source on Earth where the gravity is less, a gravitational red-shift of the solar light is observed.

This gravitation-time effect of the general theory means that clocks near the surface of the Earth should run more slowly than those at heights above the Earth, which would be in a weaker gravitational field. Although extremely small, this effect has been measured between the bottom and top floors of a building using "nuclear clocks." So if you work or live on the bottom floor of a building, your clock runs more slowly, or you age more slowly than a person on the top floor—a little general-relativity youth "tonic."

SUMMARY OF KEY TERMS

Ether a hypothetical medium for propagation of light waves.

Michelson-Morley experiment an experiment that was designed to detect the ether through velocity (vector) addition and interference. No effects were observed.

Fitzgerald contraction a proposed contraction or shortening of the length of the arm of the Michelson-Morley interferometer in the direction of the ether wind that would explain the null result of the experiment.

Special theory of relativity Einstein's theory of relativity that deals with nonaccelerating or inertial systems.

Principle of relativity all the laws of physics are the same for all observers moving at a constant velocity with respect to one another.

Constancy of light the speed of light in free space is the same for all observers regardless of the motion of the source or the motion of the observer.

Time dilation the observation (t) of a clock in a moving system (t_o) running more slowly, according to the equation

$$t = t_o/\sqrt{1 - v^2/c^2}.$$

Length contraction the observation (L) of a shortening of a length (L_o) in the direction of motion in a moving system, according to the equation $L = L_o\sqrt{1 - v^2/c^2}$.

Twin paradox the paradox of a space-traveling twin returning to Earth younger than his Earth-bound twin, which is predicted by the general theory of relativity.

Relativistic mass the mass (m) an object is observed to have in a moving system: $m = m_o/\sqrt{1 - v^2/c^2}$, where m_o is the rest mass or the mass of the object measured in its rest frame.

Mass-energy conversion the changing of mass into energy and vice versa, according to the equation $E = mc^2$. As a result, mass is considered to be a form of energy.

General theory of relativity Einstein's theory of relativity that deals with accelerated systems.

Principle of equivalence observations made in an accelerated system are indistinguishable from observations made in a gravitational field.

EXERCISES

1. How long does it take light to travel to the Earth from (a) the nearest star and (b) the next-nearest star?
2. Galileo could not get a consistent result for the speed of light because of the reaction times of his assistant and himself. Assuming the human reaction time to be 0.20 s, how far could light travel during the experimental time error due to the reaction times?
3. How might a supporter of the ether theory explain (a) a translucent object and (b) an opaque object?
4. If there really were a Fitzgerald contraction in the Michelson-Morley experiment, why not just measure the lengths of the arms and see if there was a difference?
5. When you look at a night sky, it might be said that you are looking into the past. Explain why.
6. When the driver of a constantly moving car blows the horn, he hears a frequency (pitch) of 1000 Hz, but a stationary observer in front of the oncoming car hears a frequency of 1010 Hz. Are the laws of physics different in the systems?
7. In Table 26.1, why aren't the units of γ given?
8. Two observers moving at a relative constant velocity watch the same half-hour program on their TV sets (from stations in their own reference systems). What is the proper time of the TV program? What would one observer note if he compared the program in the other system with his own?
9. A meterstick moving at a constant velocity is observed to be only one half as long as the observer's meterstick. How would the mass of the meterstick compare with that of the observer's?
10. What would happen to the length and mass of the meterstick in Problem 9 if it were turned perpendicular to the direction of the motion?
11. An observer sees a friend in a spaceship traveling by with a constant velocity. He knows his friend should be the same height as he, 1.8 m, but somewhat bigger around (fatter). How does his friend appear to him in the moving spaceship?

12. A distant star system is 100 light years away. Would it be possible for an astronaut to go in a spaceship to this star system and come back if the average life span is 70 years? Explain in terms of (a) an Earth observer and (b) the astronaut.
13. A rectangular object travels relative to an observer with $\gamma = 2$. If the long side of the object is in the direction of the motion, how does the measured *density* of the object appear to the observer as compared with a similar object in his system? How about if the long side of the object were perpendicular to the direction of motion?
14. If a moving meterstick is observed to be only 10 cm in length, how fast is the stick moving relative to the observer?
15. A student in a system moving with a speed of $0.7c$ sits through a 50-minute class period. How long does an observer in a stationary system measure the class period to be?
16. Suppose you are an astronaut and traveling away from Earth at a large constant velocity. What effects would you observe on your (a) pulse rate, (b) mass, and (c) volume?
17. Would it be possible to use the twin paradox and have a son or daughter older than the parents? Explain in terms of (a) the child being the space traveler and (b) the parents being space travelers.
18. What would you as an outside observer see for the case shown in Figure 26.15b? Show by drawing several sequential sketches.
19. Do the relativistic effects of mass, length, and time apply to ordinary observations? Explain.
20. If mass is a form of energy, why is there talk of an energy crisis when we have so much mass?
21. When a ball is thrown horizontally, it falls a distance of 4.9 m in the first second. By the principle of equivalence, a horizontal light beam also drops 4.9 m in one second. Why isn't this drop observed?
22. Laser beams are used for surveying, for example, in building tunnels. Won't the gravitational bending of a beam give rise to errors? Explain.
23. A "gedanken" or thought experiment on the effect of gravity on time is illustrated in Figure 26.19. Clocks A and B are at rest relative to each other, so they read the same time. Clock C is moving relative to B and so runs more slowly. Since A and B are equal, C also runs more slowly relative to A, even though they are in the same system. What causes this? (*Hint:* Think in terms of the principle of equivalence and simulated gravity [Chapter 4].)

Figure 26.19 See Exercise 23.

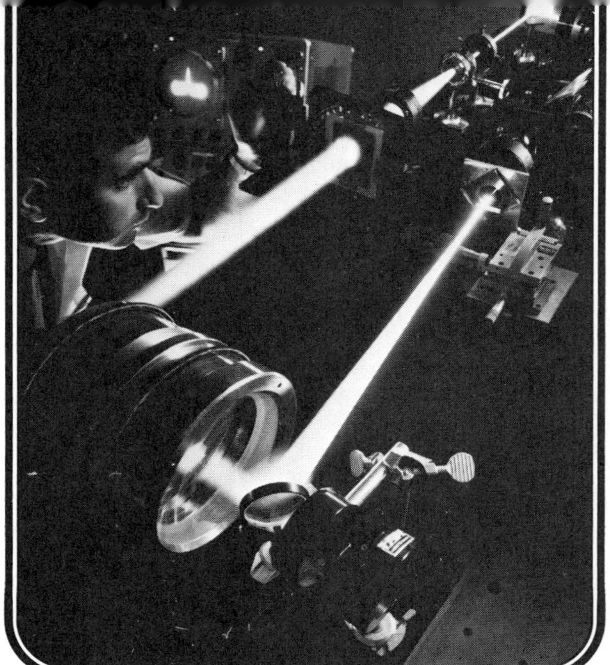

27

Quantum Physics

The question of how light propagates through the void of space or vacuum led to the theory of relativity (Chapter 26) and the nonclassical idea that light or electromagnetic waves simply propagate through empty space without requiring a medium. However, there were also problems on a more "local" level. The wave theory of light did not accurately predict other observed features of electromagnetic phenomena. It did adequately explain some things, such as interference and polarization, but it failed in other instances. In some cases, light seemed to behave as a quantum or "particle" instead of as a wave. This quantum behavior of light led to what is known as quantum theory and the "dual nature of light." In this chapter we shall review the development of quantum physics and some of its results. Let's see what was happening in the early 1900's.

The Ultraviolet Catastrophe and Quantization

Even before the turn of the century, it was well known that light of all frequencies was emitted from an incandescent solid. The relative brightness or intensity of the different frequencies, which we see as colors, depends on the temperature of the solid (Fig. 27.1). As the temperature is increased, more radiation is emitted at every frequency, and the radiation component of maximum intensity is shifted to a higher frequency. As a result, a very hot solid appears to go from a dull red to a blue-white as the temperature is increased (see Chapter 13). This would be expected, since the hotter the solid, the greater the electron vibrations and the higher the frequency of the emitted radiation.

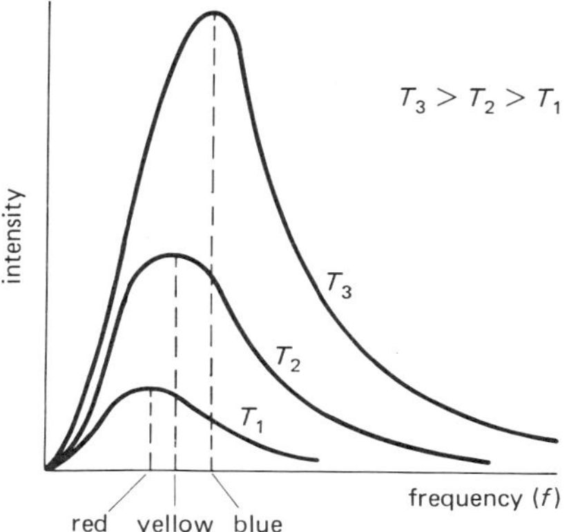

Figure 27.1 Radiation and temperature. For an incandescent solid, the relative brightness or intensity of the different frequencies depends on the temperature of the object. As a result, hot objects glow with different colors as the temperature increases.

When the emission of thermal radiation was analyzed using classical wave theory, it was predicted that the intensity [energy/(area/time)] should be proportional to the square of the frequency, $I \propto f^2$, which would be as shown in Figure 27.2. This gives rise to what was called the ultraviolet catastrophe: "ultraviolet" because the difficulty occurred at high frequencies and "catastrophe" because it predicted that the intensity of emitted energy should be overly large. Thus, the classical theory "bombed out" again, disagreeing not only with experiment but also with common sense.

The dilemma was resolved by Max Planck (pronounced "Plonk"), a German physicist, who in 1900 formulated a theory that correctly predicted the frequency distribution of thermal radiation—but only

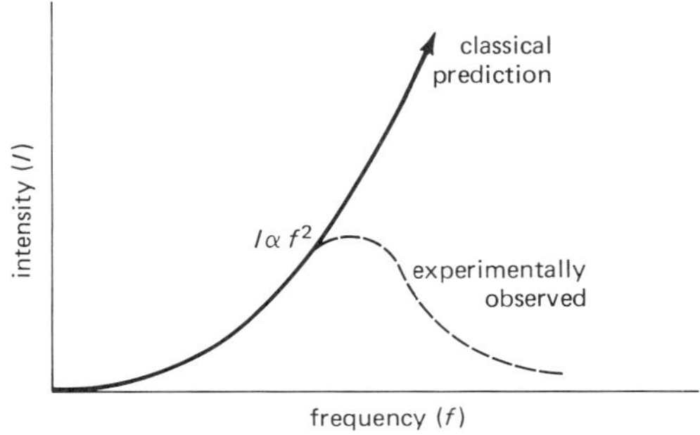

Figure 27.2 The ultraviolet catastrophe. Classical theory predicts that $I \propto f^2$ for a thermal radiator, which requires the emitted energy at higher frequencies (ultraviolet) to be infinitely large (catastrophe).

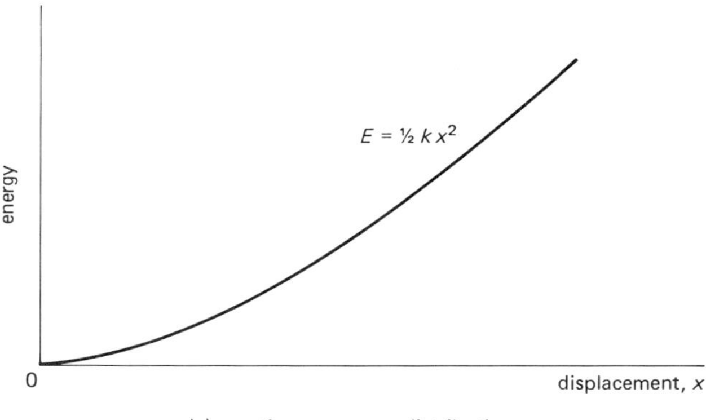

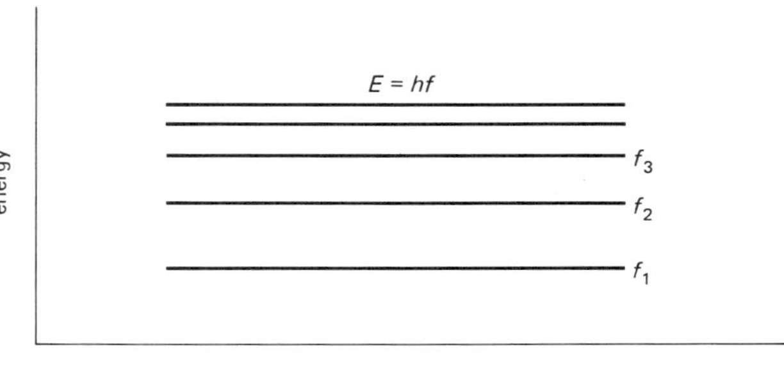

Figure 27.3 Energy distribution and quantization. (a) For a classical oscillator, such as a spring, there is a continuous energy distribution, and an oscillator can have any amount of energy from zero to some maximum value. (b) According to Planck's quantum hypothesis, a thermal oscillator could have only discrete amounts of energy or be in a particular "energy level."

with the introduction of a radically new idea. Planck considered the oscillators of the body to have only discrete energies. Classically, an ocillator is like a spring or pendulum, and it vibrates at a certain frequency. Presumably, an electron in a solid would vibrate at a certain frequency and can have any amount of energy from zero up to some maximum value. This is a continuous energy distribution (Fig. 27.3). **Planck's hypothesis** was that the energy was quantized, or that the oscillators in a thermal radiator could have only *certain* or discrete energies. Moreover, the energy of an oscillator depended on its frequency, according to the relation

$$E = hf$$

where h is a constant, called **Planck's constant** for obvious reasons. The discrete amount of energy of an oscillator was called a quantum of energy (plural, quanta; Latin *quantus,* meaning "how much").

Although the theory fit the experimental data, there was a great deal of skepticism. Even Planck himself was not overly convinced about why the energy should be quantized. However, the quantum hypothesis set the stage for the development of modern quantum physics, which is needed to explain other phenomena. For his initial work, Planck was awarded a Nobel Prize in 1918.

The Photoelectric Effect

There were some other problems with the wave theory of light in predicting observed results, and it wasn't long before the quantum theory scored some other successes. One of these was the explanation of the photoelectric effect. In the latter part of the 19th century, it was observed that certain metallic materials, such as selenium oxide and cesium oxide, were "photosensitive." That is, electrons were emitted from the surfaces of these materials when they were exposed to light. The emission of electrons from a material when exposed to light is called the **photoelectric effect.**

Light then, as a form of energy, must be doing work in freeing the electrons from the surface. You might think of the light waves as causing electrons to vibrate with greater and greater amplitudes until they break free of the material—something like the resonance vibration of a pushed swing and a person not being able to hang on at a large amplitude. Of course, electrons cannot be observed "jumping out" of a material. But if a photoelectric material is used as the cathode of a vacuum tube, a phototube is formed (Fig. 27.4). When the phototube is exposed to light, a current flows in the phototube circuit. The photoelectric effect also occurs in semiconducting diodes or photocells (Chapters 7 and 22). Such cells are commonly used in photographic light meters and in solar panels.

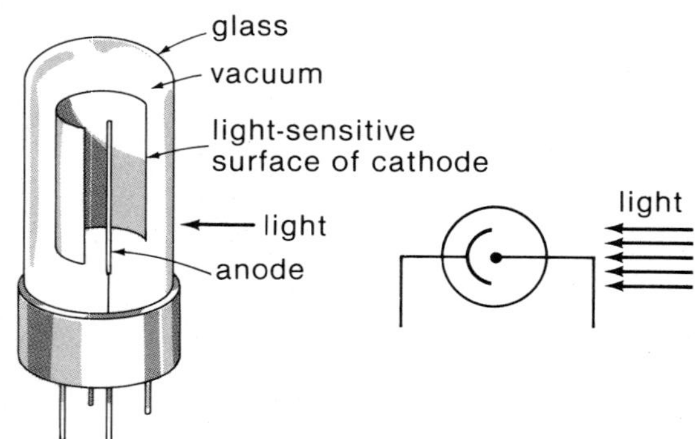

Figure 27.4 Phototube. Direct conversion of light energy to electrical energy. When the phototube is exposed to light, current flows in the circuit.

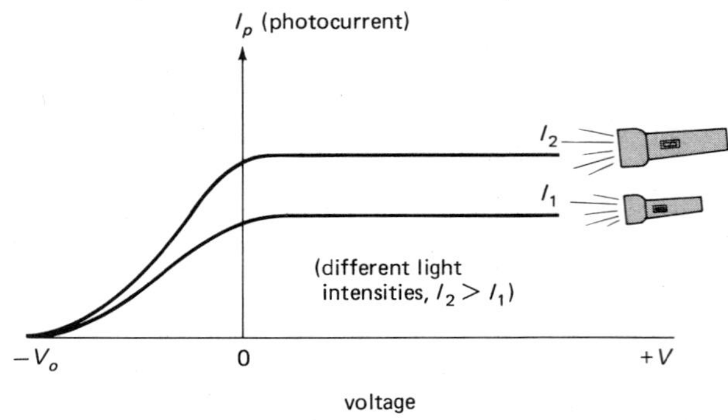

Figure 27.5 Photocurrent and light intensity. The current of a phototube or cell is proportional to the intensity of light—the greater the intensity, the greater the photocurrent. This effect is used in photographers' light meters.

In investigating the characteristics of a phototube or photomaterial, it is found that the "photocurrent" is proportional to the intensity of light (Fig. 27.5). That is, the greater the intensity or energy received, the greater the photocurrent. (This is what is used in photographers' light meters to measure the brightness or intensity of light.) Classically, one would expect this to be the case, since the greater the intensity or amplitude of the light wave vibration, the more energy there is available to free electrons from the material.

However, there are other effects that cannot be explained by classical wave theory. It is found that the kinetic energy of the emitted electrons depends on the frequency of the light but not on its intensity.* But no photoemission occurs for light having a frequency below a certain **cutoff frequency** f_o. For example, photoemission is observed for violet light but not for red light (lower f and below f_o). Also, when a photocurrent is observed, it is observed *immediately* on illumination without any delay, even though the intensity is low.

These characteristics are summarized in a "photoelectric scoreboard" in Table 27.1. Notice that classical wave theory is not having a good season in this case. Let's take a closer look at the classical losses.

Classically, the electric field of a light wave would interact with the electrons at the surface of a material. It

* The kinetic energy can be measured by applying a reverse bias or stopping potential $-V_o$ to a phototube such that no current flows (Fig. 27.5). This is similar to the cutoff voltage of a vacuum tube (Chapter 22).

Table 27.1 Photoelectric Characteristics

1. The photocurrent is proportional to the intensity of light. Predicted by classical theory?	YES
2. The kinetic energy of the emitted electrons is dependent on the light frequency, but not on the light intensity. Predicted by classical theory?	NO
3. No photoemission occurs for light having a frequency below a certain cutoff frequency f_o, regardless of the intensity. Predicted by classical theory?	NO
4. A photocurrent is observed immediately on illumination of a photomaterial when the light frequency is greater than f_o, even if the intensity is very low. Predicted by classical theory?	NO

would be thought that the stronger electric field of more intense light would interact more with the electrons so they would be emitted with higher speeds or greater kinetic energies. But this is not the case (number 2 in Table 27.1). Yet, a low-intensity or dim light of a higher frequency produces a smaller number of photoelectrons, but they have much greater kinetic energies. This is rather strange and doesn't conform to classical ideas. Also, according to wave theory, photoemission should occur for any frequency of light, provided that the intensity of the light is sufficient. Yet below a certain frequency there is no photoemission, regardless of the intensity (number 3 in Table 27.1).

The last item in Table 27.1 was very perplexing. According to wave theory calculations, the time for free electrons to acquire enough energy from a continuous light wave to have the observed kinetic energies of photoelectrons is on the order of minutes. Add to this the energy needed to do the work in freeing an electron from the surface of a material and you have quite a long time!

An explanation of the photoelectric effect was put forth by Albert Einstein in 1905, for which he received a Nobel Prize. He applied Planck's hypothesis to light. In his theory, light itself was quantized, being made up of discrete packets or quanta called **photons.** The energy of a photon of light depended on the frequency f, with a single photon having an energy of $E = hf$ †

† It is interesting to note that Newton first thought that light was made up of "corpuscles" or particles. (See the introduction to Part Six.)

Planck's constant h is a very small number (on the order of 10^{-34}), so most beams of light have many photons.

Einstein applied the conservation of energy to the situation. If an electron in a photomaterial did absorb a photon, then

$$\text{Energy of photon} = \text{energy needed to free electron} + \text{kinetic energy of emitted electron}$$

or

$$E = hf = \phi + K$$

where ϕ is called the work function of the material and is the energy needed to free an electron. Quantizing this, $\phi = hf_o$, where f_o is called the threshold frequency and is the same as the cutoff frequency, we have

$$hf = hf_o + K$$

Hence, light with a frequency f less than f_o will not have enough energy in its photons to free electrons. When f is greater than f_o, electrons will be freed, with the leftover photon energy going into the kinetic energy of an electron. Increasing the light intensity only increases the number of photons and emitted electrons, not their kinetic energy.

Also, no time delay for the emission of photoelectrons is expected in the quantum theory, since a bundle or packet of energy is "dumped on" an electron. That is, in absorbing a photon or quantum of energy, the electron receives a definite amount of energy all at once, instead of receiving it continuously as in wave theory. This is somewhat like filling a bucket by dumping another bucket (quantum) of water in it rather than filling it with a continuous stream of water from a hose as would be analogous to a continuous wave.

The Dual Nature of Light

Such developments as the photoelectric effect led to what is known as the **dual nature of light.** That is, light or electromagnetic radiation sometimes behaves as a wave and sometimes as a quantum or "particle." In order to explain classical phenomena, such as interference and polarization, light is treated as a wave. Yet for other physics phenomena, light is viewed as a packet or particle.

It is important to note that light has a single consistent nature that is described without ambiguity by modern quantum theory. This of course is beyond the scope of this text. The "dual nature" arises when we try to force it into a classical model as a "wave" or "particle." Neither fits entirely, but one *or* the other generally seems to work. A good analogy used by one of the reviewers of this text involves an airplane. Suppose a strange society knew about cars and birds but not airplanes. Upon discovery or observation of an airplane, they might attribute to it a dual nature. Under some circumstances a plane runs along the ground on wheels and carries people. At other times it soars through the sky like a bird—different from other birds perhaps, but certainly more like a bird than a car. With more knowledge and a more modern outlook, this society would describe an airplane as a single consistent entity without ambiguity.

You may wonder why we don't observe the particle or quantum nature of light in everyday observations. For example, we do not see discrete packets of light or see intermittently. This is because the energy of a single photon is very small, and many quanta must be absorbed each second to give enough energy for ordinary observations. As a result, we are macroscopically unaware of the discrete nature of the energy absorption, just as we are not aware of the individual molecules in liquid flow because they are so numerous.

Other examples of the quantum nature of light follow, but first some practical applications of the photoelectric effect are illustrated in Figure 27.6.

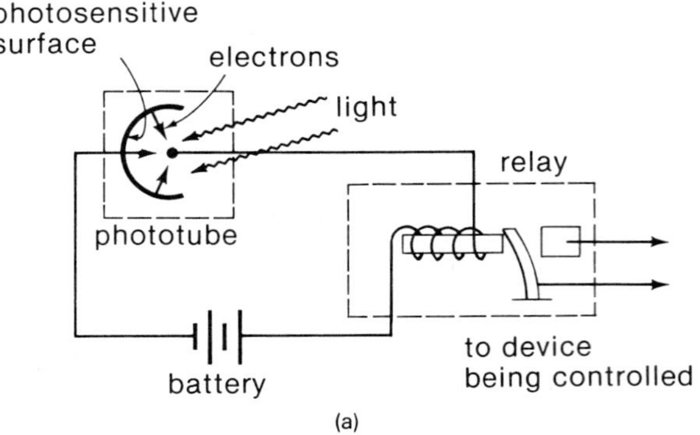

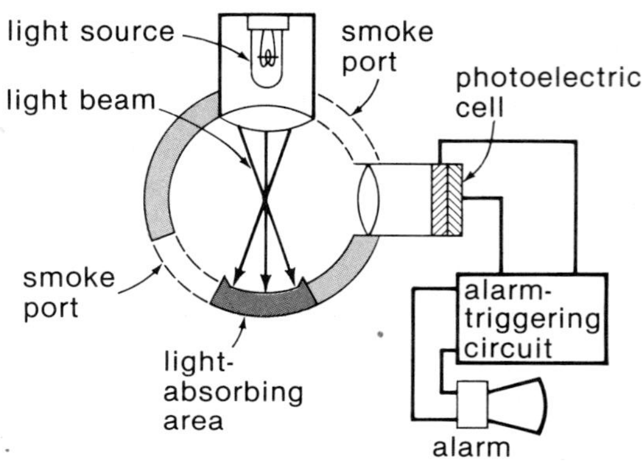

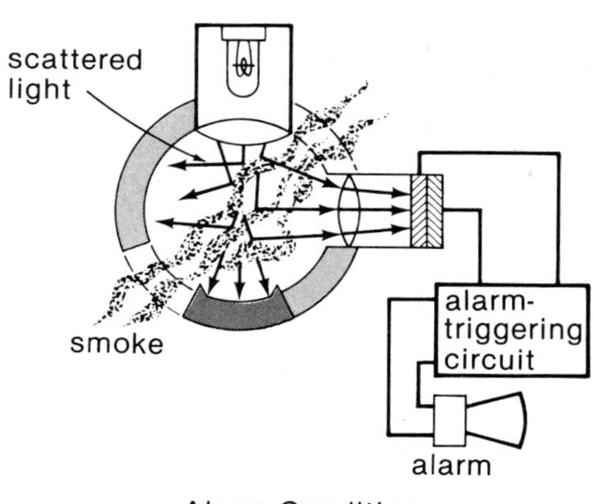

Figure 27.6 Photoelectric applications. (a) Electric eye. When light hits a phototube or cell, current flows in the circuit. Interrupting the light beam opens the circuit, and the relay (magnetic switch, Chapter 21) controls some device. (b) Smoke detector. When smoke enters the detector, the scattered light is detected by the photocell.

The Bohr Theory of the Hydrogen Atom

The quantum theory played an important role in the development of our simplified model of the atom. Recall that we visualize a planetary model of the atom, with the electrons being in orbits about the nucleus (protons and neutrons), similar to the planets in orbits about the Sun. In 1911 the British physicist Lord Ernest Rutherford showed that the protons of the atom were in a "nucleus" in the atom (Chapter 28). The idea of orbiting electrons came from another source and another consideration.

Around the turn of the century, a great deal of experimental work was being done with gas discharge tubes, for example, mercury (vapor), neon, and hydrogen. When the light from these tubes was analyzed, instead of continuous spectra as are observed from incandescent sources, discrete or line spectra were observed. That is, only light or spectral lines of certain frequencies or wavelengths were found (Fig. 27.7). Light coming directly from a tube gives an emission or bright-line spectrum. If white light is passed through a relatively cool gas, it is found that certain frequencies or wave-

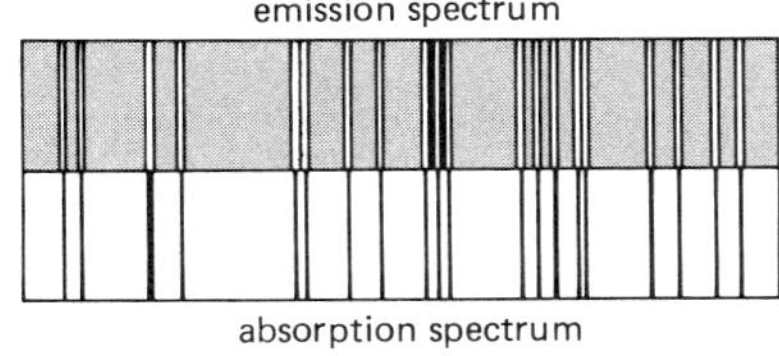

(a) emission and absorption spectra

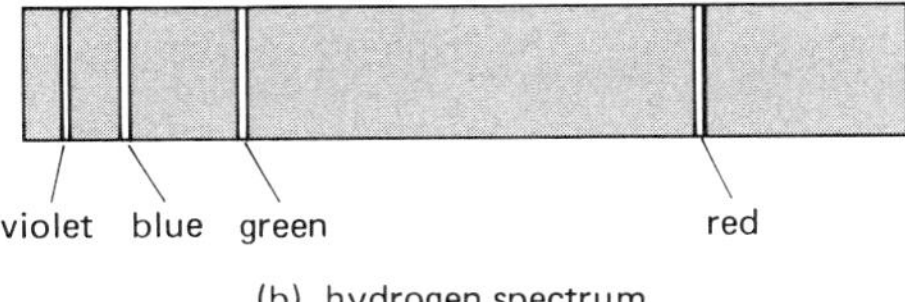

(b) hydrogen spectrum

Figure 27.7 Line spectra. (a) Light is emitted from gases only at certain frequencies, giving a discrete line emission spectrum. If white light is passed through a relatively cool gas, the same frequencies are absorbed, and these lines are missing in an absorption spectrum. (b) The visible spectrum of hydrogen.

lengths are missing in the resulting absorption spectrum, which shows dark lines. Atoms absorb light as well as emit it, and when we compare the emission and absorption spectral lines of a particular gas, we see that they occur at the same frequencies. These puzzling atomic line spectra gave the clue to the electronic structure of the atom.

The spectrum of hydrogen, the lightest element and simplest atom, received much attention because of its relative simplicity (Fig. 27.7). In the late 1800's a Swiss mathematician and physicist, J. J. Balmer, found an empirical formula that gave the positions of the spectral lines. Some new lines were found as a result, but Balmer gave no reason why the formula worked or the physics of the spectral lines.

An explanation of the spectral lines in a theory of the hydrogen atom was put forth in 1913 by a Danish physicist, Niels Bohr (Fig. 27.8). Using the results of Planck's and Einstein's quantum theories along with the Balmer formula, Bohr was able to calculate the energy states of the electron in a hydrogen atom (one electron and one proton), which corresponded to the energies of certain *discrete* circular orbits about the nuclear proton (Fig. 27.9). The possible electron orbits were characterized by a "quantum number" n, called the principal **quantum number,** that could have any whole-number value, $n = 1, 2, 3, 4, \ldots$. The *quantized* circular orbits in Bohr's theory were the result of

Figure 27.8 Niels Bohr (1885–1962) at the time he developed his theory of the hydrogen atom.

the quantization of the orbital angular momentum of the electron, which occurred in whole-number multiples of $\hbar = h/2\pi$, or $n\hbar$, where $n = 1, 2, 3, 4, \ldots$ are the principal quantum numbers. (Notice that Planck's constant comes into the picture again.)

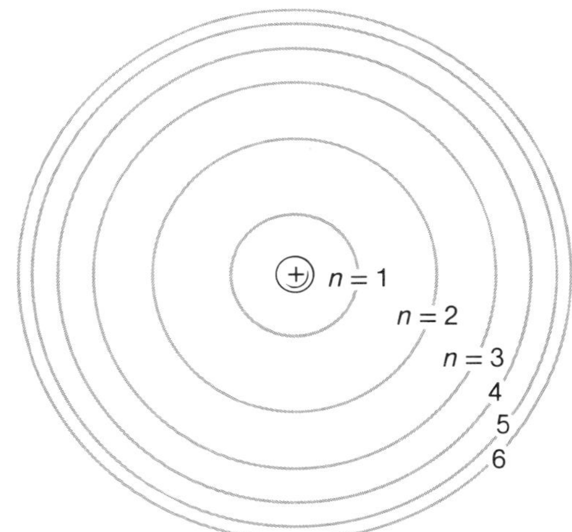

Figure 27.9 Bohr orbits. In the Bohr theory of the hydrogen atom, the electron could only be in quantized or discrete orbits.

However, there was a classical problem with Bohr's planetary theory. According to classical theory, an accelerating electron radiates electromagnetic energy, and in a circular orbit an electron is centripetally accelerating (Chapter 4). Thus, an orbiting electron should radiate energy and quickly spiral into the nucleus, similar to an Earth satellite in a decaying orbit. This doesn't happen in an atom, and so Bohr had to make some bold, nonclassical assumptions in his theory. To account for this and the observed discrete-line spectrum, Bohr postulated that the hydrogen electron *does not* radiate energy when in a bound, discrete orbit but does so only when it makes a "jump" or transition to another orbit or energy state.

The allowed orbits of the hydrogen electron are commonly expressed in terms of energy states or levels, with each state corresponding to a particular orbit (Fig. 27.10). With the electron being bound to the nuclear proton, it is in a potential energy "well," similar to a gravitational potential well (Chapter 6). The electron is normally at the bottom of the well, or in the **ground state** ($n = 1$), and must be given energy or be "excited" to raise it up in the well to an excited state. Thus, the hydrogen electron can be excited only by discrete amounts. It is sometimes convenient to consider the energy levels to be like steps on an energy "ladder." As for a real ladder, a person going up or down a ladder must do so in discrete steps on the ladder rungs. Notice that the energy "rungs" of the hydrogen atom are not evenly spaced. Given enough energy to excite the electron to the top of the well, the electron is no longer bound to the nucleus, and the atom is ionized.

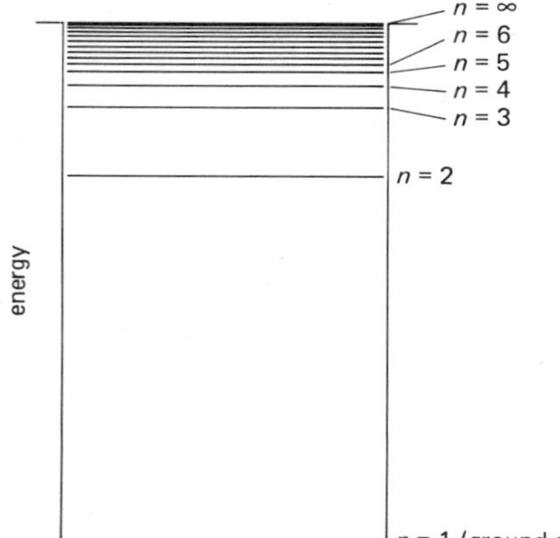

Figure 27.10 Energy levels. Each orbit corresponds to a particular energy level. The electron makes transitions between energy levels when changing orbits.

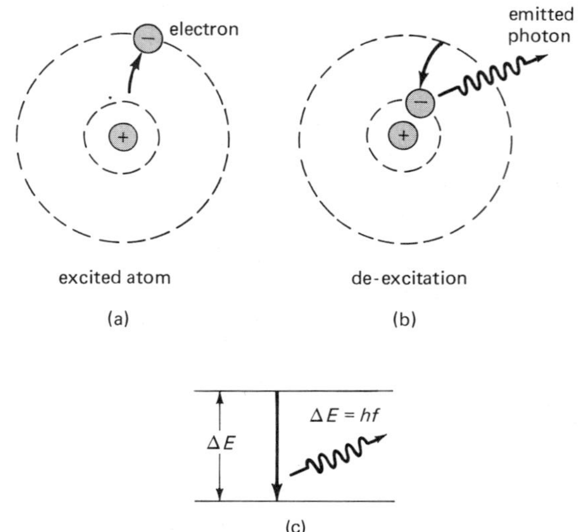

Figure 27.11 Photon emission. (a) When an electron is excited to a higher orbit or energy level, it eventually "decays" (b), with the emission of a photon or a discrete amount of energy at a certain frequency determined by the difference in the energy levels (c).

An electron does not usually remain in an excited state for long and quickly decays or makes a transition to a lower energy level. In doing so, the electron can radiate energy by the Bohr theory and does so by emitting a photon or a discrete amount of light energy (Fig. 27.11). The energy of the photon is equal to the energy difference between the two energy levels, or $\Delta E = hf$. Hence, the hydrogen atom emits only discrete frequencies of light. The electron does not always "jump" down the energy ladder one rung at a time. It can skip rungs, just as a person coming down a ladder can. In an excited hydrogen gas, as in a discharge tube, with many, many atoms, various transitions are made, and these correspond to the frequencies of the observed spectral lines (Fig. 27.12)‡

Thus, quantum physics and the quantum nature of light scored another success. The electron energy levels for other, more complex atoms are more difficult to calculate, but the line spectra of various atoms are indicative of their energy spacings and provide a characteristic "fingerprint" by which an atom may be spectrographically identified.

‡ In an incandescent solid, such as a lamp filament, the atoms interact strongly, and the electrons are not associated with a single atom. In this case, the electron energies have a continuous range and hence give rise to a continuous spectrum.

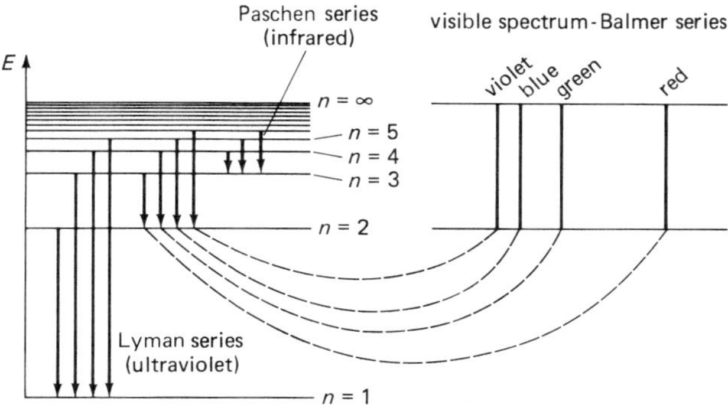

Figure 27.12 Transitions and line spectra. The visible spectrum or Balmer series of the hydrogen atom is given by transitions to $n = 2$. Other "series" are transitions to other energy levels. Bohr's theory correctly predicted the wavelengths of the observed spectral lines of hydrogen and aided in the discovery of the Lyman series.

The Laser

We hear more and more about lasers. Laser beams have been sent to the moon and have been used in eye surgery, in surveying, and to drill holes. We now hear talk of laser weapons capable of destroying satellites. Because of such publicity, the laser is sometimes thought of as an incredible source of energy. This is not the case. The laser is a light source, and like other light sources, it converts one form of energy to another—always at a loss. The uniqueness of the laser as a light source comes about because of the special properties of the light it produces.

The development of the laser was one of the crowning successes of the modern approach to science. In many instances, scientific discoveries have been made accidentally, and even though these discoveries were applied, no one fully understood how or why they worked. The discovery and use of X-rays are an example. Often, in inventing something, a trial-and-error or Edisonian approach was used. (In developing the incandescent lamp, Edison tried many filament materials until he found one that worked—a carbonized thread from his wife's sewing basket.) However, the laser was developed first "on paper" and then built as predicted. Atomic theory allowed scientists to "look" into the atom for previously unobserved and unapplied phenomena. As a result of quantum theory, we now have the laser. The theory behind the laser was developed during the middle and late 1950's by the American scientists Arthur Schalow and Charles Townes, and the first working device was constructed in 1960.

The term **laser** is an acronym for *l*ight *a*mplification by *s*timulated *e*mission of *r*adiation (*l-a-s-e-r*). The laser was the first device capable of amplifying light waves themselves. You may think that the amplification of electromagnetic radiation is not new. For example, radio waves are amplified in a radio. But here it is the amplification of the current signals *after* the electromagnetic energy has been converted to the vibrational energy of electrons in the antenna (Chapter 22). Electrical circuits are incapable of handling the frequencies of light (a million times those of radio waves), because the circuit electrons cannot oscillate at these frequencies. This limitation is overcome in the laser, in which light is amplified using energy stored in atoms.

The principle of the laser was first developed for *m*icrowave frequencies and the first device was called a *m*aser. The laser is an optical maser. The same principles are involved, but the laser involves higher frequencies. The first maser used ammonia as an amplifying medium. Other substances, such as ruby, crystals, and various gases, have been found to be suitable for higher light frequency "lasing" action or amplification. To understand this action, let's take a look at the "aser" part of a laser—*a*mplification by *s*timulated *e*mission of *r*adiation.

When an electron is excited to a higher energy level in an atom, it generally "decays" after a short time (on the order of 10^{-8} second) and returns to a lower energy level with the emission of a photon. This is called **spontaneous emission** (Fig. 27.13). However, an excited atom can be *stimulated* to emit a photon. In a **stimulated emission** process, an excited atom is struck by a photon of the same energy of the allowed transition, and the atom emits an identical photon. After emission, the two photons have the same frequency, are emitted in the same direction, and are in phase. Since one photon goes in and two come out, this is an amplification process (which was predicted by Einstein in 1917).

The monochromatic (single frequency), phase, and directional properties of the light of the lasing action give laser beams some unique properties. Light from common sources, such as an incandescent lamp, is incoherent. That is, the waves have no particular relationship to each other. For example, if you throw a handful of gravel into a pond, the resulting waves would be incoherent. In light sources, excitation occurs randomly, and atoms emit randomly and at different frequencies (different transitions). As a result, an incoherent light beam is "chaotic" (Fig. 27.14). Such a beam spreads out and becomes less intense. To get

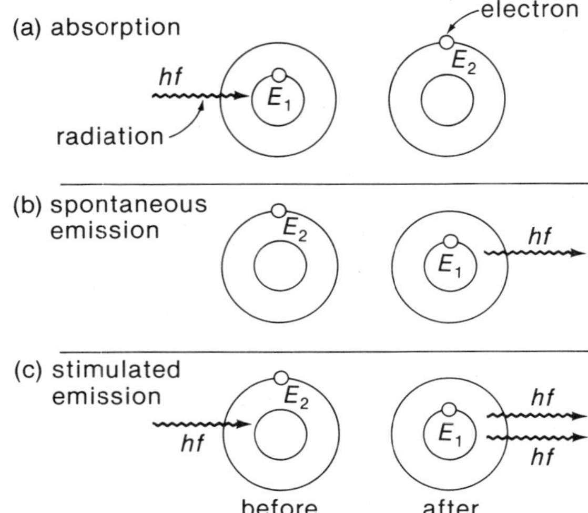

Figure 27.13 Radiation emission. (a) An atom absorbs energy and becomes excited. (b) It generally decays spontaneously after a short time, with the emission of a photon, but if another photon strikes the excited atom (c), it is stimulated into emission.

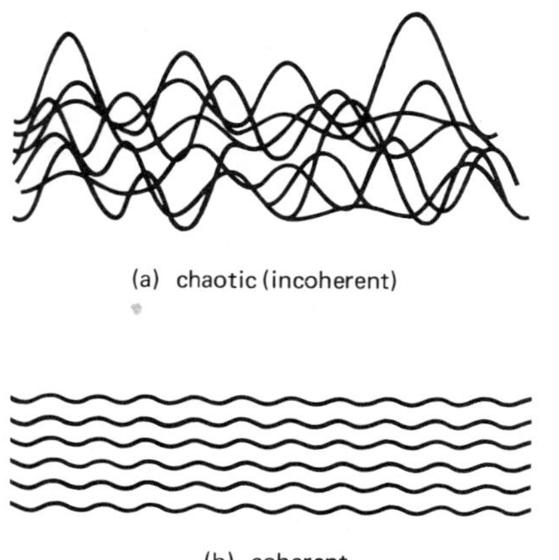

Figure 27.14 Laser light is coherent. (a) Light from common sources is chaotic or incoherent, with the waves having no particular relationship to each other. (b) Coherent laser light has the same frequency, phase, and direction.

monochromatic light, all but one frequency must be filtered out of the beam, which leaves it weak and still incoherent. When the waves (or photons) of a beam of light have the same frequency, phase, and direction, it is **coherent.** It is possible to make a coherent beam that does not spread out appreciably and that, if amplified, can be very intense.

There are pulsed lasers, which produce pulses of coherent light, and continuous lasers. Let's focus on the continuous-beam He-Ne (helium-neon) gas laser, which you are most likely to see in the classroom and in some common applications. The gas mixture is subjected to a high voltage in a laser tube, and an electrical discharge is induced (Fig. 27.15). Electron collisions efficiently excite He atoms to an excited state that has a

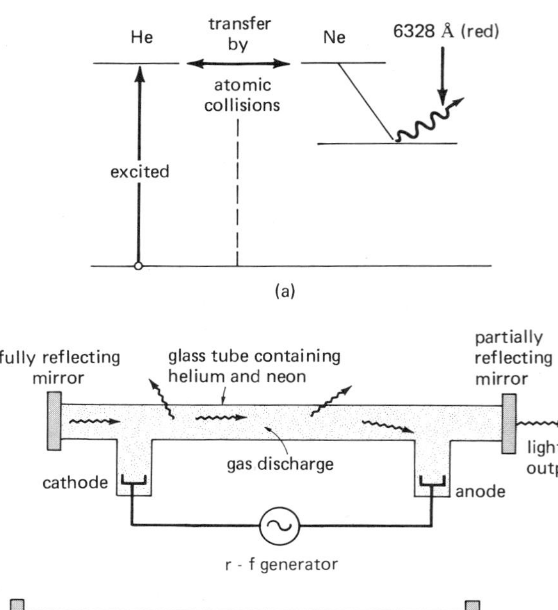

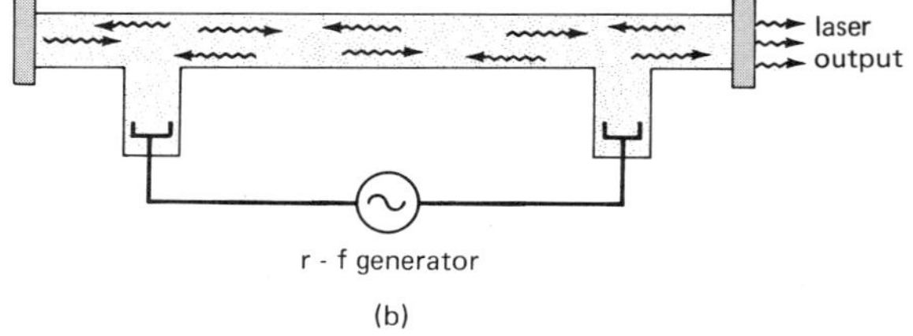

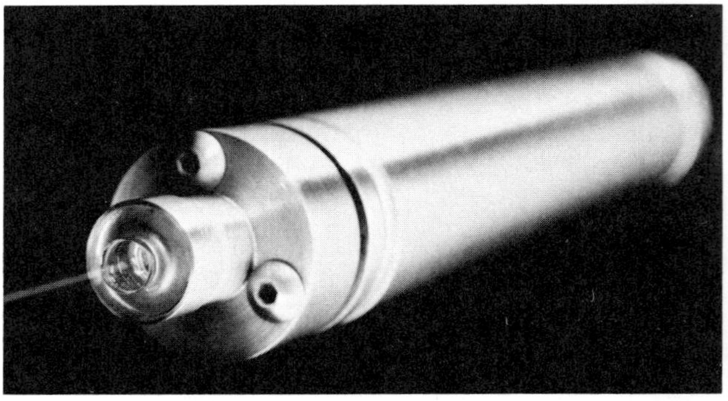

(c)

Figure 27.15 He-Ne laser. (a) The laser light comes from a neon atom transition. (b) A schematic diagram of a gas laser. (c) A laser. Steel-ceramic tubes, which are much more reliable than gas tubes, are now commonly used.

SPECIAL FEATURE 27.1

Lasers in the Supermarket

You may have noticed the pink glow of He-Ne laser light in a supermarket checkout that uses a scanner for the optical reading of the Universal Product Code (UPC) or "bar code" on prepackaged items (Fig. 27.16). The checkout person moves an item across an opening through which a laser beam is directed toward the bar code label. The dark areas of the code absorb light, and the light areas reflect. A light-sensitive detector reads the pattern of bars from the reflection pattern. This information goes into a central computer, which identifies the product and sends the programmed price for that item to the electronic cash register.

(a)

(b)

Figure 27.16 Code-bar scanning system. A He-Ne laser is used in the optical scanning system (a) that reads product codes (b) on grocery and other items. The zero to the left of the code bar indicates a grocery item. The next five digits indentify the manufacturer (13000, H. J. Heinz Co.), and the next five digits identify the product (45090, 16-oz. can of pork and beans).

relatively long lifetime ($\sim 10^{-4}$ s). The Ne atom also has an energy state at the same energy. There is a good chance that before an excited He atom emits a photon spontaneously, it will collide with an unexcited Ne atom and transfer energy to it. (There are about 10 He atoms to one Ne atom in the gas mixture.) The excited Ne atoms decay after a short time, and the photon of the transition has a frequency that is in the visible red region.

The amplification of the light emitted by the Ne atoms is accomplished by reflections from mirrors placed at each end of the laser tube (Fig. 27.15b). Reflecting back and forth in the tube, the photons cause stimulated emissions, and an intense beam of photons builds up along the direction of the tube axis. Part of the beam emerges through one of the end mirrors, which is only partially silvered. The outgoing beam is highly intense, coherent, and directional.

A common He-Ne laser application is given in Special Feature 27.1. The laser light seen in this case is scattered or indirect. An intense laser beam can be dangerous to the eye, and you should never look directly into a laser because eye damage can occur. In the medical field, such *slight* damage is used to "weld" detached retinas in eyes (Fig. 27.17).

The list of laser uses is quite long and continues to grow. The distance of the moon from the Earth has been accurately measured using a laser pulse reflected from a mirror placed on the moon by astronauts. Surveyors use laser beams to "line up" measurements, as

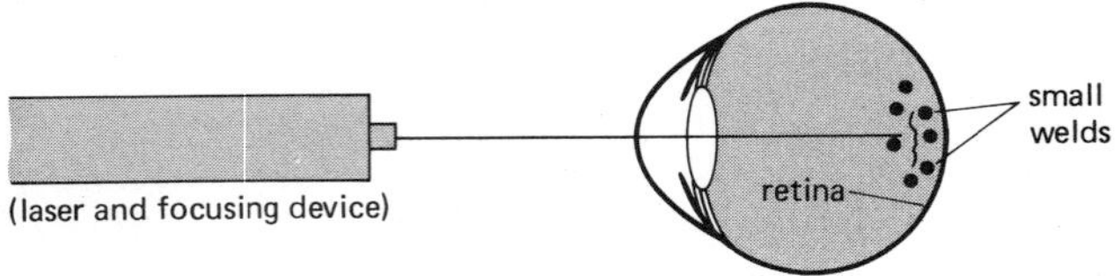

Figure 27.17 Lasers are used to "weld" detached retinas in eyes.

in drilling tunnels. Long-distance communications with laser beams in space and in optical fibers are being developed and are already in use.

High-intensity lasers can cut metals (Fig. 27.18), as well as cloth (laser knife or scissors) in commercial garment-making. Metal welding is also done with lasers. Computer printouts are now done with laser printers, and laser videodiscs are available. The videodisc is about the same size as a LP album. More than 50,000 separate images can be stored on each side of a disc, along with two-channel audio. Information is coded in tracks of microscopic indentations. As the disc spins, a laser "needle" reads the tracks for playback on a conventional TV receiver.

The coherent property of laser light has made the production of three-dimensional images possible through holography (Greek *holos,* meaning "whole"). A hologram is made by using reference and interference beams (Fig. 27.18). Their interference patterns on a photographic film record the information from "in depth" parts of the object. When developed, the interference pattern appears as a meaningless pattern of light and dark areas. However, when illuminated with a laser beam (or another light source, for some holograms), the recorded information is reproduced and perceived as a three-dimensional image. Someday, holographic, 3-D TV sets may be common.

There are a variety of new types of lasers—glass, chemical, and semiconductor. The light produced ranges from infrared through ultraviolet, and some types are "tunable" and can be adjusted to different frequencies. X-ray lasers are also being developed. This holds the promise of in-depth views, possibly three-di-

(a)

Figure 27.18 Laser uses. (a) A laser being used to cut metal. (b) An arrangement for making and viewing a hologram.

photographic
plate
object
object
beam
reference
beam
partially
reflecting
mirror
laser

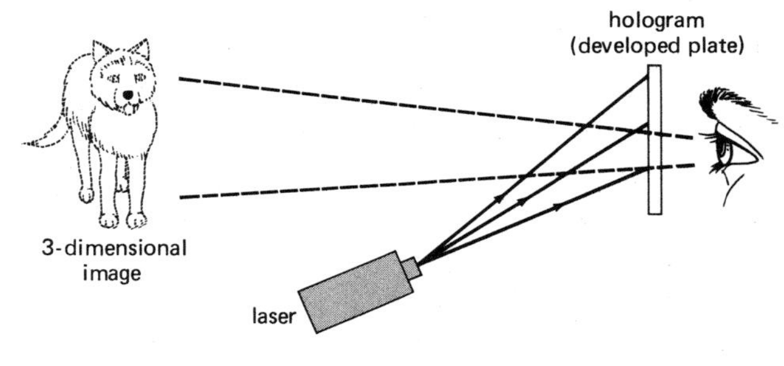

(b)

mensional (holograhic), of microscopic structures. Also, as will be learned in the next chapter, experimental work is being done in the use of lasers to induce nuclear fusion.

Matter Waves and Quantum Mechanics

The dual nature of light—what had been thought to be a wave, with evidence to that effect, sometimes behaves as a "particle," also with evidence. A logical question follows: Do mass particles have a wave nature? Since nature exhibits a great deal of symmetry, the French physicist Louis deBroglie reasoned that matter, as well as light, had properties of both particles and waves. In 1924 he postulated that any moving particle or object has a wave associated with it, with a wavelength (λ) given by

$$\lambda = h/mv$$

where m and v are the mass and velocity of the particle (recall that mv = momentum) and h is Planck's constant again.

The waves associated with moving particles were called **matter waves or deBroglie waves.** If a photon is taken to be a "massless" particle, then an electromagnetic wave is the deBroglie wave for a photon, but the deBroglie waves associated with mass particles, such as electrons and protons, are *not* electromagnetic waves.

As you might expect, deBroglie's hypothesis met with a great deal of skepticism. The idea that photons were electromagnetic waves did not seem unreasonable, but to think that a mass in motion somehow had wave properties was another matter. In support of his hypothesis, deBroglie showed how it could give an interpretation of the quantization of the electron orbits in Bohr's theory of the hydrogen atom. As the electron travels around in one of its circular orbits, the associated matter wave would be expected to be a standing or stationary wave. That is, only a certain number of wavelengths would fit into a given orbit, as in the case of a standing wave in a string (Chapter 16). DeBroglie's theory predicted this (Fig. 27.19). A wave must match or meet itself after going around one full circumference. If it didn't match or wasn't in phase with itself, the wave would interfere destructively, and the associated particle could not be in that orbit. Thus, there were only discrete, or allowed, orbits.

QUESTION: If moving masses have wave properties, why aren't these observed normally?

ANSWER: According to deBroglie's equation for the wavelength of a wave associated with a moving "particle," for normal objects and speeds the wavlengths are so small that the wave properties go unnoticed. For example, a 1000-kg car traveling at 100 km/h would have a wavelength

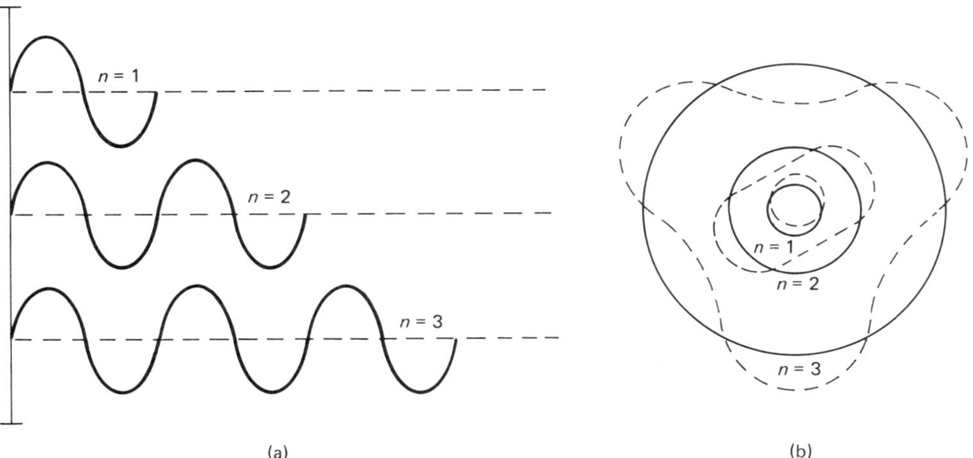

Figure 27.19 DeBroglie waves and the hydrogen atom. (a) Only a certain number of (deBroglie) wavelengths would fit in the electron orbits (b), and discrete standing waves (characteristic frequencies or wavelengths) correspond to the discrete orbits in Bohr's theory.

on the order of 10^{-38} meters. Recall that wave properties, like diffraction, are observed when the wavelength is on the same order as an object or opening. A wavelength of 10^{-38} m is just too small for wave effects to be observed.

Although deBroglie's hypothesis provided an interpretation of discrete electron orbits, this was hardly proof. The motions of particles had been studied extensively, and no wave-like behavior had ever been observed. The preceding Question and Answer tells you why. What would be needed is a particle of very small mass so that the deBroglie wavelength would be appreciable. In 1927 two physicists in the United States, C. J. Davisson and L. H. Germer, took advantage of this and showed that a beam of electrons could be diffracted in a nickel crystal, thereby demonstrating the wave-like properties of particles! (Recall that crystals had also been used to study the wavelengths of X-rays, which also have very short wavelengths.)

Another experiment carried out by G. P. Thomson in Great Britain in the same year supported this result. The diffraction pattern for an energetic electron beam passing through a thin metal film was obtained. This was compared to a transmission X-ray diffraction pattern (Fig. 27.20). The comparison of the patterns leaves little doubt about the wave-like properties of the electron particles. For an important application of the wave nature of particles, see Special Feature 27.2.

As a result of the dual natures of waves and particles, a new kind of physics based on a synthesis of wave and quantum ideas was born in the 1920's and 1930's. This new physics, called quantum mechanics, replaced the old mechanistic view that all things moved according to exact natural laws with a new concept of probability. According to this concept, all physical observations were to some degree uncertain. The idea of probability was quite different from previous scientific thought, which considered it possible to predict the exact locations of particles. In quantum mechanics the matter waves "guide" or "pilot" the particles and give an indication of the probability that a particle will be in a particular location. For example, the quantum-mechanics analysis of the hydrogen atom predicts that the electron will *most probably* be in the discrete orbits of the Bohr theory. The probability of finding the electron decreases on each side of the radii of the discrete orbits

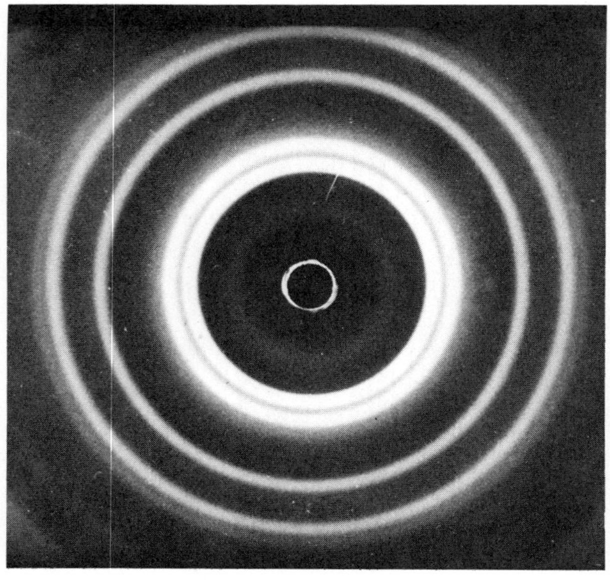

(a)

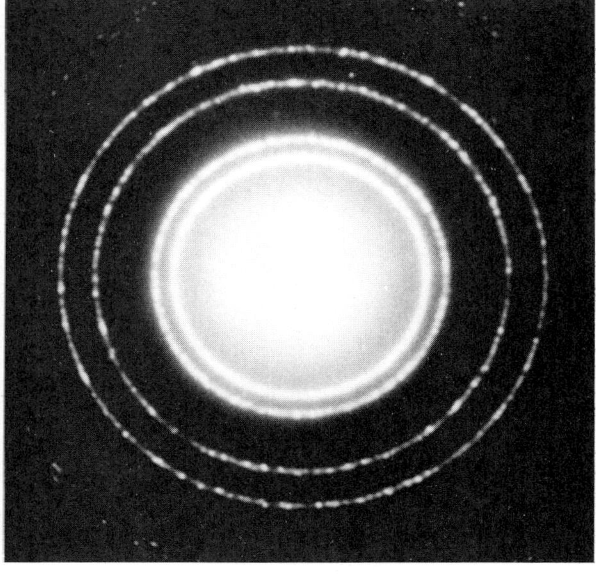

(b)

Figure 27.20 Diffraction patterns made by (a) X-rays passing through aluminum foil and (b) a beam of electrons passing through the same foil. (Center of pattern in X-ray diffraction is black because the center of the film is blocked off to prevent overexposing the film.)

and is very low in the intervening regions. This prediction gives rise to the idea of an "electron cloud" around the nucleus, where the cloud density reflects the probability that the electron is in that region.

SPECIAL FEATURE 27.2

The Electron Microscope

In some instances, electromagnetic waves have a particle-like nature — the dual nature of light. Conversely, it has been shown that particles exhibit a wave-like nature. For example, a beam of electrons can be diffracted, and the behavior is described by a wave relationship.

The "dual nature of particles" is applied in the electron microscope. A magnetic coil can focus an electron beam just as a glass lens can focus a light beam. Such magnetic "lenses" are used to focus the electron beam in an electron microscope in an arrangement similar to that of glass lenses in a light microscope (Fig. 27.21).

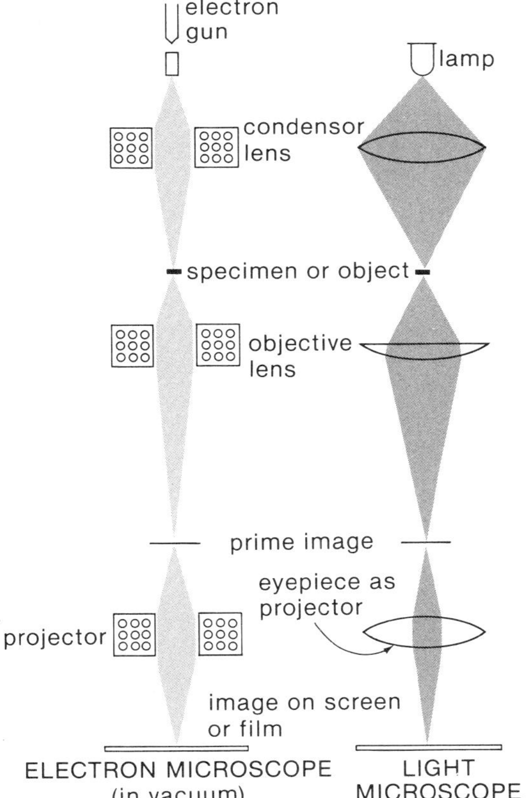

Figure 27.21 A comparison of electron and light microscopes. The electron microscope uses magnetic-coil "lenses," and the light microscope uses glass lenses.

The electron beam is directed on a very thin specimen. Different numbers of electrons pass through different parts of the specimen, depending on its structure. The transmitted beam is then brought to focus, forming an image on a film or a fluorescent screen.

Specimens that cannot be thin-sectioned can be viewed by other means. A specimen is coated with a thin layer of metal. The electron beam is then made to scan across the specimen. The surface irregularities cause variations in the intensity of the reflected electrons, and the reflected beam is focused to form an image. The metal coating makes the specimen electrically conducting. Otherwise, a nonuniform charge would build up on the specimen and cause the image to be distorted.

Both the transmission electron microscope and the scanning electron microscope are enclosed in a high-vacuum chamber so that the electrons are not deflected by air molecules. As a result, an electron microscope looks nothing like a light microscope (Fig. 27.22). However, magnifications up to 100,000 times can be achieved with an electron microscope, whereas a light microscope is limited to a magnification of about 2000 times.

A new microscope technique is being used to enhance the study of such things as living cells (Fig. 27.23). This method uses X-rays to form a replica of the cell in a photosensitive material called resist (Chapter 22). The replica is then observed in an electron microscope. Resist, which is used in fabricating electronic circuits, can record much smaller features than can the photographic films used for conventional X-ray pictures.

This method makes it possible to see cell features only 50 angstroms (50×10^{-10} m), which is about 1/100 the wavelength of visible light. The blood platelet shown in Figure 27.23 is about 3 micrometers (~ 1/10,000 inch) in diameter. The dark filaments are believed to be a kind of skeleton in the cell, probably formed of muscle filaments. This structure is of great interest to biologists and cannot be seen in conventional electron micrographs.

(Continued)

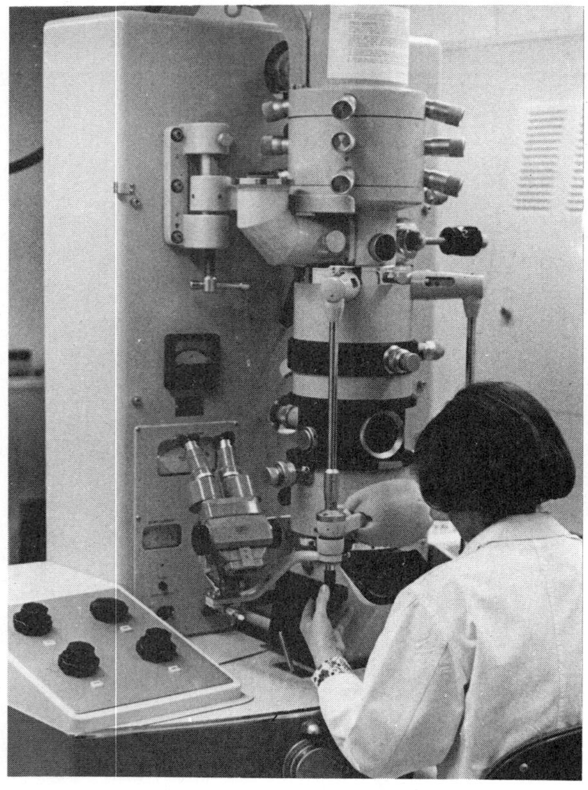

(a)

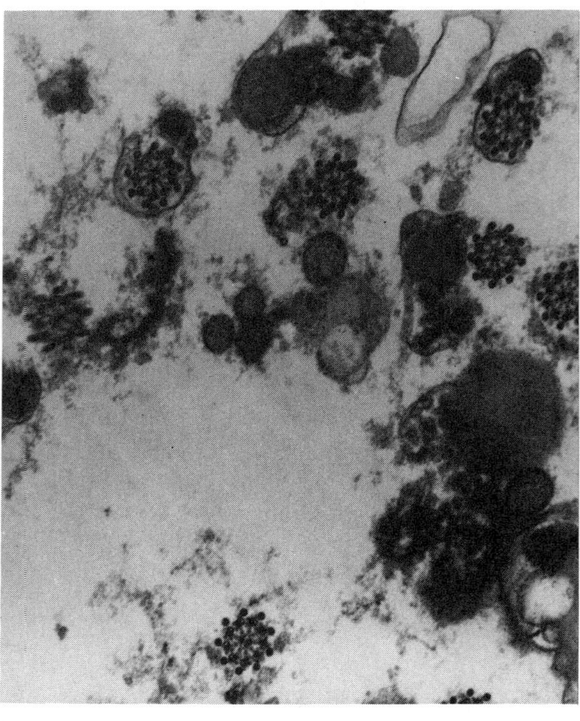

(b)

Figure 27.22 (a) An electron microscope. (b) An electron micrograph of biological tissue at 33,000× magnification.

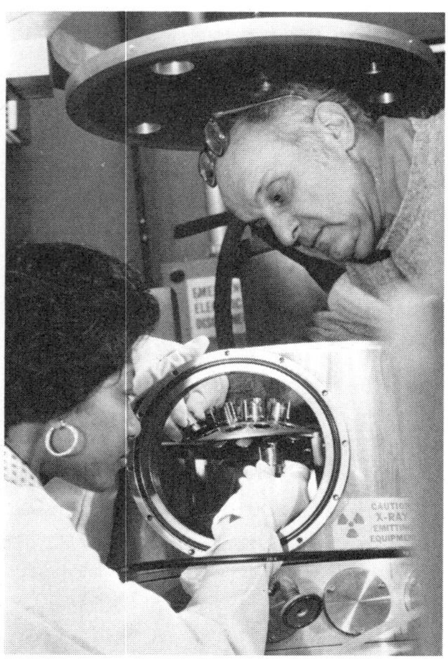

(a)

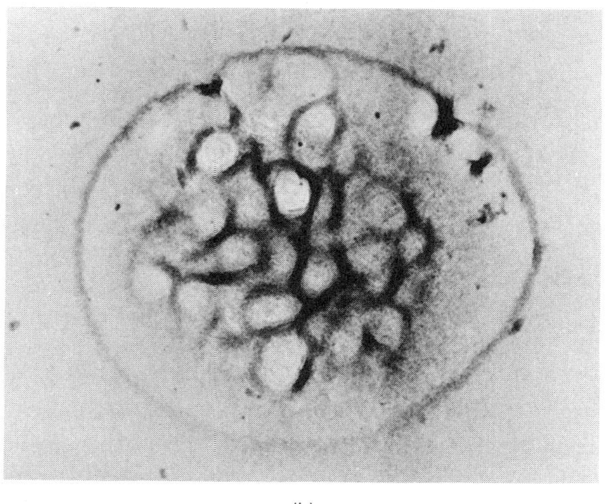

(b)

Figure 27.23 A new microscope technique. (a) X-rays are used to form a replica of an object in photosensitive material, and the replica is then observed with an electron microscope. (b) Electron micrograph of an X-ray replica of a blood platelet, which is only about 3 micrometers (3×10^{-6} m) in diameter.

SUMMARY OF KEY TERMS

Ultraviolet catastrophe the problem with the classical wave theory prediction for thermal radiation that the radiation intensity or energy at high frequencies (ultraviolet) should be infinitely large (catastrophe).

Planck's quantum hypothesis the oscillators of a thermal radiator could have only discrete energies given by $E = hf$, where h is Planck's constant and f is the frequency of the oscillator.

Photoelectric effect the emission of electrons from a material that is exposed to light.

Photon a packet or quantum of light with energy $E = hf$, where f is the frequency of the light.

Dual nature of light light or electromagnetic radiation sometimes behaves as a wave and sometimes as a quantum or "particle."

Bohr theory of the hydrogen atom the theory that quantizes the hydrogen electron in discrete orbits or energy levels. Photons are emitted when the electron makes transitions to lower energy levels.

Laser a light source that amplifies light through stimulated emisson (*l-a-s-e-r*, *l*ight *a*mplification by *s*timulated *e*mission of *r*adiation).

deBroglie hypothesis particles in motion have a wave nature, with their "matter waves" having a wavelength given by $\lambda = h/mv$.

Quantum mechanics the physics that uses deBroglie or "pilot" waves to predict the probability of a physical event.

EXERCISES

1. When a solid, such as an iron ball, is heated, why does it first glow with a dull-red color?
2. If a photoelectric material has a cutoff frequency, why does white light always give a photocurrent in a light meter?
3. A photographer checks out a modeling set that is illuminated with red light with a light meter and gets a zero reading. What is wrong? Is the meter broken?
4. In terms of the water bucket analogy of quantum absorption given in the chapter, explain why there is no photoemission for $f < f_o$.
5. Which has more energy, (a) a quantum of red light or a quantum of blue light? (b) infrared-radiation quanta or visible-light quanta?
6. Can you have a photon of white light? Explain.
7. A sunburn is caused by ultraviolet light. Why doesn't visible light cause a burn?
8. Explain how the photoelectric effect is used in the application of an electric "eye" counter (Fig. 27.24) or burglar alarm. (See Fig. 27.24 and note the light source and receiver on each side of the door.)
9. If electromagnetic radiation is made up of quanta, why don't we hear the radio intermittently with the arrival of discrete packets of energy?
10. We know that the Sun contains certain elements. How do we know this when we only receive light from the Sun?
11. Incandescent lamp filaments are made of tungsten. Does the light from these lamps exhibit a discrete tungsten spectrum? Explain.
12. Does the light from neon signs have a continuous spectrum? Explain.
13. An absorption spectrum is obtained by passing light through a cool gas. The gas atoms absorb light, but they also re-emit it, so why are there dark absorption lines?
14. Why is the electron in a Bohr orbit accelerated if it is traveling with a constant speed?

Figure 27.24 See Exercise 8.

15. Classically, why should an electron in a circular orbit emit radiation? Why was it necessary for Bohr to assume that a bound electron in an orbit did not emit radiation?
16. In which transition is the photon of greater energy emitted, $n = 3$ to $n = 1$ or $n = 2$ to $n = 1$?
17. A hydrogen electron is in the excited state $n = 4$. How many photons of different frequencies could possibly be emitted in returning to the ground state?
18. Is a satellite in circular orbit about the Earth like the hydrogen electron in circular orbit about the nuclear proton? Give differences and/or similarities.
19. Incoherent light is sometimes likened to a crowd of people. How would the people be moving in this case? Could a group of people be analogous to coherent light? How about a marching band? Explain.
20. In addition to stimulated emission in a laser, another condition for laser operation is "population inversion." That is, more atoms must be in an excited state than in the ground state. Why is this necessary?
21. Would a He-Ne laser operate if one end of the tube were covered with a transparent piece of glass? Explain.
22. Could holograms be made with an incandescent lamp? Explain.
23. Estimate your deBroglie wavelength when you are running. Recall that $h \approx 10^{-34}$ in SI units and that 1 lb is equivalent to about 0.5 kg (actually 0.45 kg). For the computation, estimate how fast you can run in m/s.
24. What is a basic difference between classical mechanics and quantum mechanics?

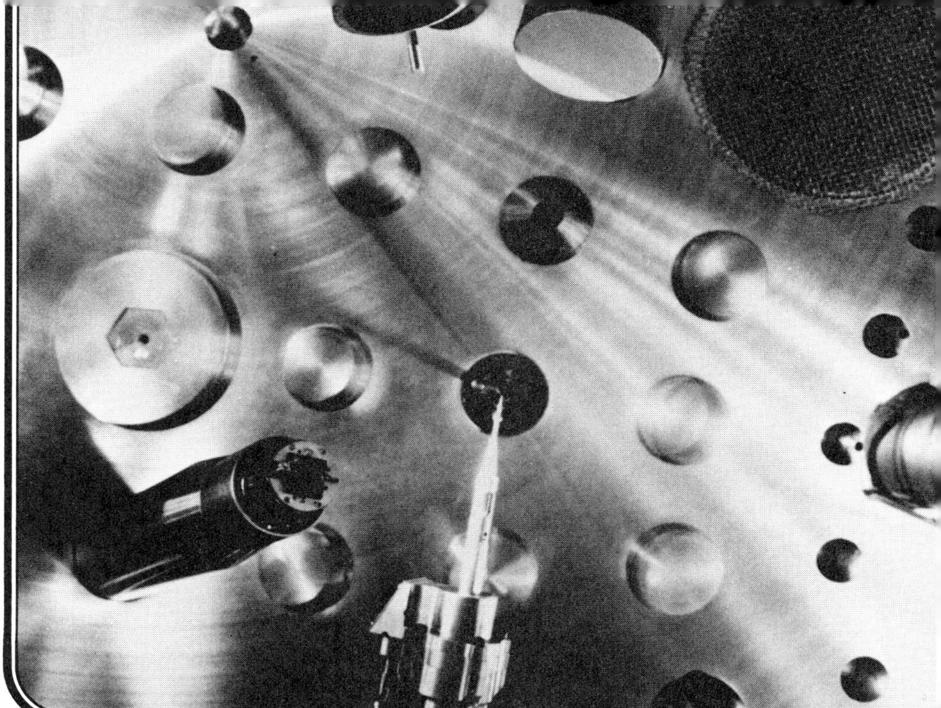

28

The Nucleus and Radioactivity

The Atomic Nucleus

Having looked at the orbiting "planet" electrons of the simplified planetary model of the atom, let's now look at the atomic "sun"—the nucleus of the atom. Another name for our model of the atom is the Rutherford-Bohr model. The Bohr contribution was discussed in the last chapter. The Rutherford part of the model is concerned with the central core or nucleus of the atom.

In the early 1900's the atom was thought to have a "plum-pudding" structure. This plum-pudding model viewed the electrons of the atom as being spread out like raisins in a sphere of positively charged "pudding." The pudding atom was thought to be about 10^{-8} cm across. But around 1910 there were experimental results that didn't agree with this model. The experiment was suggested by the British physicist Ernest Rutherford and was performed in his laboratory. Positively charged particles, called alpha particles, were directed toward a thin gold foil "target" (Fig. 28.1). These alpha particles are thousands of times more massive than electrons, and the plum-pudding model predicted that they would be deflected only slightly as a result of collisions with the electrons. However, the results were completely different. Some of the alpha particles were scattered through large angles, even scattered backward. As Rutherford put it, this was "almost as if you fired a 15-inch shell at a piece of tissue paper, and it came back and hit you."

In 1911 an explanation of the alpha-particle scattering was offered by Rutherford that gave a different view of the atom. If all the positive charges of an atom were concentrated in a central massive core or nucleus, then an alpha particle coming close to this region would experience a large deflecting force and would even be back-scattered in head-on collisions (Fig. 28.2). Theoretical calculations showed that the theory fit the data, and the atomic nucleus was "born." For his efforts,

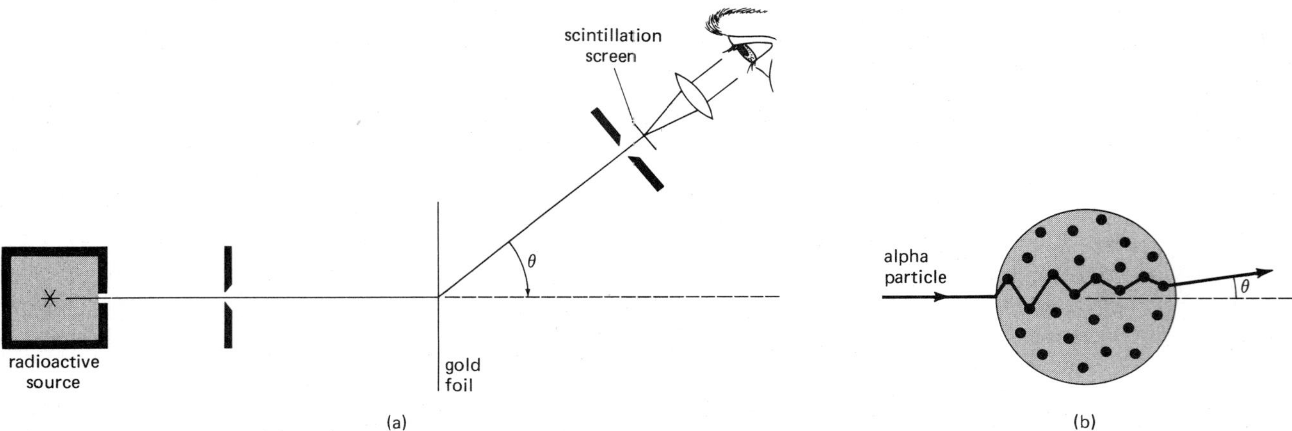

Figure 28.1 Rutherford's scattering experiment. (a) The experimental setup. (b) The plum-pudding model of the atom predicted that the alpha particles would be only slightly deflected as a result of collisions with electrons. The experimental results were different.

Rutherford was given a title (he became Lord Rutherford) and received a Nobel Prize.

The alpha scattering is an example of a "black box" experiment that is very common in nuclear physics. In general, you know what goes in, observe what comes out, and infer what happened in between. The atomic nucleus was Rutherford's "black box." Notice in Figure 28.2 that the back-scattering gives an upper limit on the nuclear size or radius. This is on the order of 10^{-14} m for a typical atom. The atomic electrons in the Rutherford-Bohr model are much farther out, on the order of 10^{-10} m (Fig. 28.3). Because of the small size of the atom, most people do not realize the extent of the "void of space" within the atom. The relative dimensions of the atom's structure have been likened to a large major-league football stadium with the nucleus being a small marble (on the 50-yard line). If the nucleus of a typical atom were expanded to the size of the Sun, then on a proportionate basis, the "inner planets" (Mercury and Venus) in Figure 28.3 would be well beyond the outer reaches of our solar system—many, many times beyond.

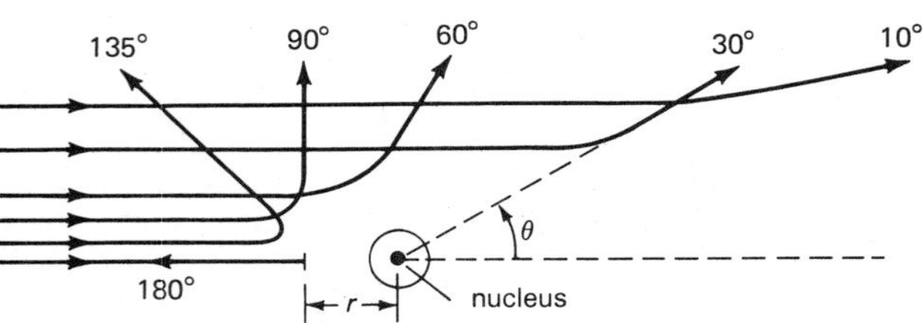

Figure 28.2 The nucleus. The scattering of the Rutherford experiment could be explained if the positive charges in the atom were concentrated in a central core or nucleus.

Nuclear Notation and Isotopes

Recall that it is the number of nuclear protons in an atom that determines what kind of atom it is (Chapter 9). The nucleus of a hydrogen atom is a single proton, helium atoms have two protons, lithium atoms have three protons, and so on. It was found that all of the nuclear mass could not be accounted for by protons. Rutherford suggested in 1920 that there may be another electrically neutral particle in the nucleus, which he called a **neutron.** The existence of the neutron was confirmed experimentally in 1932. We now know that all nuclei, with the exception of the common hydrogen nucleus (a proton), contain neutrons, which are electri-

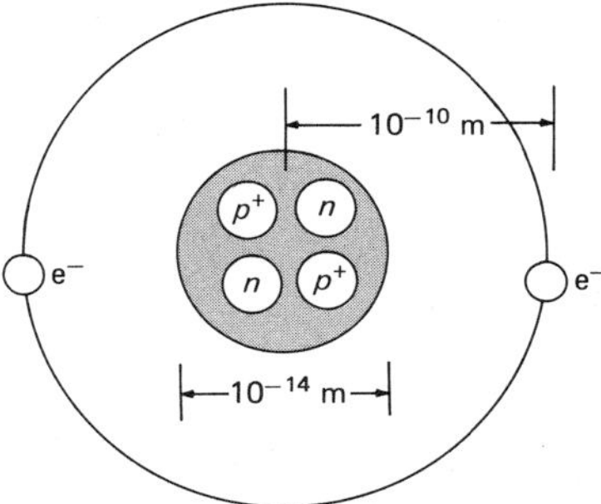

Figure 28.3 A schematic diagram of a helium atom showing atomic dimensions.

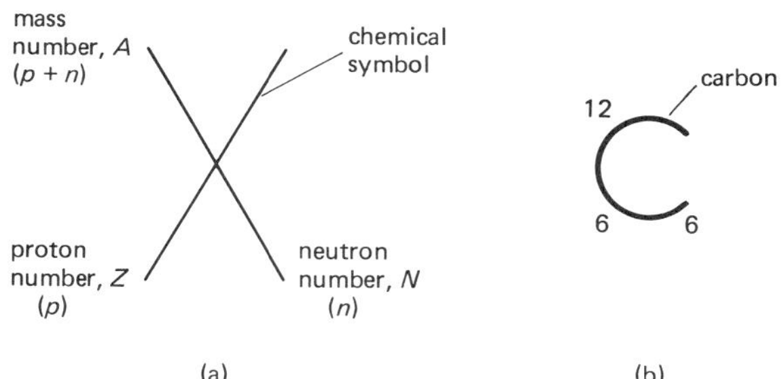

Figure 28.4 Nuclear notation for (a) a general case and (b) a carbon nucleus. The notation conveys that this carbon nuclide has six protons (as do all carbon nuclei) and six neutrons.

cally neutral particles with a mass about the same as that of a proton.

Protons and neutrons are about 2000 times more massive than electrons, so the vast majority (more than 99.9 percent) of the atomic mass lies in the nucleus. Nuclei are generally spherical, but they sometimes deviate from this shape and may resemble a watermelon or a door knob. Nuclear protons and neutrons are collectively referred to as **nucleons,** but we speak specifically of the proton number (sometimes called the atomic number) and the neutron number of a nucleus, which of course are just the numbers of protons and neutrons in a particular type of nucleus.

To designate a particular nucleus or nuclear species, which is called a **nuclide,** a special notation, shown in Figure 28.4 for a general case and a carbon nuclide, is used. Notice that the **proton number** is on the lower left of the chemical symbol of the element or atom. The proton number is also commonly called the atomic number and is designated by the letter Z. The fact that a nucleus has six protons makes it a carbon nucleus. The **neutron number** is on the lower right. Notice that if you add the number of protons and the number of neutrons $(6 + 6)$, you get the number (12) to the upper left of the symbol. This is called the **mass number** $(p + n)$. It is customary to leave off the neutron number and simply write $^{12}_{6}C$. This still gives you the same information. (Why?) In referring to this nucleus, we say that it is a carbon-12 nucleus.

If you examined a bunch of carbon nuclei (all with six protons) you would occasionally find a nucleus with more than six neutrons—sometimes seven neutrons, and on rarer occasions, eight neutrons. Nuclei or nuclides of the same element having different numbers of neutrons are called **isotopes.** Isotopes of hydrogen and carbon are illustrated in Figure 28.5. A group of iso-

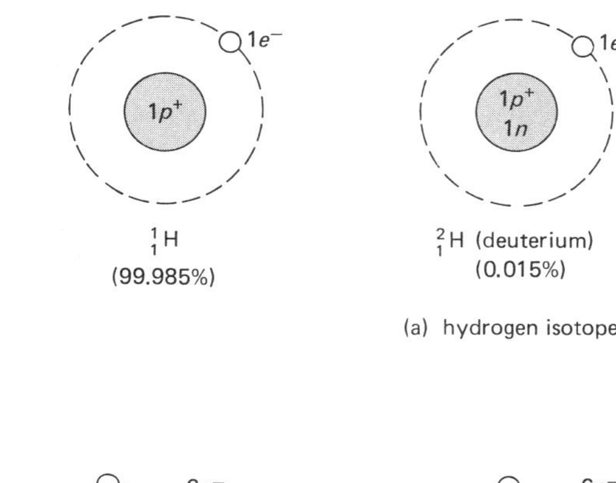

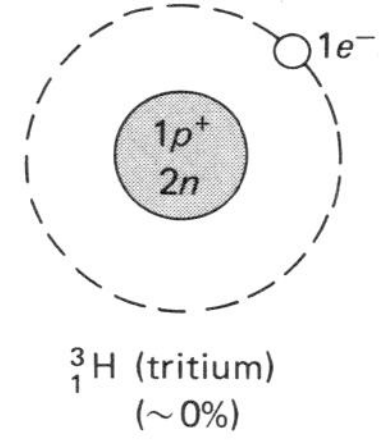

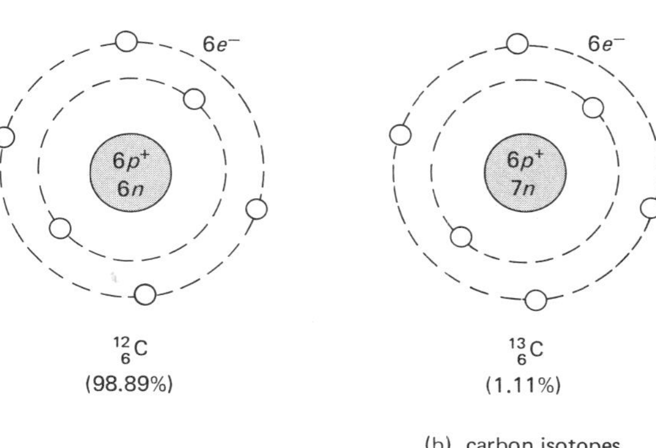

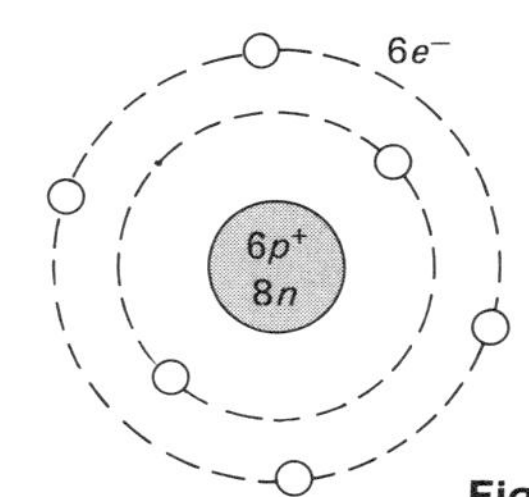

Figure 28.5 Isotopes. All hydrogen isotopes have one proton, and all carbon isotopes have six protons, but the neutron numbers are different for the isotopes in each family.

topes is somewhat like a family—they are all Joneses or Smiths (for example, hydrogen or carbon), but the individual family members are different. In the nuclear case, the "family" members differ by the number of neutrons they contain. Some elements have large nuclear isotope "families" with a dozen or more members.

Only the isotopes of hydrogen are given specific names. ^1_1H, the most common isotope, is just a proton. The other isotopes, ^2_1H and ^3_1H, are called deuterons and tritons, respectively, or deuterium and tritium in atomic form. When the more massive deuterium atom replaces the common hydrogen atom in H_2O, we have what is called "heavy water." For every 6500 or so atoms of ordinary (light) hydrogen in water, there is one atom of deuterium. The oceans contain millions of tons of deuterium. Tritium is radioactive. You will learn more about both these isotopes later in the discussion on nuclear energy. Other isotopes are referred to by their element-mass number designation, for example, carbon-12, carbon-13, and carbon-14.

The Nuclear Force

Within the small confines of the nucleus, the nucleons are clustered together, with each nucleon taking up about the same amount of space (Fig. 28.6). The nucleons are on the order of 10^{-15} m apart. This means that there are large repulsive electrical forces between the positively charged protons. What then holds the nucleus together? Since the nucleons are mass particles, the ever-present attractive gravitational force is there. However, calculations show that the gravitational force between two nucleons is a factor of 10^{-40} smaller than the electrical force between two protons. On a relative basis, with a factor of 10^{-40}, the gravitational force is so small that it is negligible, so that's out.

Obviously, there must be some other strongly attractive force acting in the nucleus that overcomes the repulsive electrical force. We call this attractive force acting between nucleons the nuclear force, or the strong interaction.* The **nuclear force** is a fundamental force like the electrical and gravitational forces, but it is more complicated and not completely understood. From experiments, we believe that this force acts between any pair of nucleons—proton-proton, neutron-neutron, and proton-neutron.

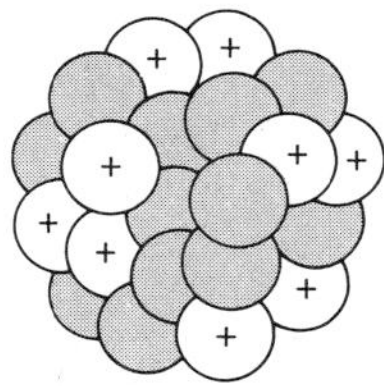

Figure 28.6 Nucleons are clustered together in the nucleus, with each nucleon taking up about the same amount of space.

The strong nuclear force is a short-range force. That is, it falls off very quickly with nucleon separation distance. Recall that the electrical force falls off as the inverse square of the distance ($1/r^2$). The strong nuclear interaction weakens much more quickly. For example, the repulsive electrical force between two protons on opposite sides of a sizable nucleus may be appreciable, but the attractive nuclear force is very small. An electrical force also exists between the nuclear protons and the orbiting atomic electrons (centripetal force), but the nuclear force does not extend outside the nucleus.

If the attractive nuclear and repulsive electrical forces acted only between protons, then there would be only small nuclei or atoms because of the short range of the nuclear force. However, there are very large stable nuclei, for example, the lead isotope $^{208}_{82}Pb$, with 82 protons and 126 neutrons. The key is the number of neutrons. Remember that the nuclear force acts between all nucleons, so the neutrons act as a nuclear "glue" to hold the nucleus together. As the number of protons increases for larger stable nuclei, the number of neutrons increases too, so that the short-range nuclear forces are greater than the long-range repulsive electrical forces. Neutrons provide attractive nuclear forces between *both* protons and other neutrons.

This effect is shown graphically in Figure 28.7. For lighter or less massive nuclides up to about calcium (proton number of 20), the stable nuclei have an equal or an approximately equal number of protons and neutrons. But for heavier nuclides, there are more neutrons than protons.

If you look at the magnitudes of the attractive nuclear forces between nucleons and the attractive electrical forces between nuclear protons and orbiting atomic electrons, the latter are about one-billionth as great. This gives a hint about why the nuclear energy of atoms is so much greater than chemical energy. In a chemical reaction, such as the burning or combustion of gasoline, there is a rearrangement of the electrons in the atoms or molecules. The nuclei of the atoms do not

* There is also a weak nuclear force of weak interaction, which will be discussed later.

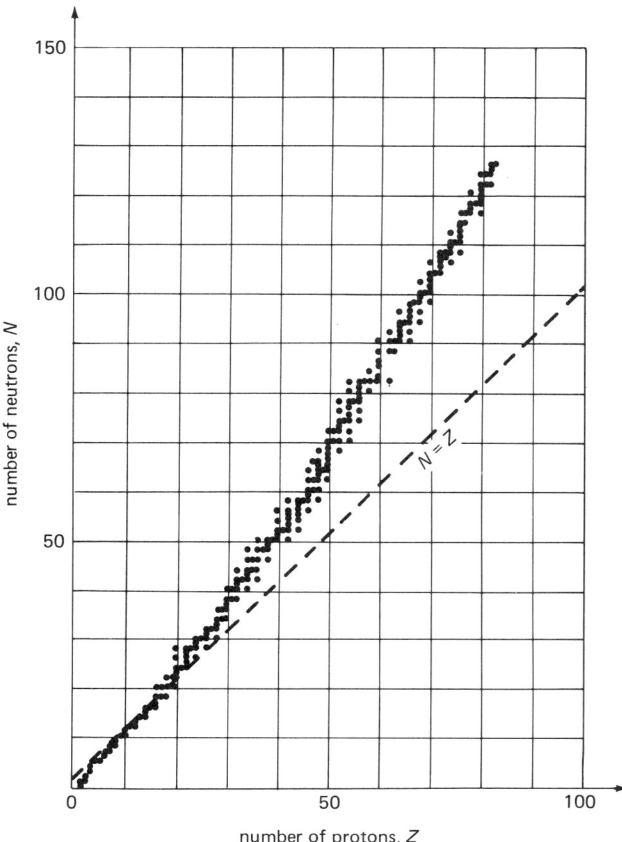

Figure 28.7 Neutron number versus proton number for nuclides. Nuclides with proton numbers greater than 20 have more neutrons than protons. The excess neutrons act as a nuclear "glue" that holds the nucleus together.

change. In a nuclear reaction, however, there is a "rearrangement" or change of nuclear particles. This involves strong forces and the release of large amounts of energy, as will be learned in the next chapter in the discussion on nuclear energy.

Radioactivity

There are more than 250 stable nuclei making up the graph in Figure 28.7. However, there are approximately 1400 known nuclides. The big difference reflects the known unstable nuclides. Unstable nuclei spontaneously "decay" with the emission of energetic particles and are said to be radioactive. The "radio–" part refers to the emitted radiation, which in the modern context may be a particle or a wave. So a radioactive nuclide or isotope is "active" in emitting radiation. Notice that the graph in Figure 28.7 terminates at proton number 83. There are no stable nuclei with proton numbers greater than 83.

Of the nearly 1200 known unstable nuclides, only a small number occur naturally. The other radioactive nuclei are made artificially. The unstable nuclides found in nature decay with the emission of alpha particles or beta particles, which are sometimes accompanied by gamma rays. Using a magnetic field, which deflects electrically charged particles, it was found that alpha "rays" were positively charged particles and beta "rays" were negatively charged particles (Fig. 28.8). Because they are deflected more, beta particles must be less massive than alpha particles. Gamma rays were not deflected at all, so they must be uncharged. Actually, a gamma ray or "particle" is a photon or quantum of energy.

Radioactivity was discovered accidentally by the French physicist Henri Becquerel in 1896. The circumstances were not unlike Roentgen's discovery of X-rays. Becquerel noticed that a photograhic plate in a light-tight wrapper that he had left in a drawer with a uranium compound showed signs of exposure when developed. Evidently, some type of radiation or "rays" from the uranium was able to penetrate the wrapper and expose the film. A couple of years later, the husband-and-wife team of Pierre and Marie Curie announced the discovery of two new radioactive elements, radium and polonium, which they had isolated

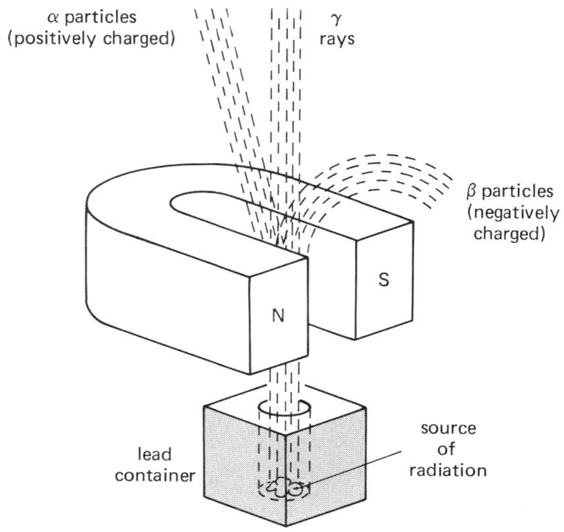

Figure 28.8 Radioactive decay particles. Passing radiation through a magnetic field shows there are positively charged alpha particles, negatively charged beta particles, and neutral gamma rays.

from uranium pitchblende ore. (See Special Feature 28.1 and Figure 28.9.) They had painstakingly isolated 10 mg of radium and a smaller amount of polonium from 8 tons of ore! The Curies and Becquerel shared the 1903 Nobel Prize in physics for their work in radioactivity. Madame Curie, as she is commonly known, also received the Nobel Prize in chemistry in 1911 for her contributions in chemically isolating radioactive materials. Carrying on the family tradition, the Curies' daughter, Irene Joliot-Curie, and her husband Frederic Joliot won the 1935 Nobel Prize in chemistry.

Let's take a closer look at the ABCs of radioactivity: alpha (α), beta (β) and gamma (γ) decays and their respective particles.

SPECIAL FEATURE 28.1

Marie and Pierre Curie

Marie Sklodowska (1867–1934) was born in Warsaw, Poland, and received her early scientific training from her father. She became involved in a students' revolutionary organization and found it advisable to leave Warsaw. In Paris she earned a science degree and in 1895 she married Pierre Curie, who was a physicist well known for his work on crystals and magnetism. One of his most important discoveries was that the magnetic properties of substances change at a certain temperature, the "Curie temperature."

After their marriage, Marie did doctoral research on radioactivity, and Pierre joined his wife in this work. In 1898 they announced their discovery of two new radioactive elements that they had isolated, radium and polonium (named after Marie Curie's native country). From eight tons of uranium pitchblende ore, they painstakingly isolated 10 mg of radium and a smaller amount of polonium. In 1903 they were awarded the Nobel Prize in physics for their discovery in radioactivity. The prize was shared with Henri Becquerel, who discovered the radioactive properties of uranium in 1896. Pierre Curie was killed in a horse-drawn vehicle accident in 1906. Marie Curie (commonly known as Madame Curie) was appointed to his professorship at the Sorbonne, the first woman to have this post.

In 1911 Mme. Curie was awarded the Nobel Prize in chemistry for her work on radium and the study of its properties. She was the first person to win two Nobel Prizes in science. The rest of her career was spent in establishing and supervising laboratories for research on radioactivity and the use of radium in the treatment of cancer. In 1921 President Harding, on behalf of the women of the United States, presented her with a gram of radium in recognition of her services to science.

Mme. Curie died in 1934 of leukemia, a form of blood cancer, which may have been caused by overexposure to radioactive substances. This occurred shortly before an event that would no doubt have made her very proud. In 1935 the Curies' daughter, Irene Joliot-Curie, and her husband, Frederic Joliot, were awarded the Nobel Prize in chemistry.

Figure 28.9 Marie and Pierre Curie.

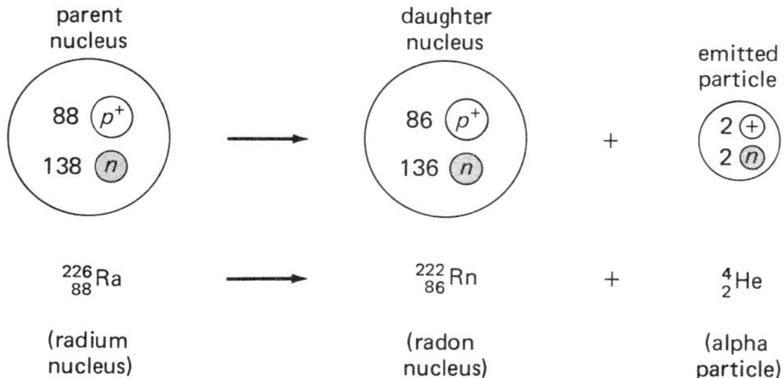

Figure 28.10 An example of alpha decay.

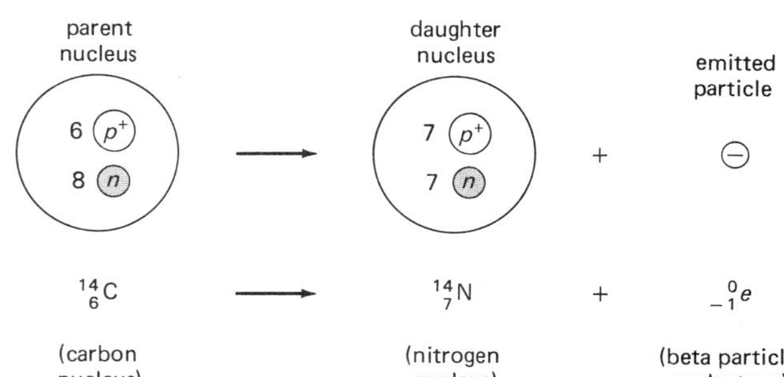

Figure 28.11 An example of beta decay.

ALPHA DECAY

An **alpha particle** consists of two protons and two neutrons. It is the same as a helium nucleus ($^4_2$He, sometimes written $^4_2\alpha$), with a 2+ electrical charge. When a nucleus undergoes alpha decay, an alpha particle is spontaneously emitted, and the nucleus is *transmuted* (changed) into the nucleus of another element, since two protons are lost. An example of alpha decay is given in Figure 28.10. The original nucleus in a decay process is commonly called the parent nucleus (radium-226 in this case) and the resulting nucleus is called the daughter nucleus (radon-222).

The decay process is written in a nuclear reaction equation (similar to a chemical reaction equation). Notice in this reaction, *as in all nuclear reactions,* that

a. the total number of nucleons remains constant (conservation of nucleons) and
b. the total charge remains constant (conservation of charge).

Condition (a) means that the sums of the superscript mass numbers on each side of the equation are equal ($226 = 222 + 4$). Condition (b) requires that the sums of the subscript proton numbers on each side of the equation are equal ($88 = 86 + 2$).

BETA DECAY

A **beta particle** is simply an electron. An example of beta decay is given in Figure 28.11. Notice in the nuclear reaction equation that the -1 subscript of the electron (its *charge* number) allows for the conservation of charge ($6 = 7 - 1$).

But where does an electron come from within the nucleus? Our nuclear model has only protons and neutrons in the nucleus. Looking at the proton and neutron numbers in the beta-decay reaction, we see that the daughter nucleus has one less neutron and one more proton than the parent nucleus. This indicates that a neutron must have been converted into a proton and an electron in the beta-decay process; i.e.,

$$^1_0n \rightarrow {}^1_1p + {}^0_{-1}e$$

Notice how a neutron is written in nuclear notation. (^1_1p, the proton, is equivalent to $^1_1$H. Why?)

A nuclear force distinct from the strong interaction is associated with beta decay. This is the weak nuclear force or weak interaction. All the particles we will study interact through the weak interaction, while some also interact electromagnetically and still fewer act through the strong interaction. For example, two nuclear protons interact by all four fundamental forces—strong, electromagnetic, weak, and gravitational interaction. The relative strengths of the fundamental interactions are given in Table 28.1. As in classical physics, we generally consider only the important forces and ignore the others or consider them negligible in a particular situation; for example, the strong and electromagnetic interactions dominate between nucleons. In the case of beta decay involving an electron, there is no strong force between the nucleons and an electron, so the weak interaction becomes important. The weak interaction is complex and is not well understood.

Table 28.1 Relative Strengths of Fundamental Interactions

Interaction	*Relative strength*	*Range*
Strong	1	Short range, about 10^{-14} m
Electromagnetic	10^{-2}	Infinite ($1/r^2$)
Weak	10^{-13}	Extremely short range
Gravitational	10^{-40}	Infinite ($1/r^2$)

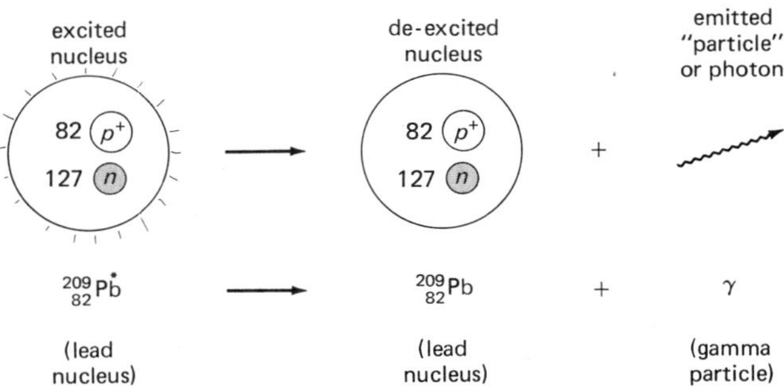

Figure 28.12 An example of gamma decay.

GAMMA DECAY

A **gamma "particle"** is a quantum of electromagnetic energy. The nucleus has energy levels similar to the atomic electron energy levels. In the de-excitation of a nucleus, a gamma ray is emitted (Fig. 28.12). The asterisk in the equation indicates that the nucleus is in an excited state. It decays with the emission of a gamma ray or particle. Gamma rays are photons, just as visible light and X-rays are photons or quanta of lower energies or frequencies ($E = hf$; see Fig. 27.11). Since it is a de-excitation process, there is no nuclear transmutation with gamma decay; that is, there is no change in the type of species of nucleus, only in its state.

Decay Rate and Half-Life

Radioactive isotopes decay at vastly different rates. However, the decay rates of the various isotopes are remarkably constant and are unaffected by environmental conditions such as temperature and pressure. In fact, we know of no way to change the rate of decay or "activity." Radioactive nuclei just "do their own thing" and decay at their own fixed rates.

The decay rate of a radioactive isotope is measured in terms of a characteristic time called the **half-life.**

> One half-life is the time it takes for one half of the nuclei of a sample of a given radioactive isotope to decay.

The radioactive decay curve has the form shown in Figure 28.13. Suppose we started with 1 gram of strontium-90 ($^{90}_{38}Sr$), which beta-decays. After 28 years, or one half-life ($t_{1/2}$), only half of the nuclei in the sample would be ^{90}Sr, or there would only be $\frac{1}{2}$ gram of ^{90}Sr. The other half of the sample or the other half of the nuclei would have decayed (into the daughter nucleus, $^{90}_{39}Y$). After another half-life or another 28 years (56 years total), half of the remaining half of the sample would have decayed and only $\frac{1}{4}$ gram of ^{90}S would be left, and so on. The half-lives of some radioactive nuclides are listed in Table 28.2.

Of course, in practice the half-life of a particular isotope is not determined by counting nuclei or weighing the sample. This is done by monitoring the activity of the sample. The activity of a radioactive sample is the number of nuclear decays per unit time. The result of radioactivity or nuclear decay is emitted particles, and these are what are counted. When the number of radioactive nuclei decreases by one half (in a time $t_{1/2}$), so does the number of emitted decay particles.

The common unit of radioactivity is named in honor of the Curies. One **curie** (Ci) is defined as follows:

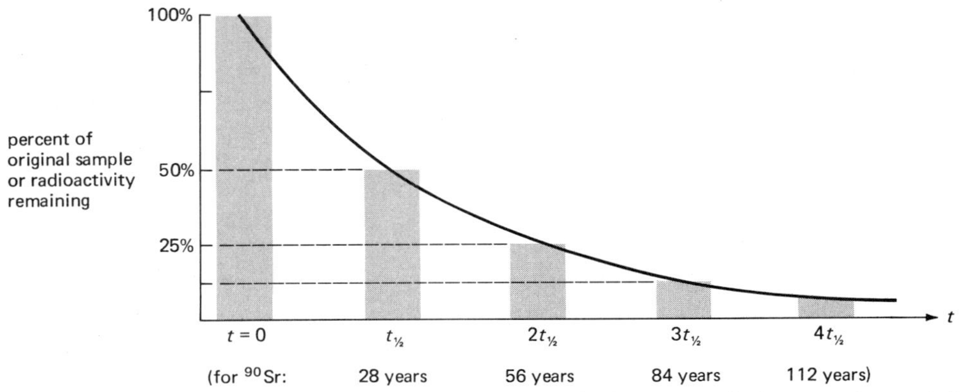

Figure 28.13 Radioactive decay curve. The percent of original sample decreases by one half during each half-life. The half-life ($t_{\frac{1}{2}}$) of strontium-90, used as an example, is 28 years.

Table 28.2 The Half-lives of Some Radioactive Nuclides

Nuclide		Decay mode	Half-life
Beryllium-8	($^{8}_{4}$Be)	α	1×10^{-16} second
Polonium-213	($^{213}_{84}$Po)	α	4×10^{-6} second
Carbon-16	($^{16}_{6}$C)	β	0.75 second
Aluminum-28	($^{28}_{13}$Al)	β	2.24 minutes
Magnesium-28	($^{28}_{12}$Mg)	β	21 hours
Iodine-131	($^{131}_{53}$I)	β	8 days
Cobalt-60	($^{60}_{27}$Co)	β	5.3 years
Strontium-90	($^{90}_{38}$Sr)	β	28 years
Radium-226	($^{226}_{88}$Ra)	α	1600 years
Carbon-14	($^{14}_{6}$C)	β	5730 years
Uranium-238	($^{238}_{92}$U)	α	4.5×10^{9} years
Rubidium-87	($^{87}_{37}$Rb)	β	4.7×10^{10} years

$$1 \text{ Ci} = 3.7 \times 10^{10} \text{ decays/sec}$$

The curie unit was originally defined as the activity of 1 gram of radium-226. It is rather a large unit, so millicuries (mCi) and microcuries (μCi) are commonly used. The SI unit of radioactivity is the **becquerel** (Bq), which is simply defined as the number of decays per second,or

$$1 \text{ Bq} = 1 \text{ decay/sec}$$

Although it is more convenient, the becquerel unit has not gained widespread use.

So a radioactive material continually decays. The activity of a sample may become so small that it is considered to be negligible; the radioisotope has "decayed away" (into stable nuclei). For a radioactive material to be found in nature, it must have a half-life that is not much smaller than the age of the Earth ($\sim$4.5 billion years), or it must be continually produced by the decay of another radioactive isotope—for example, as in the decay series shown in Figure 28.14—or it must be continually produced by some other means, such as carbon-14 decay, as discussed in Special Feature 28.2.

Uranium-238 has a half-life about equal to the age of the Earth, so it is still around or found in nature, as are the products of the decay series. There are only 25 unstable naturally occurring nuclides that are not products of the decay chains of heavy nuclides. Notice how some nuclides decay by two modes. The series terminates with a stable lead isotope.

It is believed that when the universe was formed, all nuclides, both stable and unstable, were formed in varying amounts. The unstable nuclei with half-lives much less than a billion years have long since decayed away into stable nuclei. Recall from Chapter 9 that the elements technetium and promethium are not found in nature at all because these elements have short half-lives and no stable isotopes. We will discuss how artificial isotopes are made in the next chapter.

Radiation Detectors

Since we cannot generally sense the particles of nuclear radiations when they are emitted from decaying nuclei, they must "do" something before we know they are there. For example, the detection of nuclear radiations by photographic films has been mentioned. Nuclear radiations and X-rays expose various film materials, and persons working with radiation often wear "film badges." The amount of exposure is indicated by the degree of darkening of the film after it has been developed. This is fine, but you would like to be able to detect radiations immediately, along with getting a count of the emissions or decays.

Energetic alpha, beta, and gamma rays from nuclear decay processes have the ability to ionize neutral atoms. As a result, they are commonly referred to as ionizing radiations. For example, when an energetic charged alpha or beta particle passes through a gas, it interacts with gas molecules and rips some of the loosely bound outer electrons from the gas atoms. Although some of the freed electrons might recombine or attach themselves to other atoms or molecules, most will remain free for a while. This ionizing property of nuclear radiations forms the basis of a commonly used detection instrument, the Geiger counter. Your instructor probably has one of these detectors and will demonstrate it for you. It was developed chiefly by Hans Geiger, a student and colleague of Rutherford's.

The basic principle of the Geiger counter is illustrated in Figure 28.16. A voltage is applied across the Geiger tube, which contains argon gas at a low pressure (about 5 percent of normal atmospheric pressure). When an ionizing particle enters the tube through a thin window and "strikes" one or more argon atoms, electrons are ejected. These electrons are accelerated by the applied voltage and produce further ionizations. As a result, an "avalanche" current is set up momentarily in the tube. This current pulse is amplified and counted by the counter circuitry. Hence, radiation particles are detected and counted by counting the current pulses.

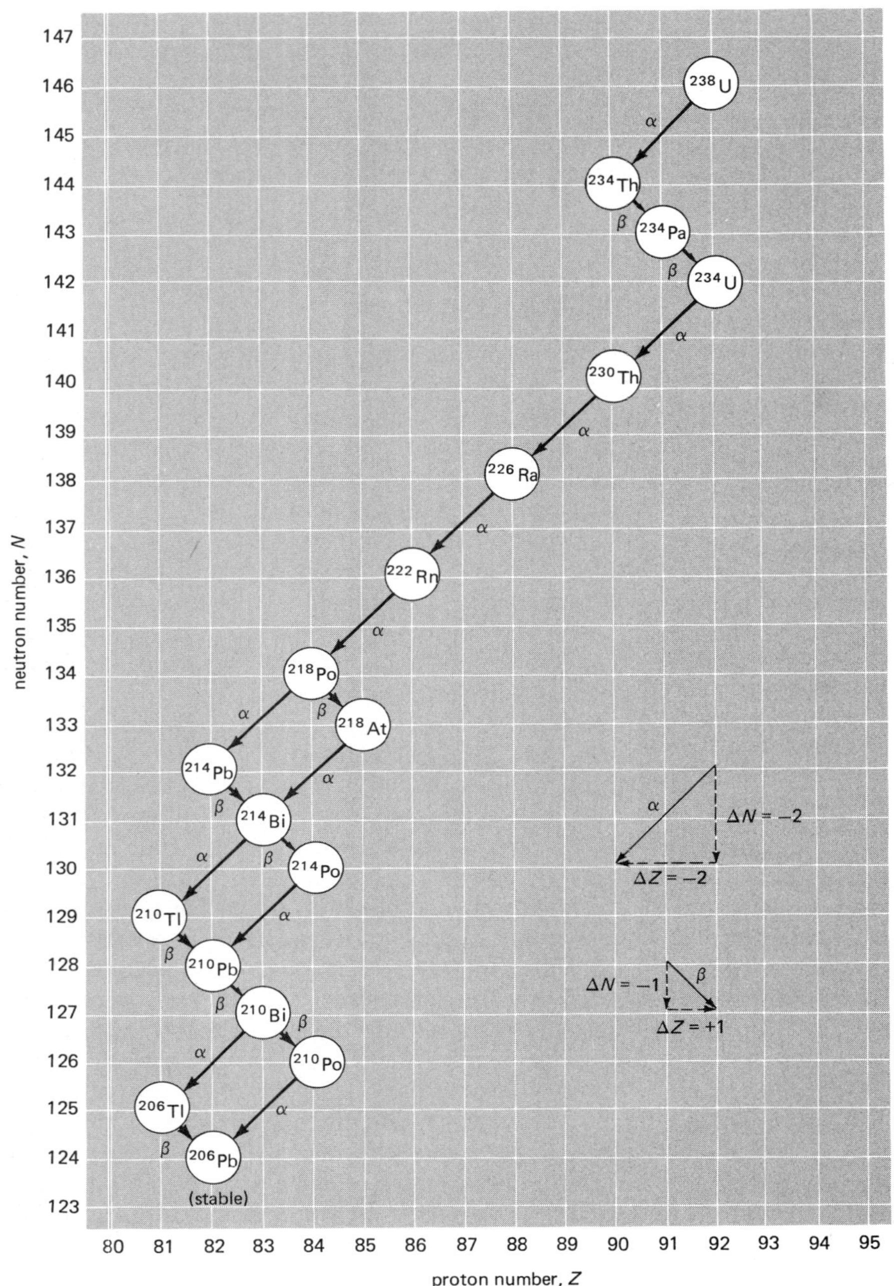

Figure 28.14 Radioactive decay series (uranium). A series of alpha and beta decays leads to a stable nuclide.

One of the main disadvantages of the Geiger counter is its "dead time." Dead time refers to the time it takes for the tube to recover or clear itself between counts. If a lot of particles enter the Geiger tube at a rapid rate, the tube will not have time to recover, and some particles will not be counted. Another method of counting gives a shorter recovery time. This is used in the scintillation counter. Certain phosphor materials have the ability to convert the energy of a particle into visible light. For example, if an alpha particle falls on a screen coated with zinc sulfide, it produces a tiny flash of light that is bright enough to be seen in the dark with the aid

SPECIAL FEATURE 28.2

Radioactive Dating

Because of their fixed decay rates, radioactive isotopes can be used as nuclear "clocks." For example, suppose you had half a gram of ^{90}Sr (Fig. 28.13). Then you know that 28 years (one half-life) have "ticked off" on the clock since there was 1 gram of ^{90}Sr. Similarly, if you had $\frac{1}{4}$ gram, you would know that it was $2t_{1/2}$ (two half-lives) or $2 \times 28 = 56$ years past the 1-gram mark on your nuclear "clock." It is this radioactive "ticking" and counting backward that forms the basis of radioactive dating, which allows scientists to determine how old various objects are.

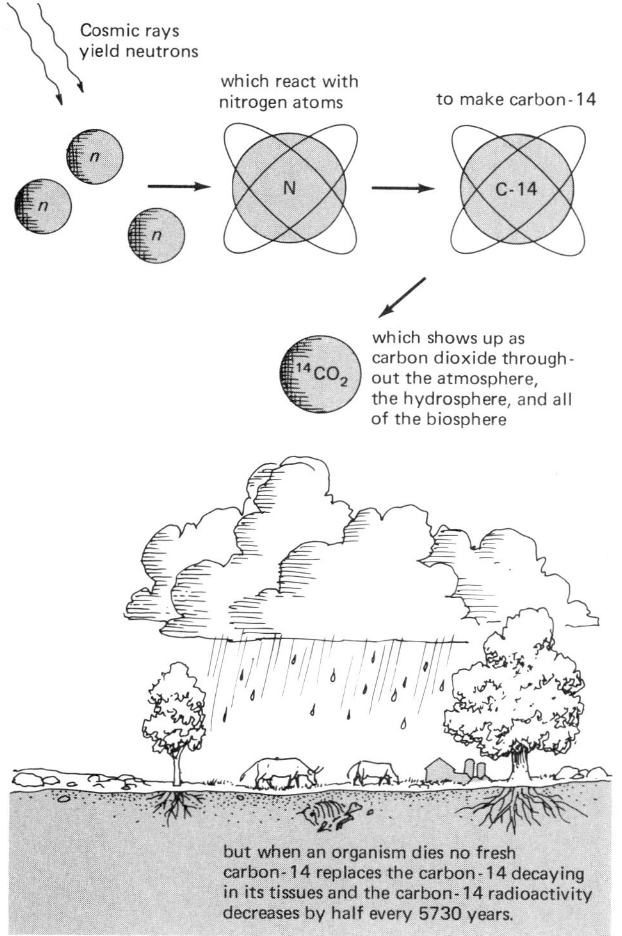

Figure 28.15 Carbon-14 in the environment.

An important radioactive-dating process uses the radioisotope carbon-14. Carbon-14 is a natural part of the environment (Fig. 28.15). It is formed in the upper atmosphere as a result of cosmic ray interactions. (Cosmic rays are particles from outer space, mostly protons.) The carbon-14 enters the biosphere, and all living things—plants and animals—obtain a certain amount of carbon-14. A radioactive isotope behaves chemically just like a stable isotope, and part of the carbon in your body is radioactive carbon-14. The activity of carbon-14 in a living organism remains fairly constant because of life processes. In each gram of carbon in a body, there is enough carbon-14 so that there are about 16 beta emissions per minute (per gram of carbon). Recall from Figure 28.11 that carbon-14 beta-decays. However, when the organism dies and the carbon-14 is not replenished by the intake in life processes, the activity decreases.

Suppose an old bone found in an archeological "dig" has a carbon-14 activity of 4 beta emissions per minute per gram of carbon. When the bone was a part of a living organism, there were 16 beta emissions per minute. So, counting backward in time ($4 \rightarrow 8 \rightarrow 16$), we know that the carbon-14 in the bone has gone through two half-lives. Carbon-14 has a half-life ($t_{1/2}$) of 5730 years (Table 28.2), and $2t_{1/2} = 2(5730) = 11{,}460$ years. Hence, the bone is about 11,500 years old.

Carbon-14 dating can be used to date objects up to about 30,000 years old. Other radioactive isotopes with longer half-lives can be used to date older objects. For example, ^{48}K and ^{87}Rb are useful in determining the age of geological rock formations.

of a magnifying lens. This was the detection device used in the Rutherford scattering experiment. In this experiment, the counting was done visually, which was slow and tiring. It has been said that whenever Rutherford himself did the actual counting, he would "damn" vigorously for a few minutes and then have one of his assistants take over. Overall, more than a million scintillations or flashes were counted in the experiment.

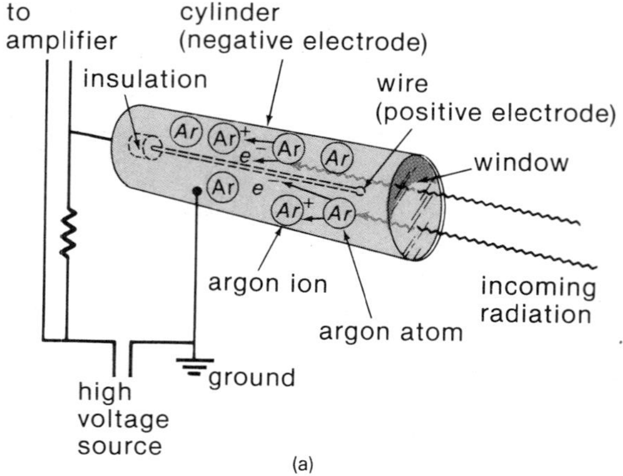

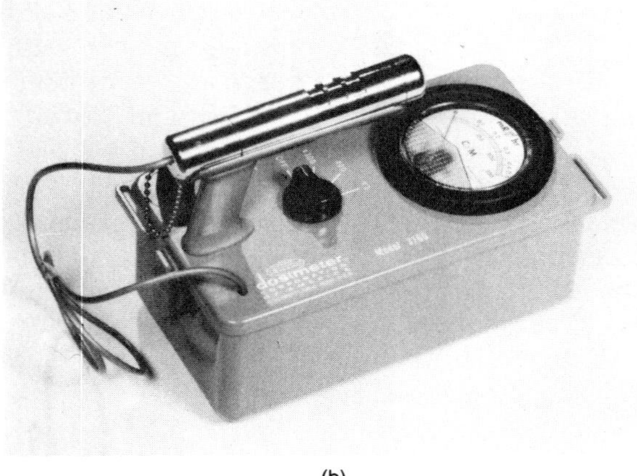

Figure 28.16 Geiger counter. (a) Radiation ionizes the gas molecules in the tube, giving rise to an electrical pulse, which is counted. (b) A portable (battery-operated) Geiger counter.

This laborious task was eliminated with the development and application of the photomultiplier tube (Fig. 28.17). Photons from the scintillator strike a photoelectric material, and photoelectrons are emitted. The photocurrent is amplified in a photomultiplier tube, which consists of a series of electrodes, called "dynodes," at successively higher potentials. The accelerated photoelectrons cause secondary electrons to be emitted when they strike a dynode, and so the current is multiplied. In this manner, weak scintillations are converted into sizable electrical pulses that can be counted electronically. Also, since the energy of the incident particle and the photon is proportionately amplified in the photomultiplier tube, the resulting output pulse gives a measure of the energy of the incident particle.

With the development of semiconducting materials, "solid-state" detectors are now available that can measure the energy of radiation very precisely and are capable of very high counting rates—very "fast" detectors (very small dead times). Charged particles passing through a semiconductor produce electron-hole pairs (solid-state "ionization"). With an applied voltage, these electron-hole pairs give rise to electrical pulses that can be amplified and counted.

Other types of detectors include cloud and bubble chambers. These are used primarily in research applications. Vapors and liquids are used in the respective chambers. When the temperature and pressure are properly adjusted, vapor condenses on the ionized molecules along the path of an energetic particle, leaving a trail of liquid droplets in a cloud chamber, and boiling or bubble formation is enhanced in the region of charged particles, leaving a trail of bubbles in a bubble chamber. Photographs may be taken of the trails in these chambers, which provide a permanent record of the particle's path and the "events" that take place along the way (Fig. 28.18). When a magnetic field is applied across a chamber, electrically charged particles are deflected, and their energies can be calculated from the radius of curvature of the particle's path, as measured from the photograph. Nuclear research has found

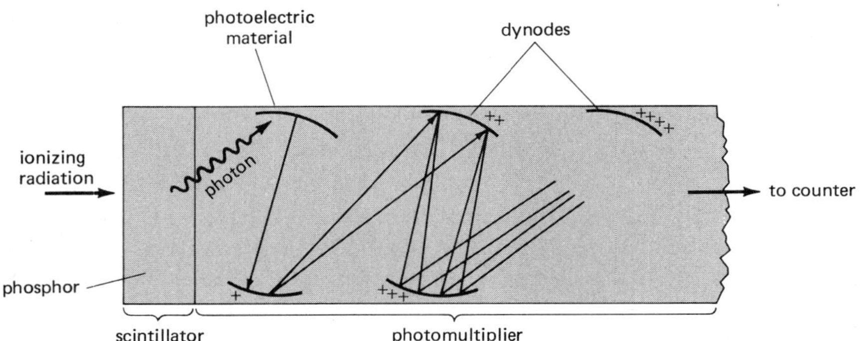

Figure 28.17 Photomultiplier tube. The photocurrent is amplified or multiplied at successive dynodes.

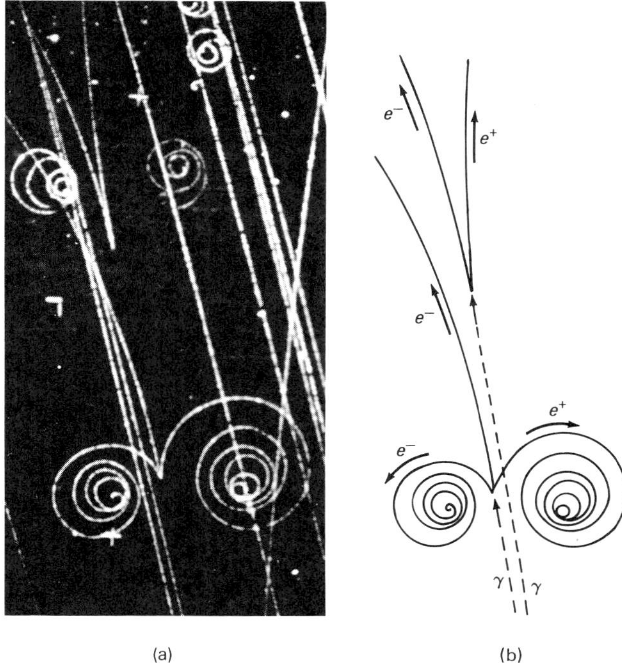

Figure 28.18 Bubble chamber photograph. (a) Particles leave a trail of bubbles in their wake. (b) A magnetic field causes the opposite deflection of oppositely charged particles.

a variety of particles in the nucleus, called elementary particles. More about these shortly.

Applications of Radioactivity

Since the discovery of radioactivity less than 100 years ago, nuclear decay has been used in a variety of applications. Let's consider a few medical and industrial applications here and the application to the production of nuclear energy in the next chapter.

The fact that radioactive isotopes behave chemically just like nonradioactive isotopes of the same element makes them useful as radioactive tracers. For example, an atom with a carbon-14 nucleus will react to form a molecule of carbon dioxide (CO_2) the same way as a nonradioactive atom of carbon-12 would. Radioactively "tagged" molecules can be traced in various processes by detecting the nuclear radiations from these molecules. For example, the location of an underground pipe or a clog can be determined by using a liquid radioactive tracer, and research in the wearing of machine parts can be carried out by tracing processes (Fig. 28.19).

Similarly, in medicine radioactive tracers allow the monitoring of body functions. For example, the iodine uptake of the thyroid gland can be measured by using radioactive iodine-131. The minimum daily requirement of iodine is 150 micrograms. If a patient is given some radioactive iodine, the thyroid gland doesn't know the difference, and a scan of the gland shows how the radioactive iodine is distributed and concentrated in the gland (Fig. 28.19).† The distribution is altered if there is some abnormality in the gland function. Various other internal parts of the body can be "observed" using radioactive tracers. A detector is used to scan the area in question. The information is stored in a computer and later put together as a scan "picture."

The penetration of radiation depends on the density of a particular material and is inversely proportional to the thickness of the material—the greater the thickness, the less penetration. This follows a given relationship, so radiation absorption can be used in thickness measurement and monitoring, for example, in metal sheet and foil manufacturing (Fig. 28.20). Another common application is discussed in Special Feature 28.3.

Elementary Particles

The search for the ultimate or *fundamental* "building blocks" of nature is as old as physics itself. The simple picture of an indivisible atom gave way to an atom with subatomic particles. By the 1930's scientists had identified four of these: the electron, the proton, the neutron, and the photon. Since then, more than 200 various particles have been discovered coming from the nucleus. They are called "elementary" particles, elementary because we are not certain which of them are really fundamental. You may have heard of some of these—neutrinos, positrons, muons, mesons, pions, lambda particles, and others.

It is believed that most elementary particles do not exist outside the nucleus, but come into being when the nucleus is disrupted. Some elementary particles are thought to be the "exchange" particles responsible for the four fundamental forces or interactions. For example, it is thought that the "exchange" particle for the strong nuclear interaction is the pi meson or pion. Nuclear particles with a strong interaction are viewed as interacting by exchanging pions back and forth (the

† Iodine-131 beta-decays into xenon-131, which emits gamma rays. It is the gamma radiation that is monitored in a thyroid scan.

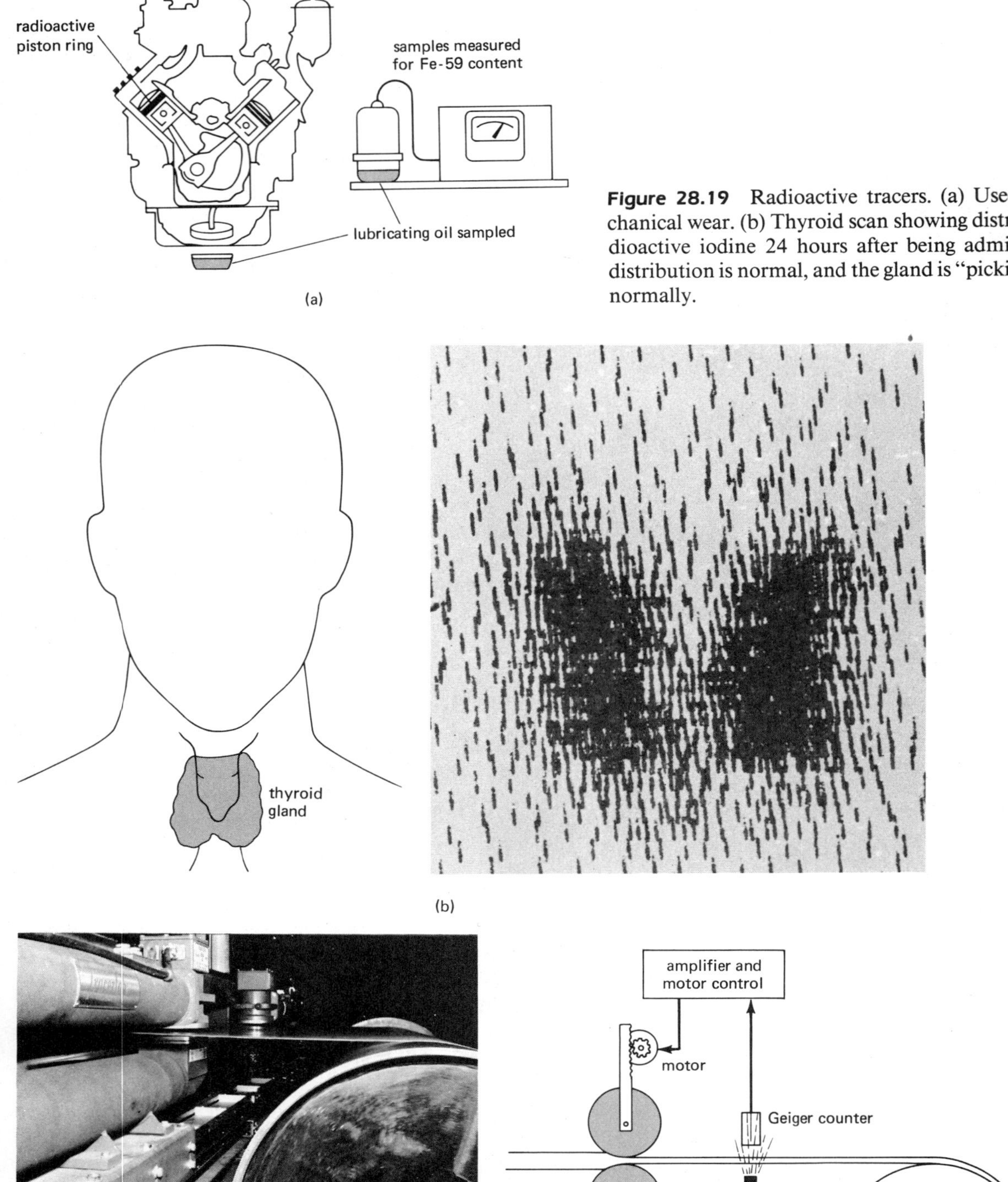

Figure 28.19 Radioactive tracers. (a) Used to test mechanical wear. (b) Thyroid scan showing distribution of radioactive iodine 24 hours after being administered. The distribution is normal, and the gland is "picking up" iodine normally.

Figure 28.20 Thickness measurement. The penetration of radiation depends on the density and hence the thickness, of a material. As a result, radioactive particles can be used in thickness monitoring and control.

SPECIAL FEATURE 28.3

The Smoke Alarm

Smoke alarms are common in many homes. The early warning of smoke is important in personal and fire protection. Smoke may be detected by several means. A photoelectric type of smoke detector is described in Chapter 27. However, more common is the ionization type of smoke detector, which uses radioactivity (Fig. 28.21). A weak radioactive source ionizes the air, setting up a small current in a detector circuit. If smoke is present, the ions become attached to the smoke particles. The slower movement of the heavier, charged smoke particles causes a reduction in the detector current. This is electronically sensed and causes an alarm to sound.

Such smoke detectors are quite sensitive. Smoke vapors from cooking (from burnt foods) often set them off.

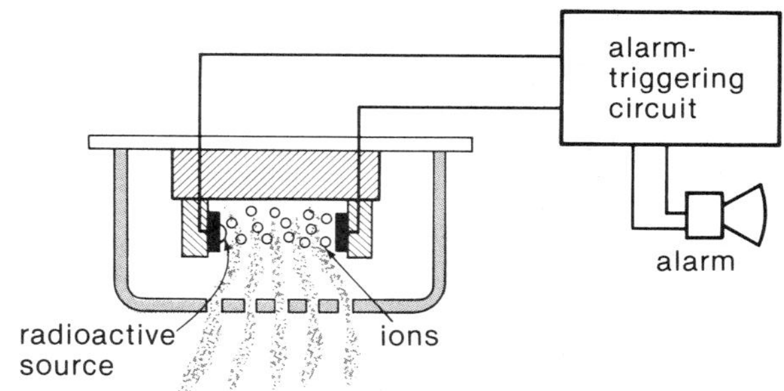

Figure 28.21 Smoke detection. In an ionization-type smoke detector, a weak radioactive source ionizes the air and sets up a small current. Smoke affects the current and causes an alarm to sound.

way you interact with someone by exchanging a beach ball back and forth). All strongly interacting particles, which include the nuclear protons and neutrons, are called hadrons.

Similarly, it is thought that the long-range electromagnetic action-at-a-distance force is due to the exchange of a special kind of photon between charged particles. The theoretical exchange particle for the weak interaction is the W-particle, and for the gravitational interaction it is the graviton. Scientists hope someday to simplify things in a unified field theory that would unite all four fundamental forces and show them to be aspects of the same force.

There have been efforts to simplify matters. In 1964 Murray Gell-Mann at the California Institute of Technology and George Zweig in Switzerland independently suggested that all hadrons, the elementary particles with the strong interaction, were made up of no more than three subparticles with fractional electronic charges, called quarks.‡ For example, a proton is thought to be made up of three quarks (two with a $+\frac{2}{3}$ electronic charge and one with a $-\frac{1}{3}$ electronic charge). Quarks apparently do not exist as free particles but are permanently bound inside the nucleus.§ There is, however, indirect experimental evidence for their existence. The theoretical exchange particle between quarks is the gluon — the "glue" that holds the world together.

At first there were only two types, or "flavors," of quarks, the u and d (for up and down, associated with spins), and their antiparticles or antiquarks. Then a new quark flavor, the strange quark, denoted by the letter s, joined the club. The discovery of a new elementary particle in 1974 added the necessity of a fourth flavor, a "charmed" or c quark. Then, to make the physics agree with experimental observations, it was assumed that each quark comes in three different "editions," which are called colors, for example, red, green, and blue u and d quarks. In the exchange of gluons between quarks, it seems that a gluon can change the color of a quark, or is "color-carrying."

The story does not stop here. The discovery of still another elementary particle in 1978 gave evidence for a new flavor, the b quark. To round out the quark symmetry (u-d, s-c), the b quark is associated with a t quark. There was some effort to call the b-t quark pair *b*eauty

‡ From a line in James Joyce's novel *Finnegans Wake,* which reads, "Three quarks for Muster Mark!" The "three quarks" denote the children of Mr. Finn, who sometimes appears as Mister (Muster) Mark. A bit strange, but there are also "strange" quarks. Read on.

§ At the time of this writing, there is one report that claims to have found fractional charges experimentally, which would indicate the discovery of free or at least unbalanced quarks. However, attempts to confirm this by other experiments have failed.

and *t*ruth, but this seems to have lost out to *b*ottom and *t*op. So we might say that quarks come in six flavors, each of which can have three colors.

The preceding description of quarks was not meant to be facetious, but rather to simply describe the enormous difficulties in trying to understand the intricacies of nature and the enormous strides scientists have made. Scientists apply the scientific method, and nature will be the final judge of their theories and understanding.

SUMMARY OF KEY TERMS

Atomic nucleus the central core of the atom in which the protons and neutrons of an atom are located.

Nucleon a nuclear proton or neutron.

Nuclide a particular nucleus or nuclear species.

Proton number the number of protons in a nucleus, which defines its atoms as being a particular element.

Mass number the sum of the protons and neutrons in a nucleus.

Isotopes nuclei or nuclides of the same element with different numbers of neutrons.

Nuclear force the strong attractive interaction that acts between nucleons.

Radioactivity the spontaneous decay of certain isotopes with the emission of energetic particles.

Alpha particle a particle consisting of two protons and two neutrons, which is the same as a helium nucleus.

Beta particle an electron.

Gamma particle a quantum or photon of energy.

Half-life the time it takes for one half of the nuclei of a sample of a given radioactive isotope to decay.

Geiger counter a common radiation detector based on the ionizing nature of radiation.

Dead time the time required for a detector to recover for another detection or count.

Scintillation counter a radiation detector based on the ability of a phosphor material to convert the energy of a particle into visible light.

EXERCISES

1. Tin ($_{50}Sn$) has 23 isotopes. Write the complete nuclear symbol (with neutron number) for the isotopes of tin with mass numbers from 114 to 120. (The complete family of isotopes have mass numbers from 108 to 130.)
2. Speculate about why the Curies named one of the new elements they discovered in 1898 "radium." (If you're wondering about polonium, it was named after Madame Curie's native country. See Special Feature 28.1.)
3. (a) Write the nuclear notation for the isotopes of hydrogen using the symbols H, D, and T. (b) Using the H, D, and T symbols, it is possible to write six chemical formulas for water. Write these, and identify which three would be radioactive forms of water.
4. Discuss how the gravitational and electrical forces might contribute to the stability or instability of an atom, taking into account nucleons *and* orbital electrons.
5. Given only the proton numbers or neutron numbers or mass numbers of several nuclei, would you *always* be able to distinguish between different nuclides? Explain.
6. Explain the role of neutrons in the nucleus in increasing nuclear stability.
7. Considering the following nuclides, show that the neutron number generally exceeds the proton number for nuclei with proton numbers greater than 20: lithium-6, bromine-80, silicon-28, titanium-48, fluorine-19, and platinum-179. (*Hint:* see Table 9.1 or 9.2.)
8. Write the complete nuclear symbol for the heaviest (most massive) stable nucleus. Speculate why heavier nuclei are not stable in terms of the forces acting in the nucleus. (*Hint:* Think about the range of the forces.)
9. Suppose an electric field were used in Figure 28.8 instead of a magnetic field to distinguish between nuclear radiations. What would be observed?
10. Write the nuclear equations for (a) the alpha decay of $^{226}_{88}Ra$, (b) the beta decay of $^{60}_{27}Co$, (c) the gamma decay of $^{210}_{84}Po$.
11. Complete the following decay reactions:

    (a) $^{64}_{29}Cu \rightarrow ^{64}_{30}Zn + $ _____

    (b) $^{190}_{78}Pt \rightarrow ^{186}_{76}Os + $ _____

    (c) $^{123}_{52}Te \rightarrow ^{123}_{52}Te + $ _____

    (d) $^{3}_{1}H \rightarrow ^{3}_{2}He + $ _____
12. Write the nuclear decay equations that show how gamma rays are used to monitor the thyroid gland in iodine-131 uptake. (*Note:* iodine-131 beta decays.)
13. A radioactive decay series called the neptunium series is shown in Figure 28.22. None of its members have half-lives comparable to the age of the Earth, so the nuclides in this series do not occur naturally. However, they can be made artificially. Identify the members of this decay series.

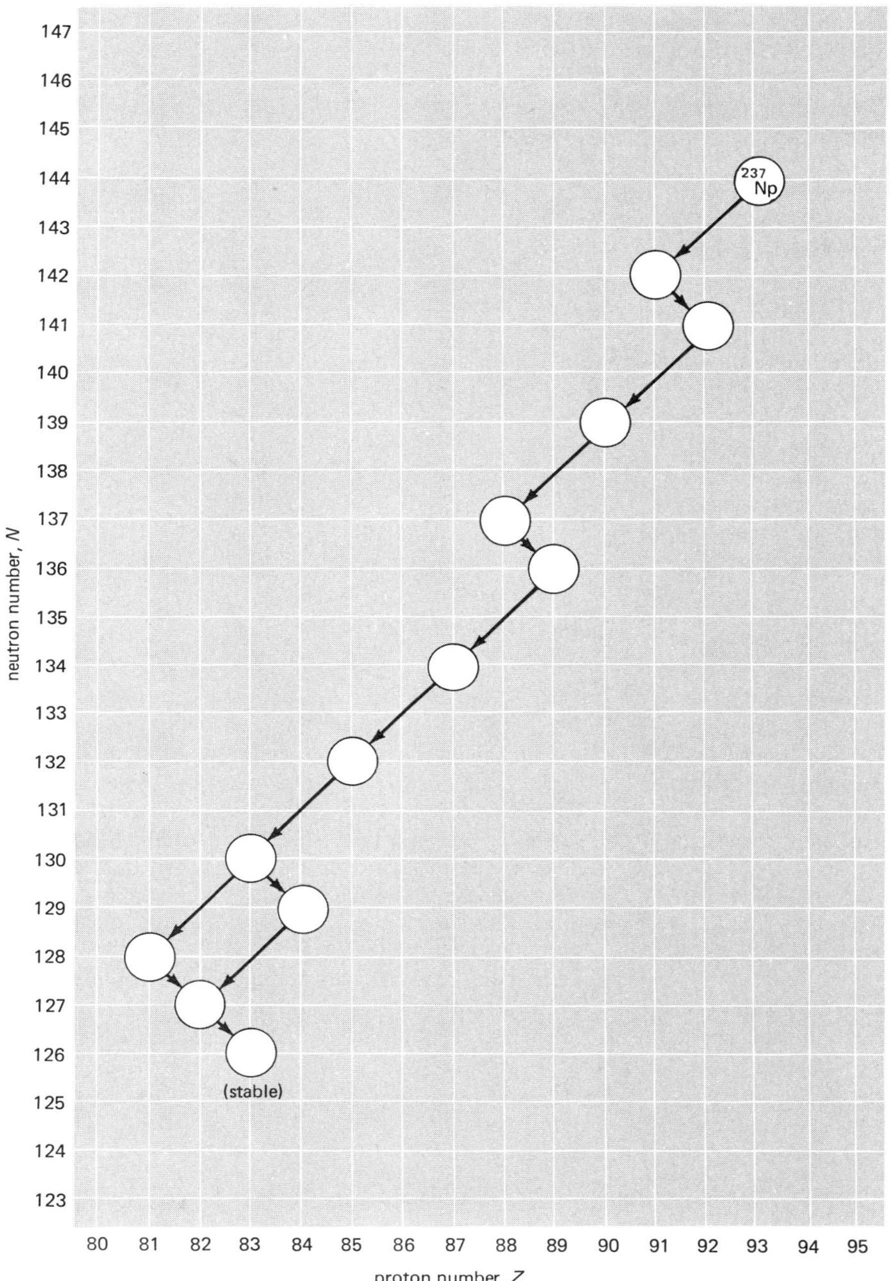

Figure 28.22 See Exercise 13.

14. Protactinium-233 ($^{233}_{91}Pa$) undergoes beta decay. (a) What is the daughter nucleus in this process? Write the equation for it. (b) The daughter nucleus then undergoes alpha decay. What is the "granddaughter" nucleus?
15. Radon-222 ($^{222}_{86}Rn$), a radioactive noble gas, undergoes alpha decay. (a) What is the daughter nucleus of this process? Write the equation for the decay. (b) The daughter nucleus undergoes both beta decay and alpha decay. What are the "granddaughter" nuclei in each case?
16. Why can't the half-life of a radioactive sample be determined by weighing the sample after periods of time?
17. Bismuth-209 ($^{209}_{83}Bi$) is said to be the heaviest stable nu-

cleus. Yet this nuclide alpha-decays with a half-life greater than or equal to 2×10^{18} years. Why is it justified to say that this bismuth isotope is "stable"?

18. Some short-lived radioisotopes are found in nature. Why? Shouldn't these have decayed away a long time ago?

19. If the bubbles "decay" in Figure 28.23 like radioactive nuclei, what is the half-life of this soap-bubble "isotope"? How many bubbles will be present when the clock reads 8 minutes?

20. How many becquerels are there in one curie?

21. A radioactive isotope has a half-life of 1 hour. If a sample contains 100 micrograms of this isotope, how much is left after (a) 1 hour, (b) 2 hours, and (c) 3 hours?

22. A sample of cobalt-60 has an activity of 120 μCi. How long will it take for the activity of the sample to decrease to 15 μCi?

23. A Geiger counter near a radioactive sample registers 1600 cps (counts per second). Twelve hours later, the count rate is 200 cps. What is the half-life of the radioactive nuclide?

24. A particular radioactive nuclide has a half-life of 10 minutes. By what percentage would the activity of a sample of this isotope decrease in 1 hour?

25. A patient in a hospital is given 100 micrograms of iodine-131. Assuming that none is lost through body functions, how much of the iodine remains after one month?

26. A petrified wood carving is found in an archeological dig. It is found that there are 2 beta emissions/g of C/min from a sample of the carving. Approximately how old is the wood carving?

27. If a large animal died today, in what year would there be 4 beta emissions/g of C/min from its bones?

28. Why can't carbon dating be used accurately for ages of 25,000–30,000 years and older?

29. The technique of carbon dating relies on the assumption that the cosmic ray intensity has been constant for the last 25,000–30,000 years. Suppose we found out that the cosmic ray intensity was much greater 10,000 years ago. How would this affect the ages of samples that have been dated by carbon-14 dating? Explain.

Figure 28.23 See Exercise 19.

30. (a) What are the names of the particles believed to "carry" the fundamental forces between particles? (b) What are the elementary particles believed to make up larger particles such as protons and neutrons? Are these particles electrically neutral?

31. How many quark flavors and colors are there? What is the total number of quarks accounted for by this scheme? (Don't forget antiquarks.) Can you name them?

29

Nuclear Energy: Fission and Fusion

The Background of the "Nuclear Age"

The dropping of the "atomic" bomb on Hiroshima, Japan, in 1945 was the devastating beginning of what is sometimes called the atomic or, more correctly, the nuclear age. It was a public announcement of the awesome power that could be released from the nucleus. Prior to this, the secret of "splitting the atom" was known to only a handful of scientists. The scenario began in the first part of this century, not long after the discovery of radioactivity and the introduction of the nuclear model of the atom. If a radioactive nucleus can decay into the nucleus of another element with the emission of a particle, how about the reverse reaction —putting a particle into a nucleus and getting a nucleus of another element? "Reverse radioactivity," so to speak.

Scientists quickly thought of this and showed that it could be done. Attempts were then made to make new, heavy elements. Uranium, the heaviest element known at the time, was bombarded with particles — in particular, neutrons — to do this. One of the unexpected things found was that the uranium nucleus sometimes ended up in fragments or broken ("split") apart.

In 1938 the German chemists Otto Hahn and Fritz Strassman identified some uranium nucleus fragments as the nuclei of lighter elements. Hahn related this discovery to Lise Meitner, a long-time coworker, who was forced to flee Nazi Germany because she was "non-Aryan" (Fig. 29.1). Ms. Meitner and her nephew, Otto Frisch, worked out the theory of this reaction in Sweden, calling it "fission" in analogy with biological fission or division of a living cell into two parts. In addition to the emission of neutrons from the fission reaction, there was also a release of energy.

They sent their results to a scientific journal in Denmark, which you may recall was the native country of Niels Bohr (Chapter 27). Bohr carried the news of the discovery of fission with him to a scientific meeting in Washington, D.C. Here, scientists realized the poten-

Figure 29.1 Lise Meitner and Otto Hahn collaborated in scientific research for 30 years. In 1938 Meitner was forced to leave Nazi Germany. She was in Sweden when she published the first report on nuclear fission with her nephew O. R. Frisch.

tial of nuclear fission in the production of energy. Albert Einstein, who now lived in America (having also fled Germany), wrote a now-famous letter in 1939 to President Roosevelt to make him aware of the potential of nuclear fission. Subsequently, the "Manhattan Project" was initiated, which resulted in the development of the "atomic" bomb.

The first fission reactions had actually been produced in 1934 by Enrico Fermi and his coworkers in Italy when they bombarded many elements with neutrons (Fig. 29.2). However, the reaction was unrecognized at the time. Even so, Fermi won the 1938 Nobel Prize in physics for his work in creating artificial isotopes by neutron bombardment. The Italian dictator Mussolini allowed Fermi and his family to go to Stockholm for the presentation ceremony, but Fermi didn't return. He came to the United States, where he was a principal figure in the Manhattan Project.

Figure 29.2 Enrico Fermi, a principal figure in the development and understanding of nuclear fission.

Although first used for military applications, nuclear energy has peaceful uses. One of the major ones is the production of nuclear energy in reactors for the generation of electricity. This has been going on since the late 1950's. However, some 20 years later in 1979, an "incident" at Three Mile Island (see opening photograph) dramatically brought to the public attention the potential dangers associated with nuclear reactors.

Now, well into the "nuclear age," we hear such terms as "meltdown," "radioactive wastes," "nuclear warheads," and "nuclear proliferation" and debates between "nuke" and "anti-nuke" groups. One of the main purposes of this chapter is to make you aware of the facts about nuclear energy so you can better understand the issues in the nuclear debate. You may be called upon to make a voting decision on one or more of these issues some day.

Nuclear Reactions

Thus far, we have considered only nuclear decay reactions in which a radioactive nucleus spontaneously changes or transmutes into the nucleus of another element with the emission of an alpha or beta particle. However, it is possible to induce a nuclear reaction by "adding" particles to the nucleus, which creates different nuclides.

Lord Rutherford produced the first induced nuclear reaction in 1919 by bombarding nitrogen (N-14) gas with alpha particles from a natural bismuth-214 source (Fig. 29.3). The particles coming from the reactions were identified as protons. Rutherford reasoned that an alpha particle colliding with a nitrogen nucleus can occasionally knock out a proton. This results in the *artificial* disintegration or transmutation of a nitrogen nucleus to an oxygen nucleus:.

$$\underset{\text{Alpha particle}}{{}^{4}_{2}\text{He}} + \underset{\text{Nitrogen}}{{}^{14}_{7}\text{N}} \rightarrow \underset{\text{Oxygen}}{{}^{17}_{8}\text{O}} + \underset{\text{Proton}}{{}^{1}_{1}\text{H}}$$

This reaction was discovered almost by accident because it took place so infrequently in Rutherford's experiment. One proton is produced for about every one million alpha particles that pass through the gas. Notice that the conservation of nucleons and the conservation of charge hold in this and all nuclear reactions. That is,

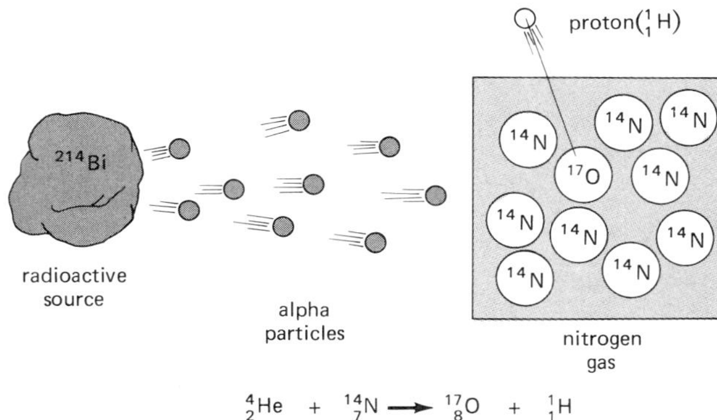

Figure 29.3 Rutherford's reaction. Alpha particles from a radioactive source cause nuclear reactions that change nitrogen to oxygen.

the sums of the mass numbers and proton numbers (symbol superscripts and subscripts) on the opposite sides of the equation are equal.

But think of the implications of this reaction. By use of such bombardment processes, one element can be changed into another. In a sense, it is the old dream of the alchemists come true. Of course, the alchemists' main concern was the changing of common elements into gold. This can be done too with nuclear reactions, for example,

$$^2_1H + ^{196}_{78}Pt \rightarrow ^{197}_{79}Au + ^1_0n$$

$$\gamma + ^{198}_{80}Hg \rightarrow ^{197}_{79}Au + ^1_1H$$

Gold (Au)-197 is the only stable isotope of gold. However, making gold by this process is not too economical. It costs more to produce the gold than the gold is worth.

The neutrons produced in nuclear reactions with charged particles (protons, deuterons, or alpha particles) can in turn be used to induce nuclear reactions to make artificial radioactive isotopes for industrial and medical applications, for example,

$$\underset{\text{Neutron}}{^1_0n} + \underset{\text{Aluminum}}{^{27}_{13}Al} \rightarrow \underset{\text{Magnesium}}{^{27}_{12}Mg} + \underset{\text{Proton}}{^1_1H}$$

Magnesium-27 is radioactive, and it beta-decays with a half-life of 9.5 minutes.

Neutrons are especially effective as nuclear "bullets" in initiating reactions because they have no electric charge. They are not subject to the repulsive electrostatic forces experienced between positively charged "bullets" and the nucleus and are more likely to penetrate nuclei than protons or alpha particles.

Examples of other important neutron-induced reactions are

$$^1_0n + ^{238}_{92}U \rightarrow (^{239}_{92}U)^* \rightarrow ^{239}_{93}Np + ^{\ 0}_{-1}e$$

$$^1_0n + ^{235}_{92}U \rightarrow (^{236}_{92}U)^* \rightarrow ^{141}_{56}Ba + ^{92}_{36}Kr + 3(^1_0n)$$

In the first reaction, a *transuranic* (above uranium) element, neptunium, is made. There are no naturally occurring elements above uranium. Neptunium beta-decays into plutonium. Some artificial transuranic elements are listed in Table 29.1. The second reaction is an example of the dividing ("splitting") or fissioning of a large nucleus. More about this shortly.

The early study of nuclear reactions was limited by the kind of particle "bullet" that could be used to bombard nuclei. Only alpha particles from natural radioactive sources could bring about nuclear reactions, and these particles were low in energy and intensity. The nucleus is a "hard nut to crack," and a particle must have enough energy to penetrate it. When heavier elements are bombarded with alpha particles, the greater repulsive electric force from the greater charge in the heavier nuclei makes it difficult for alpha particles to penetrate the nuclei. There is an advantage to using protons or deuterons, since the repulsive force on them is not as great as that on an alpha particle.

Since the 1930's many devices for accelerating charged particles have been invented. In all of these, charged particles are accelerated by electric fields, and the energetic beams can be guided or steered by magnetic fields. Particle accelerators are used to produce radioactive isotopes for military, medical, and industrial purposes, but very large accelerators are used mainly to probe the nucleus in elementary particle research (Fig. 29.4). The use of electric fields or voltage differences to accelerate charged particles in accelerators has brought about a commonly used unit of energy in nuclear physics—the electron volt (eV). An **electron volt** is simply the amount of energy received by an electron when accelerated through a potential of 1 volt. (Recall that voltage is the work or energy per charge, so eV is a unit of energy.) The electron volt is a small unit. One electron (charge) times 1 volt or $1\ eV = (1.6 \times 10^{-19}\ C)(1\ V) = 1.6 \times 10^{-19}\ J$. Accelerator energies are expressed in MeV (M, mega–, 10^6 or million electron volts) and GeV (G, giga–, 10^9 or billion electron volts).

* These intermediate, unstable nuclei are called "compound" nuclei and quickly decay. They are commonly omitted from the nuclear reaction equations.

(a)

(b)

Figure 29.4 Accelerators. (a) An early cyclotron with inventors Ernest O. Lawrence (left) and M. S. Livingston. (b) Fermi National Accelerator Laboratory near Batavia, Illinois.

Endoergic and Exoergic Reactions

If you compare the sum of the masses of the reactants of a nuclear reaction with the sum of the masses of the reaction products, you might be surprised to find that they are not the same. For example, in Rutherford's original experiment of bombarding nitrogen nuclei with alpha particles, there is more mass at the end than was started with (Fig. 29.5). This might seem strange,

Table 29.1 Artificial Transuranic Elements

Symbol	*Name*	*Named for*	*Year*
$_{93}$Np	Neptunium	The planet Neptune	1940
$_{94}$Pu	Plutonium	The planet Pluto	1940
$_{95}$Am	Americium	America	1944
$_{96}$Cm	Curium	Marie and Pierre Curie	1944
$_{97}$Bk	Berkelium	Berkeley, California	1949
$_{98}$Cf	Californium	California	1950
$_{99}$Es	Einsteinium	Albert Einstein	1952
$_{100}$Fm	Fermium	Enrico Fermi	1953
$_{101}$Md	Mendelevium	Dmitri Mendeleev	1955
$_{102}$No	Nobelium	Alfred Nobel	1957
$_{103}$Lw	Lawrencium	Ernest Lawrence	1961
$_{104}$Rf	Rutherfordium	Ernest Rutherford*	1964
$_{105}$Ha	Hahnium	Otto Hahn	1970

* Called Kurchatovium (Ku) by Russian scientists and named after Igor Kurchatov, who independently created the element in Russia at about the same time as it was created in America.

but remember that according to Einstein, mass is a form of energy ($E = mc^2$, Chapter 26). So if the total mass increases, the increase must come from energy—the energetic alpha particle has energy, some of which is converted to mass in the reaction.

This type of reaction is called an **endoergic reaction.** That is, energy (*-ergic*) must be put into (endo-) the reaction for it to occur or proceed. This is analogous to an endothermic chemical reaction in which heat (energy) must be added for the reaction to proceed, or more energy is put in than is obtained.

However, some chemical reactions are *exo*thermic. Here, you get more (heat) energy out than you put in, for example, in combustion such as burning logs in a fireplace. Similarly, for nuclear reactions there are **exoergic reactions.** An example is the fission reaction shown previously (Fig. 29.6). Here, there is less total mass after the reaction than before, on the order of 200 MeV mass equivalents less. Hence, we have a *source* of energy—nuclear energy.

Scientists were quick to realize this, as evidenced by Einstein's letter to President Roosevelt in 1939 (mentioned previously). A result was the awesome releases of energy over Hiroshima and Nagasaki, Japan, on August 6 and 9, 1945, and the world has never been the same. We had entered the "nuclear age."

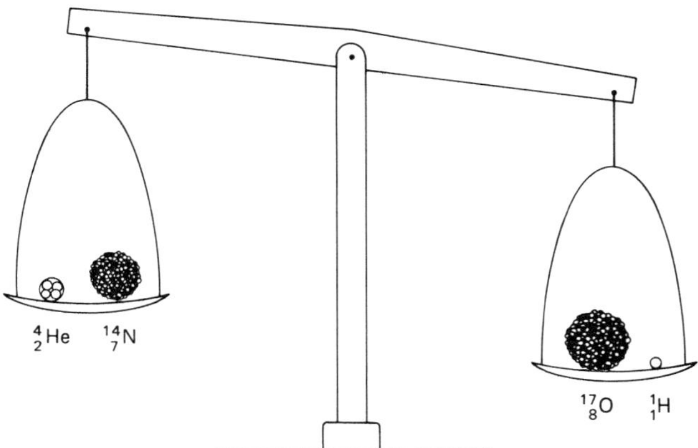

Figure 29.5 Endoergic reaction. The total mass of the products of the reaction is greater than that of the reactants. Energy is converted into mass.

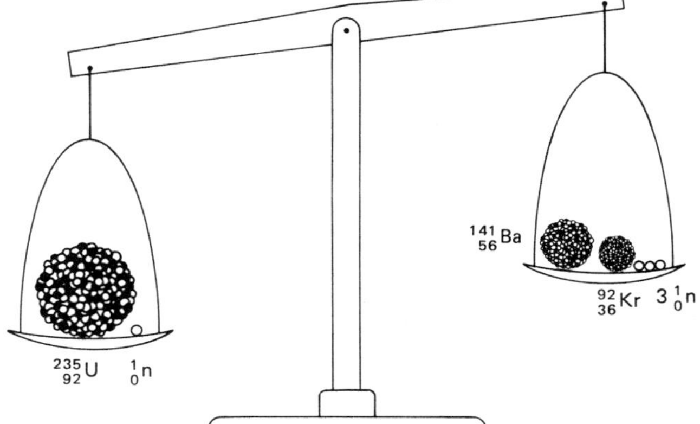

Figure 29.6 Exoergic reaction. The total mass of the products is less than that of the reactants. Mass is converted into energy.

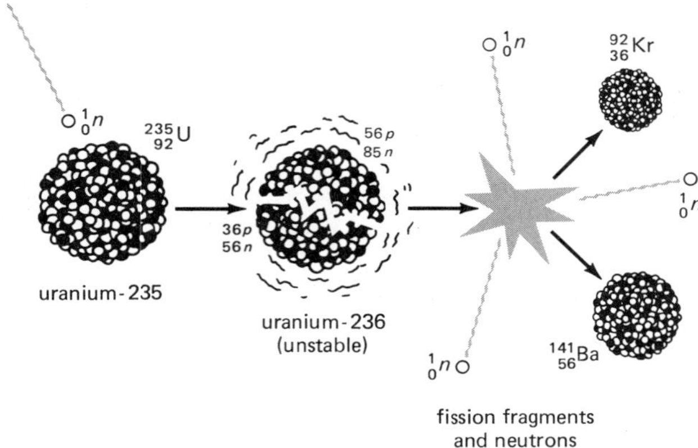

Figure 29.7 Fission. An incident neutron causes a uranium-235 nucleus to divide into two intermediate lighter nuclei with the emission of two or more energetic neutrons.

Fission

The exoergic fission reaction forms the basis for the current production of nuclear energy. In a **fission reaction,** a heavy nucleus divides into two intermediate lighter nuclei with the emission of two or more energetic neutrons (Fig. 29.7). Energy is emitted in the process, being carried off by the fission fragments and neutrons.

Only certain nuclides undergo fission, and the probability of the reaction occurring for a particular nuclide depends on the energy of the incident neutrons. For example, neutrons of any energy will cause U-235 and Pu-239 to undergo fission, but a reaction is more probable for neutrons with low energies of less than 1 eV. We say that these nuclei have a greater cross-section for "slow" neutrons. (Cross-section is a measure of the probability for a reaction to occur.) Thorium-232, on the other hand, prefers "fast" neutrons with energies of 1 MeV or more.

A fission reaction can proceed in a variety of ways. For example, some fission modes other than those given previously are

$$^1_0n + ^{235}_{92}U \rightarrow ^{140}_{54}Xe + ^{94}_{38}Sr + 2(^1_0n)$$

$$^1_0n + ^{235}_{92}U \rightarrow ^{132}_{50}Sn + ^{101}_{42}Mo + 3(^1_0n)$$

Now, back to the production or release of nuclear energy. As mentioned previously, each fission reaction produces about 200 MeV of energy. MeV (*million* electron volts) may sound like a large amount of energy, but it is not really. Recall that 1 eV = 1.6×10^{-19} J, so 200 MeV is only 3.2×10^{-11} J of energy, which isn't much. When you pick your physics book up from a table you do about 5 J of work or expend 5 J of energy. However, keep in mind that there are billions of nuclei in a small sample of fissionable material. The trick is to get enough fission reactions to produce practical amounts of energy.

This is accomplished by a **chain reaction** (Fig. 29.8). The neutrons released in a fission reaction induced other fission reactions, and so on, so that a growing "chain" reaction takes place. Energy is released with each fission reaction. The chain reaction proceeds very quickly, and there is a tremendous release of energy from the millions of nuclei undergoing fission.

For a chain reaction to occur, there must be a sufficient amount of the fissionable material present. Otherwise, neutrons would escape from the sample before reacting with a nucleus, and the chain reaction

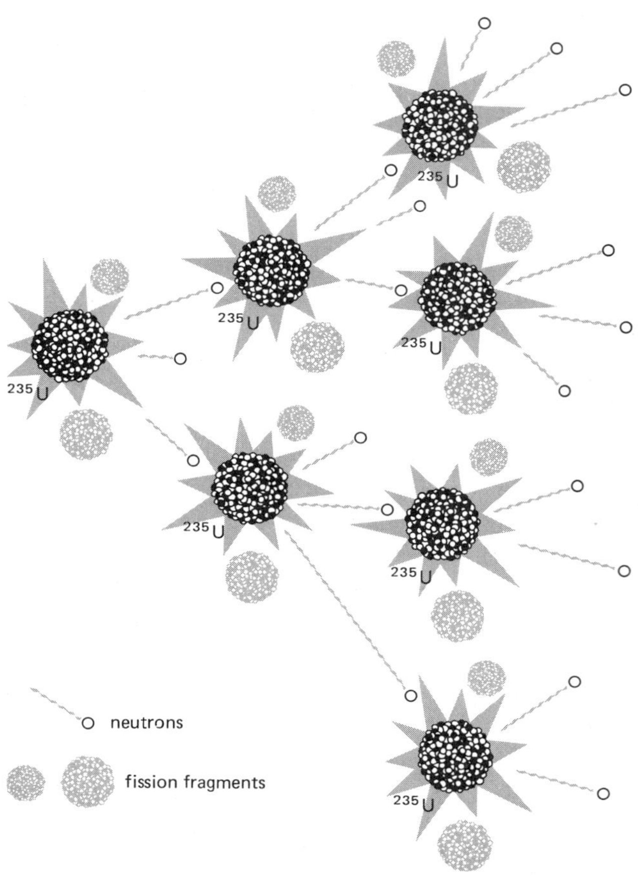

Figure 29.8 A fission chain reaction. Neutrons from one nucleus undergoing fission produce other fissions in a growing chain of reactions.

would not proceed. The chain would be "broken," so to speak. Scientists call the amount of fissionable material necessary to sustain a chain reaction a **critical mass.**

Natural uranium is composed of 99.3 percent U-238 and only 0.7 percent of the fissionable U-235. To have more fissionable U-235 nuclei present in a sample, the U-235 is concentrated or "enriched." Weapons-grade material may be enriched to 99 percent or more, whereas the enriched uranium used in nuclear reactors for the production of electricity is only 3 to 5 percent U-235. Recall from Chapter 7 that this is why a nuclear reactor cannot explode like a bomb. In a bomb, a critical mass of highly enriched material must be held together for a short time to get an explosive release of energy (Fig. 29.9). Segments of the fissionable material in an "atomic" bomb or "nuclear device" (modern terminology) are separated before explosion so there is not a critical mass for the chain reaction. A chemical explosive is used to bring the segments together, and the explosive force on interlocking configurations

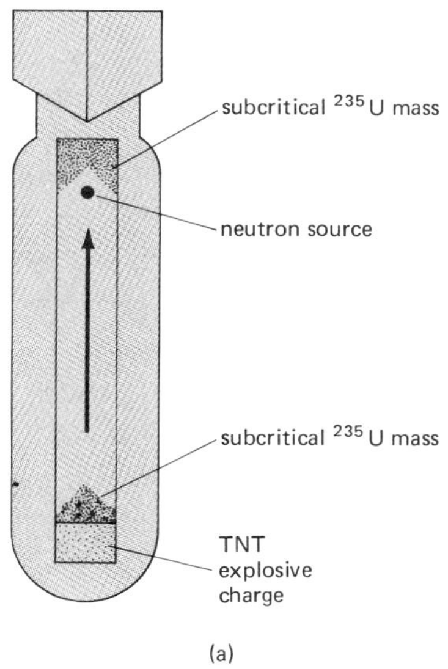

(a)

(b)

(c)

Figure 29.9 Uncontrolled nuclear fission. (a) In an "atomic" bomb, a critical mass is obtained by bringing two masses together by a small explosive charge. (b) An atomic bomb of the "Little Boy" type detonated over Hiroshima, Japan, during World War II. The bomb is 28 inches in diameter and 120 inches long, and has an equivalent yield of 20,000 tons of TNT. (c) An underwater nuclear detonation. Notice the ships in the foreground.

holds them together long enough for a large fraction of the material to undergo fission. The result is an explosive release of energy. A bomb is an example of *uncontrolled* fission, which we hope will never be used again. A nuclear reactor is an example of *controlled* fission, where we control the growth of the chain reaction and the release of energy.

FISSION REACTORS

The basic design of a fission nuclear reactor is shown in Figure 29.10. Fuel pellets in tubes form the fuel rods in the reactor core. The chain reaction and energy output are controlled by means of boron or cadmium **control rods.** These materials absorb neutrons. Hence, by insertion of the control rods between the fuel rod assemblies, the number of available neutrons for inducing fission and the chain reaction can be controlled. The control rods can be adjusted so that energy is released at a relative steady rate. This requires that, on the average, one neutron from each fission event initiates only one other event. If more energy is needed, the rods are withdrawn farther. When fully inserted, the control rods can curtail the chain reaction and shut the reactor down. A nuclear reactor can run for about four years before having to be "refueled."

Recall that a reactor is basically just an energy source and that the energy is removed by a coolant (see Fig. 7.7). Water flowing through the fuel assemblies and the reactor vessel acts not only as a coolant–heat transfer agent, but also as a **moderator.** The U-235 nuclei react best with "slow" neutrons (kinetic energies of only fractions of an electron volt). The "fast" neutrons

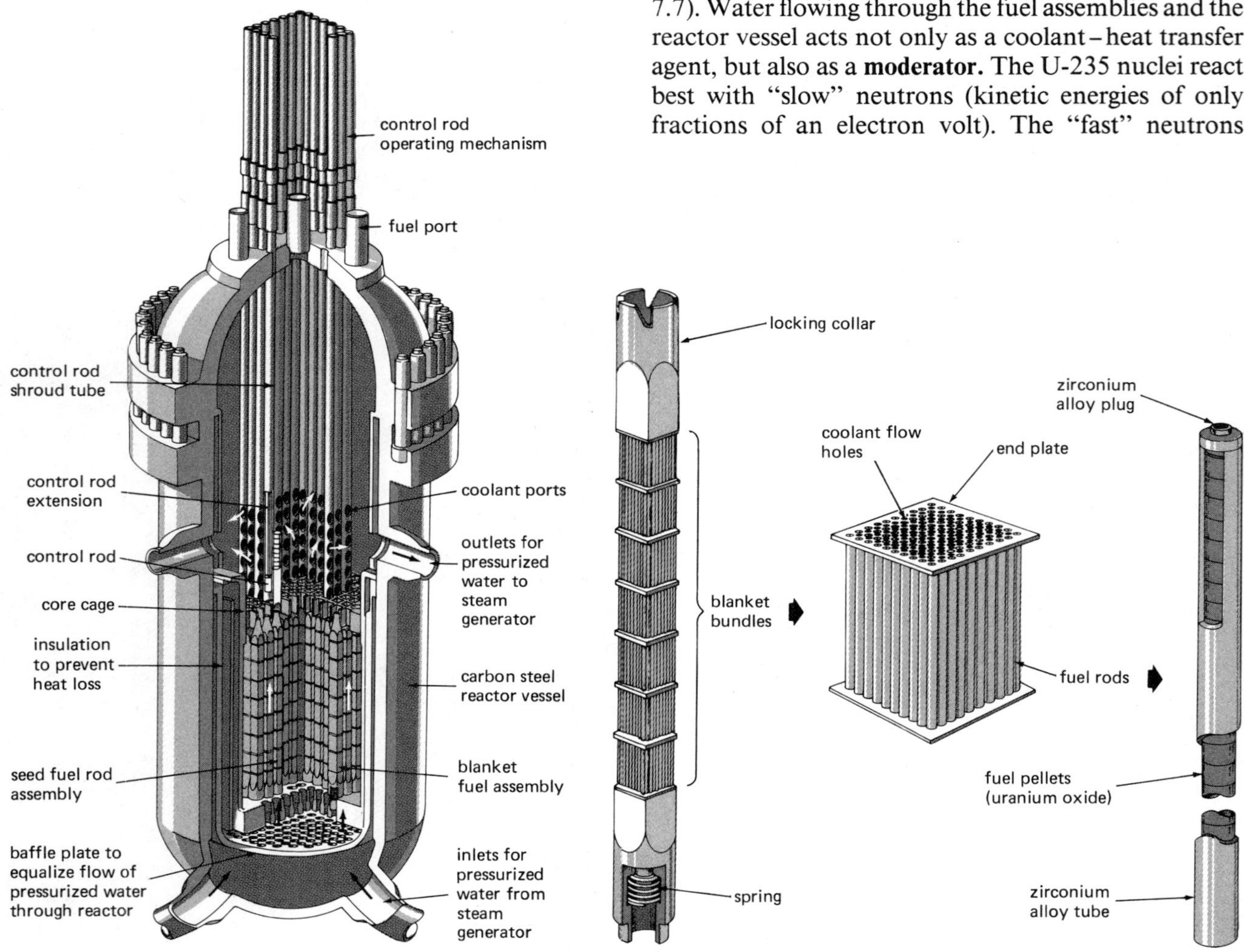

Figure 29.10 Nuclear fission reactor. (a) The basic elements of a reactor. **(Continued)**

emitted from the fission reactions (energies up to 20 MeV with an average of about 2 MeV) are slowed down or "moderated" by energy losses through collisions. Hydrogen atoms in water are very effective in slowing down neutrons because the mass of the hydrogen nucleus is about the same as that of a neutron. Recall from Chapter 3 that when particles of nearly equal masses collide, there is a large transfer of energy. It takes only about 20 collisons on the average to slow fast neutrons down to energies less than 1 eV.

Other materials, such as beryllium and graphite (carbon), have been used as moderators. Because these have heavier nuclei, not as much energy is transferred per collision. About 120 collisions with carbon atoms are needed to slow fast neutrons down to an optimal "slow" speed. Although it is not the best moderator, carbon in the form of graphite permits a chain reaction to occur in natural (unenriched) uranium fuel arranged in a large mass of graphite. The first self-sustaining fission chain reaction was accomplished in such a reactor in 1942 by a team working with Enrico Fermi (Fig. 29.11). It was called a "pile" because it essentially consisted of a pile of graphite blocks. Owing to wartime secrecy, no photographs were taken of the first reactor, or at least, none are available.

This was the first "proof" that the fission chain reaction would work. The Manhattan Project involved a large number of scientists and engineers, including refugees from Europe. They worked under the pressure of World War II to develop a nuclear weapon before the Germans, who were also working on one. The first nuclear bomb was detonated in a test explosion over

(b)

Figure 29.10 (Continued) (b) A fuel-rod assembly and a partially disassembled reactor.

(a)

(b)

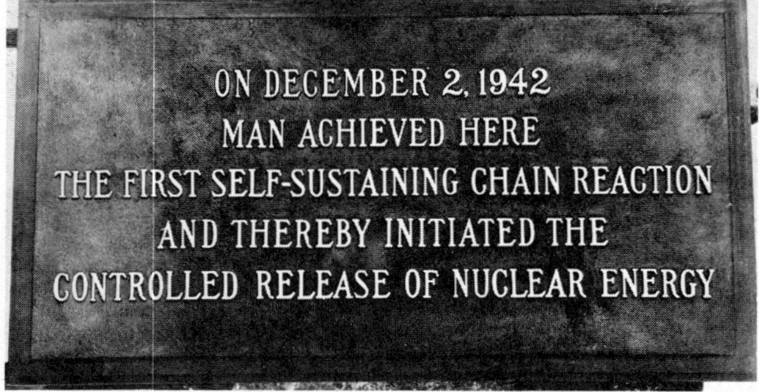

(c)

Figure 29.11 Birthplace of the atomic age. (a) The west stand of Stagg Field at the University of Chicago (being demolished in 1957). It was under this stand in a squash court that the first nuclear chain reaction was initiated on December 2, 1942. (b) and (c) The plaque installed on the wall of the stand in 1952 remains at the location. Enrico Fermi stands under the plaque (second from right) at the unveiling.

the desert in New Mexico on July 16, 1945, only $2\frac{1}{2}$ years after the first reactor was built.

Heavy water (D_2O, where D stands for deuterium) can also be used as a moderator in a reactor that uses natural uranium as fuel. The hydrogen nuclei of regular water have a high tendency to capture neutrons when colliding with them.† As a result, it is impossible to achieve a chain reaction with natural uranium and ordinary water. Ordinary water can be used only with enriched uranium. Deuterium nuclei, on the other hand, do not absorb neutrons very readily, and so heavy water can be used as a moderator in reactors with natural uranium fuel. Most of the reactors in Canada are of this type, called heavy-water reactors, whereas most of the reactors in the United States are "light"-water reactors. The first nuclear reactor for the purpose of electrical generation went into operation in 1957 at Shippingsport, Pennsylvania.

One of the major concerns in reactor safety is that a reactor may get out of control or in a "runaway condition" because of a malfunction of the control rods or a **loss-of-coolant accident** (a "LOCA"). With an uncontrolled release of energy, the coolant may not be able to handle the excess heat generated, or if there were a loss of coolant, heat could build up in the core. In either case (or both) the reactor core might get so hot that the fuel rods would melt and the fused mass would burn through the reactor vessel and the containment floor. Such a "meltdown" would cause leakage of dangerous radioactive materials into the environment. (This is the so-called "China syndrome," in which the fissioning mass might melt its way through the Earth to China. Calculations show that this cannot really happen.) The introductory photo of this chapter shows the Three Mile Island nuclear facility, where a LOCA occurred and where a meltdown almost happened in 1979. (See Special Feature 29.1 at the end of the chapter for more details.)

As pointed out earlier (Chapter 7), a nuclear reactor cannot explode like a bomb. The uranium enrichment is too low for the chain reaction to proceed fast enough for the required energy release in a short time. Moreover, to explode a system must be prevented from flying apart before this time elapses. Should the uncontrolled pressure buildup in a reactor core cause it to burst apart, the resulting separation of the fuel rods would give neutrons more opportunity to escape from the system, and the chain reaction would stop.

† ${}^{1}_{0}n + {}^{1}_{1}H \rightarrow {}^{2}_{1}H + \gamma$

The Breeder Reactor

We have about a 50-year known domestic reserve of uranium. As with oil and gas, our domestic supply of uranium will run out in the not-too-distant future. However, it is possible to extend the supply of fissionable material through breeder reactors. A breeder reactor converts nonfissionable material, such as U-238, into fissionable material, and the reactor produces or "breeds" more fuel than it consumes.

This is not a case of getting something for nothing. Rather, it is a process of converting a nonfuel into a fuel. For example, in an ordinary nuclear reactor there is a great deal of U-238 that "goes along for the ride"; the 3 percent or so of U-235 is the fissionable fuel that does the job. A U-238 nucleus can absorb a fast-moving neutron and undergo a change to plutonium-239 (Pu-239) by the reaction

$$\underset{\text{Fertile material}}{^{238}\text{U}} + {}^{1}n \rightarrow \underset{\text{Unstable products}}{^{239}\text{U} \rightarrow {}^{239}\text{Np}} \rightarrow \underset{\text{Fissionable material}}{^{239}\text{Pu}}$$

Pu-239 is a fissionable material that can do the same job as U-235. The original material is called a fertile material. Other fertile materials that can be used in a breeding reaction are Th-232 and Pu-240. These breed fissionable U-233 and Pu-241 isotopes, respectively.

In a U-238 breeder reactor, the moderation or slowing down of fast neutrons is minimized, and with the presence of an appreciable number of fast neutrons, Pu-239 is bred from U-238 while energy is obtained from U-235 fission. The breeding reaction is illustrated in Figure 29.12. France and several other countries have breeder reactors, but only experimental breeder reactors have been operational in the United States because of limited funding. One of the reasons for the curtailment of the development of the breeder reactor follows in the next discussion.

Nuclear Wastes and Proliferation

The building and planning of nuclear reactors were greatly curtailed after the Three Mile Island incident. This was not only because of public concern for safety, but also because it was realized that there was not the demand for electricity that had been projected, and new reactors were not needed. People were conserving energy. Another factor was falling fossil-fuel costs. Overall, it was cheaper to build and operate a fossil-fuel plant than a nuclear one, so economics played a role.

Even so, there are about 70 nuclear reactors operating in the United States (Fig. 29.13). (There are about 130 nuclear reactor generating facilities operating in other countries.) In doing so, they are creating large amounts of nuclear wastes. These are the "ashes" from the fission process. A large variety of fission products are formed in nuclear reactors, most of which are radioactive. Some of these products have very long half-lives, up to tens of thousands of years or more. This

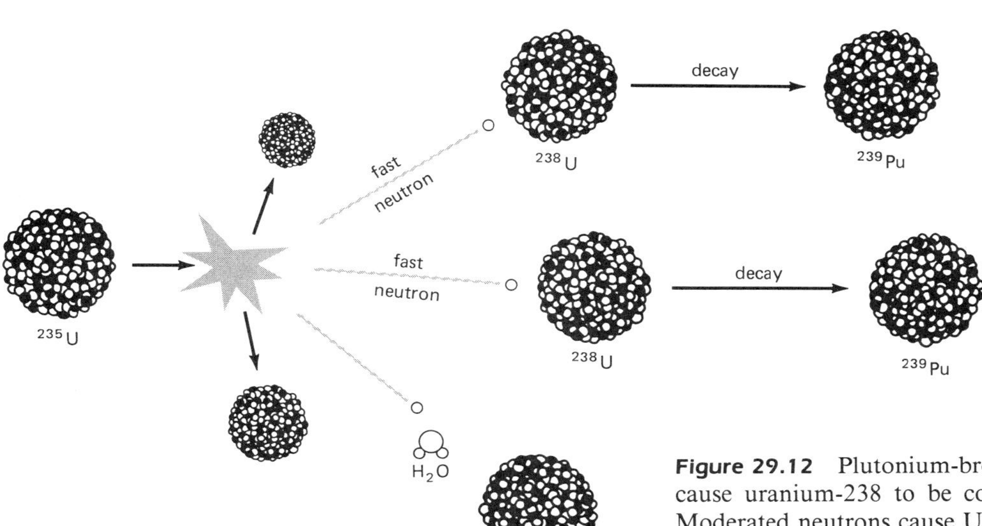

Figure 29.12 Plutonium-breeding reaction. Fast neutrons cause uranium-238 to be converted into plutonium-239. Moderated neutrons cause U-235 nuclei to undergo fission and sustain a chain reaction. In a breeding reaction, more fissionable nuclei are produced than consumed.

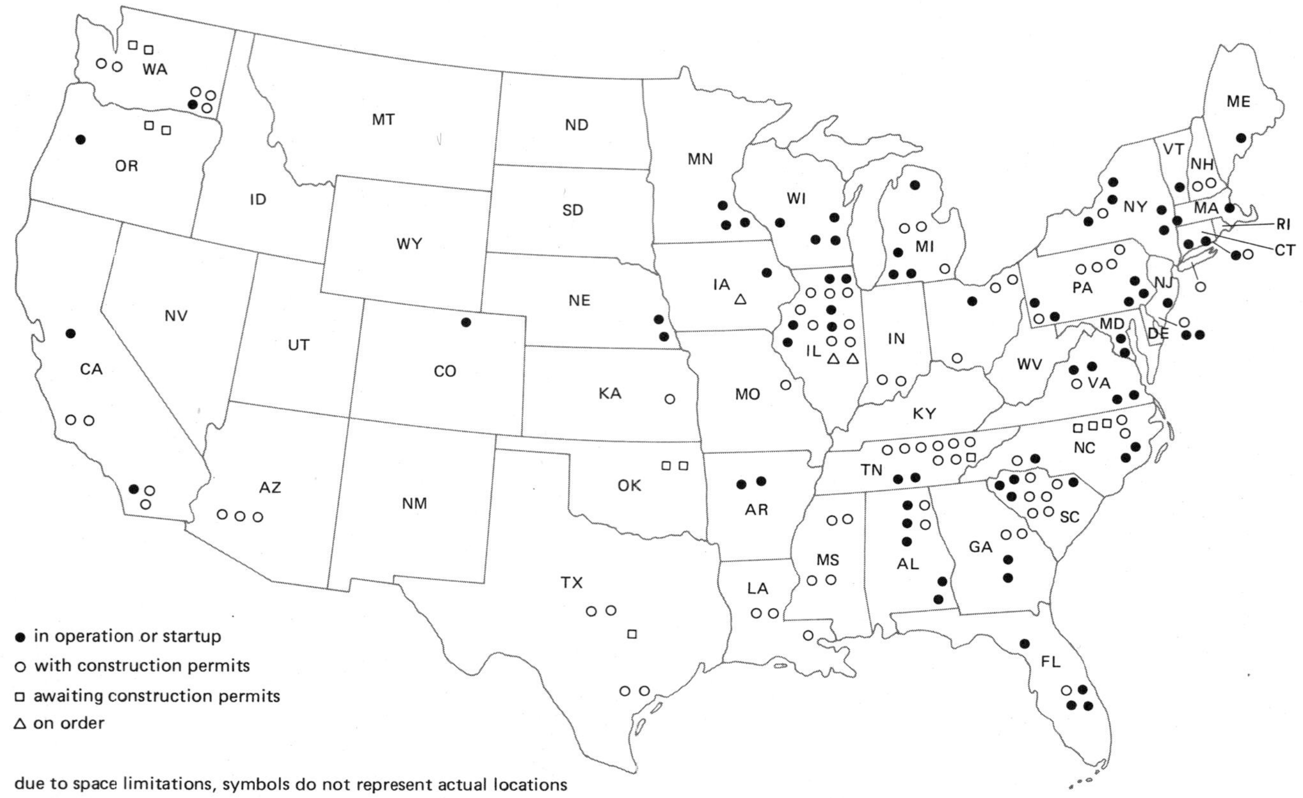

Figure 29.13 Nuclear power plant status in the United States.

means that these "high-level" wastes will be around for hundreds of years, and the concentrations make them highly dangerous.

There are still useful fissionable materials in the spent fuel assemblies, and it would be desirable to recover them. Currently, however, nuclear wastes are being stored in double-walled storage tanks that are covered with earth (Fig. 29.14). Plans call for the "ultimate" disposal of these wastes to be in underground storage in stable geologic repositories, such as salt or rock formations. A law or bill passed by the Congress calls for these repositories to be available by the year 1996. "Permanent" underground burial requires the development of long-lasting containers or the conversion of wastes into solid, insoluble form, for example, embedded in glass (Fig. 29.14). Stainless-steel containers are currently being considered. Appropriate geological formations must be found where the wastes could cause no contamination of underground water should they leak.

In addition to the radioactive fragments of the fission process, the transuranic element plutonium-239 is formed in the normal operation of a reactor. Some U-238 is converted to Pu-239 through "fast" neutron capture and two beta decays. This is one of the reactions promoted in a breeder reactor, discussed previously. Pu-239 is a fissionable material and does the same job as U-235. In the reactor operation, the freshly produced Pu-239 also undergoes fission and produces energy. As a result, it is possible to extend the normal "lifetime" of the reactor fuel elements, and they do not have to be replaced as quickly.

However, it is possible to remove the fuel elements before the Pu-239 is used up and to separate it through difficult and complicated procedures. The Pu-239 can then be processed into nuclear weapons. (The second "atomic bomb" that destroyed the city of Nagasaki, Japan, used Pu-239 instead of U-235 as in the first one dropped on Hiroshima. The Pu-239 had been made in an experimental reactor.) A nuclear power plant reactor can generate up to 200 to 300 kg of Pu-239 a year. A great deal of this is used up in the fission process, but only 5 to 10 kg of Pu-239 are needed to make a nuclear bomb capable of destroying a medium-size city. Thus, a nuclear reactor can be used to produce electricity and/or to produce Pu-239, which could be used as ma-

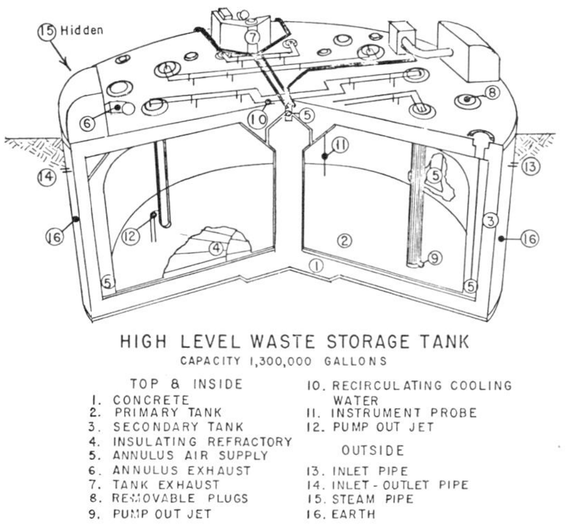

(a)

(b)

Figure 29.14 Radioactive wastes. (a) Nuclear wastes are stored in double-walled tanks that are covered with earth. (b) Embedding radioactive waste in glass is a possible method of storage.

terial for weapons. This capability gives rise to what is called the problem of "nuclear proliferation."

This was made particularly evident in 1974 when India exploded a "nuclear device" as a test for "peaceful applications," according to that government. The explosion showed that countries other than the five "nuclear powers" (those who had tested and demonstrated nuclear weapons) could produce Pu-239 and use it in nuclear weapons if they so desired.

In the early days of the "nuclear age" there were only three countries who had the bomb. The United States, Great Britain, and the Soviet Union test-exploded several bombs in the atmosphere after World War II. The large amounts of radioactive fission products from atmospheric explosions that come to Earth are called radioactive fallout. These products can be carried around the world by winds and eventually settle out or are carried to the Earth by rain and snow. As you can imagine, radioactive fallout can be dangerous. The greatest source of radioactive fallout in the absence of atmospheric nuclear explosions is from coal-fired plants. Radioactive materials in the coal are vented to the atmosphere in combustion gases and particles. (The biological effects of radiation will be discussed shortly.)

As a result of public and scientific concern, the three nuclear powers and many other countries signed the Nuclear Test Ban Treaty in 1963, by which they agreed to a moratorium on further tests in the atmosphere. Underground testing was still permitted. The most notable exceptions who did not sign the treaty were France and China. Both countries have since exploded bombs, now more politely called "nuclear devices," in the atmosphere and have joined the ever-growing nuclear weapons club.

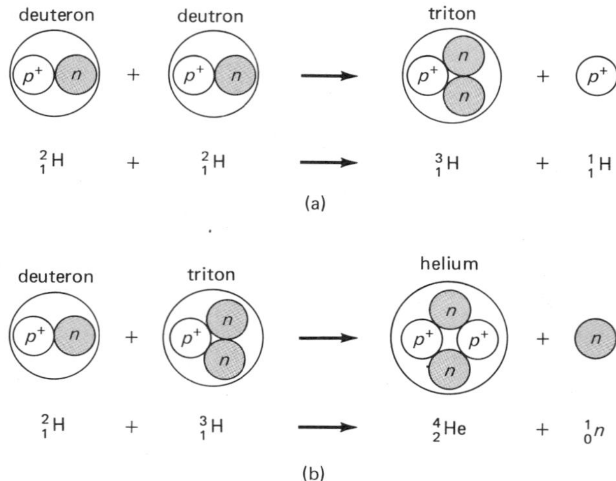

Figure 29.15 Fusion reactions. (a) D-D reaction. (b) D-T reaction.

Fusion and Fusion Reactors

Another class of nuclear reactions that release energy is called fusion reactions. In these exoergic reactions, light nuclei are "fused" together to form heavier nuclei with the release of energy, for example, hydrogen fusion, as illustrated in Figure 29.15.

Fusion is the source of energy of stars, which includes our Sun. In the Sun, the fusion process results in the production of a helium nucleus from four protons (hydrogen nuclei). The process does not take place in a single reaction, but goes through different sets of reactions with the net result of

$$4{}^{1}_{1}\mathrm{H} \rightarrow {}^{4}_{2}\mathrm{He} + 2({}^{0}_{+1}e)$$

(The symbol ${}^{0}_{+1}e$ represents a positron, which is the antiparticle of the electron.)

Many scientists look upon fusion as an ultimate and ideal source of energy for several reasons. Among these are the availability or supply of raw materials and advantages in waste disposal. "Heavy hydrogen" or deuterium (${}^{2}_{1}\mathrm{H}$) is present in all water (heavy-water molecules). For about every 6500 atoms of ordinary (light) hydrogen in water, there is one atom of deuterium. This may not seem like much, but taking into account the vast amounts of water in the oceans and other surface waters, it is estimated that there is enough deuterium to supply the world's energy needs for more than a million years!

The other heavy isotope of hydrogen, tritium (${}^{3}_{1}\mathrm{H}$) is very rare in nature. It is radioactive, with a half-life of about 12 years. As "nuclear waste" of the first or "D-D" reaction in Figure 29.15, it would be a relatively short-term disposal problem, as compared with fission wastes. Notice that the tritium formed by deuterium fusion could immediately react by the second "D-T" reaction with available deuterium. For use as a fusion "fuel," tritium is produced by the nuclear reaction of neutron capture by lithium (Li). As a raw material, lithium is widely distributed over the Earth, being estimated to be more abundant than tin. Should this supply of lithium not be adequate, it could be extracted from the oceans, which contain large amounts (but in small concentrations).

The energy releases from the D-D and D-T reactions are about 4.0 MeV and 17.6 MeV, respectively. This is significantly less than the 200-MeV energy release from a typical fission reaction. But keep in mind that, "kilogram for kilogram," a given mass of hydrogen isotopes has many, many more nuclei than an equivalent mass of a "heavy," fissionable isotope.

In the case of fusion, there is no critical mass or size, since there is no chain reaction to maintain. However, there are problems in practical fusion energy production. One of the major ones is in obtaining a self-sustaining reaction. The repulsive electric force between positively charged nuclei opposes their coming together and fusing. For hydrogen (one nuclear proton), this force is not excessive, which is why hydrogen is the fusion candidate. Fusion reactions can be produced in particle accelerators. For example, deuterons (deuterium nuclei) can be accelerated and "slammed" into a solid deuterium target to produce D-D fusion reactions. This shows fusion to be possible, but much more energy is spent in accelerating the deuteron than is produced by the small number of fusion reactions that occur.

To get an appreciable energy output, one might use a confined gas of hydrogen isotopes that are to undergo fusion. If the temperature of the gas is raised, the kinetic energy of the molecules would be increased, and hence it is only a matter of attaining a sufficiently high temperature for fusion to occur. In the confined space of the gas, the moving nuclei would collide repeatedly until fusion reactions take place. However, it is found that this requires temperatures on the order of millions of degrees (Fahrenheit, Celsius, or kelvins, take your choice).

At such high temperatures, almost all of the hydrogen-isotope atoms will be stripped of their electrons. Such a gas, consisting almost entirely of positively charged ions and free, negatively charged electrons, is

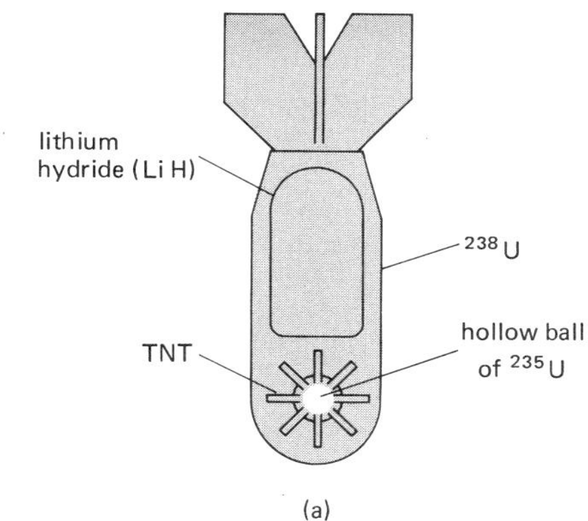

(a)

(b)

Figure 29.16 Uncontrolled fusion. (a) In a hydrogen bomb, a small fission bomb is used to "trigger" the fusion reaction. (b) A thermonuclear explosion in the Pacific in 1954.

called a **plasma.** Plasmas have a number of special properties that have caused them to be referred to as a "fourth state of matter," a term used in 1879 by William Crookes, an English chemist, who generated plasmas in gas-discharge or Crookes tubes. Some of these properties can be used to advantage in potential fusion-reactor techniques, but others create problems.

Large amounts of fusion energy have been released, but this occurred in uncontrolled fusion in the form of the hydrogen (H) bomb. In the H-bomb, a nuclear fission bomb is used to "trigger" the fusion reaction or to supply the energy to get it started (Fig. 29.16). Because of the high temperatures required to bring about fusion reactions in plasmas, they are called thermonuclear reactions, and fusion weapons are called thermonuclear weapons or "devices."

The fusionable material in the bomb is in the form of solid lithium hydride, LiH, where the H is deuterium (^{2}H or D). With the explosion of the fission bomb, a hot plasma develops in which there are D-D reactions. Neutrons from the fission bomb react with the Li to produce tritium (^{3}H or T) for D-T reactions.‡ The neutrons emitted in the D-T reactions come off with high energy (most of the 17 MeV released in the reaction). These neutrons are speedy enough to induce fission in ^{238}U, so the bomb is surrounded by a layer of uranium, which caps off the explosion with another fission energy release. This fission-fusion-fission process takes place in about a millionth of a second.

The controlled release of fusion energy might be accomplished by adding fuel in small amounts to a fusion reactor, similar to the way we control the amount of fuel in conventional energy (heating) sources, such as a gas furnace. However, fusion reactors are not yet practical. There are major problems with confinement—confinement of an ultra-hot plasma and confinement of sufficient energy in the plasma to initiate and sustain fusion. No material could be used to "hold" a plasma. Tungsten, the material with the highest melting point, melts at around 3370°C. Also, heat would be readily transferred from the confinement region. Considering all of the various factors, it is calculated that to achieve a net energy production from D-T fusion would require a reactor operating temperature of at least 100 million degrees. Even higher temperatures would be required for D-D fusion.

Two types of confinement, magnetic and inertial, are now under development.

MAGNETIC CONFINEMENT. Since the plasma is a gas of charged particles, it can be controlled and manipulated with electric and magnetic fields. In magnetic confinement, magnetic fields are used to hold the plasma in a confined space (a magnetic "bottle" or "ring"). Electric fields produce electric currents that heat and raise the temperature of the plasma (Fig. 29.17).

The energy breakeven point (getting as much out as put in) depends on two things: (1) the temperature of the plasma and (2) its density and confinement time. Temperatures of 100 million kelvins have been achieved in tokamaks (see Fig. 29.17), and in other experiments the density-confinement time requirement seems to be satisfied. The trick scientists must now perform is to achieve these requirements at the same time, which may prove more difficult than reaching them separately.

‡ $^1_0n + ^6_3\text{Li} \rightarrow ^3_1\text{H} + ^4_2\text{He}$

(a)

(b)

Figure 29.17 Magnetic confinement. (a) ORMAK fusion research device. The magnetic system used is known as a tokomak (a Russian word for toroidal magnetic chamber) and is a circular device shaped much like a donut. (b) The Doublet III magnetic fusion research machine. It uses a unique confinement concept invented by General Atomic Company for use in fusion-power reactors.

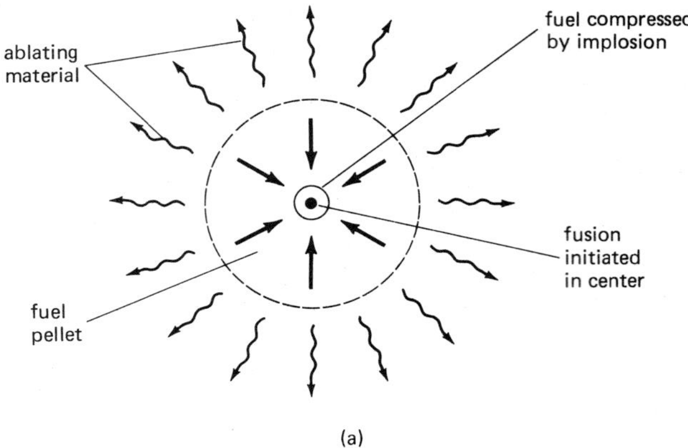

(a)

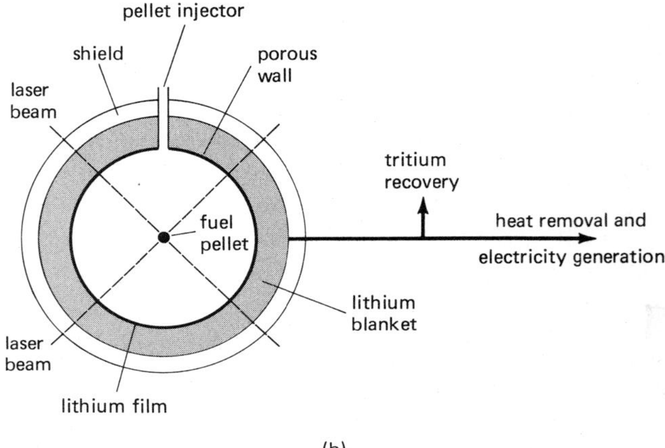

(b)

Figure 29.18 Inertial confinement and implosion. (a) The compression of a fuel pellet is enhanced by the use of an outer shell of material—for example, a plastic—that is ablated or vaporized. This drives an inner shell inward and compresses the fusion fuel. (b) A design of a laser fusion reactor cavity. The laser beams compress and heat the fuel pellet. Neutrons released from the fusion are captured in the lithium blanket, creating tritium, which can be recovered and used as a fuel.

INERTIAL CONFINEMENT. In this technique, it is hoped that laser, electron, or ion beams can be used to initiate fusion. It calls for hydrogen fuel pellets to be dropped into a reactor chamber (Fig. 29.18). Energetic beam pulses would cause a pellet to "implode," causing compression and high temperatures. Fusion could occur if the pellet stays together for a sufficient time. This depends on the inertia of the pellet (hence the name inertial confinement). At the present time, lasers and particle beams are not powerful enough to induce fusion by this means, but research has begun (Fig. 29.19).

If you think about it, the technological problems of fusion reactors are enormous, but so are the benefits. Practical energy production from fusion is not expected until well into the 21st century.

(a)

(b)

(c)

Figure 29.19 Fusion research. (a) Model of the Shiva laser system for laser fusion research at the Lawrence Livermore Laboratory. The 20 laser beams are pointed at and focused onto a tiny fusion target about the size of a grain of sand located in the center of a target chamber (b). This huge laser system (c) will deliver more than 30 trillion watts of optical power in less than one billionth of a second.

Biological Radiation Hazards

You may have seen signs like the one shown in Figure 29.20 if you have been in the vicinity of a radioactive source or a general radiation area. It is a notification or a *warning* (caution) sign. But why the warning? What is the potential danger from the improper use of radiation or radioactive materials?

The main hazard of radiation is to living cells. When "radiation," such as X-rays, gamma rays, beta particles, and alpha particles, passes through matter, it can knock electrons out of the atoms, forming electron-ion pairs. (Recall from Chapter 28 that this allows the detection of radiation, such as by Geiger counters.) Hence, we refer to such radiations as ionizing radiations. The amount of ionization and the penetration of

Figure 29.20 The symbol used to indicate the presence of a radioactive source or a general radiation area.

radiation in a material, including living tissue, depend on the radiation energy, the electrical charge, and the density of the material.

A major injury can occur to the reproductive mechanism of cells. Because of radiation ionization, living cells may receive slight radiation damage, heal, and resume their normal activities and ability to divide and reproduce. In extreme cases, the cells may be mutilated so that they cannot reproduce. If this happens to enough cells, the irradiated tissue may eventually die.

In other instances, damage to the chromosomes in a cell nucleus (genetic damage) can cause the cells to lose their identity and/or to reproduce at an abnormal rate. The condition of an abnormal reproduction rate of cells with "amnesia" is called cancer.

Cancer cells may grow slowly with little effect on the surrounding normal tissue, forming a benign tumor. When cancer cells grow at the expense of the surrounding tissue, we call it a malignant tumor. Skin cancer due to overexposure to ultraviolet radiation (too much sun) and leukemia ("cancer of the blood") are examples of the cancerous effects of radiation. (Leukemia is the unregulated production of white cells—leukocytes—in the blood.)

One of the most abundant fission products is strontium-90 ($^{90}_{38}Sr$). It is chemically similar to calcium-40 ($^{40}_{20}Ca$)—check to see in which chemical groups of the periodic table these elements are located (Chapter 9). From radioactive fallout, the strontium-90 can get into our bodies and into our bones—for example, it falls on grass, the grass is eaten by cows, it appears in milk, the milk is drunk by humans. The radiation from the strontium-90 may cause cancer, such as bone tumors and leukemia, particularly in children.

On the other hand, cell damage may serve a useful purpose. Radiations, such as gamma rays from cobalt-60, are used in medical treatment to control the growth of tumors by producing ionization in the tumor-cell nuclei so that some of the tumor cells are "killed," and the growth of the tumor is slowed.

SPECIAL FEATURE 29.1

Case History: Three Mile Island*

The Three Mile Island nuclear plant in Middletown, Pennsylvania, is a subsidiary of the General Utilities Corporation (see Chapter introductory photo). Starting at 4 AM on March 28, 1979, a series of mishaps occurred. The sequence is enormously complicated; some 40 different events have been identified during the subsequent investigations.

The operation that seems to have started the trouble was a routine one—the changing of a batch of water purifier in a piping system. For a nuclear plant, this job is considered to be as routine as, say, changing the oil filter in an automobile. But a problem developed; some air got into the pipe, causing an interruption in the flow of water. Of course, the back-up systems in a nuclear plant are designed to respond automatically to such an event, but in this instance, several other things went wrong.

· Two spare feedwater pumps were supposed to be ready to operate at all times. However, the valves that control the water from these pumps were out of service for routine maintenance; therefore, the spare pumps could not deliver water. The controls for these valves were provided with tags to indicate that they were being repaired. The tags hung down over red indicator lights that go on when the spare pumps are not feeding water. Since the lights were obscured by the tags, the operators did not realize that no water was flowing. The control of a nuclear power plant is an array of indicator lights and switches (Fig. 29.21).

· As a result, pressure built up in the reactor core. A relief valve in the primary coolant loop then opened automatically (as it should have) to let out superheated steam. But *the relief valve failed to close,* causing a dangerous drop in pressure. This malfunction is considered to be the crucial failure of equipment in the entire sequence.

· When the emergency core-cooling system came on automatically, the pressure gauges in the control

* Adapted from Turk, J., and A. Turk, *Physical Science,* Second Edition, Saunders College Publishing, Philadelphia, 1981. The purpose here is to give the general sequence of events that occurred at Three Mile Island (TMI), which came dangerously close to a "meltdown" situation. The TMI "incident" has been examined many times to determine the underlying causes of the accident and who was at fault. If you are interested in this aspect, there is a public report of the President's Commission on the Accident at Three Mile Island, issued on October 30, 1979. This report is known informally as the "Kemeny Report" because the Commission was chaired by John G. Kemeny, President of Dartmouth College. The Commission found enough fault to go around.

Figure 29.21 The control room of a nuclear power plant showing the array of indicator lights and control switches.

room gave a false reading, leading operators to think that the water level was still above the fuel rods. It wasn't. Instead, bubbles of gas from below were pushing the water up, leaving part of the core exposed.

· The primary and the emergency cooling pumps, which should have been left on, were turned off twice by operators misled by the faulty pressure gauges.

The net result of this confusing sequence of mishaps was that the nuclear core overheated. The indicators of the temperatures inside the reactor vessel climbed off the recorder charts. For $13\frac{1}{2}$ hours the situation was very unclear. It seemed that the reactor core was partially exposed above the cooling water and that there were voids or perhaps bubbles in the system. In fact, subsequent investigations shown that the entire core was exposed for some time. This means that only steam, not liquid water, was circulating through the core to remove excess heat.

The "void" or "bubble" that caused the problem was something entirely unexpected—it was a 1000-cubic-foot (28,000-liter) volume of steam and hydrogen gas. Where did the hydrogen come from? There are two possibilities, both of which probably played a part. One was radiolysis, which is chemical change produced by radiation. In other words, the radioactivity in the core chemically decomposed the water and produced hydrogen (as well as oxygen). The other possibility was the chemical reaction of water with metals, which also produces hydrogen (but not oxygen). The metal tubes holding the uranium fuel are made of zirconium, which reacts with water when the temperature gets hot enough. Pure hydrogen gas is not explosive. Mixtures of hydrogen with air or oxygen are explosive, but only in certain concentrations. Therefore radiolysis (which produces both hydrogen and oxygen) differs in explosive potential from the reaction of water with metals (which produces hydrogen but not oxygen).

The superheated steam released to the atmosphere early in the emergency was radioactive, and for this reason Pennsylvania's Governor Thornburgh ordered the evacuation of children and pregnant women from the area near the plant. Others left of their own accord. As the problem subsided, the evacuees returned, with misgivings one can only imagine.

SUMMARY OF KEY TERMS

Artificial transmutation the changing of the nuclei of one element into the nuclei of another element through an induced nuclear reaction.

Electron volt (eV) a unit of energy, the energy an electron receives when accelerated through a potential of 1 volt: $1 \text{ eV} = 1.6 \times 10^{-19}$ J.

Endoergic reaction a nuclear reaction in which energy is converted to mass, or in which there is a net energy input.

Exoergic reaction a nuclear reaction in which mass is converted to energy, or in which there is a net energy output.

Fission the dividing or "splitting" of a heavy nucleus into two intermediate lighter nuclei with the emission of two or more energetic neutrons.

Chain reaction a growing series of induced fission reactions due to the emitted neutrons of previous fission reactions.

Critical mass the amount of mass or concentration of fissionable material needed for a sustained chain reaction.

Breeder reactor a reactor that produces more fuel than it consumes by converting nonfissionable nuclei into fissionable nuclei.

Nuclear waste the unwanted radioactive by-products from the production of nuclear energy.

Nuclear proliferation the increased availability of radioactive materials that could be used to produce nuclear weapons.

Fusion a nuclear reaction in which two light nuclei are "fused" or reacted together to form a heavier nucleus, with the release of energy.

EXERCISES

1. Discuss the pros and cons of the free exchange of scientific information among scientists of different countries.
2. Identify each of the following and the role they played in the development of the atomic bomb: (a) Hahn, (b) Meitner, (c) Fermi, (d) Bohr, (e) Einstein.
3. Was the Manhattan Project carried out in Manhattan (New York City)?
4. A "mad" scientist wants to sell you a large quantity of gold-194 that he made by artificial nuclear reactions for only one fourth the current price of gold. Should you buy it? (*Hint:* If you checked a chart of nuclides, you would find that gold-194 beta-decays with a half-life of about 40 hours.)
5. In trying to initiate nuclear reactions, would there be any advantage to using protons or hydrogen nuclei instead of alpha particles?
6. Explain what is meant by the conservation of nucleons and the conservation of charge in nuclear reactions.
7. Complete the following nuclear reactions:

   (a) ${}^{45}_{21}\text{Sc} + {}^{4}_{2}\text{He} \rightarrow \_\_\_\_ + {}^{1}_{1}\text{H}$

   (b) ${}^{27}_{13}\text{Al} + {}^{1}_{0}n \rightarrow {}^{28}_{13}\text{Al} + \_\_\_\_$

   (c) $\_\_\_\_ + {}^{4}_{2}\text{He} \rightarrow {}^{35}_{17}\text{Cl} + {}^{1}_{1}\text{H}$

   (d) ${}^{1}_{0}n + {}^{235}_{92}\text{U} \rightarrow \_\_\_\_ + {}^{92}_{36}\text{Kr} + 3({}^{1}_{0}n)$

   (e) ${}^{2}_{1}\text{H} + {}^{2}_{1}\text{H} \rightarrow \_\_\_\_ + {}^{1}_{0}n$
8. Write the fission reaction for uranium-235 "splitting" into rubidium-94 and cesium-139.
9. Why are only two transuranic elements named after planets (Table 29.1)?
10. Identify the people and places in Table 29.1. (Use an encyclopedia if necessary.)
11. Why is the symbol for the transuranic element Rutherfordium Rf instead of Ru?
12. How big a unit is the electron volt? Give some comparisons with other units.
13. Is natural radioactivity an endoergic or exoergic reaction? Explain.
14. Many people think that it was not necessary to drop atomic bombs on Japan in 1945, since World War II could have been brought to an end by an invasion of the Japanese Islands. Discuss the pros and cons of dropping the bomb.
15. Give an example of a *chemical* "chain reaction."
16. Discuss the advantages and disadvantages of nuclear electrical generation as compared with fossil-fuel electrical generation.
17. Why is it impossible to have a sustained fission chain reaction in a very small piece of fissionable material?
18. Why can't a nuclear reactor blow up like a bomb?
19. Could the moderator (or lack of it) be used to control the chain reaction in a reactor? Explain.
20. Is it necessary always to use enriched uranium in a reactor? Explain.
21. What is meant by a "meltdown"? How is this related to the "China syndrome"?
22. A reactor can be shut down in a few seconds by the control rods, but a fuel rod fusion is still possible. Why?
23. Why is an A-bomb needed to start an H-bomb?
24. Tritium can be made by neutron bombardment of lithium-6. Write the nuclear equation for this reaction.
25. The energy release from each fusion reaction is less than 20 MeV, whereas about 200 MeV of energy is released from a fission reaction. Even so, discuss the advantages that fusion reactors would have over fission reactors.

26. Why would a LOCA be less serious or less likely in a fusion reactor than in a fission reactor?
27. The Sun is a big fusion reactor. How are the fusion reactions confined in this case?
28. How does fusion get started in stars?
29. Why is strontium-90 so readily absorbed in bones?
30. On the average, people normally receive a certain amount of radiation exposure or dose of radiation each year, for example, from diagnostic X-rays. About half of the radiation comes from what is called natural background radiation. What gives rise to "natural" radiation? How about "unnatural" radiation?

30

Astrophysics

When taking a long look at the sky on a clear, starry night, have you not been awed by the vastness and the beauty? The distant stars and galaxies appear to be fixed and unchanging, but actually you are observing a small portion of a dynamic and evolving universe.

This intrigue of the heavens prompted the development of one of the earliest sciences—astronomy. At first, astronomy was the study and observation of "objects of the sky" that could be seen with the unaided eye. Our "known" universe was relatively very small, since the universe is all the matter and energy in existence anywhere, observable or not (known or unknown). The invention of the telescope (Chapter 25) expanded the known universe tremendously. The most distant observable object that now denotes the boundary of the known and unknown universe is believed to be some 16 billion light years away!

Into all of this is woven astrophysics, which is physics applied to astronomical situations. One of the major applications of astrophysics is in **cosmology,** the study of the nature and structure of the universe. This term is often broadened to include the origin and evolution of the universe—its past, present, and future.* The **"cosmos"** is the world or universe regarded as an orderly system. In this chapter we'll take a look at how astrophysics is applied in various theories of cosmology, and then we will focus on a particular evolution process—the life and death of a star.

The Expanding Universe

The known universe has "expanded" as our observations and knowledge have increased. However, there is evidence that our universe is physically expanding—flying apart. During the early 1900's V. M. Slipher, an

* Technically, the study of the origin or creation of the universe is called cosmogony.

astronomer at the Lowell Observatory in Arizona, studied the spectra of light from nearby galaxies and noted that there were Doppler shifts in the spectra (Chapter 17). By 1914 a dozen or more of the brightest galaxies had been studied. Strangely enough, most of the spectra showed a red shift rather than a random mixture of red and blue shifts, as might be expected. Slipher's measurements showed that most of these galaxies were moving away from the Earth at high speeds. The most notable exception was the Andromeda galaxy, which showed a blue shift and hence is approaching.†

In the 1920's Edwin Hubble studied the spectra and measured the speeds of many galaxies at greater distances, using the largest telescope in the world at the time, the 100-in (2.5-m) reflecting telescope on Mt. Wilson in California. Without exception, all of the distant galaxies were moving away from the Earth at high speeds. After World War II, the new 200-in (5.1-m) telescope on Mt. Palomar was focused on even more distant galaxies and galactic clusters in the cosmos, and all were found to be receding from the Earth at even greater speeds (Fig. 30.1).

Figure 30.1 The 200-inch (5.1-meter) Hale telescope. Moonlight view of the dome with shutters open, showing the instrument. (See also Fig. 25.19.)

One of the results of Hubble's observations of a number of distant galaxies was that the more distant a galaxy, the faster it moved away. This was shown generally to follow a simple linear relationship between speed and distance, or $v \propto d$ (Fig. 30.2). That is, if one galaxy is twice as far from us as another, it will be moving away twice as fast; if three times as far, the more distant galaxy will be moving away three times as fast, and so on. This relationship can be expressed mathematically by an equation that is called **Hubble's law:**

$$v = Hd$$

where H is called Hubble's constant.

Are We at the Center of the Universe?

If all the galaxies are generally moving away from us, as observations indicate, the question asked in this section heading quickly follows. Is the Earth at the center of the universe as the old geocentric theory predicts (Chapter 5)? The answer is that we can't say for certain, but probably not. If you were on a planet in another galaxy, you would see the distant galaxies receding from you in the same way as we observe from Earth.

A common analogy used to demonstrate this is to picture a simplified model of the universe in which galaxies are represented by points on the surface of a balloon (Fig. 30.3). If the balloon is blown up, each point (galaxy) will move away from all other "galaxies." In addition, the speed at which any two galaxies separate will be proportional to the distance between them. In effect, an observer at any point on the balloon will see what we see from our Earth in the Milky Way galaxy "point."

Of course, we can't prove that this view represents the universe as a whole. As geocentric as we might be, it is highly improbable that we are at the center of the cosmos. Keep in mind that although the balloon model illustrates Hubble's law, the universe is not a surface like that of a balloon. A surface is "carved" out of three-dimensional space. The universe is three-dimensional, and a more realistic model must be carved out of four-dimensional space-time. A three-dimensional analogy of the expanding universe is given by an expanding raisin cake during baking (Fig. 30.4). Every raisin would "see" every other raisin moving away from it at a speed that depends on the distance from it. Thus, each raisin would be at the center of the expansion, measured from its own position.

† The Andromeda galaxy, which is relatively close to the Earth, is approaching our planet at a speed of 300 km/s. Galaxies have a certain amount of random motion, and for nearby neighboring galaxies, a small receding motion may be masked by random motions.

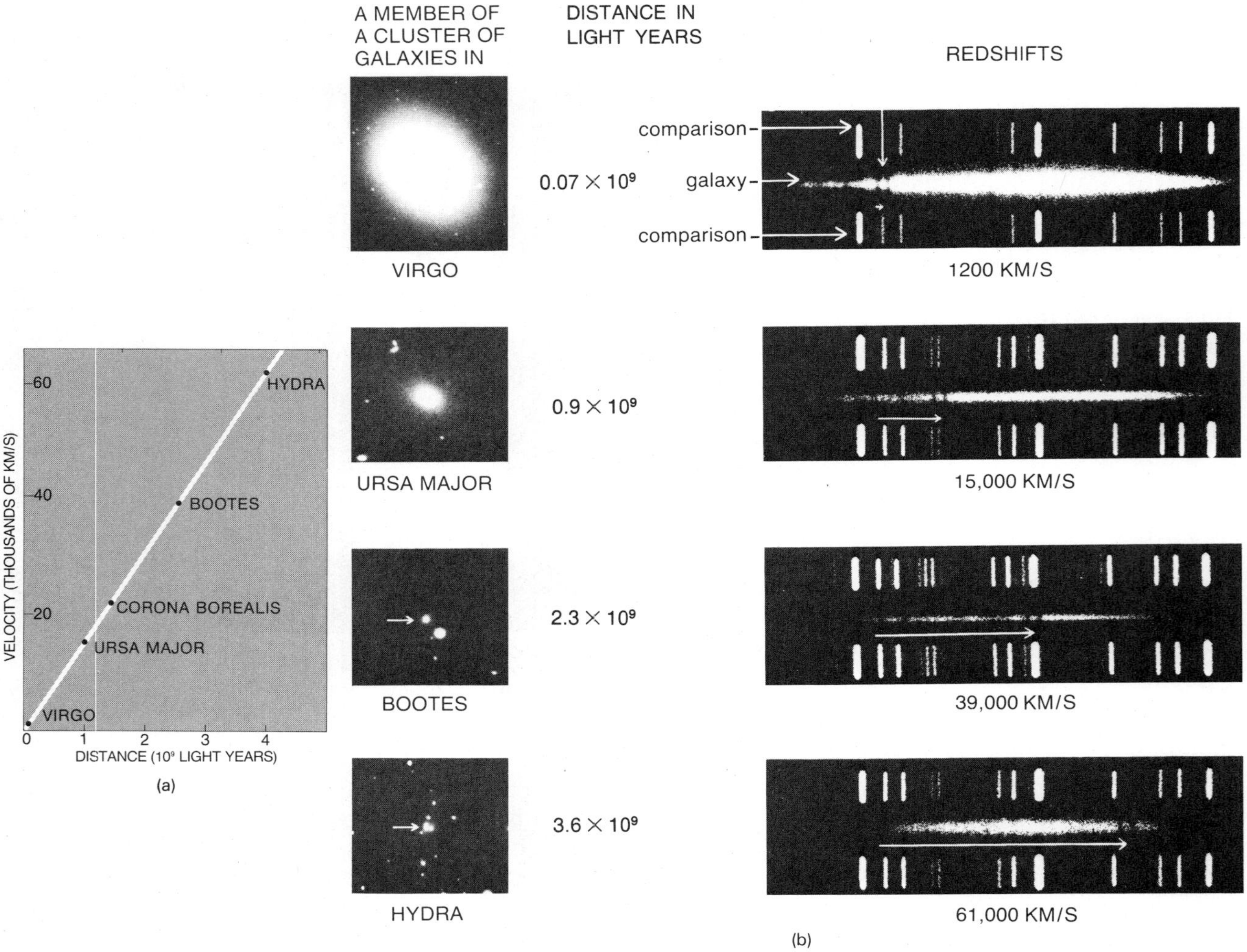

A MEMBER OF A CLUSTER OF GALAXIES IN	DISTANCE IN LIGHT YEARS	REDSHIFTS
VIRGO	0.07×10^9	1200 KM/S
URSA MAJOR	0.9×10^9	15,000 KM/S
BOOTES	2.3×10^9	39,000 KM/S
HYDRA	3.6×10^9	61,000 KM/S

comparison
galaxy
comparison

VELOCITY (THOUSANDS OF KM/S)
60
40
20
HYDRA
BOOTES
CORONA BOREALIS
URSA MAJOR
VIRGO
0 1 2 3 4
DISTANCE (10^9 LIGHT YEARS)

(a)

(b)

Figure 30.2 Hubble's constant. The red shifts of the spectra from galaxies show that their velocities are proportional to their distances, $v \propto d$. If velocity is plotted against distance, Hubble's constant H can be determined ($v = Hd$). (b) The spectra of galaxies, like the examples shown here, provided the data necessary to determine Hubble's constant. The spectra are all reproduced to the same scale. The arrow below each horizontal streak of spectrum shows how far the spectral line being considered is redshifted. The spectrum of a known source located inside the telescope building appears as vertical lines above and below each galactic spectrum to provide a comparison with a redshift known to be zero.

Figure 30.3 Expansion. When the balloon is blown up and it expands, the dots get farther apart. An observer on any dot would see the other dots moving away, analogous to what we observe for galaxies from Earth.

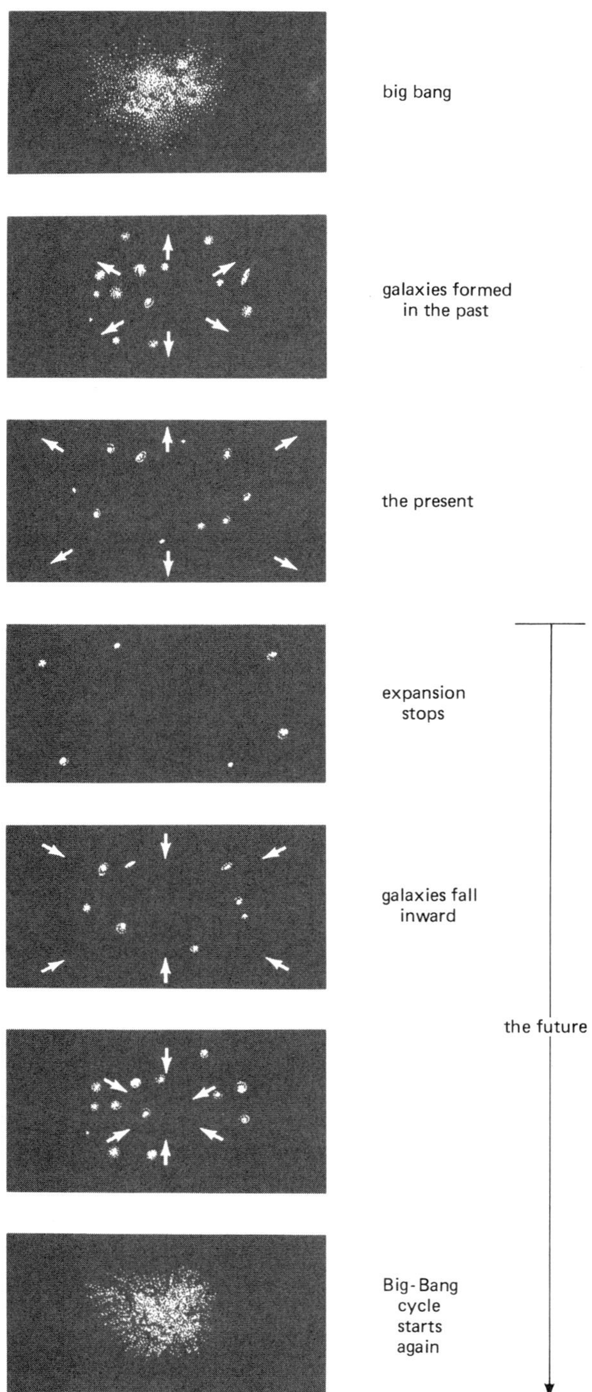

Figure 30.10 The oscillating theory of the universe. In this theory, the universe "oscillates" through never-ending cycles of birth, life, and death, with a new and different universe being formed with each new cycle.

If the average density is computed from all the visible matter in the universe, it is found that this amount of matter is only about 10 percent of that needed. There is a great deal of radiation in the universe, which has a mass equivalence by Einstein's $E = mc^2$ equation. Taking this into account only increases the amount of mass by a percent or two.

Of course, we don't see everything in the universe. Not everything is luminous, for example, burned out, dead stars, distant stars with small intensities that we cannot see, black holes, and so on. Estimates of this "invisible" mass have been made through the study of motions in a galactic cluster and the required gravitational force involved, which gives an indication of the total mass — visible and invisible. Although the results indicate that there is much more mass in the universe than we can see, the total mass is still about 10 percent short of the critical mass needed for an oscillating universe.

Recent data indicate that an elementary particle called the neutrino, which had been thought to be massless, may have a tiny mass. Neutrinos result from nuclear reactions in stars, and they pervade the universe. If this presumed mass is added, the universe's mass is near the boundary between being opened or closed. More measurements are needed for a definite answer, but we have plenty of time for further investigations.

Cosmologists distinguish between the Big Bang theory and the oscillating theory by talking about open and closed universes. An open universe is one composed of space with no boundaries, in which the expansion of the galaxies continues forever. The open-ended Big Bang theory predicts an open universe. A closed universe is one with limits or boundaries, in which galaxies can recede only a limited distance from each other before coming together. The oscillating theory predicts a closed universe.§

Recall from Chapter 5 that in Einstein's general relativity model, gravity is not treated as a "force." Instead, mass is pictured as distorting or warping four-dimensional space-time. If there is enough matter in the universe, then there is a closure of space-time. In effect, this means that a galaxy would travel around a "closed universe" and return to its starting point. This is analogous to a particle traveling on a spherical surface in three-dimensional space.

§ If the universe did not oscillate by never rebounding from the "big crunch," this would also be a closed universe.

Another way of looking at this is to consider the effect of gravity on light beams (Chapter 26). In a closed universe, two parallel light beams would eventually meet (Fig. 30.11). In an open universe, the parallel light beams would eventually diverge. Should the universe have exactly the critical amount of mass so that it is neither open nor closed, the beams would remain parallel.

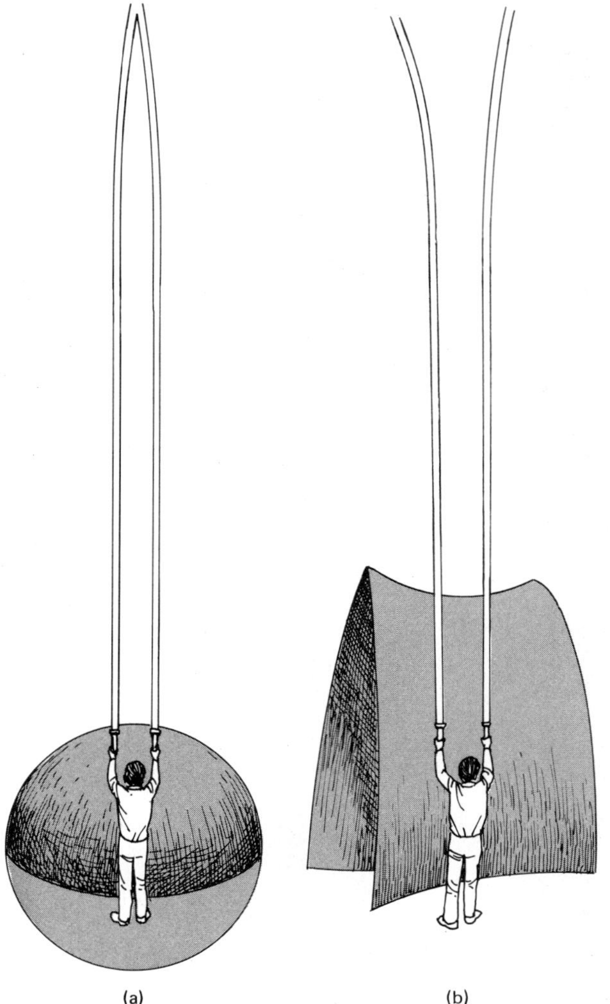

Figure 30.11 Open or closed universe? In a closed universe, two parallel light beams would eventually meet (a). In an open universe, the parallel light beams would eventually diverge (b).

Stellar Evolution — The Birth, Life, and Death of a Star

As gaseous matter expanded from the universe's explosive birth, swirling "pockets" of gas were randomly formed, being held together by gravity. In time, these gas pockets gravitationally collapsed into isolated, dense clumps of gas, called protogalaxies (galaxies in formation). Inside the protogalaxies, smaller "subclumps" of gas came together, which were protostars (stars in formation). We now see the millions of galaxies that resulted from protogalaxies. One of these is our own Milky Way galaxy, which is an aggregate of some 100 billion separate stars held together by mutual gravitational attraction (Fig. 30.12).

Protostars eventually become stars, one of which is our Sun. The Sun is about 4.6 billion years old, whereas the Milky Way sports an age of 12 to 14 billion years. This implies that not all the stars of a galaxy are formed at the time of galactic birth. In fact, stars are being born now, as well as dying. This process of stellar evolution forms one of the most interesting topics of astrophysics.

STELLAR BIRTH

A star is conceived in large clouds of interstellar gas (Fig. 30.13). By some random fluctuation, enough gas atoms or molecules come together in a clump of gas. This cluster of predominantly hydrogen atoms, held together by the grip of its own gravity, is a protostar.

The continual pull of gravity within the clump of gas causes the protostar to shrink in size. That is, the protostar collapses under its own gravitational attraction. When it does so, the density and temperature of the protostar increase. The protons or hydrogen nuclei falling inward gain speed, and hence the temperature of the gas increases. After about ten million years of collapse, the interior temperature of the protostar reaches about 10 million kelvin. At this critical temperature, protons have enough speed to overcome their mutual electrical repulsion, and nuclear fusion is initiated at the center of the contracting mass — a star is born! The release of nuclear energy from the fusion of hydrogen nuclei into helium nuclei generates an outward pressure that balances the inward pressure due to the force of gravity, the gravitational collapse stops, and the shining, newborn star is stable.

Small pockets of contracting gas may not have enough mass and gravity for the collapse to result in the initiation of fusion and become a star. These objects are sometimes called **brown dwarfs.** They are faint, small objects radiating the heat of compression in infrared radiation.

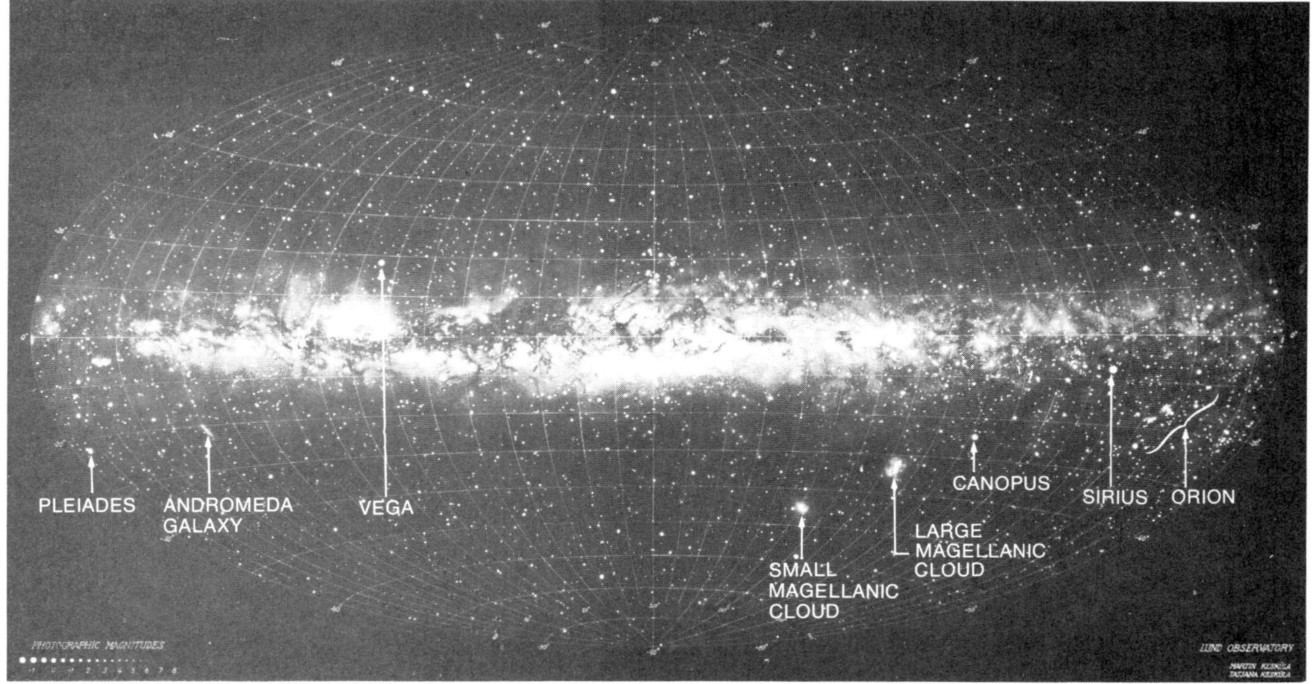

Figure 30.12 The Milky Way. A drawing showing an equatorial view of the Milky Way Galaxy. It is a vast, rotating, wheel-shaped system of approximately 100 billion stars, with a diameter on the order of 100,000 light years. Where are we located? The Sun is about two thirds of the way from the center out to the rim of the wheel in an orbital revolution around the galactic center that takes about 200 million years. Unlike a wheel, the Galaxy does not rotate as a solid body.

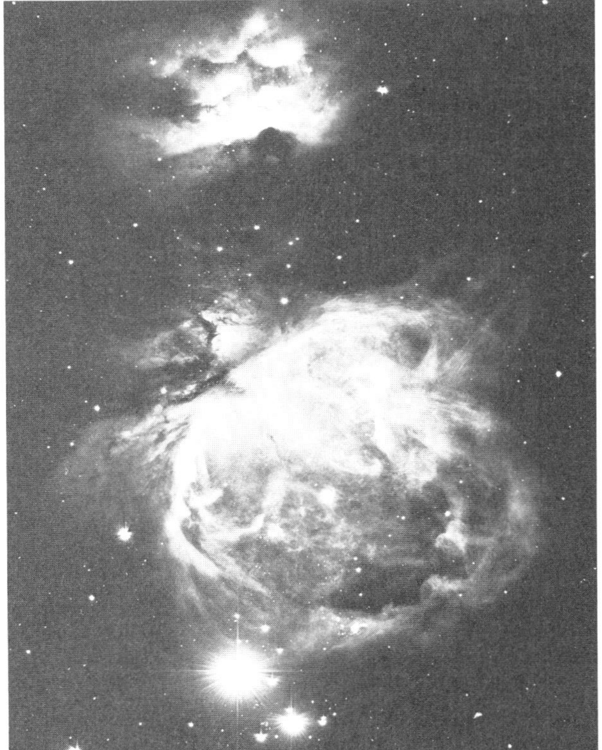

Figure 30.13 Interstellar gas. The gaseous nebula in the constellation Orion.

STELLAR LIFE

The hydrogen "burning" stage of a star, such as the Sun, accounts for about 90 percent of its lifetime. The Sun is "middle-aged." Being about 4.6 billion years old, it has four to five billion years to live.

As with humans, some stars have relatively short lifetimes, whereas others exist considerably longer. The lifetime of a star is related to its mass. Surprisingly, the largest stars have the shortest lifetimes. They have more fuel to burn, but they burn it faster than smaller stars. This is due to higher temperatures resulting from greater gravitational attraction, which increases the reaction rate. A star 15 times more massive than the Sun may live only 10 million years (a lifetime 1000 times shorter than that of the Sun). On the other hand, a star with only a fraction of the Sun's mass may live 100 million years or even a trillion years.

STELLAR OLD AGE

The helium produced by the fusion or burning of a star's hydrogen accumulates at the center of the star, where most of the nuclear reactions take place. When a star's hydrogen fuel starts to run low and it is predominantly composed of a helium core, it begins to show signs of old age.

The temperature of the helium core is not high enough for its fusion, so with no release of energy the core begins to collapse under the influence of gravity. The shrinking helium core gets hotter, which causes the inner portion of the surrounding hydrogen shell to "burn" faster. It might be thought that the star would become brighter, but much of the energy is absorbed by the gaseous hydrogen shell, and it expands outward. With a greater surface area to emit the energy that escapes, the surface temperature of the star drops, and it appears distinctly red in color. When the expanded hydrogen shell cannot absorb any more energy, the energy escapes as radiation, and the brightness or luminosity of the star soars. In this condition of being an expanded, brilliant red star, it is called a **red giant.**

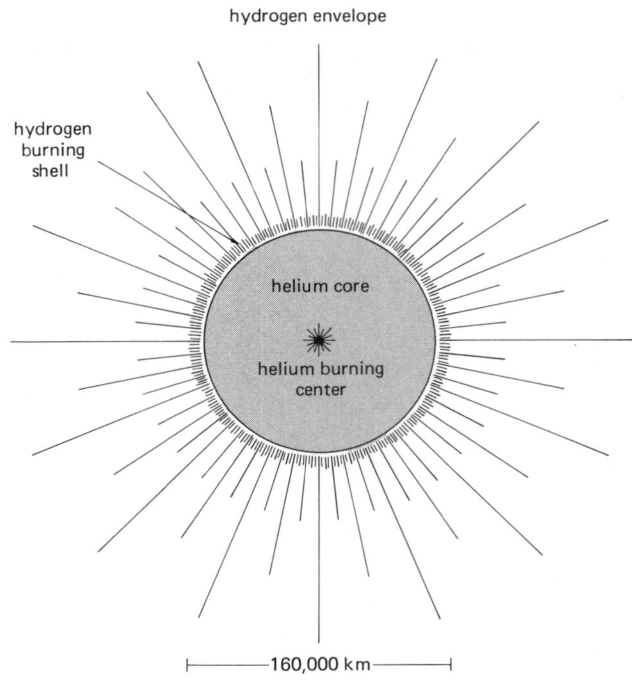

Figure 30.14 Red giant star. With a huge hydrogen envelope at a relatively low temperature, the star appears red in color. Further gravitational collapse causes helium fusion to be initiated in the core.

STELLAR DEATH

Any star exhausts its original hydrogen fuel eventually and then heads toward demise. How this happens depends on the mass of the star. Let's first consider the death of a relatively small star.

As the red-giant stage of any star progresses, the helium core continues to collapse. The critical temperature for stellar hydrogen fusion is about 10 million kelvin, but getting helium nuclei to fuse requires temperatures on the order of 100 million kelvin. When the temperature of the star's core grows large enough for helium fusion, there is a sudden burst of liberated energy, called the helium flash. The star then continues to fuse helium nuclei into carbon, and the burning of helium prevents further collapse (Fig. 30.14).

It might be thought that this cycle would continue with carbon fusion, and so on, but this does not happen for a small star that does not have sufficient mass for gravitational compression to raise the temperature of the core to about 600 million kelvin. Eventually, as the star's helium fuel nears its end, its outer layers expand outwardly, forming a planetary nebula (Fig. 30.15). So-called because they appeared similar to the outer planets when first discovered with small telescopes, these fuzzy (nebular) forms have nothing to do with planets, but are signs of aging stars.

As a relatively small star approaches the end of its energy generation and life, its shrunken core becomes extremely dense, with a volume about as big as that of a good-sized planet. Although such a star is very faint, its surface is quite hot, in fact, white-hot, and we call it a **white dwarf.** Calculations show that if a star's mass is less than about 1.4 solar masses at this critical point in

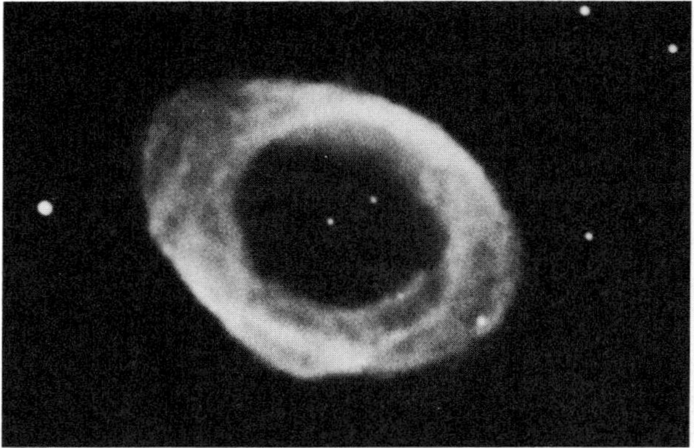

Figure 30.15 The Ring Nebula, a planetary nebula in the constellation Lyra.

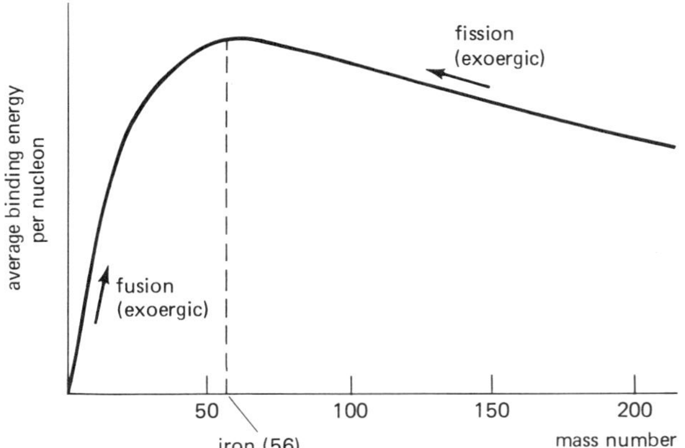

Figure 30.16 Binding energy. The nuclides form a general curve, as shown here. The higher their position on the curve, the more tightly bound together are the nucleons in a nucleus. Iron is a particularly stable nucleus.

its life, it will become a white dwarf. The Sun will almost certainly become a white dwarf, which quietly and increasingly slowly radiates its energy into space and cools. Eventually ceasing to "shine" after billions of years, a white dwarf becomes a **black dwarf,** or a cold hulk in the "graveyard" of stars.

A different fate awaits a more massive star—an explosive death. For stars a few times more massive than the Sun, this occurs with a "carbon flash" as soon as the carbon burning begins. In still more massive stars, carbon fusion sets in and continues. Successive contractions and fusion reactions produce heavier elements until iron is formed. Iron is very special, inasmuch as it marks the limit of exoergic fusion. The formation of heavier elements would require energy input (endoergic). That is, iron is the demarcation point between fusion and fission. Elements heavier than iron would not release energy if their nuclei underwent or could undergo fission (Fig. 30.16). With no more fusion, the massive star gravitationally collapses, then explosively rebounds like a giant compressed spring.

Such an exploding star is called a **supernova.**‖ It is the most energetic of all stellar explosions. They occur somewhat frequently on an astronomical time scale. The earliest reported supernova was observed by Chinese astronomers in A.D. 1054. This "guest star," as they called it, exceeded the brightness of any object in the night sky for several weeks. The remnant of this supernova is now observed as a great expanding cloud of gas called the Crab Nebula (Fig. 30.17). Its expansion rate is on the order of 1600 km/s (1000 mi/s). The last observed supernova in our galaxy was seen in 1604. Many supernovae have been observed in other galaxies (Fig. 30.17).

‖ A nova or "new" star is a brilliant exploding star that blows off a small amount of mass. This is believed to be associated with an exchange of mass in a binary star (a two-star system) that has white dwarf and red giant components.

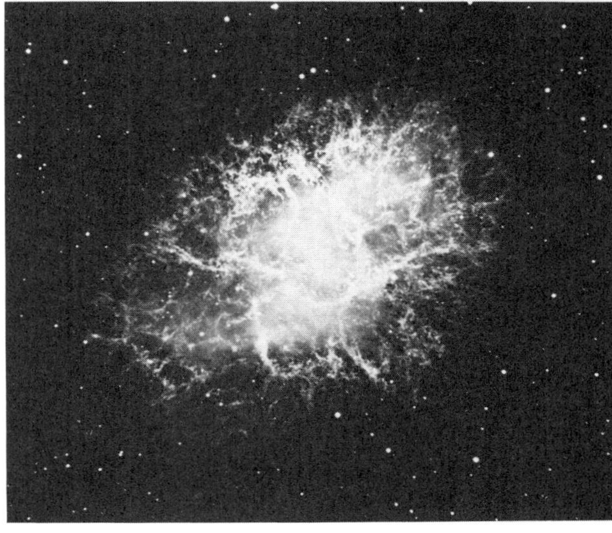

(a)

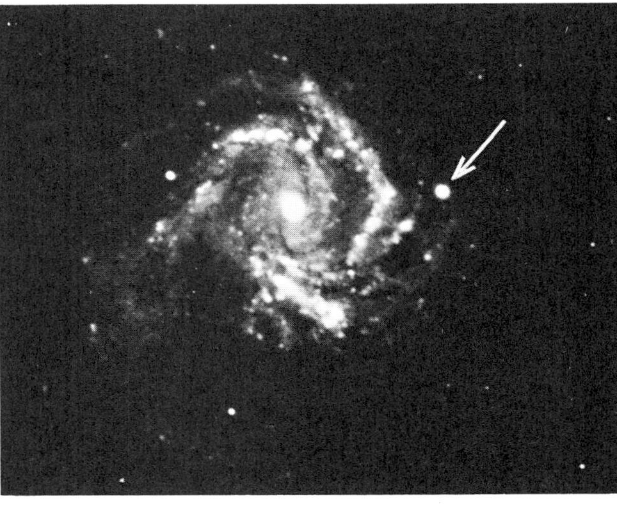

(b)

Figure 30.17 (a) The Crab Nebula is a great expanding cloud of gas, the remnant of a supernova. (b) The supernova that appeared in the galaxy NGC 4303 in 1961 shone for several weeks with an absolute brightness 300 million times that of the Sun.

In the cataclysmic explosions of supernovae, heavier elements are made by nuclear reactions. In the final moments of these stars' lives, it is assumed that all of the elements found in nature are made, even those heavier than iron, for example, gold, lead, and uranium. The heavier elements above iron are relatively rare in abundance compared with the lighter elements, as might be expected. All the elements are scattered into space and become mixed with interstellar hydrogen, which will eventually be condensed into protostars. With the evolution of stars and, presumably, of planet "spin-offs" such as our Earth, the elements are recycled. According to this theory, both the Earth and you are made up in large part of "stardust."

Pulsars, Neutron Stars, and Black Holes

In late 1967 at a radio telescope installation in England, a graduate student, Jocelyn Bell, noticed that a celestial radio source emitted regular pulses or "beeps" of radio waves every 1.33733 seconds. It was at first thought that these might be signals from an intelligent life form in space. Several other beeping celestial "radio stations" were soon found. Not surprisingly, the sources were at first dubbed LGMs (for Little Green Men). However, subsequent investigations excluded this possibility, and the term **pulsar** replaced LGM.

An explanation of pulsars came from a previous prediction. In the explosion of a supernova, a tremendous inward pressure would compress the core of the star. This would be so great that individual electrons and protons would combine to form neutrons. Hence, it was predicted that an incredibly dense ball of pure neutrons, or **"neutron star,"** would be left at the center of a supernova explosion. A comparison of some stellar densities is shown in Figure 30.18. In 1968 a pulsar was discovered at the center of the Crab Nebula, right where a neutron star would be expected to be located.

But what causes the regular pulses or beeping? Scientists believe that violent, localized disturbances occur on the surfaces of neutron stars (which are only about 25 to 30 km in diameter), similar to those on the surface of the Sun. These localized storms emit radiation in well-defined directions—like beams of light. If a neutron star were spinning rapidly, the beam would be observed as a series of pulses or beeps, much the same as from a rotating flasher seen on police cars. Hence, pulsars are believed to be rotating neutron stars that resulted from supernova explosions.

Calculations show that a neutron star cannot exist with a mass three to five times greater than that of the Sun. When a very massive star becomes a supernova, it is speculated that the large core continues to collapse right past the density of a neutron star and forms a **black hole.** Recall that the gravitational attraction of a black hole is so great that it prevents light or anything else from leaving it. Hence, black holes must be invisible, and we can only search for them by indirect means, as described in Chapter 5. The fascinating subject of black holes is treated in more detail in Special Feature 30.1 at the end of the chapter.

Quasars

Before leaving our study of astrophysics, let's take a brief look at something you may have heard and wondered about—quasars. When radio telescopes became a common astronomical tool (see Chapter 25), astronomers busied themselves viewing the "radio sky." In addition to many galaxies and nebulae being radio sources, some "stars" were also found to emit strong radio signals—ordinary stars that were assumed to be in our galaxy. It was not understood how an ordinary star could emit such large amounts of radio waves, but for the moment, the strange objects were dubbed "radio stars."

Investigations of the visible spectra of these radio stars gave rise to even more intrigue. No spectra like these had been seen before coming from ordinary stars or galaxies. It was finally realized that the problem of nonrecognition occurred because the spectra were extremely shifted toward the red end of the spectrum. If this is a Doppler shift, it indicates that the "radio stars" are receding from us at very high speeds. The first two "stars" studied showed recession speeds of about 15 and 30 percent of the speed of light. Others have red shifts that indicate they must be moving at speeds greater than 90 percent of the speed of light! Using Hubble's law, we find that this puts these "stars" at distances of 3, 6, and 16 billion light years, respectively. The latter would be the most distant object known. Thus, it would seem that quasars are on the forefront of the expanding universe.

The fact that we can see these objects at such tremendous distances means that they must have enormous energy outputs. In fact, they can't be stars as we know them, since their energy outputs are comparable to those of galaxies. Hence, it appears that the "radio stars" are not stars at all, but distant galaxies. Since these strange objects looked somewhat like stars, they

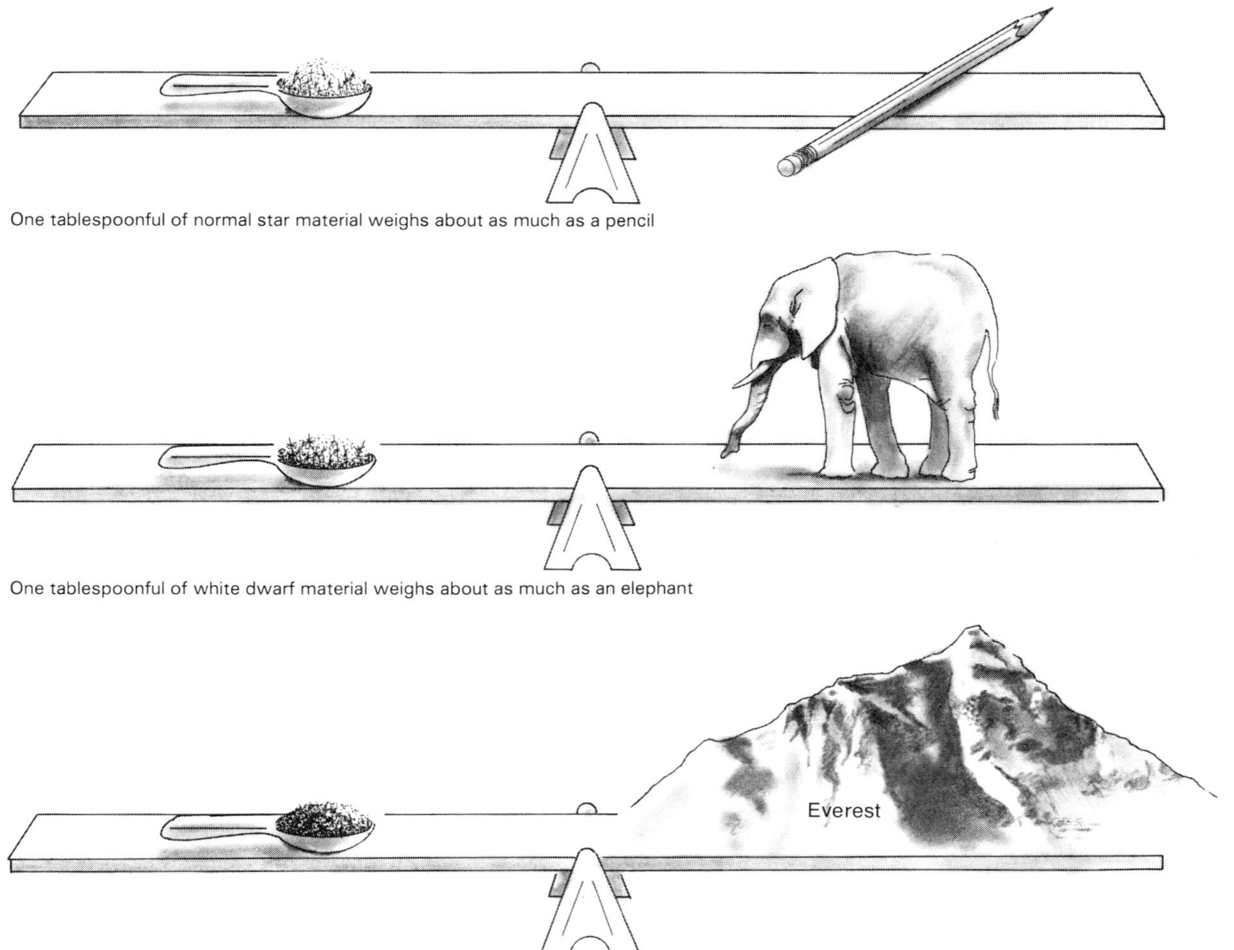

Figure 30.18 A comparison of star densities.

were given the name "quasi-stellar radio sources." The *quasi-stellar* part of the name means "starlike." The name was quickly shortened to **quasar.**

Quasars at present are an astronomical puzzle or mystery. If they are at such great distances, then their energy outputs must be up to 100 times greater than those of large galaxies such as the Andromeda Galaxy and our own Milky Way Galaxy. Furthermore, data indicate that quasars are very small — only light days to light weeks in diameter. (The Milky Way Galaxy is about 100,000 light years across.) Such an energy source is difficult to imagine.

Perhaps the red shift is not really a Doppler shift. In Einstein's general theory of relativity, (light) photons are affected by gravity, and the photons lose a fraction of their energy in escaping from a star. If gravity robs the photons of energy, then by the energy-frequency relationship, $E = hf$, the frequency becomes smaller and the wavelength becomes longer — a red shift. This is called a gravitational red shift. However, to radiate as much energy and to produce the observed large red shift of light from quasars, an object would have to collapse quickly into a black hole. This makes the gravitational mimicking of the effects of a Doppler shift unlikely.

Other questions arise: Is Hubble's law valid for large distances? Could quasars really be nearby objects, which would then put them more in line with galactic energy outputs? Could the energy of a distant quasar result from the supernova of two colliding stars in a distant galaxy? All these questions have been asked, with no definite (proven) answers. The riddle of the quasars is just one more example of the unknown that makes science so interesting and fascinating.

SPECIAL FEATURE 30.1

Black Holes

The idea of black holes has caught the public's fancy. Imagine—something so dense that neither light nor anything else can escape from it. Black holes are becoming increasingly common in science fiction, but in actuality they are still largely theoretical. Any object, even you, could become a black hole if it shrank enough, but in most cases this would be highly unlikely. The most favored candidates to become (or that may have become) black holes are collapsing stars that have too much mass to become neutron stars. Such stars have three to five times the mass of the Sun.

As we said, any object could theoretically become a black hole if it collapsed enough to a sufficiently small size or radius. This of course would depend on the mass of the object. As a star collapses under the influence of gravity, the important point to focus on is the gravitational force at the *surface* of the shrinking star. The force gets larger because the distance from the surface to the center becomes less (recall that $F \propto 1/r^2$, Chapter 5). If a star's radius shrinks by a factor of $\frac{1}{2}$, then the attractive gravitational force at the surface would *increase* by a factor of four. Similarly, if the star shrinks so that its radius is ten times smaller, the gravitational force at the surface is 100 times stronger, and so on.

Recall from the chapter that a photon leaving a star loses some energy as a result of gravity, which gives rise to a gravitational red shift. For a photon leaving the Sun, this energy loss is about 0.0005 percent. If the Sun should collapse, the increased gravity at the surface would rob more energy from the photons. Finally, if the Sun shrank to a radius of about 3 km, the escaping photons would lose 100 percent of their energy. In other words, they would not escape at all, and the Sun would become a black hole. Of course, the Sun is not expected to do this, as mentioned previously. If the Earth could shrink to become a black hole, its gravitational force at the surface would not let light escape when the Earth had a radius of 1 cm! (On a relative scale, what would be your size when your mass became a black hole?)

The idea of light rays escaping at various stages from a shrinking star is illustrated in Figure 30.19. At the size of a neutron star, the light rays (or photons) still escape, but they are bent considerably, owing to the strong gravitational field at the surface. As the star shrinks further, the rays are bent more, and finally some of the

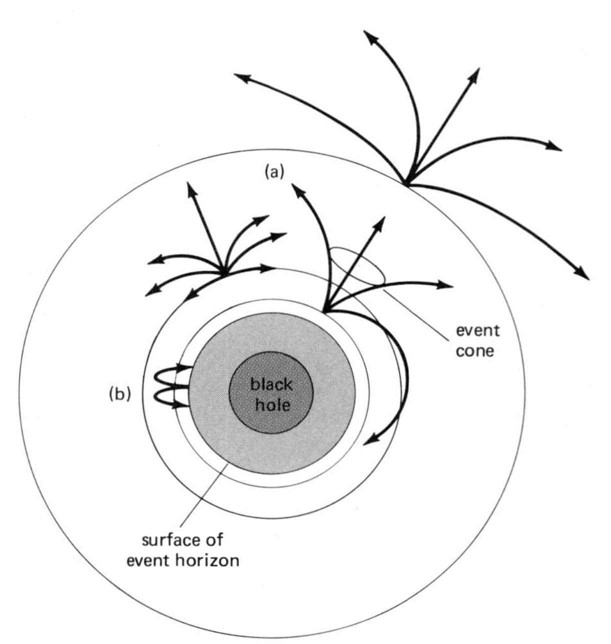

Figure 30.19 Light and a shrinking star. At (a) light is bent by dense stars. With further contraction, only light within an event cone escapes from the surface. At a critical (Schwarzchild) radius, which defines the surface of the event horizon (b), light can no longer escape, and a black hole is born. The mass of a black hole may continue to shrink.

more horizontal rays are bent back into the star and do not escape. Only the rays emitted within a certain **event cone** can escape. As the star contracts still further past a certain critical point or radius, light and everything else is trapped inside—a black hole is born!

The critical size or radius at which a shrinking star becomes a black hole is obtained by setting the escape velocity at the surface equal to the speed of light. The surface at this point is called the **event horizon,** and the critical radius is called the **Schwarzchild radius.*** As noted previously, this would be 3 km for a star the mass of the Sun. For a star with 10 solar masses, the Schwarzchild would be about 30 km, making the density of a black hole incredibly large.

* Named for Karl Schwarzchild, a German astronomer and mathematician. The escape velocity (Chapter 5) from a body of mass M and radius R is given by $v_e = \sqrt{2GM/R}$. Setting this equal to the speed of light c and solving for R, we have $R = 2GM/c^2$ (Schwarzchild radius).

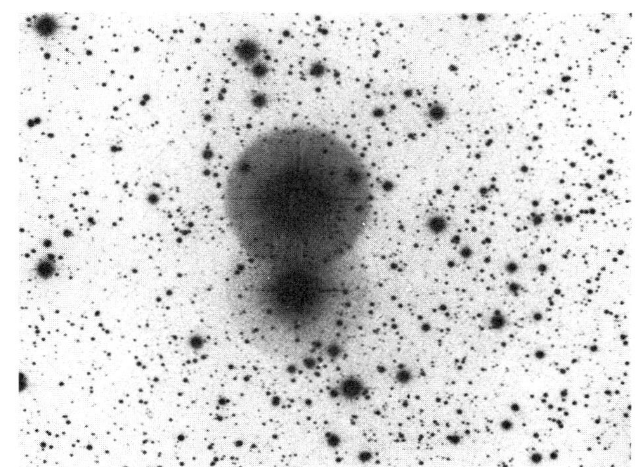

Figure 30.20 Cygnus X-1. The overexposed large dark area at the center is a giant star believed to be a companion of the X-ray source and possible black hole, Cygnus X-1.

The boundary of a black hole is the surface of the event horizon at its Schwarzchild radius. The matter within could continue to contract, but the force of gravity at the event horizon would still be the same. Thus, the boundary of a black hole is not a sphere of matter, but the radius at which the gravitational force is sufficiently strong to keep light from escaping. The actual radius of the matter of a black hole may be much less. What form matter takes inside a black hole is unknown. If we ever determine the location of a black hole in space, any probe we might send to check this out could never report back when it arrived. Think about it.

The experimental proof of the existence of black holes is another thing. How do you observe a black hole? The most likely possibility comes from binary star systems, which are quite common. Cygnus X-1, the first X-ray source discovered in the constellation Cygnus, provides the best evidence so far for a black hole in our Milky Way galaxy. (Approximately 100 or more X-ray sources have been discovered by observations from rockets and satellites. X-rays do not penetrate the Earth's atmosphere.) If we look at the location of the Cygnus X-ray source visibly (Fig. 30.20), we can also observe a giant star whose spectra show Doppler shifts, indicating an orbital motion.

It is speculated that one member of a binary star system has become a black hole. Matter drawn from the other member would cause an **accretion disk** of spiraling matter around the black hole (Fig. 30.21). The matter falling into the disk would be accelerated and heated. Collisions would give rise to X-rays, and the black hole would be an X-ray source.

So, maybe black holes exist and maybe they don't. Time and investigations (the scientific method) will tell. In the meantime, there are a lot of speculations. Matter directed toward a black hole would be gone forever. Perhaps black holes are "vacuum cleaners" in space, gobbling up any matter that comes within its event horizon. As a black hole gains more matter, its horizon would expand. However, the consumption of matter does not set well with our idea of conservation of mass. Where does the matter go? This question has given rise to the speculation that perhaps matter going into a black hole "pops out" somewhere in a parallel universe from what has been termed a **"white hole."** This may sound like a support of the steady-state theory of the universe, but the idea is questionable in terms of a black hole being the ultimate result of a collapsing star. Perhaps the situation can be summed up in the words of the scientist John Haldane: "My suspicion is that the universe is not only queerer than we suppose, but queerer than we can suppose."

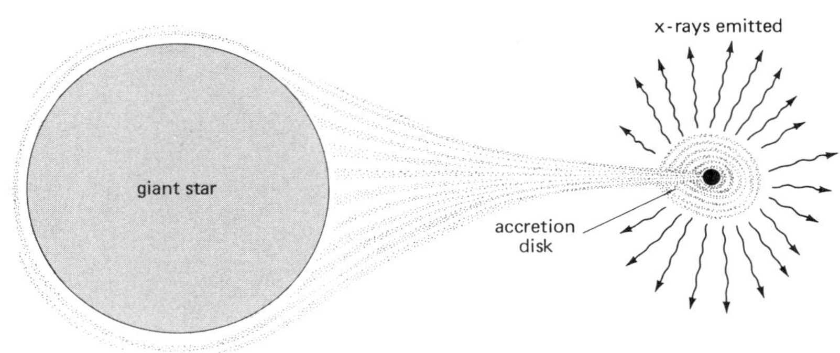

Figure 30.21 X-rays and black holes. Matter drawn from the other member of a binary star system forms a spiraling accretion disk around the hole. Matter falling into the disk is accelerated, and collisions give rise to X-rays.

SUMMARY OF KEY TERMS

Astrophysics physics applied to astronomical situations.

Cosmology the study of the nature, structure, and origin of the universe.

Cosmos the world or universe regarded as an orderly system.

Hubble's law a relationship between the speed and distance of receding objects, $v = Hd$, where H is Hubble's constant.

Big Bang theory a theory of cosmology tht views the universe in the process of expanding from an explosive beginning. The expansion is indefinite, and the universe will eventually fade away into darkness.

Steady-state theory a theory of cosmology that views the universe as having no beginning or end, but being replenished by the appearance or creation of matter out of nothing.

3-K background radiation radiation permeating space that is believed to have come from the original fireball of the Big Bang beginning of the universe.

Oscillating universe theory a theory of cosmology that predicts that the expanding universe will stop, contract, and be reborn again with another explosion. The process continues indefinitely, and the universe oscillates between expansion and contraction.

Open universe a universe with no boundaries, in which the expansion of the galaxies continues forever.

Closed universe a universe with limits or boundaries in which galaxies can recede only a limited distance from each other before coming together.

Red giant a star near the end of its hydrogen burning that has an expanded hydrogen shell that appears red.

White dwarf a star at the end of its helium burning that is white-hot due to core contraction.

Black dwarf the remnant of a white dwarf after it has radiated its energy into space and cooled.

Supernova an exploding star.

Pulsars pulsating radio sources believed to be rotating neutron stars that are formed and located at the center of supernovae.

Black hole a gravitationally collapsed star that is so dense that light cannot escape from it.

Quasar (quasi-stellar radio source) a starlike object thought to be a galaxy, whose spectrum shows large red shifts. This indicates a large speed of recession, presumably because of the object's enormous distance from us.

EXERCISES

1. Considering the definition of the cosmos, explain the term used to describe particles called *cosmic* rays.
2. Describe what the known universe consisted of for early civilizations. What did the invention of the telescope provide? What is the present-day boundary of the known universe?
3. Are we at the center of the universe? Can we determine exactly where the Big Bang took place? Explain. (Suppose that three cars started from the same place and then moved along a straight road at different rates. Considering yourself to be stationary in the middle car or at the center of things, what would the other cars appear to be doing? What would the drivers of the other cars see?)
4. Discuss the average density of the universe in terms of the (a) Big Bang theory, (b) steady-state theory, and (c) oscillating universe theory.
5. Is there a violation of the conservation of mass-energy in any of the theories of cosmology? Explain.
6. Is it possible that some of the stars we see no longer exist? Explain.
7. If you plotted values of the Hubble constant determined from stars at various distances versus time, would you get a horizontal straight line? Explain.
8. Explain how Hubble's constant would be expected to be affected according to the oscillating universe theory.
9. How is the appearance of the railroad tracks in Figure 30.22 analogous to the behavior of parallel light beams in a closed universe?
10. Does the steady-state theory predict an open or closed

Figure 30.22 See Exercise 9.

universe? How would the continual creation of matter required by the theory affect this?

11. When does a protogalaxy become a galaxy? A protostar become a star?
12. Explain why a star has such a long gestation period.
13. How does stellar evolution support the Big Bang theory over the steady-state theory?
14. Will the Sun ever be able to have carbon fusion? Explain.
15. Stars with masses up to 4 solar masses can become white dwarfs, but a white dwarf itself cannot have more than 1.4 solar masses. How is the mass lost?
16. When the Sun becomes a red giant, it will expand to a diameter of about 100 million miles. Considering this, what can you say about the future of the Earth?
17. What determines whether a star becomes a white dwarf, a neutron star, or a black hole?
18. Give a brief history of what the matter in the universe has been doing for the last 15 billion years and what it will be doing in the future.
19. Are there any similarities between the lifespans of very massive stars and those of overweight persons?
20. Assuming that a star collapsed directly into a black hole, the black hole would have the same amount of mass as the star. Why would the gravitational effects be any different?
21. Determining the distances to quasars using Hubble's constant is equivalent to linearly extrapolating Hubble's law in Figure 30.2 to many more billions of light years in distance. Is this really justified? (*Hint:* Think in terms of extrapolating Hooke's law for a spring, $F = kx$.)
22. Quasars are believed to be only light days to light weeks in diameter. What are these lengths in kilometers and miles? ($c = 3 \times 10^5$ km/s $= 1.86 \times 10^5$ mi/s). The diameter of the Sun is on the order of 10^6 km. What is this length in an appropriate light-distance unit such as light seconds?

Illustration Credits

Various photographs and line drawings are courtesy of:

Abell, G., *Exploration of the Universe,* Fourth Edition, Saunders College Publishing, Philadelphia, 1982.
Figures 5.5, 5.7, and 26.18

Adams, T.
Accoustic photograph color plate

AIP Niels Bohr Library
Chapter 26 Introductory photo, Figures 26.6 (top), 27.8, and 28.9. W.F. Meggers Collection.

AiResearch Manufacturing Company
Figure 8.10(b)

Bethlehem Steel Corporation
Chapter 13 Introductory photo

Black, Lloyd, Saunders College Publishing
Figures 1.3(a), 10.4(c), 10.7(a), 19.8, 19.11, and 24.23

Bureau of Reclamation
Chapter 21 Introductory photo

Cabri, Charlotte, Lander College
Figure 21.5(b)

California Institute of Technology, Courtesy of the Archives
Figure 26.6

Cedar Point photos by Dan Feicht
Figure 4.17 and Chapter 8 Introductory photo

Clemson University
The Athletic Department: Chapter 2 Introductory photo
Department of Information and Public Services: Figure 8.23

Corning Glass Works
Chapter 14 Introductory photo

Cousteau Society, Inc., 930 W. 21st St., Norfolk, Virginia 23517, a membership-supported environmental organization.
Figure 12.20

Duke Power Company
Figure 7.7(b)

Ealing Corporation
Figure 9.4. Courtesy of the Ealing Corporation.

Edgerton, Harold E., Massachusetts Institute of Technology
Chapter 3 Introductory photo and Figure 3.2

Educational Development Center, Newton, Massachusetts
Figures 4.9(c), 20.6, and 23.10

Environmental Science Services Administration
Figures 14.11 and 20.19

Freedman, Jay, Saunders College Publishing
Chapter 16 Introductory photo

Greenberg, L., *Physics for Biology and Pre-Med Students,* W.B. Saunders Co., Philadelphia, 1975.
Figures 23.14(b), 24.8, and 28.19(b)

Hahn, Otto, *A Scientific Autobiography,* Charles Scribner's Sons, New York, 1966. Courtesy of AIP Niels Bohr Library.
Figure 29.1

Highsmith, P., *Physics, Energy, and Our World,* W.B. Saunders Co., Philadelphia, 1975.
Figures 3.5 and 6.13

Highsmith P., and A. Howard, *Adventures in Physics,* W.B. Saunders Co., Philadelphia, 1972.
Figure 3.22

Holton, G., F. Rutherford, and F. Watson, *Project Physics,* Holt, Rinehart and Winston, New York, 1980.
Figures 2.1, 2.9, and 3.8, Chapter 16 Introductory photo, Figures 17.7(b), 24.11(b), 27.20, 28.18, 28.20, and 29.4

IBM
Figure 27.23

Jones, M., et al., *Chemistry, Man, and Society,* Fourth Edition, Saunders College Publishing, Philadelphia, 1983.
Figure 10.13

Kelly, Lee, Lander College
Figures 6.4, 7.22, 8.11, 8.24, 8.25, 9.10(b), 11.6(b), 14.1, Chapter 15 Introductory photo, Figures 18.21(b), 18.22, and 22.15(a)

Kuhn, K., and J. Faughn, *Physics In Your World,* Second Edition, Saunders College Publishing, Philadelphia, 1980.
Figures 4.22, 6.18, 7.2, 7.4, 7.11, 8.5(b), and 8.27, Chapter 9 Introductory photo, Figures 9.12, 10.3, 10.10(b), 12.4, 12.19, and 13.7, Chapter 14 Introductory photo, Figures 16.13, 17.10, 17.13(b), 18.12 (photo), and 20.1, Chapter 21 Introductory photo, Figures 21.8(b) and 21.15, Chapter 23 Introductory photo, Figures 24.15(a) and 25.8.

Lick Observatory
Figures 30.13 and 30.17(b)

Lund Observatory, Sweden
Figure 30.12

Luray Caverns Corporation
Chapter 10 Introductory photo

Masterton, W.L., et al., *Chemical Principles,* Fifth Edition, Saunders College Publishing, Philadelphia, 1981.
Figures 9.3(a), 9.7, 9.9(b), 10.9, and 10.10(a)

Merken, M., *Physical Science,* Third Edition, Saunders College Publishing, Philadelphia, 1984.
Figures 1.3, 13.1, and 18.21(a), and Chapter 20 Introductory photo

Metrologic Instruments, Inc., 143 Harding Ave., Bellmawr, New Jersey, 08031. Used by permission.
Figure 27.15(c)

National Aeronautics and Space Administration (NASA)
Figures 2.18, 3.19, Chapter 5 Introductory photo, Figures 5.13, 5.16, 5.20, 5.21, and space color plates

Nave, C., and B. Nave, *Physics for the Health Sciences,* W.B. Saunders Co., Philadelphia, 1975.
Figure 12.13

New York Times Pictures
Figure 18.13
Palomar Observatory Photographs
Figure 25.19, Chapter 30 Introductory photo, Figures 30.1, 30.2, 30.7, 30.15, 30.17(a), and 30.20
Pasachoff, J., *Contemporary Astronomy,* Third Edition, Saunders College Publishing, Philadelphia, 1981.
Figures 30.4 and 30.8
Photo Researchers, Inc.
Figure 16.9(b) and Chapter 29 Introductory photo
Sanders, Jocelyn, Lander College
Figure 11.2
Sandia Laboratories
Figure 7.15
Serway, R., *Physics for Scientists and Engineers,* Saunders College Publishing, Philadelphia, 1982.
Chapter 4 Introductory photo, Figures 4.6, 4.19, 4.21, 16.3, 16.16(b), and Chapter 19 Introductory photo
Stampf, Edward, Lander College
Chapter 12 Introductory photo and Figure 12.10
Stevenson, R., and R.B. Moore, *Theory of Physics,* W.B. Saunders Co., Philadelphia, 1967.
Figures 4.3(b), 4.10(a), 24.2 (photo), and 24.3 (photo)
Turk, J., and A. Turk, *Environmental Science,* Third Edition, Saunders College Publishing, Philadelphia, 1984.
Figures 7.5 and 7.9(a)
Turk, J., and A. Turk, *Physical Science,* Second Edition, Saunders College Publishing, Philadelphia, 1981.
Figures 2.22, 2.23, 3.15(a), 3.20, 4.1(b), and 6.18, Chapter 7 Introductory photo, Figures 7.8, 7.13(a), 7.20(b), 7.23, 8.10(a), 8.26, 9.11, 9.13, 10.16, and 10.18, Chapter 13 Introductory photo, Figures 15.13, 15.14, 15.16, and 17.23, Chapter 24 Introductory photo, Figures 24.10(b), 24.24, 25.15, 27.14, 28.23, 29.21, and 30.18
University of Chicago
Figures 29.2 and 29.11
U.S. Coast Guard
Figure 11.10
U.S. Department of Commerce, National Bureau of Standards
Figures 1.4(b), 1.7(b), 1.11, and 1.15, and Table 1.2
National Atmospheric and Oceanic Administration (NOAA) Figure 14.11, Chapter 18 Introductory photo, Figure 20.19 (Vic Hessler, University of Alaska)
U.S. Department of Energy
Figures 7.1, 7.6, 7.9(b), 7.10, 7.13(b), 7.17, 7.18, 21.12, 28.15, 29.10, 29.13, 29.14, 29.16(b), 29.17, 29.18, 29.19, and energy research color plates
Welch Scientific Co.
Spectrum chart adaption, color plate
Wicks Organ Company
Figure 17.20 (photo)
Williams, J., F. Trinklein, and H. Metcalfe, *Modern Physics,* Holt, Rinehart and Winston, New York, 1980.
Figures 2.21(b), 8.20(b), and 9.3(b), Figures 10.2, 11.13, 11.14, 13.19, 17.17(b), 17.19, 22.14, 23.11, and 28.19(a)
Wilson, J., *Technical College Physics,* Saunders College Publishing, Philadelphia, 1982.
Figures 4.11, 4.16, and 5.3, Chapter 6 Introductory photo, Figures 6.17, 7.14, 7.16(b), 8.10(b), 8.22, and 10.15, Chapter 11 Introductory photo, Figures 11.8, 11.9(b), 11.11, 11.16, 11.20, 11.22, 11.23, 12.6, 12.7, 12.23, 13.2(b), 13.4, 13.6, 13.11, 13.12, 13.13, 13.14, 13.15, 13.17, 14.18, 15.5, 15.6, 15.7, 15.8, 15.10, 15.11, 15.12, 16.15, 16.18, 17.3, 17.4, 17.5, 17.11, 17.13(c), 17.16, 18.5, 18.16, 18.17, 18.20, 19.5, 19.17, 19.18, 19.21, 20.3, 20.9, 20.10, 20.11(a) and (b), 20.13, 20.14, 20.15, 20.16, 21.8(a), 21.11 (photos), and 21.14, Chapter 22 Introductory photo, Figures 22.4, 22.7, 22.8, 22.9, 22.13, 22.21, 22.22, 22.23, 22.24, 23.2, 23.3, 23.5(b), 23.8, 23.9, 23.15, 24.6, 24.7(a), 24.14(a), 24.18, 24.21, 25.1, 25.13, 25.18 (photos), 25.20, and 25.21, Chapter 27 Introductory photo, Figures 27.6, 27.7(b), 27.16, 27.18, 27.22, and 29.9
Yerkes Observatory
Figure 25.17(b)

Index

(Numbers followed by *t* and *n* refer to pages with items in tables and footnotes, respectively.)

N

O

T

U

V